Zinc Oxide Bulk, Thin Films and Nanostructures

Zinc Oxide Bulk, Thin Films and Nanostructures

Processing, Properties and Applications

Edited by

Chennupati Jagadish

and

Stephen Pearton

2006

ELSEVIER

AMSTERDAM • BOSTON • HEIDELBERG • LONDON • NEW YORK • OXFORD
PARIS • SAN DIEGO • SAN FRANCISCO • SINGAPORE • SYDNEY • TOKYO

Elsevier
The Boulevard, Langford Lane, Kidlington, Oxford OX5 1GB, UK
Radarweg 29, PO Box 211, 1000 AE Amsterdam, The Netherlands

First edition 2006
Reprinted 2007

British Library Cataloguing in Publication Data
A catalogue record for this book is available from the British Library

Library of Congress Cataloging-in-Publication Data
A catalog record for this book is available from the Library of Congress

ISBN–13: 978-0-08-044722-3
ISBN–10: 0-08-044722-8

Printed and bound by CPI Group (UK) Ltd, Croydon, CR0 4YY

Transferred to Digital Print 2011

Contents

Foreword

Since the invention of the first semiconductor transistor in 1947 by the scientists of Bell Labs, the semiconductor industry has grown at an incredible pace, fabricating faster, smaller, more powerful devices while manufacturing in larger volume at lower costs. Even though the very first semiconductor transistor was made from germanium (Ge), silicon (Si) became the semiconductor of choice as a result of the low melting point of Ge that limits high temperature processes and the lack of a natural occurring germanium oxide to prevent the surface from electrical leakage. Due to the maturity of its fabrication technology, silicon continues to dominate the present commercial market in discrete devices and integrated circuits for computing, power switching, data storage and communication. For high-speed and optoelectronic devices such as high-speed integrated circuits and laser diodes, gallium arsenide (GaAs) is the material of choice. It exhibits superior electron transport properties and special optical properties. GaAs has higher carrier mobility and higher effective carrier velocity than Si, which translate to faster devices. GaAs is a direct bandgap semiconductor, whereas Si is indirect, hence making GaAs better suited for optoelectronic devices. However, physical properties required for high power, high temperature electronics and UV/blue light emitter applications are beyond the limits of Si and GaAs. It is essential to investigate alternative materials and their growth and processing techniques in order to achieve these devices. Wide bandgap semiconductors exhibit inherent properties such as larger bandgap, higher electron mobility and higher breakdown field strength. Therefore, they are suitable for high power, high temperature electronic devices and short wavelength optoelectronics.

Zinc oxide is a direct, wide bandgap semiconductor material with many promising properties for blue/UV optoelectronics, transparent electronics, spintronic devices and sensor applications. ZnO has been commonly used in its polycrystalline form for over a hundred years in a wide range of applications: facial powders, ointments, sunscreens, catalysts, lubricant additives, paint pigmentation, piezoelectric transducers, varistors, and as transparent conducting electrodes. Its research interest has waxed and waned as new prospective applications revive interest in the material, but the applications have been limited by the technology available at the time. ZnO has numerous attractive characteristics for electronics and optoelectronics devices. It has direct bandgap energy of 3.37 eV, which makes it transparent in visible light and operates in the UV to blue wavelengths. The exciton binding energy is ~60 meV for ZnO, as compared to GaN ~25 meV; the higher exciton binding energy enhances the luminescence efficiency of light emission. The room temperature electron Hall mobility in single crystal ZnO is ~200 $cm^2 V^{-1}$, slightly lower than that of GaN, but ZnO has higher saturation velocity. ZnO has exhibited better radiation resistance than GaN for possible devices used in space and nuclear applications. ZnO can be grown on inexpensive substrate, such as glass, at relatively low temperatures. Nanostructures, such as nanowires and nanorods, have been demonstrated. These structures are ideal for detection applications due to its large surface area to volume ratio. Recent work on ZnO has shown ferromagnetism in ZnO by doping with transition metal, e.g. Mn, with practical Curie temperatures for spintronic devices. One main attractive feature of ZnO is the ability to bandgap tuning via divalent substitution on the cation site to form heterostructures. Bandgap energy of ~3.0 eV can be achieved by doping with Cd^{2+}, while Mg^{2+} increases the bandgap energy to ~4.0 eV. ZnO has a hexagonal wurtzite crystal structure, with lattice parameters a = 3.25 Å and c = 5.12 Å. The Zn atoms are tetrahedrally coordinated with four O atoms, where the Zn d-electrons hybridize with the O p-electrons. The bonding between the Zn atoms and O atoms are highly ionic, due to the large difference in their electronegative values (1.65 for Zn and 3.44 for O). Alternating Zn and O layers form the crystal structure.

The first utilization of ZnO for its semiconductor properties was detectors in build-your-own radio sets in the 1920s. A thin copper wire, known as "cat's whisker," is placed in contact to sensitive spots on a ZnO crystal. The metal/semiconductor junction allows current to flow only in one direction, converting the incoming radio waves from alternating current to direct current in the radio circuit. In 1957, the New Jersey Zinc Company published a book entitled "Zinc Oxide Rediscovered" to promote the material's "frontier" properties (semiconductor, luminescent, catalytic, ferrite, photoconductive, and photochemical properties) and illustrative applications. Research focused mainly on growth, characterization and applications that do not require single crystals such as varistors, surface acoustic wave devices and transparent conductive films.

Recent improvements in the growth of high quality, single crystalline ZnO in both bulk and epitaxial forms has renewed interest in this material. Originally, research efforts in ZnO growth were intended for gallium nitride (GaN) epitaxy. GaN is another wide, direct bandgap semiconductor that has been the focus of intensive research for high power, high frequency electronics that can operate at elevated temperatures and UV/blue optoelectronics. The lack of a native substrate has led to a search for suitable choices of substrate in other materials, including sapphire, silicon carbide (SiC) and ZnO. The work of Look and his colleagues played a major role in the revival of interest in ZnO research. They also organized the First Zinc Oxide Workshop in 1999 that brought together researchers from all over the world to disseminate their findings and to exchange ideas. Furthermore, Look et al. were the first to publish convincing results of carefully characterized p-type ZnO homoepitaxial film grown by molecular beam epitaxy (MBE), a critical step in achieving p-n junctions for light emitting devices. Subsequent ZnO Workshops, in 2002 and 2004, continue to encourage research efforts on ZnO.

Significant efforts in the last few years have been aimed at controlling conductivity and improving crystal quality. However, in order to fully realize ZnO devices, additional material and process development issues must be overcome. The purpose of this book is to provide an overview of recent progress in ZnO research and identify future areas that need work.

Chennupati Jagadish
Canberra, ACT, Australia

Stephen Pearton
Gainesville, FL, USA

Chapter 1

Basic Properties and Applications of ZnO

V. A. Coleman and C. Jagadish

Department of Electronic Materials Engineering, Research School of Physical Sciences and Engineering, The Australian National University, Canberra, ACT 0200, Australia

1.1 Introduction

Recently, zinc oxide (ZnO) has attracted much attention within the scientific community as a 'future material'. This is however, somewhat of a misnomer, as ZnO has been widely studied since 1935 [1], with much of our current industry and day-to-day lives critically reliant upon this compound. The renewed interest in this material has arisen out of the development of growth technologies for the fabrication of high quality single crystals and epitaxial layers, allowing for the realization of ZnO-based electronic and optoelectronic devices.

With a wide bandgap of 3.4 eV and a large exciton binding energy of 60 meV at room temperature, ZnO, like GaN, will be important for blue and ultra-violet optical devices. ZnO has several advantages over GaN in this application range however, the most important being its larger exciton binding energy and the ability to grow single crystal substrates. Other favorable aspects of ZnO include its broad chemistry leading to many opportunities for wet chemical etching, low power threshold for optical pumping, radiation hardness and biocompatibility. Together, these properties of ZnO make it an ideal candidate for a variety of devices ranging from sensors through to ultra-violet laser diodes and nanotechnology-based devices such as displays.

As fervent research into ZnO continues, difficulties such as the fabrication of *p*-type ZnO that have so far stalled the development of devices are being overcome [2]. We are thus moving ever closer to the future in which ZnO will be a viable and integral part of many functional and exotic devices.

In this chapter, an overview of the basic properties of ZnO, including the crystal structure, energy band structure and thermal properties is presented, as well as an

Zinc Oxide Bulk, Thin Films and Nanostructures
C. Jagadish and S. Pearton (Editors)

introduction to the mechanical properties, basic electronic and optical properties and potential applications of ZnO.

1.2 Crystal structure and lattice parameters

At ambient pressure and temperature, ZnO crystallizes in the wurtzite (B4 type) structure, as shown in figure 1.1. This is a hexagonal lattice, belonging to the space group $P6_3mc$, and is characterized by two interconnecting sublattices of Zn^{2+} and O^{2-}, such that each Zn ion is surrounded by a tetrahedra of O ions, and vice-versa.

This tetrahedral coordination gives rise to polar symmetry along the hexagonal axis. This polarity is responsible for a number of the properties of ZnO, including its piezoelectricity and spontaneous polarization, and is also a key factor in crystal growth, etching and defect generation. The four most common face terminations of wurtzite ZnO are the polar Zn terminated (0001) and O terminated $(000\bar{1})$ faces (*c*-axis oriented), and the non-polar $(11\bar{2}0)$ (*a*-axis) and $(10\bar{1}0)$ faces which both contain an equal number of Zn and O atoms. The polar faces are known to posses different chemical and physical properties, and the O-terminated face possess a

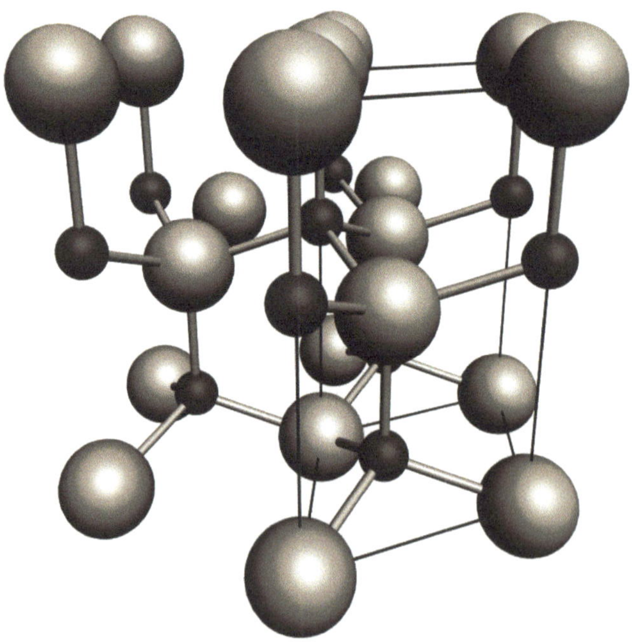

Figure 1.1: The hexagonal wurtzite structure of ZnO. O atoms are shown as large white spheres, Zn atoms as smaller black spheres. One unit cell is outlined for clarity.

slightly different electronic structure to the other three faces [3]. Additionally, the polar surfaces and the (1010) surface are found to be stable, however the $(11\bar{2}0)$ face is less stable and generally has a higher level of surface roughness than its counterparts. The (0001) plane is also basal.

Aside from causing the inherent polarity in the ZnO crystal, the tetrahedral coordination of this compound is also a common indicator of sp^3 covalent bonding. However, the Zn–O bond also possesses very strong ionic character, and thus ZnO lies on the borderline between being classed as a covalent and ionic compound, with an ionicity of $f_i = 0.616$ on the Phillips ionicity scale [4]. The lattice parameters of the hexagonal unit cell are $a = 3.2495$ Å and $c = 5.2069$ Å, and the density is 5.605 g cm^{-3} [5].

In an ideal wurtzite crystal, the axial ratio c/a and the u parameter (which is a measure of the amount by which each atom is displaced with respect to the next along the c-axis) are correlated by the relationship $uc/a = (3/8)^{1/2}$, where $c/a = (8/3)^{1/2}$ and $u = 3/8$ for an ideal crystal. ZnO crystals deviate from this ideal arrangement by changing both of these values. This deviation occurs such that the tetrahedral distances are kept roughly constant in the lattice. Experimentally, for wurtzite ZnO, the real values of u and c/a were determined in the range $u = 0.3817$–0.3856 and $c/a = 1.593$–1.6035 [6–8].

Additional to the wurtzite phase, ZnO is also known to crystallize in the cubic zincblende and rocksalt (NaCl) structures, which are illustrated in figure 1.2.

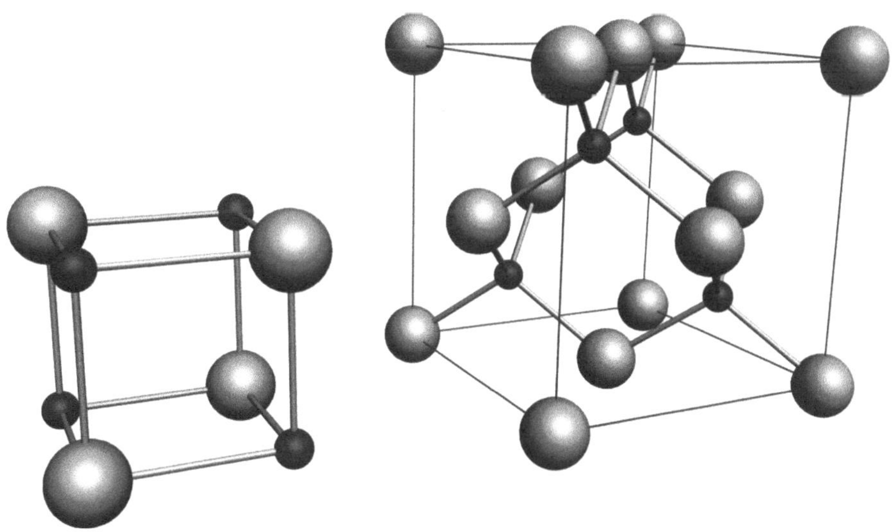

Figure 1.2: The rock salt (left) and zincblende (right) phases of ZnO. O atoms are shown as white spheres, Zn atoms as black spheres. Only one unit cell is illustrated for clarity.

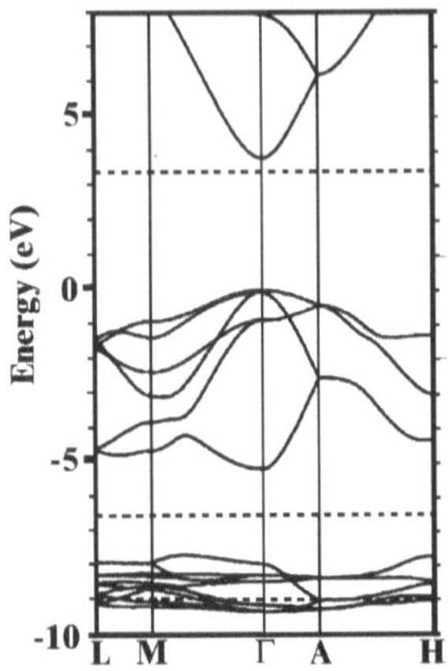

Figure 1.3: The LDA band structure of bulk wurtzite ZnO calculated using dominant atomic self-interaction-corrected pseudopotentials (SIC-PP). This method is much more efficient at treating the d-bands than the standard LDA method. [Reprinted with permission from D. Vogel, P. Krüger and J. Pollmann, Phys. Rev. B **52**, R14316 (1995). Copyright 1995 by the American Physical Society.]

Zincblende ZnO is stable only by growth on cubic structures [9–11], whilst the rocksalt structure is a high-pressure metastable phase forming at ~10 GPa, and can not be epitaxially stabilized [12]. Theoretical calculations indicate that a fourth phase, cubic cesium chloride, may be possible at extremely high temperatures, however, this phase has yet to be experimentally observed [13].

1.3 Energy band gap

The electronic band structure of ZnO has been calculated by a number of groups [13–19]. The results of a band structure calculation using the Local Density Approximation (LDA) and incorporating atomic self-interaction corrected pseudopotentials (SIC-PP) to accurately account for the Zn $3d$ electrons is shown in figure 1.3 [19]. The band structure is shown along high symmetry lines in the hexagonal Brillouin zone. Both the valence band maxima and the lowest conduction band minima occur at the Γ point $k = 0$ indicating that ZnO is a direct band gap semiconductor. The bottom 10 bands (occurring around −9 eV) correspond to Zn $3d$ levels. The next 6 bands from −5 eV to 0 eV correspond to O $2p$ bonding states. The first two conduction band states are strongly Zn localized and correspond to empty Zn $3s$ levels. The higher conduction bands (not illustrated here) are free-electron-like. The O $2s$ bands (also not illustrated here) associated with core-like energy states, occur around −20 eV. The band gap as determined from this calculation is 3.77 eV. This

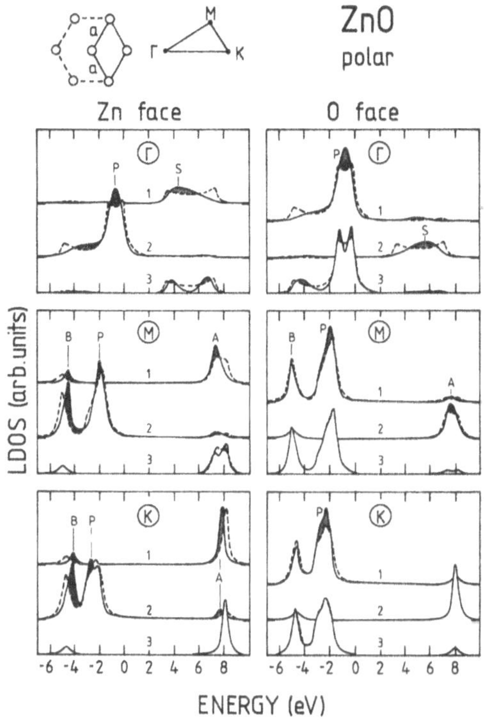

Figure 1.4: Wave-vector-resolved LDOS's on the first three layers of the (0001)-Zn (left panel) and (000$\bar{1}$)-O (right panel) surfaces. The bulk LDOS is given by the dashed lines and surface induced positive changes to the LDOS are shown as hatched. The letters A, B, P and S represent anti-back bonds, back bonds, P resonances and S resonances respectively. [Reprinted with permission from I. Ivanov and J. Pollmann, Phys. Rev. B **24**, 7275 (1981). Copyright 1981 by the American Physical Society.]

correlates reasonably well with the experimental value of 3.4 eV, and is much closer than the value obtained from standard LDA calculations, which tend to underestimate the band gap by ~3 eV due to its failure in accurately modeling the Zn 3*d* electrons.

In addition to calculations for the band structure of bulk ZnO, Ivanov and Pollmann have also carried out an extensive study on the electronic structure of the surfaces of wurtzite ZnO [18]. Using the empirical tight-binding method (ETBM) to determine a Hamiltonian for the bulk states, the scattering theoretical method was applied to determine the nature of the surface states. The calculated data was found to be in very good agreement with experimental data obtained from electron-energy-loss spectroscopy (EELS) and ultra-violet photoelectron spectroscopy (UPS). Figure 1.4 shows the wave-vector-resolved local density of states

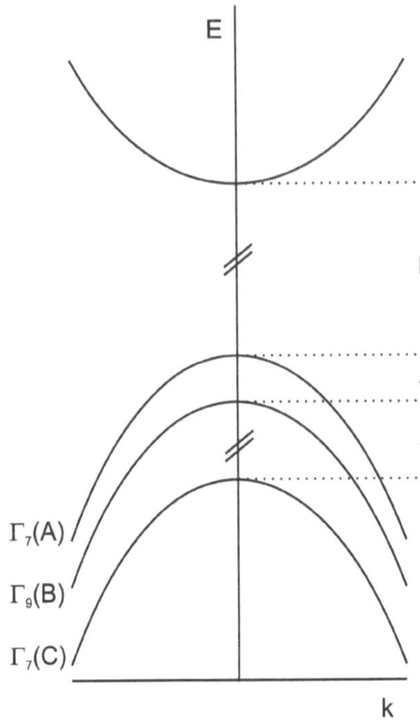

Figure 1.5: Schematic diagram representing the crystal-field and spin-orbit splitting of the valence band of ZnO into 3 subband states A, B and C at 4.2 K.

(LDOSs) on the first three layers of the (0001)-Zn (left panel) and $(000\bar{1})$-O (right panel) surfaces, for the Γ, M and K points of the surface Brillouin zone. The bulk LDOS (calculated using the ETBM) is given by the dashed lines. Surface induced positive changes to the LDOS are shown as hatched. As discussed in section 1.2, the properties of the two polar faces are expected to be different, and this is reflected in this data. Whilst it indicates that no surface states are present in the band gap, the Zn surface shows an increase in back bonds (denoted by B in fig. 1.4) and anti-back bonds (denoted by A) surface states, while the O face simply shows an increase in P resonances and states. This result suggests that the Zn face possesses more covalent character, arising from the Zn 4*s*–O 2*p* states, whilst the O face is more ionic.

Experimentally, the ZnO valence band is split into three band states, A, B and C by spin-orbit and crystal-field splitting. This splitting is schematically illustrated in figure 1.5. The A and C subbands are known to posses Γ_7 symmetry, whilst the middle band, B, has Γ_9 symmetry [20]. The band gap has a temperature dependence

Table 1.1: Key properties of the binary II-VI oxides [35,57].

	ZnO	MgO	CdO
E_g (eV)	3.4	7.8	2.2
m_e^* (m_0)	0.28	0.35	–
m_{hh}^* (m_0)	0.78	1.60 [001], 2.77 [111]	–
m_{lh}^* (m_0)	–	0.35 [001], 0.31 [111]	–
a (Å)	3.2	4.2	4.7
c (Å)	5.2	–	–
Stable crystal structure	Wurtzite	Rocksalt	Rocksalt

up to 300 K given by the relationship:

$$E_g(T) = E_g(T = 0) - \frac{5.05 \times 10^{-4} T^2}{900 - T} \tag{1.1}$$

These properties, combined with the lattice dynamics (discussed in section 1.5) of ZnO give rise to interesting optical properties which will be discussed in section 1.8.

1.3.1 Opportunities for band gap engineering

For a semiconductor to be useful, particularly in reference to optoelectronic devices; band gap engineering is a crucial step in device development. By alloying the starting semiconductor with another material of different band gap, the band gap of the resultant alloy material can be fine tuned, thus affecting the wavelength of exciton emissions. In the case of ZnO, alloying with MgO and CdO is an effective means of increasing or decreasing the energy band gap respectively [21–23]. Some of the key properties of ZnO, MgO and CdO are shown in table 1.1 whilst available relevant parameters for the alloy semiconductors $Zn_{(1-x)}Mg_xO$ and $Zn_{(1-y)}Cd_yO$ are shown in table 1.2.

Currently however, due to the relative newness of the field, only limited experimental and theoretical work has been done for these materials, and thus the information available is both incomplete and not well verified.

1.4 Mechanical properties

Table 1.3 gives a brief overview of the well accepted and experimentally useful parameters describing the mechanical properties of ZnO. As seen from the table, ZnO is a relatively soft material, with a hardness of ~5 GPa at a plastic penetration

Table 1.2: Important parameters for the alloy semiconductors $Zn_{(1-x)}Mg_xO$ and $Zn_{(1-y)}Cd_yO$.

Parameter	$Zn_{(1-x)}Mg_xO$	$Zn_{(1-y)}Cd_yO$
Typical energy range (eV)	3.37–4	2.9–3.37
Maximum percentage alloy incorporation	43% [58]	70% [59]
a-axis length expansion (Å)	$3.250 + 0.036x$ [60]	$3.252 + 0.143y - 0.147y^2$ [61]
c-axis length expansion (Å)	$3.34 - 0.063x$ [60]	$5.204 + 0.956y - 5.42y^2$ [61]
E_g (eV)	$3.37 + 2.51x$ [58]	$3.29 - 4.40y + 5.93y^2$ [61]
$\Delta E_c/\Delta E_v$	70/30 [62]	–

Table 1.3: Key mechanical properties of *c*-axis oriented wurtzite ZnO, as determined by experiment and theory.

Parameter	Experimental	Theoretical
Bulk Young's modulus, E (GPa)	111.2 ± 4.7^a	
Bulk hardness, H (GPa)	5.0 ± 0.1^a	
Epitaxial Young's modulus, E (GPa)	310 ± 40^b	
Epitaxial hardness, H (GPa)	5.75 ± 0.8^b	
Bulk modulus, B (GPa)	142.4^c	156.8^d
dB/dP	3.6^c	3.6^d
e_{33} (C m^{-2})	0.96^e	1.19^f
e_{31} (C m^{-2})	-0.62^e	-0.55^f
e_{15} (C m^{-2})	-0.37^e	-0.46^f
Spontaneous polarization (C m^{-2})		-0.047^g
c_{11} (GPa)	209^h	246^f
c_{33} (GPa)	216^h	246^f
c_{12} (GPa)	120^h	127^f
c_{13} (GPa)	104^h	105^f
c_{44} (GPa)	44^h	56^f
Born effective charge, Z^*		2.1^g

[a] Spherical indentation on bulk ZnO (ref. [24]).
[b] Spherical indentation on epitaxial ZnO (ref. [26]).
[c] See ref. [63].
[d] Ab initio Hartree Fock calculation (ref. [7]).
[e] See refs. [30,31].
[f] Ab initio Hartree Fock calculation (ref. [29]).
[g] Calculation using LDA and Hartree Fock (ref. [28]).
[h] See ref. [64].

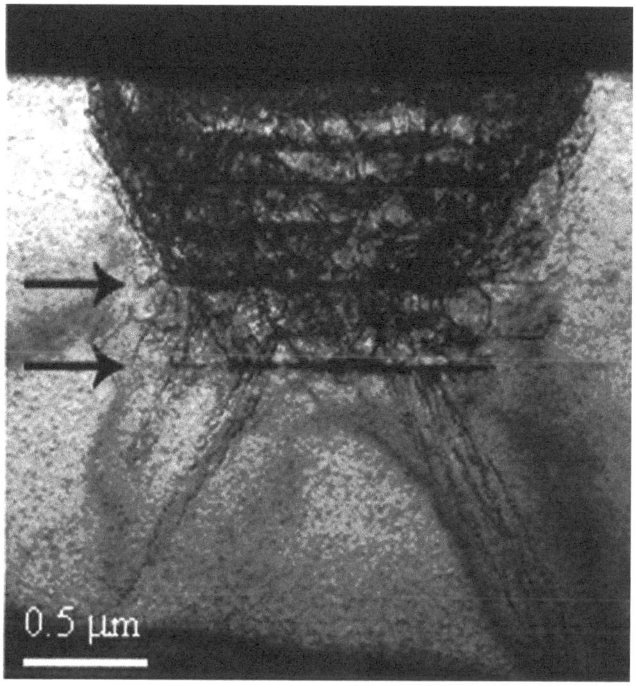

Figure 1.6: A bright-field XTEM image of a spherical indent in bulk, *c*-axis oriented ZnO created using a ~4.2 µm indenter tip. The maximum load is 15 mN. Extensive damage can be seen in the region directly under the indent, extending well beyond the volume under contact. Slip bands occurring along the basal planes can also be clearly seen and are indicated by arrows.

depth of 300 nm (for *c*-axis oriented bulk ZnO) [24]. This needs to be taken into consideration when processing and designing ZnO-based devices. Nanoindentation studies conducted by Bradby et al. [25] of ZnO with a spherical indenter of radius ~4.2 µm have shown that the primary mechanism for deformation in this semiconductor is the nucleation of slip on the basal and pyramidal planes. For loads of up to 50 mN, there has been no observation of phase transformations or cracking in this material. Nanoindentation is a useful technique for probing the mechanical properties of a material, whilst also providing information on the behavior of a material under contact induced damage, such as that experienced during device processing. For ZnO, indentation results in significant quenching of the excitonic luminescence. Additionally, extensive damage is created in the ZnO material, with defects propagating far beyond the volume under contact [25]. Figure 1.6 shows a bright field cross sectional transmission electron microscopy image of indented ZnO. From this image, the extent to which damage propagates well beyond the volume under

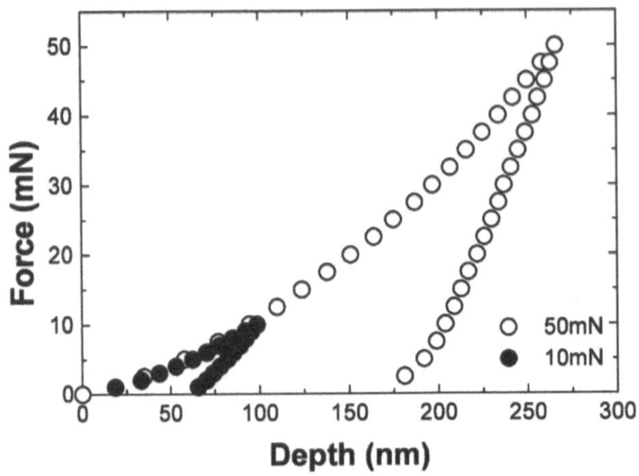

Figure 1.7: Load–unload curves of ~500 nm thick *c*-axis oriented ZnO grown on sapphire. The indents were made using a ~4.2 μm spherical indenter. Closed circles indicate a load of 10 mN whilst open circles are for a 50 mN load. No 'pop-in' events are seen, suggesting that slip along the basal planes has been suppressed in this system.

contact can be clearly seen. Slip bands along the basal planes which give rise to so called 'pop-in' events during indentation [25] loading can also be clearly seen.

There is also some indication that the crystal orientation of ZnO influences the mechanical properties due to the orientation of the basal planes [26,27]. A-axis oriented bulk ZnO is significantly softer than *c*-axis material, with a hardness of ~2 GPa at a plastic penetration depth of 50 nm below contact. In *a*-axis material, the basal planes lie perpendicular to the surface and are thus more susceptible to slip. This is reflected in the occurrence of only one large pop-in event for indentation in this material [26]. As with *c*-axis ZnO, *a*-axis material does not show any evidence of phase transformation or cracking under indentation, however the excitonic luminescence is quenched and very large propagation of defects is seen [27].

In addition to bulk material, epitaxial ZnO will be important for ZnO devices, and thus it is important to understand differences in the mechanical properties of this system. Studies show that epitaxial ZnO grown on sapphire is slightly harder than its bulk counterpart, with a hardness of ~5.7 GPa for *c*-axis epitaxial layers [26]. This increase in hardness is due to the presence of the underlying layer which inhibits the slip mechanism along the basal planes. This is evidenced by the absence of 'pop-in' events during indentation loading in epitaxial material. This is illustrated in figure 1.7 which shows an indentation load–unload curve made in a ~500 nm thick *c*-axis oriented ZnO layer grown on sapphire. As-grown threading dislocations in epitaxial material also increase its hardness by inhibiting via strain compensation [26].

Table 1.4: Experimentally determined principal phonon modes of wurtzite ZnO at 300 K [20].

Phonon mode	Value (cm^{-1})
E_2^{low}	101
E_2^{high}	437
TO (A_1)	380
LO (A_1)	574
TO (E_1)	591

Piezoelectricity is also an important mechanical property. ZnO is believed to have a piezoelectric tensor equal to or even greater than that of GaN and AlN which means that ZnO is a suitable candidate for device applications requiring a large electromechanical coupling [28]. Many studies, both theoretical and experimental, have been made to determine the three piezoelectric stress coefficients, e_{ik} for wurtzite ZnO [28–33], and a selection of these values are also listed in table 1.3, along with the elastic constants c_{hk}.

1.5 Lattice dynamics

In single crystal wurtzite ZnO, there are 4 atoms per unit cell, giving rise to 12 phonon modes. These modes are important for understanding the thermal, electrical and optical properties of the crystal, and are as follows: one longitudinal-acoustic (LA), two transverse-acoustic (TA), three longitudinal-optical (LO) and six transverse-optical (TO) branches. The A_1 and E_1 branches are Raman and infra-red active, while the two E_2 branches (non-polar) are only Raman active. The E_2^{low} mode is associated with the vibrations of the Zn sub-lattice, whilst the E_2^{high} mode is associated with the oxygen atoms only. The B_1 branches are always inactive. The phonon modes of ZnO have been extensively studied and modelled (see ref. [34] and references therein). Table 1.4 gives a list of the experimental values for the most common phonon modes visible at 300 K [20].

1.6 Thermal properties

In this section, the thermal properties of wurtzite ZnO will be presented, including the thermal expansion coefficients, thermal conductivity and specific heat.

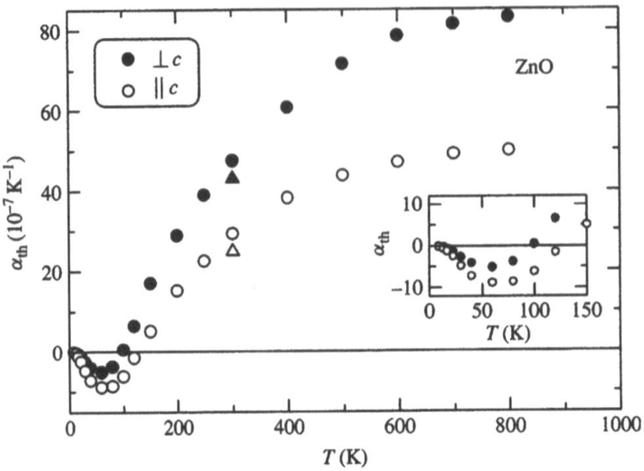

Figure 1.8: Graph of the ZnO thermal expansion coefficient α_{th} as a function of temperature. [Reprinted with permission from H. Ibach, Phys. Stat. Sol. (b) **33**, 257 (1969). Copyright 1969 by WILEY-VCH.]

1.6.1 Thermal expansion coefficients

The thermal expansion coefficient of a material describes lattice deformation as a function of temperature. For ZnO, these coefficients are given as $\alpha_a = 4.31 \times 10^{-6}\,\mathrm{K}^{-1}$ and $\alpha_c = 2.49 \times 10^{-6}\,\mathrm{K}^{-1}$ at 300 K [35]. Additionally, figure 1.8 shows a plot of the ZnO thermal expansion coefficient, α_{th}, as a function of temperature.

1.6.2 Thermal conductivity

The thermal conductivity, κ (W cm^{-1} K^{-1}) of a semiconductor is an important property when considering high-power/high temperature devices. It is a kinetic property influenced by the vibrational, rotational and electronic degrees of freedom, and is predominately limited by phonon-phonon scattering in a pure crystal. ZnO, like most other semiconductors, contains a large number of point defects, and these have a significant effect on the thermal conductivity. The highest measured values of thermal conductivity come from a study done on vapour-phase grown samples which measured the conductivity on the polar faces of ZnO [36]. This study gives the values of $\kappa = 1.02 \pm 0.07$ and 1.16 ± 0.08 W cm^{-1} K^{-1} from the Zn face of two different samples, and $\kappa = 1.10 \pm 0.09$ and 0.98 ± 0.08 W cm^{-1} K^{-1} from the O

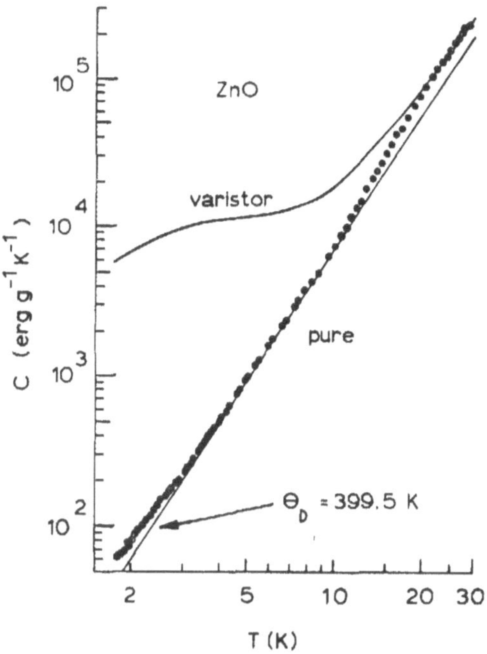

Figure 1.9: Specific heat data for pure (bulk) and varistor ZnO measured between 1.7 and 25 K. [Reprinted with permission from W. N. Lawless and T. K. Gupta, J. Appl. Phys. **60**, 607 (1986). Copyright 1986, American Institute of Physics.]

face of the same two samples. These values are considerably higher than other values measured from ZnO which typically fall in the range $\kappa = 0.6$ 1 W cm^{-1} K^{-1} [34].

1.6.3 Specific heat

The specific heat of a material is influenced by the lattice vibrations, free carriers and defects within the material. In crystals of high quality and purity, the specific heat is mainly influenced by the lattice vibrations. Unfortunately, there is very limited data available in the literature for specific heat measurements on ZnO. The Handbook of Chemistry and Physics ([5]) gives a value of the specific heat capacity of ZnO at constant pressure as $C_p = 40.3$ J mol^{-1} K^{-1}. A study by Lawless and Gupta [37] measures the specific heat for pure and varistor ZnO between 1.7 and 25 K. The data obtained in this study is shown in figure 1.9. From this figure, it can be seen that the specific heat of pure ZnO diverges from that of the varistor below 20 K as a result of impurities and defects incorporated in the varistor grain boundaries.

Additionally, the specific heat data presented here for ZnO is unlikely to be accurate for modern ZnO crystals, as dramatic improvements in growth technology have taken place since this study was undertaken, which in turn will influence the specific heat measurements. However, at the time of publication, no further specific heat measurements on ZnO single crystals could be found in the literature.

1.7 Electrical properties

The electrical properties of ZnO are hard to quantify due to large variance of the quality of samples available. The background carrier concentration varies a lot according to the quality of the layers but is usually $\sim 10^{16}$ cm^{-3}. The largest reported n-type doping is $\sim 10^{20}$ electrons cm^{-3} and largest reported p-type doping is $\sim 10^{19}$ holes cm^{-3}, however such high levels of p-conductivity are questionable and have not been experimentally verified [38]. The exciton binding energy is 60 meV at 300 K, and is one of the reasons why ZnO is so attractive for optoelectronic device applications. As mentioned in section 1.3.1, the electron effective mass is $0.24m_0$, and the hole effective mass is $0.59m_0$. The corresponding electron Hall mobility at 300 K for low n-type conductivity is $\mu = 200\,\text{cm}^2\,\text{V}^{-1}\,\text{s}^{-1}$, and for low p-type conductivity is 5–50 cm^2 V^{-1} s^{-1} [39].

1.8 Optical properties

As mentioned in section 1.3, the optical properties of ZnO are heavily influenced by the energy band structure and lattice dynamics. For a comprehensive review of the optical properties of excitonic recombinations in bulk, n-type ZnO, please refer to the work of B. K. Meyer et al. [20]. This work gives a comprehensive treatment and analysis of the excitonic spectra obtained from ZnO, and assigns many defect related spectral features, as well as donor–acceptor pair (DAP) emission. A broad defect related peak extending from ~1.9 to ~2.8 eV is also a common optical feature of ZnO. Known as the green band, the origin of its luminescence is still not well understood and has in the past been attributed to a variety of different impurities and defects. Figure 1.10 shows a typical photoluminescence spectra of n-type ZnO measured at 4.2 K. The excitonic, DAP and extended green band emission can all be clearly seen, as can the phonon replicas produced from the longitudinal optical phonons (LO). Due to the lack of available data on p-type ZnO, a corresponding spectrum is not shown here.

In terms of the more fundamental optical properties of ZnO, there have been a number of comprehensive studies to determine the refractive index and dielectric

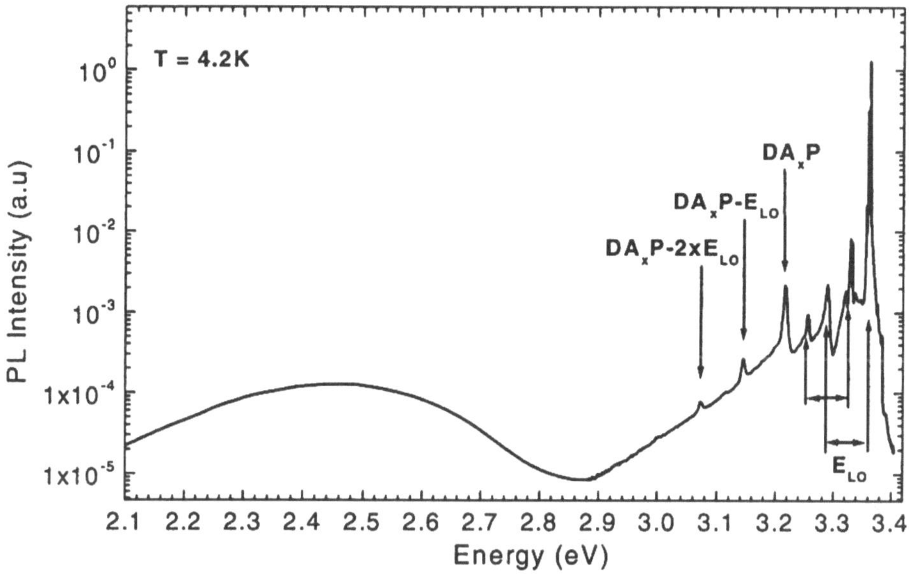

Figure 1.10: Photoluminescence spectrum of n-type bulk ZnO (HeCd excitation) showing excitonic, donor acceptor pair and green-band emission. The longitudinal optical phonons with the corresponding phonon replicas are indicated on the figure. [Reprinted with permission from B. K. Meyer, H. Alves, D. M. Hofmann, W. Kriegseis, D. Forster, F. Bertram, J. Christen, A. Hoffmann, M. Straßburg, M. Dworzak, U. Haboeck and A. V. Rodina, Phys. Stat. Sol. (b) **241**, 231 (2004). Copyright 2004 by WILEY-VCH.]

Table 1.5: Static (ϵ_0) and high frequency dielectric constant (ϵ_∞) data for ZnO [40,41]

		Film [41]	Bulk [41]	Bulk [40]
ϵ_0	$E \perp c$	7.46	7.77	
	$E \parallel c$	8.59	8.91	
ϵ_∞	$E \perp c$	3.7	3.6	3.68
	$E \parallel c$	3.78	3.66	3.72

constants of this material [40–42]. The measurements were all carried out using spectroscopic ellipsometry. The values determined for the dielectric constants of ZnO are shown in table 1.5 whilst the refractive index dispersion for both $E \perp c$ and $E \parallel c$ as measured and calculated by Yoshikawa and Adachi [40] is shown in figure 1.11. The refractive index of wurtzite ZnO is commonly given as $n_\omega = 2.008$ and $n_e = 2.029$ [39].

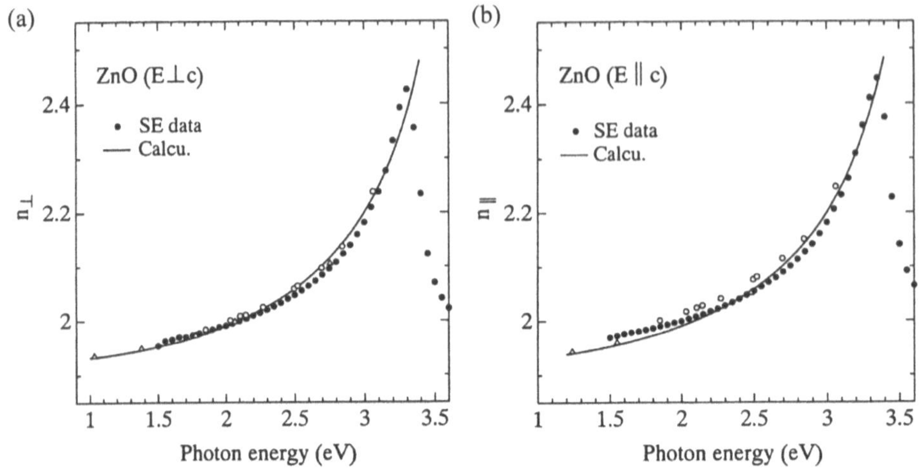

Figure 1.11: Refractive index dispersion of ZnO for E ⊥ *c* and E ∥ *c* below the fundamental absorption edge. The solid circles represent the spectroscopic ellipsometry data whilst the solid line is calculated data. [Reprinted with permission from H. Yoshikawa and S. Adachi, Jpn. J. Appl. Phys. **36**, 6237 (1997). Copyright 2004 by Jpn. J. Appl. Phys.]

1.9 Issues

The main issue currently limiting the production of ZnO-based devices is that of the achievement of *p*-type ZnO. This has been the case for most wide-band gap semiconductors, including GaN. ZnO is predicted to be an intrinsic semiconductor, however it almost exclusively occurs naturally as *n*-type. The cause of this inherent doping, combined with the difficulty in achieving *p*-type material is still not directly understood. Native defects such as Zn interstitials and O vacancies [43] are thought to compensate the donors and give rise to the native conductivity. Background impurities such as H introduced during growth could also play a role, as could other deep impurity levels [44]. Finally, it has been suggested that low solubility of dopants within this material could also be responsible [45]. Recent improvements in the as-grown quality of ZnO material as well as successes with dopant atoms indicate that *p*-type doping of ZnO is an achievable goal.

Another issue important for the realization of ZnO devices is that of contacts to ZnO. While Ohmic contacts to *n*-type ZnO are relatively easily obtained, the production of good reliable Schottky diodes still remains an issue [39]. A recent report by Grossner et al. [46] showed that Pd contacts on clean ZnO surfaces produced Schottky behavior with a barrier height of 0.83 eV, which is in good agreement with the value for the barrier height of Pd on ZnO predicted by theory.

Whilst other issues within ZnO exist, these two are by far the most important and currently play a limiting role in the realization of ZnO devices.

1.10 Applications

In chapters 10–16, some of the principal applications for ZnO will be discussed in detail, however a brief overview of these applications is provided here. As mentioned previously, ZnO is already widely used in our society, and indeed it is a key element in many industrial manufacturing processes including paints, cosmetics, pharmaceuticals, plastics, batteries, electrical equipment, rubber, soap, textiles, floor coverings to name just a few. With improvements in growth technology of ZnO nanostructures, epitaxial layers, single crystals and nanoparticles, we are now moving into an era where ZnO devices will become increasingly functional and exotic.

ZnO-based nanostructures including nanowire arrays hold a host of opportunities for flat screen displays, field emission sources, gas, chemical [47] and biological sensors, and as UV light emitters and switches [47–50].

Epitaxial layers and single crystals will be important for the development of optoelectronic (blue and ultraviolet light emitters and detectors) [51], piezoelectric [52] and spintronic [53] devices, and together with GaN may form the light source of the 21st century [54]. Epitaxial ZnO also holds much promise as a semi-conducting transparent thin film [55], which again will be important for solar cells, gas sensors, displays and wavelength selective applications.

Existing technologies are also being revolutionized with ZnO nanoparticles, which have led to the development of improved sunscreens, paints and coatings to name just a few.

Additionally, the radiation hardness of ZnO to MeV proton irradiation makes it an ideal candidate for space applications [56].

Thus ZnO whilst already possessing a wide application base, has enormous opportunities for society and industry alike due to its unique properties which are now being explored and applied. The future in which ZnO devices become part of our everyday lives is already approaching reality.

Acknowledgments

We would like to thank Dr. Jodie Bradby for helpful discussions, Mr. Aert van de Hulsbeek for his assistance with the figures presented in this chapter, and the Australian Research Council for financial support.

References

[1] C. W. Bunn, Proc. Phys. Soc. London 47 (1935) 835.

[2] A. Tsukazaki, A. Ohtomo, T. Onuma, M. Ohtani, T. Makino, M. Smiya, K. Ohtani, S. F. Chichibu, S. Fuke, Y. Segawa, H. Ohno, K. Koinuma, M. Kawasaki, Nat. Mater. 4 (2005) 42.

[3] O. Dulub, L. A. Boatner, U. Diebold, Surf. Sci. 519 (2002) 201.

[4] J. C. Phillips, Bonds and Bands in Semiconductors, Academic, New York, 1973.

[5] D. R. Lide (Ed.), CRC Handbook of Chemistry and Physics, 73rd Edition, CRC Press, New York, 1992.

[6] E. H. Kisi, M. M. Elcombe, Acta Cryst. C45 (1989) 1867.

[7] J. E. Jaffe, A. C. Hess, Phys. Rev. B 48 (1993) 7903.

[8] L. Gerward, J. S. Olsen, J. Synchrotron Radiat. 2 (1995) 233.

[9] T. Kogure, Y. Bando, J. Electron Microsc. 47 (1993) 7903.

[10] A. B. M. A. Ashrafi, A. Ueta, A. Avramescu, H. Kumano, I. Suemune, Y. W. Ok, T. Y. Seong, Appl. Phys. Lett. 76 (2000) 550.

[11] S. K. Kim, S. Y. Seong, C. R. Cho, Appl. Phys. Lett. 82 (2003) 562.

[12] C. H. Bates, W. B. White, R. Roy, Science 137 (1962) 993.

[13] J. E. Jaffe, J. A. Snyder, Z. Lin, A. C. Hess, Phys. Rev. B 62 (2000) 1660.

[14] J. R. Chelikowsky, Solid State Commun. 22 (1977) 351.

[15] U. Rossler, Phys. Rev. 184 (1969) 733.

[16] S. Bloom, I. Ortenburger, Phys. Stat. Sol. (b) 58 (1973) 561.

[17] M. Usuda, N. Hamada, T. Kotani, M. van Schilfgaarde, Phys. Rev. B 66 (2002) 125101.

[18] I. Ivanov, J. Pollmann, Phys. Rev. B 24 (1981) 7275.

[19] D. Vogel, P. Krüger, J. Pollmann, Phys. Rev. B 52 (1995) R14316.

[20] B. K. Meyer, H. Alves, D. M. Hofmann, W. Kriegseis, D. Forster, F. Bertram, J. Christen, A. Hoffmann, M. Straßburg, M. Dworzak, U. Haboeck, A. V. Rodina, Phys. Stat. Sol. (b) 241 (2004) 231.

[21] T. Makino, C. H. Chia, N. T. Tuan, H. D. Sun, Y. Segawa, M. Kawasaki, A. Ohtomo, K. Tamura, H. Koinuma, Appl. Phys. Lett. 77 (2000) 975.

[22] K. Ogata, K. Koike, T. Tanite, T. Komuro, F. Yan, S. Sasa, M. Inoue, M. Yano, J. Cryst. Growth 251 (2003) 623.

[23] T. Makino, Y. Segawa, M. Kawasaki, A. Ohtomo, R. Shiroki, K. Tamura, T. Yasuda, H. Koinuma, Appl. Phys. Lett. 78 (9) (2001) 1237.

[24] S. O. Kucheyev, J. E. Bradby, J. S. Williams, C. Jagadish, M. V. Swain, Appl. Phys. Lett. 80 (2002) 956.

[25] J. E. Bradby, S. O. Kucheyev, J. S. Williams, C. Jagadish, M. V. Swain, P. Munroe, M. R. Philips, Appl. Phys. Lett. 80 (2002) 4537.

[26] V. A. Coleman, J. E. Bradby, C. Jagadish, P. Munroe, Y. W. Heo, S. J. Pearton, D. P. Norton, M. Inoue, M. Yano, Appl. Phys. Lett. 86 (2005) 203105.

[27] V. A. Coleman, J. E. Bradby, C. Jagadish, P. Munroe, M. R. Phillips (*unpublished*).

[28] A. D. Corso, M. Posternak, R. Resta, A. Baldereschi, Phys. Rev. B 50 (1994) 10715.

[29] M. Catti, Y. Noel, R. Dovesi, J. Phys. Chem. Solids 64 (2003) 2183.

[30] Landolt-Börnstein, O. Madelung (Eds.), New Series, Group III: Solid State Physics, Low Frequency Properties of Dielectric Crystals: Piezoelectric, Pyroelectric and Related Constants, vol. 29b, Springer, Berlin, 1993.
[31] E. E. Tokarev, I. B. Kobyakov, I. P. Kuz'mina, A. N. Lobachev, G. S. Pavo, Sov. Phys. Soild State 17 (1975) 629.
[32] A. R. Hutson, Phys. Rev. Lett. 4 (1960) 505.
[33] Y. Noel, M. Llunell, R. Orlando, P. D'Arco, R. Dovesi, Phys. Rev. B 66 (2002) 214107.
[34] U. Özgür, Y. I. Alivov, C. Liu, A. Teke, M. A. Reshchikov, S. Doŭan, V. Avrutin, S. J. Cho, H. Morkoç, J. Appl. Phys. 98 (2005) 041301.
[35] S. Adachi, Properties of Group-IV, III-V and II-VI Semiconductors, John Wiley and Sons, Ltd, West Sussex, England, 2005.
[36] D. I. Florescu, L. G. Mourokh, F. H. Pollak, D. C. Look, G. Cantwell, X. Li, J. Appl. Phys. 91 (2002) 890.
[37] W. N. Lawless, T. K. Gupta, J. Appl. Phys. 60 (1986) 607.
[38] D. C. Look, B. Clafin, Y. I. Alivov, S. J. Park, Phys. Stat. Sol. (a) 201 (2004) 2203.
[39] S. J. Pearton, D. P. Norton, K. Ip, Y. W. Heo, T. Steiner, Prog. Mater. Sci. 50 (2005) 293.
[40] H. Yoshikawa, S. Adachi, Jpn. J. Appl. Phys. 36 (1997) 6237.
[41] N. Ashkenov, B. M. Mbenkum, C. Bundesmann, V. Riede, M. Lorenz, D. Spemann, E. M. Kaidashev, A. Kasic, M. Schubert, M. Grundmann, G. Wanger, H. Neumann, V. Darakchieva, H. Arwin, B. Monemar, J. Appl. Phys. 93 (2003) 126.
[42] X. W. Sun, H. S. Kwok, J. Appl. Phys. 86 (1999) 408.
[43] W. Walukiewicz, Phys. Rev. B 50 (1994) 5221.
[44] C. G. van de Walle, Phys. Rev. Lett. 85 (2000) 1012.
[45] C. G. van de Walle, D. B. Laks, G. F. Neumark, S. T. Pantelides, Phys. Rev. B 47 (1993) 9425.
[46] U. Grossner, S. Gabrielsen, T. M. Borseth, J. Grillenberger, A. Y. Kuznetsov, B. G. Svensson, Appl. Phys. Lett. 85 (2004) 2259.
[47] L. Liao, J. C. Li, D. F. Wang, C. Liu, C. S. Liu, Q. Fu, L. X. Fan, Nanotech. 16 (2005) 985.
[48] Y. Xia, P. Yang, Y. Sun, Y. Wu, B. Mayers, B. Gates, Y. Yin, F. Kim, H. Yan, Adv. Mater. 15 (2003) 353.
[49] Z. L. Wang, J. Phys. Cond. Matter 16 (2004) R829.
[50] D. J. Sirbuly, M. Law, H. Q. Yan, P. D. Yang, J. Phys. Chem. B 109 (2005) 15190.
[51] D. C. Look, Mater. Sci. Eng. (b) 80 (2001) 383.
[52] J. G. E. Gardeniers, Z. M. Rittersma, G. J. Burger, J. Appl. Phys. 83 (1998) 7844.
[53] D. P. Norton, S. J. Pearton, A. F. Hebard, N. Theodoropoulou, L. A. Boatner, R. G. Wilson, Appl. Phys. Lett. 82 (2003) 239.
[54] J. Nause, Comp. Semicond. 11 (2005) 29.
[55] H. Hartnagel, A. L. Dawar, A. K. Jain, C. Jagadish, Semiconducting transparent thin films, Institute of Physics Publishing, Bristol and Philadelphia, 1995.
[56] D. C. Look, D. C. Reynolds, J. W. Hemsky, R. L. Jones, J. R. Sizelove, Appl. Phys. Lett. 75 (1999) 811.

[57] T. Makino, Y. Segawa, M. Kawasaki, H. Koinuma, Semicond. Sci. Technol. 20 (2005) S78.
[58] K. Koike, K. Hama, I. Nakashima, G. Takada, K. Ogata, S. Sasa, M. Inoue, M. Yano, J. Cryst. Growth 278 (2005) 288.
[59] S. Shigemori, A. Nakamura, T. Aoki, J. Temmyo, Jpn. J. Appl. Phys. 43 (2004) L1088.
[60] A. Ohtomo, M. Kawasaki, T. Koida, K. Masubuchi, H. Koinuma, Y. Sakurai, Y. Yoshida, T. Yasuda, Y. Segawa, Appl. Phys. Lett. 72 (1998) 2466.
[61] T. Makino, Y. Segawa, M. Kawasaki, A. Ohtomo, R. Shiroki, K. Tamura, T. Yasuda, H. Koinuma, Appl. Phys. Lett. 78 (2001) 1237.
[62] G. Coli, K. K. Bajaj, Appl. Phys. Lett. 78 (2001) 2861.
[63] S. Desgreniers, Phys. Rev. B 58 (1996) 11425.
[64] Landolt-Börnstein, O. Madelung (Eds.), New Series, Group III: Solid State Physics, Low Frequency Properties of Dielectric Crystals: Elastic Constants, vol. 29a, Springer, Berlin, 1993.

Chapter 2

Doping and Defects in ZnO

David C. Look

Semiconductor Research Center, Wright State University, Dayton, OH 45435, USA;
Materials and Manufacturing Directorate, Air Force Research Laboratory,
Wright-Patterson Air Force Base, OH 45433, USA

Abstract: ZnO is a wide bandgap semiconductor material with numerous present applications, such as varistors and surface acoustic wave devices, and future applications, including UV light-emitting diodes and transparent field-effect transistors. However, all of these applications are either dependent upon, or are affected by, impurities and defects. We will consider donor-type impurities H, Al, Ga, and In; and acceptor-type impurities N, P, As, and Sb. Among defects, we will concentrate on Zn interstitials, Zn vacancies, O vacancies, and complexes of each. The main experimental techniques discussed here include temperature-dependent Hall-effect and low-temperature photoluminescence measurements, because they alone can provide donor and acceptor concentrations, and donor energies. Surface conduction is important in the Hall-effect analysis of annealed samples, and is included by means of a two-layer analysis. The important topic of *p*-type ZnO is considered in some detail, especially in connection with the most useful acceptor dopants.

2.1 Introduction

The last decade has witnessed an enormous growth of ZnO related research, largely because of the possibilities of new or improved types of electronic and photonic devices [1–3]. One device that has great commercial potential is a UV light-emitting diode (LED), which could be combined with phosphors to produce solid-state white lighting [4–6]. Another is a transparent field-effect transistor, which could serve as an active element in large-area displays [7–9]. In each of these cases, ZnO possesses fundamental properties that give it advantages over competitive materials, such as GaN. For example, the free exciton is ZnO has a binding energy of about 60 meV, whereas that of GaN is about 24 meV. Thus, efficient excitonic emission processes can persist in ZnO at room temperature and higher, and in fact, prototype ZnO-based LEDs have been shown to operate at nearly 400°C. ZnO also has many other advantages, including the availability of large-area native substrates, amenability

Zinc Oxide Bulk, Thin Films and Nanostructures
C. Jagadish and S. Pearton (Editors)

to low-temperature growth and wet chemical etching, and high radiation resistance. But in regard to LED development, ZnO also has one major disadvantage, namely, the lack of a reliable technology for producing *p*-type material. Fortunately, in recent years, this situation has improved, and now several groups have reported *p*-type ZnO [6,10–43], and a few have even produced light-emitting p-n junctions [4–6,44]. (For a review of ZnO-based LEDs, 2001–2003, see Ref. 45.) However, the optical and electrical properties of ZnO are still not well understood, and this means that the roles of various impurities and defects in ZnO are not well understood, because as is well known, impurities and defects generally control such properties. Thus, in this chapter, we will attempt to summarize our present knowledge, both experimental and theoretical, of the roles of defects and impurities in controlling the optical and electrical properties of ZnO. Indeed, our understanding of some of these matters has undergone major changes in the last few years.

It is useful to summarize some of these changes. As far as we know, as-grown ZnO has always been found to be *n*-type, and temperature-dependent Hall-effect (T-Hall) measurements have usually yielded donor energies of 30 to 70 meV [46]. Since most samples are grown under Zn-rich conditions, it was always natural in the past to assume that the dominant donor was either the O vacancy V_O, or the Zn interstitial Zn_I. In the year 2000, this conclusion was strongly challenged by Kohan et al. [47] who showed theoretically that both V_O and Zn_I have high formation energies in *n*-type ZnO, and that furthermore, they are deep, not shallow, donors. Thus, it was concluded that neither V_O nor Zn_I would exist in measurable quantities, and that even if one or the other were present, its ionization energy would be too high to produce free electrons. Other theoretical analyses have concluded that Zn_I is actually a shallow donor, rather than deep [48,49], as has also been shown by electron-irradiation experiments [50]; however, the high formation energy of Zn_I mentioned earlier would still limit its ability to influence the conductivity of *n*-type material. Also in the year 2000, the assignments of V_O and Zn_I as the dominant donors in as-grown ZnO was further challenged by Van de Walle's theoretical result that H is *always* a donor in ZnO, that it is easily ionized, and that it has a low enough formation energy to be abundant; thus, Van de Walle suggested that it was likely to be a dominant background donor in ZnO materials that were exposed to H during growth [51]. This proposal has been subjected to testing, because H-containing, high-quality, bulk ZnO, grown by a seeded chemical vapor transport (SCVT) technique, has been commercially available for the last few years [46]. In general, these tests have confirmed that a shallow donor due to H exists in SCVT ZnO [52–57] and can contribute significantly to its conductivity. This fact, coupled with the theoretical evidence of high formation energies for the native donors has led to a prevailing opinion that native donors do not play a significant role in the conductivity of as-grown ZnO. However, we will challenge this conclusion

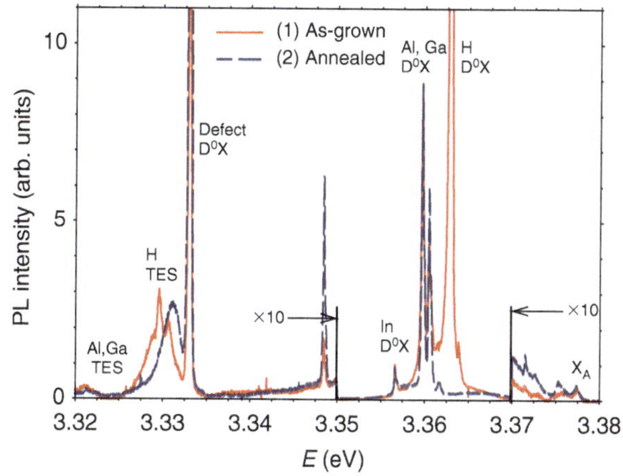

Figure 2.1: Photoluminscence at 4 K for a SCVT-grown ZnO sample, #1: (1) as-grown; and (2) annealed at 715°C for $\frac{1}{2}$ hr in flowing N_2 gas.

and show that native donors, especially Zn_I, do indeed contribute significantly to conduction in ZnO, but as complexes, rather than isolated elements. It should be pointed out that there is really no conflict with the theoretical results of Kohan et al. and Van de Walle, because those theories deal with isolated defects, not complexes.

Besides H, several other donor impurities are known to be important in ZnO. In particular, the Group III elements Al, Ga, and In, substitute easily for Zn, and each can be incorporated to very high concentrations, $>10^{20}$ cm^{-3} [58,59]. In fact, photoluminescence (PL) spectral features assigned to Al, Ga, or In, are seen in almost every ZnO sample (cf. Fig. 2.1), although of course, PL is not quantitative. The Group VII elements, in particular F and Cl, also are known to have donor activity in ZnO [60]. However, there is not much electrical or optical evidence that they are significant *background* dopants, so we will not discuss them in detail here. It is also possible that the Group IV elements, such as C, Si, and Ge could be electrically active as either donors or acceptors. However, there is no strong evidence that any of these elements forms a shallow donor or acceptor [61], so we will also not consider them here.

Acceptor dopants are of great interest in ZnO, because the realization of high-conductivity *p*-type ZnO could lead to a viable p-n-junction UV LED, which could have an enormous impact on the solid-state white-lighting industry. In this regard, the Group I elements, substituting for Zn, and Group V elements, substituting for O, all must be considered. Interestingly, theory predicts shallow acceptor levels for both Li_{Zn} and Na_{Zn} [62,63], but neither dopant produces high-conductivity, *p*-type material [64]. The reason, also suggested by theory, is that Li and Na can be incorporated either as donors or as acceptors. In contrast, theory predicts relatively deep levels for P_O, As_O, and Sb_O [62], but each of these dopants has been used to produce *good* *p*-type ZnO [12,24,32,65]. Again, theory offers an explanation [66], which we will discuss. Among defect-type acceptors, the Zn vacancy is definitely present in much of the ZnO being grown today, and in fact is the *dominant* acceptor in some cases [67].

Up to now, all of the defects mentioned, such as V_{Zn}, V_O, or Zn_I-X, are quite small and qualify as point defects. However, there are also much larger defects that are present and important. For example, many ZnO samples contain line defects, such as threading dislocations; and surface defects, such as stacking faults. We will not consider such extended defects here, because their optical and electrical properties are not well known as yet. Much more information on dislocations and stacking faults is available in the GaN literature, and it is likely that their respective behaviors in ZnO will be similar.

The main experimental techniques discussed here will be T-Hall and low-temperature PL measurements. The reasons are: (1) these are the main characterization methods that have been applied to ZnO in the author's laboratory; and (2) they are the only common techniques that can deliver quantitative information about donor and acceptor concentrations and energies. Many of the impurities and defects mentioned above have had "fingerprints" established, or at least suggested, by T-Hall and PL measurements. Other characterization techniques, such as transmission electron microscopy (TEM), x-ray diffraction (XPD), secondary-ion mass spectroscopy (SIMS), and optical absorption spectroscopy (OAS) have also contributed much to our understanding of ZnO, and will be mentioned here as needed. However, detailed discussions of these and other techniques will have to be sought in other chapters in this volume or in the literature.

The establishment of T-Hall and PL fingerprints of impurity donors and acceptors has benefited from doping experiments, and we will refer to some of those reported in the literature. However, concentrations of impurities can often be changed by high-temperature annealing, and we will employ that technique here. For *defect-related* donors and acceptors, high-energy electron irradiation is by far the best way to introduce such species, and we also make extensive use of this technique. Thus, our experimental program here involves irradiation and annealing steps, and T-Hall and PL measurements after each step.

2.2 Samples and apparatus

The samples discussed here are $5 \times 5 \times 0.5$-mm squares cut from larger wafers that had been sliced from (0001) boules grown by the seeded chemical vapor transport (SCVT) method at ZN Technology [68]. The SCVT technique produces very high quality material. Ohmic In dots were soldered on the corners of the samples, and van der Pauw Hall-effect measurements were performed with a LakeShore Model 7507 apparatus, including a closed-cycle He cooling system operating from 15 to 320 K. From measurements of Hall coefficient R and conductivity σ, the Hall mobility $\mu_H = R\sigma$ and the Hall concentration $n_H = 1/eR$ could be calculated at each temperature. The true carrier concentration n is related to n_H by $n = rn_H$, where r is the so-called Hall r-factor [46,69]. The r-factor was calculated at each temperature, but typically was so close to unity that it was ignored in the plots. Thus, we present only n_H and μ_H data in the plots.

Photoluminescence measurements were performed at 4.2 K. Excitation, dispersion, and detection were accomplished, respectively, with a 45-mW HeCd laser, a 1.25-m spectrometer, and a photomultiplier detector. Resolution was better than 0.01 meV in the spectral range important for this study.

Electron irradiations were carried out at room temperature with a 2-MeV van de Graaff accelerator. A 1.0-MeV beam, of current about 2 μA/cm^2, was directed onto the (0001) Zn-face. Calculations show that a 1-MeV electron beam in the [0001] direction should produce O and Zn displacements at rates of about 0.2 and 0.3 cm^{-1}, respectively [70]. Thus, e.g., a fluence of 3×10^{17} cm^{-2} would be expected to produce about 6×10^{16} cm^{-3} O displacements, and 9×10^{16} cm^{-3} Zn displacements.

2.3 Hall-effect theory

The basic equations for Hall-effect analysis, allowing for the energy dependence of the electrons, are as follows [69]:

$$j_x = \frac{ne^2\langle\tau\rangle}{m^*}E_x \equiv -ne\mu_c E_x \tag{2.1}$$

$$R_H = \frac{E_y}{j_x B} = -\frac{1}{ne}\frac{\langle\tau^2\rangle}{\langle\tau\rangle^2} = -\frac{r}{ne} \tag{2.2}$$

where electric E and magnetic B fields are imposed in the x and z directions, respectively, and

$$\langle\tau^n(E)\rangle = \frac{\int_0^\infty \tau^n(E)E^{3/2}\frac{\partial f_0}{\partial E}\,dE}{\int_0^\infty E^{3/2}\frac{\partial f_0}{\partial E}\,dE} \rightarrow \frac{\int_0^\infty \tau^n(E)E^{3/2}e^{-E/kT}\,dE}{\int_0^\infty E^{3/2}e^{-E/kT}\,dE} \tag{2.3}$$

where the integration is over energy E. (Although these equations are written for electron current, they hold equally well for hole current with the simple transformation $n \rightarrow p$ and $e \rightarrow -e$.) This formulation is called the relaxation-time approximation (RTA) to the Boltzmann Transport Equation (BTE). Here f_0 is the Fermi-Dirac distribution function and the second equality in Eqn. 2.3 holds for non-degenerate electrons, i.e., those describable by Boltzmann statistics. The relaxation time, $\tau(E)$, depends upon how the electrons interact with the lattice vibrations as well as with extrinsic elements, such as charged impurities and defects. For example, acoustical-mode lattice vibrations scatter electrons through the deformation potential ($\tau_{\rm ac}$) and piezoelectric potential ($\tau_{\rm pe}$); optical-mode vibrations through the polar potential ($\tau_{\rm po}$); ionized impurities and defects through the screened coulomb potential ($\tau_{\rm ii}$); and charged dislocations, also through the coulomb potential ($\tau_{\rm dis}$). The strengths of these various scattering mechanisms depend upon certain lattice parameters, such as dielectric constants and deformation potentials, and extrinsic factors, such as donor, acceptor, and dislocation concentrations, N_D, N_A, and $N_{\rm dis}$, respectively. The total momentum scattering rate, or inverse relaxation time, is

$$\tau^{-1}(E) = \tau_{\rm ac}^{-1}(E) + \tau_{\rm pe}^{-1}(E) + \tau_{\rm po}^{-1}(E) + \tau_{\rm ii}^{-1}(E) + \tau_{\rm dis}^{-1}(E) \tag{2.4}$$

and this expression is then used to determine $\langle \tau^n(E) \rangle$ via Eqn. 2.3, and thence, $\mu_H = e\langle \tau^2 \rangle / m^* \langle \tau \rangle$. Formulas for $\tau_{\rm ac}$, $\tau_{\rm pe}$, $\tau_{\rm po}$, $\tau_{\rm ii}$, and $\tau_{\rm dis}$, can be found in the literature [69,71]. Note that μ_H, the Hall mobility, is related to μ_c, the conductivity mobility (Eqn. 2.1), by $\mu_H = r\mu_c$.

The fitting of μ_H vs. T data, described above, should be carried out in conjunction with the fitting of n vs. T, which is derived from the charge-balance equation (CBE):

$$n + N_A = \sum_i \frac{N_{Di}}{1 + n/\phi_{Di}} \tag{2.5}$$

where the subscript i denotes a particular donor and where

$$\phi_{Di} = \frac{g_{0i}}{g_{1i}} e^{\alpha_{Di}/k} N_C' T^{3/2} e^{-E_{D0i}/kT} \tag{2.6}$$

For each donor, g_0/g_1 is a degeneracy factor, $N_C' = 2(2\pi m_n * k)^{3/2}/h^3$ is the effective conduction-band density of states at 1 K, h is Planck's constant, E_D is the donor energy, and E_{D0} and α_D are defined by $E_D = E_{D0} - \alpha_D T$. Eqn. 2.5 describes the simplest type of charge balance, in which each of the one or more donors has only one charge-state transition within a few kT of the Fermi energy. An example of such a donor is Ga on a Zn site in ZnO. If there are double or triple donors, or more than one acceptor, proper variations of Eqn. 2.5 can be found in the literature [69].

For a p-type sample, the simple CBE becomes

$$p + N_D = \sum_i \frac{N_{Ai}}{1 + n/\phi_{Ai}} \tag{2.7}$$

where $N_V' = 2(2\pi m_p * k)^{3/2}/h^3$, $\phi_A = (g_1/g_0)N_V' \exp(\alpha_A/k)T^{3/2} \exp(-E_{A0}/kT)$, and $E_A = E_{A0} - \alpha_A T$, for each acceptor. Note that ϕ_A has the term "g_1/g_0", which is inverted from the first term in Eqn. 2.6 because the degeneracies still refer to electrons, not holes. From Eqn. 2.7, it can be shown that the hole concentration in a nondegenerate, single-donor/single-acceptor model, will be given by [69]

$$p = \frac{1}{2}(\phi_A + N_D)\left\{\left[1 + \frac{4\phi_A(N_A - N_D)}{(\phi_A + N_D)^2}\right]^{1/2} - 1\right\} \tag{2.8}$$

where the various symbols were defined earlier.

2.4 Multi-layer analysis

There is extensive recent evidence that the ZnO near-surface region can be highly conductive, possibly due to H donors in this region [72–74]. As we will show later, the samples discussed in this work can be reasonably modeled as having two layers: (1) a bulk layer, comprising most of the sample; and (2) a surface layer, much thinner than the bulk layer, but containing a high sheet concentration of electrons, high enough that they are degenerate (temperature-independent). Under these conditions, the total mobility and carrier concentration can be written as [74]:

$$\mu = \frac{n_b\mu_b^2 + n_s\mu_s^2}{n_b\mu_b + n_s\mu_s} \tag{2.9}$$

$$n = \frac{(n_b\mu_b + n_s\mu_s)^2}{n_b\mu_b^2 + n_s\mu_s^2} \tag{2.10}$$

Here b denotes a bulk quantity, and s, a surface quantity. However, it must be noted that we have written n_s as a volume concentration, when in reality we can measure only the sheet concentration n_s. The *true* volume concentration of the near-surface electrons would be n_s/d_s, but we don't know d_s either, in general. Thus, to simplify matters, we normalize n_s to the bulk thickness d_b; i.e., $n_s = n_s/d_b$. In this way, we can plot n_b and n_s on the same axis; however, it must be remembered that the true volume concentration of the near-surface electrons is $n_s d_b/d_s$, a much higher quantity than the plotted n_s values presented later.

2.5 Hydrogen and Group I impurities

It has long been known that H incorporates as a donor in ZnO, even in *n*-type ZnO [75,76]. This fact is somewhat surprising, because H is amphoteric in most other semiconductor materials and thus should incorporate mainly as an acceptor in *n*-type ZnO. Van de Walle explained this paradox by showing theoretically that H^- never has a lower formation energy than H^0 or H^+, so that the donor state is always the preferred state in thermodynamic equilibrium [51]. Subsequent experiments have confirmed this assertion [52–57], and in some cases H is even the dominant donor. For example, in Fig. 2.1, the strongest photoluminescence (PL) line is one at 3.36270 eV, arising from an exciton bound to interstitial H^0 [77]. (We will designate such a complex particle as a neutral donor-bound exciton, D^0X.) Note also in Fig. 2.1, and even Fig. 2.2, this line virtually disappears during a $\frac{1}{2}$-hour anneal at 715°C in N_2 gas, because a large share of the H is known to leave the sample at this temperature [55,78]. We obtain further information about the H donor from its two-electron satellite (TES) line at 3.32961 eV. A TES line occurs when the donor is left in an excited ($n = 2$) state upon the collapse of the exciton. From a hydrogenic model, the donor ground state ($n = 1$) energy can then be obtained from $E_D = 4/3(E_{D^0X} - E_{TES})$, giving 44 meV for the H donor energy.

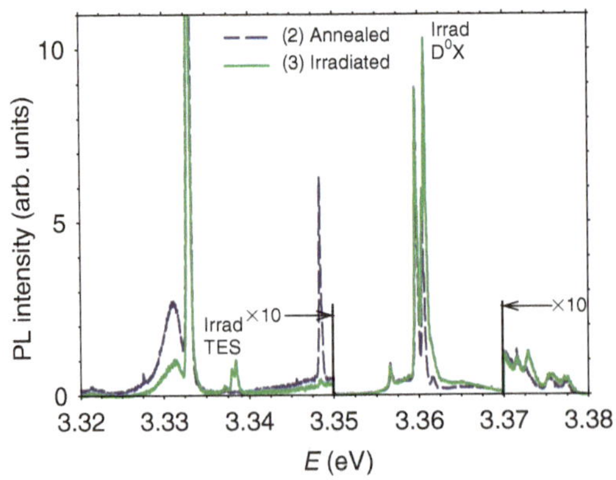

Figure 2.2: Photoluminscence at 4 K for sample #1: (2) annealed at 715°C for $\frac{1}{2}$ hr in flowing N_2 gas; and (3) irradiated with 1-MeV electrons to a fluence of 3×10^{17} cm^{-2}.

To determine the concentration of the H donor, we turn to T-Hall measurements, presented in Figs 2.3–2.6. The solid lines in these figures are theoretical fits using a three-donor model (cf. Eqn. 2.5, above), with the energies of the three donors chosen as 30, 44, and 75 meV. Although we cannot claim complete uniqueness for this choice of energies, they are justified from the following considerations: (1) they produce excellent fits of n vs. T, in most cases; and (2) the 30 and 44 meV values are computed from PL TES lines. The 75 meV choice does not have an obvious

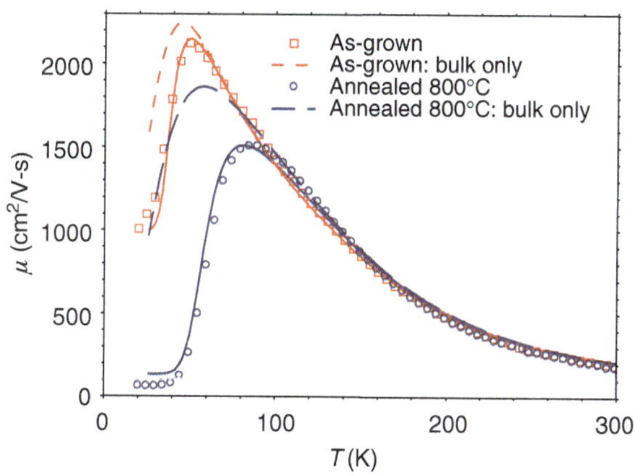

Figure 2.3: Mobility vs. temperature for a SCVT-grown ZnO sample, #2, as-grown, and annealed at 800°C for $\frac{1}{2}$ hr in flowing N_2 gas. The solid lines are theoretical fits to the data, and the dashed lines are the *predicted* mobilities in the absence of surface conduction.

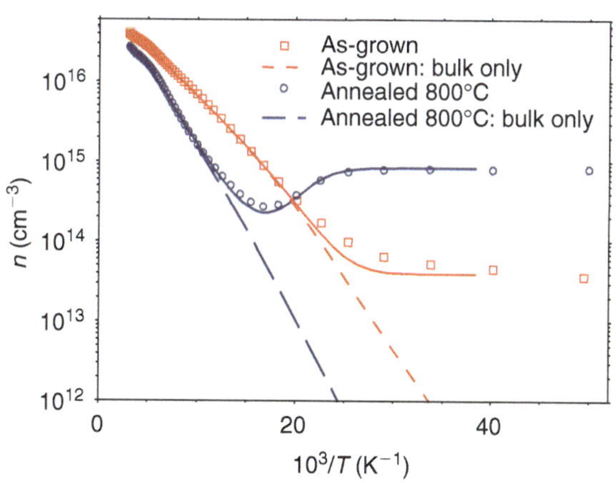

Figure 2.4: Carrier concentration vs. inverse temperature for #2, as-grown, and annealed at 800°C. The solid lines are theoretical fits to the data, and the dashed lines are the *predicted* carrier concentrations in the absence of surface conduction.

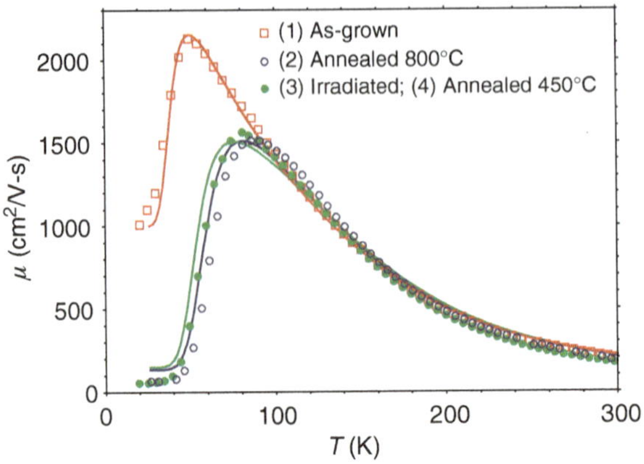

Figure 2.5: Mobility vs. temperature for sample #2: (1) as-grown; (2) annealed at 800°C; (3) irradiated with 1 MeV electrons; and (4) annealed at 450°C. The solid lines are theoretical fits to the data.

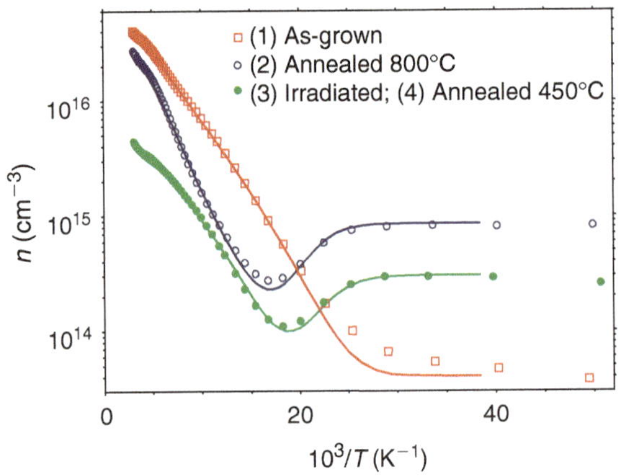

Figure 2.6: Carrier concentration vs. inverse temperature for sample #2: (1) as-grown; (2) annealed at 800°C; (3) irradiated with 1 MeV electrons; and (4) annealed at 450°C. The solid lines are theoretical fits to the data.

TES origin; instead, there are lines near 3.32 eV, which have been assigned to the TES lines of Al and Ga [77], and which would then give about 55 meV for $E_{D,Al}$ and $E_{D,Ga}$. Unfortunately, the fits seem to be better with 75 than 55 meV, so there is some uncertainty connected with the origin of the 75 meV donor. However, most of our evidence suggests that it is indeed associated with Al and/or Ga.

Before going on, we note that the solid lines in Figs 2.3 and 2.4 are the fits to the actual experimental data, including the contributions of surface conduction, represented in Eqns 2.9 and 2.10 by parameters μ_s and n_s. On the other hand, the

Table 2.1: Donors and acceptor concentrations in sample #1 after various treatments. All anneal times were $\frac{1}{2}$ hr. The 400°C anneal was preceeded by anneals at 250, 300, and 350°C. The 1-MeV-irradiation fluence was 3×10^{17} cm^{-2}. The concentration unit is 10^{16} cm^{-3}.

Treatment	N_{D1} (30 meV)	N_{D2} (44 meV)	N_{D3} (75 meV)	N_A
As-grown	0.45	3.0	2.0	0.13
715°C anneal	0.74	0.1	3.2	0.7
1-MeV irrad.	~20	0.1	0.13	~20
400°C anneal	1.35	0.1	0.4	1.2

Table 2.2: Donors and acceptor concentrations in sample #2 after various treatments. The 800°C anneal was preceeded by anneals at 500, 600, and 700°C. The 450°C anneal was preceeded by anneals at 300, 350, and 400°C. The 1-MeV-irradiation fluence was 3×10^{17} cm^{-2}. The concentration unit is 10^{16} cm^{-3}.

Treatment	N_{D1} (30 meV)	N_{D2} (44 meV)	N_{D3} (75 meV)	N_A
As-grown	0.35	1.9	3.1	0.13
800°C anneal	0.16	0.1	2.7	0.19
1-MeV irrad.	~30		0.1	~30
450°C anneal	0.26	0.1	0.27	0.27
500°C anneal	0.22	0.1	0.70	0.23

dashed lines are the predictions of what the solid curves would look like in the *absence* of surface conduction, i.e., with n_s set to 0. It is clear that the surface conduction produces large changes in the mobility and carrier concentration curves at low temperatures. This phenomenon has been investigated in detail only recently, and probably arises from H adsorbing on the surface from the ambient, or from H diffusing to the surface from the interior of the sample [74]. In any case, it is easy to include the effects of surface conduction in the analysis (Eqns 2.9 and 2.10), because the values of μ_s and n_s come directly from the low-temperature μ and n data, respectively.

Returning to the 44-meV donor concentration, the T-Hall fitting parameters are summarized in Tables 2.1 and 2.2. Here it is seen that this concentration drops precipitously as a result of 800 or 715°C anneals, and the same happens to the 3.36270-eV D^0X and 3.32961-eV TES lines. Coupled with the studies of H diffusion and effusion as a function of temperature, it is entirely reasonable to assign these T-Hall and PL fingerprints to H.

Another valence-1 element of high importance is Li. Although theory predicts that Li on the Ga site Li_{Ga} should be a relatively shallow acceptor [62,63], it turns out that doping with Li almost always produces semi-insulating (SI) material [17,64]. The reason evidently involves the fact that the Li interstitial Li_I has a lower formation energy than Li_{Zn} in *p*-type ZnO, and Li_I is a donor. Thus, a sample with high concentrations of Li would attain a balance between Li_{Ga} and Li_I and the Fermi level would end up somewhere close to midgap, producing SI material. It should be noted that Li also produces a deep state in the gap, as determined by EPR analysis [79], but this state may arise from a complex [63]. The same considerations evidently hold for Na, also, although not nearly as much experimental work has been carried out for Na.

2.6 Group II elements

The group II elements Mg and Cd form solid solutions with Zn, and give a potential range of bandgaps from about 2 eV to 8 eV. Heterostructures of ZnO with MgZnO and CdZnO have already been demonstrated, and will undoubtedly be important in future optical and electronic devices based on ZnO. In fact, this subject is emphasized in Chapter 3, and so will not be discussed further here.

2.7 Group III elements

The Group III elements Al, Ga, and In, are well-known donor dopants in ZnO, and each can produce carrier concentrations in the $>10^{20}$ cm^{-3} range [58,59]. In fact, these types of ZnO produce some of the highest-conductivity transparent material available today. However, as *background* impurities, their roles are somewhat of a mystery. From doping experiments, the D^0X lines at 3.36042, 3.35964, and 3.35664 have been assigned to Ga_{Zn}, Al_{Zn}, and In_{Zn}, respectively, as seen in Fig. 2.1 [77]. Indeed, as noted in Tables 2.1 and 2.2, they are the dominant D^0X lines in sample #1, and have intensities second only to that of H in the *unannealed* version of sample #2. However, analytical measurements, such as secondary-ion mass spectroscopy (SIMS), and glow-discharge mass spectroscopy (GDMS), often give much lower concentrations of these elements than that of the 75-meV donor found from T-Hall measurements [80]. In fact, after annealing, the 75-meV donor is always dominant, so either the 75-meV donor is not related to Group-III elements, or the analytical measurements are inaccurate. This latter solution is most likely, because concentrations in the low-10^{16}-cm^{-3} range are not easy to measure by either SIMS or GDMS. More studies on this matter will have to be carried out, but our belief is that the 75 meV donor is indeed associated with Group-III impurities.

2.8 Group IV elements

Very little is known of the electrical and optical activity of the Group-IV elements, such as C and Si. It is difficult to get accurate concentrations of C from mass spectroscopic techniques because of CO in the spectrometer background and also because of surface contamination from polishing and etching processes. According to density functional theory (DFT), the split-interstitial complex $(CO)_O$, should be a donor [81]; however, its energy is questionable because of the crude DFT treatment of electron correlation, known as the local density approximation (LDA). If N is also present, then the complex $(NC)_O$ is expected to be a shallower donor. With respect to Si, this element is often present at the 1×10^{16}-cm^{-3} level in SCVT-grown ZnO, according to GDMS analysis [80]; however, there is no indication that Si produces shallow levels in the bandgap.

2.9 Group V elements

This group, comprising N, P, As, and Sb, are extremely important for the realization of *p*-type ZnO. A glance at the periodic table would suggest that N would make a good acceptor dopant, because it has an electronic core structure and ionic radius similar to those of O and so should readily substitute for O. Indeed, this is the case, with concentrations of $10^{20}\,cm^{-3}$ having been measured in many N-doped ZnO samples. In our SCVT samples, SIMS measurements show an N concentration of about $10^{17}\,cm^{-3}$ [17], and EPR measurements find that it it primarily goes onto the O site, as an acceptor [82,83]. Interestingly, however, the total, *active* acceptor concentration in as-grown SCVT samples is only in the low 10^{15}-cm^{-3} range [46], leading to a possible discrepancy. Evidently, either (1) the SIMS measurements are wrong, or (2) most of the N is passivated. Indeed, there is direct evidence that at least some of the N in SCVT ZnO is passivated with H, because H-N vibrational modes have been seen in the IR absorption spectra; furthermore, other evidence suggests that most of the H in hydrothermally-grown ZnO is associated with the passivation of acceptors (such as N_O), rather than the creation of shallow donors [55,84,85].

Many different growth techniques have been used to produce N-doped ZnO, including chemical vapour deposition (CVD), pulsed laser deposition (PLD), molecular-beam epitaxy (MBE), metal-organic chemical vapour deposition

(MOCVD), and rf or dc sputtering [86]. Recently, a MBE scheme using temperature (T) modulation has been employed to produce good *p*-type material, which has been used to fabricate homoepitaxial light emitting diodes (LEDs) [6]. In this T-modulation method, a thin layer of N-rich ZnO is grown at low temperature, so that the N will incorporate more readily, and then a thin *undoped* layer is grown at higher temperatures, so that the crystal quality will be better. This process is repeated until the full layer is grown. There are reports that this growth method is being commercialised in Japan, and only time will tell if it is the preferred way to produce UV LEDs.

Several other attempts to dope ZnO with N have also produced *p*-type material, but in some cases the *p*-type nature was weak, and seemed to be unstable and susceptible to type-change under light. In this connection, great care has to be exercised when measuring the present generation of *p*-type ZnO samples [86]. The hole mobilities are small, so that the Hall voltages tend to be small; also, the contacts in *p*-type ZnO can be noisy, sometimes leading to "*n*-type" spikes, as the contact configurations are switched during the van der Pauw analysis. Also, even room light can sometimes produce persistent electron-dominated photoconductivity, and render the sample "*n*-type", because the electron mobility is so much larger than the hole mobility. These effects are a nuisance, but they are not the same as an actual *lattice* instability, in which the sample *permanently* goes from *p*-type to *n*-type in time. This latter phenomenon is very rare, in our experience, although it may occur in some cases [42].

Conventional wisdom suggests that if it is difficult to make *p*-type ZnO by N doping, then it should be nearly impossible to do so by P, As, or Sb doping. However, the exact opposite is true; that is, good *p*-type material has been made using all three dopants [12,24,32,65]. The conventional wisdom is based on acceptor energy and solubility, which are functions of ionic size. Of the Group V elements, N_O should have the smallest acceptor transition energy and highest solubility, and P_O, As_O, and Sb_O, should be deeper and more difficult to incorporate. To explain the experimental disagreement with this scenario, Limpijumnong et al. examined what would happen if As actually went on the Zn site, where it would fit better, rather than the O site. Using DFT, they indeed found that a complex, As_{Zn}-$2V_{Zn}$, has a rather low formation energy (high solubility), and a relatively shallow acceptor transition energy [66]. Two recent studies are consistent with this finding: (1) *implanted* As tends mainly to go on the Zn site [87]; and (2) As-doped ZnO contains high quantities of Zn vacancies [88]. These results don't prove the As_{Zn}-$2V_{Zn}$ conjecture, but do lend credence to it.

Photoluminescence and T-Hall results on *p*-type ZnO layers are presented in Figs 2.7 and 2.8, respectively. In Fig. 2.7, a N-doped layer grown by MBE, and P- and As-doped layers grown by rf sputtering, are compared with a bulk, *n*-type SCVT

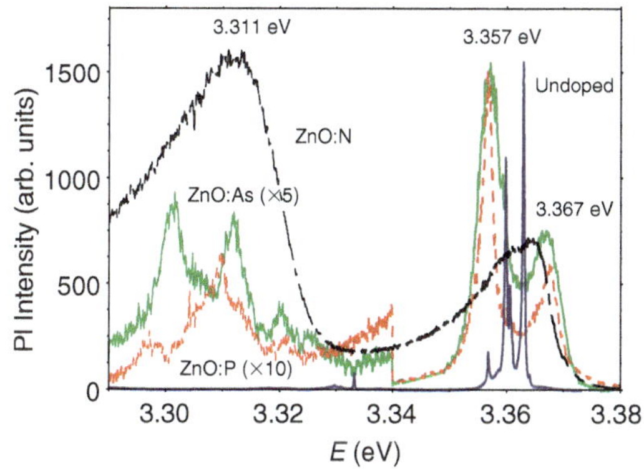

Figure 2.7: A comparison of photoluminscence at 4 K for a SCVT-grown, undoped, *n*-type ZnO sample; a MBE-grown, N-doped, *p*-type ZnO layer; a rf-sputtered, P-doped, *p*-type ZnO layer; and a rf-sputtered, As-doped, *p*-type ZnO layer.

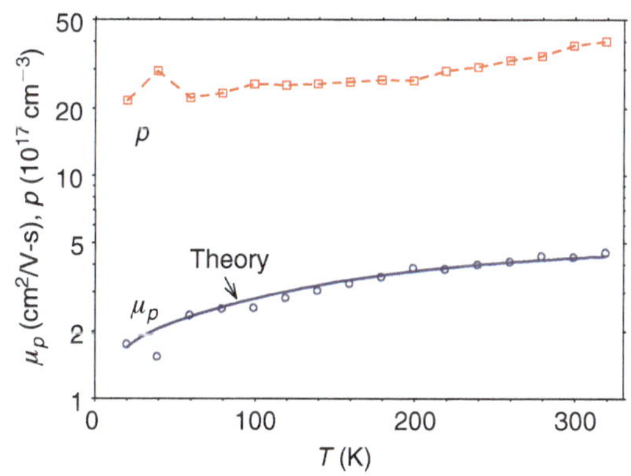

Figure 2.8: Mobility and carrier concentration vs. temperature for a rf-sputtered, As-doped, *p*-type ZnO layer.

sample. It is seen that PL lines at 3.31, 3.357, and 3.367 eV are relatively strong in all of the *p*-type materials. The 3.31-eV line has been variously conjectured to arise from acceptor-bound excitons, donor-acceptor pairs, or isolated acceptors ($e^- + A^0 \rightarrow A^-$), but at this point, none of these models is certain. The 3.357-eV line is often attributed to In, but another, broader line often appears in the same vicinity after 800°C annealing [89]. The 3.367-eV line has been attributed to an *ionized-donor*-bound exciton [77], which indeed might be reasonable in *p*-type material. But at this stage, none of these lines has been unambiguously identified.

The T-Hall results in Fig. 2.8 involve the As-doped sample shown in Fig. 2.7 [32]. The 300 K mobility is about 5 cm^2/V-s, a good value for present-day *p*-type ZnO, and close to what is also usually found in p-type GaN. A simple, one-band fit to the hole-mobility data gives donor and acceptor concentrations in the high $10^{19}\,cm^{-3}$ range, close to the As concentration found by SIMS. These results are consistent with the p-type nature being due to As acceptors, whether As_O, or As_{Zn}-$2V_{Zn}$. Of course, some possible secondary phases, such as As metal or Zn_3As_2, can also be p-type, and must be considered in the analysis of a sample such as this one. Neither As metal nor Zn_3As_2 can explain the data of Fig. 2.8, but potential secondary phases must always be in mind when doping with P, As, or Sb.

2.10 Group VI elements

Although some of the Zn-VI ternary compounds, such as ZnS_xSe_{1-x}, have been studied in great detail, only a few reports exist on compounds of the type $ZnVI_xO_{1-x}$ [90]. Thus, we will not review mixtures of ZnO with Group VI elements.

2.11 Group VII elements

The elements F, Cl, Br, and I should substitute for O and act as donors in ZnO. Indeed, studies have found that F doping greatly decreases the resistivity in ZnO films grown by chemical spray techniques [60,91]. However, the Group III elements are preferred as donor dopants, and they also seem to be more prevalent as background impurities than the Group VI elements. In any case, there is little recent literature on Group-VII doping, and we will not consider this class of elements further.

2.12 Donor-type point defects

As mentioned earlier, the prevalent n-type nature of ZnO has usually, in the past, been attributed to native defect donors, either the Zn interstitial Zn_I or the O vacancy V_O. Nowadays it is known that H, along with the Group III elements Al and Ga, are also background donors that can contribute significantly to the conductivity of as-grown ZnO, and we have discussed their optical and electrical fingerprints above. To study the point-defect-related donors, the best way is to create them by high-energy electron irradiation [50,92], and in Fig. 2.2 we present PL data on sample #1, grown by the SCVT method and irradiated with 1 MeV electrons to a fluence of $3 \times 10^{17}\,cm^{-2}$. A new, sharp D^0X line pops up at 3.3607 eV, and it also has TES lines in the region of 3.338 eV. Thus, the defect donor energy is approximately $E_D = 4/3(E_{D^0X} - E_{TES}) = 30\,meV$. In a separate work, we argue that these lines are associated with a Zn_I complex, probably Zn_I-N_O [93]. The changes in mobility and carrier concentration due to irradiation are shown in Figs 2.5 and 2.6, respectively, and the fits to these curves give the donor and acceptor concentrations shown in Tables 2.1 and 2.2. The main results of irradiation are an increase in the 30-meV donor, and decrease in the 75-meV donor. We believe that the increase in the 30-meV donor is due to the creation of Zn interstitials, and the decrease of the 75-meV donor may be due to the complexing of O interstitials (acceptor-like defects) with Group-III donors. However, this model will have to be tested further.

The O vacancy produces a deep donor state, according to theory [47–49]. In agreement, EPR experiments have seen deep states associated with V_O [94]. On the other hand, positron annihilation spectroscopy (PAS) suggests that a shallow (~100 meV) V_O-related state may also exist [95]. Thus, more research on this matter is needed.

Besides Zn_I and V_O, the antisite Zn_O should also be a donor, although there is some disagreement among theoreticians on whether it is shallow or deep [47–49]. In any case, right now there is little strong evidence for any optical or electrical activity caused by Zn_O.

2.13 Acceptor-type point defects

According to theory, the O interstitial O_I and Zn vacancy V_{Zn} should behave as acceptors [47–49], and V_{Zn}, at least, should be prevalent in *n*-type ZnO, especially in material created under O-rich conditions. Experimentally, little is known about O_I, which can exist in both tetrahedral and octahedral positions. However, much is known about V_{Zn}, because as a negatively charged vacancy, it can easily trap positrons. Here again, electron irradiation has been used to create a large concentration of V_{Zn} and thereby facilitate the establishment of its PAS fingerprints. Then, by comparing the PAS signals with acceptor concentrations, as determined by T-Hall measurements, it can be shown that the small acceptor concentration of $1–2 \times 10^{15}\,cm^{-3}$ in as-grown SCVT ZnO (cf. Tables 2.1 and 2.2) can be entirely explained by V_{Zn}-related defects [67]. As mentioned before, there are also N_O-related acceptors in SCVT ZnO, but they are likely neutralized by H unless the sample has been subjected to annealing or irradiation [55,84,85].

2.14 Summary

Temperature-dependent Hall-effect (T-Hall) and low-temperature photoluminescence (PL) measurements have been carried out on as-grown, annealed, and electron-irradiated ZnO samples, and have yielded the energies and concentrations of several donors and acceptors. Donors which have been identified by T-Hall, PL, and other techniques, include interstitial H, substitutional Al and Ga, a Zn-interstitial complex, and the O vacancy. For the acceptors, substitutional N, P, As, and Sb, and the Zn vacancy seem to be the most important species. Although ZnO research and technology has expanded greatly in the last few years, a major problem is still the development of a reliable means of producing high-conductivity *p*-type material. A critical aspect of this quest is to understand how P, As, and Sb form acceptors in ZnO, and, in particular, which lattice site each occupies. Future research will undoubtedly address these and other issues.

Acknowledgments

We wish to thank T.A. Cooper and W. Rice for technical assistance, and B.B. Claflin for helpful discussions. Support was provided by US Air Force Contract F33615-00-C-5402, and subcontracts with Structured Materials Industries, Inc., and SVTA, Inc. Much of the work was performed at the Air Force Research Laboratory, Wright-Patterson AFB, Ohio.

References

[1] D.C. Look, Mater. Sci. Eng. B **80**, 383 (2001).
[2] S.J. Pearton, D.P. Norton, K. Ip, Y.W. Heo, and T. Steiner, Prog. Mater. Sci. **50**, 293 (2005).
[3] G-C. Yi, C. Wang, and W.I. Park, Semicond. Sci. Technol. **20**, S22 (2005).
[4] Ya.I. Alivov, E.V. Kalinina, A.E. Cherenkov, D.C. Look, B.M. Ataev, A.K. Omaev, M.V. Chukichev, and D.M. Bagnall, *Appl. Phys. Lett.* **83**, 4719 (2003).
[5] A. Osinsky, J.W. Dong, M.Z. Kauser, B. Hertog, A.M. Dabiran, P.P. Chow, S.J. Pearton, O. Lopatiuk, and L. Chernyak, *Appl. Phys. Lett.* **85**, 4272 (2004).
[6] Tsukazaki, A. Ohtomo, T. Onuma, M. Ohtani, T. Makino, M. Sumiya, K. Ohtani, S.F. Chichibu, S. Fuke, Y. Segawa, H. Ohno, H. Koinuma, and M. Kawasaki, Nature Mater. **4**, 42 (2005).
[7] J.F. Wager, Science **300**, 1245 (2003).
[8] R.L. Hoffman, J. Appl. Phys. **95**, 5813 (2004).
[9] E.M.C. Fortunato, P.M.C. Barquinha, A.C.M.P.G. Pimental, A.M.F. Gonçalves, A.J.S. Marques, L.M.N. Pereira, and R.F.P. Martins, Advanced Materials **17**, 590 (2005).
[10] K. Minegishi, Y. Koiwai, and K. Kikuchi, Jpn. J. Appl. Phys. **36**, L1453 (1997).
[11] M. Joseph, H. Tabata, and T. Kawai, Jpn. J. Appl. Phys. **38**, L1205 (1999).

[12] Y.R. Ryu, S. Zhu, D.C. Look, J.M. Wrobel, H.M. Jeong, and H.W. White, J. Crystal Growth **216**, 330 (2000).

[13] X-L. Guo, H. Tabata, and T. Kawai, J. Crystal Growth **223**, 135 (2001).

[14] T.V. Butkhuzi, M.M. Sharvashidze, N.M. Gamkrelidze, Kh.V. Gelovani, T.G. Khulordava, N.P. Kekelidze, and E.E. Kekelidze, Semicond. Sci. Technol. **16**, 575 (2001).

[15] A.B.M.A. Ashrafi, I. Suemune, H. Kumano, and S. Tanaka, Jpn. J. Appl. Phys. **41**, L1281 (2002).

[16] G. Xiong, J. Wilkinson, B. Mischuck, S. Tuzemen, K.B. Ucer, and R.T. Williams, Appl. Phys. Lett. **80**, 1195 (2002).

[17] D.C. Look, D.C. Reynolds, C.W. Litton, R.L. Jones, D.B. Eason, and G. Cantwell, Appl. Phys. Lett. **81**, 1830 (2002).

[18] X. Li, Y. Yan, T.A. Gessert, C.L. Perkins, D. Young, C. DeHart, M. Young, and T.J. Coutts, J. Vac. Sci. Technol. A **21**, 1342 (2003).

[19] K.H. Bang, D-K. Hwang, M-C. Park, Y-D. Ko, I. Yun, and J-M. Myoung, Appl. Surf. Sci. **210**, 177 (2003).

[20] B.S. Li, Y.C. Liu, Z.Z. Zhi, D.Z. Shen, Y.M. Lu, J.Y. Zhang, X.W. Fan, R.X. Mu, and D.O. Henderson, J. Mater. Res. **18**, 8 (2003).

[21] A.V. Singh, R.M. Mehra, A. Wakahara, and A. Yoshida, J. Appl. Phys. **93**, 396 (2003).

[22] J. Huang, Z. Ye, H. Chen, B. Zhao, and L. Wang, J. Mater. Sci. Lett. **22**, 249 (2003).

[23] Y.R. Ryu, T.S. Lee, and H.W. White, Appl. Phys. Lett. **83**, 87 (2003).

[24] K-K. Kim, H-S. Kim, D-K. Hwang, J-H. Lim, and S-J. Park, Appl. Phys. Lett. **83**, 63 (2003).

[25] J. Lu, Y. Zhang, Z. Ye, L. Wang, B. Zhao, and J. Huang, Mater. Lett. **57**, 3311 (2003).

[26] J. Wang, G. Du, B. Zhao, X. Yang, Y. Zhang, Y. Ma, D. Liu, Y. Chang, H. Wang, H. Yang, and S. Yang, J. Cryst. Growth **255**, 293 (2003).

[27] C. Wang, Z. Ji, K. Liu, Y. Xiang, and Z. Ye, J. Crystal Growth **259**, 279 (2003).

[28] J.F. Rommeluère, L. Svob, F. Jomard, J. Mimila-Arroyo, A. Lusson, V. Sallet, and Y. Marfaing, Appl. Phys. Lett. **83**, 287 (2003).

[29] Y. Ma, G.T. Du, S.R. Yang, Z.T. Li, B.J. Zhao, X.T. Yang, T.P. Yang, Y.T. Zhang, and D.L. Liu, J. Appl. Phys. **95**, 6268 (2004).

[30] C-C. Lin, S-Y. Chen, S-Y. Cheng, and H-Y. Lee, Appl. Phys. Lett. **84**, 5040 (2004).

[31] Z-Z. Ye, F-Z. Ge, J-G. Lu, Z-H. Zhang, L-P. Zhu, B-H. Zhao, and J-Y. Huang, J. Cryst. Growth **265**, 127 (2004).

[32] D.C. Look, G.M. Renlund, R.H. Burgener, II, and J.R. Sizelove, Appl. Phys. Lett. **85**, 5269 (2004).

[33] S. Kim, B.S. Kang, F. Ren, Y.W. Heo, K. Ip, D.P. Norton, and S.J. Pearton, Appl. Phys. Lett. **84**, 1904 (2004).

[34] S.Y. Lee, E.S. Shim, H.S. Kang, S.S. Pang, and J.S. Kang, Thin Solid Films **473**, 31 (2005).

[35] J-H. Lim, K-K. Kim, D-K. Hwang, H-S. Kim, J-Y. Oh, and S-J. Park, J. Electrochem. Soc. **152**, G179 (2005).

[36] V. Vaithianathan, B-T. Lee, and S.S. Kim, Appl. Phys. Lett. **86**, 062101 (2005).

[37] G.D. Yuan, Z.Z. Ye, L.P. Zhu, Q. Qian, B.H. Zhao, R.X. Fan, C.L. Perkins, and S.B. Zhang, Appl. Phys. Lett. **86**, 202106 (2005).
[38] X. Li, B. Keyes, S. Asher, S.B. Zhang, S-H. Wei, T.J. Coutts, S. Limpijumnong, and C.G. Van de Walle, Appl. Phys. Lett. **86**, 122107 (2005).
[39] C-C. Lin, H-P. Chen, and S-Y. Chen, Chem. Phys. Lett. **404**, 30 (2005).
[40] H.W. Liang, Y.M. Lu, D.Z. Shen, Y.C. Liu, J.F. Yan, C.X. Shan, B.H. Li, Z.Z. Zhang, J.Y. Zhang, and X.W. Fan, Phys. Stat. Sol. (a) **202**, 1060 (2005).
[41] T.S. Jeong, M.S. Han, J.H. Kim, C.J. Youn, Y.R. Ryu, and H.W. White, J. Cryst. Growth **275**, 541 (2005).
[42] T.M. Barnes, K. Olson, and C.A. Wolden, Appl. Phys. Lett. **86**, 112112 (2005).
[43] H.Y. Xu, Y.C. Liu, Y.X. Liu, C.S. Xu, C.L. Shao, and R. Mu, Appl. Phys. B **80**, 871 (2005).
[44] T Aoki, Y. Hatanaka, and D.C. Look, Appl. Phys. Lett. **76**, 3257 (2000).
[45] D.C. Look and B. Claflin, Phys. Stat. Sol. (b) **241**, 624 (2004).
[46] D.C. Look, D.C. Reynolds, J.R. Sizelove, R.L. Jones, C.W. Litton, G. Cantwell, and W.C. Harsch, Solid State Commun. **105**, 399 (1998).
[47] A.F. Kohan, G. Ceder, D. Morgan, and C.G. Van de Walle, Phys. Rev. B **61**, 15019 (2000).
[48] S.B. Zhang, S-H. Wei, and A. Zunger, Phys. Rev. B **63**, 075205 (2001).
[49] F. Oba, S.R. Nishitani, S. Isotani, H. Adachi, and I. Tanaka, J. Appl. Phys. **90**, 824 (2001).
[50] D.C. Look and J.R. Sizelove, Phys. Rev. Lett. **82**, 2552 (1999).
[51] C.G. Van de Walle, Phys. Rev. Lett. **85**, 1012 (2000).
[52] S.F.J. Cox, E.A. Davis, S.P. Cottrell, P.J.C. King, J.S. Lord, J.M. Gil, H.V. Alberto, R.C. Vilao, J.P. Duarte, N.A. de Campos, A. Weidinger, R.L. Lichti, and S.J.C. Irvine, Phys. Rev. Lett. **86**, 2601 (2001).
[53] D.M. Hofmann, A. Hofstaetter, F. Leiter, H.J. Zhou, F. Henecker, B.K. Meyer, S.B. Orlinskii, J. Schmidt, and P.G. Baranov, Phys. Rev. Lett. **88**, 045504 (2001).
[54] K. Shimomura, K. Nishiyama, and R. Kadono, Phys. Rev. Lett. **89**, 255505 (2002).
[55] N.H. Nickel and K. Fleischer, Phys. Rev. Lett. **90**, 197402 (2003).
[56] K. Ip, M.E. Overberg, Y.W. Heo, D.P. Norton, S.J. Pearton, C.E. Stutz, B. Luo, F. Ren, D.C. Look, and J.M. Zavada, Appl. Phys. Lett. **82**, 385 (2003).
[57] Y.M. Strzhemechny, H.L. Mosbacker, D.C. Look, D.C. Reynolds, C.W. Litton, N.Y. Garces, N.C. Giles, L.E. Halliburton, S. Niki, and L.J. Brillson, Appl. Phys. Lett. **84**, 2545 (2004).
[58] K-K. Kim, S. Niki, J-Y. Oh, J-O. Song, T-Y. Seong, S-J. Park, S. Fujita, and S-W. Kim, J. Appl. Phys. **97**, 066013 (2005).
[59] T. Makino, Y. Segawa, S. Yoshida, A. Tsukazaki, A. Ohtomo, and M. Kawasaki, Appl. Phys. Lett. **85**, 759 (2004).
[60] A. Guillén-Santiago, M. de la L. Olvera, A. Maldonado, R. Asomoza, and D.R. Acosta, Phys. Stat. Sol. (a) **201**, 952 (2004).
[61] D.H. Fan, Z.Y. Ning, and M.F. Jiang, Appl. Surf. Sci. **245**, 414 (2005).

[62] C.H. Park, S.B. Zhang, and S.H. Wei, Phys. Rev. B **66**, 073202 (2002).
[63] M.G. Wardle, J.P. Goss, and P.R. Briddon, Phys. Rev. B **71**, 155205 (2005).
[64] X.S. Wang, Z.C. Wu, J.F. Webb, and Z.G. Liu, Appl. Phys. A **77**, 561 (2003).
[65] F.X. Xiu, Z. Yang, L.J. Mandalapu, D.T. Zhao, and J.L. Liu, Appl. Phys. Lett. **87**, 152101 (2005).
[66] S. Limpijumnong, S.B. Zhang, S-H. Wei, and C.H. Park, Phys. Rev. Lett. **92**, 155504 (2004).
[67] F. Tuomisto, K. Saarinen, and D.C. Look, Phys. Rev. Lett. **91**, 205502 (2003).
[68] ZN Technology, 910 Columbia Street, Brea, CA 92821.
[69] D.C. Look, Electrical Characterization of GaAs Materials and Devices (Wiley, New York, 1989), Ch. 1.
[70] P. Erhart, K. Albe, N. Juslin, and K. Nordlund, unpublished.
[71] D.C. Look and J.R. Sizelove, Phys. Rev. Lett. **82**, 1237 (1999).
[72] O. Schmidt, P. Kiesel, C.G. Van de Walle, N.M. Johnson, J. Nause, and G.H. Döhler, Jpn. J. Appl. Phys. Part 1 **44**, 7271 (2005).
[73] O. Schmidt, A. Geis, P. Kiesel, C.G. Van de Walle, N.M. Johnson, A. Bakin, A. Waag, and G.H. Döhler, Superlattices and Microstructures **39**, 8 (2006).
[74] D.C. Look, H.L. Mosbacker, Y.M. Strzhemechny, and L.J. Brillson, Superlattices and Microstructures **38**, 406 (2005).
[75] H.E. Brown, "ZnO Rediscovered" (The New Jersey Zinc Co., New York, 1957), p. 31.
[76] Y-S. Kang, H-Y. Kim, and J-Y. Lee, J. Electrochem. Soc. **147**, 4625 (2000).
[77] B.K. Meyer, H. Alves, D.M. Hofmann, W. Kriegseis, D. Forster, F. Bertram, J. Christen, A. Hoffmann, M. Strassburg, M. Dworzak, U. Haboeck, and A.V. Rodina, Phys. Stat. Sol. (b) **241**, 231 (2004).
[78] D.C. Look, R.L. Jones, J.R. Sizelove, N.Y. Garces, N.C. Giles, and L.E. Halliburton, Phys. Stat. Sol. (a) **195**, 171 (2003).
[79] O.F. Schirmer and D. Zwingel, Solid State Commun. **8**, 1559 (1970).
[80] G. Cantwell, ZN Technology, private communication.
[81] S. Limpijumnong, X. Li, S-H. Wei, and S.B. Zhang, Appl. Phys. Lett. **86**, 211910 (2005).
[82] W.E. Carlos, E.R. Glaser, and D.C. Look, Physica B **308–310**, 976 (2001).
[83] N.Y. Garces, N.C. Giles, L.E. Halliburton, G. Cantwell, D.B. Eason, D.C. Reynolds, and D.C. Look, Appl. Phys. Lett. **80**, 1334 (2002).
[84] C.H. Seager and S.M. Myers, J. Appl. Phys. **94**, 2888 (2003).
[85] L.E. Halliburton, L. Wang, L. Bai, N.Y. Garces, N.C. Giles, M.J. Callahan, and B. Wang, J. Appl. Phys. **96**, 7168 (2004).
[86] D.C. Look, B. Claflin, Ya.I. Alivov, and S.J. Park, Phys. Stat. Sol. (a) **201**, 2203 (2004).
[87] U. Wahl, E. Rita, J.G. Correia, A.C. Marques, E. Alves, and J.C. Soares, Phys. Rev. Lett. **95**, 215503 (2005).
[88] F. Tuomisto, I. Makkonen, M.J. Puska, K. Saarinen, D.C. Look, G.M. Renlund, and R.H. Burgener II, Superlattices and Microstructures, in press.

[89] D.C. Reynolds, D.C. Look, B. Jogai, C.W. Litton, T.C. Collins, W. Harsch, and G. Cantwell, Phys. Rev. B **57**, 12151 (1998).
[90] Y-Z. Yoo, Z-W. Jin, T. Chikyow, T. Fukumura, M. Kawasaki, and H. Koinuma, Appl. Phys. Lett. **81**, 3798 (2002).
[91] P.M. Ratheesh Kumar, C. Sudha Kartha, K.P. Vijayakumar, F. Singh, and D.K. Avasthi, Mater. Sci. Eng. B **117**, 307 (2005).
[92] D.C. Look, D.C. Reynolds, J.W. Hemsky, R.L. Jones, and J.R. Sizelove, Appl. Phys. Lett. **75**, 811 (1999).
[93] D.C. Look, G.C. Farlow, S. Limpijumnong, S.B. Zhang, and K. Nordlund, Phys. Rev. Lett. **95**, 225502 (2005).
[94] P. Kasai, Phys. Rev. **130**, 989 (1963).
[95] F. Tuomisto, K. Saarinen, D.C. Look, and G.C. Farlow, Phys. Rev. B **72**, 085206 (2005).

Chapter 3

Synthesis and Characterization of Nitrogen-Doped ZnO Films Grown by MOCVD

Timothy J. Coutts, Xiaonan Li, Teresa M. Barnes, Brian M. Keyes, Craig L. Perkins, Sally E. Asher, S. B. Zhang and Su-Huai Wei

National Renewable Energy Laboratory, Golden, CO 80401, USA

Sukit Limpijumnong

School of Physics, Institute of Science, Suranaree University of Technology and National Synchrotron Research Center, Nakhon Ratchasima, Thailand

3.1 Introduction

The reasons behind the current worldwide interest in zinc oxide (ZnO) are well known and have been discussed at length in this book, making it unnecessary to repeat them here. However, the reasons for the relevance of ZnO, in the context of energy efficiency and renewable energy, are less well known and may be of interest to the reader. We study ZnO at the National Renewable Energy Laboratory (NREL) because: i) *n*-ZnO (both doped and undoped) is currently used in the copper indium gallium diselenide (CIGS, or Cu(In,Ga)Se_2) thin-film solar cell; ii) the added design flexibility that *p*-ZnO may afford new design of thin-film solar cells; iii) the potential importance of ZnO (and its alloys) to solid-state lighting; and iv) the potential of ZnO devices to the photoelectrochemical splitting of water to generate hydrogen.[1–4]

This chapter is concerned with *p*-ZnO, and item i) above is not immediately relevant. However, there is a growing worldwide shortage of indium, and a severe inflation of its price has occurred during the last two years.[5] Of course, this has taken place at a time when the flat-panel display industry is expanding very rapidly and demand is increasing.[6] Industries that use indium tin oxide (ITO) appreciate the need to develop an alternative, and *n*-ZnO is one serious possibility.

p-ZnO could also be important to thin-film solar cells because additional designs to those presently studied could possibly be developed. For example, the structure of

Zinc Oxide Bulk, Thin Films and Nanostructures
C. Jagadish and S. Pearton (Editors)

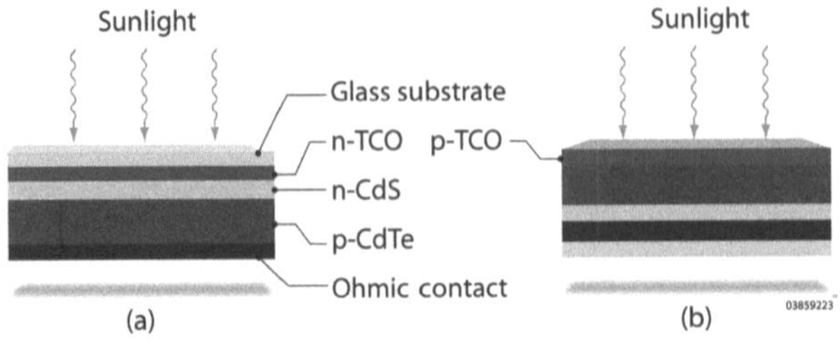

Figure 3.1: Schematic of single-junction, thin-film solar cells. Figure 3.1(a) illustrates a solar cell for which the illumination passes first through the glass substrate and the TCO window layer (typically SnO_2), then the CdS buffer layer, and is then absorbed by the CdTe absorber. Figure 3.1(b) shows an inverted configuration in which the light passes first through the *p*-type TCO, then the CdTe absorber, and then the CdS that forms the electrostatic junction.

the cadmium telluride (CdTe) solar cell, as illustrated in Fig. 3.1(a), shows that the *p*-CdTe layer is deposited as the second-to-last film, with only the opaque contact being deposited afterward.[7] In this device, the light passes first through the glass substrate, then through the *n*-TCO (transparent conducting oxide), after which it passes through the active layers of the device. It may be advantageous to construct this device so that the light passes first through a transparent *p*-type TCO that makes contact to the *p*-CdTe and then through the rest of the device layers. A schematic of the suggested structure, which, to our knowledge, has never been demonstrated, is shown in Fig. 3.1(b). *P*-type TCOs are equally relevant to the development and, possibly, the construction of tandem thin-film solar cells.[8] In such devices, the individual subcells may be connected in series with the subcell currents being matched at their maximum power points, or in parallel, with the subcell voltages being matched at their maximum power points. The devices may be grown as a monolithic stack or as discrete cells in a mechanically stacked configuration.[9] In both connectivities, the solar radiation first passes through the top cell, and the bottom cell uses photons with energies greater than its own bandgap but less than that of the top cell. These two structures are shown in Fig. 3.2. In Fig. 3.2(a), the two subcells are connected in series. Figure 3.2(b) shows a mechanically stacked tandem solar cell in which we have assumed that the top subcell is based on a CdTe absorber. This, of course, is not a prerequisite. However, in both configurations, an interconnect that is highly transparent to the lower-energy photons is required and a high-performance *p*-TCO could be useful, although other approaches have also been considered.[7]

According to the US Department of Energy (DOE), more than 20% of the electricity consumed in the United States (i.e., more than 8% of total energy consumption)

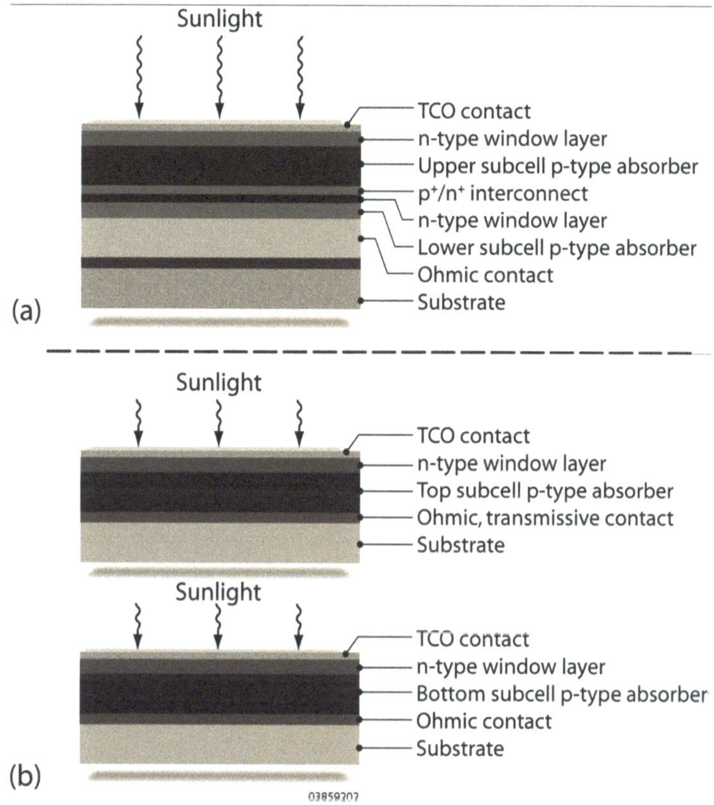

Figure 3.2: Schematics of tandem, thin-film solar cells. The photons with energies less than that of the top-subcell bandgap are transmitted, to be used by the bottom subcell (if their energy is greater than its bandgap). To achieve this, a transparent interconnect is needed and this may require a p^+-TCO, such as ZnO. Figure 3.2(b) shows a mechanically stacked device in which the two subcells are grown separately on their own substrates.

is used in domestic and industrial lighting.[2] Most of the electricity consumed by incandescent lights, for example, is actually used in heating and the creation of unusable infrared radiation, with only about 10% of the input electrical energy being converted to visible light. Clearly, there would be great advantages in developing a more-efficient form of lighting. Most of us are aware that solid-state lighting (light-emitting diodes, or LEDs) is increasingly found in everyday life – for example, in toys, electronics, traffic lights, and brake lights in cars – but the longer-term goal is to develop white-light-emitting systems with an output of 150–200 lumens per watt and a cost of less than 1 cent per lumen.[2] Such a development could lead to enormous financial and environmental savings – the cumulative savings in dollars being about $100 billion over the period 2000–2020 and reduced carbon emissions being about 28 million tons per year. ZnO has a bandgap of about 3.2 eV and, if it can be used as a light emitter, would radiate at about 388 nm, which is in the blue/near-ultraviolet portion of the electromagnetic spectrum. If efficient lasers or LEDs can be made of ZnO, the near-ultraviolet photons that would be emitted could be used to excite phosphors in suitable combinations to produce white light. Alternatively, alloying with cadmium would reduce the bandgap, and the emitted photons would be of longer wavelength. Hence, ZnO may have a role to play in the solid-state lighting project that is being vigorously sponsored by the DOE.[2]

Photoelectrochemical (PEC) cells may be used for the solar splitting of water into hydrogen and oxygen, and this is one of the approaches being considered in the DOE's hydrogen initiative.[10] The goal of the initiative is to develop a PEC cell, driven only by solar radiation, with a solar-to-hydrogen conversion efficiency of 10% and a system lifetime of at least ten years. It should be emphasized, however, that there are other forms of renewable energy that could be used to split water.[11] A PEC cell consists of three components: a semiconductor with a bandgap of at least 1.6–1.7 eV, an electrolyte (water), and a counter electrode (typically platinum or calomel). In reality, a somewhat larger bandgap is needed to drive the reaction. The conduction and valence band edges of the semiconductor must straddle the standard redox potentials of the H_2O/H_2 and H_2O/O_2 couples. The semiconductor must also be stable in water. For water, with a pH $\sim$10, ZnO meets these requirements. Although ZnO has a bandgap that is too large to permit efficient absorption of solar radiation, the promise of modern bandgap-engineering techniques has allowed ZnO to be seriously considered for this application. The corrosion problem remains an issue and must be solved if ZnO is to be used for this application.

In this chapter, we shall begin by reviewing the literature that deals with the fabrication, characterization, and analysis of ZnO:N. We shall then discuss our own work on the growth of ZnO:N by metal-organic chemical-vapor deposition (MOCVD), characterization of the material, and theoretical analysis of our results to date.

3.2 Background to nitrogen-doped ZnO

Efforts to make thin films of *p*-ZnO have involved several deposition methods (e.g., molecular-beam epitaxy, chemical-vapor deposition, sputtering, spray deposition, diffusion, oxidation of Zn_3N_2), and the use of different dopants (e.g., nitrogen, phosphorous, arsenic, antimony, lithium, etc.). In addition, co-doping with Ga-N, Al-N, and In-N has been reported as having varying degrees of success. In some cases, post-deposition annealing has been used in an attempt to optimize the material properties. The material is far from being fully understood, and results are rather inconsistent. Devices have also been made, and strongly diode-like behavior has been reported, as has the emission of near-band-edge light emission.

However, low-resistivity *p*-type ZnO has not yet been obtained using nitrogen (N) doping. For N-doped ZnO, the typical carrier concentrations are only $10^{15}\sim10^{18}\,cm^{-3}$, although concentrations of N as high as $10^{21}\,cm^{-3}$ have been detected. The reasons for the low electrical activity of incorporated N are under investigation and will be discussed extensively in this chapter. This could be due, in part, to the ionization energy of the substitutional nitrogen (N_O) acceptor level. However, some experiments have reported this quantity to be about 160 meV, from which one may still expect about 1/500th of the N impurities to be ionized at room temperature. Given the concentrations of N observed in the ZnO samples, it is therefore apparent that at least one mechanism of compensation must exist. Possible mechanisms could include donors such as hydrogen (H), and defect complexes involving H, carbon (C), and N, the latter in the form of N_2.

Although H is known to be amphoteric in most semiconductors, it is predicted to act only as a donor in ZnO.[12] H is generally a component of the precursors used in MOCVD, it is likely that it will be incorporated in the film and may passivate N acceptors. The same comment may be made of C, which is also always present in one or more of the growth precursors. Later in this chapter, we shall discuss the possible roles of these impurities, including the possible role of defect complexes that may compensate N-doping.

3.2.1 Theory

ZnO has asymmetric doping limitations that strongly favor *n*-type conductivity. It is difficult to form a preponderance of acceptor-like defects because of the abundance of compensating donor defects.[13,14] Zhang et al. discussed intrinsic ZnO and showed that it cannot be doped *p*-type with native defects (O_i, V_{Zn}).[14] Although these form shallow acceptors, compensating donor-like defects, such as Zn_i, V_O, and Zn_O, form much more easily. Growth conditions for *p*-type films are also favorable for the formation of native donors when attempting anionic doping.[15] In 1983, Kobayashi et al. made the first theoretical prediction that N may be a suitable *p*-type dopant.[16] This work analyzed all group-V elements as potentially suitable *p*-type dopants in wide-bandgap II-VI materials. Even at this early stage, the challenges of *p*-type doping, such as self-compensation by native donor defects and relatively deep acceptor levels, were appreciated. According to the calculations performed,

N is the only group-V element that could act as a relatively shallow substitutional anionic acceptor in ZnO. Later work by Park et al. also suggested that N may be the best candidate as a *p*-type dopant.[17]

Although theoretical work predicted that N would be the best *p*-type dopant for ZnO, the calculations also identified three distinct sets of challenges to successfully realizing high-quality *p*-type material. These challenges are: i) the solubility and efficiency of incorporation of the N;[18,19] ii) the fact that the acceptor level is relatively far above the valence-band maximum;[17] and iii) the existence of compensating effects.[17,18,20] These problems would all hinder the production of highly conductive *p*-type ZnO. The first set of problems is caused by thermodynamic obstacles to the incorporation of N into ZnO. Nitrogen forms an acceptor in ZnO only if a single atom occupies an oxygen (O) site (N_O) in the lattice. However, the formation energy of this defect (which is dominated by the energy required to break the strong N_2 bond) is relatively high. Diatomic N will form a double shallow donor if it substitutes for O, $(N_2)_O$, and compared with N_O, its formation energy is expected to be relatively low.[18,19] This difference in defect formation energies is partially caused by the extremely strong N_2 bond that must be broken to liberate atomic N from several of the gas-phase precursors, such as N_2 or N_2O, that have been used. Yan suggested that the use of NO or NO_2 should yield *p*-type material with higher hole concentrations and mobilities, because the defect formation energies are not dominated by the thermodynamic constraint of breaking the N_2 bond. Instead, they are controlled by the kinetics of incorporation and reactions to form N_2.[19]

Various co-doping approaches have also been proposed to facilitate the formation of N_O defects. Co-doping relies on doping with N and a donor species, such as excess Zn,[17] H,[21] Ga, Al, or In,[22] to increase the number of N_O defects and, possibly, reduce the ionization energy of the acceptor. Although some encouragement can be taken from the results obtained from these efforts, the acceptor concentrations achieved were not notably higher than those achieved using N alone.

Once the N_O defect is formed, it does not necessarily contribute a hole to the valence band, because the ionization energy of the acceptor is rather high. Several theorists have estimated the ionization energy to be as high as 0.40–0.45 eV,[17] which would certainly cause a low ionization efficiency. Additionally, and perhaps more significantly, holes can easily be compensated by a number of defects. Intrinsic donor defects, such as Zn interstitials and O vacancies, form easily in the Zn-rich conditions that are generally used to grow *p*-type ZnO, when attempting to dope the O site with a group-V impurity.[14] Unintentionally incorporated hydrogen forms a shallow donor and is essentially impossible to exclude from the films.[12] Hydrogen is plentiful in the atmosphere, cannot be completely removed from any vacuum system, and can also be incorporated from incompletely reacted organometallic precursors.[23] Carbon may also have a detrimental compensation effect. Compensation by both of these elements will be discussed in greater detail later in the chapter. Nitrogen itself can introduce compensating donors by forming donor-like $(N_2)_O$ defects.[18,19] Consequently, it is necessary to minimize the concentration of all these compensating defects, while simultaneously creating a high hole concentration. In addition to all the difficulties described, it has also been appreciated that N_O may be metastable in the ZnO lattice. N_O is bonded to four Zn atoms, and these bonds are higher in energy than Zn-O bonds. This excess energy may provide a driving force for N to move in the lattice, possibly forming the double donor $(N_2)_O$.[20]

3.2.2 Experimental results for ZnO:N

3.2.2.1 Deposition techniques

N-doped ZnO has been synthesized by a variety of techniques, including an assortment of chemical-vapor deposition (CVD) processes, molecular-beam epitaxy (MBE), pulsed-laser deposition (PLD), annealing of Zn_3N_2, and sputtering. No technique has been able to synthesize stable *p*-type ZnO:N, reliably, reproducibly, and with the required properties: all have confirmed the difficulties suggested by theoretical calculations.

The first report of *p*-type ZnO:N, prepared by CVD from elemental Zn, ZnO powder, H_2, and NH_3 on sapphire substrates, was made by Minegishi in 1997.[24] Excess Zn is required for N incorporation by this technique, and *p*-type conductivity was only seen in films grown at high temperatures. In addition, high post-deposition temperatures were required to drive off incorporated H.

MOCVD work at NREL combined diethyl zinc (DEZn) and NO in a thermally driven reaction to form *p*-type ZnO:N.[25] NO was chosen as the oxidizer because of the theoretical work of Yan et al.[19] described above. The N-incorporation level achieved in the ZnO films was as high as 2.6×10^{21} cm^{-3}. However, hole concentrations were several orders lower – in the range of $1.0 \times 10^{15}{\sim}1.0 \times 10^{18}$ cm^{-3}. The minimum film resistivity achieved was ~20 Ω-cm. Secondary-ion mass spectrometry (SIMS) analysis indicated high levels of H and C impurities. The *p*-type conductivity was found to be unstable.[26] X-ray photoelectron spectroscopy (XPS) and Fourier transform infrared spectroscopy (FTIR) also showed high levels of H. The associated characterization work will be discussed throughout this chapter.

In the work of Barnes et al. the ZnO films were produced using high-vacuum plasma-assisted CVD (HVP-CVD), a technique that combines a remote plasma source with a high-vacuum deposition environment.[27] Oxygen and nitrous oxide (N_2O) were combined and dissociated in a high-density inductively coupled plasma source. The cracked products then diffused through a small orifice into the

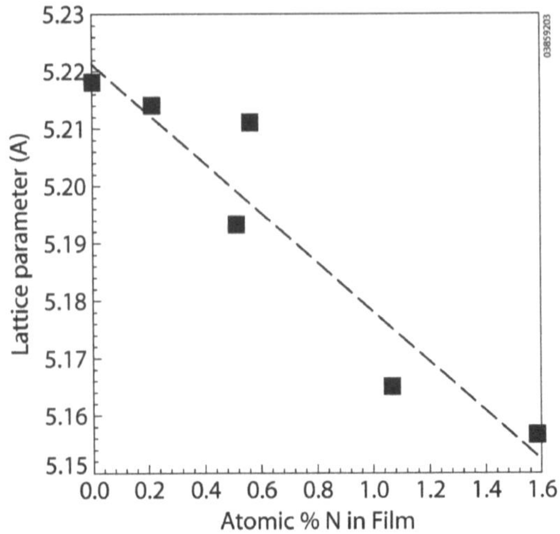

Figure 3.3: Lattice constant of ZnO:N films as a function of the atomic percentage of N. The lattice constant was measured using XRD and the amount of N using XPS.[29]

high-vacuum environment containing a background pressure of DMZ of about 10^{-5} torr. ZnO films were grown in high vacuum on both silicon and glass at 500 K. Nitrogen incorporation was controlled by adjusting the ratio of the O to N_2O in the plasma source. In contrast to the observation reported by Li et al.[25] using MOCVD growth, high concentrations of N were found in films deposited in the presence of O_2 in HVP-CVD.[28]

A series of films was made by HVP-CVD with different amounts of N, and both the structural and electrical properties were characterized.[29] Figure 3.3 shows that the lattice parameter decreased systematically with the quantity of N incorporated, and the authors concluded that this was most likely due to strain rather than to the formation of Zn-N bonds. The films shown in Fig. 3.3 were all *n*-type or highly resistive when measured several days or weeks after deposition.[29]

The conductivity type of the films was assessed using Hall and Seebeck probes, and films that were initially *p*-type were noted to change over time. The hole concentration and mobility were initially about 6×10^{17} cm^{-3} and 0.5–1 cm^2 V^{-1} s^{-1} respectively. Figure 3.4 illustrates the change in the sign of the Seebeck coefficient, indicating a change in carrier type as a function of time after deposition for several films.[29]

As the films changed in conductivity type, their structural properties were also measured. Figure 3.5 shows the change in lattice constant with time after deposition.[29] The lattice constants gradually increased to values similar to those for

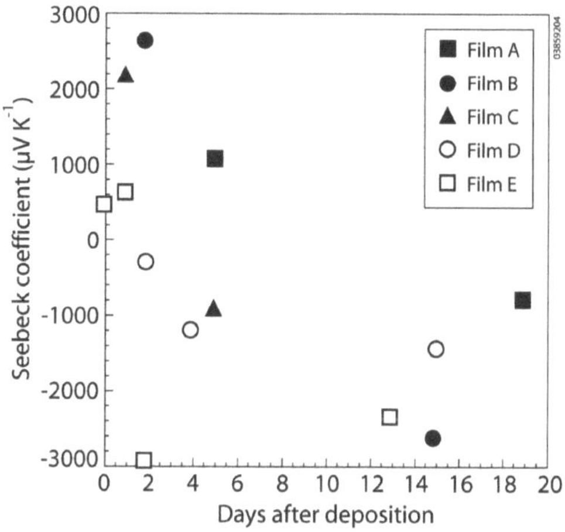

Figure 3.4: Variation of Seebeck coefficient with time after deposition for five ZnO:N films. In all cases, the Seebeck coefficient changed from positive, indicating *p*-type conductivity, to negative, indicating *n*-type conductivity.[29]

undoped HVP-CVD-grown ZnO. As with the work of Li et al. SIMS and XPS were used to show that N remained in the films for a long period after deposition. The authors commented that adsorbed water vapor or atmospheric H could contribute to the incorporation of H-related donors and possibly explain the change in conductivity type. However, they also pointed out that this mechanism could not explain the change in lattice constant.[29]

The observed increase in lattice constant, combined with the large N concentration measured by XPS, indicates that N may be moving through the lattice.[29] Nitrogen may migrate from a substitutional to an interstitial site to form a N_i-V_O or $(N_2)_O$ donor-like defect complex. This migration could possibly be driven by the compressive strain in the film.

MBE techniques have been used to make *p*-type ZnO:N using many different precursors. Ashrafi and coworkers combined DEZn, H_2O, and mono-methyl hydrazine (MMH) to make reproducible *p*-type ZnO:N by MBE.[30] The electrical properties strongly depended on the Zn/O ratio and MMH partial pressure. They found that films grown with plasma-cracked O_2 instead of H_2O were generally *n*-type.[30] Plasma-assisted (P-MBE) processes have produced ZnO:N from the reaction of elemental Zn with active O- and N-containing species generated from an N_2/O_2 plasma source. A film fabricated by P-MBE was characterized as *p*-type by Ashrafi,[30] whereas those grown by Iwata and coworkers[31] were not. Many other groups have been unsuccessful at producing *p*-type ZnO by P-MBE. It is not clear

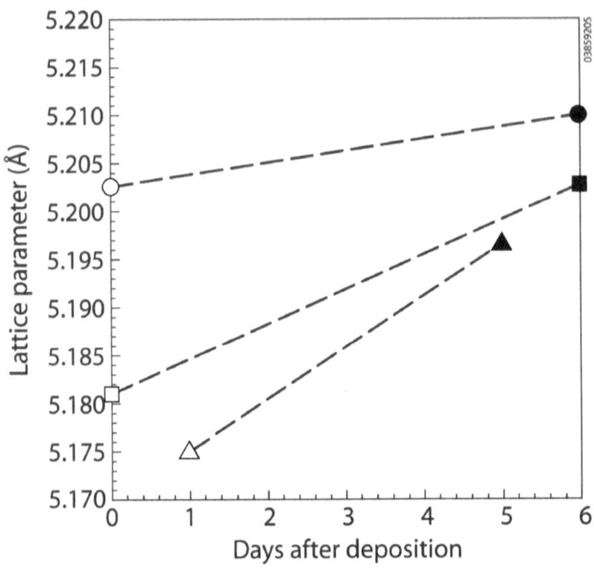

Figure 3.5: Lattice constant of ZnO:N films as a function of time after deposition for three films. In all three cases, the lattice constant relaxed from its initial value (empty symbols) to a value nearer that of undoped ZnO (5.2069 Å).[29]

if these films were unstable or if the *p*-type films of Ashrafi et al. could not be repeated.[30]

Repeated temperature-modulation epitaxy (RTM) is an exceptionally promising MBE technique for depositing *p*-type ZnO:N. RTM has been used to make a blue LED composed entirely of ZnO.[32] N-doped films with hole densities of 2×10^{16} cm^{-3} were produced by alternating 10–15-nm-layer growth steps at 450°C with high-temperature annealing steps at 950°C to make a 150–400-nm film. The low-temperature growth facilitates N incorporation, and the rapid ramp to high temperature quickly recovers the surface smoothness and quality during the annealing step.[33] It is possible that this type of temperature modulation may be applicable to CVD processes, with the additional potential benefit of driving off incorporated C and H.

Most PLD processes for *p*-type ZnO:N have relied on irradiating a ZnO target in the presence of an N_2O or N_2 plasma. However, only a small fraction of films grown this way have been reported to exhibit *p*-type conductivity.[34,35] A recent study compared ZnO:N synthesis by PLD using three plasma-activated sources (N_2, N_2O, and NO).[36] All three sources provided a high level of N incorporation, as determined by SIMS, but the films remained *n*-type. Sputtering is similar to PLD in that a ZnO target is ablated in a reactive atmosphere. In co-doping experiments using N and Ga, the carrier type and concentration were typically found to strongly depend on the O partial pressure. *P*-type conductivity was only realized with a sputtering gas mixture of 60% O_2 and 40% N_2.[37]

Several post-deposition film treatments have been employed, either to synthesize or enhance *p*-type ZnO:N. Li et al. grew Zn_3N_2 by PECVD on fused-silica substrates and then thermally oxidized them at 700°C. The resulting ZnO:N film was reported to be *p*-type, with a hole concentration of 4.2×10^{17} cm^{-3}.[38] This work, again, identified a narrow process window for producing *p*-type ZnO:N.

N^+ ion implantation, followed by annealing at 850°C, of sputtered ZnO films has yielded films with carrier concentrations around 10^{17} cm^{-3} and mobilities of 2–6 cm^2 V^{-1} s^{-1}.[39] However, ion implantation in ZnO is not well understood and generally results in large regions of *n*-type residual damage.[40]

Despite the wide variety of deposition techniques used to make ZnO:N, none has demonstrated systematic control over the electrical properties of the material. Very little is known about the deposition chemistry leading to the formation of *p*-type ZnO:N by techniques other than HVP-CVD. Consequently, the relationships between processing conditions and film properties are poorly understood. The typical observation is of a narrow processing window for producing *p*-type ZnO:N material. Stability may also play a major role in the apparent narrow processing windows. If the films are stable for only days,[29] weeks, or months,[26] it is possible that electrical measurements are not being performed before they become *n*-type. The electrical data presented so far also indicate high resistivity, making a definitive assignment of carrier type difficult.

3.2.2.2 Determination of p*-type conductivity*

Experimentally confirming *p*-type conductivity in ZnO is generally not an easy task. Even the best films are highly resistive, making electrical measurements difficult. Measurement of the Seebeck and/or Hall effects may be used to test the carrier type. To measure the Seebeck voltage, a temperature gradient is established along the length of a sample, and the sign of the resulting voltage (in the same direction as the temperature gradient) gives the carrier type. If $p\mu_p > n\mu_n$, the sign of the Seebeck voltage is positive.

Measurement of the Hall effect is a more stringent test of *p*-type conductivity, and it enables one to determine both carrier concentration and mobility. However, to conclude that a sample is *p*-type in a two-carrier system, it is necessary that $p\mu_p^2 > n\mu_n^2$. This can lead to ambiguities because hole mobilities are generally ~1 cm^2 V^{-1} s^{-1}. The formation of ohmic contacts to *p*-type material is also challenging, and contact-induced noise may contribute to the somewhat inconsistent Hall effect results observed for high-resistivity, nominally *p*-type, material. Plausible Hall effect data have been reported for ZnO:N samples with hole concentrations on the order of 10^{15}–10^{18} cm^{-3} and mobilities on the order of 0.1–1 $cm^2 V^{-1} s^{-1}$, i.e., resistivities of about 40–100 Ω-cm.[26,29,35,41] A recent paper has suggested that unreliable Hall measurements are the cause for the variability in conductivity type reported by many researchers.[41] However, none of this adequately explains the conversion of *p*-type films to semi-insulating or *n*-type, as seen for films grown by MOCVD or HVP-CVD.

Capacitance-voltage (C-V) measurements can also be used to determine the carrier type and concentration. Ashrafi et al. used capacitance/voltage profiling and concluded that their samples were *p*-type with hole concentrations of 10^{15}–10^{16} cm^{-3}.[30] It has consistently been observed that *p*-type ZnO:N is difficult to characterize electrically, and it is advisable to combine several techniques to ensure an unambiguous result.

Photoelectron spectroscopy using either ultraviolet light (UPS) or X-rays (XPS) can be used to determine the position of a material's valence-band maximum relative to the Fermi energy, thereby indicating the conductivity type. Perkins et al. used UPS to determine that the conduction-band minima of several different types of N-doped ZnO samples were close to the Fermi energy, implying that the samples were *n*-type under the conditions of the experiment.[42] Complications of using photoemission for determining conductivity type include the possible preferential excitation of donors with the high-energy photons used in these experiments and the general requirement of sputter-cleaning samples prior to analysis. Sputtering oxides usually causes some degree of reduction and should generate donors in ZnO.[43]

Optical techniques have been used to determine the presence and nature of N-related defects. Using photoluminescence, the N_O acceptor level was determined to be about 0.15–0.165 eV above the valence-band maximum, using a photoluminescence line at 3.315 eV, which was attributed to an acceptor/bound exciton.[41,44,45] This result is in good agreement with the theoretical estimate of ionization energy discussed earlier.[46] Hall measurements indicate the N_O acceptor level to be 0.09 eV above the valence-band maximum, but this has not yet been duplicated by other workers.[41]

3.3 NREL research on the synthesizing and characterizing of nitrogen-doped ZnO films

3.3.1 Theoretical basis of NREL research into p*-ZnO*

In this section, we shall discuss our efforts to synthesize *p*-ZnO films doped with N. This work was motivated by the points discussed in the introductory section and guided by the theoretical work of Yan et al.[19] The latter suggested that nitric oxide (NO) should be a more-effective dopant than other N-containing molecules. Consequently, we shall begin by reviewing some of the key points made in this paper because of its relevance to our work and because they are also relevant to several of the papers reviewed above. We shall then give a brief description of the approach used to calculate the various defects that form in ZnO, most of which act as donors. In later sections, we shall discuss the defects in much greater detail and provide more information about the theoretical approaches used.

The solubility of dopants depends on the energy of formation of the defects with which they are associated. The formation energies depend, in turn, on the chemical potentials of the host atoms (Zn and O) and of the impurity itself (N). The formation energy of charged defects also depends on the position of the Fermi level in the bandgap. And this position can cause the formation of compensating defects to become energetically more likely than the incorporation of further amounts of the impurity. Yan et al. suggested that NO, which is the source gas we used, is a better dopant to dope ZnO with N than N_2.[19] They also pointed out that it is important to ensure that the NO molecule arrives intact at the growing surface. By using molecular doping (with NO), the likelihood of N incorporation is greatly increased, and Yan et al. suggested that high hole concentrations should be obtained.

Density functional theory was the main tool used, and the assumptions and approximations used to solve the Schröedinger equation were very similar to those described later to calculate the energetics of more-complicated defects. In equilibrium conditions, the concentrations of defects were defined in terms of the chemical potentials of the host and impurity atoms, in relation to appropriate elemental and molecular reservoirs. The main defects considered were substitutional N atoms (N_O) and substitutional N molecules $(N_2)_O$. The former is an acceptor and the latter is a donor, the relative numbers of these defects ultimately govern the conductivity. However, as discussed later, it is now known that several other defect complexes act as donors, and the analysis presently being discussed was probably oversimplified.

Yan et al. considered four molecules containing N and O and concluded that the two having single N atoms (NO and NO_2) were inherently superior to the two having two N atoms (N_2 and N_2O) because energy must be supplied to break the N-N bonds of the latter. Hence, the energy of formation of the desired acceptor is significantly lower than it would have been had either of the two diatomic N gases been used. This means that the monatomic N gases may be incorporated directly into the growing film to form the acceptor-like defect. On the other hand, the diatomic N gases (N_2 or N_2O) would be more likely to form the donor-like defect.

Even if monatomic N is incorporated into the film, it is possible that two such atoms could diffuse to form $(N_2)_O$. However, provided that the concentration of N is much less than that of O, the probability of this is very small. Figure 3.6 shows

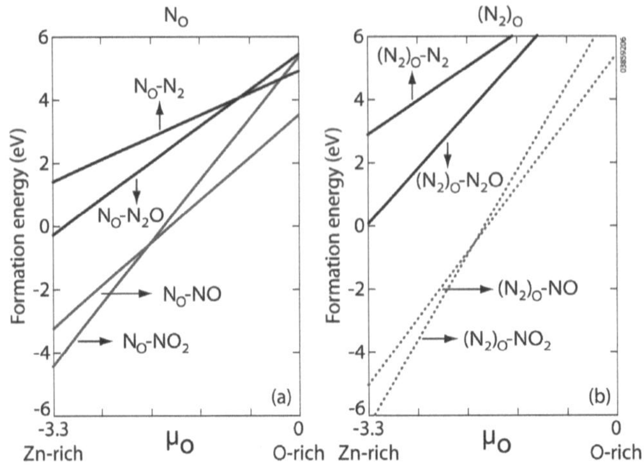

Figure 3.6: Calculated formation energies of N_O and $(N_2)_O$ defects, as functions of the oxygen chemical potential. The N_O acceptor defect is supplied from N_2, N_2O, NO, or NO_2 gases (Fig. 3.6(a)). In Fig. 3.6(b), the calculations are repeated for $(N_2)_O$ donor defects. $\mu_O = -3.3$ eV is the Zn-rich limit condition, and $\mu_O = 0$ eV is the O-rich limit condition.[19]

the formation energies for the N_O and $(N_2)_O$ defects as a function of the ambient O condition (in terms of the O chemical potential). The ordinate $\mu_O = -3.3$ eV represents the zinc-rich limit, whereas the ordinate $\mu_O = 0$ eV represents the O-rich limit. Figure 3.6 shows that the monatomic N molecules are much more likely to produce the desired substitutional acceptor than are the diatomic species, and the formation energies at the Zn-rich limit are significantly lower than those at the O-rich limit. The same monatomic N-doping sources, however, also favor the formation of unwanted $(N_2)_O$ donor defects, as shown in Fig. 3.6. One may overcome this shortcoming by carefully controlling the growth kinetics, because the formation of $(N_2)_O$ from monatomic N sources requires two N atoms to meet and bind to each other. The probability for such reaction is low, provided that the N concentration is significantly smaller than that of O. In this kinetic limit case, the formation of a secondary phase such as Zn_3N_2 is also not considered. At higher N concentrations, however, one would have to take into account the Zn_3N_2 phase and recalculate the available chemical-potential range for μ_O and μ_{Zn}. In such cases, negative formation energies in Fig. 3.6 are impossible because μ_O and μ_{Zn} are now confined to a significantly smaller energy range near the O-rich limit in Fig. 3.6.[47]

Yan and Zhang then used their qualitative conclusions to explain the results of Joseph et al. in which the pulsed lasser deposition growth conditions used N_2O

as a background gas. Their equipment included electron cyclotron resonance as an optional means of cracking the nitrous oxide. In addition, co-doping with gallium was attempted. The predictions of Yan et al. were able to account for all the observations of Joseph.[34]

Hydrogen is of particular relevance to our work because of its inevitable inclusion in our films. Work by van de Walle et al. showed that H acts as a donor in ZnO, and we have confirmed this result in our own work, as well as calculating its most likely locations in the ZnO lattice.[12] We calculated the possible H-related vibration frequencies, and this enabled us to make comparisons with the local vibrational modes measured using FTIR spectroscopy. This work has allowed us to make semiquantitative comparisons between the theoretically predicted spectra and the features observed using FTIR. We shall discuss our results and give further details of the methods used later.

3.3.2 Synthesis and characterization

Films were fabricated using low-pressure (LP-MOCVD) and, for fabricating undoped material, the precursors were diethylzinc (DEZ, or $(C_2H_5)_2Zn$) and O. DEZ and dilute NO were used to fabricate N-doped ZnO. The latter was used as both the oxidizer and dopant, and the dilution was 2 wt.% NO/Ar. The carrier gas in all cases was N_2. The substrate temperature was varied between 200 and 550°C and, for most depositions, both Si and Corning 1737 glass substrates were used. The ZnO samples on Si substrates were mainly used for FTIR analysis, whereas the samples on glass substrates were used for most of the other characterizations, such as crystal structure, composition, optical, and electronic property analyses.

We used atomic force microscopy (AFMAutoprobe LS from Park Scientific Instruments with Si cantilevers) to assess the topography of the films and X-ray diffraction (Scintag Model PTS) to assess their crystal structure. We investigated the total transmittance and reflectance over the wavelength range of 200–2000 nm using a Cary 5G spectrophotometer equipped with an integrating-sphere detector. We studied the electrical properties using a Hall probe (BioRad Model HL5500) and capacitance-voltage profiling (Keithley 590) using a double-Schottky surface-contact mercury probe (MSI Electronics Hg-2C-5). We analyzed the N, H, and C concentrations using SIMS (CAMECA IMS 5f), XPS (Physical Electronics Model 5600), and Auger electron spectroscopy (AES, Physical Electronics 670 scanning Auger nanoprobe). We used standard sensitivity factors to determine absolute values of the N concentration in the ZnO films, and the XPS spectrometer binding-energy scale was calibrated with the Au $4f_{7/2}$ (83.98-eV) and Cu $2p_{3/2}$ (932.67-eV) peaks.

Finally, the signatures of the various defects suggested by our theoretical analysis were studied by room-temperature FTIR transmittance measurements of films grown on parallelogram-shaped, single-crystal silicon substrates mounted in an attenuated total-reflectance (ATR) accessory attached to a Nicolet Magna-IR 550 spectrometer.

3.3.3 Results

Given that our intention is to substitute N for O, it follows that an O-deficient growth ambient is required. Our work has shown that if we use O as the oxidizer and NO gas as the dopant, we achieve no detectable N doping.[26] However, when we use NO as both oxidizer and dopant, N concentration levels up to 2.60×10^{21} cm^{-3} can be achieved. Although a high N concentration can be achieved for a wide range of deposition parameters, the window for fabricating *p*-type films is rather narrow and, in particular, we were only able to fabricate *p*-type material for substrate temperatures between 400 and 440°C.

Figure 3.7 shows the topography of a typical ZnO:N film on a glass substrate, as assessed by AFM, clearly indicating polycrystallinity. The surface roughness is about ±10–20 nm. XRD analysis indicated that the ZnO:N films had a wurtzite structure, similar to that of an undoped ZnO film. However, we found that the undoped material had a strong (002) orientation, whereas the ZnO:N was randomly oriented. In Fig. 3.8, we show the transmittances of undoped and ZnO:N films. The undoped ZnO is colorless and very transmissive because of the wide fundamental optical bandgap. On the other hand, the N-doped material showed increased absorption near the band edge, which gave the film a yellowish appearance. These properties of polycrystallinity and high transmittance are typical of transparent conducting oxide films. Clearly, the addition of N causes increased absorption losses, possibly due to near-band-edge defects.

Table 3.1 shows the results of Hall probe analysis of several ZnO samples. Undoped ZnO is *n*-type, with a carrier concentration in the range of $10^{16}{\sim}10^{18}$ cm^{-3}. The N-doped ZnO samples are no longer *n*-type, but are either insulating or weakly *p*-type, with a hole concentration of about 10^{13}–10^{18} cm^{-3}. Sample L3358.2 is highly resistive, resulting in large uncertainties in the mobility and carrier concentration measurements. We confirmed that the films were *p*-type using C-V profiling and by measuring the Seebeck coefficient, which was positive, again indicating *p*-type behavior. We used the C-V technique to characterize a ZnO:N sample that had previously been measured with the Hall probe. The latter had indicated a hole concentration of 6.6×10^{13} cm^{-3}, whereas C-V indicated *p*-type conductivity and a carrier concentration of 10^{14} cm^{-3}. The relatively good agreement between these techniques lends confidence to our measurements.

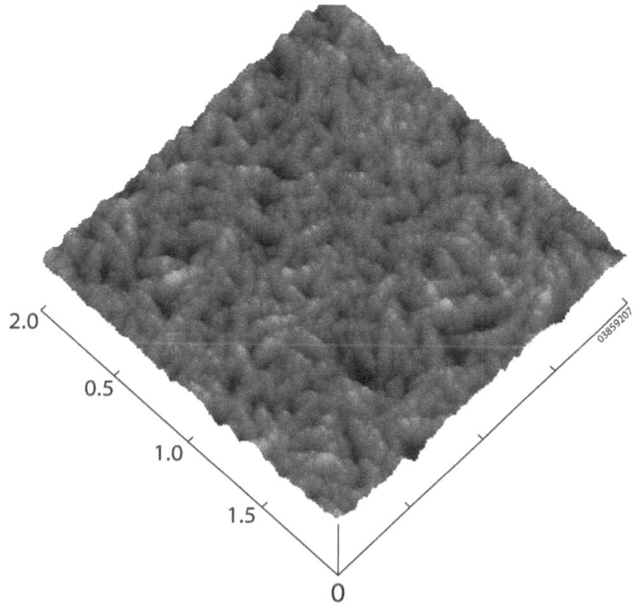

Figure 3.7: The topography of a ZnO:N film grown on a glass substrate. The growth temperature was 400°C, and the thickness of the film was 0.74 μm.

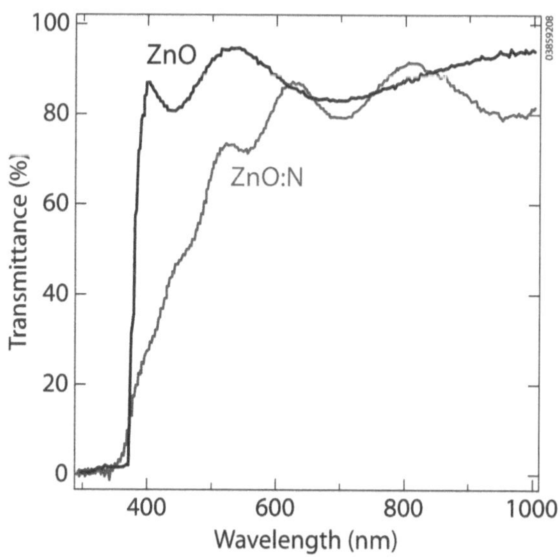

Figure 3.8: Optical transmittance of ZnO and ZnO:N films. Both films were deposited on glass substrate at 400°C. The effect of the incorporated impurities can be seen in the region near the band edge.

Table 3.1: Electrical properties of ZnO films deposited at a substrate temperature of 400°C.

Doping condition	Sample ID	Carrier conc. (cm^{-3})	Mobility ($cm^2\ V^{-1}\ s^{-1}$)	Resistivity (Ω-cm)	Conductivity type
Undoped	L3353.2	-8.42×10^{18}	3.2	0.235	*n*
Undoped	L4084.1	-8.38×10^{16}	6.3	11.8	*n*
N-doped	L3358.2	9.24×10^{13}	236	286	*p*
N-doped	L4098.2	8.36×10^{17}	4.55	1.64	*p*

We found *p*-type behavior only for growth temperatures of about 400–450°C. For temperatures less than 400°C, the film deposition rate was low and the films were *n*-type. The best electronic properties achieved in this temperature range for a single film were obtained using a sapphire substrate. The film had a carrier concentration of 8.36×10^{17} cm^{-3}, Hall mobility of 4.55 $cm^2\ V^{-1}\ s^{-1}$, and resistivity of 1.64 Ω-cm. If we used a substrate temperature greater than 500°C, we again observed *n*-type conductivity.

3.3.4 Impurity analysis

The typical N concentration in MOCVD-formed ZnO:N films is about 1–2.60×10^{21} cm^{-3}. The N concentration of ZnO:N samples was analyzed using AES, XPS, and SIMS. Measurement of the N concentration by SIMS is difficult, but a detailed discussion is not appropriate here. We shall discuss only the concentration data obtained using AES and XPS in this chapter. Although both techniques can provide an estimate of N concentrations in the 10^{21} cm^{-3} range, in general, it was found that XPS yielded more consistent values for N concentrations than AES, perhaps due to electron-beam-stimulated desorption of N during AES experiments. In addition, it was found that high-resolution XPS was able to differentiate between different chemical states of N in ZnO:N. Initial determinations of N concentrations were conducted via AES, whereas subsequent work was conducted with XPS. By using low primary-beam intensities we were able to ensure and reproduce N concentrations using AES analysis. Table 3.2 lists the N concentration data, obtained by AES, for several ZnO:N samples and various NO flow rates. This set of data indicates that the N concentration is affected by the partial pressure of NO in the growth ambient, although it does not increase linearly with flow rate because NO functions as both dopant and oxidizer. As the NO rate was increased, with fixed DEZ flow rate, the ratio of O to Zn must also have increased, leading to a more

Table 3.2: Nitrogen concentration data varies as the function of the NO gas flow rates. The ZnO:N films were deposited at 400°C on glass substrate, and the N concentration was measured using AES.

ID	NO flow rate (sccm)	F_{DEZn} (sccm)	NO/DEZ	Nitrogen concentration ($\times 10^{21}$ cm^{-3})
ZnO:NA113	40	0.4	80	1.66
ZnO:NB1	40	1	40	1.75
ZnO:NB2	36	1	36	2.21
ZnO:NB3	30	1	30	2.32
ZnO:NB4	20	1	20	1.87

O-rich environment. Therefore, even though the available N in the film growth ambient increased with NO flow rate, the N concentration in the films did not.

Other than N, H and C can also have an impact on the ZnO properties. H is present in ZnO, regardless of deposition technique. An isolated interstitial H impurity is predicted to act only as a donor in ZnO, i.e., it always exists in the form of H^+ and tends to bond with the host O^-.[12] Carbon impurities are more specific to MOCVD because C is contained in all metal-organic precursors. A recent publication indicated that C could be a donor or an acceptor, depending on the position of the Fermi level in the bandgap.[48] For *n*-type ZnO, C may substitute on O vacancies and act as an acceptor. When ZnO is insulating or *p*-type, the C may bond with O or N and form a donor state. Our experience has shown that H and C are always incorporated during growth and may originate from the metal-organic precursor. The extent of the incorporation of C and H depends strongly on the growth conditions. We shall discuss these issues in the next section.

We used SIMS depth profile analysis to investigate the H and C impurity levels in the undoped ZnO films, and C was also studied using XPS, as we shall discuss later. Of these two techniques, only SIMS can detect H directly and, therefore, we shall make comparisons between C and H using this technique. The dependences of the H and C concentrations on the growth temperature for undoped ZnO are shown in Fig. 3.9. This figure indicates that in undoped material, H and C levels are high at low deposition temperature, but as the deposition temperature increases, the concentrations of both decrease and stabilize at about 400°C.

The similar dependence of the C and H concentrations on the deposition temperature suggest that they may exist in a bound state – a theory that is strongly supported by FTIR data.[49] The FTIR absorbance spectra of undoped films in the wavenumber range of 2500–4000 cm^{-1} were studied. Several important local

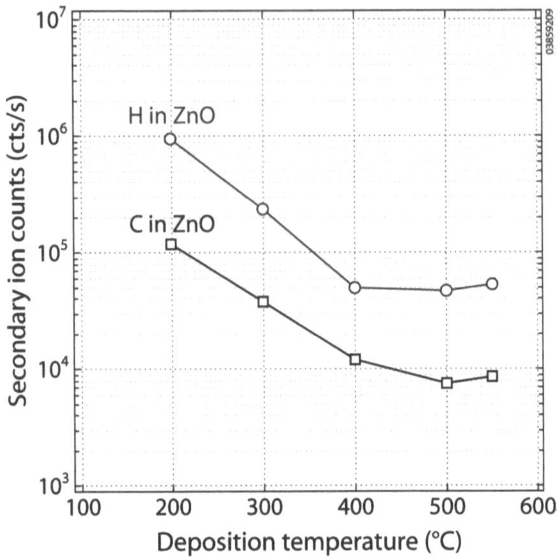

Figure 3.9: SIMS analysis of H and C in MOCVD-fabricated, undoped ZnO samples. The films were deposited at substrate temperatures between 200 and 550°C using flow rates of DEZ and O_2 of 1 and 100 sccm, respectively.

stretch modes involving H bonded to C, N, and O are located in this spectral region. FTIR absorbance spectra show a series of sharper absorption peaks in the range of the CH_x stretch-mode frequencies for the undoped film. These specific peaks correlate well with the observed frequencies of the CH_x stretch modes. Combining both SIMS and FTIR data leads us to conclude that C and H in the films are very likely bonded together as CH_x defect complexes. Therefore, they almost certainly originate from the metal-organic precursor. At low deposition temperature, as products of the decomposition of the DEZ, the various CH_x species can be expected to be buried in the growing film. However, as the deposition temperature increases, the CH_x molecules may obtain enough thermal energy to desorb from the substrate surface, leading to a reduced concentration of CH_x in the ZnO film.[27] This has also been observed in GaN.[50,51]

The dependences of H and C concentrations on growth temperature are different for the N-doped materials. Figure 3.10 shows that the H and C concentrations in the ZnO:N do not simply decrease with increasing deposition temperature. Films grown at 400°C have the lowest H and C concentrations, but as the deposition temperature is increased to greater than 400°C, the concentrations increase again. The N concentration has the opposite behavior and decreases for deposition temperatures

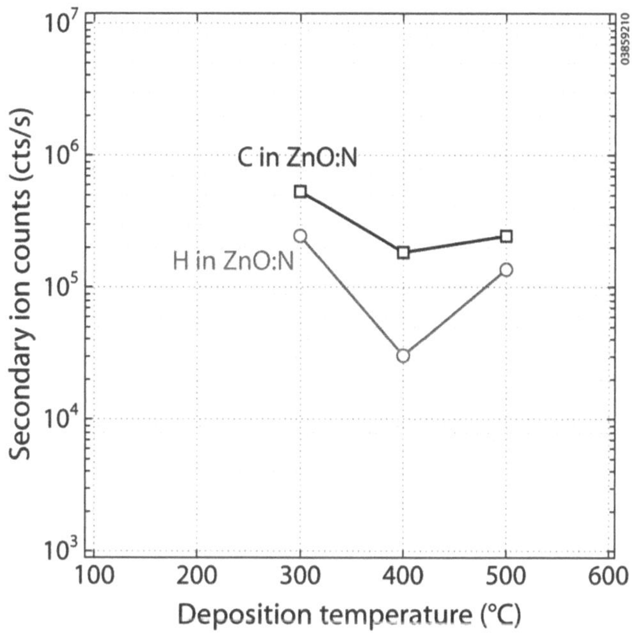

Figure 3.10: SIMS analysis of H, C, and N concentration on the MOCVD-fabricated ZnO:N samples. The ZnO:N films were deposited at temperatures between 200 and 550°C with DEZ and NO gas flow rates of 1 and 36 sccm, respectively.

greater than 400°C. In addition, when we compare the H and C concentrations over the entire range of growth temperatures, we find that both H and C concentrations are higher for N-doped samples than for undoped samples. This suggests that there must be a chemical reaction between NO and the DEZ precursor. However, at this stage, we are unable to determine whether this reaction occurs in the gas phase, on the surface of the ZnO film, or via both mechanisms. Nevertheless, it is clear that when NO is used instead of O_2 as the oxidizer, the C and H concentrations in the ZnO are increased. We are presently attempting to clarify this issue.

The detailed FTIR analysis indicates that with incorporation of N, the broad asymmetrical band around 3400 cm^{-1} becomes narrower and more symmetrical at 3410 cm^{-1}. This could be due to a surface mode of an O-H bond.[23] Furthermore, the peaks around 2900 cm^{-1} mostly disappeared. Instead, a new peak formed at 3007 cm^{-1}, which can be fit by individual Gaussian peaks at 3020 and 2970 cm^{-1}. The peak at 2970 cm^{-1} is possibly due to the remaining C-H_x bonds, and the peak at 3020 cm^{-1} is new.

We do not believe that the increased C and H concentrations necessarily indicate that the concentration of C-H_x complexes also increases. The FTIR absorbance spectrum indicated that, with N doping, the relatively well-understood and well-characterized C-H_x modes are strongly reduced at a deposition temperature of 400°C and a new N_O-H peak appears. This indicates that either the NO assists the breaking of C-H_x bonds in the gas phase or that the N_O-H complex in the film is energetically more favorable (i.e., more stable) than the C-H_x defect.

As mentioned above, the C concentrations in the ZnO:N films are much higher than in the undoped films, perhaps suggesting that bonding occurs between C and N. Using FTIR, we identified very strong infrared absorption peaks in the range of frequencies consistent with NC, NO, NN, and CO modes.[49] XPS data suggest that C-N bonds are likely present in MOCVD-grown ZnO:N films.

Many of these issues will be discussed in greater detail in the coming section. Although we treat the N, H, and C impurities separately, one must appreciate that there is considerable interaction between them.

3.3.5 Compensation and passivation effects in ZnO:N films

3.3.5.1 Nitrogen-related defects

Although we have found that the acceptor concentration increases with N concentration, the concentration of N in the films is much greater than that of the holes. One explanation for the low carrier concentration is that the acceptors must be heavily compensated. In previous work, we showed that H^+ could passivate the N_O acceptors. Here, we will discuss some other possible compensation mechanisms in MOCVD-grown, ZnO:N films.

First, we shall discuss the possible roles of N. In ZnO, N can act as a donor as well as an acceptor, and its particular role depends on several factors, including the conditions under which the film was made and the position of the Fermi level in the bandgap. For example, when the Fermi level is in the lower half of the bandgap, i.e., when the material is *p*-type, an interstitial N (N_i) defect could be a donor with any one of three charge states (3^+, 2^+, 1^+). If a single N atom substitutes for O (N_O), it is a single acceptor (1^-). For *p*-type material, when N_2 substitutes for an O atom forming $((N_2)_O)$[1], it is expected to act as a double donor (2^+). Therefore, knowledge of the N concentration in the ZnO film is not sufficient to understand the net doping of the material. For this, we require some knowledge of the relative concentrations of each of the possible defects. In general, the concentration of N_O defects must exceed that of the other two configurations. However, the ratio of the N_O to N_i or to $(N_2)_O$ depends on the formation energy of each of these defects and

[1] This can also be represented by N_i-N_O, and it is also known as a split interstitial.

on the position of the Fermi level. Note that the formation energy also depends on the source materials. Because the isolated N_i defect has a large formation energy, we shall assume that its concentration will be very low.

However, isolated interstitial and substitutional defects experience a strong Coulombic force of attraction and are likely to form a strongly bound defect, i.e., the split interstitial, N_i-N_O, which, as mentioned above, acts as double donor. Hence, under N-rich fabrication conditions, in which interstitials are likely to form, there is a high probability that they will bind to substitutional N to form the $(N_2)_O$ donor-like defect. The $(N_2)_O$ complex is stable in the charge states 2^+ and 1^+. Thus, for N concentration levels as high as 10^{21} cm^{-3}, one of the concerns is that N will form $(N_2)_O$, instead of N_O.[18] Hence, it is to be expected that some of the N would have been incorporated in its molecular form. The results showed that it is very difficult to achieve a high acceptor concentration in ZnO:N films when N_2 is used as the sputtering gas, presumably because of self-compensation. This observation is consistent with the theoretical ideas above.

We have used XPS to help understand the role of N in films. We have found that, with appropriate care, it should be possible to identify the $(N_2)_O$ and N_O defect states in ZnO. Perkins et al. studied the role of N in ZnO films by using XPS.[42] To aid in the assignments of the N 1s spectra of the N-doped ZnO films, N-containing standards were prepared within the XPS analysis chamber by N_2^+ ion bombardment of the following: i) Zn metal foil, ii) C-containing MOCVD-grown ZnO, and iii) C-free ZnO (single-crystal and ZnO films grown by reactive sputtering from a Zn target in various mixtures of N and O). Comparisons of data from the standards and the N-doped ZnO films allowed conclusions to be drawn about the types and quantities of N chemical states in the ZnO:N films.

Prior to the *in situ* implantation of N, the ZnO (0001) substrates (supplied by Princeton Scientific Corporation) were sputter-cleaned and annealed at 400°C. The foils were cleaned by *ex situ* scraping and *in situ* sputtering. Nitrogen was implanted at an energy of 3 keV.

The core-level spectra for both the sputtered ZnO:N films and the implanted foil were strikingly similar, as can be seen in Fig. 3.11. In the case of the implanted foil, peaks were observed at 396.5 and 404.9 eV. These energies were very close to those observed for the ZnO:N film. In the foil, the lower-energy peak was believed to be due to N^{3-}, whereas in the sputtered film it probably corresponded to the desired N_O substitutional defect. It was further concluded that the higher-energy peak in the spectrum of the foil was probably associated with a diatomic N species. Assuming this is also the case for the ZnO film, it is probable that this corresponded to the donor-like $(N_2)_O$ defect. Figure 3.12 shows the 1s spectrum for a MOCVD-grown NO-doped ZnO film. The data were taken in ultra-high vacuum after 120 seconds of 3-keV Ar^+ sputter cleaning (20 μA cm^{-2}). As discussed above, the N 1s peaks at 396

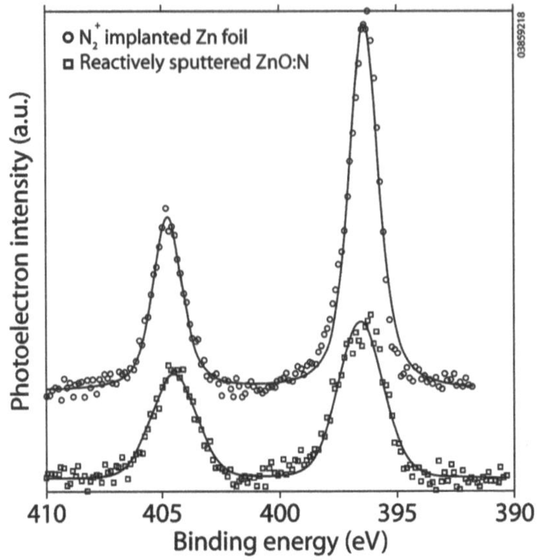

Figure 3.11: XPS spectra of a ZnO:N film grown by reactive sputtering and of a N-implanted Zn foil.[42]

and 404 eV are believed to be due to N_O and N_2, respectively. The areas under the two N-related peaks in the XPS spectrum, corrected for the appropriate sensitivity factors, provide the estimated N_O and N_2 concentration produced by N doping. The N_O:$(N_2)_O$ ratio estimated in the MOCVD-grown ZnO film is much higher than the ratio in a film grown by reactive sputtering in N_2. This result indicated that, by changing the N source, the N self-compensation could be reduced, and it added to the mounting evidence that using NO has a significant advantage over N_2. This is in agreement with the prediction of Yan et al.[19]

For an isolated N_O acceptor, we calculated the ionization energy to be about 0.2 eV, which is in good agreement with recent experimental estimates of 0.165 eV.[45] The microscopic structure of N_O in the 1^- charge state is illustrated in Fig. 3.13. The N atom draws the nearest-neighbor atoms closer by about 3% with respect to the Zn-O bond in bulk ZnO. In the neutral charge state, the defect geometry is almost identical to the 1^- charge state, but the N atom draws the nearest-neighbor Zn atoms closer by about 2% with respect to the Zn-O bond in bulk ZnO.

The $(N_2)_O$ is stable in 2^+ and 1^+ charge state under *p*-type conditions. The corresponding vibrational frequency of 2^+ and 1^+ charge state are calculated as 2108 and 1852 cm^{-1}, respectively. Figure 3.14 shows that there are several peaks in this wavelength range, as observed using FTIR spectroscopy. Although the theoretical calculations predict bond lengths close to those of the corresponding free

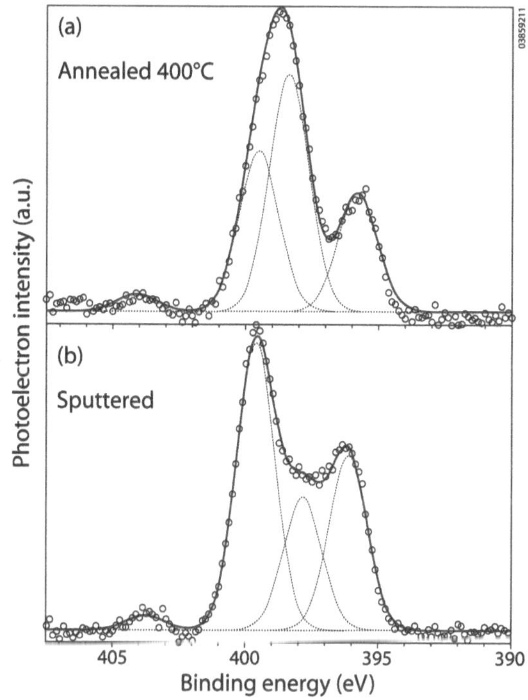

Figure 3.12: XPS spectrum of a ZnO:N film that was grown by MOCVD. The film was subjected to cycles of annealing in ultra-high vacuum and sputtering. The spectra were reversible for many cycles. The dotted line shows the deconvoluted spectrum indicating the individual peaks.[42]

diatomic molecule, the existence of sufficient asymmetry might result in a nonzero absorption coefficient. The observed peaks centered at 2000 and 1710 cm^{-1} are in reasonable qualitative agreement with the calculated positions of the $(N_2)_O^{2+}$ and $(N_2)_O^{1+}$ defects, respectively.

3.3.5.2 Hydrogen-related defects

As mentioned already, H itself is always a donor in ZnO, and we have devoted significant effort to understanding its various bonding configurations. In this section, we shall discuss some H-related defect complexes. The theoretical calculation of possible H-related vibrational modes was performed following the approach used by Limpijumnong et al.[52] Due to the light mass of the H atom, it is imperative to construct a potential vs. displacement curve that includes vibrational amplitudes of

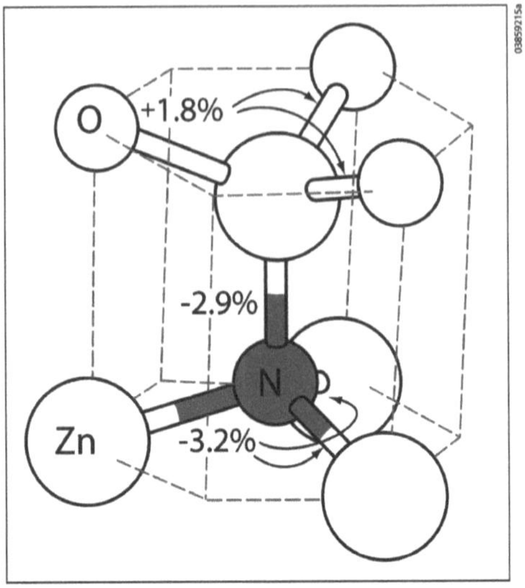

Figure 3.13: Atomic structure of the N_O acceptor in ZnO in the 1^- charge state. The large white spheres are the Zn atoms, the small white spheres are the O atoms, and the small shaded spheres are the N atoms. The bond lengths are given in terms of the percentage differences from bulk ZnO bond.[23]

at least 0.2–0.3 Å, in which region anharmonicity is important. The potential vs. displacement curve can be fitted by a 4th degree polynomial, i.e.,

$$V(x) = \frac{k}{2}x^2 + \alpha x^3 + \beta x^4. \tag{3.1}$$

The harmonic approximation, i.e., $V(x) = kx^2/2$, represents the potential vs. displacement curve inadequately. Next, we used perturbation theory to obtain an approximate analytical solution to the Schrödinger equation, using this potential.[53] We used the reduced mass (between the H atom and its oscillator pair – in this case, the N atom) for the oscillator mass in the Schrödinger equation.

To estimate the uncertainty in the calculated vibrational frequencies associated with the approximations in the computational approach (e.g., pseudopotentials, local density approximation, density functional theory), which should be mostly systematic in nature, we calculated the symmetric stretch-mode frequency of a free ammonia (NH_3) molecule. We obtained a value of $3194\,cm^{-1}$.[54] In comparison with the measured value ($3337\,cm^{-1}$), our calculation underestimated the actual value by $143\,cm^{-1}$, or 4.2%. We assume that this underestimation is systematic for

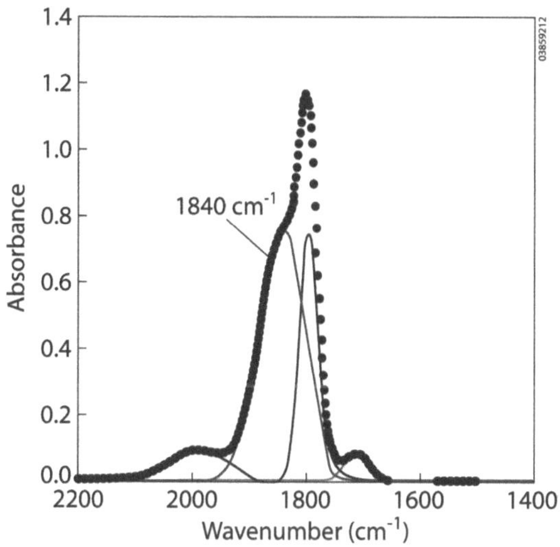

Figure 3.14: FTIR absorbance spectra of N-doped ZnO films from 1400 to 2200 cm^{-1}. The solid lines show the deconvoluted data.

other H-containing molecules and that our calculated N-H stretch frequencies in ZnO are consistently lower than the measured values by 143 cm^{-1}.

The FTIR absorbance spectra of undoped and N-doped ZnO material are shown in Fig. 3.15. The vibrational frequencies of several important H-related stretch modes involving C, N, and O are found within this spectral region.[55] On inspection, we see that there is an asymmetrical band at 3400 cm^{-1} for the undoped films, most likely due to an O-H bond. A series of sharper absorption peaks in the range of the CH_x stretch-mode frequencies was also observed for the undoped film. Fitting this sharper absorbance spectrum near 3000 cm^{-1} with Gaussian peaks yields individual absorbance modes centered at 2967, 2926, and 2856 cm^{-1}.

The presence of both C and H in the DEZn precursor used in this work, and the detection of both impurities in the bulk of the thin film, are reasons to expect CH modes in the FTIR spectra of this material.[23] The peaks centered at 2967, 2926, and 2856 cm^{-1} are typical of CH stretch frequencies.[55] We shall discuss the role of C more extensively later.

With incorporation of N, the broad asymmetrical band around 3400 cm^{-1} becomes a narrower and more symmetrical peak at 3410 cm^{-1}, which is in a region typically associated with N-H and O-H vibrations.[55] Theoretical calculations predict O-H vibrations in ZnO ranging from 3216 to 3644 cm^{-1}, depending on the configuration and number of H impurities in the complex.[56] Experimentally observed peaks in single-crystal ZnO assigned to O-H have been observed in as-grown and H-treated material.[56–59] We tentatively assign the peak at 3410 cm^{-1} to O-H.

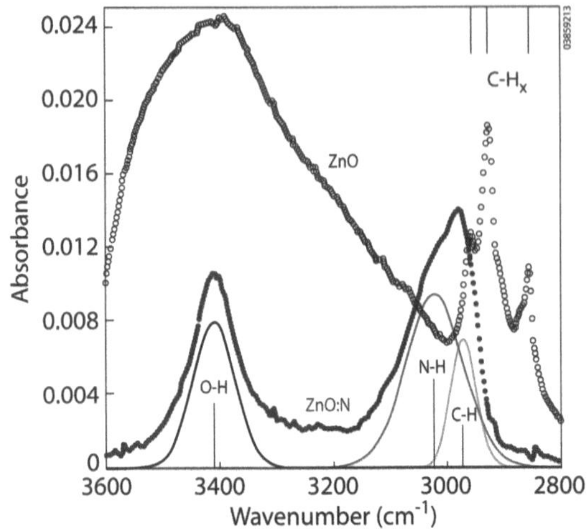

Figure 3.15: The curve fitting for N-doped sample provides Gaussian peaks (solid lines) located at 3410, 3020, and 2970 cm^{-1}. The peak positions of the C-H_x local vibrational mode found from the undoped ZnO film are indicated on the top of the plot.[23]

In addition, and as mentioned above, the addition of N caused the C-H_x stretch modes (in the vicinity of 2900 cm^{-1}) to become much smaller. A new peak that appeared at 3007 cm^{-1} could be deconvoluted into two peaks, one of which (at 2970 cm^{-1}) may have been due to additional C-H_x modes. The other, centered at 3020 cm^{-1}, is higher in frequency than any expected C-H stretch mode and lower in frequency than any observed or calculated O-H stretch mode in ZnO.[56–59] This peak is in the spectral region associated with N-H bonds in III-V semiconductors, which makes the N-H bond a likely candidate for this observed mode.[60,61] The N-H bond has been associated with a vibrational line at 3096 cm^{-1}, in unintentionally N-doped single-crystal ZnO.[62] It is interesting that none of the peaks in the latter work is assigned to O-H, even though the O concentration is significantly higher than the N concentration. This suggests H may preferentially bond to N.

To understand the possible HN bonds in additional depth, we calculated the formation energies of an isolated interstitial H impurity, a N acceptor N_O, and a N_O-H complex (see Fig. 3.16). Zn-rich growth conditions are assumed, and the chemical potentials for N and H atoms are based on N_2 and H_2 precipitation limits. For an isolated H^+ location, there are four low-energy sites surrounding an O atom, with the so-called $BC_{||}$[2] site having the lowest energy.[56] The other three sites

[2] AB signifies anti-bonding. BC signifies bond center.

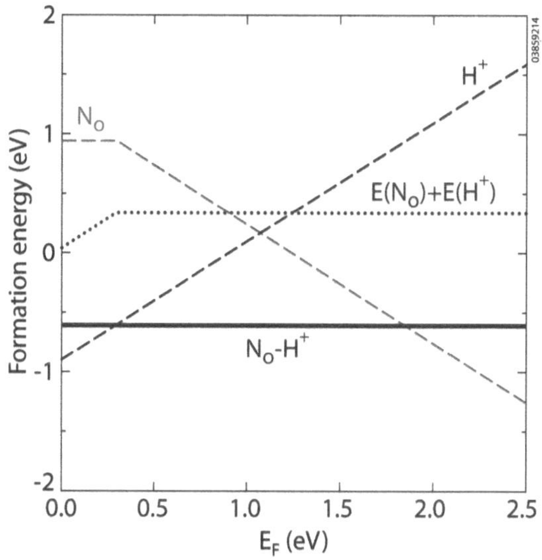

Figure 3.16: Calculated formation energy of interstitial H, N_O (dashed lines) and N_O-H (solid line) complexes as a function of electron Fermi energy. The Fermi energy is referenced to the top of the valence-band maximum of ZnO. The dotted line shows the sum of the formation energies of an isolated interstitial hydrogen and an isolated N_O. The Zn-rich condition, N_2 phase precipitation limits (for nitrogen chemical potential), and H_2 phase precipitation limits (for H chemical potential) were used in the calculations. The slopes of the curves reflect the charge states of the defects.[23]

have slightly larger energies, by about 0.2 eV. Only the formation energy of the lowest-energy configuration $BC_{||}$ is used for the plot in Fig. 3.16.

For almost the entire Fermi energy range – that is, the bandgap – the isolated N_O is stable in a 1^- charge state, whereas an isolated interstitial H atom is always stable as a 1^+ charge state. The two defects, therefore, experience a Coulombic attraction that results in a strong tendency to form a neutral defect complex. We found the formation energy of the N_O-H complex (solid line in Fig. 3.16) to be lower than the sum of the formation energies of the individual defects (dotted line) by 0.95 eV. This implies that the complex has a binding energy of 0.95 eV. An important consequence of this, for the interpretation of FTIR data, is that, in the presence of N acceptors, H atoms are energetically likely to passivate the acceptors and to form N-H bonds, instead of existing as isolated interstitial H^+ and forming O-H bonds with host O atoms. If the sample were strongly *p*-type, then the latter statement may not be correct.

In a N_O-H defect complex, H is more likely to occupy the sites surrounding the N atom. Similar to the case of an isolated interstitial H – and due to wurtzite

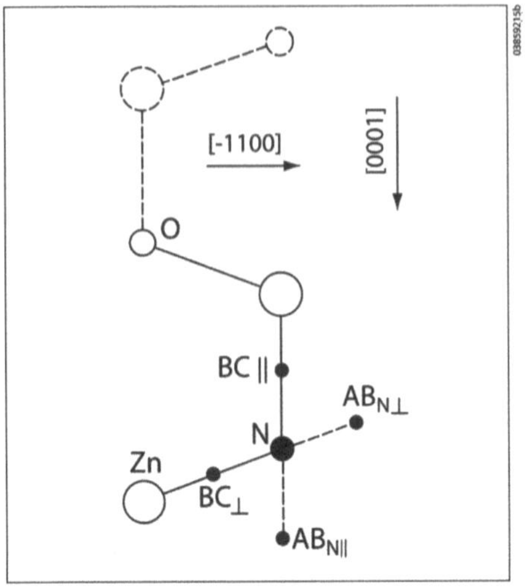

Figure 3.17: Atomic structure of the N_O acceptor in ZnO in the 1^- charge state and possible H sites surrounding the N_O acceptor in the $(11\bar{2}0)$ plane of wurtzite ZnO. The large white spheres are the Zn atoms, the small white spheres are the O atoms, and the small shaded spheres are the N atoms. The small solid circles represent the possible sites for H.[23]

Table 3.3: Calculated relative formation energy, N-H bond distance, and vibration frequency for N_O-H defect complex in four lowest-energy configurations in wurtzite ZnO. ω^0 is the harmonic component of the vibrational frequency, $\Delta\omega$ is the anharmonic contribution, and ω is the total frequency: $\omega = \omega^0 + \Delta\omega$.

Site	ΔE (eV)	d_{N-H} (Å)	ω^0 (cm^{-1})	$\Delta\omega$ (cm^{-1})	ω (cm^{-1})
$AB_{N\perp}$	0.00	1.050	3144	−232	2912
$AB_{N\|\|}$	0.08	1.047	3173	−232	2941
$BC_{\perp}$	0.32	1.027	3395	−88	3307
$BC_{\|\|}$	0.18	1.029	3478	−147	3331

symmetry – there are four possible high-symmetry sites surrounding the N atom: two BC and two AB_N, as seen in Fig. 3.17. We have calculated the energetics of all four sites and found the $AB_{N\perp}$ configuration to have the lowest formation energy. The relative formation energies, stretch-mode vibrational frequencies, and calculated N_O-H bond distances for all four configurations are tabulated in Table 3.3, and the geometry of all four configurations is illustrated in Fig. 3.18. The vibrational frequencies were calculated as described earlier in this section.

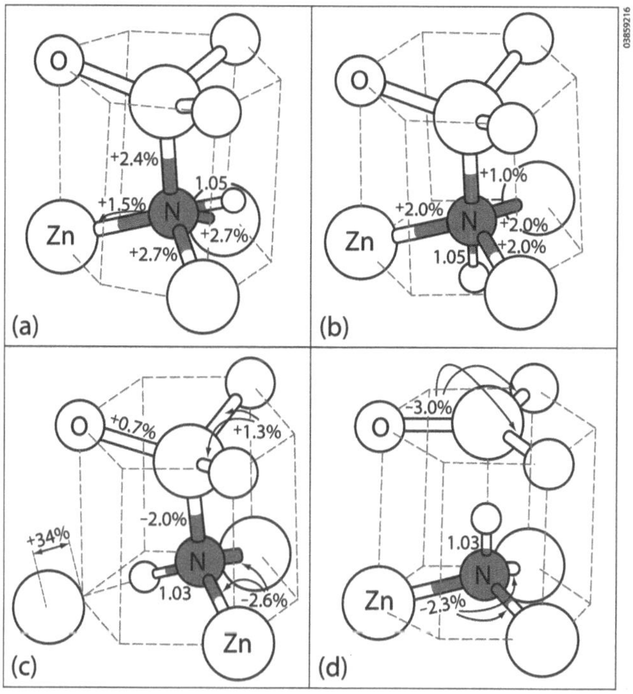

Figure 3.18: Atomic structures of N_O-H complexes: (a) $AB_{N\perp}$ (b) $AB_{N||}$ (c) $BC_{\perp}$, and (d) $BC_{||}$ configuration. The large white spheres are the Zn atoms, the medium white spheres are O atoms, the medium shaded spheres are N atoms, and the small white spheres are H atoms. The distance between the H and N atoms are given in Å; some Zn-N and Zn-O bond lengths are given in terms of the percentage differences from the bulk ZnO bond.

Both AB_N configurations have similar N_O-H stretch-mode frequencies with an average value of 2927 cm^{-1}, and the anharmonic contributions are large, i.e., a 8%. Both BC configurations also have similar N_O-H stretch-mode frequencies with an average value of 3319 cm^{-1}, i.e., significantly higher than those of the AB_N configurations. The anharmonic contribution associated with the BC configurations is less than half that associated with the AB_N configurations. The smaller anharmonic contribution was also found for H in GaN.[52] To compare the calculated frequencies with the observed values, we applied the systematic computational error of 143 cm^{-1} (obtained for the free NH_3 molecule) to the calculated N-H vibrational frequencies. Therefore, the systematic-error-adjusted vibration frequencies associated with the AB_N and BC configurations are 3070 cm^{-1} (2927 + 143) and 3462 cm^{-1} (3319 + 143), respectively. However, the AB_N configurations are energetically more stable than the BC by about 0.2–0.3 eV. With this energy difference,

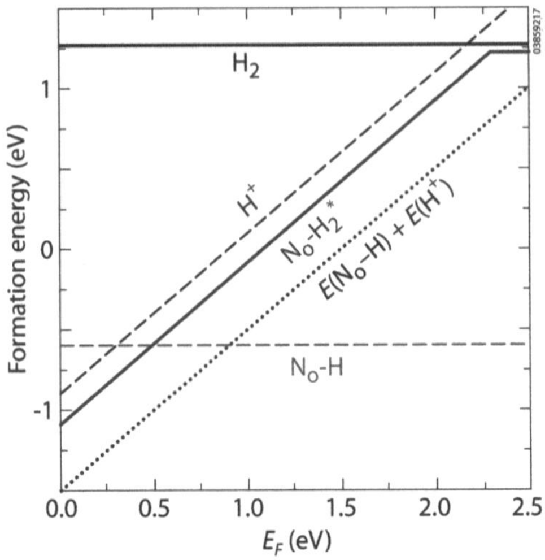

Figure 3.19: The energetics of formation of the N_O-H_2^* complex. Dashed lines represent the formation energies of N_O-H complex and an isolated H^+ defect. The solid lines represent the formation energy of the N_O-H_2^* complex. The dotted lines show the sum of the formation energies of N_O-H complex and an isolated H^+ defect. The N_O-H_2^* complex has a total formation energy higher than the sum of individual defects, which means the complex is unstable. The Zn-rich condition, the NO_2 phase precipitation limits (for N chemical potential), and H_2 phase precipitation limits (for H chemical potential) were used in the calculations. The formation energy of H_2 molecule in ZnO is also shown.

the possibility of forming the BC configuration is negligible. From the analysis above, we are therefore confident in assigning the observed absorption band at 3020 cm^{-1} to N_O-H in the AB_N configuration.

An additional test calculation using slightly different computational details yields results in qualitative agreement with calculations in the last section.[23] For example, the $AB_{N\perp}$ configuration is also found to be the site with the lowest formation energy (the other sites are slightly higher by about 0.2 eV) and with a vibrational frequency of 3117 cm^{-1}.

We also explored some other H-related defect complexes such as H_2, H_2^*, and N_O-H_2^*. The calculations predict that an interstitial H_2 molecule is unstable in ZnO, except in extreme *n*-type conditions (i.e., the Fermi level close to the conduction-band minimum). The calculated formation energy (plotted per H atom) is shown in Fig. 3.19 in comparison with the formation energy of an isolated H^+ and N_O-H complex in ZnO. Under Zn-rich conditions, the formation energy of H_2 molecules in ZnO is larger than that of an isolated H^+ and N_O-H complex, for almost the entire Fermi energy range. Again, only the formation energy of the lowest-energy configuration (in this case AB_N) is used in the plot.

The H_2^* complexes, one of the possible H-related configurations that has been predicted to be stable in GaP:N.[63] and metastable in GaN,[52] have a high formation energy in ZnO and are thus unlikely to exist. The reason that the H_2^* complexes are unstable in ZnO, as opposed to most III-V and group-IV semiconductors, is that H^- is not stable in ZnO. A H_2^* complex can be viewed as a complex between H^+ bonded with O, and H^- bonded with Zn.

The calculated formation energy of the N_O-H_2^* molecule in ZnO is shown by the solid line in Fig. 3.19, in which it is compared with the sum of the formation energies of an isolated H^+ and a N_O-H complex (dotted line). The role of N in ZnO is very different from its role in dilute GaP:N and GaAs:N systems. In these III-V semiconductors, N is an isovalent impurity when substituted for the anion, whereas it is an acceptor when substituted for O in ZnO. When N substitutes for phosphorus or arsenic, the lattice mismatch is large compared to the mismatch when substituting for O in ZnO. N_O is a single acceptor in ZnO, when it forms a complex with an H^+, the complex has a neutral charge state. There is no Coulombic driving force for the complex to bind with another H^+ to form a N_O-H_2^* complex. Therefore, it is even more difficult to form a N_O-H_2^* complex than an isolated H_2^*, which is also not stable in ZnO. On the other hand, in an isolated N_O defect, the N atom causes a slight inverse relaxation of its neighboring atoms, as shown Fig. 3.13, that slightly expands empty space around the Zn atoms. Although, in general, larger empty space helps to lower the formation energy of H^-, it is still insufficient to stabilize H^-. Thus, N_O-H_2^* complexes will not be formed by N_O-H binding with H^-.

In summary, it is probable that the inevitable presence of H has a negative influence when the goal is to achieve *p*-type material. Hydrogen has several possible configurations in ZnO. An isolated interstitial H is a donor in ZnO, and an interstitial H_2 molecule is unstable in ZnO except in extreme *n*-type conditions. Therefore, of various possible H-related defects, we need only consider the N_O-H complex for NO-doped ZnO:N films. As we have previously stated, the formation of the neutral N_O-H complex should increase the solubility of H in ZnO:N. Therefore, with N doping, the H concentration should be higher than that in undoped material.

3.3.5.3 Carbon-related defects

As seen above, there are theoretical grounds for believing that defect complexes may play a significant role in ZnO and may be responsible for the low hole concentrations observed in practice. It is also reasonable to expect that the reaction between the precursors would have led to the formation of various CN compounds. In particular, Perkins et al. used XPS to study the role of N and C in films grown by thermal CVD, from DEZ and NO precursors, and by reactive sputtering.[42]

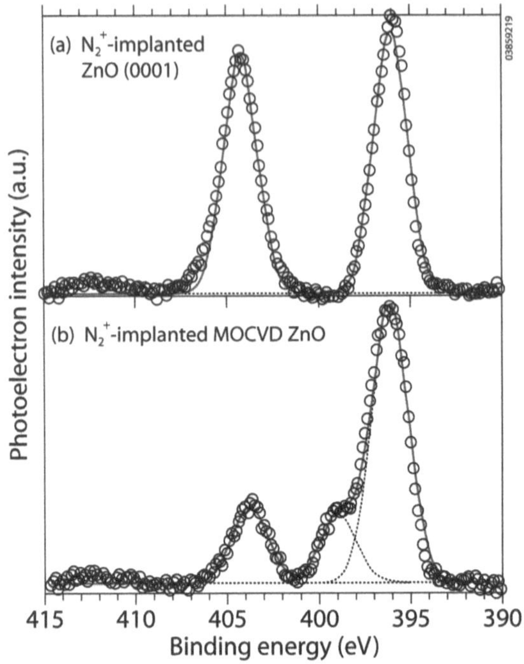

Figure 3.20: XPS spectra of a N-implanted single-crystal of ZnO, (a), and a N-implanted film of undoped ZnO grown by MOCVD. The dotted line in (b) shows the deconvolution of the spectrum, indicating an additional peak not present in the single-crystal.

XPS data showed substantial differences between ZnO:N films grown by reactive sputtering and by MOCVD. Carbon was found to be a bulk impurity in the MOCVD films, whereas the sputter-grown films had only slight surface C contamination. The N 1s spectra of films prepared by MOCVD indicated that nitrogen occupied a total of four different chemical states, two more than in the sputter-grown ZnO:N films. The two extra chemical states manifested themselves as peaks at 398.3 eV (probably corresponding to CN bonds) and 399.5 eV (in the binding energy range of amines, i.e., NH_x, and CN compounds).

Further evidence that the reaction between precursors led to the formation of various CN compounds was provided by implantation of N into i) a C-free single-crystal of ZnO (0001) and a sputter-grown, C-free ZnO film and ii) a C-containing, N-free, MOCVD-grown ZnO film. The latter was grown using DEZ and O_2, rather than NO. As can be seen in Fig. 3.20a, the XPS spectrum of the single-crystal showed the same peaks that had previously been seen in Fig. 3.11 for the N-implanted Zn foil and for the reactively sputtered ZnO:N. However, in the case of the MOCVD-grown, C-containing film, an additional peak was observed at an energy of 398.9 eV, as seen in Fig. 3.20b. The spectra shown in Fig. 3.12 were found to be reversible for many cycles of annealing (Fig. 3.12a) and sputter cleaning (Fig. 3.12b), during which H in the film would have been expected to be removed. Hence, the conclusion was that the additional peak seen in Fig 3.12 was probably not due to a N-H bond, but was more likely due to a CN species.

Ultraviolet photoelectron spectroscopy was also used to characterize the films, and it was concluded, from the position of the Fermi energy with respect to the valence-band maximum, that all films studied were *n*-type. However, it is known that there are changes in the conductivity type of *p*-ZnO films if they are stored in illuminated conditions, and it is possible that the ultraviolet photons may have accelerated this effect. The work described has shown that there are several point defects present in the films discussed. Theoretical considerations lead us to believe that these are all likely to act as donors and to compensate the single acceptor-like defect.

First-principles calculations indicate that C has several possible defect configurations and charge states in ZnO. Interstitial C (C_i) is a donor and substitutional C (C_O) is either neutral or is a deep acceptor when the Fermi level lies in the lower half of the bandgap. Both C_i and C_O have a high formation energy, and the concentration of interstitial and substitutional C is expected to be correspondingly low. However, the N_O and C_i could bond to form an $(NC)_O$ defect complex. The binding energy of this complex is large; therefore, it is likely to form. This prediction is strongly supported by the experimental observation that the C solubility increased in the N-doped material and by the possible existence of NC bonds, as discussed above. The NC defect complex is predicted to be a donor over most of the bandgap (except very close to the conduction-band minimum), thus the increased C will compensate N_O doping. Therefore, the observed high C concentration could be one of the important factors that causes the MOCVD-grown ZnO to be *n*-type. Especially for N-doped ZnO, when N_O concentration is low, the C impurity could be a dominant effect. We suggest that this may explain the very narrow process window for making *p*-type ZnO by MOCVD.

First-principles calculations predict that, under *p*-type conditions, the $(NC)_O$ defect complex is stable in a 1^+ charge state and has a vibrational frequency of 1995 cm^{-1}. This is in reasonable qualitative agreement with the observed peaks in the vicinity of 2000 cm^{-1} observed by FTIR. If the Fermi level moves into the top half of the bandgap, the $(NC)_O$ complex could change its charge state. The $(NC)_O^0$ complex is expected to have a lower vibrational frequency of 1688 cm^{-1}. Our FTIR experiment observed no peak in this wavelength region, and this observation could be well explained by the theoretical prediction. The ZnO:N samples we studied are either insulator or *p*-type (the Fermi level is not in the top half of the bandgap). Thus, the $(NC)_O^0$ complex is not stable. From the above discussion, we have clearly shown that it is very important to reduce the C concentration to reduce the compensation effect and achieve a high acceptor concentration. This may be difficult to achieve in films grown via MOCVD.

3.4 Discussions and summary

It may also be useful to discuss our targets for the electrical and optical properties of *p*-ZnO. What must we achieve to define our project at NREL as successful? In the context of solar cells, *p*-type ZnO TCO would be potentially useful if it could be used to provide additional design flexibility via, for example, enabling device designers to reconfigure the CdTe cell. However, for the *p*-ZnO window layer contact to be used for this purpose, a hole concentration of about $10^{20}\,cm^{-3}$ would be needed. Clearly, we are several orders of magnitude distant from this goal! In addition, the charge mobility, which is usually about $0.1–1\,cm^2\,V^{-1}\,s^{-1}$ for *p*-ZnO, must be increased substantially. The effective mass of holes is about $0.6\,m_e$, whereas that of electrons is about $0.4\,m_e$.[64] The typical mobility of *n*-ZnO is in the range of $20–30\,cm^2\,V^{-1}\,s^{-1}$, i.e., at least 20× greater than that of *p*-ZnO. Hence, the ratio of the effective masses does not account for this large difference, and it is possible that the mobility is limited by a very short hole relaxation time caused by additional scattering due to the very high N concentrations that we have observed. The N may be molecular or in the form of complexes. This, of course, does not apply to other approaches discussed in this book, although similar arguments may. For an *n*-type window layer with a carrier concentration of $5 \times 10^{20}\,cm^{-3}$ and a mobility of $20\,cm^2\,V^{-1}\,s^{-1}$, the resistivity is about $6 \times 10^{-4}\,\Omega$-cm. If *p*-ZnO were to be used in an inverted configuration solar cell, as shown in Fig. 3.1, a similar resistivity would be necessary. In addition to these electrical requirements, the window layer must also be highly transmissive, i.e., at least 80%. The transmittance is closely linked to the mobility of the charge carriers in a TCO, with both the height and width of the free-carrier absorption band increasing as mobility decreases. Hence, a high hole concentration alone is a necessary, but not sufficient, property to meet the needs of a window layer.

For the monolithically grown, series-connected tandem thin-film solar cell discussed earlier, an integral tunnel-junction interconnect is required. For *p*-ZnO to be a partner in a tunnel-junction interconnect, it must be degenerate – meaning that the Fermi level must lie in the valence band. If, for the purposes of making a semiquantitative point, we assume that the Fermi surface is spherical, it is straightforward to show that the Fermi level aligns with the top of the valence band when the hole concentration is about $3 \times 10^{19}\,cm^{-3}$. In this application, the optical and electrical properties are not as critical as those described for the window layer above. The two sides of the tunnel junction are each very thin (on the order of 10–30 nm), and neither their optical nor their electrical properties would significantly affect the properties of the tandem cell. In this sense, the application may be somewhat simpler in principle, although the practical problems may be much more difficult, given the thermal sensitivity of the stack of thin-film layers.

In typical III-V LEDs, each side of the junction has a doping concentration on the order of $3–5 \times 10^{18}\,cm^{-3}$.[65] Hence, this is the most-promising near-term application, and it may be important to solid-state lighting and other applications involving LEDs. As is true with the other device applications, even if it proves possible to achieve hole concentrations of this level, growth of the second of the ZnO layers must not disrupt the first layer. ZnO also has the attraction that higher- and lower-bandgap alloys may be fabricated using Mg and Cd, respectively.

In this chapter, we have discussed work at the National Renewable Energy Laboratory, and collaborating institutions, concerned with developing MOCVD-grown N-doped *p*-type ZnO. To date, the carrier concentrations we have achieved have been relatively low, but it must be noted that this is similar to the results of other workers using different dopants and growth techniques. By using the fabrication, characterization, and theoretical capabilities available at NREL, we have developed a clear understanding of the reasons why our films have such low hole concentrations.

It appears that films, prepared using MOCVD and reactive sputtering, are inevitably highly compensated due to the inclusion of H, C, and self-compensating N-related donor-like defects. However, it is certainly not inevitable that the concentrations of these defects outnumber the N-related acceptors. In particular, it has been shown by others that heating the films at high temperature after deposition can cause out-diffusion of H.

We have also seen preliminary evidence that by careful choice of the growth parameters, it should be possible to minimize the concentration of C in the films. There are several C-related donors, and this step could be critical to our success. Increasing the flow rate of NO may increase compensation, but we have also seen a reduction in C concentration by using higher flow rates. Hence, for each change made, there appears to be several possible consequences, and this is the reason for the rather narrow fabrication window for making *p*-type material.

We are adding PECVD and MBE to our ZnO deposition capabilities in an effort to deposit higher-quality ZnO films and widen the processing window. Temperature modulation, similar to the RTM process, shows great promise for producing *p*-type ZnO. Plasma processing may also help reduce C incorporation. We have not yet attempted to use co-doping, which is a technique that other authors have reported as promising. Incorporating this technique along with the above fabrication changes are among our short-term plans.

References

[1] M. A. Contreras, K. Ramanathan, J. AbuShama, D. L. Young, B. Egaas, and R. Noufi, Progress in Photovoltaics: Research and Applications, 13, 209–216, Diode characteristics in state-of-the-art $ZnO/CdS/Cu(In_{1-x}Ga_x)Se_2$ solar cells (2005).
[2] A. Bergh, G. Craford, A. Duggal, and R. Haitz, Phys. Today, 54, 42–49, The promise and challenge of solid-state lighting (2001).
[3] A. J. Nozik, Ann. Rev. Physical Chem., 29, 189–222, Photoelectrochemistry – applications to solar-energy conversion (1978).
[4] M. Gräetzel, Nature, 414, 338–344, Photoelectrochemical cells (2001).
[5] B. O'Neill, in Indium: Markets, Applications and Alternatives, Minor Metals 2005, Lisbon, Portugal, 2005.
[6] C. Sang-Hun, Electronics makers race to dominate hot market for flat-screen televisions, International Herald Tribune, Wednesday, April 27, 2005.
[7] X. Wu, J. Zhou, A. Duda, J. C. Keane, T. A. Gessert, Y. Yan, and R. Noufi, Progress in Photovoltaics: Research and Applications, In press, 13.9%-efficient CdTe polycrystalline thin-film solar cells with an infrared transmission of ~50% (2005).
[8] T. J. Coutts, J. S. Ward, D. L. Young, K. A. Emery, T. A. Gessert, and R. Noufi, Progress in Photovoltaics: Research and Applications, 11, 359–375, Critical issues in the design of polycrystalline, thin-film tandem solar cells (2003).
[9] L. M. Fraas, J. E. Avery, V. S. Sundaram, V. T. Dinh, T. M. Davenport, J. W. Yerkes, J. M. Gee, and K. A. Emery, in Over 35% GaAs/GaSb stacked concentrator cell assemblies for terrestrial aplications, Twenty First IEEE Photovoltaic Specialists Conference, Kissimmee, FL, 1990 (IEEE, Piscataway, NJ), p. 190–195.
[10] M. Dresselhaus, 2003, Basic Research Needs for the Hydrogen Economy, DOE. Office of Science, Basic EnergySciences, www.sc.doe.gov/bes/hydrogen.pdf
[11] B. Kroposki, 2004, Hydrogen and Fuel Cell Research, http://www.nrel.gov/hydrogen/proj_production_delivery.html
[12] C. G. Van de Walle, Phys. Rev. Lett., 85, 1012–1015, Hydrogen as a cause of doping in zinc oxide (2000).
[13] T. Yamamoto and H. Katayama-Yoshida, J. Cryst. Growth, 214/215, 552–555, Unipolarity of ZnO with a wide-band gap and its solution using codoping method (2000).
[14] S. B. Zhang, S.-H. Wei, and A. Zunger, Phys. Rev. B, 63, 075205, Intrinsic n-type versus p-type doping asymmetry and the defect physics of ZnO (2001).
[15] F. Oba, S. R. Nishitani, S. Iostani, H. Adachi, and I. Tanaka, J. Appl. Phys., 90, 824–828, Energetics of native defects in ZnO (2001).
[16] A. Kobayashi, O. F. Sankey, and J. D. Dow, Phys. Rev. B, 28, 946–956, Deep energy levels of defects in the wurtzite semiconductors AlN, CdS, CdSe, ZnS, and ZnO (1983).
[17] C. H. Park, S. B. Zhang, and S.-H. Wei, Phys. Rev. B, 66, 073202-1-3, Origin of p-type doping difficulty in ZnO: The impurity perspective (2002).

[18] E. C. Lee, Y. S. Kim, Y. G. Jin, and K. J. Chang, Physica B–Condensed Matter, 308, 912–915, First-principles study of the compensation mechanism in N-doped ZnO (2001).
[19] Y. F. Yan, S. B. Zhang, and S. T. Pantelides, Phys. Rev. Lett., 86, 5723–5726, Control of doping by impurity chemical potentials: Predictions for p-type ZnO (2001).
[20] L. G. Wang and A. Zunger, Phys. Rev. Lett., 90, 256401, Cluster-doping approach for wide-gap semiconductors: The case of p-type ZnO (2003).
[21] C. G. Van de Walle, Phys. Stat. Sol. (b), 229, 221–228, Strategies for controlling the conductivity of wide-band-gap semiconductors (2002).
[22] T. Yamamoto and H. Katayama-Yoshida, Jpn. J. Appl. Phys., 38, L166–L169, Solution using a codoping method to unipolarity for the fabrication of p-type ZnO (1999).
[23] X. N. Li, B. Keyes, S. Asher, S. B. Zhang, S. H. Wei, T. J. Coutts, S. Limpijumnong, and C. G. Van de Walle, Appl. Phys. Lett., 86, 122107–122109, Hydrogen passivation effect in nitrogen-doped ZnO thin films (2005).
[24] K. Minegishi, Y. Korwai, Y. Kikuchi, K. Yang, M. Kasuga, and A. Shimizu, Jpn. J. Appl. Phys. Pt. 2, 36, L1453–L1455, Growth of p-type zinc oxide films by chemical vapor deposition (1997).
[25] X. Li, Y. Yan, T. A. Gessert, C. DeHart, C. L. Perkins, D. L. Young, and T. J. Coutts, Electrochemical and Solid State Letters, 6, C56–C58, p-type ZnO thin films formed by CVD reaction of diethylzinc and NO gas (2003).
[26] X. Li, Y. Yan, T. A. Gessert, C. L. Perkins, D. L. Young, C. DeHart, M. Young, and T. J. Coutts, J. Vac. Sci. Technol. A, 21, 1342–1346, Chemical vapor deposition-formed p-type ZnO thin films (2003).
[27] T. M. Barnes, S. Hand, J. Leaf, and C. A. Wolden, J. Vac. Sci. Technol. A, 22, 2118–2125, An investigation of the surface chemistry occurring in ZnO chemical vapor deposition from dimethylzinc and atomic oxygen (2004).
[28] T. M. Barnes, J. Leaf, S. Hand, C. Fry, and C. A. Wolden, J. Appl. Phys., 96, 7036–7044, A comparison of plasma-activated N_2/O_2 and N_2O/O_2 mixtures for use in ZnO:N synthesis by chemical vapor deposition (2004).
[29] T. M. Barnes, K. Olson, and C. A. Wolden, Appl. Phys. Lett., 86, 112112–112114, On the formation and stability of p-type conductivity in nitrogen-doped zinc oxide (2005).
[30] A. B. M. A. Ashrafi, I. Suemune, H. Kumano, and S. Tanaka, Jpn. J. Appl. Phys., 41, L1281–L1284, Nitrogen doped p-type ZnO layers prepared with H_2O vapor assisted metalorganic molcular beam epitaxy (2002).
[31] K. Iwata, P. Fons, A. Yamada, K. Matsubara, and S. Niki, J. Cryst. Growth, 209, 526–531, Nitrogen-induced defects in ZnO:N grown on sapphire substrate by gas source MBE (2000).
[32] A. Tsukazaki, M. Kubota, A. Ohtomo, T. Onuma, K. Ohtani, H. Ohno, S. F. Chichibu, and M. Kawasaki, Jpn. J. Appl. Phys., 44, L643–L645, Blue light-emitting diode based on ZnO (2005).
[33] A. Tsukazaki, A. Ohtomo, T. Onuma, M. Ohtani, T. Makino, M. Sumiya, K. Ohtani, S. F. Chichibu, S. Fuke, Y. Segawa, H. Ohno, H. Koinuma, and M. Kawasaki, Nature

Materials, 4, 42–46, Repeated temperature modulation epitaxy for p-type doping and light-emitting diode based on ZnO (2005).
[34] M. Joseph, H. Tabata, and T. Kawai, Jpn. J. Appl. Phys., 38, L1205–L1207, p-type electrical conduction on ZnO thin films by Ga and N codoping (1999).
[35] X.-L. Guo, J.-H. Choi, H. Tabata, and T. Kawai, Jpn. J. Appl. Phys., 40, L177–L180, Fabrication and optoelectronic properties of a transparent ZnO homostructural light emitting diode (2001).
[36] H. Matsui, H. Saeki, T. Kawai, H. Tabata, and B. Mizobuchi, J. Appl. Phys., 95, 5882–5888, N doping using N_2O and NO sources: From the viewpoint of ZnO (2004).
[37] A. V. Singh, R. M. Mehra, A. Wakahara, and A. Yoshida, J. Appl. Phys., 93, 396–399, p-type conduction in codoped ZnO thin films (2003).
[38] B. S. Li, Y. C. Liu, Z. Z. Zhi, D. Z. Shen, Y. M. Lu, J. Y. Zhang, X. W. Fan, R. X. Mu, and D. O. Henderson, J. Mater Res., 18, 8–13, Optical properties and electrical characterization of ZnO thin films prepared by thermally oxidizing Zn_3N_2 thin films (2003).
[39] C. C. Lin, S.-Y. Chen, S.-Y. Cheng, and H.-Y. Lee, Appl. Phys. Lett., 84, 5040–5042, Properties of nitrogen-implanted p-type ZnO films grown on Si_3N_4/Si by radio frequency magnetron sputtering (2004).
[40] S. J. Pearton, D. P. Norton, K. Ip, and Y. W. Heo, J. Vac. Sci. Technol. B, 22, 932–948, Recent advances in processing of ZnO (2004).
[41] D. C. Look and B. Claflin, Phys. Stat. Sol. (b), 241, 624–630, P-type doping and devices based on ZnO (2004).
[42] C. L. Perkins, S. H. Lee, X. N. Li, S. E. Asher, and T. J. Coutts, J. Appl. Phys., 97, 034907-1, Identification of nitrogen chemical states in N-doped ZnO via x-ray photoelectron spectroscopy (2005).
[43] V. E. Henrich and P. A. Cox, The Surface Science of Metal Oxides (Cambridge University Press, Cambridge, UK, 1994).
[44] B. K. Meyer, H. Alves, D. M. Hoffmann, W. Kriegseis, D. Forster, F. Bertram, J. Christen, A. Hoffmann, M. Strassburg, M. Dworzak, U. Haboeck, and A. V. Rodina, Phys. Stat. Sol. (b), 224, 231–260, Bound exciton and donor-acceptor pair recombination in ZnO (2004).
[45] A. Zeuner, H. R. Alves, D. M. Hofmann, B. K. Meyer, A. Hoffmann, U. Haboeck, M. Strassburg, and M. Dworzak, Phys. Stat. Sol. (b), 234, R7–R9, Optical properties of the nitrogen acceptor in epitaxial ZnO (2002).
[46] S. B. Zhang, S.-H. Wei, and Y. Yan, Physica B, 302–303, 135–139, The thermodynamics of codoping: how does it work? (2001).
[47] S.-H. Wei, Compu. Mater. Sci., 30, 337–348, Overcoming the doping bottleneck in semiconductors (2004).
[48] S. Limpijumnong, X. Li, S.-H. Wei, and S. B. Zhang, Appl. Phys. Lett., 86, 211910, Substitutional diatomic molecules NO, NC, CO, N_2 and O_2: Their vibrational frequencies and effects on p doping of ZnO (2005).
[49] B. Keyes, X. Li, and T. J. Coutts, Appl. Phys. Lett., in press, FTIR study of MOCVD ZnO:N thin films (2005).

[50] G.-C. Yi and B. W. Wessels, Appl. Phys. Lett., 70, 357–359, Carbon-hydrogen complexes in vapor phase epitaxial GaN (1997).
[51] M. O. Manasreh, J. M. Baranowski, K. Pakula, H. X. Jiang, and J. Lin, Appl. Phys. Lett., 75, 659–661, Localized vibrational modes of carbon-hydrogen complexes in GaN (1999).
[52] S. Limpijumnong, J. E. Northrup, and C. G. Van de Walle, Phys. Rev. B, 68, 075206, Identification of hydrogen configurations in p-type GaN through first-principles calculations of vibrational frequencies (2003).
[53] L. D. Landau and E. M. Lifshitz, Quantum Mechanics, 3rd ed. (Pergammon, Oxford, 1977).
[54] S. Limpijumnong, J. E. Northrup, and C. G. van de Walle, in Anharmonicity in the Vibrational Modes Associated with H-H, N-H, O-H, and C-H Bonds, Materials Research Society Symp., Pittsburgh, 2004, p. H3.6.
[55] B. Smith, Infrared Spectral Interpretation (CRC Press LLC, Boca Raton, FL, 1999).
[56] E. V. Lavrov, J. Weber, F. Borrnert, C. G. Van de Walle, and R. Helbig, Phys. Rev. B, 66, 165205-1–165205-7, Hydrogen-related defects in ZnO studied by infrared absorption spectroscopy (2002).
[57] S. J. Jokela, M. D. McCluskey, and K. G. Lynn, Physica B, 340–342, 221–224, Infrared spectroscopy of hydrogen in annealed zinc oxide (2003).
[58] M. D. McCluskey, S. J. Jokela, K. K. Zhuravlev, P. J. Simpson, and K. G. Lynn, Appl. Phys. Lett., 81, 3807–3809, Infrared spectroscopy of hydrogen in ZnO (2002).
[59] C. H. Seager and S. M. Myers, J. Appl. Phys., 94, 2888–2894, Quantitative comparisons of dissolved hydrogen density and the electrical and optical properties of ZnO (2003).
[60] B. Clerjaud, D. Cote, W. S. Hahn, A. Lebkiri, W. Ulrici, and D. Wasik, Phys. Stat. Sol. (a) – Appl. Res., 159, 121–131, On the way to the investigation of hydrogen in GaN: Hydrogen in nitrogen doped GaP and GaAs (1997).
[61] F. Jiang, M. Stavola, M. Capizzi, A. Polimeni, A. A. Bonapasta, and F. Filippone, Phys. Rev. B, 69, Vibrational spectroscopy of hydrogenated $GaAs_{1-y}N_y$: A structure-sensitive test of an H-2(*)(N) model (2004).
[62] N. H. Nickel and K. Fleischer, Phys. Rev. Lett, 90, 197402, Hydrogen local vibrational modes in zinc oxide (2003).
[63] A. Janotti, S.-B. Zhang, and S.-H. Wei, Phys. Rev. Lett., 88, 122506, Hydrogen vibration modes in GaP:N: The pivotal role of nitrogen in stabilizing the H-2* complex (2002).
[64] D. P. Norton, Y.-W. Heo, M. P. Ivill, K. Ip, S. J. Pearton, M. F. Chisholm, and T. Steiner, Materials Today, 7, 34–40, ZnO: growth, doping & processing (2004).
[65] S. E. Miller and A. G. Chynoweth, Optical Fiber Communications (Academic Press, New York, 1979).

Chapter 4

Pulsed Laser Deposition of Zinc Oxide (ZnO)

Vinay Gupta and K. Sreenivas

Department of Physics and Astrophysics University of Delhi, Delhi–110 007, India

4.1 Introduction

Zinc oxide (ZnO) exhibits an interesting combination of multifunctional properties, including optical, semiconducting, piezoelectric, electrooptic, and optoelectronic properties, and in thin film form ZnO films find immense applications in many electronic devices including sensors, actuators, transducers, and high frequency surface acoustic wave (SAW) devices.[1] Currently world-wide efforts are focused in both the basic and applied areas, and with the advent of modern thin film deposition techniques for oxide films, it is now realized that better preparation methods would lead to direct integration of ZnO onto a variety of substrates. These developments in the meantime demand good reproducibility and a clear understanding of the stability of thin film layered structures employed in the device design. Research and development efforts on ZnO thin film growth have been underway since the last four decades, and device quality films have been deposited successfully by a variety of deposition techniques. Primarily the interest in ZnO was due to its direct wide band gap and its optical properties (refractive index), and initially it found immense applications as a transparent conducting electrode and as an anti-reflection coating for solar cell applications.[2] Subsequent improvements in the growth techniques offering a better control over the crystallographic properties and the electrical conductivity led to a variety of applications based on the functional properties of ZnO as a piezoelectric layer for thin film surface acoustic wave (SAW) devices,[3] thin film actuators,[4] and conductive gas sensors.[5]

Zinc oxide is a direct band gap semiconductor with $\boldsymbol{E_g} = 3.37$ eV having hexagonal wurtzite structure with lattice parameters $\boldsymbol{a} = 0.325$ nm and $\boldsymbol{c} = 0.512$ nm. Stoichiometric films are insulating and when grown with a preferred $\boldsymbol{c}$-axis

Zinc Oxide Bulk, Thin Films and Nanostructures
C. Jagadish and S. Pearton (Editors)

orientation exhibit strong piezoelectric coupling coefficients,[6] whereas non-stoichiometric and oxygen deficient films are very good as transparent semiconducting oxides or conductors.[7] In the last 10 years intense research on ZnO thin films was triggered by the observation of excitonic lasing, when excitonic stimulated emission from epitaxial ZnO film at room temperature was first demonstrated,[8] and since then the focus has been on the development of epitaxial films on lattice matched substrates for optoelectronic device applications. Influence of various dopants in ZnO revealed potential applications in the area of UV light detection,[9] spin functional devices,[10] optical waveguides[11] and acousto-optic media.[12] Lately with the availability of epitaxial quality ZnO films, single crystal ZnO wafers, and the large exciton binding energy of 60 meV in ZnO compared with 25 meV for wide band gap GaN, the prospects for using ZnO in optoelectronic device applications has increased enormously.[13] Significant improvements in the processing technologies and a better understanding of the semiconducting properties of ZnO have now intensified the efforts for possibly achieving *p*-type doping in ZnO to develop novel device structures useful for transparent opto-electronic devices.

The immense applications of ZnO films have boosted research activity on a variety of thin film and thick film growth techniques including screen printing,[14] thermal oxidation of metallic zinc films,[15] spray pyrolysis,[16] sol gel synthesis,[17] ion beam assisted reactive deposition,[18] sputtering,[19] modification with electron cyclotron resonance (ECR) source,[20] chemical vapor deposition[21] (CVD), pulsed laser deposition[22] (PLD), and molecular beam epitaxy[23] (MBE). Amongst all these techniques sputtering, MOCVD MBE, and PLD are the foremost that have been able to produce device quality films. Much of the earlier work on sputtered films resulted in polycrystalline films, and with the advent of (dc and rf magnetron sputtering) growth of ZnO films with a preferred ***c***-axis orientation at high deposition rates and epitaxial growth were established. Initial efforts at Motorola in USA and by Murata manufacturing Co., in Japan led to the commercial development of thin film surface acoustic wave devices operating over a wide frequency range (30 MHz to 1.5 GHz) for military and civilian applications.[24–25] However a few issues still remain with the sputtering technologies relating to the growth of defect free optical quality films, reducing the *insitu* bombardment effects by neutrals and energetic oxygen ions during film growth and capability for developing doped compositions. Combining an electron cyclotron resonance (ECR) source with sputtering has enabled the deposition of highly dense and insulating ZnO films with properties approaching those of the bulk for thin film piezoelectric devices.[26] Other sophisticated techniques such as MBE allow layer-by-layer growth with excellent control of the purity and crystalline quality of the resulting film.[27] High quality films grown by MOCVD have been obtained by homoepitaxy, and to a large extent the growth is found to depend on substrate quality and surface preparation. Premature reactions between the Zn metalorganic compounds and the oxidants yield some unwanted deposits and demand improvements in the reactor design.[28] Above all the interest in MOCVD techniques is primarily from the industrial point of view for large-scale production, and the technology for ZnO growth is strongly driven by the choice of precursors, their availability, and the growth rates that can be achieved.

Successful growth of high T_c superconducting thin films by pulsed laser deposition (PLD) has led to the developments of several other multi-component oxide compositions that exhibit a wide range of properties including superconductors,[29] ferroelectrics,[30] magnetic and non-linear optical properties.[31] It is a simple deposition technique and provides many advantages for oxide epitaxy. When compared to other thin film deposition techniques, PLD demands minimum level of optimization to establish thin film growth of a given material composition, whereas sputtering, MBE, and MOCVD demand extensive optimization for achieving a good control for a complex composition. In the PLD technique oxidation occurs primarily in the ZnO ablation plume and this alleviates the difficulties encountered with other techniques where oxidation proceeds via surface reactions. Conceptually, pulsed laser deposition (PLD) technique is probably the simplest among all thin film growth techniques.[32] Thin films are obtained by vaporizing a material using high-energy laser pulses. The targets used are relatively small (25 mm in diameter), and it is very convenient to carry out compositional variation studies. It is generally easier to obtain the desired stoichiometry in the deposited films than with other deposition technologies.

Developments in ZnO thin processing by various techniques have been covered in many excellent reviews published by Ohtomo and Tsukazaki,[33] Triboulet and Perriere,[34] Norton *et al.*,[35] Pearton *et al.*,[36] Ozgur *et al.*[37] and Ogale.[38] In this chapter we present a detailed overview on the progress achieved with ZnO thin films grown by pulsed laser deposition and summarize the recent trends in the processing technologies. The practical advantages of PLD are described and the present state of art is reviewed. The emphasis is on the processing conditions, source materials and substrates used for thin film deposition. We highlight the properties achieved relating to various applications, and discuss the opportunities for research and development where the PLD grown ZnO films can make an impact.

4.2 Pulsed laser deposition

In a typical pulsed laser deposition process, film growth occurs in three different stages: (1) Firstly the laser beam strikes a solid target of known composition and produces a highly forward directed plume of gas phase material, (2) secondly the plume interacts (both chemically and physically) with a background ambient, and (3) finally the ablated material condenses onto a suitable substrate where the thin film nucleates and grows. The adjustable experimental parameters are the laser fluence, substrate to target distance, background oxygen pressure, and substrate temperature. Film quality is determined by the selected laser wavelength, target composition, its structure and density, and the identity of the substrate (amorphous, or single crystal, and the lattice matching). Understanding how these experimental conditions affect the nucleation and the growth of film deposited by PLD is an area of intense research activity.

The schematic of a typical pulsed laser deposition system used for the growth of thin films is shown in Fig. 4.1, and consists of the following main parts.

a) A spherical or cylindrical electro-polished stainless steel chamber. A target carrousel that can hold a number of targets is useful for multi layer deposition.
b) Target rotation during ablation is necessary. Facility for changing the targets allows multi layer deposition, and can be achieved either manually, or through a stepper motor drive. After laser ablation deep grooves (v-shaped or rectangular) are formed depending on the laser spot diameter. Formation of such grooves affects the film quality in subsequent deposition runs, and it becomes necessary to remove the targets, scrape them to a smooth finish and remount them for the next deposition. Alternately the target assembly can house an (x, y) translation stage as well. A stepper motor can be used so that the target is stepped in both the x and y directions while the laser interaction spot is held stationary. Such an arrangement offers a few advantages; (i) that the target surface area can be used more effectively for a longer duration without groove formation, and (ii) the target assembly can be rotated (about the vertical axis) relative to the incoming laser beam.
c) A stable substrate heater and substrate holder assembly, with capability to heat the substrates up (800°C) in oxygen environment (50 to 300 mTorr). The arrangement is placed parallel to the target surface at a suitable distance.
d) A pumping station (preferably oil free) to evacuate the chamber, feed-throughs for gas flow and gauges for pressure control.
e) Tiny single crystal substrates (up to 1 cm × 1 cm) can be easily fixed onto the substrate table using conducting silver epoxy with a slight heat treatment (150°C for 30 min in air), and for heavier substrates the substrate table can be designed with a suitable holder and clamps, and care should be taken that substrates do not fall down at higher substrate temperatures. It is preferable to mount the substrates at the proper place on the substrate table so that the forward directed plume from the target is right in front of the substrate and interacts with the deposited film.

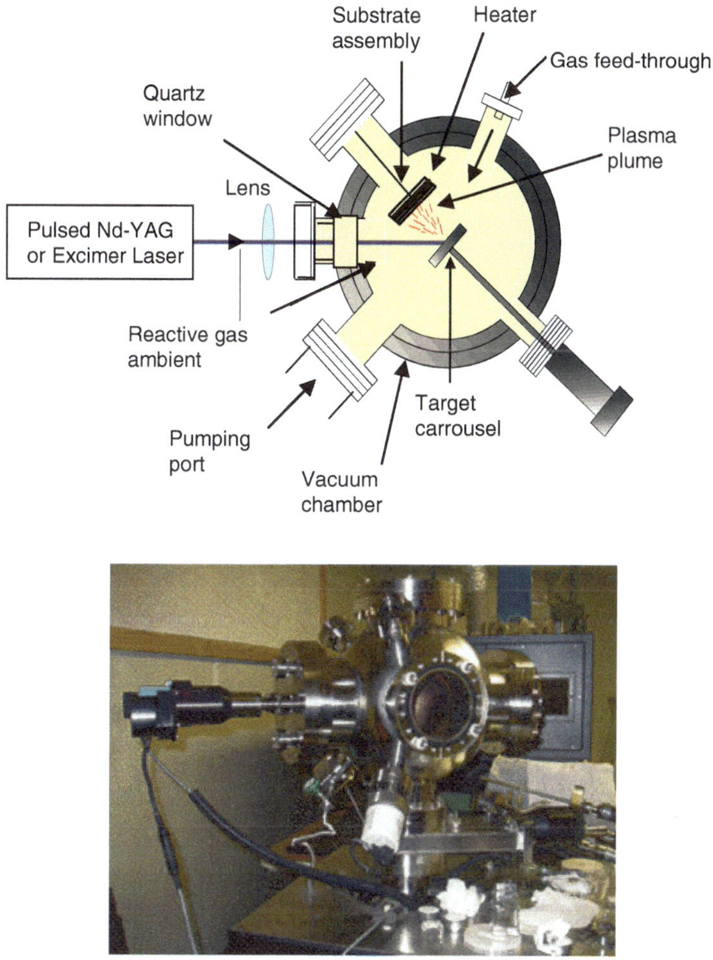

Figure 4.1: Schematic of a pulsed laser deposition system (Source: http://speclab.upr.edu, used with permission of Prof. R.S. Katiyar, Speclab, University of Puerto Rico).

Pulsed-laser deposition (PLD) involves a very simple experimental set up and uses pulsed laser radiation to vaporize the target material to be deposited as a thin film:

1. Intense laser pulses of (5 to 7 ns duration) are focused with an energy density (1 to 5 J/cm^2) onto the target surface. The pulses are directed into the vacuum chamber through a non-absorbing window onto the target material where they are absorbed. The laser pulses above a threshold power density depending upon the target material heat up the target surface locally and vaporize the material

from the surface region of the target in the form of neutrals, or ionic atoms and molecules, and is visible as a luminous cloud usually called the plume. How well the plume interacts with the substrate is a crucial factor and determines the film quality. The deposition rate is usually of the order of 1 to 10 nm per pulse, and has a strong dependence on the pulse energy, ambient gas and pressure, the repetition rate, and the target to substrate distance. The final composition in the deposited film, its crystalline quality and surface morphology can be influenced by the deposition rate.

2. Film growth by pulsed laser deposition occurs in three stages. Firstly the interaction of the laser pulse with the target matter leads to the melting of the target surface and instantaneous vaporization produces a highly forward directed plume from the thin upper layer of the molten surface. Secondly the plume interacts (both physically and chemically) with the ambient gas, and finally the ablated material condenses onto a suitable substrate where the thin film nucleates and grows.

In comparison to some of the other existing thin film deposition techniques used for depositing ZnO films pulsed laser deposition has its own limitations and advantages.

1. Complex oxide compositions with high melting points can be easily deposited provided the target materials absorb the laser energy.
2. PLD allows the growth of films under a highly reactive gas ambient over a wide range of pressure.
3. The target composition is transferred instantaneously leading to a stoichiometric deposition and as such the turnaround time is much faster for initial optimization of the growth conditions with this technique. This is a major advantage whenever there is a need for experimenting with different target compositions.
4. The PLD technique is found to be very effective and well suited for developing epitaxial films, and allows fabrication of multilayers, hetero-structures and superlattices on lattice matched substrates.
5. In the PLD process during film growth suitable kinetic energy in the range 10–100 eV, and photochemical excitation exist in comparison to other deposition techniques, and is found to be beneficial for improving the quality of the deposited film.
6. The main practical limitation of PLD is its relatively low duty cycle, incorporation of particulates in the deposited films, although this is not unique to PLD, because particulate problem exists in the case sputtering and MOCVD as well. Up scaling for industrial production and deposition on large area substrates is still a major concern. The technique is ideally suitable for deposition on small

Table 4.1: Performance features of excimer and Nd:YAG lasers.

Pulsed UV-Laser	Advantages	Disadvantages
Excimer laser	– High power output – Good stability – Flexibility for tuning laser output parameters	– Short operation lifetime – Complicated maintenance – Expensive and high purity gases, constant refilling – Space consuming
Nd:YAG laser	– Output energy sufficient for laser ablation – Simple maintenance – Compact systems	– Large energy drop for the 3rd harmonic mode

substrates, and yields high quality films useful for research and allows fabrication of discrete devices.

4.3 Lasers and target materials used for ZnO growth

ZnO thin film growth by PLD has been reported using a variety of lasers. First reports on ZnO thin film growth by PLD appeared in the year 1983.[39–40] Excimer lasers which operate in the ultraviolet region (193 to 351 nm) in mixtures of rare gases with halide molecules have been used commonly for the deposition of the multi component oxides. Bulk of the reported literature on ZnO thin film growth is focused on the use of UV excimer lasers (KrF: $\lambda = 248$ nm and ArF: $\lambda = 193$ nm); and Nd:yttrium aluminium garnet (YAG) pulsed lasers ($\lambda = 355$ nm).[41] Okoshi *et al.*[42] have used a Ti:sapphire laser at ($\lambda = 790$ nm) for depositing ZnO films on silicon, and the films were ***c***-axis oriented and highly conducting. Though visible light from a Cu-vapor[43] laser (emitting at $\lambda = 510$ and 578 nm with a 2:1 intensity ratio) with 50 ns pulse-length operating at 4400 Hz, and an energy density of 14.5 J/cm^2 per pulse has also been used for ZnO film growth, and cathodoluminescence under electron bombardment was observed, but the use of this laser has been scarcely reported. The performance features of (Nd:YAG and the Excimer laser) that are most commonly used for film deposition are compared in Table 4.1.

4.3.1 Target materials

Since the composition obtained in a deposited film is directly related to the target composition it is very important to begin the target fabrication with low impurity chemicals. For ZnO thin film deposition, a variety of target materials including sintered ceramic discs prepared from pressed powders,[44] single crystals of ZnO,[45] and the ablation of a pure metallic zinc (Zn) target in reactive oxygen atmosphere[46–47] have been reported, and sintered ceramic targets are most commonly used. The target size is typically (1 to 2 inch dia.), however in special cases when the available amount of powder is very small, it is possible to prepare and use targets as small as (1 cm × 1 cm), but a minimum thickness of 2 mm is desirable with smooth parallel surfaces, and the shape of the target is immaterial. Care should be taken to note the impurity content in single crystal targets and the oxide powders used for preparing the ceramic sintered targets[33] and sintering conditions of target can also influence the film quality. Surface impurities in sintered ceramic targets should be avoided. It is a normal practice to scrape the target before use, or intermittently after a few deposition runs, and it is customary to pre-ablate the target by shielding the substrates prior to film deposition on the substrate.

Droplet formation in the growing film is a significant problem in pulsed laser deposited films, and their density and size distribution is linked to the ejection of liquid droplets by the target during irradiations, and leads to the formation of macroscopic particulates on the surface of the deposited films.[48] Although such droplets do not pose a problem when the bulk physical properties of the films are being measured, but they tend to degrade the optical or optoelectronic properties, and the poor surface morphology limits device fabrication on the film surface. A number of studies have been reported to understand the origin for such droplet formation. Although the problem has not been completely eliminated but ways to minimize and reduce their density have emerged from the abundant literature. Laser ablation at a longer wavelength laser is found to generate more droplets.[49–50] For example, in the case of ZnO film growth a Nd:YAG laser operating at $\lambda = 532$ nm resulted in the presence of many droplets, whereas a KrF excimer laser ($\lambda = 248$ nm, 20 ns pulse width) significantly reduced the droplet formation. Droplet formation is a severe problem at higher laser fluence. Although setting the fluence just above the threshold ensures reduced droplet formation, but such an adjustment often leads to a trade off between the deposition rate and the quality of the film. Normally during the initial optimization runs the influence of the deposition parameters on the film surface morphology is carefully examined, and one arrives at a good compromise on the deposition parameter that ensures the deposition of smooth films.

Ablating highly dense ZnO targets at a lower wavelength with appropriate laser fluence, and ensuring a uniform power density across the laser spot tend to reduce the problem of droplet formation.[51] A number of alternate approaches have been investigated to avoid the problem of droplet ejection, such as (i) crossed beam laser ablation technique,[52] (ii) off axis deposition,[53] (iii) magnetic field assisted ablation,[54] (iv) using a mechanical velocity filter,[55–56] and (v) the use of a second laser[57] for preheating the target up to the molten state just before ablation.[58] The origin for the development of such particulate matter during the ablation process is still not well understood, and a few suggestions based on phase explosions, and splashing effects have not been well established.[59–60] The presence of droplets on the film surface is more or less a direct consequence of the physical nature of the target and the related thermal effects occurring within the target during the laser-matter interaction. In a way ZnO appears to be a convenient material to produce films by PLD that are droplet free because it melts and sublimes at 2248 K without a phase transformation from the molten state to the evaporative plume. However, experiments have shown[61] that its molten state does exist for a finite duration, despite the fact that ZnO sublimes at its melting point, and this could be the possible source for liquid droplet ejection when ablation is done at a higher wavelength.

Use of a femto second laser effectively reduces droplet density,[62] because the laser pulse energy is consumed instantaneously to break the chemical bonds, and minimize the thermal effects and a stoichiometric film could be grown in vacuum,[42] but the film quality is found to be poor in comparison to a nanosecond laser. The result was explained in terms of the difference in the kinetic energy of the species ejected from the target (1 keV with the femto second laser and a few hundreds of eV with the nanosecond laser).[63] Interaction of a laser pulse with a solid material can be divided into two sub categories, depending on the pulse duration. Laser-solid interaction with nanosecond or longer laser pulses of sufficient duration couple not only into the electronic but also the vibration wave function of the material, but with femto/picosecond pulses the duration of the excitation is too short to couple directly into the vibrational wave function. Thus different ablation behavior could be induced in solids after changing the pulse duration of interacting laser.

In order to reduce the droplet density investigations have been carried out on a number of materials using a dual-laser ablation method as well.[61,64–65] Laser irradiation from a CO_2 laser (10.6 μm, 200 ns) and a KrF laser strikes alternately (or partially overlap) on the target, and are temporally adjusted. The laser system and operation are complicated and the experiment demands precise temporal synchronization between the two laser pulses.[61] The infrared laser initially melts the target surface, but does not ablate and the UV laser pulse immediately ablates the molten layer to remove any large-size droplets, and additional heating up in the plume with an extended infrared laser prevents the gaseous specie to form small particles.

Reported results on laser ablated thin films from wide band gap insulators, indicate[66] that the surface morphology depends on the optical properties of the target material. The higher the optical absorption coefficient (α), the lower is the droplet density on the surface of the deposited films. Since the optical absorption of ZnO increases with decreasing wavelength it has been clearly seen that films deposited with ArF ($\lambda = 193$ nm), KrF ($\lambda = 248$ nm) excimer lasers, and frequency quadrupled

($\lambda = 266$ nm) Nd:YAG lasers yield a superior morphology in comparison to films deposited at the second harmonic mode ($\lambda = 532$ nm) and the fundamental frequency ($\lambda = 1064$ nm) of Nd:YAG lasers.[67] Currently for pulsed laser deposition of ZnO films the ArF or KrF excimer lasers, or the frequency quadrupled Nd:YAG lasers are commonly used.

The laser energy fluence, has a significant influence on the growth kinetics of the ZnO film.[68] At lower levels of laser fluence, the deposition rate is low, and this allows the growth of ZnO via a 3D-island mode, and since the kinetic energy of the species is low their reduced surface mobility leads to an island growth mode.[69] On the other hand, too high a laser fluence quickly degrades the film properties due to the bombardment of the growing film by energetic species. Laser fluence in the range 1.5 to 2.5 J/cm^2 is found to be the optimum for obtaining good quality ZnO films. Wei *et al.*[70] studied effects of laser influence in great detail while depositing ZnO films on an ITO coated glass substrate. The plasma plume has a highly forward scattered and non-symmetric flux distribution. Upon arriving at the substrate surface the high-speed particles of the plume tend to sputter atoms from the surface of the film on the substrate and produce microholes or defects. If the laser fluence is decreased the number of defects is decreased, and the crystallinity of the film improves. When the laser fluence is set just below the turn-on value of producing defects, the quality of films is found to improve significantly.

4.4 ZnO thin film processing by PLD

4.4.1 Effect of oxygen pressure and substrate temperature

The processing parameters that play an important role during the deposition process of textured ZnO films include background gas pressure, substrate temperature, laser energy density, target to substrate distance, and the repetition rate. Oxygen incorporation into a growing film originates both from the oxide target and the background oxygen gas.[71] PLD makes the composition of a stoichiometric ZnO target surface appear as $ZnO_{0.2}$ besides substantial loss of target surface material under successive laser ablation at low oxygen pressure.[72] Excess amount of zinc observed on the surface of the oxide target is related to the high volatility and the loss of oxygen due to vacuum pumping,[72] and thus a high background oxygen pressure becomes necessary to maintain the equilibrium composition of the target, and to avoid the oxygen deficiency in the deposited film. In some studies a small concentration (~8%) of ozone gas mixed with oxygen as background gas has been used because of its strong ability to oxidize.[73] Attractive flexibility of the PLD technique is that it allows film growth over a wide pressure range beginning from ultra high vacuum to a high value of ~100 Pa. This particular feature has been extremely useful for the preparation of nanostructures of ZnO material, and for fabricating heterostructures using ZnO films with many other materials that require processing at varying oxygen pressure to form a metastable crystalline layered structure.

At low oxygen pressure (<0.5 mbar), the emitted atoms and ions from the target surface after laser interaction condense on the substrate surface to form a continuous film. The mean free path of the ablated species depends upon the processing pressure and increases with decreasing pressure. The processing of film at very low oxygen pressures is generally known as laser molecular-beam epitaxy (LMBE) where the mean free path of the plume species is much larger than the target to substrate distance.[74] When films are processed under high pressures (>1 mbar) the ablated species undergo a large number of collisions with background gas molecules and the condensation of species in the gas phase results in the formation of nano-size particles. The settlement of these nanoparticles on the surface of the substrate promotes the formation of a film that is dense and smooth. The nucleation of nanoparticles on the substrate is controlled either by the processing conditions, or by using a catalytic layer, and can lead to the fabrication of interesting nanostructures. Films deposited under very low pressure are often oxygen deficient, and thereby affect the microstructural and electrical properties,[75] and therefore demand a careful optimization. Moreover films deposited at low oxygen pressure (10^{-5} to 10^{-4} Torr) are reported to be in a state of compressive strain exhibiting ***c***-axis lattice parameter larger than the bulk value, and deposition at higher oxygen pressures (10^{-2} Torr) leads to reduction in the strain with minimal surface roughness.[76] The semiconducting transport properties are also seen to be affected, and an increase in oxygen pressure is reported to reduce the electron mobility, and a maximum hall mobility of ($72\,cm^2\,V^{-1}\,s^{-1}$) was measured on films deposited at 10^{-4} to 10^{-3} Torr on a sapphire substrate. The quality of the ZnO film at high pressure is found to improve with the pre-coating of an ultrathin nucleating layer grown at low pressure (10^{-4} to 10^{-3} Torr). The optimized value of oxygen pressure for a desired property is found to be different in the reported literature and such variations can arise due to differences in the growth kinetics resulting from the varying nature of the target (oxide or metal; ceramic or single crystal), target to substrate distance, and substrate temperature etc. Oxygen pressure influences both the deposition rate and the kinetic energy of ejected species. The kinetic energy of the ablated species reduces with the increase in pressure due to large number of collisions with background gas molecules and it is commonly observed that the size of ablated plume decreases with increase in oxygen pressure. Therefore it is recommended to reduce the target to substrate distance while working at higher pressures to maintain the optimum energy of the ablated species. The oxygen pressure and the target-substrate distance are correlated by an expression Pd^{γ} = constant for the processing of a good quality oxide film by PLD,[77] where γ is the ratio of the specific heats of the elements in the plume, and depends upon the oxide target composition, P is background pressure, and d is the plume length. If the substrate is located beyond a plume size d, the adhesion coefficient for the ejected species that arrive at the substrate surface is substantially decreased, and therefore the crystal quality of ZnO films becomes worse as the ambient pressure increases beyond an optimum value.[44,78]

The substrate temperature in the PLD process is very crucial because it activates the surface mobility of the ejected species deposited onto the substrate surface and since the kinetic energy of ablated species in PLD is relatively higher than that observed in films deposited by other physical vapor techniques, a crystalline ZnO film with low surface roughness is easily obtained at low temperature.[41,69,79–80] The c-axis orientation of ZnO film degrades at substrate temperature ($<200°C$) due to reduced surface mobility and migration of species on the substrate surface, and the texture quality normally improves with increasing substrate temperatures.[80] Re-evaporation of atoms from the surface of a deposited film at very high substrate temperature ($>600°C$) has been observed resulting in the formation of an off-stoichiometric thin film.

The nucleation process depends on the interfacial energies between the substrate surface and the condensing species, and is governed by the substrate temperature. The surface mobility of the condensing species defines the crystallinity of the film. Generally the species will diffuse through several atomic distances before sticking to a stable position within the newly formed film. A high substrate temperature favors rapid and defect free growth of crystallites due to full oxidation of Zn atoms and optimum surface diffusion of the species, whereas a low substrate temperature results in the growth of a disordered or a poorly crystallized structure.[47] With the increase of substrate temperature and oxygen pressure, the crystallinity of the film is enhanced and allows the development of a smooth, dense and uniform microstructure with good adhesion to the substrate.[81] Although bulk of the reported literature on PLD deposition of ZnO films lays emphasis on relatively high substrate temperatures for growing high quality films,[82] one particular modification in the PLD technique namely "Aurora PLD method" (Fig. 4.2) has shown some promising results, and prospects for film growth at reduced substrate temperatures is yet to be explored.[83–86] It has been proposed as a way of enhancing activation and/or ionization of the ablated particles during transport from the target to substrate with the application of a magnetic field behind the substrate. In comparison to conventional PLD, epitaxy of NiO on MgO has been demonstrated at room temperature. Similarly in the case of ZnO films grown by "Aurora PLD" the photoluminescence is significantly enhanced in comparison to the conventional PLD grown films.[86]

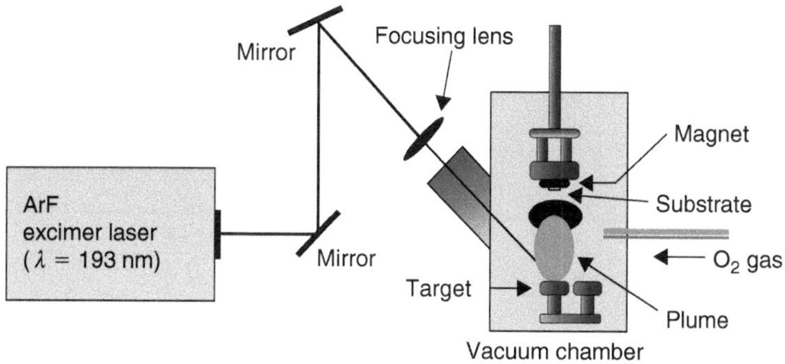

Figure 4.2: A schematic of the Aurora PLD system.[86] [Reprinted from Thin Solid Films, Vol. No. 445, S. Yata, Y. Nakashima and T. Kobayashi, "Improved crystallinity of ZnO thin films grown by the 'Aurora PLD method' ", pp. 259–262, copyright (2003), with permission from Elsevier.]

4.4.2 Substrates for ZnO deposition

A variety of substrates have been used for growing ZnO films by the PLD process and are listed in Table 4.2 and results obtained on amorphous glass, single crystal Si, GaAs and a number of oxide single crystal substrates are discussed.

4.4.2.1 Amorphous glass substrates

Generally on the amorphous substrates like glass the formation energy for the crystallization of ZnO is quite large in comparison to its growth on crystalline substrates that promote nucleation very quickly, and hence growth of high quality ultrathin ZnO films on glass has been quite difficult. On glass substrates ZnO film is normally amorphous in nature initially up to at least 5 nm thickness, and with increase in film thickness the deposited layer provides the nucleating centers and allows the formation of a crystalline layer due to the reduction in the formation energy, and highly ***c***-axis oriented film can be obtained at higher thickness, and the crystallinity is found to depend strongly on film thickness, the deposition rate, and the substrate temperature.

Zinc oxide films deposited on three different substrates, corning glass, passivated silicon, and fused quartz were characterized to study the changes in the crystallographic orientation of the film. A typical x-ray diffractrogram of the ZnO film grown on a glass substrate is shown in Fig. 4.3.

A single dominant x-ray diffraction (XRD) peak *(002)* is observed at $2\theta = 34.46°$ for ZnO film deposited at a substrate temperature of 400°C. High deposition

Table 4.2: Substrates used for ZnO thin film growth by PLD.

Substrate	Ref.
Glass	40, 73, 80, 87–91
ITO coated glass	70
Quartz	81, 92–93
(0001) Al_2O_3	94
(001) NaCl	95
(111) CaF_2	96
(100) Si	40, 41, 80, 90, 97–102
(111) Si	40, 90, 102–103
(100) GaAs	40, 100–102, 104–106
(001) GaAs	44
(111) GaAs	40
(100) InP	89, 100–101,107–108
(0001) Sapphire	40, 44, 96–97, 109–110
($1\bar{1}20$) Sapphire	111–112
(0001) $ScAlMgO_4$	45, 113–115
(111) $SrTiO_3$:La	116
(0001) $LiNbO_3$	117–118
(0001) $LiTaO_3$	118
(100) $SrTiO_3$	119
(110) $SrTiO_3$	120
(001) Mica	96, 121
(111) YSZ	122
(100) γ-$LiAlO_2$	123

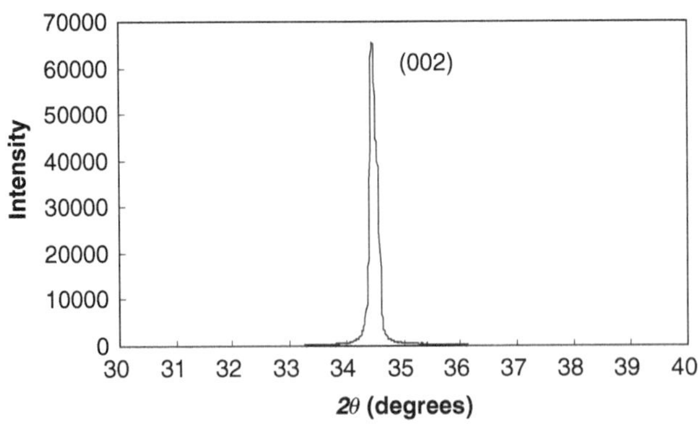

Figure 4.3: X-ray diffraction pattern of ZnO film deposited on Corning 7059 glass at 400°C by laser ablation showing a preferred ***c***-axis orientation at $2\theta = 34.46°$.

pressures are necessary for the growth of ZnO thin films with laser ablation technique in comparison to sputtering, because PLD films tend to be deficient in oxygen when the ablation experiment is done at low pressures. The substrate temperature, at which deposition takes place, is known to influence the crystallographic orientation of ZnO films. In order to understand the crystallization of ZnO films both ultrathin and thick films were deposited. Ultrathin (≈20 nm) ZnO films were deposited on corning glass and fused quartz substrate for 1 min. at a target to substrate distance of 55 mm, using a Nd:YAG laser ($\lambda = 355$ nm, 10 Hz, 1 to 1.5 J/cm^2). Likewise, thick films (≈1 μm) were deposited at a relatively lesser target to substrate distance of 30 mm for the deposition time of 45 minutes, and the substrate temperature was varied between 300 and 500°C.[87]

The structural characterization of ultrathin and thick ZnO films deposited at all the substrate temperatures revealed a preferred *(002)* crystallite orientation. However the peak intensity and full width at half maximum (FWHM) showed a variation with substrate temperature. The ultra thin films showed a decrease in FWHM with increasing substrate temperature [Fig. 4.4(a)], while the thicker films showed a decrease up to a substrate temperature of 400°C and then increased slightly.

The *(002)* peak position shifted slightly with increasing substrate temperature. In case of thick films, Fig. 4.4(b) indicates the *(002)* peak position at $2\theta = 34.46°$ for a substrate temperature of 400°C is close to the bulk value 34.43°. At all other temperatures, there is a shift in the peak position towards the higher side. This shift is attributed to the presence of compressive strain in the film. Though substrate temperature up to 450°C had a marginal influence on peak position (2θ) in the case of ultrathin films, but the films are highly stressed because the peak position is around 34.5°, which is greater than bulk single crystal value of ZnO, and at

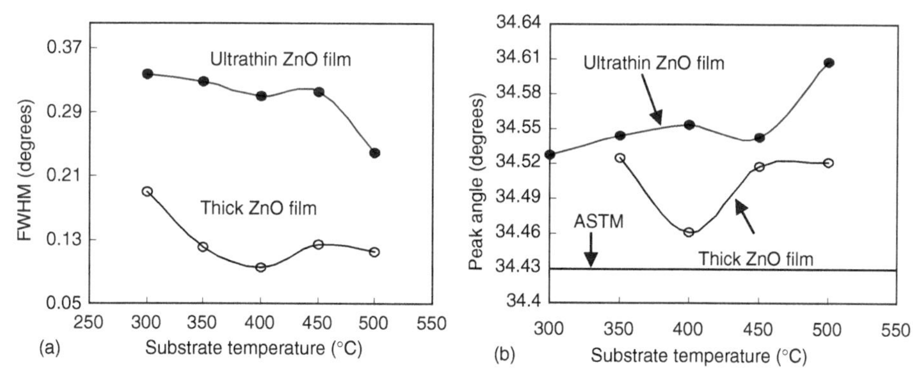

Figure 4.4: Effect of substrate temperature on (a) full width at half maxima (FWHM), (b) peak angle for the ***c***-axis (002) oriented ZnO films grown by pulsed laser deposition on glass substrates.

temperatures higher than 500°C, a significant shift in the peak position is noticed [Fig. 4.4(b)]. Surface morphology of ZnO deposited on glass substrates for varying deposition times indicated a granular structure. Ultrathin ZnO film deposited for 1 minute is found to be continuous with a uniform spread of small size grains and with increasing film thickness (or deposition time) a marked increase in the grain size is observed. The strong violet emission at ~413–424 nm was observed in the room temperature photoluminescence spectra of ZnO films fabricated on amorphous quartz substrate by the oxidation of PLD grown zinc films.[93] The intensity of emission increased and a blue shift in peak position was reported with increase in oxygen pressure and attributed to the electron transition from valance band to interstitial zinc level. Cao *et al.*[92] demonstrated the self formation of laser cavities in PLD grown ZnO film on amorphous fused silica substrates due to strong optical scattering in semiconducting polycrystalline film. The discrete cavity modes of random lasers have been observed, and open up the possibility of using disordered semiconductor microstructures as alternative sources of coherent light emission.

4.4.2.2 Single crystal substrates

Amongst the single crystal substrates used for the epitaxial growth of ZnO, (0001) sapphire has been used most frequently for epitaxial growth partly because high-quality, large-size single crystal wafers are easily available. It has been commonly observed that ***c***-axis oriented ZnO films having in-plane epitaxial relationship can be grown on (0001) sapphire and on similar hexagonal crystals, e.g., (0001) $LiNbO_3$ and $LiTaO_3$, (0001) CaF_2, and (0001) $ScAlMgO_4$. ZnO films grown on (100) and (111) planes of cubic semiconductors (Si, GaAs, InP), amorphous glass and passivated silicon (SiO_2/Si) substrates are generally found to be ***c***-axis oriented, but they possess random orientation or in-plane rotation domains. Single in-plane orientation has been reported for films grown on $(11\bar{2}0)$ sapphire.[111] The ***a***-axis oriented film was obtained for ZnO films grown on the $(\bar{1}102)$ plane of sapphire.[110,115] Most of the substrates that have been used have a large lattice mismatch with ZnO, and the deposited films contain large size crystallites separated by dense grain boundaries, and these are often detrimental, generating defects and conduction barriers for high-performance semiconductor devices. For example, Hall mobility measured for ZnO films grown on sapphire is found to be less than $100\,cm^2\,V^{-1}\,s^{-1}$, in comparison to the high values measured in bulk[124] ZnO single crystals (~$230\,cm^2\,V^{-1}\,s^{-1}$), and this shortcoming has been attributed to the dominant carrier scattering at the negatively charged grain boundaries.[125–126]

Amongst the oxide single crystal substrates listed in Table 4.2 (0001) $ScAlMgO_4$ (SCAM) is the best choice because of the small lattice mismatch (0.09%). Epitaxial growth and electrical properties of ZnO films grown on (0001) sapphire substrates, and the (0001) SCAM substrates are discussed separately in detail in next section.

4.4.2.3 ZnO growth on (0001) sapphire substrate

Ohtomo *et al.*[127] suggest that though the ZnO thin films grown on lattice matched ($\sim$18%) sapphire (0001) substrates are epitaxial but they tend to maintain a multi-crystalline structure having incoherent grain boundaries. Actually these grain boundaries appear to be useful to some extent for observing excitonic laser action. When the nanocrystal is tuned to a small size ($\sim$50 nm), excitonic lasing is most efficient, and makes use of the grain boundaries as mirrors of longitudinal cavity. However the electronic properties of such films appear to be very poor as represented by the measured electron concentration $\boldsymbol{n} \sim 10^{17}\,\mathrm{cm}^{-3}$ and Hall mobility $\boldsymbol{\mu} \sim 10\,\mathrm{cm}^2\,\mathrm{V}^{-1}\,\mathrm{s}^{-1}$ at room temperature compared to the desired values of $\boldsymbol{n}$ $\sim 10^{15}\,\mathrm{cm}^{-3}$ and $\boldsymbol{\mu} \sim 200\,\mathrm{cm}^2\,\mathrm{V}^{-1}\,\mathrm{s}^{-1}$ measured on bulk single crystals. Mobilities of 130 and $137\,\mathrm{cm}^2\,\mathrm{V}^{-1}\,\mathrm{s}^{-1}$ along with minimum carrier concentrations of 1.2×10^{17} and $7.4 \times 10^{16}\,\mathrm{cm}^{-3}$ have been obtained in ZnO films on ***c***-plane and ***a***-plane sapphire substrates respectively. Kaidashev *et al.*[128] reported the highest value of hall mobility for films grown on the (0001) sapphire by PLD. This was attained by a multistep growth process, in which a low-temperature interlayer growth followed by high-temperature annealing and overgrowth were successively repeated several times to improve the crystallinity. Kaidashev *et al.*[128] reported high electron mobilities from 115 up to $155\,\mathrm{cm}^2\,\mathrm{V}^{-1}\,\mathrm{s}^{-1}$ at 300 K in a narrower concentration range from 2 to $5 \times 10^{16}\,\mathrm{cm}^{-3}$ (Fig. 4.5), and their detailed studies suggest that the insertion of a 30 nm thin ZnO relaxation layer deposited at reduced substrate temperature is the most critical step in the multistep PLD deposition process. The high mobility epitaxial ZnO layer shows atomically flat surfaces with a large grain size of 0.5 to 1 μm, whereas the surfaces of the low mobility films revealed clearly resolved hexagonally faceted columnar grains 200 nm in size.

Generally for the growth of ZnO thin film by PLD on sapphire substrates, low pulse repetition rates (1 to 20 Hz) and high energy per pulse ($\sim$100 mJ/pulse) have been used. Hence the amount of species arriving at the substraes could be instantaneously large, causing the condensation of undesired particulates in the films and introduce various types of intrinsic defects.[76,111] The film quality tends to be poor and requires careful control of the growth parameters. Lin *et al.*[129] recently have developed a new method of growing ZnO films using a fast-pulsed laser deposition (FPLD) method. A diode pumped solid-state laser (355 nm wavelength, 15 ns pulse width) running at 10 KHz was used to ablate the ZnO target that was maintained at a distance of 50 mm from the substrate. The energy per pulse was fixed at 200 μJ, and the laser beam was tightly focused so that the fluence was maintained at approximately $2\,\mathrm{J/cm}^2$. Atomic oxygen was introduced into the growth chamber using

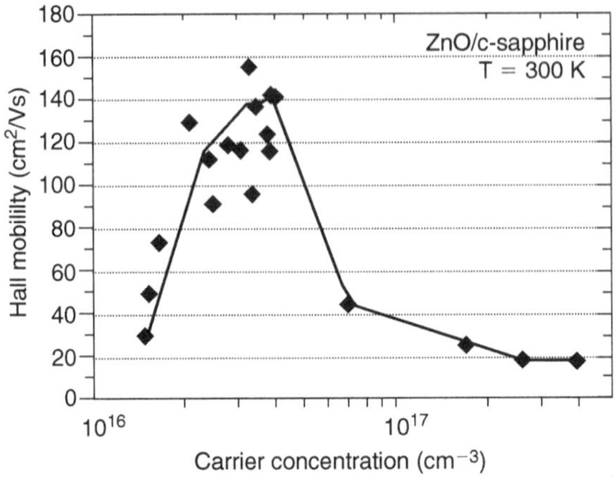

Figure 4.5: Room temperature electron mobility and carrier concentration for multi-step PLD grown ZnO film on (0001) sapphire.[128] [Reused with permission from E.M. Kaidashev, M. Lorenz, H.V. Wenckstern, A. Rahm, H.-C. Semmelhack, K.-H. Han, G. Benndorf, C. Bundesmann, H. Hochmuth and M. Grundmann, Applied Physics Letters, 82, 3901 (2003). Copyright 2003, American Institute of Physics.]

an RF atom source to assist the film growth at 800°C. It was possible to achieve a high mobility of $100\,cm^2\,V^{-1}\,s^{-1}$ at 300 K with a carrier concentration in the range $(2 \text{ to } 5) \times 10^{16}\,cm^{-3}$ in epitaxial ZnO films grown by FPLD method without a buffer layer on sapphire in comparison to the two-step approach discussed earlier.

4.4.2.4 ZnO growth on (0001) $ScAlMgO_4$ single crystal substrate

To overcome the problems associated with the use of sapphire substrates for ZnO film growth, hexagonal $ScAlMgO_4$ (0001) is the best alternate[130] which is lattice matched to ZnO more closely with lattice constant of $\boldsymbol{a} = 0.3246$ nm and $\boldsymbol{c} = 2.5195$ nm and provides in-plane lattice mismatch as small as 0.09% and promotes the growth of high quality ZnO films exhibiting superior electrical properties. $ScAlMgO_4$ is considered to be a natural supperlattice composed of alternating stacking layers of wurtzite (0001)–face $(Mg,Al)_x$ and rocksalt (111)–face ScO_y layers, and has a strong cleavage habit along the (0001) plane. The substrate is insulating, and is suitable for characterization of optical and electronic properties because of the large band-gap of 6.2 eV, and a recent study confirms that (0001) SCAM substrate surface actually terminates with the ScO_y layer,[131] and high quality single crystals can be grown by the Czochralski method.

ZnO films grown by PLD are found to be highly crystalline, and a MgZnO buffer layer on the substrate further improves the crystallinity exhibiting distinct RHEED intensity oscillations.[132] The defect density in the films is very low, and a high mobility (μ_H) of 440 $cm^2 V^{-1} s^{-1}$ that exceeds the value measured on bulk ZnO single crystals has been measured, and improved structural and optical properties have been obtained with the ZnO/MgZnO superlattices on (0001) SCAM substrates.[133] Tsukazaki *et al.*[45] deposited undoped *n*-type ZnO epitaxial films by laser–MBE on a smooth ZnO buffer layer formed on a $ScAlMgO_4$ substrates at a temperature of 950°C that yielded a carrier concentration of 10^{16} cm^{-3} at room temperature, and reported a Hall mobility of 300 $cm^2 V^{-1} s^{-1}$ and 5000 $cm^2 V^{-1} s^{-1}$ at room temperature 300 K and 100 K respectively, thus surpassing the best value for a ZnO bulk single crystal. However the reported growth temperatures in the range of 700 to 950°C are still on the higher side for obtaining high quality films on the (0001) SCAM substrates.

4.5 Doped and alloyed ZnO films

Transparent conducting oxides (TCO) with low electrical resistivity ($\sim 10^{-4}$ Ω cm), high carrier concentration density ($>10^{21}$ cm^{-3}) and high mobility are becoming very important for a variety of applications including solar cells, opto-electronics devices and displays. However the unintentional electrical conductivity and mobility of pure ZnO film is found to be very low. Potential applications for ZnO films as an active channel layer in transparent thin film transistor (TTFT) have been identified for display panel devices, and the key issues for further development are related to (i) achieving higher mobility, (ii) lowering the background electron concentration, and a good control on the carrier concentration.[134]

4.5.1 n-*type ZnO for TCO and TOS applications*

The properties of ZnO can be tuned via the substitution of intentional impurity element on either the cation site or the anion site. It is important to have a precise control over the dopant concentration, and PLD has been extensively used to study the influence of a variety of dopant materials. ZnO is a well-known transparent conducting oxide (TCO) material and its electrical conductivity is *n*-type due to the presence of a large concentration of native defects in the form of oxygen vacancies, or zinc interstitials that contribute deep and shallow states respectively. The doping of ZnO with heavily doped donor impurities such as Al, Ga, Zr or In by PLD showed excellent properties with improved *n*-type conductivity in comparison to those prepared by other deposition techniques.[135–137] Reported results on PLD grown ZnO doped films exhibiting *n*-type or *p*-type conductivity are summarized in Table 4.3. The

Table 4.3: Doped ZnO thin films prepared by PLD with *n*-type/*p*-type conductivity.

Dopant	Dopant source in ZnO target	Concentration	Substrate	Applications and studies	Ref.
Al^{3+}/Ga^{3+}	M_2O_3 (M = Al, Ga)	1–5 wt% Al 1–7 wt% Ga	Quartz/Si (100)	*n*-type TCO	137
$Al^{3+}/Ga^{3+}/B$	M_2O_3 (M = Al, Ga, B)	1–5%	Corning glass	*n*-type TCO	138
P/Mg^{2+}	P_2O_5/MgO	2%/10%	ITO/glass	TTFT & LCD	134,152
Ga^{3+}	Ga_2O_3	5%	Corning glass	*n*-type TCO	144
Zr	ZrO_2	1–10 wt%	Corning glass	TCO	144–146
Al^{3+}	Al_2O_3	2–3 wt%	Glass/quartz	TCO	135,137,139, 144,147–148, 154–156,195
Al^{3+}	Al_2O_3	1 & 10 wt%	$SrTiO_3$ (001)	Epitaxy	149
Al^{3+}/H_2	Al_2O_3/H_2 gas	1–3 wt%	Glass	*n*-type TCO	141
Al^{3+}	Al_2O_3		γ-Al_2O_3/Si (111)	Epitaxy	150
Sn/In	In_2O_3/SnO_2	1/50 wt%	Glass	*n*-type TCO	142
Ga^{3+}	Ga_2O_3	5 at.%	Glass	*n*-type TCO	136
Al^{3+}/Mg^{2+}	Al_2O_3, MgO	2.3/≤40 at.%	Glass	*n*-type TCO	157
P^{3-}	P_2O_5	≤1–5 at.%	(0001) Al_2O_3	*n*-type	158–159
N^{3-}	N	$\leq 10^{19}$ cm^{-3}	(0001) $ScAlMgO_4$	*n*-type, PL	160
N^{3-}	N_2O, NO	$\leq 10^{21}$ cm^{-3}	Glass	*n*-type, SIMS	161
N^{3-}	N	$\sim 2 \times 10^{20}$ cm^{-3}	(0001) $ScAlMgO_4$	*p*-type	45
N^{3-}	N_2O + plasma 5×10^{-4} mbar		Glass	*p*-type	162–163

N^{3-}	$N_2O + ECR$		$(11\bar{2}0)$ Al_2O_3	*p*-type	164
N^{3-}	$N_2O + ECR$		Fused silica, glass	*p*-type	111,165
N^{3-}	N2, N_2O, NO	1 at.%	MgO/Al_2O_3	Scintillator	159,166
N^{3-}/Ga^{3+}	N, Ga_2O_3	$\leq 10^{21}$ cm^{-3}	(0001) $ScAlMgO_4$	*n*-type	167–168
N^{3-}/Ga^{3+}	NO, Ga_2O_3	≤5 at.%	(0001) Al_2O_3, glass	*n*-type	169
N^{3-}/Ga^{3+}	N_2O, Ga_2O_3	≤5 at.%	Glass	*p*-type	111,162,170,222
N^{3-}/Ga^{3+}	$N_2O + Ga_2O_3$ + ECR Ga (0.1 at.%)		Glass	*p*-type	162
N^{3-}/Ga^{3+}	$N_2 + Ga_2O_3$ + RF		(0001) $ScAlMgO_4$	*n*-type	167
N^{3-}/Ga^{3+}	NO + Ga_2O_3 + ECR 5 wt%		(0001) Al_2O_3	*n*-type	169
Ga^{3+}	Ga_2O_3			TCO	155,167,171,172
P^{3-}	Zn_3P_2	Laser annealing	*n*-type ZnO	*p*-type	173
P^{3-}/Mg^{2+}	P_2O_5, MgO	2/10 at.%	Glass, sapphire	p-type	174–175
P^{3-}/Mg^{2+}	P_2O_5, MgO	2/10 at.%	ZnO:Sn/sapphire	*p*-type	176–178
Al^{3+}	Al_2O_3	5 wt%	Steel disk	Tribology	179–180
As^{3-}	GaAs	$\leq 10^{21}$ cm^{-3}	(001) GaAs	*p*-type	181
As^{3-}	(As) mol. beam		n-ZnO/SiC	*p*-type,	182–183
As^{3-}	(As) mol. beam		Sapphire	*n*-type	182
As^{3-}	Zn_3As_2 (RTA)	1 mol%	*c*-sapphire	*p*-type	184
As^{3-}	Zn_3As_2	1 mol%	*c*-sapphire	*n*-type Ω cm	184
Li^+	Li_3N	≤17 at.%	(0001) Al_2O_3	Raman	155
Li^+	Li_2O	≤30 at.%	(100) Si	Ferroelectricity	185–186
Li^+/Mg^{2+}	Li_2O, MgO	10/2.5 at.%	(100) Si	Electro-optic	187–188
Mn	Mn_2O_3	5 at.%	(0001) Al_2O_3	Electro-optic	189
Sb^{3-}	Sb_2O_3	0.5 at.%	(0001) Al_2O_3	Raman	155

room temperature electrical resistivity for Al-doped ZnO film is reported to be a minimum (1.4×10^{-4} Ω cm) with a carrier concentration of the order of 10^{21} cm^{-3} for 1 wt% of the dopant concentration.[137–140] The low resistivity of TCO films is affected by target sintering conditions, and the processing parameters. The resistivity is found to decrease at an optimum substrate temperature of 300°C, and increases at higher temperatures due to increase in oxygen content in the deposited film.[141] PLD of Al doped ZnO films has also been attempted using hydrogen atmosphere to suppress the oxidation. On glass substrates the resistivity is found to decrease slightly up to a hydrogen pressure of 1.2×10^{-2} Pa and the Hall mobility increases to 50 cm^2 V^{-1} s^{-1}. Tin doping in indium-zinc oxide thin film deposited on glass shows improved properties at an optimum composition of $Zn_2In_{1.9}Sn_{0.1}O_{5+\delta}$ and yields electrical resistivity, mobility and carrier concentration of 4×10^{-4} Ω cm, 30 cm^2/Vs and 5×10^{20} cm^{-3} respectively.[142] Pure and Sn-doped ZnO films have been deposited using a new technique named as metronome doping.[143] This technique allows continuous introduction of a dopant during film growth, with varying concentration of the dopant in a single deposition sequence. Sn concentrations up to 16% have been successfully introduced into ZnO films using the metronome doping technique, and controlled doping is achieved in a simple way by arranging a thin rod to pass through the laser beam regularly in a plane just above the target surface. Adjusting the time the rod spends under the laser beam could control the doping rate in the films. A minimum value of resistivity $\sim 2.5 \times 10^{-4}$ Ω cm with a 2.7 wt% Ga-doped ZnO target was reported in films deposited on Corning glass substrate.[138] Kim *et al.*[144] studied the effect of different dopants (Al, Ga and Zr) on the properties of ZnO film deposited on glass substrate and obtained the resistivity of the order of 10^{-4} Ω cm in all the films. Zr doped ZnO films are amorphous when grown at room temperature, and become crystalline at 150°C. Their resistivity is slightly higher (5.6×10^{-4} Ω cm) and optical transmission (~88%) is low. A sheet resistance of 18 Ω/sq with 5 wt% Zr in ZnO is found to be comparable to commercially available ITO films.[145] However the observed power efficiency (~0.9 lm/W) of an organic LED device with doped ZnO film as anode material at a luminance of 100 cd/m^2 is found to be high with Zr dopant in comparison to the Al or Ga dopant in ZnO, and the performance is comparable to devices fabricated on commercial ITO coated substrates.[144,146] Boron doped ZnO films have also been studied but the films show a higher resistivity in comparison to Ga and Al doped ZnO and have not attracted much attention for TCO applications.[138] Attempt have been made to grow Al-doped ZnO films (AZO) at room temperature using a modified oxygen radical assisted PLD technique.[147] The system consists of an oxygen radical source, made of SiO_2 and is surrounded by a helical coil, and can be fitted into a conventional PLD chamber. The molecular oxygen after entering into the radical source is decomposed into radicals and ions by the applied RF field. A low resistivity (5×10^{-4} Ω cm) with 86% optical transmission at room temperature has been obtained, and allows the flexibility for depositing transparent conducting layers of ZnO on thermally sensitive substrates (e.g. plastic) useful for lightweight flat panel displays or solar cells.[147] The low value of mobility observed in the doped ZnO films on amorphous substrates is attributed to the presence of grain boundaries and dislocations. The growth of epitaxial film delineates the effect of grain boundaries and crystallinity on the electrical properties of TCOs. Single crystal sapphire (0001) substrate has been used to grow Al doped ZnO epitaxial film by PLD at 680°C with in-plane epitaxial relationship of ZnO:Al(0001)/Al_2O_3(0001).[148] The resistivity and carrier mobility of the epitaxial film is about 2.2×10^{-4} Ω cm and 32 $cm^2 V^{-1} s^{-1}$ respectively with 90% optical transmission. Epitaxial growth of Al-doped ZnO on $SrTiO_3$ (001) substrate has been reported by PLD at 650°C.[149] A lattice mismatch of about 3% is reported between the (110) direction of the $SrTiO_3$ substrate and the ***c***-axis of highly oriented (1120) Al-doped ZnO film. Reflection high energy electron diffraction (RHEED) study indicates that the epitaxy of doped film is benefited by an initial layer-by-layer growth mode that vanishes after three monolayers. Epitaxial growth of Al-doped ZnO on epi-γ-Al_2O_3/Si(111) substrate by rf sputtering and PLD has been compared. RHEED studies indicate a 30° rotation of PLD deposited Al doped ZnO film with respect to γ-Al_2O_3/Si(111) substrate, and the quality of epitaxial film grown by PLD was reported to be far superior to rf sputtered film.[150] Doped TCO films deposited by PLD have also been tested for dry etching under CH_4/H_2 gas plasma and studies show that dry patterning techniques can be applied to ZnO based TCO films.[137] Besides the chemical and thermal stability, the main advantages of doped ZnO as an alternative to ITO for commercial applications in solar cell and flat panel displays are the low cost and the absence of toxicity.

Successful attempts have been made to fabricate transparent thin film transistors (TTFTs) using ZnO film as a transparent semiconducting oxide (TSO) deposited by PLD as an active channel layer.[151] The electrical property of ZnO film has been optimized by varying PLD processing parameters and a minimum value of carrier concentration of about 3.3×10^{16} cm^{-3} and a maximum Hall mobility 0.70 cm^2/Vs at 450°C substrate temperature has been achieved by depositing ZnO films at 3 mTorr in (oxygen with 8% O_3). A layer of SiN_x (>50 nm thick) was introduced between the ZnO and the gate oxide SiO_2 to prevent the diffusion of Zn into SiO_2. The I_{on}/I_{off} ratio of ZnO-TTFTs fabricated on Si wafer was more than 10^5. The optical transmission on glass substrate was more than 80%, and the devices could be operated in visible light. For transparent thin film transistor (TTFT) applications acceptor impurities in ZnO films are of special interest for controlling the carrier concentration in the moderate range. Ultrathin ZnO films (20 nm thick) as channel layers in TTFT devices doped with phosphorus and Mg have been deposited using PLD. Phosphorous doping in ZnO creates deep acceptors for reducing the

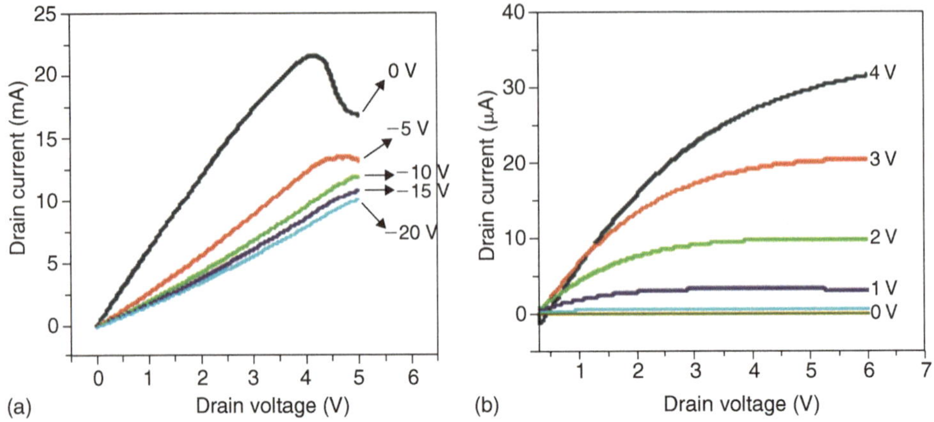

Figure 4.6: The output characteristics of the TFT with alternative channel layer (a) undoped ZnO and (b) P doped (Mg, ZnO).[134] [Reused with permission from Y.J. Li, Y.W. Kwon, M. Jones, Y.W. Heo, J. Zhou, S.C. Luo, P.H. Holloway, E. Douglas, D.P. Norton, Z. Park and S. Li, "Progress in semiconducting oxide-based thin-film transistors for displays", Semiconductor Science Technology 20, 720–725 (2005), IOP publishing house, Copyright 2005, Institute of Physics.]

carrier density for depletion mode operation, whereas Mg doping further reduces the carrier concentration, and makes doped ZnO layer more suitable for enhancement mode operation.[134] Insulating and amorphous coatings of (Ce,Tb)$MgAl_{11}O_{19}$ (CTMA) or HfO_2 have been used as an oxide gate due to their large band gap in comparison to ZnO.[134,152] The doped ZnO film was annealed at 600°C in oxygen ambient to activate the phosphorous doping. TFT using undoped ZnO and P doped ZnO with CTMA as gate oxide showed depletion-mode operation (Fig. 4.6a). The operation of a TFT device in the enhancement mode has also been successfully demonstrated using P doped (Mg, ZnO) as the channel with an on/off ratio of 10^3 (Fig. 4.6b). The carrier concentration of the doped ZnO channel is estimated to be about $3.9 \times 10^{16}\,cm^{-3}$, and the field effect mobility was ($5.32\,cm^2\,V^{-1}\,s^{-1}$) as determined from the linear portion of the drain I-V characteristics.[134] TTFTs fabricated to date using conventional transparent oxide semiconductors (TOSs) such as SnO_2 have exhibited poor performance with on-to-off current ratio of the order of 10^3 and field effect mobilities as low as $10\,cm^2\,V^{-1}\,s^{-1}$.[153] Grain-boundary potentials are thought to limit the performance. Hence it is becoming important to select materials where the carrier concentration can be controlled down to the intrinsic level, and to develop suitable methods for growing high quality crystalline thin films. Keeping in view some of these requirements Nomiura *et al.*[122] reported the fabrication of a TTFT using a single crystal thin film of a complex oxide $InGaO_3(ZnO)_5$ as an electron channel with amorphous HfO_2 as a gate insulator on a YSZ(111) substrate. The device exhibits an on-to-off ratio of 10^6 with a field effect mobility of $\sim 80\,cm^{-2}\,V^{-1}\,s^{-1}$ at room temperature and is insensitive to visible light.

4.5.2 Efforts towards p*-type ZnO*

The ability to achieve *p*-type doping in ZnO is of great interest for the fabrication of transparent two-terminal and three-terminal junction structures for device applications such as ultraviolet light emitting diodes (UV LED) and diode lasers. However the prospects for achieving *p*-type doping in ZnO has been quite difficult in comparison to *n*-type doping due to the carrier compensation by native defects such as zinc interstitial and oxygen vacancies. Group-I elements (Li, Na and K) and group-V elements (N, P and As) as substitutional dopants for Zn and O respectively have been used as *p*-type dopants. However, many of these dopants introduce deep acceptor levels, or occupy the interstitial sites instead of lattice site and cause problem for *p*-type doping. Incorporation of group-I elements results in the formation of semi-insulating film instead of producing *p*-type conductivity in ZnO due to the creation of donor type defects in the form of compensating interstitials rather than substitutional placement.[185–190] Lithium (Li) doped ZnO films have been reported to exhibit ferroelectric behavior and electro-optic property[185–188] and need further confirmation. However, these findings slowly shifted the focus onto group-V elements as effective acceptor impurities in ZnO. Efforts are also continuing towards achieving *p*-type conductivity in ZnO thin films doped with N, As, P, and co-doping methods for simultaneous introduction of donor and acceptor impurities. PLD has been used extensively to achieve *p*-type ZnO film using various kinds of dopant elements, and several modifications have been proposed during the processing for introducing the dopant element effectively into the ZnO lattice.

The most promising acceptor dopant from the group-V elements for *p*-type ZnO is nitrogen (N) because of its similar ionic radius with oxygen, and introduces an acceptor level at 0.4 eV,[199] despite the fact that the solubility of N is low in ZnO. Flowing N_2O gas at 5×10^{-4} mbar through ECR plasma source has been reported[162] to yield *p*-type ZnO film on corning glass substrate by PLD at 400°C having a high resistivity ($5 \times 10^5\ \Omega\,m$; carrier conc. $= 2 \times 10^{10}\,cm^{-3}$), and the type of conduction was confirmed by Hall and Seebeck coefficient measurements. Epitaxial ZnO films with smooth surfaces have been fabricated on a $(11\bar{2}0)$ Al_2O_3 substrate using PLD, and the use of a reactive N_2O plasma enhanced by an electron cyclotron resonance (ECR) source was found to enhance the doping of N, and strongly influence the electrical properties.[111,164] The films were highly *c*-axis oriented with an optical transmission of about 90%, and *p*-type conduction with improved resistivity (2–5 Ω cm), hole concentration (6×10^{18} cm^{-3}) and hole mobility of 0.1–0.4 cm^2/Vs were measured indicating that nitrogen (N) impurities act as an effective acceptor for producing shallow *p*-type dopant.[111,164] *P*-type ZnO films on corning glass substrate with improved electrical properties have been obtained by PLD (resistivity ~2 Ω cm and carrier conc. ~4×10^{19} cm^{-3}) by co-doping of Ga (0.1 wt% and 5 wt%) as donor, and nitrogen as acceptor impurities (via N_2O gas through a plasma source). The donor impurity (Ga) is not a *p*-type compensator, but a good reactive codopant, and deliberate codoping of donors seems essential for enhancement of acceptor dopant (N) incorporation, and it is observed that the solubility limit of N concentration is increased by simultaneous incorporation of Ga atoms in the epitaxial ZnO fabricated on *a*-face sapphire.[191] It is also suggested that an increase in nitrogen (N) concentration minimizes the formation of compensating defects created by Ga-complexes such as Ga_{Zn}-$O_{interstitial}$ and Ga_{Zn}-Zinc vacancy (V_{Zn}).[191] The use of N_2O gas through ECR plasma source is reported to be more effective in comparison to N_2 gas as a source for incorporating the nitrogen (N) dopant into ZnO for achieving *p*-type behavior.[162] The dissociation energies (N-*n*: 5.16 eV and *n*-O:6.37 eV) and ionization potential (12.9 eV) for N_2O gas are lower than the corresponding values (N-*n*: 9.76 eV) and (15.65 eV) respectively for N_2 gas, and therefore N_2O is found to be more useful in preventing the formation of native defects such as oxygen vacancies in ZnO film besides introducing N as acceptor impurities.[37,163] It is interesting to note that ZnO films prepared by co-doping of Ga and N without an ECR source exhibit *n*-type conductivity, and *p*-type conductivity is reported only when the plasma source is used.[162–163] However the reproducibility of these results is reported to be uncertain due to the sensitive dependence on processing conditions, and the doping parameters.[192] Matsui *et al.*[161] compared the incorporation of nitrogen content (N) into ZnO films deposited by PLD using three different gases N_2O, NO and N_2 as sources of nitrogen in the presence and absence of plasma discharge. The introduction of a high concentration of N ($>10^{20}$ cm^{-3}) into the ZnO lattice was difficult using a gas flow of N_2O and N_2 without plasma, and clearly established that N_2O gas flowing through a plasma discharge was more effective for enhanced incorporation of N by means of radical N_2 species that are produced. However, free radical characteristics of NO gas do allow the effective incorporation of N dopant into ZnO using a simple gas flow, but only when introduced specifically without any assistance from a plasma discharge. For example the use of plasma discharge with NO gas was ineffective to achieve *p*-type conduction when ZnO films were deposited on (0001) sapphire substrate by simultaneous doping of Ga (5 wt%) and N, using NO gas through a nitrogen ion gun and an ECR nitrogen plasma source.[169] Nitrogen concentration in the deposited ZnO layers was found to be low even when ($N_2 + O_2$) gas mixtures were used for depositing N-doped ZnO layers by PLD with RF discharges. Films were always *n*-type despite the use of sub-picosecond and nanosecond lasers, however the mobility values were higher with the sub-picosecond laser in comparison to the nanosecond laser.[193] Although the use of NO gas for nitrogen (N) incorporation has been preferred because of its lower ionization potential in comparison to N_2 and N_2O gases, but its effectiveness for nitrogen (N) incorporation into ZnO is yet to be recognized. Subsequently systematic efforts have been directed to achieve *p*-type conductivity in ZnO film deposited on lattice matched (0001)$ScAlMgO_4$ substrate by PLD with simultaneous doping of Ga and N.[167] RF radical source gun with a flow of N_2 gas (3×10^{-5} Torr) has been employed for the incorporation of nitrogen. ZnO single crystal targets and Ga doped ZnO ceramic targets were used to control the Ga concentration in the film over a wide range (10^{18} to 10^{20} cm^{-3}), but no *p*-type conduction was detected in any of the prepared samples. The failure was attributed to the probable formation of the more stable Ga-N pair with excess N dopant serving as a self-compensation center in ZnO instead of quasi-stable N-Ga-N complex which is essentially required for co-doping.[167] It is understood that the deposition of high quality undoped ZnO film (*n*-type) is very important for a reliable introduction of doping to achieve *p*-type conductivity. Laser MBE has been successfully used to achieve *p*-type conductivity.[45] Firstly an epitaxial ZnO film (undoped) was grown at 950°C on a lattice matched $ScAlMgO_4$ substrate.[33,45] Deposition of a buffer layer of semi-insulating $Mg_{0.15}Zn_{0.85}O$ and annealing were necessary to obtain a smooth surface on (0001) $ScAlMgO_4$ substrate prior to the deposition of ZnO. An electron carrier concentration of $\sim 10^{16}$ cm^{-3} and mobility $\sim$300 cm^2/Vs were measured in the deposited undoped ZnO film, and these films were further utilized for the nitrogen doping using a radical source operated at 350 W with a nitrogen flow at a pressure of 6.7×10^{-4} Pa, and a repeated temperature modulation procedure to achieve *p*-type ZnO films.[45] The incorporation of nitrogen into ZnO is difficult at high temperatures and a special deposition scheme was adopted in which a nitrogen doped ZnO film (10–15 nm) is deposited at a low temperature (400°C) followed by the growth of ultrathin (1 nm) film rapidly at a high temperature of 950°C, and the sequence was repeated successively until the desired film thickness was achieved.

The repeated temperature modulation technique is found to be advantageous to put back together any degradation in the surface roughness, and the formation of non-equilibrium defects due to N incorporation. Such modifications during processing have resulted in high quality of *p*-type ZnO:N film showing room temperature hall mobility $\sim$8 cm^2/Vs and carrier concentration $\sim 10^{16}$ cm^{-3} (Fig. 4.7).[33,45] The activation energy estimated from the temperature dependence of hole concentration was 100 meV and compensation ratio (N_D/N_A) $\sim$ 0.8 was higher than the reported values of 0.1 for *p*-type ZnO grown by MBE.[45]

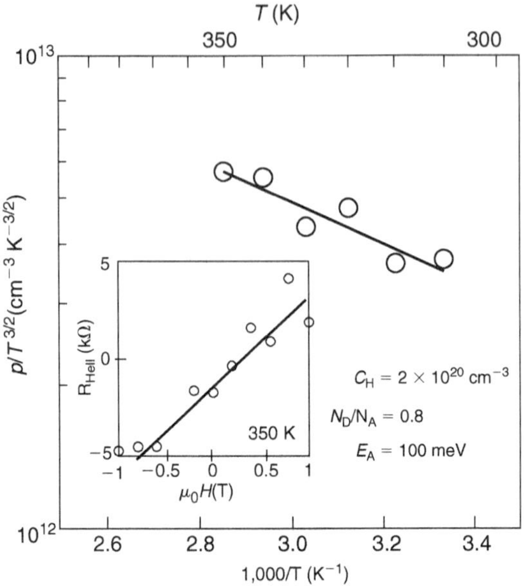

Figure 4.7: Variation of hole concentration as a function of reciprocal temperature for *p*-type ZnO film doped with nitrogen (Nitrogen concentration $\sim 2 \times 10^{20}$ cm^{-3}).[45] [Reprinted by permission from Macmillan Publishers Ltd: Nature Materials, A. Tsukazaki, A. Ohtomo, T. Onuma, M. Ohtani, T. Makino, M. Sukiya, K. Ohtani, S.F. Chichibu, S. Fuke, Y. Segawa, H. Ohno, H. Koinuma and M. Kawasaki, "Repeated temperature modulation epitaxy for *p*-type doping and light-emitting diode based ZnO", Vol. 4, (2005), pp. 42–46, copyright 2005.]

It is predicted that the substitution of other group V elements such as phosphorous and arsenic (P or As) at the O site in ZnO creates a deep acceptor level at 0.93 eV and 1.15 eV respectively due to mismatch between the ionic sizes with O, and cause resistance to the formation of *p*-type conductivity.[190] Both P and As act as amphoretic impurities, and substitution at anti-Zn site instead of O site is probable to avoid the local strain because of their larger bond lengths (2.18 Å and 2.23 Å, respectively) in comparison to Zn-O bond length (1.93 Å), and act as donor impurities. Therefore the incorporation of these elements is not expected to be reliable as acceptor dopants. However, research activities towards the incorporation of group-V elements for obtaining *p*-type ZnO is still in progress. A ZnO based diode structure was fabricated by few groups by integrating P doped *p*-type ZnO film with *n*-type ZnO.[158,173] A laser doping technique was used to introduce *p*-dopant for making *p*-type ZnO film. Initially a Zn_3P_2 thin film was deposited on the *n*-type ZnO layer by PLD, and the fabricated layered structure was subjected to annealing at high pressure by irradiation with a Nd:YAG laser. No residual layer of Zn_3P_2 was observed from surface analysis after laser irradiation, and a phosphorous doped *p*-type ZnO was obtained at the surface after diffusion of the decomposed atoms (Zn and P) into the under lying *n*-type ZnO layer.[173] The structure showed typical I-V characteristics of a diode confirming the formation of a *p-n* junction at the interface of *n*-type ZnO and *p*-type ZnO:P. Similar diode structures on *n*-type ZnO single crystal have been obtained by excimer laser annealing ($\lambda = 248$ nm, pulse width $= 20$ ns, power density $= 150$ mJ/cm^2) using Zn_3P_2 film.[158] Weak light emission of white-violet color was observed from the fabricated *p-n* junction. However, Hall or Seebeck coefficient measurements were not performed to confirm the *p*-type conductivity. A similar attempt for the doping of Sb in ZnO has been made using a 30 nm thick Sb film deposited on ZnO film surface and annealing with a KrF excimer laser under high pressure oxygen.[194] Hall measurements show the formation of *p*-type Sb doped ZnO film with resistivity $8 \times 10^{-3}\ \Omega$ cm and Hall mobility of ~ 1.5 cm^2 V^{-1} s^{-1}. One approach used for the incorporation of phosphorus (P) in ZnO film by PLD using a sintered ceramic target ($ZnO + P_2O_5$) was however not so successful. Films grown on (0001) sapphire substrate over a temperature range 300–500°C with varying concentration of P dopant from 0 to 5 at.% when annealed at 400 to 700°C in 100 Torr O_2 ambient exhibited *n*-type conductivity.[158] Subsequently, a modified ceramic composition consisting of (2 wt% P doped $Zn_{0.9}Mg_{0.1}O$) was used to suppress the residual charge carriers due to native defects,[174] and films deposited on glass and sapphire substrates had to be annealed at 400°C in 20 mTorr oxygen pressure to compensate for the loss of oxygen. Further annealing at 650°C in flowing oxygen for 1 hr produced films with a resistivity of 35 Ω cm, hole concentration $\sim 10^{16}$ cm^{-3} and mobility of 8.2 cm^2 V^{-1} s^{-1} exhibiting *p*-type conductivity.[174,178] Temperature dependence of current-voltage characteristics of *p-n* junction diode of P doped $Zn_{0.9}Mg_{0.1}O$/*n*-type ZnO grown on bulk ZnO crystal exhibited negative temperature coefficients for reverse breakdown voltage.[175,177] The growth of *p-n* junction at temperatures <500°C on tin doped (1000) ZnO single crystal substrate without post deposition annealing has been attempted using P doped $Zn_{0.9}Mg_{0.1}O$ layer having hole concentration $\sim 10^{16}$ cm^{-3}, mobility ~ 6 cm^2/V, and rectifying behavior up to 200°C has been observed, with a room temperature forward turn-on-voltage of ~ 6.5 V.[176] Arsenic (As) and antimony (Sb) have also been used to obtain *p*-type ZnO films.[181,184,194] Ryu *et al.*[181] reported the first result on *p*-type ZnO film using As dopant. The film deposited in ultra pure oxygen (35 mTorr) at 300–450°C on (001) GaAs substrate by PLD was *p*-type. In this study the dopant atoms (As) were diffused into *n*-type ZnO film grown on a GaAs substrate at a moderate temperature, and acted as suitable acceptor impurities ($\sim 10^{17}$–10^{21} atoms/cm^3) resulting in a *p*-type ZnO film at the interface with Hall mobility ~ 0.1–50 cm^2 V^{-1} s^{-1}. Crystal quality of the deposited *p*-type ZnO film could be improved after modifying the procedure for As doping during the processing of ZnO film by PLD using an *insitu* evaporator (molecular beam of dopant source) along with rf oxygen plasma in the deposition chamber, and the procedure was termed as a hybrid beam deposition (HBD) technique.[182,183,196] ZnO films after incorporating As atoms *insitu* at the O site in ZnO lattice were deposited by the HBD technique at 550°C on different substrates (ZnO, SiC and sapphire). However *p*-type conductivity was observed only on the O-face ZnO and Si-face SiC substrate with an electrical resistivity ~2 Ω cm, Hall mobility ~35 $cm^2 V^{-1} s^{-1}$ and hole concentration ~$10^{18} cm^{-3}$ at room temperature, indicating the crucial role of substrate surface polarity.[182] Homostructural ZnO *p-n* junction has been fabricated by integrating *p*-type ZnO:As film with undoped *n*-type ZnO film (electron concentration ~$10^{17} cm^{-3}$ and hall mobility ~100 $cm^2 V^{-1} s^{-1}$) on a SiC substrate.[183] The turn-on voltage of *p-n* junction was in the range 6–8 V and could be reduced by either improving the metal contacts, or increasing As incorporation in ZnO. Arsenic (As) doped ceramic targets prepared by mixing 1 mol% Zn_3As with ZnO powder have been used as a *p*-type dopant to prepare films on a ***c***-plane sapphire substrate at 600°C and 50 mTorr oxygen pressure. Highly ***c***-axis oriented single phase ZnO films with *p*-type conductivity were formed after rapid thermal treatment at 200°C in N_2 ambient for 2 minutes. The resistivity, Hall mobility and hole concentration were about 2.2–6.7 Ω cm and 0.83–11.4 $cm^2 V^{-1} s^{-1}$ and between 2.48×10^{17} and $1.18 \times 10^{18} cm^{-3}$, respectively.[184] So far, the success achieved in developing *p*-type ZnO film using group-V elements as acceptor impurities is commendable, but the origin is still unclear. The observed low value of hole mobility and conductivity in *p*-type ZnO, and its reproducibility is still a major concern. Reliable techniques for dopant introduction to achieve device quality *p*-type ZnO film is still far from being recognized and there is a need for a systematic and careful study with precise control of concentration and substitution of dopant at the lattice site in ZnO.

4.5.3 Band gap engineering

ZnO has been identified as a promising candidate for UV opto-electronic devices and the main emphasis is on band-gap engineering for the design of ZnO-based short wavelength transparent opto-electronic devices. The flexibility of depositing multilayers of different material with precise thickness using a multi-target carrousel controlled by computer without breaking vacuum gives PLD an edge over other techniques, and films can also be grown with a wide compositional variation. Furthermore PLD is a non-equilibrium process and allows the alloying with other compounds with large thermodynamic limit of solubility. Different approaches have been followed to prepare ternary alloyed films by PLD, and involve the alloying of two or more oxide systems, or deposit hetero-structures for quantum confinement, and develop superlattice structures. Majority of the work has been reported using sintered targets prepared by conventional solid state reaction of ZnO powders and the alloy material. In some cases a multi-target carousel has been used to deposit alternate layers of ZnO and the alloy compound, one over the other using two separate ceramic targets, and the overall composition of the alloy film is controlled by varying the thickness of sub-layers of both the materials.[197] The most promising use of the multi-target carousel is in the combinatorial pulsed laser deposition that was used to fabricate high quality alloyed films with compositional variations over a wide range.[198] A multi-target assembly assisted by the automated switching of a mask/shutter that moves back and forth over the substrate is used. This allows the flexibility to vary the content of alloyed material in the deposited film without changing the target having a new composition. An alloy film with a compositional gradient could be prepared, and reduces the time for preparing targets of varying composition, and such an arrangement allows the synthesis of complex compositions in thin film form, whose bulk materials are sometimes difficult to synthesize.

The preferred choice for band-gap engineering in ZnO is alloying with the same group (II-VI) oxide compound and attempts have been made to deposit the ternary alloyed semiconductors on various substrates using MgO, CdO and ZnS. It is important to note that all these compounds are not isostructural to wurtzite ZnO, and the ionic radii of Mg^{2+} (0.65 Å) is relatively close to that of Zn^{2+} (0.74 Å), whereas Cd and *S* have much larger radii. Therefore a small variation in lattice parameter with low deformation in the structure is expected in ternary alloy film using MgO in comparison to the host wurtzite ZnO. The tailored band gap of alloy compositions that are lattice matched to ZnO is useful for fabricating functional heterostructures for optoelectronic device applications. Alloying of MgO in ZnO has been studied extensively in comparison to other compounds. Table 4.4 summaries the major findings collected from the reported literature on ZnO based ternary alloy films deposited by PLD.

Ohtomo *et al.*[199] deposited high-quality, single-phase, wurtzite $Mg_xZn_{1-x}O$ (MZO) thin films on sapphire substrate at 600°C with Mg content up to 33 wt%, and reported a linear increase in band gap ($\boldsymbol{E_g}$) with Mg content from 3.3 to 4.0 eV. A much higher content of Mg in the deposited film was reported in comparison to target composition and was attributed to the difference in the vapor pressure of the ablated Mg and Zn species. Similar observations have been made in a related study where a higher Mg content ~0.34 is seen in the alloyed film using a target of composition $Mg_{0.20}Zn_{0.80}O$ and an increase in the band gap to 4.05 eV was noted, in comparison to the band gap of 3.3 eV for pure ZnO.[200–201]

Phase segregation of MgO in MZO alloyed films is found to occur for $x > 0.33$, whereas, theoretical solubility limit is 0.04 in the bulk form. The phase segregation

Table 4.4: Growth conditions and dopants used for band gap tailoring of ZnO.

Substrate and growth temperature	Targets composition	Crystallographic structure of $A_xZn_{1-x}O$ alloy film (A = Mg, Cd)	Main observations	Ref.
$SrTiO_3$ (100)/ Si (100) 400–800°C	$Mg_{0.40}Zn_{0.60}O$	Epitaxial Cubic $x > 0.62$	Tuning of $\boldsymbol{E_g}$ by growth-induced desorption. $\boldsymbol{E_g} = 5.51$ eV at $x = 0.62$. $Mg_{0.62}Zn_{0.38}O$ detector at $\boldsymbol{\lambda} = 225$ nm shows a large UV-to-visible rejection ratio.	208
(0001) Sapphire 600°C	ZnO + MgO combinatorial	Wurtzite ($x < 0.37$) Phase-separation ($0.37 < x < 0.60$) Cubic ($x > 0.60$)	Cubic MZO segregated in the form of nano-grains embedded in a matrix of wurtzite MZO for $0.37 < x < 0.6$. $\boldsymbol{E_g} = 3.27$ to 4.28 eV for x varying from 0 to 0.37 in wurtzite MZO.	198
(0001) Sapphire 550°C Annealing up to 1000°C	$Mg_xZn_{1-x}O$ $x = 0$ to 0.18 LMBE	Wurtzite ($x = 0$ to 0.23)	Band gap ~3.56 eV for $x = 0.15$, and ~3.72 eV for $x = 0.22$. Solubility limit ~0.15 at 1000°C for MgO in alloyed epitaxial film.	209
(0001) Sapphire 600°C	$Mg_xZn_{1-x}O$ $x = 0$ to 0.18	Wurtzite ($x > 0.33$) Metastable phase ($x > 0.33$)	$\boldsymbol{E_g}$ increases to 3.99 eV with $x = 0.33$. Decrease in $\boldsymbol{a}$, and increase in $\boldsymbol{c}$ lattice parameters with x. More Mg content in film. PL at 4.2 K gives peaks at 3.36 eV($x = 0$), 3.87 eV ($x = 0.33$).	199
α-sapphire 700 to 750°C	$Mg_xZn_{1-x}O$	Wurtzite ($x < 0.36$)	Decrease in $\boldsymbol{a}$, and increase in $\boldsymbol{c}$ lattice parameter with x.	202

Annealing at 750°C			$\boldsymbol{E_g}$ increases to 4.19 eV with $x = 0.36$. PL at 295 K from 3.36 to 4.0 eV	
(0001) Sapphire 700 to 750°C	$Mg_xZn_{1-x}O$	Wurtzite ($x < 0.36$)	$\boldsymbol{E_g}$ increases to 4.19 eV with $x = 0.36$. Refractive index measured by prism coupling decreases with x.	203
(0001) Sapphire 25 to 750°C	$Mg_{0.50}Zn_{0.50}O$	Cubic ($x = 0.50$ to 1.0)	Linear increase in Mg content and $\boldsymbol{E_g}$ with growth temperature. Tunable $\boldsymbol{E_g}$ (5 to 6 eV).	210
(0001) Sapphire 750°C	$Mg_{0.20}Zn_{0.80}O$	Wurtzite $Mg_{0.34}Zn_{0.66}O$	$\boldsymbol{E_g}$ increases to 4.05 eV with $x = 0.34$. High responsivity ~1200 A/W, fast rise time ~8 ns, fall time ~1.4 μs noted ($\lambda = 337$ nm), UV detector application.	200
(0001) Sapphire 750°C	$Mg_{0.10}Zn_{0.90}O$	Wurtzite	Band gap of the alloyed film increases from 3.45 to 3.78 eV with decrease in oxygen pressure from 10^{-2} to 10^{-5} Torr.	215
(0001) Sapphire 750°C	$Mg_xZn_{1-x}O$	Wurtzite ($x \leq 0.29$)	$\boldsymbol{E_g}$ increases up to 4.10 eV. Exciton binding energy decreases from 61 meV (for $x = 0$) to ~50 meV (for $x = 0.17$) and increase to ~58 meV for $x = 0.29$.	204
(0001) Sapphire 450 to 750°C	ZnO + MgO $Mg_xZn_{1-x}O$ $x = 0$ to 0.5	WZ ($x < 0.4$) Metastable ($0.4 < x < 0.6$) Cubic ($x > 0.6$)	$\boldsymbol{E_g}$ increases with decrease in ZnO sublayer thickness in multilayer system. Variation in growth temperature and thickness do not give stable MZO in $\boldsymbol{E_g}$ range 4 to 6 eV.	197 207

(Continued)

Table 4.4: (*Continued*)

Substrate and growth temperature	Targets composition	Crystallographic structure of $A_xZn_{1-x}O$ alloy film (A = Mg, Cd)	Main observations	Ref.
Glass 400°C	$Mg_xZn_{1-x}O$ $x = 0.005$–0.05	Wurtzite	Increase in ***a***, and decrease in ***c*** lattice parameters with x. $\boldsymbol{E_g}$ increases linearly from 3.274 to 3.43 eV for $x = 0$ to 0.05. Near band emission PL peak shift from 3.28 eV (for $x = 0$) to 3.39 eV (for $x = 0.05$).	205
Glass 200 to 500°C	$Mg_xZn_{1-x}O$ $x = 0.1, 0.3, 0.5$	Amorphous – 500°C Wurtzite 400°C Cubic ($x = 0.05$)	Films are amorphous at 500°C. At 400°C, ***c***-axis oriented wurtzite film at $x = 0.1, 0.3$, and cubic at $x = 0.05$.	221
(100) MgO (110) MgO (0001) Al_2O_3 Glassy SiO_2 100°C	$Mg_xZn_{1-x}O$ $x = 0, 0.1, 0.15$	Cubic rock-salt on MgO (110) $x = 0.10, 0.15$	Rock-salt structure of MZO at low Mg content ($x = 0.15$) with very low tetragonal distortion (***c/a***) ~1.02 is grown on MgO (110) substrate. Wurtzite phase on other substrates.	212
MgO (100) Al_2O_3 (0001) TiN/Si (100) 650 to 750°C	$Mg_xZn_{1-x}O$ $x = 0.75$–0.85	Cubic $x = 0.82$–1.00	Epitaxial MZO film on MgO (100) and TiN/Si (100). $\boldsymbol{E_g}$ for $x = 0.82$ was ~6.7 eV.	219

(0001) Al_2O_3 750°C	$Mg_xZn_{1-x}O$ $x = 0.025–0.2$	Wurtzite	Mg content in film is 1 to 2 times of target composition. A slight degradation in crystallinity. $\boldsymbol{E_g}$ increases linearly from 3.28 to 4.5 eV for $x = 0$ to 0.52.	201 214
(0001) Al_2O_3 627°C	$Mg_xZn_{1-x}O$ + ZnO, $x = 0.01$ combinatorial	Wurtzite $X = 0.02–0.19$	Decrease in ***c*** lattice parameter with x. $\boldsymbol{E_g}$ increases linearly from 3.29 to 3.65 eV for $x = 0.02$ to 0.19.	206
(0001) Al_2O_3 (0001) $ScAlMgO_4$ 400°C	$Cd_xZn_{1-x}O$ $x = 0$ to 0.07	Wurtzite $x = 0$ to 0.07	Increase in ***c***, ***a***, ***c/a***, cell volume with Cd content (<0.07). Degradation in quality at $x \geq 0.05$. Loss of Cd in alloy film with post-annealing 800°C in air. $\boldsymbol{E_g}$ decrease to 2.99 eV with $x = 0.07$. Near band emission PL peak at 3.19 eV at 4.2 K.	213
(0001) Al_2O_3 750°C	$Cd_xZn_{1-x}O$ $x = 0.03$ to 0.075	Wurtzite	Cd content in film is 5 to 20% less compared to target composition. A slight degradation in crystallinity. $\boldsymbol{E_g}$ decreases from 3.28 to 3.18 eV for $x = 0$ to 0.017.	201 214
(0001) Al_2O_3 700°C	ZnO + ZnS	Wurtzite ($x_s = 0–0.03$) Mixed phase ($x_s \geq 0.13$)	Shrinkage in ***c*** lattice parameter and increase in ***a***-axis length at $x_s = 0.01$. $\boldsymbol{E_g}$ decreases with increase in x_s.	216

region was evident from the appearance of two optical absorption edges corresponding to the hexagonal and the cubic phases in the alloyed films, resulting into a metastable condition. Sharma *et al.*[202] reported the solubility of MgO in ZnO to be 36 wt% by PLD at a substrate temperature 700–750°C. An increase in band gap from 3.40 to 4.19 eV was observed with x varying from 0 to 0.36, and the refractive index estimated from prism-coupled waveguide measurement showed a systematic decrease with increase in x up to 0.36 over a wavelength range 400 to 950 nm.[203]

Schmidt *et al.*[204] studied the optical properties of wurtzite MZO film ($x < 0.29$) using ellipsometry for photon energies (1 to 5 eV) and reported an increase in band gap up to 4.10 eV. The exciton binding energy was found to decrease from 61 meV (for $x = 0$) to ~50 meV (for $x = 0.17$) and again increase to ~58 meV for $x = 0.29$. A linear increase in the band gap of ***c***-axis oriented wurtzite MZO film deposited on an amorphous substrate at 400°C was observed from 3.274 to 3.43 eV for $x = 0$ to 0.05.[205] The near band emission peak in the photoluminescence measurements of MZO film was found to shift to higher energy from 3.28 eV for $x = 0$ to 3.39 eV for $x = 0.05$, and are coincident with the values of band gaps estimated from UV-VIS spectroscopy.[205] The epitaxial alloy film grown on (0001) sapphire substrate by combinatorial PLD system at 600°C exhibits an enhanced limit of solubility of MgO up to 37 mol%, and the maximum achievable band gap for the wurtzite phase MZO ($x = 0.37$) is 4.28 eV.[206] Appearance of a mixed phase corresponding to wurtzite and cubic MZO with undefined band gap is observed at $x > 0.37$. At a higher value of x (>0.62), MZO films in pure cubic phase with band gap in the range ~5.4 to 7.8.

The band gap engineering in ZnO has also been studied by depositing multilayers of ZnO and MgO thin film sequentially on (0001) sapphire substrate by PLD at different substrate temperatures.[207] The thickness of ZnO layer was varied from 0.75 to 2.5 nm while maintaining a constant thickness (1 nm) for the MgO sub-layer. The absorption edge in the optical transmission spectra (Fig. 4.8) shifted to lower wavelengths from 350 to 200 nm as the thickness of ZnO sublayer decreases from 2.5 to 0.75 nm. The band gap increased with decrease in ZnO layer thickness, and a jump in the measured band gap from 3.7 to 6.2 eV was observed with change in ZnO thickness from 2.0 to 1.25 nm (Inset of Fig. 4.8a). The solubility limit of MgO in alloy film was found to be nearly 40 at.%, and stable MZO alloy films with a continuous variation in $\boldsymbol{E_g}$ from 4.0 to 6.0 eV is reported to be very tedious even with a much smaller variation of ZnO sub-layer thickness.[207] All films exhibit a single peak in the XRD spectra (Fig. 4.8b) corresponding to MZO besides the substrate (sapphire), and a change in the structure from (002) wurtzite phase to (111) orientation of cubic phase is observed in deposited MZO film with decrease in ZnO sublayer thickness.

A variation in the solubility limit reported in the published literature is due to a variation in the nature of the substrate, or the processing conditions such as

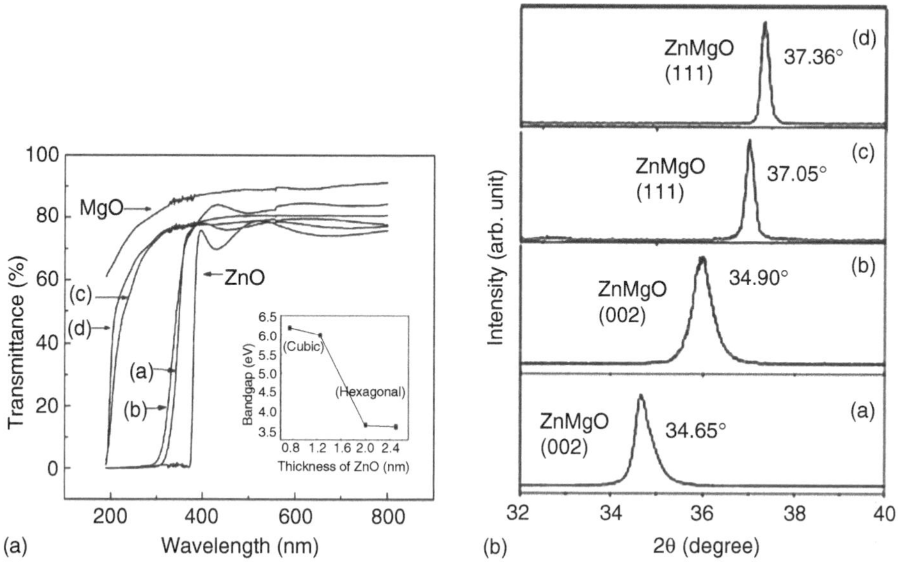

Figure 4.8: The influence of ZnO sublayer thickness in ZnO/MgO multilayer on the (a) optical transmission, Inset – corresponding band gap values; and (b) XRD patterns. [(a) 2.5 nm, (b) 2.0 nm, (c) 1.25 nm and (d) 0.75 nm)].[207] [Reprinted from Thin Solid Films, Vol. No. 447–448, P. Bhattacharya, R.R. Das, and R.S. Katiyar, "Comparative study of Mg doped ZnO and multilayer ZnO/MgO thin films", pp. 564–567, copyright (2004), with permission from Elsevier.]

substrate temperature and oxygen pressure. At higher substrate temperatures, the desorption rate of Zn on the substrate surface is much higher than that of Mg due to differences in the vapor pressure of Mg and Zn, and results in Mg rich films with larger band gap.[208] The thermal stability of supersaturated $Mg_xZn_{1-x}O$ alloy films and its interface with pure ZnO film was investigated systematically by Ohtomo *et al.*[209] as a function of post deposition annealing temperature in oxygen ambient up to 1000°C. The films were grown on (0001) sapphire substrate at 550°C by laser molecular beam epitaxy using sintered targets. The as deposited MZO films had higher Mg content (more than twice) in comparison to ceramic target composition. No significant degradation in the structural property of alloy film ($x = 0.22$) was observed after annealing treatment up to 850°C, and the appearance of a secondary phase due to MgO segregation in deposited film was reported at higher temperature. However no such segregation and inter-diffusion of impurities was observed in the deposited film with $x = 0.15$ even after annealing at 1000°C, indicating a lower value of thermodynamic solubility limit of ~0.15 at 1000°C for MgO in $Mg_xZn_{1-x}O$ epitaxial film.

The crystallographic structure of $Mg_xZn_{1-x}O$ alloy film can be tuned from wurtzite phase of ZnO for a low value of x (<0.37) to the cubic phase of MgO (for $x > 0.60$) by varying the Mg content. The MZO alloy films in the intermediate composition range ($0.37 < x < 0.6$) are in the metastable cubic phase having a mixed crystallographic structure (both wurtzite and cubic) with band gap >5.0 eV.[210] A small variation (1%) in the lattice parameters i.e., an increase in ***c***-axis and a decrease in ***a***-axis with Mg content is observed, and an insignificant change in cell volume has been reported. A complete mapping of the phase evolution in alloyed MZO system has been reported on samples with a lateral composition gradient with one edge at $x = 0$ (pure ZnO) and other at $x = 1$ (pure MgO) prepared by combinatorial PLD on (0001) sapphire substrate.[198] Coexistence of wurtzite and cubic phases were identified for intermediate values of x and defined as phase-separation region between the wurtzite phase (for $x < 0.37$) and the cubic phase ($x > 0.62$). In the phase-separation region, precipitation of a cubic MZO phase segregated in the form of nano-grains embedded in a host matrix of wurtzite MZO phase is observed, and a unique growth relationship is found to exist between cubic and hexagonal MZO where their basal planes coincide.[198] The band gap is not well defined in this region due to phase separation and is termed as a metastable phase region. A linear shift in the band gap from 3.27 to 4.28 eV is observed with increase in Mg (from $x = 0$ to 0.37) in the wurtzite phase region[198] and, a non-linear increase in band gap from 5.4 to 7.8 eV with Mg ($x > 0.62$) is reported in the cubic phase region.[208]

The dependence of Mg content on the growth temperature has also been utilized successfully to tune the composition (or the band gap) of epitaxial MZO alloy film over a wide range using a single ceramic target of specific composition.[208,210] Films were grown on (0001) sapphire substrate using $Mg_{0.5}Zn_{0.5}O$ ceramic target at different substrate temperatures ranging from room temperature to 750°C. A linear dependence of the processing temperature on the both the Mg content in the alloyed film and the band gap was observed. At room temperature the composition in the deposited alloy films was very close to the target composition, and with increasing growth temperature the Mg content increased linearly due to the large vapor pressure of Zn in comparison to Mg and desorption of Zn occurred from the substrate surface. The tunable range of band gap by such a growth-induced desorption process as a function of substrate temperature (RT to 750°C) was 5.0 to 6.0 eV, and subsequently the increase in Mg atomic fraction in the deposited films was from 0.50 to 0.86, respectively. These results indicated that the solubility of ZnO in MgO is about 50% using PLD.

As discussed earlier, the oxygen pressure strongly influences the ablated species arriving at the substrate. Therefore a large variation in the band gap of MZO alloy film can be expected with oxygen pressure during film growth, and was studied in great detail by Misra *et al.*[211] for MZO films in the wurtzite phase using $Mg_{0.1}Zn_{0.9}O$

ceramic target. The band gap of the alloy film was found to increase from 3.45 eV to 3.78 eV with decrease in oxygen pressure from 10^{-2} to 10^{-5} Torr. All deposited films were found to be ***c***-axis oriented corresponding to the wurtzite phase. The increase in oxygen pressure tends to reduce the Mg content in the alloyed film due to effective scattering of the lighter atom (Mg) compared to that of heavier atom (Zn) in oxygen ambient. The reduced content of Mg at higher pressure leads to a lower value of band gap. Similar studies have been reported towards the fabrication of a rock-salt (cubic) type MZO film after alloying with very low content of MgO.[212] A sintered ceramic target of $Mg_xZn_{1-x}O$ ($x = 0$, 0.10, 0.15) was used for the growth of MZO film on various substrates such as MgO(100), MgO(110), Al_2O_3(0001) and glassy SiO_2 with an oxygen backfill pressure of 0.1 Pa. All films deposited with $x = 0$ target were wurtzite ZnO when deposited under identical growth conditions, and films with the various target compositions on different substrate [except MgO(100)] exhibited wurtzite phase. MZO film grown on MgO(100) single crystal substrate at 100°C with targets having $x = 0.10$ and 0.15 compositions exhibited a pure cubic phase indicating the importance of substrate selection for tuning the crystallographic structure. It may be noted that the lattice mismatch between the MgO single crystal substrate and alloyed MgO film in rock-salt phase is very low ~1.6%. The solubility of ZnO in MgO is significantly enhanced to 85 wt% on MgO(100) substrate in comparison to a maximum value of 50% reported earlier on (0001) sapphire substrate.[210]

To tune the band gap of ZnO based alloy film towards the lower wavelengths (λ) corresponding to near visible region, alloying with CdO has been the preferred choice because of its small direct $E_g \sim 2.3$ eV. Growth of epitaxial quality films of the alloy composition $Cd_xZn_{1-x}O$ (CZO) on (0001) sapphire and (0001) $ScAlMgO_4$ single crystal substrates has been achieved by PLD at 400°C, using ceramic targets with $x = 0.07$ which is much larger than the thermodynamic solubility limit of 2 wt% of CdO in ZnO.[213] Contrary to MZO alloy, the lattice parameters (***c*** and ***a***), the unit cell volume, and the ***c/a*** ratio in CZO alloy films show an increase with increase in Cd content. A degradation in the crystal quality of the alloy film is noted with $x \geq 0.05$, and post deposition annealing of the deposited CZO films results in the reduction of the Cd content, and the incorporation of Cd^{2+} leads to band gap narrowing down to 2.99 eV for $x = 0.07$.[213]

Comparison of target and film composition has revealed a significant decrease in the concentration of Cd by (5 to 20%) in the alloyed film.[214] The alloyed films were deposited using two step PLD process on (0001) sapphire substrate where initially an ultra thin nucleating layer was deposited at low pressure followed by fabrication of homoepitaxial alloyed film. A slight degradation in crystallinity was reported with the incorporation of Cd^{2+}, and the band gap of the CZO alloy film is found to decrease from 3.28 to 3.18 eV for a variation in x from 0 to 0.017.[214] Alloy

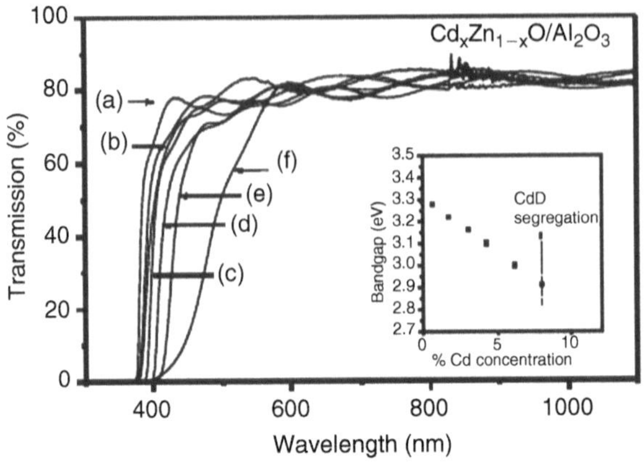

Figure 4.9: Optical transmission spectra of $Cd_xZn_{1-x}O$ thin films with $x \sim$ (a) 0, (b) 0.018, (c) 0.031, (d) 0.06, (e) 0.08 and (f) $x > 0.08$. Inset shows the corresponding band gap values below CdO segregation.[215] [Reprinted from Applied Physics A: Material Science & Processing, Vol. No. 78, (2004), 37–40, "Sequential pulsed laser deposition of $Cd_xZn_{1-x}O$ alloy thin film for engineering ZnO band gap", P. Misra, P.K. Sahoo, P. Tripathi, V.N. Kulkarni, R.V. Nandedkar and I.M. Kukreja, Figure No. 4.5, copyright (2004), with kind permission of Springer Science and Business Media.]

thin films of $Cd_xZn_{1-x}O$ with different Cd concentrations have been grown by sequential ablation of ZnO and CdO targets alternately.[215] The high vapor pressure of CdO compared to ZnO presents a serious problem, and the Cd concentration in the film is found to be around four times less than that in the target. The films are highly crystalline with wurtzite phase only up to a Cd concentration of 8% and an optical transmission higher than 80% is found in the visible region for all the samples (Fig. 4.9). A continuous decrease in the band gap ($\boldsymbol{E_g} = 3.28$ to 2.9 eV) in CZO alloy films is observed with a variation in the Cd content from $x = 0$ to 0.53 (Inset of Fig. 4.9). At higher Cd content (8%), a large broadening in the optical transmission spectrum is observed due to segregation of CdO phase.[215]

ZnS has also been used for studying the band gap engineering in ZnO using PLD.[216] Sintered ceramic targets of high purity ZnO and ZnS were alternatively ablated to introduce S in ZnO. The content of S (x_s) in the ZnO:S film deposited at 700°C on (0001) sapphire substrate was controlled by changing the number of laser pulses for each target. The deposited films were in single phase in the range ($x_s = 0$ to 0.03) with $\boldsymbol{c}$-axis orientation corresponding to wurtzite structure. The $\boldsymbol{a}$-axis length of ZnO:S film is found to increase by 0.5% at $x_s = 0.01$ and an abrupt shrinkage in $\boldsymbol{c}$-axis length is noted. At higher value of $x_s = 0.13$, a mixed phase begins to appear, and the incorporation of S at interstitial site becomes probable.

A continuous decrease in the band gap was reported with increase in x_s, despite the fact that the band gap of ZnS is higher (3.74 eV) than ZnO.[216]

In comparison to the nitrides, it is clearly observed that the single-phase alloy films of II–VI oxides such as $A_xZn_{1-x}O$ (with **A** = Mg, Cd) can tailor the band gap over a wide range with a small variation in lattice parameter, while maintaining the hexagonal wurtzite structure. The content of **A** in the alloyed ternary $A_xZn_{1-x}O$ compound allows the band gap engineering of ZnO, and the new value of band gap can estimated as follows[217]

$$\boldsymbol{E_g}(x) = \boldsymbol{b}(x-1)x + [\boldsymbol{E_g}(\mathrm{AO}) - \boldsymbol{E_g}(\mathrm{ZnO})] + \boldsymbol{E_g}(\mathrm{ZnO})$$

where, $\boldsymbol{b}$ is the bowing parameter, $\boldsymbol{E_g}$(ZnO) and $\boldsymbol{E_g}$(AO) are the band gap values of ZnO and AO compound respectively. The band gap of ZnO after alloying with MgO increases with increasing content of Mg, whereas, alloying with other compounds (CdO and ZnS) results in a decrease in $\boldsymbol{E_g}$. The linear increase in band gap with Mg content as discussed earlier shows that the bowing parameter $\boldsymbol{b}$ is zero with MgO alloying in ZnO. However, in the case of ZnO:S alloyed films, a decrease in $\boldsymbol{E_g}$ with increase in sulphur (S) content is noted despite the higher value of the band gap of pure ZnS (3.74 eV) over that of pure ZnO (3.3 eV), and suggests a positive value for the bowing parameter.[216] The influence on the bowing parameter is still not properly understood, and it is still difficult to draw any clear conclusions from such preliminary studies, and the analysis need further refinement with carefully measured data. There is a need for extensive studies on other compatible systems with reliable data to establish the fact, because even in the case of ZnO:S alloy films, there is a need to stabilize *S* and to suppress the $ZnSO_4$ formation at high temperature.[216] For example in the case of epitaxial ZnSeO alloy films grown by molecular beam epitaxy on GaAs substrate using rf plasma it was noted that the band-gap energy decreases with increasing oxygen composition and indicated the occurrence of large band gap bowing. Bang gap energy data was fitted to a quadratic function of composition and a bowing parameter of 8 eV was estimated.[218]

A buffer layer of $SrTiO_3$ or TiN has been used to overcome the large lattice mismatch between MZO film and Si (100) substrate to grow epitaxial MZO films with tuned composition on silicon.[217,219] The $SrTiO_3$ (STO) buffer layer was deposited by a three step process on Si substrate to avoid the formation of SiO_x interfacial layer, and the undesirable (110) orientation. This was achieved by growing a 3 nm SrO followed by a 10 nm STO grown on Si in vacuum at room temperature, and the bilayer was annealed at 850°C to deposit a STO film of 10 nm and 50 nm in vacuum and oxygen ambient respectively. Compositionally tuned epitaxial films of MZO were prepared on the $SrTiO_3$ buffered silicon substrate at different substrate temperatures using $Mg_{0.4}Zn_{0.6}O$ target. However all the deposited MZO films were in cubic phase and possessed higher Mg content relative to target composition. The cubic MZO film and STO follow cube-on-cube alignment with 45° rotation of the unit cell with respect to Si lattice, and lattice mismatch between STO and Si(100) is reduced from –28.2% (without lattice rotation) to +1.6% (with lattice rotation).[208] Heterostructures with epitaxial films of cubic MZO on MgO(100) and TiN/Si(100) substrates have been grown by PLD using a sintered MZO ceramic target with $x = 0.75$ to 0.85.[219] X-ray diffraction and transmission electron microscopy confirmed the epitaxial nature of MZO film, and showed the formation of a spinel $Mg_2Zn_xTiO_4$ thin layer at the interface with TiN. The band gap of epitaxial cubic $Mg_{0.82}Zn_{0.18}O$ films was reported to be ~6.7 eV.[219]

The wide tunability of band gap in alloyed ZnO films opens the door for the realization of novel optoelectronic devices especially short wavelength light emitters and photodetectors. Tunability of E_g with Mg composition from 3.3 eV to 7.8 eV for wurtzite and cubic MZO alloy covers the UV-A, UV-B and UV-C regions with wavelength varying from 200 nm to 400 nm, and has been successfully exploited to sense mid and deep ultraviolet (UV) light.[220] UV photodetectors based on metal-semiconductor-metal (MSM) structure have been demonstrated using MZO films.[198,200,208] Fig. 4.10 shows the normalized spectral photoresponse of an array of compositional spread MZO photodetectors with interdigital electrodes of gold (Inset of Fig. 4.10) having a finger width and separation of ~3 μm.[198] Photo detectors operating at different λ have been fabricated with MZO films. Photo detectors using wurtzite epitaxial MZO films (formed at lower x) give a peak response at $\lambda = 308$ nm.[200] The visible blind UV detector using $Mg_{0.34}Zn_{0.66}O$ alloy film exhibits a high responsivity of 1200 A/W with a fast response speed of ~8 ns rise time and ~1.4 μs fall time at $\lambda = 337.1$ nm.[200] The spectral responsivity of a photo detector fabricated using epitaxial cubic MZO alloy film in a metal-insulator-metal device structure is found to be in the deep UV region with a peak at $\lambda = 225$ nm (corresponding to $E_g = 5.51$ eV) and shows an UV-to-VIS rejection ratio of more than one order of magnitude.[208] Also by varying the Mg content (62 to 100%) in the cubic MZO alloy films, it was possible to use the detector over a wide range of the UV spectrum ($\lambda = 157$ to 230 nm).[208]

4.5.4 ZnO based dilute magnetic semiconductors (DMS)

Currently there is a lot of interest in exploiting spin of charge carriers (electrons or holes) in semiconductors to provide functionality for microelectronic devices. Transition metal (TM) doped ZnO has been identified as a promising dilute magnetic semiconductor (DMS) for its possible use in solid-state spintronic devices.

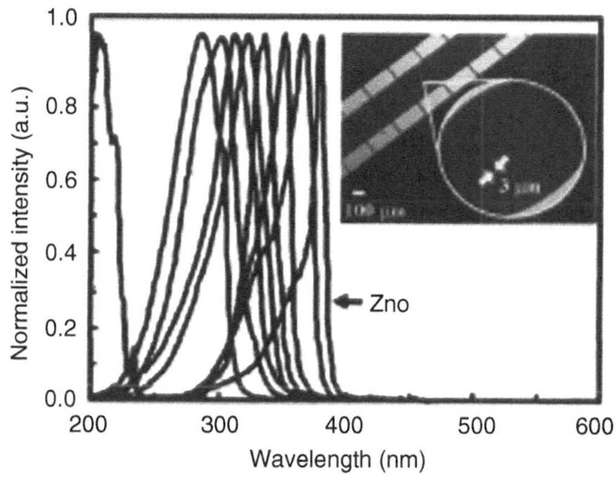

Figure 4.10: Normalized spectral response of an array of compositional spread MZO photodetector. Inset shows an enlarged view of IDT electrodes.[198] [Reused with permission from I. Takeuchi, W. Yang, K.-S. Chang, M.A. Aronova, T. Venkatesan, R.D. Vispute and L.A. Bendersky, Journal Applied Physics, 94, 7336 (2003). Copyright 2003, American Institute of Physics.]

The theoretical prediction of Dietl *et al.*[223] regarding possibility of room temperature magnetism in ZnO containing 5% of Mn with a high hole concentration (3.5×10^{20} cm^{-3}), resulted in a number of experimental investigations on transparent ZnO films doped with ferromagnetic TM. Besides Mn many other TM dopant elements including Co, Cu, V, Cr, Fe and Ni have been studied extensively in host ZnO for making a room temperature thin film DMS by the PLD technique. The experimental work on the growth of transition metal doped ZnO film was started by Fukumura *et al.*[224] using PLD in 1999 and reported a higher solubility ~35 at.wt% in ZnO with wurtzite structure which is much higher than the thermal equilibrium limit ~13 at.% at 600°C. The solubility of Mn and Co is expected to be higher in ZnO due to closeness of atomic radii and the same valence state as Zn^{2+}. Epitaxial film of ZnO films doped with Mn in the range (0 to 0.36 at.%) were deposited on (0001) sapphire substrate at 600°C at an oxygen pressure of $\sim 10^{-3}$ Torr. All films were ***c***-axis oriented with wurtzite structure having homogenous distribution of Mn. Both ***a***- and ***c***-lattice parameters expand monotonously with the increase in Mn content up to $x = 0.35$, and ferromagnetism was not observed at any temperature.[223] Subsequently a lot of work has been reported on the investigation of magnetic properties of TM-doped ZnO films by PLD in ZnO using ceramic targets of varying compositions, and targets sintered at various temperatures. Attempts have also been made to prepare ZnO based DMS thin films by ablating two oxide targets, or metal targets, and the deposited layers were mostly *n*-type.

Fukumura *et al.*[225] deposited $Zn_{0.64}Mn_{0.36}O$ film on sapphire substrate and found a spin-glass behavior with spin-freezing temperature ~13 K. A number of transition metals including (Co, Mn, Cr, and Ni) were studied as dopants in $Zn_{1-x}TM_xO$ films ($x = 0.05$–0.25) deposited on **α**-sapphire (1120), however, no ferromagnetism was observed in any of the films, except the films doped with cobalt (Co).[226] All deposited films were single phase and possessed a preferred ***c***-axis orientation with a solubility limit < 50%. The ***c***-axis parameter increased linearly with Co content, and $Zn_{0.85}Co_{0.15}O$ film showed ferromagnetic behavior with $\mathbf{T_c} \sim 300$ K, and a saturation magnetization ($\mathbf{M_s}$) ~ $2\mu_B$/Co as noted from the M-H hysteresis curve. However the reproducibility of the results was less than 10%. The magneto-optic properties of $Zn_{1-x}TM_xO$ (TM = Sc, Ti, V, Cr, Mn, Co, Ni, Cu) films deposited on (0001) sapphire were investigated using magnetic circular dichroism (MCD) at 5 K.[227] Only Mn, Co, Ni and Cu doped ZnO films exhibited pronounced negative MCD peak near 3.4 eV corresponding to the band gap of host ZnO and is due to the sizable sp-d exchange interaction indicating the fabrication of a DMS thin film. The amplitude of MCD effect ~−2.7°/cm Oe in Co-doped ZnO with $x = 0.012$ was reported at 5 K and disappears at room temperature.[228] The Faraday effect measured using a polarization modulation method at 5 K was also huge near 3.4 eV in $Zn_{1-x}Co_xO$ ($x = 0.012$). Combinatorial laser MBE was used to grow epitaxial film doped with TM (Mn, Co, Cr and Fe) on (0001) sapphire substrate using two oxide targets (ZnO and TMO). The solubility limit of Mn and Co in comparison to other TM was reported to be large.[229] A strong MCD signal was observed for Mn- and Co-doped films near the optical band edge. However, ferromagnetism was not observed when studied with a scanning superconducting quantum interference microscopy down to 3 K. Yoo *et al.*[230] observed that an alternate method of doping using ablation of two oxide targets gives better crystallinity and surface morphology for Co-doped ZnO film. Some efforts for observing ferromagnetism with vanadium (V) and nickel (Ni) doping was also carried out, and the solubility of Ni was reported to be 25 at.%.[231–232] ZnO doped films with Ni ($x = 0.03$ to 0.25) and V (0.05 to 0.15) do exhibit ferromagnetic behavior at 2 K and 350 K respectively. Ferromagnetism in V-doped samples was observed only at a high carrier concentrations (~10^{18} cm^{-3}), and Ni-doped ZnO films show a super-paramagnetic behaviour between 30–300 K.[231] $Zn_{0.85}V_{0.15}O$ films show a hysteresis loop at 300 K with $\mathbf{M_s} \sim 0.5\,\mu_B$/V and coercive field ($\mathbf{H_c}$) ~ 200 Oe.[163] Co-doped ZnO films ($x = 0.05$–0.15) deposited on (0001) sapphire substrate show the ferromagnetism. The reported value of $\mathbf{M_s} \sim 2\mu_B$/Co was obtained in the film ($x = 0.15$) with Curie temperature of 300 K. However, the reproducibility of the results was claimed to be less than 10%.[163] $Zn_{0.9}Mn_{0.1}O$ films deposited by Laser-MBE on a (0001) sapphire substrate in an ultra high vacuum $\sim 10^{-9}$ Torr with $x = 0.1$ showed M-H hysteresis loop at 5 K with $\mathbf{M_s} \sim 0.15\,\mu_B$/Mn and $\mathbf{H_c} \sim 300$ Oe.[233] The band gap of the film increases from 3.27 to 4.2 eV with x from 0 to 0.3. Kim *et al.*[234] investigated the influence of PLD processing conditions on Co doped ZnO films, and reported on the paramagnetic property of homogeneous films (grown at low substrate temperatures $\leq$600°C) exhibiting spin glass behavior with freezing temperature $\sim$8 K, whereas, room temperature ferromagnetism was observed in inhomogeneous films of wurtzite ZnO phase, mixed with rock-salt CoO and hexagonal Co phases formed at a high substrate temperature and low pressure $\leq 10^{-5}$ Torr, and the observed results were attributed to the presence of clustering.

Despite the contradictory results in the reported magnetic properties of PLD $Zn_{1-x}TM_xO$ films, and lack of reproducibility, efforts continued to modify the processing conditions and the source material to achieve room temperature ferromagnetism. Prellier *et al.*[46] prepared high quality Co-doped ZnO films using two metal targets (Zn and Co) in oxidizing ambient. The purpose for introducing alternate deposition from metal targets was to allow a homogeneous repartition of the Co into wurtzite ZnO. Films doped with 5% Co showed ferromagnetism in the temperature range from 5 to 300 K using a superconducting quantum interference device magnetometer. A clear transition from ferromagnetic state to paramagnetic state was observed, and ruled out the possibility of its origin due to clustering because the transition temperature of a metallic Co cluster is >1000 K. The growth of Co doped films at low oxygen partial pressure ($<10^{-6}$ Torr) leads to a DMS film and room temperature ferromagnetism was observed in $Zn_{0.075}Co_{0.25}O$ film with an out-of-plane easy magnetization axis and $\mathbf{M_s} = 1.3\,\mu_B$/Co.[235] Film growth at low pressures induced oxygen vacancies and thereby free electrons mediate the ferromagnetic exchange interactions. $Zn_{1-x}Co_xO$ film grown under low oxygen pressure exhibited a giant positive magneto resistance (MR) for $x > 0.15$.[236] The MR ratio at 5 K in 70 KOe field was $\sim$10% for $x = 0.15$ and reaches up to 60% for $x = 0.20$ indicating representative features of DMS. Epitaxial $Zn_{0.9}Co_{0.1}O$ film deposited by PLD at 500°C on ***c***-sapphire with a 30° or 90° rotation relative to the substrate shows M-H hysteresis loop, and the coercive field increases from 35 to 50 Oe with a decrease in temperature from 300 K to 10 K.[237] A linear decrease in magnetic moment of $\sim$3% is observed over the entire temperature range at 500 G. The substitution of Co at Zn site gives the characteristic absorption edges at 658, 616 and 568 nm and was correlated with d-d transitions of tetrahedrally co-coordinated Co^{2+} ions. Fouchet *et al.*[238] investigated the effect of Co content on the structural and magnetic properties of $Zn_{1-x}Co_xO$ thin films deposited by ablating Zn and Co metal targets alternately, and showed the ferromagnetic state in epitaxial films grown on (0001) sapphire with a curie temperature of 150 K and 300 K for $x = 0.016$ and 0.066 respectively. The origin for ferromagnetism at 300 K with saturation moment

0.7 μ_B/mole Co was not due to clustering of Co metal, since transition from ferromagnetic to paramagnetic state was clearly observed. The Co doped ZnO with $x = 0.03$ deposited by PLD at low oxygen pressure $\sim 5 \times 10^{-5}$ Pa and substrate temperature 400°C on (0001) sapphire without forming second phase exhibited ferromagnetism at 290 K with $\mathbf{M_s} \sim 0.17\ \mu_B$/Co measured using VSM at a magnetic field ±1 T.[239] However, phase segregation and appearance of second phase of Co or Co_2O_3 was evident at higher growth temperatures (>600°C) and background oxygen pressure. $Zn_{0.85}Co_{0.15}O$ films with 1 at.% Al and 15% Mn incorporation exhibited $\mathbf{M_s} \sim 0.08\ \mu_B$/Co and $0.19\ \mu_B/(0.5Co + 0.5\,Mn)$ respectively. The Co doped films were metallic in nature, and incorporation of Mn leads to semiconducting property.[239] Venkatesan *et al.*[240] investigated the origin of magnetism at high temperatures in PLD films as a function of (i) different dopant elements (TM = Sc, Ti, V, Fe, Co, Ni, Cr, Mn, Cu), (ii) content of Co in ZnO and (iii) oxygen pressure. The (110) oriented ZnO films on R-cut sapphire substrate were deposited by PLD using ceramic targets having 5 at.% of Sc, Ti, V, Fe, Co, or Ni, and room temperature ferromagnetism was observed. A large value of $\mathbf{M_s} \sim 2.6\ \mu_B$/Co, 0.5 μ_B/Ti, 0.3 μ_B/Sc were reported for Co-, Ti- and Sc-doped-ZnO films respectively. A large value of magnetization was measured in the Co doped ZnO film when the applied field is parallel to the edge of the substrate, and a low value along the ***c***-axis of the ZnO i.e. along the diagonal (Fig. 4.11).[240] The Co-doped ZnO films grown at low oxygen pressure (<0.1 mbar) was *n*-type conductor, and becomes an insulator when deposited at a higher pressure. A weak ferromagnetism in epitaxially strain relaxed $Zn_{1-x}Co_xO$ films ($x = 0.1$) ablated on ZnO (00.$\bar{1}$) single crystal substrate using two oxide targets was observed only after introduction of 1.6 at.% Al, and ferromagnetism with an average magnetization of 0.06 μ_B/Co is reported for a low value of $x = 0.06$.[241]

Room temperature ferromagnetism in transparent Mn doped ZnO has been obtained by optimizing the quality of the source target with sintering temperature.[242–243] $Zn_{1-x}Mn_xO$ thin films were deposited on amorphous quartz substrate by PLD using ceramic target ($x = 0.1$) sintered at 500°C at a substrate temperature of 400°C in an ambient pressure of 0.3 mbar. The deposited films were ***c***-axis oriented and had a single phase with homogenous distribution of Mn, and the content of Mn was found to be 2.21 at.%. Room temperature ferromagnetism was observed in $Zn_{0.978}Mn_{0.022}O$ film using SQUID (Fig. 4.12).

It is reported that the crystalline quality and the photoluminescence properties of PLD Mn-doped ZnO films become better at higher oxygen pressure. Post deposition annealing is found to improve the crystalline quality of Mn doped films deposited on sapphire substrate and the Mn content in the doped film was reported to be the same as that of the source target (13 at.%).[244] Mn doping was found responsible for the strong quenching of blue band emission in PL. The effect of initial carrier

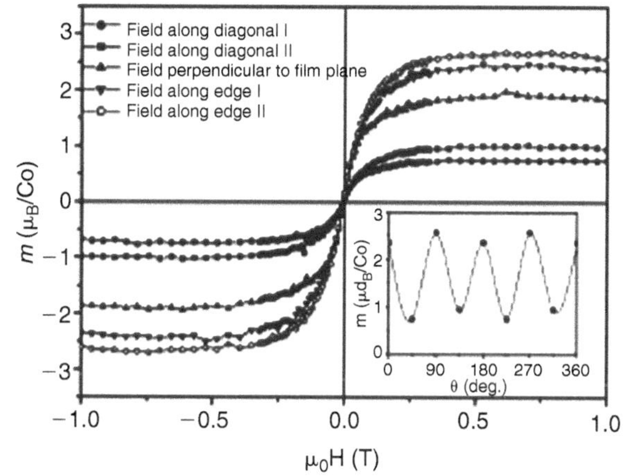

Figure 4.11: Magnetization curves of $Zn_{0.93}Co_{0.07}O$ film with magnetic field applied in different directions. The curve was obtained after correcting the diamagnetic contribution of the substrate (-0.20×10^{-12} m^3). Inset shows the in-plane anisotropy of results.[240] [Reprinted figure (No. 2) with permission from M. Venkatesan, C.B. Fitzgerald, J.G. Lunney and J.M.D. Coey, Physical Review Letters, 93, 177206 (2004). Copyright 2004 by American Physical Society.]

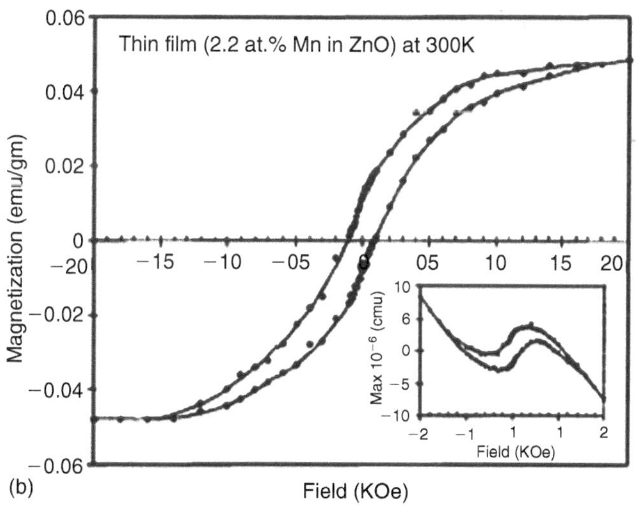

Figure 4.12: Magnetization curve at room temperature for PLD $Zn_{0.978}Mn_{0.022}O$ thin film on amorphous quartz substrate. The curve was obtained after correcting the diamagnetic contribution of the substrate. Inset shows the as-obtained data from the SQUID measurement.[242] [Reprinted from Journal of Magnetism and Magnetic Materials, Vol. No. 282, P. Sharma, A. Gupta, F.J. Owens, A. Inoue and K.V. Rao, 'Room temperature spintronic material-Mn doped ZnO revisited', pp. 115–121, copyright (2004), with permission from Elsevier.]

concentration on the magnetic properties of Mn doped ZnO film grown on sapphire substrate by PLD was investigated after high dose Mn implantation and annealing treatment.[245]

ZnO films doped with 2 at.% phosphorous have been obtained after annealing for one hour in oxygen ambient (100 mTorr) in the temperature range 425–600°C, with a wide range of carrier concentration from 7.5×10^{15} to 1.5×10^{20} cm^{-3} and corresponding mobility in the range (16 to 6 cm^2/Vs). No secondary phase formation was evident in the films after implantation with 250 KeV Mn^+ ions to a dose of 3×10^{16}. However, a remarkable change in the values of saturation magnetization and coercivity from the room temperature B-H loop have been observed on various samples, and clearly indicates that the crystallinity and defects present in the film do influence the magnetic properties besides the carrier concentration.[245] Incorporation of Mn ($x = 0.05$) in PLD ZnO films is found to improve the epitaxial growth on (0001) sapphire with 30° in-plane rotated domain, and suppress the hexagonal-on-hexagonal growth whereas, a deterioration in crystallinity is reported for higher value of x.[246] The lattice parameters (***a*** and ***c***) increase and surface roughness decreases with increase in x (0 to 0.35), and no evidence of ferromagnetism is observed with the variation in the composition. Effect of sintering temperature (500–900°C) for the preparation of PLD targets was studied to see its effect on the occurrence of ferromagnetism in the deposited films.[247] Mn doped ZnO films prepared from targets sintered at 500°C alone exhibited the ferromagnetism and a high value of curie temperature ~375 K was reported. A maximum value of $\mathbf{M_s}$ was observed at $x = 0.09$, but a good hysteresis loop with highest $\boldsymbol{E_c}$ and a large ratio between remanence was obtained at $x = 0.065$. The value of $\boldsymbol{E_c}$ was found to depend on grain size and a maximum value ~234 Oe was reported for films with a grain size of 140 nm, and Mn doping quenched the PL peaks as well.[247] Improved ferromagnetic properties were observed in $Zn_{1-x}Mn_xO$ films grown by PLD using targets sintered at 500°C.[248] Formation of clusters at higher growth temperature (>500°C) tends to degrade the ferromagnetic property. A grain size of (50–60 nm) is reported to increase dramatically with substrate temperature (500–600°C) along with a decrease in the surface roughness. Films deposited at 500°C shows pronounced ferromagnetic property, whereas at higher temperature (~650°C), measurements indicate antiferromagnetic, or paramagnetic behavior.[248] Ivill *et al.*[249] established the inverse correlation between the magnetization of Mn doped film with carrier density. They identified the Mn oxide precipitation even at $x = 0.03$ in the doped films prepared at very high oxygen pressure (20 mTorr), and controlled the resistivity from 195 to 0.185 Ω cm by introducing Sn (0–1 at.%) as a co-dopant. Films possessing minimal Mn oxide segregation exhibit a good hysteresis loop after introduction of Sn, and a decrease in magnetization and increase in $\boldsymbol{E_c}$ are noted with increase in Sn content i.e. charge carriers.

Subsequently, Hong *et al.*[250] investigated the role of Cu as a co-dopant on the magnetic properties of $Zn_{1-x}Mn_xO$ (ZMO) films grown by PLD. Highly (110) oriented $Zn_{0.9}Mn_{0.1}O$ (ZMO) and $Zn_{0.85}Mn_{0.10}Cu_{0.05}O$ films were grown on R-cut sapphire substrate. ZMO films deposited at 650°C at an oxygen pressure ~0.1 Torr are ferromagnetic with a Curie temperature higher >400 K and $\mathbf{M_s} \sim 0.075\ \mu_B$/Mn. Incorporation of Cu enhanced the value of $\mathbf{M_s}$ to 0.1 μ_B/Mn, and did not affect the magnetic properties. Films deposited at 400°C and 10^{-6} Torr pressure show very weak ferromagnetism below 75 K. ZnO films doped with V (5 at.%) grown at 650°C on R-cut sapphire show a hysteresis loop with $\mathbf{M_s} \sim 0.06\ \mu_B$/V and a Curie temperature higher than 400 K.[251] Single phase films were in wurtzite structure but polycrystalline in nature, and antiferromagnetic phase is seen to dominate at higher substrate temperature (~700°C)

Results discussed so far have been mainly on doped ZnO films that exhibited *n*-type electrical conductivity. However theoretical studies indicate ferromagnetic ordering in doped-ZnO is favored by *p*-type ZnO. Attempts have been made to develop room temperature DMS by introducing a precise mount of Cu into *p*-type ZnO by PLD.[252] Two targets of ZnO and metallic Cu were used and laser pulses were swept cyclically across the target radius to prevent localized heating. The targets were ablated alternatively using a controlled shutter, and ultra thin layers (<one monolayer) were deposited on (0001) sapphire substrate to insure a uniform solid solution of ZnO-Cu. Films deposited over a range of oxygen pressure (10^{-2}–10^{-5} Torr) and substrate temperature (400–700°C) with Cu content in the range 1–4 at.% were found to be *n*-type and in the paramagnetic state. However pure ZnO film deposited at 400°C in N_2O ambient pressure $\sim 5 \times 10^{-5}$ Torr showed a *p*-type conductivity with a resistivity ~28 Ω cm, hall mobility ~0.8 cm²/Vs and carrier density $\sim 3 \times 10^{17}$ cm^{-3}. Introduction of Cu (1 at.%) in *p*-type ZnO exhibited ferromagnetism with Curie temperature ~390 K at $E_c \sim 40$ Oe.[252] The magnetic moment at 200 K was found to be ~0.4 μ_B/Cu atom, and the magnetization was reported to be independent of temperature (10–390 K) for a higher Cu content (7 to 16 at.%). The reported results on TM doped ZnO films are presented in Table 4.5.

4.6 Multilayers, heterostructures and superlattices

Layered structures attract a lot of attention for potential applications in integrated devices with enhanced or new physical properties, and sometimes combine the functional properties of two different materials. In the case of ZnO thin films initially its piezoelectric properties were of immense interest for acoustic excitation in acousto-optic layered structures, and the studies were mostly focused on the integration of

Table 4.5: Summary on PLD grown TM doped ZnO for DMS applications.

Target composition and target sintering temperature	Substrate/growth temperature and pressure	$\mathbf{T_c}$ (K)	Main observations	Ref.
$Zn_{1-x}Mn_xO$ (1 at.% Al_2O_3) $x = 0$ to 0.35 900°C	(0001) Sapphire 600°C, 5×10^{-5} Torr	–	Film having localized Mn spin 5/2. ***a***- and ***c***-lattice length increases with x. Films showed large Magnetoresistance. No evidence of ferromagnetism.	224
$Zn_{1-x}Mn_xO$ (1 at.% Al_2O_3) $x = 0.36$ 900°C	(0001) Sapphire 600°C, 5×10^{-5} Torr	–	Spin glass behavior.	225
$Zn_{1-x}TM_xO$ (1 at.% Al_2O_3) TM = Co, Cr, Ni, Mn $x = 0.05$ to 0.25	(1120) Sapphire 350 to 600°C, 2 to 4 mTorr	300 K	No ferromagnetism for TM = Mn, Cr and Ni. ***c***-axis length increases with Co content. Co ($x = 0.15$) doped film with high carrier conc. shows ferromagnetism ($\mathbf{M_s} = 2\,\mu_B$/Co).	226
$Zn_{1-x}TM_xO$ TM = Sc, Ti, V, Cr, Mn, Fe, Co, Ni, and Cu	(0001) Sapphire 600°C 10^{-6} Torr	–	Synthesis by Cobinatorial laser MBE. Among all the doped 3d ions, Mn^{2+} and Co^{2+} show the highest solubility.	259
$Zn_{1-x}TM_xO$ TM = Sc, Ti, V, Cr, Mn, Co, Ni, Cu $x = 0$ to 0.016	(0001) Sapphire	5 K	(Mn, Co, Ni and Cu) doped ZnO films show negative MCD peak near 3.4 eV. The amplitude of MCD effect ∼−2.7°/cm Oe in Co-doped ZnO with $x = 0.012$ at 5 K.	227
$Zn_{1-x}Co_xO$ $x = 0$ to 0.049	(0001) Sapphire 600°C	200 K 1×10^{-9} Torr	$Zn_{0.84}Co_{0.16}O$ showed MCD signal up to 200 K Faraday effect was huge near 3.4 eV at 5 K.	228
$Zn_{1-x}Co_xO$ $x = 0.05$ to 0.25	(0001) Sapphire 280 to 400°C	300 K	Ferromagnetism in films with $x = 0.05$–0.15. $\mathbf{M_s} \sim 2\,\mu_B$/Co for $x = 0.15$ with $\mathbf{T_c} = 300$ K. $\mathbf{M_s} \sim 1.8\,\mu_B$/Co & $\mathbf{H_c}$ ∼50 Oe for $x = 0.05$. Reproducibility of results was poor (<10%).	163

$Zn_{1-x}Co_xO$ $x = 0.06$–0.30	(0001) Sapphire 600°C 1×10^{-4} Torr	–	Spin glass property at low temperature ($x > 0.15$). Antiferromagnetic super-exchange interaction. A linear increase in band gap with Co content.	262
$Zn_{1-x}V_xO$ $x = 0.05$ to 0.15	(1120) Sapphire 300°C, 600°C 10^{-3} to 10^{-5} Torr	>350 K	Ferromagnetism in film ($x = 0.05$–0.15). $\mathbf{M_s} \sim 0.5\ \mu_B$/V & $\mathbf{E_c} \sim 200$ Oe at 300 K for $x = 0.15$. Magnetic property depends on x and carrier concentration.	163 232
$Zn_{1-x}Ni_xO$ $x = 0$ to 0.25	(0001) Sapphire 300–700°C 1×10^{-5} Torr	2 K	Solubility of Ni in ZnO is 25 at.%. ***c***-axis increases with x (<0.05) and decreases thereafter. Films ($x = 0.003$–0.25) show ferromagnetism at 2 K. Super-paramagnetic behaviour from 30 to 300 K.	231
ZnO + TMO (1 at.% Al_2O_3) TM = Mn, Co, Cr, Fe	(001) Sapphire 600°C 10^{-6} to 10^{-9} Torr	–	Strong MCD signal observed for Mn- and Co-doped films near optical band edge. Ferromagnetism was not observed down to 3 K.	229
ZnO + $Zn_{0.95}Co_{0.05}O$ (1 at.% Al_2O_3)	(0001) Sapphire 600°C 5×10^{-5} Torr	–	Crystallinity improved by alternate doping. Solubility of Co was more using AD method.	230
$Zn_{1-x}Mn_xO$ $x = 0$ to 0.30	(0001) Sapphire 600°C 10^{-9} Torr	45 K	***c***-lattice length increases with x (0 to 0.3). Band gap increases from 3.27 to 4.2 eV with x. Films with $x = 0.1$ & 0.3 show ferromagnetism. $\mathbf{M_s} \sim 0.15\ \mu_B$/Mn, $\mathbf{H_c} \sim 300$ Oe for $x = 0.1$.	233
$Zn_{1-x}Mn_xO$ $x = 0$ to 0.36	(0001) Sapphire 610°C 5×10^{-5} Torr	–	Solubility of Mn in ZnO up to 36 at.% ***c***-axis length increases with x. Band gap increases with Mn atomic fraction. No evidence of ferromagnetism in any film.	253

(Continued)

Table 4.5: (*Continued*)

Target composition and target sintering temperature	Substrate/growth temperature and pressure	T_c (K)	Main observations	Ref.
$Zn_{1-x}Co_xO$ $x = 0.02$ to 0.049 950°C	(0001) Sapphire 300 to 700°C 10^{-1} to 10^{-6} Torr	300 K	Spin glass behavior in homogeneous film at ≤600°C. Ferromagnetism in inhomogeneous film (wurtzite ZnO phase + rock-salt CoO + hexagonal Co phases) at high substrate temp. (700°C) and low pressure $\leq 10^{-5}$ Torr.	234 255
Zn + Co (5 to 9% Co doping)	(0001) Sapphire 500 to 750°C 10^{-1} Torr	300 K	Homogenous film with wurtzite structure. Ferromagnetism in film ($x = 0.05$–0.08). Transition observed from ferromagnetic state to paramagnetic state, and rule out clustering.	46 254
ZnO, $Zn_{0.75}Co_{0.25}O$ $Zn_{0.85}Co_{0.15}O$ (1 at.% Al_2O_3) 1100°C	(0001) Sapphire 300 to 600°C 5×10^{-7} to 10^{-6} Torr	>300 K	Growth at $<10^{-6}$ Torr gives DMS film. Ferromagnetic $Zn_{0.075}Co_{0.25}O$ films are insulating with an out-of-plane easy magnetization axis and $\mathbf{M_s} = 1.3\ \mu_B$/Co.	235
$Zn_{1-x}Co_xO$ (1 at.% Al_2O_3) $x = 0.02$ to 0.25	(0001) Sapphire 600°C 10^{-5} Torr	–	Giant positive magnetoresistance at $x > 0.15$. MR ratio ~10–60% at 5 K in 70 K Oe.	236
$Zn_{0.87}Mn_{0.13}O$ 900°C	(0001) Sapphire 400°C 0.3 to 0.4 mbar	–	Mn content in film was 13 at.%. Annealing for 20 min improves crystallinity. Strong quenching of blue band emission in PL observed with Mn doping.	244
$Zn_{0.9}Mn_{0.1}O$ 500°C	Quartz 400°C 0.3 mbar	>300 K	The content of Mn in film was ~2.21 at.% with homogenous distribution. Ferromagnetism in $Zn_{0.978}Mn_{0.022}O$ film.	242 243

$Zn_{0.9}Co_{0.1}O$	(0001) Sapphire 500 to 650°C 10^{-6} Torr	>300 K	Epitaxial $Zn_{0.9}Co_{0.1}O$ film was ferromagnetic with $\boldsymbol{E_c} = 35$ Oe, Moment $= 2.88 \times 10^{-5}$. Co substitution at Zn site gives characteristic absorption edges at 658, 616 and 568 nm. No evidence of clustering formation.	237
Zn + Co ($x = 0$ to 0.066)	(0001) Sapphire 500 to 750°C 10^{-1} Torr	300 K	Ferromagnetic state (FM) with $x = 0.016$–0.066. No M-H loop was found at any temperature. FM to paramagnetic transition was observed. $\mathbf{M_s} = 0.7\ \mu_B$/Co was not due to clustering of Co.	238
$Zn_{1-x}Co_xO$ (1 at.% Al_2O_3), $Zn_{1-x}Co_xO$ ($x = 0$ to 0.3) $Zn_{10.70}Mn_{0.15}Co_{0.15}O$ 1000°C	(0001) Sapphire 400 to 600°C 5×10^{-5} Pa	290 K	Growth at low temperature and high pressures gives $\mathbf{M_s} \sim 0.08\ \mu_B$/Co for $Zn_{0.85}Co_{0.15}Al_{0.01}O$, $0.17\ \mu_B$/Co for $Zn_{0.7}Co_{0.3}O$ and $0.19\ \mu_B/(0.5Co + 0.5Mn)$ for $Zn_{0.7}(Mn_{0.15}Co_{0.15})$ Other phases found at high temperature and pressure.	239
$Zn_{1-x}TM_xO$ TM = Sc, Ti, V, Fe, Co, Ni, Cr, Mn, Cu $x = 0.05$	R-cut sapphire $(1\bar{1}02)$ 600°C 10°–10^{-4} mbar	300 K	Ferromagnetism in doped ZnO films found to depend only on TM = Sc, Ti, V, Fe, Co, or Ni. $\mathbf{M_s} \sim 2.6\ \mu_B$/Co, $0.5\ \mu_B$/Ti, $0.3\ \mu_B$/Sc. **H** parallel to substrate edge gives large $\mathbf{M_s}$ and lower value along ***c***-axis of ZnO (diagonal). $Zn_{0.95}Co_{0.05}O$ at <0.1 mbar was conductor (*n*-type) and becomes insulator at higher pressure.	240
ZnO:$P_{0.02}$ Mn^+ ion implant at 250 KeV	(0001) Sapphire 400°C 20 mTorr	300 K	Annealing treatment enhances the crystallinity. $\mathbf{M_s}$ and $\boldsymbol{E_c}$ values depend on initial carrier concentration of the as deposited film.	245
$Zn_{1-x}Cr_xO$	(0001) ZnO (0001) Sapphire	400 K	Magnetization of conducting samples decreases with temperature (20 K to 400 K).	116

(Continued)

Table 4.5: (*Continued*)

Target composition and target sintering temperature	Substrate/growth temperature and pressure	T_c (K)	Main observations	Ref.
	350 to 600°C 10^{-6} to 10^{-3} Torr		Magnetization is constant for non-conductive sample at 200 K.	
$Zn_{1-x}Mn_xO$ $x = 0$ to 0.35 1100°C	(0001) Sapphire 600°C 5×10^{-4} Torr	–	Epitaxy improves with Mn doping ($x = 0.05$) and deteriorates at higher $x > 0.05$. Lattice parameters $\boldsymbol{a}, \boldsymbol{c}$ increase and roughness decreases with x. All films in paramagnetic state.	246
$Zn_{1-x}TM_xO$ TM = Co and V	$(11\bar{2}0)$ Sapphire	–	Co and V doped ZnO films formed homogenous solid solution up to $x \sim 5$ at.%. Segregation of secondary phases at $x > 10$ at.% Impurity mode technique found sensitive to probe lattice defect induced by dopant impurities.	256
$Zn_{1-x}TM_xO$ TM = Mn, Fe, V 1000°C	Glass	–	Raman spectra do not show any additional modes for Mn and Fe doping. V:ZnO film shows an additional mode for $x \geq 5\%$.	257
$Zn_{1-x}M_xO$ (M = Fe, Sb, Ga) $x = 0.005$ to 0.08	(0001) Sapphire	–	Raman scattering studies in doped ZnO films. Additional modes at 531, 631 and 720 cm^{-1} seen specific for Sb, Ga, and Fe dopants.	258
$Zn_{1-x}Mn_xO$ $x = 0$ to 0.22 500°C, 900°C	(0001) Sapphire 500°C >0.2 mbar	>375 K	Mn content in films was same as in target. Target sintering at (500°C) yields ferromagnetic film with large $\boldsymbol{E_c} \sim 234$ Oe at $x = 0.065$. Film grown at >0.2 mbar shows magnetic domain. Mn doping quenched the PL peaks.	247

$Zn_{1-x}Mn_xO$ $x = 0.06$ 500°C	(0001) Sapphire 500 to 650°C 1 mTorr	>300 K	Films grown at 500°C show ferromagnetism. Room temperature hysteresis shrinks with $\mathbf{T_s}$ The field for $\mathbf{M_s}$ decreases from 980 G to 780 G with temperature (5 K to 300 K). Antiferromagnetism/paramagnetism at 600°C due to clustering.	248
$Zn_{1-x}Mn_xO$ $x = 0.001$ and 0.091 Mn	(0001) Sapphire	–	X-band EPR spectra recorded. Fine & hyperfine structure splitting parameters determined. Establish relation between EPR spectra and crystallinity of the films.	260
$Zn_{1-x}M_xO$, M = Co, Mn	Hydrothermal ZnO (00.1) 500°C	–	Weak ferromagnetic bahavior observed in $Zn_{1-x}Co_xO$ with few Al atoms (1.6%).	241
$Zn_{0.95}TM_{0.05}$ O TM = Sc, Ti, V, Cr, Mn, Fe, Co, Ni, or Cu, and $Ga_{1-x}Mn_xN$ ($x = 0.005$–0.1)	***c***-cut (001) Al_2O_3 600 to 900°C 10^{-4} to 10^{-1} mbar	300 K	In $Zn_{0.95}TM_{0.05}O$ films there is a systematic variation in magnetic moment with TM dopant. Co shows the highest moment, whereas Cr, Mn, Cu or pure ZnO show no magnetic moment. $Ga_{1-x}Mn_xN$ is ferromagnetic at room temperature	261
$Zn_{1-x}Mn_xO$ (0 to 1 at.% SnO_2) $x = 0.03$ 1000°C	(0001) Sapphire 400 to 600°C 20 mTorr	>300 K	Segregation in film at high pressure (20 mTorr). Electron density is controlled by Sn co-doping. Films with ($x = 0.3$) exhibit hysteresis loop. $\mathbf{M_s} \sim 0.5$ Bohr magnetons/Mn decreases with Sn, and $\boldsymbol{E_c}$ increases (50–100 G) with Sn (0.1–1 at.%).	249

(Continued)

Table 4.5: (*Continued*)

Target composition and target sintering temperature	Substrate/growth temperature and pressure	T_c (K)	Main observations	Ref.
$Zn_{0.9}Mn_{0.1}O$ and $Zn_{0.85}Mn_{0.10}Cu_{0.05}O$	R-cut sapphire (1102) 400 to 600°C 10^{-1} to 10^{-6} Torr	>400 K	Highly (110) oriented films in wurtzite phase. Films grown at 650°C and ~0.1 Torr in FM state. $\mathbf{M_s}$ increases (0.075–0.1 μ_B/Mn) with Cu doping, and films deposited at 400°C and 10^{-6} Torr show weak ferromagnetism below 75 K.	250
$ZnO + TMO + Al_2O_3$ TM = Co, Mn (1 to 2 at.% Al with Co)	ZnO (00.**1**)-O face 550°C 10^{-6} Torr	10 K	Epitaxial strains partially relaxed through dislocations in the (10.0) glide planes. ***c***-lattice length increases with Co content. Average **M** for Co(0.06) is 0.06 μ_B/Co, $\boldsymbol{E_c} \sim 90$ Oe. Paramagnetic with more Co content ($x > 0.06$). Weak ferromagnetism observed in Co-doped film ($x = 0.1$) with 1.6% Al.	241
$Zn_{0.95}V_{0.05}O$	R-cut sapphire (1102) 600 to 700°C 10^{-1} Torr	>400 K	Room temperature ferromagnetism in film at >600°C. $\mathbf{M_s} \sim 0.6\,\mu_B$/V in the film grown at $\mathbf{T_s} = 650$°C. At $\mathbf{T_s} > 650$°C, antiferromagnetic phase dominates below 200 K.	251
ZnO + Cu (1 to 14 at.% of Cu)	(0001) sapphire 400 to 700°C 10^{-2}–10^{-5} Torr (O_2) $\sim 5 \times 10^{-5}$ Torr (N_2O)	>390 K	ZnO films doped with Cu (1–4 at.%) grown at 400–700°C in O_2 pressure (10^{-2}–10^{-5} Torr) are *n*-type, and are in the paramagnetic state. P-type ZnO grown in N_2O ambient is in FM state with (1 at.%) Cu doping, and shows $\boldsymbol{E_c} \sim 40$ Oe and $\mathbf{M} \sim 0.4\,\mu_B$/Cu at 200 K.	252

sputtered piezoelectric ZnO films with SiO_2, Si_3N_4 and Al_2O_3, and epitaxial layers of GaAs which assist in optical wave guiding. Although many layered structures consisting of ZnO/$Al_{0.08}Ga_{0.92}As$/$Al_{0.25}Ga_{0.75}As$ on GaAs, ZnO/TeO_2 on SiO_2-Si, TeO_2/ZnO on silicon and ZnO/SiO_2 on GaAs were investigated theoretically for acousto-optic Bragg deflector applications, but the promising features of these layered structures have not been fully realized due to complications arising during processing by conventional sputtering methods.[12,263–264]

A few simple structures consisting of ZnO/SiO_2/Si, ZnO/Si_3N_4/SiO_2/Si and SiO_2/ZnO/Al_2O_3 have been studied by depositing sputtered ZnO films, and in most of these studies a textured film of ***c***-axis oriented ZnO was obtained suitable enough for low frequency ($<$60 MHz) device applications, particularly thin film surface acoustic wave (SAW) devices.[265–267] However for high frequency SAW device applications, and the optoelectronic devices, the need for high quality epitaxial ZnO films on matched substrates has been clearly identified. As a result the chemical and structural compatibility of ZnO with underlying layers/substrates is a major concern, and efforts are directed in two specific areas: (i) the development of smooth interfaces for building high quality abrupt interfaces with alternate materials, and (ii) appropriate selection of materials that ensure chemical stability and structural uniformity with minimum inter-diffusion. The main objective is to develop thin film layered structures with periodicity in structure and composition, and fabrication of multilayered structures and superlattice structures often leads to a number of constraints while optimizing the growth conditions by any thin film growth technique. PLD in particular is found to be quite advantageous because with a multi target carrousel, a number of materials can be deposited alternately, and can be processed over a wide range of oxygen pressure to form interesting heterostructures and superlattices. In this section some of the important developments achieved by PLD are reviewed. The structure of various multilayers, heterostructures, and superlattices formed with ZnO films are summarized in Table 4.6, and important considerations during the processing of layered structures and their related applications are discussed.

A periodic multilayer consisting of the (0001) textured ZnO and amorphous alumina layers were fabricated at a low temperature to demonstrate resonant piezoelectric attenuation of a microwave signal,[269] and ZnO quantum dots embedded in amorphous alumina have been synthesized by PLD and these are of interest for exploring the nonlinear optical response, and open up new avenues with various other material combinations.[292] Heterostructures consisting of $BaTiO_3$ layers embedded between conductive ZnO layers with appropriate crystallographic orientation envision interesting optoelectronic applications.[302] Transparent optical thin film memory sandwich structures can be prepared, and the ferroelectric polarization can be used to control the charge carrier density formed at the

Table 4.6: ZnO multilayer, heterostructures and superlattices prepared by PLD.

Structure	Substrate	Applications, experiments & new results	Ref.
Pt-$BaTiO_3$-ZnO-Pt	(001) Si	Thin film capacitor for ferroelectric memory	268
ZnO-$BaTiO_3$-ZnO	(0001) sapphire	Dielectric and electro-optic studies, phase transition of $BaTiO_3$ observed	302
$[ZnO/Al_2O_3]_n$	Si (001) α-Al_2O_3	Multilayers with periodically modified piezoelectric coefficient modulated for high frequency resonators	269
Al:ZnO/Cu_2O	Cu_2O sheet	Heterojunction solar cell applications	270
AlN/ZnO	(0001) Al_2O_3	Epitaxial growth of AlN on ZnO films	271–272
GaN/ZnO	(0001) Al_2O_3	Epitaxial growth of GaN on ZnO buffered sapphire	273–277
ZnO/$Mg_{0.2}Zn_{0.8}O$	(0001) Al_2O_3	Quantum confinement effects, superlattices, MQW	127,279
ZnO/$Mg_xZn_{1-x}O$	(0001) $ScAlMgO_4$	Superlattices, Stimulated emission of excitons Excited stimulation observed up to 373 K	133
ZnO/$Zn_{0.74}Mg_{0.26}$	(0001) $ScAlMgO_4$	Multiple quantum wells (MQW), enhanced biexciton binding energy	280
$Zn_{1-x}Cd_xO/Mg_xZn_{1-x}O$		Lattice matched MQW structure	278,281
ZnO/$Zn_{.9}Mn_{0.1}O$/ZnO	(0001) Al_2O_3	Modulated doping, and magneto transport behavior	282
(111)ZnO/(111)MgO	(0001) Al_2O_3	Short period superlattice, interface alloying	207
ZnO/$Zn_{0.8}Mg_{0.2}O$	(0001) sapphire	Double barrier resonant tunneling device (DBRTD)	283
$HfO_2/InGaO_3$(ZnO)	(111) YSZ	Transparent field effect transistor (FET)	122

p-$SrCu_2O_2$/n-ZnO	(111) ITO/(111) YSZ	Transparent *p-n* junction, near UV emission	289
n-ZnO/p-NiO/ITO	(111) YSZ	Photo response, transparent *p-n* junction	290
p-ZnO/n-ZnO	n-ZnO	Phosphorous doped p-ZnO used for diode fabrication, light emission observed at 110 K	284
ZnO/MgZnO	(0001) sapphire	Superlattice on sapphire, density of stacking faults high, PL blue shifted, and enhanced emission	285
p-$Zn_{0.9}Mg_{0.1}O$/n-ZnO	ZnO crystal	ZnO based *p-n* junction, *n*-type buffer necessary for acceptable rectification	177
ZnO/$Zn_{0.74}Mg_{0.26}O$	(0001) $ScAlMgO_4$	Multi quantum wells with Laser MBE, strong I-order, II-order LO phonon replicas observed	286
MgZnO/ZnO	(0001) $ScAlMgO_4$	Free exciton emission observed only in MZO capped ZnO layers. Burstein Moss shift confirmed	287
ZnO/$La_{0.7}Sr_{0.3}MnO_3$	(0001) Al_2O_3	Integrate magneto resistive properties with non linear opto-electronic applications of ZnO	291
ZnO in Al_2O_3		ZnO quantum dot structure in amorphous matrix of alumina	292
p-ZnO/i-ZnO/n-ZnO	(0001) $ScAlMgO_4$	Light emitting diode based on ZnO	45
p-LSMO/i-STO/n-ZCO	(111) STO:La	*p-i-n* heterostructure with rectifying *I-V* curves	116
ZnO/AlN	(111) Si	epitaxial growth and bright UV luminescence	293
ZnO/MgO/TiN	(111) Si	Domain matching of TiN allow epitaxial ZnO, MgO	293

interface. ZnO-$BaTiO_3$-ZnO structures on **c**-plane sapphire have shown interesting electrooptic properties, and similarly Pt-$BaTiO_3$-ZnO-Pt thin film capacitor structures have exhibited rectifying semiconductor-ferroelectric polarization hysteresis loops.[268,302]

Rapid progress on III-V nitride materials that enabled the fabrication of blue light emitting diodes in fact triggered the initial research on the effectiveness of ZnO films in multi-layered structures for optoelectronic device applications. For example GaN films that are required on a low cost substrate such as sapphire normally contain a high density of defects due to the poor structural and thermal mismatch with sapphire. It is in this context that the effectiveness of ZnO as a buffer layer has been recognized. ZnO that has a relatively small lattice mismatch to group III-nitrides such as AlN and GaN has been found quite useful as a buffer layer on sapphire, and it is observed that the nitrides can be epitaxially grown on ZnO with a very low defect content.[271–272,274–277] These studies were in good agreement with the earlier efforts suggesting the growth of high quality GaN films on sputtered ZnO films and ZnO single crystals, and the findings could be successfully extended for establishing the suitability of GaN layers as a buffer for the growth of ZnO.[273] Heterojunctions of *p*-type nitrides and *n*-type ZnO have been fabricated using other techniques demonstrating LEDs operable by current injection.[294–295] The feasibility for growing high quality ZnO/GaN and GaN/ZnO heterostructures on sapphire provides immense opportunities for developing hybrid ZnO/GaN optoelectronic devices.[296]

The $Mg_xZn_{1-x}O$ (MZO) alloy has manifested itself as one of the most convenient material composition for realizing band-gap modulation in ZnO-based heterostructures, its lattice mismatch with ZnO is small, and the growth conditions of these alloy layers are compatible to those of ZnO epitaxial layers. Owing to their immense application potential in the field of optoelectronics, optical properties of ZnO/$Mg_xZn_{1-x}O$ multiple quantum well structures and superlattices on the (0001) sapphire substrate are being studied extensively, and indicated a clear shift of the photoluminescence peak to higher energies with a decrease in the ZnO layer thickness.[127,279] The MZO alloy composition with $x = 0.2$ in the ZnO/$Mg_{0.2}Zn_{0.8}O$ structure presents a very low in-plane lattice mismatch of 0.13% and corresponds to 0.5 eV difference in the energy band gap, and the superlattice structure thus formed exhibited quantum confinement effects, but necessitated the insertion of an ultra thin ZnO layer to achieve smooth alloy layers.[279] Further improvement has been reported with the use of lattice matched (0001) $ScAlMgO_4$ substrate that yielded high quality ZnO/$Mg_xZn_{1-x}O$ superlattices with negligibly small lattice mismatch (0.08%) at the superlattice-substrate interface. In comparison to ZnO epitaxial layers, improved structural and optical properties have been reported in ZnO/(Mg, Zn)O superlattices grown with laser molecular beam epitaxy (LMBE) with a lower threshold ($\sim$11 KW/cm^2 for $x = 0.12$) for the stimulated

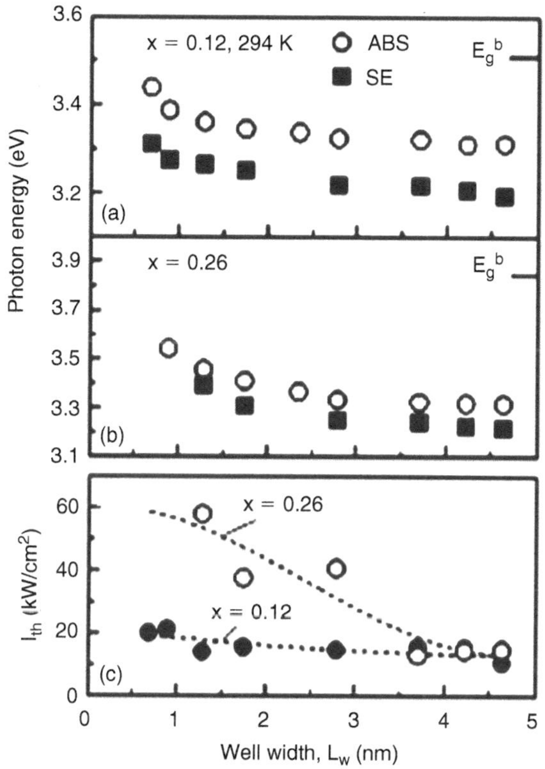

Figure 4.13: Optical transmission energies of subband absorption (open circles) and stimulated emission (closed squares) as a function of well layer thickness of ZnO in $Mg_xZn_{1-x}O$ superlattices with (a) $x = 0.12$, (b) $x = 0.26$; and (c) shows the variation of the threshold for stimulated emission with well layer thickness.[133] [Reused with permission from A. Ohtomo, K. Tamura, M. Kawasaki, T. Makino, Y. Segawa, Z.K. Tang, G.K.L. Wong, Y. Matsumoto and H. Koinuma, Applied Physics Letters, 77, 2204 (2000). Copyright 2000, American Institute of Physics.]

emission, and the excited stimulated emission was observed up to 373 K.[133] The confined exciton states could be clearly observed by optical absorption spectroscopy,[133,298] and the emission energy could be tuned between 3.2 and 3.4 eV by varying the well thickness and/or the Mg content in the barrier layer as shown in Fig. 4.13.

Some of the developments with ZnO-based multi quantum well structures[297] have been so exciting that low cost diode lasers in violet and UV region can now be envisaged with further improvements in the fabrication techniques and establishing good reproducibility. Fabrication of perfectly lattice-matched $Zn_{1-x}Cd_xO/Mg_xZn_{1-x}O$ multi-quantum wells has also been demonstrated,[278,281] and the most interesting feature of these two ZnO alloys is that $Zn_{1-x}Cd_xO$ exhibits a decreasing band gap, whereas $Mg_xZn_{1-x}O$ exhibits an increase in the band gap with varying Cd and Mg content in the respective alloys. A modulated doping structure can be prepared with ZnO/$Mg_xZn_{1-x}O$ combination, and also with ZnO/$Zn_{0.9}Mn_{0.1}O$ combination.[282] With the availability of $Zn_xMn_{1-x}O$ alloy film which has a wide band-gap and is lattice matched to ZnO, many innovative heterostructures can be fabricated.[224]

A large number of reports indicate the enhancement of band gap of ZnO by alloying it with different concentrations of MgO,[298–299] and it has been shown that with higher Mg concentration, the MZO alloy forms a metastable material The metastable phase and the degree of metastability are the limiting factors for the realistic applications of ZnO based devices. ***C***-axis oriented metastable hexagonal ZnO films with 30% MgO content made by PLD were reported by Ohtomo *et al.*[199] The binary phase diagram of ZnO and cubic MgO shows the formation of a perfect solid solution up to 40% of Mg.[300] Choopun *et al.*[210] reported a wide band gap metastable MZO alloy (~6 eV) with 50% Mg insertion at Zn sites using the PLD technique, and a wide variation in Mg concentration (50% to 85%) was shown by varying the substrate temperature. Surprisingly such films after rapid thermal annealing for 1 min at 750°C suddenly degraded and exhibited a band gap of 3 eV. However Bhattacharya *et al.*[207] showed that MZO alloy film formation by PLD can be stabilized by preparing ZnO/MgO multilayers in a short period superlattice, despite the large lattice mismatch between (111) MgO and (0001) ZnO layers on sapphire substrates. A stable wide band gap MZO alloy with (~6 eV) using ZnO/MgO multilayer could be formed using a constant MgO thickness of 1 nm, and by varying the thickness of the ZnO layers in the range 0.75 to 2.5 nm. Sanati *et al.*[301] suggest the possible presence of nontrivial atomic ordering in the vicinity of the phase separation. Highly transparent and oriented hexagonal and cubic thin films were obtained by changing the ZnO sublayer thickness inside the MgO host, and post deposition annealing at 750°C does not influence the structural and optical properties. Alloying at the hetero interface was found to occur at a substrate temperature of ~650°C, and these processing temperatures are considerably lower in comparison to the results on the ZnO/$Mg_xZn_{1-x}O$ heterostructure.

Heterostructures with *n*-type ZnO and a number of other materials exhibiting *p*-type semiconducting behavior have been studied in view of their potential applications for fabricating *p-n* heterojunctions. Initially due to the non availability of *p*-type ZnO, the advantages of *n*-ZnO as an active layer have been explored by depositing *n*-ZnO, or *n*-MgZnO on various *p*-type semiconductor layers such as ZnTe, Cu_2O, $SrCu_2O_2$, GaN, and AlGaN grown by a variety of growth techniques, and have been well covered in an earlier review.[288] YSZ (111) single crystal substrate has been commonly used because it allows the epitaxial growth of an indium-tin-oxide (ITO) layer as a transparent conducting electrode, however the large lattice mismatch in some of these systems and the ionicity of the oxides (e.g., *p*-type $SrCu_2O_2$) was a major drawback. Epitaxial growth of *n*-type ZnO on p-NiO film grown on ITO coated YSZ (111) substrates exhibit efficient UV responsivity and are comparable to those of commercial GaN UV detectors,[290] and *p*-*n* junctions based on *p*-type manganite ($La_{0.7}Sr_{0.3}MnO_3$) and *n*-type ZnO layers grown by PLD on ***c***-cut sapphire substrates provide interesting systems to integrate magnetic properties with non linear opto-electronic applications of ZnO.[291] Latest advances are related to the development of *p*-type ZnO using laser-MBE to produce high quality ZnO *p*-*i*-*n* homojunction for light emitting diode (LED) application, where nitrogen was used as the *p*-type dopant and a repeated temperature modulation epitaxy was specially developed for growing high quality superlattice structures (shown in Fig. 4.14a) on a lattice matched (0001) SCAM substrate.[45] The current-voltage characteristics of *p*-*i*-*n* device show good rectification with threshold voltage of ~7V (Fig. 14b).[45] The aforementioned experiments clearly suggest the immense application potential of ZnO based multi layers and super lattices, and ultimately these developments now demand a high level of reproducibility and repeatability.

4.7 Nanostructures by PLD

ZnO is a promising wide band gap material that has gained considerable attention for the growth of nanostructures whose configurations are much richer than any known materials including carbon nanotubes.[303] A variety of ZnO nanostructures including nanowires, nano needles, nano rods, nano helices, and nano belts have been synthesized by a variety of techniques such as MBE, thermal evaporation, vapor phase transport, sputtering and laser ablation techniques.[303–310] Preparation of defect free nanostructures with pure and doped ZnO having a controlled orientation and morphology is the major focus of nanoscience research. They are useful for short wavelength opto-electronic applications such as UV light emitters, detectors, gas sensing elements and nano heterojunctions. PLD as discussed earlier has been extensively used for the growth of epitaxial films and heterostructures. However the fabrication of ZnO nanostructures with PLD has been given less attention. The ability to ablate target materials over a wide range of pressure is one of the major advantages of PLD over the other techniques, and allows a precise control on morphology, size and composition of the fabricated nano-structures. A variety of ZnO nanostructures including an array of nano-rods/wire/needles etc. have been possible by PLD by carefully varying the deposition conditions and modifications in the processing chamber.

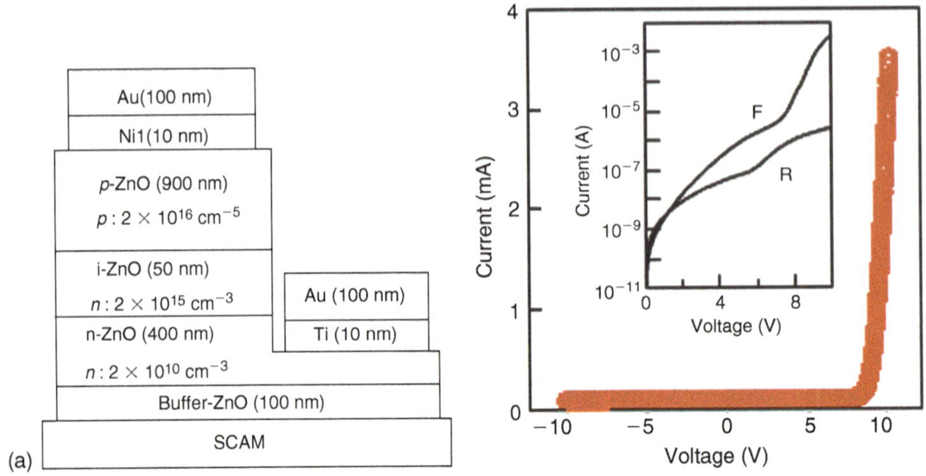

Figure 4.14: (a) PLD grown ZnO homostructural *p-i-n* junction LED structure, (b) rectifying current–voltage characteristics of device at 300 K. The inset has logarithmic scale in current with F and R denoting forward and reverse bias conditions respectively.[45] [Reprinted by permission from Macmillan Publishers Ltd: Nature Materials, A. Tsukazaki, A. Ohtomo, T. Onuma, M. Ohtani, T. Makino, M. Sukiya, K. Ohtani, S.F. Chichibu, S. Fuke, Y. Segawa, H. Ohno, H. Koinuma and M. Kawasaki, 'Repeated temperature modulation epitaxy for *p*-type doping and light-emitting diode based ZnO', Vol. 4, (2005), pp. 42–46, copyright 2005.]

An array of well-aligned ZnO nano rods were grown on single crystal (0001) sapphire substrate without a catalyst under a high oxygen pressure (5–10 Torr) ambient[309] at a target-to-substrate distance of 2 cm. The array of nanorods with hexagonal cross section was obtained at substrate temperatures of (600 to 700°C) and their length was along the ***c***-axis. Isolated ZnO nanorods with uniform diameter ~300 nm and up to 6 μm long were obtained at 700°C, and samples prepared at 400°C turn into a polycrystalline film with a rough microstructure. The fabrication of nanorod structures was attributed to the formation of nanoparticles of ZnO in the gas phase, and subsequent transportation to the substrate surface. The surface reaction of the nanoparticles on the substrate is found to be very crucial for the growth of nanorods, and a high substrate temperature of ~700°C was necessary for the melting and fusing of the nanoparticles sticking on the substrate surface for the nucleation of a rod. The characteristic luminescence spectra was observed from nanorods near the ZnO band gap ~390 nm. Gupta *et al.*[311] have synthesized an array of vertically aligned and well-isolated ZnO nanorods on Si (100) substrate at a relatively low substrate temperature of 450°C without using any catalyst. The target-to-substrate distance was ~2.5 cm, and the diameter of nanorods was found

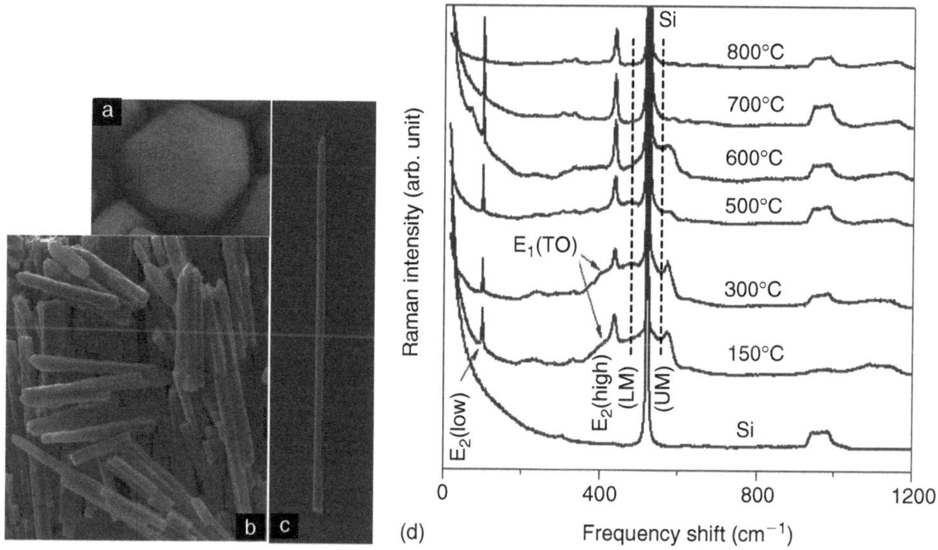

Figure 4.15: Scanning electron micro-photographs of ZnO nanorods: (a) hexagonal cross section, (b) fractured nano-rods, (c) single nano rod with fine tip, (d) Raman scattering spectra in back scattering configuration with incident light parallel to the length of rod, showing activation of surface phonon modes [after Gupta *et al.* (ref. 311)].

to depend strongly on pressure, and a uniform diameter was obtained at a minimum pressure of 5 Torr. Under optimum conditions ZnO nanorods with a diameter in the range ~120 to 200 nm, and length up to 12 μm have been obtained (Fig. 4.15a–c). At high temperature (>600°C) the nucleating centers due to the sticking of nanoparticles begin to coalesce and form a continuous film with densely packed large size grains.[311] The presence of a thin catalytic layer of gold on Si substrate is found to influence the growth significantly and ZnO rods with much larger diameter under similar processing conditions are obtained, in comparison to nanorods fabricated without a catalyst. Raman spectroscopy confirmed the formation of well-isolated single crystalline ZnO nanorods, and the activation of two surface phonon modes (LM and UM) was observed (Fig. 4.15d). High quality ZnO nanowires have also been synthesized by laser ablation with a slight modification in the growth procedure.[312] A Zn metal target is placed in a quartz tube furnace outside the hot zone, and SiO_2 passivated Si substrates covered with 20 nm gold clusters were placed at the center of the tube. The target was ablated with Nd:YAG laser in a reactive ambient consisting of 0.02% oxygen mixed with argon gas at 400 Torr pressure inside the furnace that was heated to 900°C. Large-scale nanowires were synthesized with average diameter ~20 nm and length ~10 μm. The formation of nanowires is due to diffusion of Zn vapors into gold particles, and when the zinc-gold alloy supersaturates the growth of nanowires is activated. X-ray diffraction (XRD) studies on these nanowires confirmed the formation of ZnO with wurtzite structure.[312]

Nobis *et al.*[313] investigated the inhomogeneous impurity distribution in ZnO micro-pillars using spatially and spectrally resolved cathodoluminescence (CL) microscopy. ZnO micro crystals were grown at 950°C on ***a***-plane sapphire using high pressure PLD in argon ambient. The pillars with hexagonal cross section of different diameter (1–5 μm) and height (3–20 μm) were grown using thin gold films of a few nanometers thickness as a catalyst. The room temperature CL of micro-pillars shows two peaks at 3.23 eV associated with exciton emission, and at 2.35 eV due to recombination at deep levels. The relative intensities of these peaks are found to vary spatially on pillar cross section. Excitonic emission was very strong at the center of the hexagonal cross section, whereas deep level emission dominates near the borders of the pillars.

ZnO nano-needles with high aspect ratio have been grown on ***a***-plane sapphire substrate using gold as a catalyst at high pressure PLD.[310] The observed needles were isolated and were found to be ***c***-axis oriented along the length of the needle. The cross-section of the fabricated needles was a regular hexagonal. The diameter of the needle was continuously decreasing from bottom (substrate side) to the top (needle tip) as shown in Fig. 4.16b. Spatially resolved cathodoluminescense studies of a single tapered ZnO nano-needle showed a continuous shift in the spectral maxima towards higher energies (in the region 1.8 to 2.9 eV) while approaching towards the needle tip (Fig. 4.16a), because the radius decreases along the longitudinal axis. For the first time the whispering gallery modes in hexagonal ZnO nano-needles for mode number in the range 1 to 6 were analyzed. The target-to-substrate distance was reported to be extremely important in all these studies, since the length of the laser plume decreases with increase in oxygen pressure.

Nanostructures of ZnO doped with Mg or Ga have also been attempted using PLD.[314–315] An array of vertically aligned Ga doped ZnO nano-rods have been fabricated on (0002) GaN coated sapphire ($11\bar{2}0$) single crystal substrate by alternate ablation of two oxide targets (ZnO and Ga_2O_3).[314] The samples were prepared at different oxygen pressures (10^{-1} to 10^{-3} Torr) and at growth temperatures (500–700°C) with a varying Ga dopant content in the range 0–5 at.% by controlling the number of pulses per cycle for each target. GaN was used due to its small lattice mismatch with ZnO, and the formation of uniformly distributed well-aligned nano-needle like ZnO:Ga (1%) nanorods with a diameter of ~60 nm and length ~300 nm formed around 750°C at a pressure of 10^{-2} Torr. The growth of nanorods is attributed to the initial formation of a wetting layer that act as the nucleating center for the formation of nanorods assisted by electric field produced due

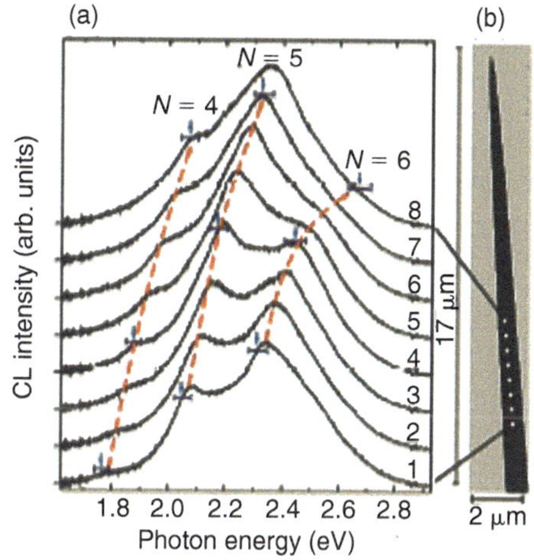

Figure 4.16: (a) Spatially resolved cathodoluminescence (CL) spectra of a single tapered ZnO nano-needle. (b) Tapered nanoneedles with indicated marked by white dots from where the CL data has been collected.[310] [Reprinted figure with permission from T. Nobis, E.M. Kaidashev, A. Rahm, M. Lorenz and M. Grundmann, Physical Review Letters, 93, 103903 (2004). Copyright 2004 by American Physical Society.]

to the interaction of target surface with the laser pulses of high power density. In contrast a continuous film without any evidence of nanorod formation is observed for pure ZnO and could be due to the low oxygen pressure used in the processing chamber. The nanorod formation is observed specifically with 1–5% Ga content in ZnO, and increase in Ga content increased the diameter, and decreased the density of nanorods.

Lorenz *et al.*[314] developed an array of $Mg_xZn_{1-x}O$ ($x = 0$–0.2) nanowires on (0001) sapphire substrate at a high pressure using PLD with the assistance of an ultra thin (~1 nm) gold catalytic layer. KrF excimer laser was focused on the edges of one of the two targets (ZnO and $Mg_{0.26}Zn_{0.74}O$) rotating continuously about a common axis inside a T-shape quartz chamber that was designed specially for the processing of materials under high pressures (25–200 mbar) and high temperatures (870–950°C). The composition of the film was controlled by varying the background pressure and the number of pulses on the edges of each target. Two substrates (***a***-plane and ***c***-plane sapphire) were placed parallel to the plume length, and the target to substrate distance (front edge) was varied from 0.5 to 3.5 cm. Growth of nanowires is observed only on the gold-coated area on the substrate. Transmission electron microscopy cross sections examined with energy dispersive x-rays indicated the absence of gold particles near the edges of the nanowire, and the role of gold has been identified as a nucleating centre for the growth of nano-wires.

The variation in diameter of the pure ZnO nano-wire across the length was found to be more on ***a***-plane sapphire in comparison to ***c***-plane sapphire, and becomes more homogenous with increase in Mg content.[314] Mg content was found to decrease continuously in the fabricated nanowires from $x = 0.16$ to 0.05 for a lateral movement on the substrate from the front edge to 8 mm. The diameter of $Mg_xZn_{1-x}O$ nanowires increases from 150 nm to 2.5 μm with decreasing substrate-to-target distance from 3.0 to 1.0 cm, and the length varied from 1.5 to 20 μm. Alternately, the use of a NiO nucleation layer instead of gold resulted in a thin thread-like nanostructure. However, CeO_2 buffer layer on the substrate produced hexagonal microdots with nanowires on top. The growth conditions of different kinds of ZnO nano structures prepared by pulsed laser ablation and major highlights of the reported work are summarized in Table 4.7.

4.8 Other studies on PLD grown ZnO films

ZnO thin films by PLD have been investigated for a variety of other applications as well. Verardi *et al.*[319] reported on the piezoelectric properties and the fabrication of a high frequency bulk acoustic wave transducer (BAW) transducer on silicon (Si) operating in the range 1 to 3 GHz. The estimated value of piezoelectric coefficient d_{33} (8×10^{-12} C/N) is high and close to the bulk value, and the measured electro mechanical coupling constant of about 0.25 confirmed the excellent quality of the deposited films. A piezoelectric modulated multilayered structure comprising of a piezoelectrically active ZnO and inactive Al_2O_3 layers on Si substrate has been developed to generate acoustic waves and a high frequency resonance at 10.6 GHz with a return loss of about 15 dB.[269] In this novel ZnO/Al_2O_3/Si multilayered structure films with (001) orientation generate the piezoelectric coefficient perpendicular to the substrate for transducer application and Al_2O_3 is used as a medium for microwave and supersonic waves. The bandwidth of the structure can be tuned by varying the thickness of the two layers and is useful for high frequency transducers with broad bandwidth. Phase velocity of the bulk acoustic phonons of ZnO film deposited on (100) InP substrate is reported to be about 6100 m/s.[107] Seabury *et al.*[320] deposited stress free and highly ***c***-axis oriented ZnO films by PLD and microwave resonators with Q values of about 300 and a coupling coefficient of 0.22 at 1.5 GHz have been fabricated, and the operation of a multipole band pass filter with −28 dB of rejection and −7 dB insertion loss has been demonstrated.

Some recent studies have pointed out that ZnO films exhibit strong non-linear second order optical susceptibility $\chi^{(2)}$, and efficient second harmonic (SH) has been generated in thin films of ZnO.[321–323] Considering the fact that most of the non-linear optical devices use higher order susceptibility $\chi^{(3)}$, the high susceptibility

Table 4.7: ZnO nanostructures prepared by pulsed laser deposition.

Structure	Material/ dopant	Substrate and catalyst used (thickness)	Growth temperature and pressure	Diameter and length	Main observations	Ref.
Nanorods	ZnO	Sapphire (0001)	600–700°C 1 to 5 Torr 20 mm	300 nm 6 μm	Array of oriented nanorods. Nanoparticles act as nucleation center. PL peaks at ~390 nm.	309
Nanorods	ZnO	Si (100)	450–500°C 5 Torr 25 mm	120–200 nm 12 μm	Array of isolated nanorods. Activation of surface phonon modes in Raman spectra.	311
Nano-needles	ZnO	*a*-sapphire gold thin film (<3 nm)	–	~50–900 nm	Isolated and tapered ***c***-axis oriented nano-needles with high aspect ratio. Whispering gallery modes were studied.	310
Nanorods	ZnO:Ga (1–5%)	Sapphire (1120) GaN (0002) layer (3.2 μm)	750°C 10^{-2} Torr	40–80 nm	Array of ZnO:Ga nanorods. Dia. ~60 nm & length ~300 nm. Ga content increases diameter and decreases density of rods.	315
Nanowire	ZnO:Mg (0–20%)	*a*-Sapphire *c*-Sapphire gold layer (1 nm)	870–950°C 25–200 mbar 5–35 mm	150–2500 nm 1.5–20.0 μm	Array of nanowires with uniform dia. across length. Mg content decreases with lateral position on substrate. NiO nucleation layer gives a thin thread-like nanostructure. CeO_2 layer gives micro-dots.	314

(*Continued*)

Table 4.7: (*Continued*)

Structure	Material/ dopant	Substrate and catalyst used (thickness)	Growth temperature and pressure	Diameter and length	Main observations	Ref.
Nanowire	ZnO	SiO_2/Si gold clusters (20 nm)	900°C 400 Torr	20 nm 10 μm	Zn target ablated in 0.02% O_2 + Ar ambient at 400 Torr give large quantity nanowires with high aspect ratio ~500:1.	312
Micro-pillars	ZnO	*a*-Sapphire Gold layer (few nm)	950°C 500 Torr Ar	1.0–5.0 μm 3.0–20.0 μm	Array of micro-pillars with hexagonal cross-section. CL peaks at 3.23 and 2.35 eV. Strong excitonic emission from center of pillar cross-section, and deep level emission dominates at border.	313
Nanorods	ZnO	Sapphire (0001)	600°C 5 Torr 20 mm	300 nm 6 μm	Array of oriented nanorods. Stimulated emission at ~388 nm with pump energy >50 mJ/cm^2.	316
Nano-cluster	ZnO:Co (1–10%): Al (10^{19}–10^{20} cm^{-3})	SiO_2	380–450°C (4 mbar O_2 + 1.5 mbar He)	40–100 nm	No secondary phase in highly Co doped nanostructure film. Average size of clusters decreases from 100 to 40 nm with Co (1–10%) $\boldsymbol{E_g}$ increases (3.24–3.28 eV) with Co.	317 318

values measured for ZnO films [$\chi^{(2)} = 2.3 \times 10^{-12}$ e.s.u. at 1.064 μm], and $\chi^{(3)} = 1.2 \times 10^{-13}$ e.s.u. qualify ZnO as one of the potential materials for non linear optical device applications.[324] There is a significant demand for thin film non-linear optical materials that can be integrated with opto-electronic devices. Second harmonic signal from ZnO thin films deposited on sapphire substrate by PLD has been measured at the fundamental mode of Q-switched Nd:YAG laser, and found to exhibit a strong non-linear second order optical susceptibility [$X^{(2)}_{zzz}$] as large as 8.8 pm/V.[321] It was concluded that a significant part of the efficient second harmonic has been generated at grain boundaries and interfaces. Very high conversion efficiency to the third-harmonic radiation has been reported as shown in Fig. 4.17 in sub-micron thick ***c***-axis oriented nano-crystalline ZnO films deposited on fused silica substrate under a mixed O_2-He gas atmosphere for an unamplified near-IR radiation of femtosecond Cr^{4+}:forsterite laser (1200–1300 nm).[31] A strong third order non-linear optical coefficient of ZnO makes its suitable for third order autocorrelation measurements of weak pulses.

Mazingue *et al.*[325] developed the waveguide structure using PLD grown ZnO coatings on SiO_2 (001) substrates for the detection of different concentrations (100 to 1000 ppm) of butane in dry air and nitrogen. The *m*-line technique ($\lambda = 633$ nm) was used for the measurements of refractive index and gas detection. ZnO film showed well-contrasted mode lines with slight anisotropy. The refractive indices of

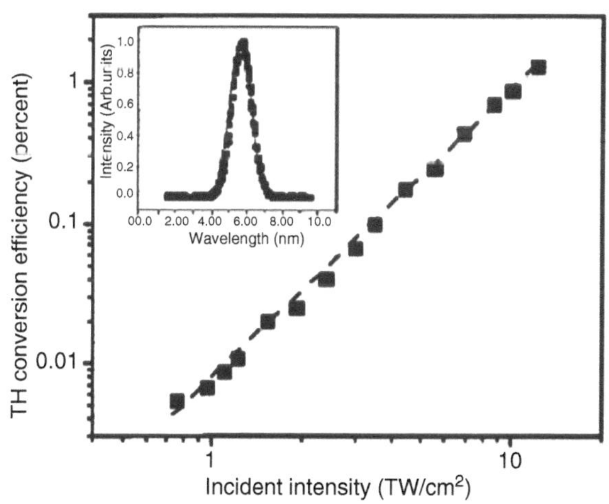

Figure 4.17: Experimentally measured conversion efficiency $P_{3\omega} = P_\omega$, where $P_{3\omega}$ is the power of the generated third harmonics (TH), and P_ω is the incident power of the fundamental beam. Inset shows the TH spectrum at maximum available input power.[31] [Reused with permission from G.I. Petrov, V. Shcheslavskiy, V.V. Yakovlev, I. Ozerov, E. Chelnokov and W. Marine, Applied Physics Letters, 83, 3993 (2003). Copyright 2003, American Institute of Physics.]

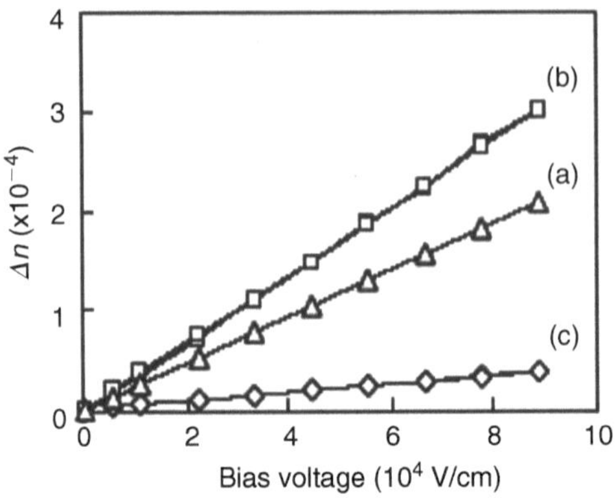

Figure 4.18: Bireferingence shift (**Δn**) of 5 at.% Mn doped ZnO film as a function of applied field at (a) 100 Hz, (b) 1 KHz and (c) 10 KHz.[189] [Reused with permission from T. Nagata, A. Ashida, N. Fujimura and T. Ito, Jpn. J. Appl. Phys. Vol. 41, 6916 (2002). Copyright 2002, The Institute of Pure and Applied Physics, Japan.]

the deposited films were in the range 1.95–1.98. The physical interactions between the butane gas and the ZnO layer leads to the variations in refractive index at the air-ZnO interface, and effective index variation down to 0.005 was detected during butane exposure at room temperature. Attempts have also been made to study the electro-optic effect by measuring the electric field induced birefringence in PLD grown pure and doped ZnO film.[188,189] Pure ZnO film grown by PLD show a low electro-optic effect due to the presence of large leakage current. The electro-optic response in Li and Mg doped ZnO has been demonstrated successfully with a dc electric field[12] however the presence of mobile Li ions is influenced by polarization behavior. In these films a linear increase in the birefringence (**Δn**) shift was observed with increase in applied bias voltage (Fig. 4.18). Linear electro-optic coefficients $r_c = -(2.\Delta n)/(n^3 \cdot E)$ of 7.8, 11.7 and 1.5 pm/V were estimated at frequencies of 100 Hz, 1 KHz and 10 KHz respectively, which are much larger than that of single crystal ZnO (~2.6 pm/V).[189] An investigation of the frequency dependent dielectric properties and electro-optic effects revealed that a rearrangement of the space charge affects the electro-optic response at < 10 KHz. Doping with Mn (5 at.%) in ZnO exhibits a linear electro-optic (EO) response with an ac bias voltage, and a linear EO coefficient originating only from the dipolar polarizability is found to be 0.5 pm/V.[189]

Holographic grating formation has been demonstrated with PLD grown ZnO films doped with 2 wt% Al. Two laser beams (TE polarized) from Kr-ion laser (356.4 nm) were incident symmetrically on film surface at 21.6° with respect to the surface normal. The diffraction grating was created on the film surface in the vicinity of point where two beams are interfering with an efficiency of 3×10^{-6} in a 200 nm thick film, and the optical response of Al doped ZnO film is reported to be superior to that of ITO.[154] Li doped ZnO films grown by PLD are reported to exhibit ferroelectric and electrochromic properties.[185,326] $Zn_{1-x}Li_xO$ ($x = 0.1$–0.17) films grown on Si (100) substrate exhibited hysteresis characteristics in the capacitance-voltage measurements, and the coercive field (~0.044 μC/cm^2) was maximum for $x = 0.17$ with a memory window of 1.2 V.[185] The electrochromic properties of nanocrystalline ZnO thin films have been studied using cyclic voltametery and *insitu* visible spectrochemical measurements.[326] ZnO films deposited on indium tin oxide coated glass substrate by PLD at 20 Pa oxygen pressure and 300°C were used as a working electrode along with two pure Li foils as reference and counter electrodes. ZnO electrode grown by PLD was reported to be stable for at least 500 cycles with a charge capacity of 5.6 μC/cm^2. ZnO film showed coloration in the visible region due to insertion of e^- and Li^+ ions and the efficiency was ~16 to 27 cm^2/C.[326] Onodera *et al.*[327] reported ferroelectric properties with Li^+ and Mg^+ dopants in ZnO, and reported ferroelectric hysteresis loops and measured phase transition in the dielectric properties which were strongly dependent on the Li and Mg content in ZnO.

Influence of several dopants such as Er, Ga, W, etc. have been studied in ZnO films grown by PLD to modify the luminescence properties for various applications in optoelectronics. Er was incorporated to enhance the 1.54 μm emission for its possible application in optical communication system, since the emission is within the wavelength range of a minimum loss for silica optical fibers.[328–330] As deposited ZnO:Er (0.5 at.%) films deposited on quartz substrate were ***c***-axis oriented with low electrical conductivity $\sim 6.4 \times 10^{-3}$ Ω cm.[328] The room temperature photoluminescence (PL) spectra exhibit a sharp and intense PL peak at 1.54 μm besides a peak in the UV region corresponding to ZnO. The peak at 1.54 μm originates from the intra-4f shell transition in Er^{3+} ions, and no change is observed under different excitation conditions. However, a shorter rise time is noted for indirect excitation indicating superior excitation efficiency. Photo luminescence studies on heavily doped ZnO films with Er = 0.6 to 2.8 at.% deposited on SiO_2/Si substrate indicate the probable incorporation of Er ions at the grain boundaries along with the lattice site, and exhibit a strong luminescence quenching.[330] The emission corresponding to 1.54 mm could be enhanced by avoiding the segregation of Er at the grain boundaries via epitaxial growth on a lattice matched substrate such as sapphire, or $ScAlMgO_4$. Lee *et al.*[331] deposited polycrystalline and epitaxial Ga doped ZnO ($ZnGa_2O_4$) film on amorphous glass and (100) MgO single crystal substrates respectively by PLD. The epitaxial film on MgO showed enhanced blue-white luminescence intensity with a broad emission band (415–700 nm) under UV excitation in comparison to the polycrystalline films on glass. The broad band was expected to be composed of multiple peaks. The blue-white light is emitted from the film having Zn/Ga ratio in the range 0.3 to 0.4. Hayashi *et al.*[332] identified the composition in the ZnO-WO_3 binary system for blue phosphor material using combinatorial PLD system with two ceramic targets. A film with a compositional spread was achieved by depositing a combination of layers (WO_3/ZnO) that were piled up by successive ablations on a platinum coated Si substrate. A chemically stable phase of the deposited film was obtained after annealing at 750°C for 3 hours in oxygen ambient, and the brightest blue emission in the CL spectra was observed in the regions having WO_3 composition of ~40 to 60 mol%.[332]

ZnO films grown by PLD have been extensively used as an effective buffer layer to promote the growth of high quality films of other complex materials useful for electronic device applications.[274] The epi-ZnO layer deposited on sapphire is useful as a lattice matched semiconducting substrate for the epitaxial growth of GaN because ZnO is an isomorphic material (elements from same row of periodic table) with a small lattice mismatch (2.2%) between the basal planes of GaN and ZnO in comparison to large lattice mismatch of (14%) between sapphire and GaN.[271] Sun *et al.*[275] showed that the GaN films grown on a bare sapphire substrate by liquid target PLD possesses a mixed orientation, and exhibits the epitaxial relationship only after introducing a ZnO buffer layer as in the structure (0001)GaN/(0001)ZnO/(0001)Al_2O_3. Two dimensional growth of GaN is greatly enhanced by using a ZnO buffer layer deposited by PLD in comparison to earlier attempts with rf sputtered ZnO buffer layers.[276] The epi-ZnO/sapphire has also been used for the epitaxial growth of AlN film exhibiting ω-rocking curve FWHM of 0.23°, which is lower by about 40–50% than the AlN film grown directly on sapphire under identical conditions. The lattice mismatch between the AlN and ZnO is within 4%. After the growth of an epitaxial AlN film on ZnO buffer layer, it is found that a 3–6 nm thick layer of spinel $ZnAl_2O_4$ is formed at the interface of ZnO and AlN due to inter-diffusion, but the initial epitaxial growth of AlN remains unaffected.[272] Thus a buffer layer of ZnO provides a suitable host lattice for the promotion of the epitaxial growth of nitride materials (GaN and AlN). Highly oriented and single-phase $Ba_2Zn_2Fe_{12}O_{22}$ (Zn_2Y) ferrite films have been obtained by introducing a ZnO (~40 nm thick) buffer layer on single-crystal (0001) sapphire substrate. The underlying ZnO layer acts as a source of Zn during Zn_2Y film growth resulting in single phase formation exhibiting epitaxial relationship of (0001)Zn_2Y/(0001)ZnO/(0001)Al_2O_3.[333] Pure and Al doped ZnO films grown by PLD have been studied for tribological behavior, and ***c***-axis oriented doped film deposited in vacuum exhibit a friction coefficient of about 0.15. Lateral force microscopy measurement on the surface of Al doped ZnO film shows the presence of a lubricant phase on wear surface.[179]

ZnO buffer layers are also found beneficial for obtaining oriented GaN films on amorphous or Si (100) substrates.[276,334] The surface morphology and optical

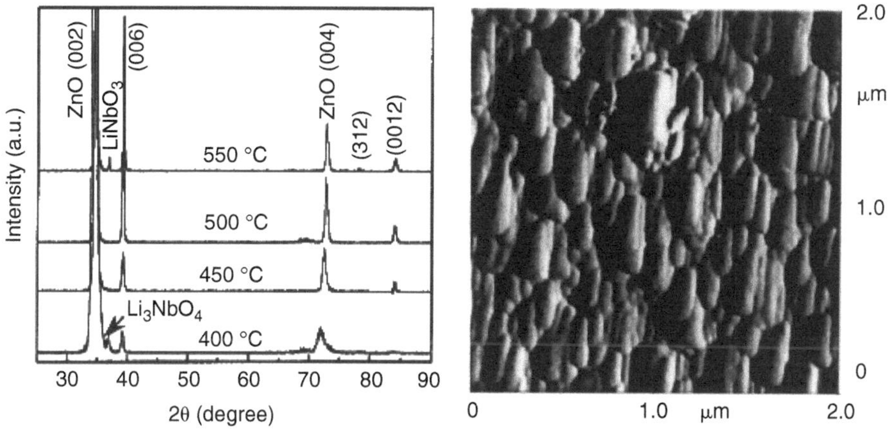

Figure 4.19: (a) Effect of growth temperature on the crystallization of oriented $LiNbO_3$ film grown on transparent conducting Al:ZnO coated Si substrate. (b) Surface morphology of ***c***-axis oriented $LiNbO_3$ film.[99] [Reused with permission from V. Gupta, P. Bhattacharya, Y.I. Yuzuk, R.S. Katiyar, M. Tomar and K. Sreenivas, J. Mater. Res., Vol. 19, No. 8 (2003). Copyright 2004, Materials Research Society.]

transparency of the GaN films is greatly improved by the ZnO buffer layer. Similarly the growth of highly oriented ***c***-axis $LiNbO_3$ thin film has been achieved on an amorphous quartz substrate using an ultra thin buffer layer of ZnO grown by PLD. A textured $LiNbO_3$ film with preferred orientation has been deposited on a Si (100) substrate using an interlayer of Al (1%) doped ZnO (TCO) thin film (Fig. 4.19). The Raman scattering and dielectric studies confirm the growth of highly ***c***-axis oriented and phase pure $LiNbO_3$ film under optimum growth conditions of 100 mTorr and 500°C.[99] Electrical measurements on such oriented $LiNbO_3$ films was difficult earlier due to non-availability of suitable conducting electrodes, however, with development of Al doped ZnO film grown by PLD with low electrical resistivity and high optical transmission the electro optic effects can be measured conveniently.

Piezoelectric activity on the surface of a ***c***-axis oriented $LiNbO_3$ film deposited on ZnO coated glass substrate by PLD has been imaged using atomic force microscopy.[87] The influence of ZnO buffer layer on the crystalline quality of $LiNbO_3$ is found to be significant and an optimum thickness ~20 nm is shown to promote the ***c***-axis growth of $LiNbO_3$ film on glass and SiO_2 passivated Si substrates. Strong piezoelectric activity in a laser-ablated $LiNbO_3$/ZnO multilayer has been observed. The AFM tip displacement in the presence of an in-plane, low-frequency (0.1–5.0 Hz) ac electric field was monitored and imaged. The AFM tip displacement normal to the film surface with applied field was much larger in the $LiNbO_3$/ZnO multilayer structure in comparison to ZnO film alone, and was attributed to the high electro mechanical coupling coefficient of $LiNbO_3$.[87]

4.9 Conclusions

In comparison to other thin film deposition techniques, the turn around time is much faster with the pulsed laser deposition technique right from film deposition for a specific material composition, to the optimization of the properties desired for device applications. Suitable modifications in the deposition techniques have enabled thin film materials scientists to meet the challenges required for the engineering of the ZnO material with tailored properties by incorporating various dopants and alloying with other oxide materials successfully. Present understanding of ZnO thin film processing by PLD has increased to a high level and researchers have become more aware of the potential offered by ZnO for optoelectronic device applications. Deposition of high quality epitaxial films of doped and undoped ZnO and fabrication of lattice matched superlattice structures has been achieved, however, there is still room for improvement for reducing the processing temperatures, and for establishing reproducible methods for achieving *p*-type doping in ZnO. In summary, the continued rapid expansion of the field, and overcoming some of the earlier limitations herald new opportunities. The potential applications of ZnO thin films in optoelectronic devices and the development of epitaxial superlattices have slowly transformed the present "art" of thin film fabrication into an established science.

Acknowledgements

The authors wish to thanks all the publishers and the authors who gave the permission to reuse some figures from their published work. We would like to specially acknowledge the contributions of Prof. R.S. Katiyar, Univ. of Puerto Rico, Puerto Rico and Dr. S.P. Singh, National Physical Laboratory, New Delhi, India for fruitful discussions, and the Defence Research and Development Organization (DRDO), Govt. of India for support.

References

[1] F.S. Hickernell, Intl. J. of High Speed Electronics and Systems, 10 (2000) 603.
[2] H.L. Hartnagel, A.L. Dawar, A.K. Jain and C. Jagadish, "Semiconducting transparent thin films", IOP, Bristol (1995), ISBN No. 0–7503–0322–0.

[3] M. Tomar, V. Gupta, K. Sreenivas and A. Mansingh, IEEE Trans. on Device and Materials Reliability, 5 (2005) 494.
[4] G.A. Racine, P. Muralt and M.A. Dubois, Smart Mater. Struct. 7 (1998) 404.
[5] K.D. Mitzner, J. Sternhagen and D.W. Galipeau, Sens. Actuators B 93 (2003) 92.
[6] T. Mitsuyu, O. Omazaki, K. Ohji and K. Wasa, Ferroelectrics 42 (1982) 233.
[7] T. Minami, H. Nanto and S. Takata, Thin Solid Films 124 (1985) 43.
[8] P. Zu, Z.K. Tang, G.K.L. Wong, M. Kawasaki, A. Ohtomo, H. Koinuma and Y. Segawa, Solid State Commun. 103 (1997) 459.
[9] P. Sharma, K. Sreenivas and K.V. Rao, J. Appl. Phys. 93 (2003) 3963.
[10] S.J. Pearton, W.H. Heo, M. Ivill, D.P. Norton and T. Steiner, Semicond. Sci. Technol. 19 (2004) R59.
[11] N. Mehan, V. Gupta, K. Sreenivas and A. Mansingh, J. Appl. Phys. 96 (2004) 3134.
[12] S. Jain and A. Mansingh, J. Phys. D: Applied Physics 25 (1992) 1116.
[13] I.-S Jeong, J.-H. Kim and S. Im, Appl. Phys. Lett. 83 (2003) 2946.
[14] V. Craciun, R.K. Singh, J. Perriere, J. Spear and D. Cracium, J. Electrochem. Soc. 147 (2000) 1077.
[15] S.J. Chen, Y.C. Liu, J.G. Ma, D.X. Zhao, Z.Z. Zhi, Y.M. Lu, J.Y. Zhang, D.Z. Shen and X.W. Fan, J. Cryst. Growth 240 (2002) 467.
[16] R. Ayouchi, F. Martin, D. Leinen and J.R. Ramos-Barrado, J. Crystal. Growth 247 (2003) 497.
[17] W. Tang and D.C. Cameron, Thin Solid Films 238 (1994) 83.
[18] Y. Takahashi, M. Kanamori, A. Kondoh, H. Minoura and Y. Ohya, Jpn. J. Appl. Phys. 33 (1994) 6611.
[19] V. Gupta and A. Mansingh, J. Appl. Phys. 80 (1996) 1063.
[20] M. Kadota and M. Minakata, IEEE Trans. Ultrasonics Ferro. and Freq. Control 42 (1995) 345.
[21] M. Aoki, K. Tada and T. Murai, Thin Solid Films 83 (1981) 283.
[22] X.W. Sun and H.S. Kwok, J. Appl. Phys. 86 (1999) 408.
[23] H.B. Kang, K. Nakamura, S.H. Lim and D. Shindo, Jpn. J. Appl. Phys. 37 (1998) 371.
[24] F.S. Hickernell, J. Appl. Phys. 44 (1973) 1061.
[25] K. Higaki, H. Nakahata, H. Kitabayashi, S. Fuji, K. Tanabe, Y. Seki and S. Shikata, IEEE Trans. Ultrason. Ferro. and Freq. Control 44 (1997) 1395.
[26] M. Kadota, Jpn, J. Appl. Phys. 36 (1997) 3076.
[27] K. Iwata, P. Fons, A. Yamada, K. Matsubara and S. Niki, J. Cryst. Growth 209 (2000) 526.
[28] C.R. Gorla, N.W. Emanetoglu, S. Liang, W.E. Mayo and Y.J. Lu, J. Appl. Phys. 85 (1999) 2595.
[29] T. Koga, B.M. Moon, I. Zakharchenko, K.V. Rao and H. Medelius, IEEE Trans. Magnetics 29 (1993) 3601.
[30] C. Bjormander, K. Sreenivas, M. Duan, A.M. Grishin and K.V. Rao, Appl. Phys. Lett. 66 (1995) 2493.
[31] G.I. Petrov, V. Shcheslavskiy, V.V. Yakovlev, I. Ozerov, E. Chelnokov and W. Marine, Appl. Phys. Lett. 83 (2003) 3993.

[32] D.B. Christey and G.K. Hubler, "Pulsed laser deposition of thin films", John Wiley & Sons Inc. (1994).
[33] A. Ohtomo and A. Tsukazaki, Semicond. Sci. and Technol. 20 (2005) S1–S12.
[34] R. Triboulet and J. Perriiere, Progress in crystal growth and characterization of materials, Elsevier, 47 (2003) 65.
[35] D.P. Norton, Y.W. Heo, M.P. Ivill, K. Ip, S.J. Pearton, M.F. Chisholm and T. Steiner, Materials Today, June 2004, pp.34.
[36] S.J. Pearton, D.P. Norton, K. Ip and Y.W. Heo, J. Vac. Sci. Technol B 22 (2004) 932.
[37] U. Ozgur, Y.I. Alivov, C. Liu, A. Teke, M.A. Reshchikov, S. Dogan, V. Avrutin, S.-J. Cho and H. Morkoc, J. Appl. Phys. 98 (2005) 041301.
[38] S.B. Ogale, Thin films and heterostructures for oxide electronics, Springer Science 2005, New York.
[39] T. Nakayama, Surf. Sci. 133 (1983) 101.
[40] H. Sankur and J.T. Cheung, J. Vac. Sci. Technol A 1 (1983) 1806.
[41] V. Craciun, J. Elders, J.G.E. Gardeniers and I.W. Boyd, Appl. Phys. Lett. 65 (1994) 2963.
[42] M. Okoshi, K. Higashikawa and M. Hanabusa, Appl. Surf. Sci. 154 (2000) 424.
[43] L.N. Dinh, M.A. Schildbach, M. Balooch and W. Mclean II, J. Appl. Phys. 86 (1999) 1149.
[44] Y.R. Ryu, S. Zhu, J.D. Budai, H.R. Chandrasekhar, P.F. Miceli and H.W. White, Appl. Phys. Lett. 88 (2000) 201.
[45] A. Tsukazaki, A. Ohtomo, T. Onuma, M. Ohtani, T. Makino, M. Sukiya, K. Ohtani, S.F. Chichibu, S. Fuke, Y. Segawa, H. Ohno, H. Koinuma and M. Kawasaki, Nature Mater. 4 (2005) 42.
[46] W. Prellier, A. Fouchet, B. Mercey, C. Simon and B. Raveau, Appl. Phys. Lett. 82 (2003) 3490.
[47] A. Fouchet, W. Prellier, B. Mercey, L. Mechin, V.N. Kulkarni and T. Venktesan, J. Appl. Phys. 96 (2004) 3228.
[48] C. Belouet, Appl. Surf. Sci. 96–98 (1996) 630.
[49] V. Craciun, D. Craciun, M.C. Bunescu, R. Dabu and I.W. Boyd, Appl. Surf. Sci. 109–110 (1997) 354.
[50] V. Craciun, D. Craciun, M.C. Bunescu, R. Dabu and I.W. Boyd, J. Phys. D: Appl. Phys. 32 (1999) 1306.
[51] V. Craciun, N. Basim, R.K. Singh, D. Craciun, J. Herman, C.-B. Leborgne, Appl. Surf. Sci. 186 (2002) 288.
[52] A. Tselev, A. Gorbunov and W. Pompe, Rev. Sci. Instrum. 72 (2001) 2665.
[53] S.G. Lee, D.S. Hwang, Y.K. Park and J.C. Park, Appl. Phys. Lett. 65 (1994) 764.
[54] H. Minami, D. Manage, Y.Y. Tsui, R. Fedosejevs, M. Malac and R. Egerton, Appl. Phys. A 73 (2001) 531.
[55] T. Yoshitake, G.G. Shiraishi and K. Nagayama, Appl. Surf. Sci. 197–198 (2002) 379.
[56] W.P. Barr, J. Phys. E 2 (1969) 112.

[57] A.A. Gorbunov, W. Pompe, A. Sewing, S.V. Gapanov, A.D. Akhsakhalyan, I.A. Zabrodin, I.A. Kaskov, E.B. Klyenkov, A.P. Morozov, N.N. Salaschenko, R. Dietsch, H. Mai and S. Vollmar, Appl. Surf. Sci. 96–98 (1996) 649.
[58] S. Witanachchi, K. Ahmed, P. Sakthivel and P. Mukherjee, Appl. Phys. Lett. 66 (1995) 1469.
[59] V. Craciun, D. Craciun, M.C. Bunescu, C.-B. Leborgne and J. Hermann, Phys. Rev. B 58 (1998) 6787.
[60] R. Kelly and A. Miotello, Phys. Rev. E 60 (1999) 2616.
[61] P. Mukherjee, S. Chen, J.B. Cuff, P. Sakthivel and S. Witanachchi, J. Appl. Phys. 91 (2002) 1828.
[62] E. Millon, O. Albert, J.C. Loulergue, J. Etchepare, D. Hulin, W. Selier and J. Perriere, J. Appl. Phys. 88 (2000) 6937.
[63] J. Perriere, E. Millon, W. Seiler, C.-B. Leborgne, V. Craciun, O. Albert, J.C. Loulergue and J. Etchepare, J. Appl. Phys. 91 (2002) 690.
[64] P. Mukherjee, S. Chen and S. Witanachchi, Appl. Phys. Lett. 74 (1999) 1546.
[65] P. Mukherjee, S. Chen, J.B. Cuff, P. Sakthivel and S. Witanachchi, J. Appl. Phys. 91 (2002) 1837.
[66] O.-G. Noël, R.G. Roman, J. Perrière, J. Hermann, V. Craciun, C.-B. Leborgne and P. Barboux, J. Appl. Phys. 80 (1996) 1803.
[67] V. Craciun, S. Amirhaghi, D. Craciun, J. Elders, J.G.E. Gardeniers and I.W. Boyd, Appl. Surf. Sci. 86 (1995) 99.
[68] R.D. Vispute, S. Choopun, R. Enck, A. Patel, V. Talyansky, R.P. Sharma, T. Venkatesan, W.L. Sarney, L. Salamanca-Riba, S.N. Andronescu, A.A. Illiadis and K.A. Jones, J. Electron. Mater. 28 (1999) 275.
[69] S.G. Mayr, M. Moske, K. Samwer, M.E. Taylor and H.A. Atwater, Appl. Phys. Lett. 75 (1999) 4091.
[70] J. Wei, B. Zhang, N. Yao, X. Wang, H. Ma and S. Wang, J. Vac. Sci. Technol B 19 (2001) 1082.
[71] R.G. Roman, R. Perez Casero, C. Marechal, J.P. Enard and J. Perriere, J. Appl. Phys. 80 (1996) 1787.
[72] F. Claeyssens, A. Cheesman, S.J. Henley and M.N.R. Ashfold, J. Appl. Phys. 92 (2002) 6886.
[73] S. Hayamizu, H. Tabata, H. Tanaka and T. Kawai, J. Appl. Phys. 80 (1996) 787.
[74] H. Koinuma, MRS Bull. 19 (1994) 21.
[75] S. Im, B.J. Lin and S. Yi, J. Appl. Phys. 87 (2002) 4558.
[76] S. Choopun, R.D. Vispute, W. Nolch, A. Balasamo, R.P. Sharma, T. Venkatesan, A. Illiadis and D.C. Look, Appl. Phys. Lett. 75 (1999) 3947.
[77] H.S. Kwok, H.S. Kim, D.H. Kim, W.P. Shen, X.W. Sun and R.F. Xiao, Appl. Surf. Sci. 109–110 (1997) 595.
[78] P.E. Dyer, A. Issa and P.H. Key, Appl. Phys. Lett. 57 (1990) 186.
[79] P.W. Wilmott and J.R. Huber, Rev. Mod. Phys. 72 (2000) 315.
[80] S.L. King, J.G.E. Gardeniers and I.W. Boyd, Appl. Surf. Sci. 96–98 (1996) 811.

[81] J.L. Zhao, X.M. Li, J.M. Bian, W.D. Yu and X.D. Gao, J. Cryst. Growth 276 (2005) 507.
[82] J. Mass, P. Bhattacharya and R.S. Katiyar, Mat. Sci. Engg. B 103 (2003) 9.
[83] T. Kobayashi, H. Akiyoshi and M. Tachiki, Appl. Surf. Sci. 197–198 (2002) 294.
[84] M. Tachiki and T. Kobayashi, Jpn. J. Appl. Phys. 38 (1999) 3642.
[85] M. Tachiki, T. Hosomi and T. Kobayashi, Jpn. J. Appl. Phys. 39 (2000) 1817.
[86] S. Yata, Y. Nakashima and T. Kobayashi, Thin Solid Films 445 (2003) 259.
[87] P. Sharma, K. Sreenivas, L.M. Belova and K.V. Rao, J. Mater. Res. 18 (2003) 2025.
[88] M. Ogasawara, M. Shimizu and T. Shiosaki, Jpn. J. Appl. Phys. 31 (1992) 2971.
[89] K. Ramamoorthy, C. Sanjeevraja, M. Jayachandran, K. Sankarnarayanan, P. Bhattacharya and L.M. Kukreja, J. Cryst. Growth 226 (2001) 281.
[90] J.H. Choi, H. Tabata and H. Kawai, J. Cryst. Growth 226 (2001) 493.
[91] F.K. Shan, G.X. Liu, W.J. Lee, I.S. Kim, B.C. Shin and Y.C. Kim, J. Cryst. Growth 277 (2005) 284.
[92] H. Cao, Y.G. Zhao, H.C. Ong, S.T. Ho, J.Y. Dai, J.Y. Wu and R.P.H. Chang, Appl. Phys. Lett. 73 (1998) 3656.
[93] X.M. Fan, J.S. Lian, L. Zhao and Y.H. Liu, Appl. Surf. Sci. (2005) (in press).
[94] T. Ohshima, R.K. Thareja, T. Ikegami and K. Ebihara, Surf. and Coat. Technol. 169–170 (2003) 517–520.
[95] S.J. Henley, M.N.R. Ashfold and D. Cherns, Thin Solid Films 422 (2002) 69.
[96] J.A. Sans, A. Segura, M. Mollar and B. Mari, Thin Solid Films 453–454 (2004) 251.
[97] S.H. Bae, S.Y. Lee, H.Y. Kim and S. Im, Appl. Surf. Sci. 168 (2000) 332.
[98] S.S. Kim and B.T. Lee, Thin Solid Films 446 (2004) 307.
[99] V. Gupta, P. Bhattacharya, Y.I. Yuzuk, R.S. Katiyar, M. Tomar and K. Sreenivas, J. Mater. Res. 19 (2004) 2235.
[100] K. Ramamoorthy, M. Jayachandran, K. Sankaranarayanan, P. Misra, L.M. Kukreja and C. Sanjeeviraja, Solar Energy 77 (2004) 193.
[101] K. Ramamoorthy, C. Sanjeeviraja, M. Jayachandran, K. Sankaranarayanan, P. Misra and L.M. Kukreja, Materials in Semiconductor Processing 6 (2003) 219.
[102] F.K. Shan, B.C. Shin, S.W. Jang and Y.S. Yu, J. Eur. Ceram. Soc. 24 (2004) 1015.
[103] A. Tiwari, M. Park, C. Jin, H. Wang, D. Kumar and J. Narayan, J. Mater. Res. 17 (2002) 2480.
[104] V. Cracium, J. Elders, J.G.E. Gardeniers, J. Geretovsky and I.W. Boyd, Thin Solid Films 259 (1995) 1.
[105] Y.R. Ryu, S. Zhu, S.W. Han, H.W. White, P.F. Miceli and H.R. Chandrasekhar, Appl. Surf. Sci. 127–129 (1998) 496.
[106] Y.R. Ryu, S. Zhu, J.M. Wrobel, H.M. Jeong, P.F. Miceli and H.W. White, J. Cryst. Growth 216 (2000) 326.
[107] E. Vasco, J.-R. Zuazo, L.V'azquez, C. Prieto and C. Zaldo, J. Vac. Sci. Technol. B 19 (2001) 224.
[108] S.E. Shim, H.S. Kang, S.S. Pang, J.S. Kang, I. Yun and S.Y. Lee, Mater. Sci. Engg. B 102 (2003) 366.

[109] Y. Segawa, A. Ohtomo, M. Kawasaki, H. Koinuma, Z.K. Tang, P. Yu and G.K.L. Wong, Phys. Status Solid B 202 (1997) 669.
[110] V. Srikant and D.R. Clarke, J. Appl. Phys. 81 (1997) 6357.
[111] X.L. Guo, H. Tabata and T. Kawai, J. Cryst. Growth 223 (2001) 135.
[112] Z.Q. Chen, S. Yamamoto, A. Kawasuso, Y. Xu and T. Sekiguchi, Appl. Surf. Sci. 244 (2005) 377.
[113] A. Ohtomo, K. Tamura, K. Saikusa, K. Takahashi, T. Makino, Y. Segawa, H. Koinuma and M. Kawasaki, Appl. Phys. Lett. 75 (1999) 2635.
[114] K. Tamura, A. Ohtomo, K. Saikusa, Y. Osaka, T. Makino, Y. Segawa, M. Sumiya, S. Fluke, H. Koinuma and M. Kawasaki, J. Cryst. Growth 214/215 (2000) 59.
[115] T. Koida, S.F. Chichibu, A. Uedona, T. Sota, A. Tsukazaki and M. Kawasaki, Appl. Phys. Lett. 84 (2004) 1079.
[116] I. Satoh and T. Kobayashi, Appl. Surf. Sci. 216 (203) 603.
[117] J. Yin, Z.G. Liu, H. Liu, X.S. Wang, T. Zhu and J.M. Liu, J. Cryst. Growth 220 (2000) 281.
[118] G.H. Lee, Solid State Commun. 128 (2003) 351.
[119] M. Suguira, Y. Nakashima, T. Nakasaka and T. Kobayashi, Appl. Surf. Sci. 197–198 (2002) 472.
[120] E. Bellingeri, D. Marre, I. Pallechi, L. Pellegrino and A.S. Siri, Appl. Phys. Lett. 86 (2005) 012109
[121] A. Segura, J.A. Sans, F.J. Manjon, A. Munoz and M.J.-H. Cabrerano, Appl. Phys. Lett. 83 (2003) 278.
[122] K. Nomura, H. Ohta, K. Ueda, T. Kamiya, M. Hirano and H. Hosno, Science 300 (2003) 1269.
[123] J. Zou, S. Zhou, C. Xia, Y. Hang, J. Xu, S. Gu and R. Zhang, J. Cryst. Growth 280 (2005) 185.
[124] D.C. Look, J.W. Hemsky and J.R. Sizelove, Phys. Rev. Lett. 82 (1999) 2552.
[125] G.E. Pike and H. Seager, J. Appl. Phys. 50 (1979) 3414.
[126] G. Blatter and G. Greutr, Phys. Rev. B 33 (1986) 3952.
[127] A. Ohtomo, M. Kawasaki, I. Ohkubo, H. Koinuma, T. Yasuda and Y. Segawa, Appl. Phys. Lett. 75 (1999) 980.
[128] E.M. Kaidashev, M. Lorenz, H.V. Wenckstern, A. Rahm, H.C. Semmelhack, K.H. Han G. Benndorf, C. Bundesmann, H. Hockmuth and M.Grundmann, Appl. Phys. Lett. 82 (2003) 3901.
[129] M.Z. Lin, C.T. Su, H.C. Yan and M.Y. Chern, Jpn. J. Appl. Phys. 44 (2005) L995.
[130] A. Ohtomo, H. Kimura, K. Saito, T. Makino, Y. Segawa, H. Koinuma and M. Kawasaki, J. Cryst. Growth 214–215 (2000) 284.
[131] B. Wessler, A. Steinecker and W. Mader, J. Cryst. Growth 242 (2002) 283.
[132] A. Tsukazaki, A. Ohtomo, M. Kawasaki, T. Makino, C.H. Chia, Y. Segawa and H. Koinuma, Appl. Phys. Lett. 83 (2003) 2784.
[133] A. Ohtomo, K. Tamura, M. Kawasaki, T. Makino, Y. Segawa, Z.K. Tang, G.K.L. Wong, Y. Matsumoto and H. Koinuma, Appl. Phys. Lett. 77 (2000) 2204.

[134] Y.J. Li, Y.W. Kwon, M. Jones, Y.W. Heo, J. Zhou, S.C. Luo, P.H. Holloway, E. Douglas, D.P. Norton, Z. Park and S. Li, Semicond. Sci. Tech. 20 (2005) 720.
[135] A. Suzuki, T. Matushita, N. Wada, Y. Sakamoto and M. Okuda, Jpn. J. Appl. Phys. 35 (1996) L56.
[136] G.A. Hirata, J. McKittrick, J. Siqueiros, O.A. Lopez, T. Cheeks, O. Contreras and J.Y. Yi, J. Vac. Sci. Technol. A 14 (1996) 791.
[137] M. Hiramatsu, K. Imaeda, N. Horio and M. Nawata, J. Vac. Sci. Technol. A 16 (1998) 669.
[138] K. Matsubara, P. Fons, K. Iwata, A. Yamada, K. Sakurai, H. Tampo and S. Niki, Thin Solid Films 431–432 (2003) 369.
[139] A.V. Singh, R.M. Mehra, N. Buthrath, A. Wakahara and A. Yoshida, J. Appl. Phys. 90 (2001) 5661.
[140] K. Ellmer, J. Phys. D: Appl. Phys. 34 (2001) 3097.
[141] H. Tanaka, K. Ihara, T. Miyata, H. Sato and T. Minami, J. Vac. Sci. Technol. A 22 (2004) 1757.
[142] N. Naghavi, C. Marcel, L. Dupont, C. Guery, C. Maugy and J.M. Tarascon, Thin Solid Films 419 (2002) 160.
[143] E. Holmelund, J. Schou, S. Tougard and N.B. Larsen, Appl. Surf. Sci. 197–198 (2002) 467.
[144] H. Kim, J.S. Horwitz, W.H. Kim, A.J. Makinen, Z.H. Kafafi and D.B. Chrisey, Thin Solid Films 420–421 (2002) 539.
[145] S.B. Qadri, H. Kim, J.S. Horwitz and D.B. Chrisey, J. Appl. Phys. 88 (2000) 664.
[146] H. Kim, J.S. Horwitz, W.H. Kim, S.B. Qadri and Z.H. Kafafi, Appl. Phys. Lett. 83 (2003) 3809.
[147] K. Matsubara, P. Fons, K. Iwata, A. Yamada and S. Niki, Thin Solid Films 422 (2002) 176.
[148] H. Kim, J.S. Horwitz, S.B. Qadri and D.B. Chrisey, Thin Solid Films 420–421 (2002) 107.
[149] M. Karger and M. Schilling, Phys. Rev. B 71 (2005) 075304.
[150] M. Kumar, R.M. Mehra, A. Wakahara, M. Ishida and A. Yoshida, J. Appl. Phys. 93 (2003) 3837.
[151] S. Masuda, K. Kitamura, Y. Okumura, S. Miyatake, H. Tabata and T. Kawai, J. Appl. Phys. 93 (2003) 1624.
[152] Y. Kwon, Y. Li, Y.W. Heo, M. Jones, P.H. Holloway, D.P. Norton, Z.V. Park and S. Li, Appl. Phys. Lett. 84 (2004) 2685.
[153] M.J. Prins, S.E. Zinnemers, J.F.M. Cillesen and J.B. Giesbers, Appl. Phys. Lett. 70 (1997) 458.
[154] B. Thestrup, C.-D. Hansen, J. Schou and P.M. Johansen, J. Opt. A: Pure Appl. Opt. 2 (2002) 196.
[155] C. Bundesmann, N. Ashekenov, M. Schubert, D. Spemann, T. Butz, E.M. Kaidashev, M. Lorenz and M. Grundmann, Appl. Phys. Lett. 83 (2003) 1974.
[156] F.K. Shan and Y.S. Yu, Thin Solid Films 435 (2003) 174.

[157] K. Matsubara, H. Tampo, H. Shibata, A. Yamada, P. Fons, K. Iwata and S. Niki, Appl. Phys. Lett. 85 (2004) 1374.
[158] Y.W. Heo, S.J. Park, K. Ip, S.J. Pearton and D.P. Norton, Appl. Phys. Lett. 83 (2003) 1128.
[159] K. Ip, Y.W. Heo, K.H. Baik, D.P. Norton, S.J. Pearton and F. Ren, J. Vac. Sci. Technol. B 22 (2004) 171.
[160] K. Tamura, T. Makino, A. Tsukazaki, M. Sumiya, S. Fuke, T. Furumochi, M. Lippmaa, C.H. Chia, Y. Segawa, H. Koinuma and M. Kawasaki, Solid State Commun. 127 (2003) 265.
[161] H. Matsui, H. Saeki, T. Kawai, H. Tabata and B. Mizobuchi, J. Appl. Phys. 95 (2004) 5882.
[162] M. Joseph, H. Tabat and T. Kawai, Jpn. J. Appl. Phys. 38 (1999) L1205.
[163] H. Tabata, M. Saeki, S.L. Guo, J.H. Choi and T. Kawai, Physica B 308 (2001) 993.
[164] X.L. Guo, H. Tabata and T. Kawai, J. Cryst. Growth 237–239 (2002) 544.
[165] X.L. Guo, H. Tabata and T. Kawai, Optical Mater. 19 (2002) 229.
[166] M. Lorentz, H. Hochmuth, J. Lenzer, T. Nobis, G. Zimmermann, M. Diaconu, H. Schmidt, H.V. Wenckstern and M. Grundmann, Thin Solid Films 486 (2005) 205.
[167] A. Tsukazaki, H. Saito, K. Tamura, M. Ohtani, H. Koinuma, M. Sumiya, S. Fuke, T. Fukumura and M. Kawasaki, Appl. Phys. Lett. 81 (2002) 235.
[168] M. Sumiya, S. Fuke, A. Tsukazaki, K. Tamura, A. Ohtomo, M. Kawasaki and H. Koinuma, J. Appl. Phys. 93 (2003) 2562.
[169] T. Ohshima, T. Ikegami, K. Ebihara, J. Asmussen and R.K. Thareja, Thin Solid Films 435 (2003) 49.
[170] M. Joseph, H. Tabata, H. Saeki, K. Ueda and T. Kawai, Phys. B 302–303 (2001) 140.
[171] A. Suzuki, T. Matsushita, Y. Sakamoto, N. Wada, T. Fukuda, H. Fujiwara and M. Okuda, Jpn. J. Appl. Phys. 35 (1996) 5457.
[172] T. Makino, Y. Segawa, S. Yoshida, A. Tsukazaki, A. Ohtomo and M. Kawasaki, Appl. Phys. Lett. 85 (2004) 759.
[173] S.Y. Lee, E.S. Shim, H.S. Kang, S.S. Pang, and J.S. Kang, Thin Solid Films 473 (2005) 31.
[174] S. Kim, B.S. Kang, F. Ren, Y.W. Heo, K. Ip, D.P. Norton and S.J. Pearton, Appl. Phys. Lett. 84 (2004) 1904.
[175] Y.W. Heo, Y.W. Kwon, Y. Li, S.J. Pearton and D.P. Norton, Appl. Phys. Lett. 84 (2000) 3474.
[176] H. Yang, Y. Li, D.P. Norton, S.J. Pearton, S. Jung, F. Ren and L.A. Boatner, Appl. Phys. Lett. 86 (2005) 172103.
[177] K. Ip, Y.W. Heo, D.P. Norton, S.J. Pearton, J.R. Laroche and F. Ren, Appl. Phys. Lett. 85 (2004) 1169.
[178] Y.J. Li, Y.W. Heo, Y. Kwon, K. Ip, S.J. Pearton and D.P. Norton, Appl. Phys. Lett. 87 (2005) 072101.
[179] S.V. Prasad, J.J. Nainaparampil, J.S. Zabinski, J. Vac. Sci. Technol. A 20 (2002) 1738.
[180] S.V. Prasad and J.S. Zabinski, Wear 203 (1997) 498.

[181] Y.R. Ryu, S. Zhu, D.C. Look, J.M. Wrobel, H.M. Jeong and H.W. White, J. Cryst. Growth 216 (2000) 330.
[182] Y.R. Ryu, T.S. Lee and H.W. White, Appl. Phys. Lett. 83 (2003) 87.
[183] Y.R. Ryu, T.S. Lee, J.H. Leem and H.W. White, Appl. Phys. Lett. 83 (2003) 4032.
[184] V. Vaithianathan, B.T. Lee and S.S. Kim, Appl. Phys. Lett. 86 (2005) 062101.
[185] M. Joseph, H. Tabata and T. Kawai, Appl. Phys. Lett. 74 (1999) 2534.
[186] X.S. Wang, C.Z. Wu, J.F. Webb and Z.G. Liu, Appl. Phys A 77 (2003) 561.
[187] T. Nagata, T. Shimura, A. Ashida, N. Fujimura and T. Ito, J. Cryst. Growth 237–239 (2002) 533.
[188] T. Nagata, Y. Nakano, A. Ashida, N. Fujimura, T. Ito, Jpn. J. Appl. Phys. 40 (2001) 5615.
[189] T. Nagata, A. Ashida, N. Fujimura and T. Ito, Jpn. J. Appl. Phys. 41 (2002) 6916.
[190] C.H. Park, S.B. Zhang and S.H. Wei, Phys. Rev. B 66 (2002) 073202.
[191] H. Matsui, H. Saeki, H. Tabata and T. Kawai, Jpn. J. Appl. Phys. 42 (2003) 5494.
[192] S.J. Pearton, C.B. Abernathy, M.E. Overberg, G.T. Thaler, D.P. Norton, N. Theodoropulou, A.F. Hebarard, Y.D. Park, J. Kim and L.A. Boatner, J. Appl. Phys. 93 (2003) 1.
[193] M. Jelinek, A. Klini, T. Kocourek, R. Zeipl, A. Santoni, C. Fotakis and E. Kaminska, Surface and Coat. Technol. 200 (2005) 418.
[194] T. Aoki, Y. Shimizu, A. Miyake, A. Nakamura, Y. Nakanishi and Y. Hatanaka, Phys. Stat. Sol. (b) 229 (2002) 911.
[195] H. Kim, C.M. Gilmore, J.S. Horwitz, A. Pique, H. Murata, G.P. Kushto, R. Schlaf, Z.H. Kafafi and D.B. Chrisey, Appl. Phys. Lett. 76 (2000) 259.
[196] Y.R. Ryu, T.S. Lee and H.W. White, J. Cryst. Growth 261 (2004) 502.
[197] P. Bhattacharya, R.R. Das and R.S. Katiyar, Thin Solid Films 447 (2004) 564.
[198] I. Takeuchi, W. Yang, K.-S. Chang, M.A. Aronova, T. Venkatesan, R.D. Vispute and L.A. Bendersky, J. Appl. Phys. 94 (2003) 7336.
[199] A. Ohtomo, M. Kawasaki, T. Koida, K. Masubuchi, H. Koinuma, Y. Sakurai, Y. Yoshida, T. Yasuda and Y. Segawa, Appl. Phys. Lett. 72 (1998) 2466.
[200] W. Yang, R.D.Yispute, S. Choopun, R.P. Sharma, T. Venkatesan and H. Shen, Appl. Phys. Lett. 78 (2001) 2787.
[201] D. Spemann, E.M. Kaidashev, M. Lorenz, J. Vogt and T. Butz, Nuc. Instrum. and Methods in Phys. Res. B 219–220 (2004) 89.
[202] A.K. Sharma, J. Narayan, J.F. Muth, C.W. Teng, C. Jin, A. Kvit, R.M. Kolbas and O.W. Holland, Appl. Phys. Lett. 75 (1999) 3327.
[203] C.W. Teng, J.F. Muth, U. Ozgur, M.J. Bergmann, H.O. Everitt, A.K. Sharma, C. Jin, J. Narayan, Appl. Phys. Lett. 76 (2000) 979.
[204] R. Schmidt, B. Rheinlander, M. Schubert, D. Spemann, T. Butz, J. Lenzner, E.M. Kaidashev, M. Lorenz, A. Rahm, H.C. Semmelhack, M. Grundmann, Appl. Phys. Lett. 82 (2003) 2260.
[205] F.K. Shan, B.I. Kim, G.X. Liu, Z.F. Liu, J.Y. Sohn, W.J. Lee, B.C. Shin and Y.S. Yu, J. Appl. Phys. 95 (2004) 4772.

[206] Y. Matsumoto, M. Murakami, Z. Jin, A. Ohtomo, M. Lippmaa, M. Kawasaki and H. Koinuma, Jpn. J. Appl. Phys. 38 (1999) L603.
[207] P. Bhattacharya, R.R. Das and R.S. Katiyar, Appl. Phys. Lett. 83 (2003) 2010.
[208] W. Yang, S.S. Hullavarad, B. Nagaraj, I. Takeuchi, R.P. Sharma, T. Venkatesan, R.D. Vispute and H. Shen, Appl. Phys. Lett. 82 (2003) 3424.
[209] A. Ohtomo, R. Shiroki, I. Ohkubo, H. Koinuma and M. Kawasaki, Appl. Phys. Lett. 75 (1999) 4088.
[210] S. Choopun, R.D. Vispute, W. Yang, R.P. Sharma, T. Venkatesan and H. Shen, Appl. Phys. Lett. 80 (2002) 1529.
[211] P. Misra, P. Bhattacharya, K. Mallik, S. Rajagopalan, L.M. Kukreja and K.C. Rastogi, Solid State Commun. 117 (2001) 673.
[212] M. Kunisu, I. Tanaka, T. Yamamoto, T. Suga and T. Mizoguchi, J. Phys: Cond. Matter. 16 (2004) 3801.
[213] T. Makino, Y. Segawa, M. Kawasaki, A. Ohtomo, R. Shiroki, K. Tamura, T. Yasuda and H. Koinuma, Appl. Phys. Lett. 78 (2001) 1237.
[214] M. Lorenz, E.M. Kaidashev, H.V. Wenckstern, V. Riede, C. Bundesmann, D. Spemann, G. Benedorf, H. Hochmuth, A. Rahm, J.-C. Semmelhack and M. Grundmann, Solid State Electron. 47 (2003) 2205.
[215] P. Misra, P.K. Sahoo, P. Tripathi, V.N. Kulkarni, R.V. Nandekar and L.M. Kukreja, Appl. Phys. A 78 (2004) 37.
[216] Y.Z. Yoo, Z.W. Jin, T. Chikyow, F. Fukumura, M. Kawasaki and H. Koinuma, Appl. Phys. Lett. 81 (2002) 3798.
[217] J.A.V. Vechten and T.K. Bergstresse, Phys. Rev. B 1 (1970) 3351.
[218] Y. Nabetani, T. Mukawa, Y. Ito, T. Kato and T. Matsumato, Appl. Phys. Lett. 83 (2003) 1148.
[219] J. Narayan, A.K. Sharma, A. Kvit, C. Jin, J.F. Muth and O.W. Holland, Solid State Commun. 121 (2001) 9.
[220] K. Sakurai, T. Takagi, T. Kubo, D. Kajita, T. Tanabe, H. Takasu, Shizuo Fujita and Shigeo Fujita, J. Cryst. Growth 237–239 (2002) 514.
[221] R.K. Thareja, H. Saxena and V. Narayanan, J. Appl. Phys. 98 (2005) 034908.
[222] H. Matsui, H. Saeki, H. Tabata and T. Kawai, J. Electrochem. Soc. 159 (2003) G508.
[223] T. Dietl, H. Ohno, F. Matsukura, J. Cibert and D. Ferrand, Science 287 (2000) 1019.
[224] T. Fukumura, Z. Jin, A. Ohtomo, H. Koinuma and M. Kawasaki, Appl. Phys. Lett. 75 (1999) 3366.
[225] T. Fukumura, Z. Jin, M. Kawasaki, T. Shono, T. Hasegawa, S. Koshihara and H. Koinuma, Appl. Phys. Lett. 78 (2001) 958.
[226] K. Ueda, H. Tabata and T. Kawai, Appl. Phys. Lett. 79 (2001) 988.
[227] K. Ando, H. Saito, Z. Jin, T. Fukumura, M. Kawasaki, Y. Matsumoto and H. Koinuma, J. Appl. Phys. 89 (2001) 7284.
[228] K. Ando, H. Saito, Z. Jin, T. Fukumura, M. Kawasaki, Y. Matsumoto and H. Koinuma, Appl. Phys. Lett. 78 (2001) 2700.

[229] Z. Jin, T. Fukumura, M. Kawasaki, K. Ando, H. Saito, T. Sekiguchi, Y.Z. Zoo, M. Murakami, Y. Matsumoto, T. Hasegawa and H. Koinuma, Appl. Phys. Lett. 78 (2001) 3824.
[230] Y.Z. Yoo, T. Fukumura, Z. Jin, K. Hasegawa, M. Kawasaki, P. Ahmet, T. Chikyow, H. Koinuma, J. Appl. Phys. 90 (2001) 4246.
[231] T. Wakano, N. Fujimura, Y. Morinaga, N. Abe, A. Ashida and T. Ito, Physica E 10 (2001) 260.
[232] H. Saeki, H. Tabata and T. Kawai, Solid State Commun. 120 (2001) 439.
[233] S.W. Jung, S.J. An, G.C. Yi, C.U. Jung, S.I. Lee and S. Cho, Appl. Phys. Lett. 80 (2002) 4561.
[234] J.H. Kim, H. Kim, D. Kim, Y.E. Ihm, W.K. Choo, J. Appl. Phys. 92 (2002) 6066.
[235] K. Rode, A. Anane, R. Mattana, J.P. Contour, O. Durand, R. LeBourgeois, Appl. Phys. Lett. 93 (2003) 7676.
[236] J.H. Kim, H. Kim, D. Kim, Y.E. Ihm and W.K. Choo, Physica B 327 (2003) 304.
[237] S. Ramachandran, A. Tiwari and J. Narayan, Appl. Phys. Lett. 84 (2004) 5255.
[238] A. Fouchet, W. Prellier, P. Padhan, C. Simon, B. Mercey, V.N. Kulkarni and T. Venkatesan, Appl. Phys. Lett. 95 (2004) 7187.
[239] L. Yan, C.K. Ong and X.S. Rao, Appl. Phys. Lett. 96 (2004) 508.
[240] M. Venkatesan, C.B. Fitzgerald, J.G. Lunney and J.D.M. Coey, Phys. Rev. Lett. 93 (2004) 177206.
[241] Y. Zheng, J.C. Boulliard, D. Demaille, Y. Bernard and J.F. Petroff, J. Cryst. Growth 274 (2005) 156.
[242] P. Sharma, A. Gupta, K.V. Rao, F.J. Owens, R. Sharma, R. Ahuja, J.M.O. Guillen, B. Johansson and G.A. Gehring, Nature Material 2 (2003) 673.
[243] P. Sharma, A. Gupta, F.J. Owens, A. Inouea and K.V. Rao, J. Magnetism and Magnetic Mater. 282 (2004) 115.
[244] E.D. Posada, G. Tobin, E. McGlynn and J.G. Lunney, Appl. Surf. Sci. 208–209 (2003) 589.
[245] Y.W. Heo, M.P. Ivill, K. Ip, D.P. Norton, S.J. Pearton, J.G. Kelly, R. Rairigh, A.F. Hebard and T. Steiner, Appl. Phys. Lett. 84 (2004) 2292.
[246] S.S. Kim, J.H. Moon, B.T. Lee, O.S. Song and J.H. Je, J. Appl. Phys. 95 (2004) 454.
[247] M. Diaconu, H. Schmidt, H. Hochmuth, M. Lorenz, G. Benndorf, J. Lenzner, D. Spemann, A. Setzer, K.W. Nielsen, P. Esquinazi and M. Grundmann, Thin Solid Films 486 (2005) 117.
[248] A.K. Pradhan, K. Zhang, S. Mohanty, J.B. Dadson, D. Hunter, J. Zhang, D.J. Sellmyer, U.N. Roy, Y. Cui, A. Burger, S. Mathews, B. Joseph, B.R. Sekhar and B.K. Roul, Appl. Phys. Lett. 86 (2005) 152511.
[249] M. Ivill, S.J. Pearton, D.P. Norton, J. Kelly and A.F. Hebard, J. Appl. Phys. 97 (2005) 53904.
[250] N.H. Hong, V. Brize and J. Sakai, Appl. Phys. Lett. 86 (2005) 82505.
[251] N.H. Hong, J. Sakai and A. Hassini, J. Phys: Cond. Matter 17 (2005) 199.

[252] D.B. Buchholz, R.P.H. Chang, J.H. Song and J.B. Ketterson, Appl. Phys. Lett. 87 (2005) 82504.
[253] A. Tiwari, C. Jin, A. Kvit, D. Kumar, J.F. Muth and J. Narayan, Solid State Commun. 121 (2002) 371.
[254] W. Prellier, A. Fouchet, C. Simon and B. Mercey, Solid State Commun. 109 (2004) 192.
[255] J.H. Kim, H. Kim, D. Kim, Y.E. Ihm and W.K. Choo, J. Eur. Ceram. Soc. 24 (2004) 1847.
[256] H. Harima, J. Phys: Cond. Matter 16 (2004) S5653.
[257] N. Awasthi, P. Bhattacharya and R.S. Katiyar, Proc. Mat. Res. Soc. Symp. Vol. 825 (2004) G2.7.1.
[258] C. Bundesmann, N. Ashkenov, M. Schubert, D. Speman, T. Butz, E.M. Kaidashev, M. Lorenz and M. Grundmann, Appl. Phys. Lett. 83 (2003) 1974.
[259] Z. Jin, M. Murakami, T. Fukumura, Y. Matsumoto, A. Ohtomo, M. Kawasaki and H. Koinuma, J. Cryst. Growth 214–215 (2000) 55.
[260] M. Diaconu, H. Schmidt, A. Poppl, R. Bottcher, J. Hoentsch, A. Klunker, D. Spemann, H. Hochmuth, M. Lorenz and M. Grundmann, Phys. Rev. B 72 (2005) 085214.
[261] L.S. Dorneles, D.O. Mahony, C.B. Fitzerald, F. McGee, M. Venkatesan, I. Stanca, J.G. Lunney and J.M.D. Coey, Appl. Surf. Sci. 248 (2005) 406.
[262] J.H. Kim, H. Kim, D. Kim, S.G. Yoon, W.K. Choo, Solid State Commun. 131 (2004) 677.
[263] M.N. Armenise, A.M. Matteo and V.M.N. Passaro, Proc. SPIE 1583 (1991) 289.
[264] W.C. Shin, S.M. Wu and T. Shiosaki, IEEE Trans. Ultrasons. Ferro. and Freq. Control 40 (1993) 642.
[265] T. Shiosaki, N. Kitamura and A. Kawabata, IEEE 7th Int. Symp. on Applications of Ferroelectrics (Urbana Champaign, IL, 6–8 June 1990), New York: IEEE, p.296.
[266] S. Valette, J. Lizet, P. Mottier, J.P. Jadot, S. Renard, A. Fournier, A.M. Grouitlet, P. Gidon and H. Denis, Electron. Lett. 19 (1983) 883.
[267] M. Kamauchi, T. Shiosaki and A. Kawabata, IEEE 7th Int. Symp. on Applications of Ferroelectrics (Urbana Champaign, IL, 6–8 June 1990), New York: IEEE, p.733.
[268] N. Ashkenov, M. Schubert, E. Twerdowski, H.V. Wenckstern, B.N. Mbenkum, H. Hochmuth, M. Lorenz, W. Grill and M. Grundmann, Thin Solid Films 486 (2005) 153.
[269] W.S. Hu, Z.G. Liu, R.X. Wu, Y.F. Chen, W. Ji, T. Yu and D. Feng, Appl. Phys. Lett. 71 (1997) 548.
[270] H. Tanaka, T. Shimakawa, T. Miyata, H. Sato and T. Minami, Appl. Surf. Sci. 244 (2005) 568.
[271] R.D. Vispute, V. Talyansky, Z. Trajanovic, S. Choopun, M. Downes, R.P. Sharma, T. Venkatesan, M.C. Woods, R.T. Lareau, K.A. Jones and A.A. Iliadis, Appl. Phys. Lett. 70 (1997) 2735.
[272] J. Narayan, K. Dovidenko, A.K. Sharma and S. Oktyabrsky, J. Appl. Phys. 84 (1998) 2597.

[273] R.D. Vispute, V. Talyansky, R.P. Sharma, S. Choopun, M. Downes, T. Venkatesan, Y.X. Li, L.G. Salamanca-Riba, A.A. Iliadis, K.A. Jones and J. McGarrity, Appl. Surf. Sci. 127–129 (1998) 431.
[274] R.P. Wang, H. Muto Y. Yamada and T. Kusumori, Thin Solid Films 411 (2002) 69.
[275] X.W. Sun, R.F. Xiao and H.S. Kwok, J. Appl. Phys. 84 (1998) 5776.
[276] T.F. Huang, T. Ueda, S. Spruytte and J.S. Harris Jr., IEEE Intl. Symp. on Compound Semiconductor, Sept. 1997, pp.11, Digital object identifier 10.1109/ISCS.1998.711531.
[277] R.F. Xiao, X.W. Sun, Z.F. Li, N. Cue, H.S. Kwok, Q.Z. Liu and S.S. Lau, J. Vac. Sci. Technol. A 15 (1997) 2207.
[278] Makino, N.T. Tuan, C.H. Chia, Y. Segawa, M. Kawasaki, A. Ohtomo, K. Tamura and H. Koinuma, Int. J. Mod. Phys. B 15 (2001) 3853.
[279] A. Ohtomo, M. Kawasaki, Y. Sakurai, I. Ohkubo, R. Shiroki, Y. Yoshida, T. Yasuda, Y. Segawa and H. Koinuma, Mater. Sci. Eng. B 56 (1998) 263.
[280] C.H. Chia, T. Makino, K. Tamura, Y. Segawa, M. Kawasaki, A. Ohtomo and H. Koinuma, Appl. Phys. Lett. 82 (2003) 1848.
[281] T. Makino, C.H. Chia, N.T. Tuan, Y. Segawa, M. Kawasaki, A. Ohtomo, K. Tamura and H. Koinuma, Appl. Phys. Lett. 77 (2000) 1632.
[282] T. Edahiro, N. Fujimura and T. Ito, J. Appl. Phys. 93 (2003) 7673.
[283] S. Krishnamoorthy, A.A. Iliadis, A. Inumpudi, S. Choopun, D. Ratnakar, D. Vispute and T. Venkatesan, Solid State Electronics 46 (2002) 1633.
[284] T. Aoki, Y. Hatanaka and D.C. Look, Appl. Phys. Lett. 76 (2000) 3257.
[285] J.F. Muth, C.W. Teng, A.K. Sharma, A. Kvit, R.M. Kolbas and J. Narayan, Mat. Res. Soc. Symp. 617 (2000) j6.7.1.
[286] H.D. Sun, Y. Segawa, M. Kawasaki, A. Ohtomo, K. Tamura and H. Koinuma, J. Appl. Phys. 91 (2002) 6457.
[287] T. Makino, K. Tamura, C.H. Chia, Y. Segawa, M. Kawasaki, A. Ohtomo and H. Koinuma, Phys. Stat. Sol. (b) 229 (2002) 853.
[288] D.C. Look, B. Claffin, Y.I. Alivov and S.J. Park, Phys. Stat. Sol. (a) 201 (2004) 2203.
[289] H. Ohta, K. Kawamura, M. Orita, M. Hirano, N. Sarukura and H. Hosono, Appl. Phys. Lett. 77 (2000) 475.
[290] H. Ohta, M. Hirano, K. Nakahara, H. Maruta, T. Tanabe, M. Kamiya, T. Kamiya and H. Hosono, Appl. Phys. Lett. 83 (2003) 1029.
[291] A. Tiwari, C. Jin, D. Kumar and J. Narayan, Appl. Phys. Lett. 83 (2003) 1773.
[292] S. Barik, A.K. Srivastava, P. Misra, R.V. Nandedkar and L.M. Kukreja, Solid State Commun. 127 (2003) 463.
[293] C. Jin, R. Narayan, A. Tiwari, H. Zhou, A. Kvit and J. Narayan, J. Mater. Sci. Eng. B 117 (2005) 348.
[294] Y.I. Alivov, J.E.V. Nostrand, D.C. Look, M.V. Chukichev and B.M. Ataev, Appl. Phys. Lett. 83 (2003) 2943.
[295] Y.I. Alivov, E.V. Kalinina, A.E. Cherenkov, D.C. Look, B.M. Ataev, A.K. Omaev, M.V. Chukichev and D.M. Bagnall, Appl. Phys. Lett. 83 (2003) 4719.

[296] R.D. Vispute, V. Talyansky, S. Choopun, R.P. Sharma, T. Venkatesan, M. He, X. Tang, J.B. Halpern, M.G. Spencer, Y.X. Li, L.G. Salamanca-Riba, A.A. Lliadis and K.A. Jones, Appl. Phys. Lett. 73 (1998) 348.
[297] H.D. Sun, T. Makino, N.T. Tuan, Y. Segawa, Z.K. Tang, G.K.L. Wong, M. Kawasaki, A. Ohtomo, K. Tamura and H. Koinuma, Appl. Phys. Lett. 77 (2000) 4250.
[298] T. Makino, C.H. Chia, N.T. Tuan, H.D. Sun, Y. Segawa, M. Kawasaki, A. Ohtomo, K. Tamura and H. Koinuma, Appl. Phys. Lett. 77 (2000) 975.
[299] T. Makino, N.T. Tuan, H.D. Sun, C.H. China, Y. Segawa, M. Kawasaki, A. Ohtomo, K. Tamura, T. Suemoto, H. Akiyama, M. Baba, T. Tomita and H. Koinuma, Appl. Phys. Lett. 78 (2000) 1979.
[300] E.R. Segnit and A.E. Holland, J. Am. Ceram. Soc. 48 (1965) 412.
[301] M. Sanati, G.L.W. Hartand and A. Zunger, Phys. Rev. B 68 (2003) 155210.
[302] B.N. Mbenkum, N. Ashkenov, M. Schubert, M. Lorenz, H. Hochmuth, D. Michel, M. Grundmann and G. Wagner, Appl. Phys. Lett. 86 (2005) 091904.
[303] Z.L. Wang, Mater. Today 7 (2004) 26.
[304] G.-C. Yi, C.W. Wang and W. Park II, Semicond. Sci. Technol. 20 (2005) S22.
[305] M.H. Huang, S. Mao, H. Feick, H. Yan, Y. Wu, H. Kind, E. Weber, R. Russo and P. Yang, Science 292 (2001) 1897.
[306] P.X. Gao, Y. Ding, W. Mai, W.L. Hughes, C. Lao, Z.L. Wang, Science 309 (2005) 1700.
[307] S. Kumar, V. Gupta and K. Sreenivas, Nanotechnology 16 (2005) 1167.
[308] H.J. Fan, R. Scholz, A. Dadgar, A. Krost and M. Zacharias, Appl. Phys. A 80 (2005) 457.
[309] M. Kawakami, A.B. Hartano, Y. Nakata and T. Okada, Jpn. J. Appl. Phys. 42 (2003) L33.
[310] T. Nobis, E.M. Kaidashev, A. Rahm, M. Lorentz and M. Grundmann, Phys. Rev. Lett. 93 (2004) 103903.
[311] V. Gupta, P. Bhattacharya, Y.I. Yuzuk and R.S. Katiyar, Mat. Res. Soc. Symp. Proc. Vol. 818 (2004) M8.26.1.
[312] Z. Liu, D. Zhang, C. Li and C. Zhou, Proc. 3rd IEEE Conf. on Nanotechnology, 2003, Vol. 2, pp. 592, Digital object identifier 10.1109/NANO.2003.1230980.
[313] T. Nobis, E.M. Kaidashev, A. Rahm, M. Lorentz, J. Lenzner and M. Grundman, Nano Letters 4 (2004) 797.
[314] M. Lorentz, E.M. Kaidashev, A. Rahm, T. Nobis, J. Lenzner, G. Wagner, D. Specmann, H. Hochmuth and M. Grundmann, Appl. Phys. Lett. 86 (2005) 143113
[315] M. Yan, H.T. Zhang, E.J. Widjaja and R.P.H. Chang, J. Appl. Phys. 94 (2003) 5240.
[316] B.H. Agung, M. Kawakami, Y. Nakata, X. Ning and T. Okada, The 5th Pacific Rim Conference on Laser and Electro-optics, 2003, Vol. 2, pp. 667, Digital object identifier 10.1109/CLEOPR.2003.1277210.
[317] I. Ozerov, D. Nelson, A.V. Bulgakov, W. Marine and M. Sentis, Appl. Surf. Sci 212–213 (2003) 349.
[318] I. Ozerov, F. Chabre and W. Marine, Mater. Sci. Engg. C (2005) (in press).

[319] P. Verardi and M. Dinescu, Proc. of IEEE Ultrason. Symp. 1995, Vol. 2, pp. 1015. Digital object identifier 10.1109/ULTSYM.1995.495735.
[320] C.W. Seabury, J.T. Cheung, P.H. Kobrin and R. Addison, Proc. IEEE Ultrason. Symp. Proc. 1995, pp. 909.
[321] H. Cao, J.Y. Yu, H.C. Ong, J.Y. Dai and R.P.H. Change, Appl. Phys. Lett. 73 (1998) 572.
[322] G. Wang, G.T. Kiehne, G.K.L. Wong, J.B. Ketterson, X. Liu and R.P.H. Change, Appl. Phys. Lett. 80 (2002) 401.
[323] G.P. Agarwal, Nonlinear fiber optics (Academic, New York, 2001).
[324] R. Adair, L.L. Chase and S.A. Payne, Phys. Rev. B 39 (1989) 3337.
[325] T. Mazingue, L. Escoubas, L. Spalluto, F. Flory, G. Socol, C. Ristoscu, E. Axente, S. Grigorescu, I.N. Mihailescu and N.A. Vainos, J. Appl. Phys. 98 (2005) 074312.
[326] F. Ding, Z. Fu and Q. Qin, Electrochemical and Solid-State Letters 2 (1099) 418.
[327] A. Onodera, N. Tamaki, K. Jin and H. Yamashita, Jpn. J. Appl. Phys. 36 (1997) 6008.
[328] S. Komura, T. Katsumata, T. Morikawa, X. Zhao, H. Isshiki and Y. Aoyagi, J. Appl. Phys. 88 (2000) 7129.
[329] M. Ishil, S. Komuro, T. Morikawa and Y. Aoyagi, J. Appl. Phys. 89 (2001) 3679.
[330] R.P. Casero, A.G. Llorente, O.P.Y. Moll, W. Seiler, R.M. Defoureau, D. Defourneau, E. Millon, J. Perriere, P. Goldner and B. Viana, J. Appl. Phys. 97 (2005) 54905.
[331] Y.E. Lee, D.P. Norton, C. Park and C.M. Rouleau, J. Appl. Phys. 89 (2001) 1653.
[332] H. Hayashi, A. Ishizaka, M. Haemori and H. Koinuma, Appl. Phys. Lett. 82 (2003) 1365.
[333] H. Kim, J.S. Horwitz, A. Piqué, H.S. Newman, P. Lubitz, M.M. Miller, D.L. Knies, and D.B. Chrisey, J. Vac. Sci. Technol. A 17 (1999) 3111.
[334] R.F. Xiao, H.B. Liao, N. Cue, X.W. Sun and H.S. Kwok, J. Appl. Phys. 80 (1996) 4226.

Chapter 5

Optical Properties of ZnO and Related Alloys

Ü. Özgür and H. Morkoç

Department of Electrical Engineering and Physics Department, Virginia Commonwealth University, Richmond, VA 23284-3072, USA

5.1 Introduction

Optical properties and processes in ZnO as well as its refractive index were extensively studied many decades ago.[1,2,3,4,5,6,7,8,9,10,11,12,13,14,15,16,17,18,19] The renewed interest in ZnO is fuelled and fanned by its prospects in optoelectronics applications owing to its direct wide band gap ($E_g \sim 3.3$ eV at 300 K) and efficient radiative recombination. The large exciton binding energy of ~60 meV, references 20 and 21, paves the way for an intense near band edge excitonic emission at room and even higher temperatures, because this value is 2.4 times that of the room temperature (RT) thermal energy ($k_BT = 25$ meV). Therefore, laser operation based on excitonic transitions, as opposed to electron-hole plasma, is expected. In this vein, there have also been a number of reports on laser emission from ZnO-based structures at RT and beyond.

The optical properties of a semiconductor have their genesis in both intrinsic and extrinsic effects. Intrinsic optical transitions take place between the electrons in the conduction band and holes in the valence band, including excitonic effects due to the Coulomb interaction. Excitons are classified into free and bound excitons. In high-quality samples with low impurity concentrations, the free excitons can also exhibit excited states, in addition to their ground-state transitions. Extrinsic properties are related to dopants/impurities or point defects and complexes, which usually create electronic states in the bandgap, and therefore influence both optical absorption and emission processes. The electronic states of the bound excitons, which may be bound to neutral or charged donors and acceptors, depend strongly on the semiconductor material, in particular the band structure. For a shallow neutral donor

Zinc Oxide Bulk, Thin Films and Nanostructures
C. Jagadish and S. Pearton (Editors)

bound exciton (DBE), for example, the two electrons in the bound exciton state are assumed to pair off into a two-electron state with zero spin. The additional hole is then weakly bound in the net hole-attractive Coulomb potential set up by this bound two-electron aggregate. Similarly, neutral shallow acceptor bound excitons (ABE) are expected to have a two-hole state derived from the topmost valence band and one electron interaction. Other extrinsic transitions could be seen in optical spectra such as free-to-bound (electron-acceptor), bound-to-bound (donor-acceptor) and the so-called yellow/green luminescence. The well-known green band in ZnO luminescence spectra (manifesting itself as a broad peak around 500–530 nm), observed nearly in all samples regardless of growth conditions, is related to singly ionized oxygen vacancies by some and to residual copper impurities by others. Simply, a requisite consensus on this issue is still lacking.

5.2 Optical transitions in ZnO

Optical transitions in ZnO have been studied by a variety of experimental techniques such as optical absorption, transmission, reflection, photoreflection, spectroscopic ellipsometry, photoluminescence (PL), cathodoluminescence (CL), calorimetric spectroscopy, etc. The luminescence from bulk ZnO extends from the band edge to the green/orange spectral range with a common broad band centered around 2.45 eV. The sharp lines dominating the band edge region of the spectra originate from various bound exciton recombinations (excitons bound to neutral donors, D^0X, and/or acceptors, A^0X) followed by longitudinal optical (LO) phonon replicas with an energy separation of 72 meV. On the high energy side of the bound excitons, free exciton transitions appear with the A-valence band (FX_A) positioned at 3.377 eV. At lower energies from 3.34 to 3.31 eV two-electron satellite (TES) recombination lines of the neutral donor bound excitons are observed. In some samples a donor-acceptor-pair (DAP) transition at ~3.22 eV that is again followed by phonon replicas is found even though the chemical identity of the acceptor is unknown. In this chapter, a review of all these transitions is given. Additionally, the refractive index of ZnO and MgZnO and stimulated emission and carrier dynamics in ZnO are discussed.

5.2.1 *Free excitons and polaritons*

The wurtzite ZnO conduction band is mainly constructed from the s-like state having Γ_7^c symmetry, whereas the valence band is a p-like state, which is split into three bands due to the influence of crystal field and spin-orbit interactions.[22,23] The near bandgap intrinsic absorption and emission spectrum is therefore dominated by transitions involving these three valence bands and the conduction band. The related free-exciton transitions from the conduction band to these three valence bands or vice versa are usually denoted by A (also referred to as the heavy hole), B (also referred to as the light hole), and C (also referred to as the crystal-field split band). By treating the wurtzite energy levels as a perturbation of those in zincblende, Hopfield[3] has derived the formulae for the valence band mixing, the extent of which is controlled by the relative magnitudes of the spin-orbit and crystal field splittings. If the energy of the top most valence band is taken as zero ($E_A = 0$), the energies of the two other exciton lines in this formalism are given by

$$E_{B,C} = -\frac{\delta + \Delta}{2} \pm \sqrt{\left[\left(\frac{\delta + \Delta}{2}\right)^2 - \frac{2}{3}\delta\Delta\right]} \tag{5.1}$$

where δ and Δ represent the spin orbit and crystal field parameters, respectively, which can be calculated by using the measured energy level splittings E_{AB} and E_{BC}.

Figure 5.1 shows the band diagram for the zincblende and the wurtzite ZnO structures derived by Birman[23] who suggests the A-Γ_9, B-Γ_7, and C-Γ_7 valence band ordering for wurtzitic ZnO. However, the ordering of the crystal-field and spin-orbit coupling split states of the valence-band maximum in wurtzite ZnO has been a subject of controversy since 1960s.[2,3,4,5,6,21,22,24,25] Thomas[2] was the first to propose, based on the polarization dependence of the absorption and reflectivity spectra, that the valence-band ordering of ZnO is anomalous compared to the usual one in other II-VI wurtzite materials, and that the top of the valence band has A-Γ_7 symmetry, which can be understood in terms of an effective negative spin-orbit splitting. These results were later confirmed by Liang and Yoffe,[12] however, challenged by Park *et al.*[4] The essential difference has its genesis in the interpretation of the spectral line, which was initially assigned to the intrinsic ground state A-exciton transition by Thomas,[2] and in contrast to extrinsic, ionized donor-bound exciton complex transition by Park *et al.*[4] The free-exciton nature of this absorption line was later confirmed by Segall,[5] by a study of the onset of phonon-assisted absorption. Usually, the polarization dependence of absorption, reflection, or PL by individually does not provide sufficient information to determine the valence band symmetry. In contrast, angular dependent magneto-optical spectroscopy is accepted as a more reliable tool.[21,24,26] Recent availability of high quality ZnO single crystals has paved the way to observe intrinsic exciton transitions in low temperature

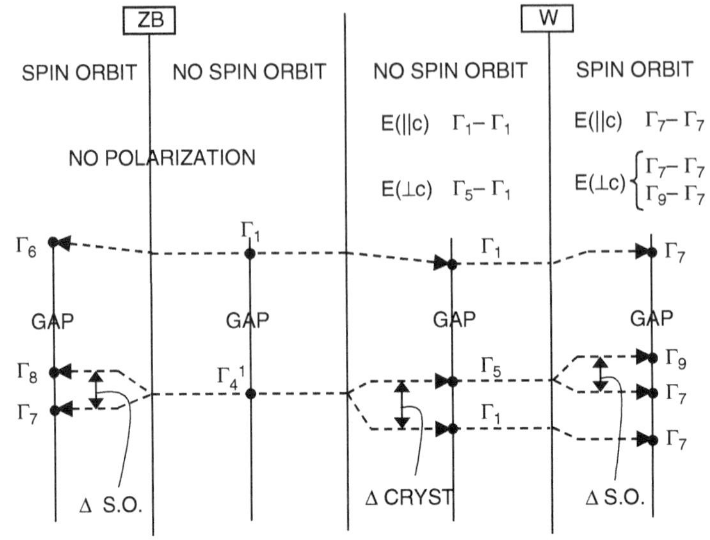

Figure 5.1: Band structure and selection rules for ZB and W structures. Crystal field and spin-orbit splittings are indicated schematically. Transitions which are allowed for various polarizations of photon electric field vector with respect to the *c*-axis are indicated. (After Birman[23])

PL, magneto-luminescence, and reflectance measurements.[21,24,26,27,28] Reynolds *et al.*[21] readdressed this band ordering issue using polarized magnetoluminescence and second-order PL spectra, which helped resolve the additional fine structure of the excitons. They concluded that the spectral line identified as an ionized DBE complex transition by Park *et al.* was indeed a free-exciton line, but that, nevertheless, the valence-band symmetry ordering in ZnO is not reversed but the same as that observed in most other wurtzitic II-VI structures and GaN (A-Γ_9, B-Γ_7, and C-Γ_7). Supporting these conclusions were calculations of energy positions and oscillator strengths of respective excitons by Gil,[29] who showed that the reversal of this ordering would require ZnO films with high biaxial tension. However, Lambrecht *et al.*[24] argued that the analysis of Reynolds *et al.* was not well taken owing to the assumptions (by Reynolds *et al.*) about the sign of the Landé *g* factor for holes and that the valence band ordering is in fact A-Γ_7, B-Γ_9, and C-Γ_7. The valence band ordering issue still remains controversial even today although most papers adopted the interpretation of Thomas.[2]

Group theory arguments and the direct product of the group representations of the band symmetries (Γ_7 for the conduction band, Γ_9 for the A valence band, upper Γ_7 for B valence band, and lower Γ_7 for C valence band) will results in the following intrinsic exciton ground states symmetries:

$$\Gamma_7 x \Gamma_9 \rightarrow \Gamma_5 + \Gamma_6,$$
$$\Gamma_7 x \Gamma_7 \rightarrow \Gamma_5 + \Gamma_1 + \Gamma_2.$$

The Γ_5 and Γ_6 exciton ground states are both doubly degenerate, whereas Γ_1 and Γ_2 are both singly degenerate. In ideal, i.e. strain free, wurtzite crystals free excitons should obey the selection rules in optical one-photon processes; Γ_5 and Γ_1 are allowed transitions with E ⊥ c and E || c polarizations, respectively, but the Γ_6 and Γ_2 are not allowed. Therefore, all three excitons are allowed in the σ polarization (E ⊥ c and k ⊥ c), but the C-exciton is quite weak. The C-exciton is strongly allowed in the π polarization (E || c and k ⊥ c). However, the A-exciton is forbidden and the B-exciton is only weakly observable in this geometry. In the α polarization (E ⊥ c and k || c) all three transitions are clearly observable. Each of these fundamental excitonic states are expected to have a fine structure due both to exciton-polariton longitudinal-transverse splitting and the splitting caused by electron-hole exchange interaction, which are on the order of a few meV.[16,30]

The transition energies of the intrinsic excitons were measured by employing the low temperature absorption,[12,31] reflection,[3,21,32] photoreflectance (PR),[33,34] and PL[21,27,28,32] spectroscopy techniques. These measurements are useful for the determination of exciton binding energies, exciton Bohr radii, the dielectric constant, and with the aid of the quasi-cubic model, spin-orbit and crystal field parameters. Using the ground state and excited state energies and assuming that the exciton has a hydrogen-like set of energy levels, the exciton binding energy can be calculated from

$$E_n = E_{gap} - \frac{E_B}{n^2}, \tag{5.2}$$

where n is the main quantum number and $E_B = 4/3(E_2 - E_1)$ is the exciton binding energy. E_B can also be expressed in terms of the low-frequency dielectric constant ε_0 and the reduced exciton mass μ^*_{ex} as

$$E_B = \frac{e^4 \mu^*_{ex}}{2\hbar^2 \varepsilon_0^2} = 13.6 \frac{\mu^*_{ex}}{\varepsilon_0^2} (\text{eV}). \tag{5.3}$$

Using μ^*_{ex}, one can also estimate the effective exciton Bohr radius from

$$a_0 = \frac{\hbar^2 \varepsilon_0}{\mu^*_{ex} e^2} n^2. \tag{5.4}$$

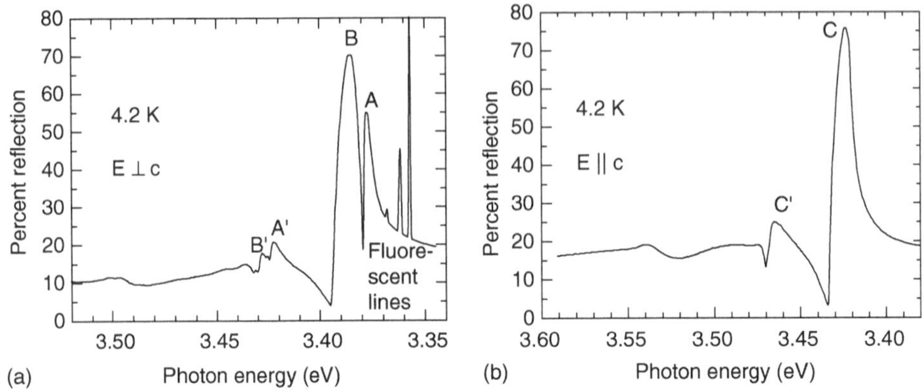

Figure 5.2: Reflection from ZnO at 4.2 K for (a) E ⊥ c and (b) E || c. Notice the PL lines for E ⊥ c data. (After Thomas[2])

The fundamental absorption edge and the exciton structure of ZnO was first studied by Thomas[2] in 1960 who measured the low temperature absorption and reflectance spectra of polished and etched thin crystals grown from the vapor phase. Thomas analyzed the reflectance data by means of the Kramers-Kronig relation and three peaks arising from $n = 1$ exciton transitions involving holes from each one of the *three* valence bands. At 4.2 K the $n = 2$ exciton states were apparent in reflection, allowing estimates to be made of the exciton binding energy and reduced mass. An example of reflectance spectra obtained in the σ and π polarizations is given in Figure 5.2 for a single crystal ZnO sample.

Park *et al.*[4] reported the fundamental spectra of ZnO and successfully explained the exciton spectrum on the basis of Birman's model[23] of the wurtzite structure analyzing the energy and polarization of the fine structure around the fundamental absorption region. Motivated by these results, Hopfield and Thomas[16] studied the polariton absorption lines in ZnO using a magnetic field in appropriate geometry to allow couplings at polariton crossings. A satisfactory agreement was attained when the crossing frequencies of the polariton dispersion and light with E || c and E ⊥ c polarizations were compared with the measured absorption line energies. Based on the results Hopfield and Thomas also argued that the B-exciton in fact belongs to the Γ_9 symmetry group.

Among these reports, there exists a discrepancy in the quoted values for the spin-orbit interaction energy and the exciton binding energies, which most likely may have its genesis in the difficulty of interpreting the complex spectra near the band edge. Table 5.1 lists some of the excitonic parameters for ZnO from various references. The features associated with A-, B-, and C-exciton transitions, their excited states, exciton-polaritons, and exciton-LO phonon complexes are

Table 5.1: Some excitonic parameters for ZnO.

Parameter	
E_{gap} (eV)	3.436[a], 3.435[b], 3.438[e], 3.437[h], 3.445[i]
A-exciton binding energy, E_B^A (eV)	0.0609[a], 0.042[b], 0.060[c], 0.064[e], 0.0604[f], 0.0595[g], 0.060[h], 0.067[i]
B-exciton binding energy, E_B^B (eV)	0.0593[a], 0.053[c], 0.053[e], 0.0575[f], 0.0566[g], 0.062[i]
C-exciton binding energy, E_B^C (eV)	0.058[a], 0.049[c], 0.0561[f], 0.0563[g], 0.061[i]
E_{AB} (eV)	0.0062[a], 0.040[b], 0.0095[d], 0.006[e], 0.0127[h], 0.0075[i]
E_{BC} (eV)	0.0385[a], 0.037[b], 0.0397[d], 0.0408[i]
Spin-orbit parameter, δ (eV)	−0.0087[a], 0.014[c], 0.016[d]
Crystal field parameter, Δ (eV)	0.0410[a], 0.048[c], 0.043[d]
Exciton effective mass, μ_{ex}^* (m_0)	0.31[a], 0.20[b]
Exciton Bohr radius, a_0 (Å)	14[a], 24[b]
Biexciton binding energy, E_B^{XX} (eV)	14.7[j], 15[k]

[a] Thomas[2] (needles grown from vapor phase).
[b] Park *et al.*[4] (platelets and bulk crystals grown from the vapor phase).
[c] Chichibu *et al.*[33] (bulk crystals grown by chemical vapor deposition).
[d] Reynolds *et al.*[21] (bulk crystals grown by a seeded physical vapor transport method).
[e] Thonke *et al.*[27] (bulk crystals grown by seeded chemical vapor transport).
[f] Reynolds *et al.*[25] (melt grown bulk crystals).
[g] Reynolds *et al.*[25] (vapor grown bulk crystals)
[h] Teke *et al.*[32] (melt grown bulk crystals).
[i] Liang and Yoffe[12] (platelets grown by vapor transport).
[j] Hvam.[120]
[j] Yamamoto *et al.*[119] (epitaxial films grown by plasma-MBE).

sometimes misinterpreted resulting in reports of different energy values. To shed additional light, Liang and Yoffe[12] performed polarization dependent transmission measurements on single crystal ZnO thin platelets grown by the vapor transport method. In addition to the transitions associated with the ground, the first, and the second excited states of intrinsic excitons, the exciton-phonon complexes were also observed in the transmission spectrum. In the same vein but later on, Reynolds *et al.*[35] reported the shift in intrinsic exciton transitions analyzing the reflection data from strained ZnO single crystals grown by the vapor transport method. During growth, defect complexes (probably pairs) are preferentially incorporated in certain crystallographic orientations producing compressive strain in the vicinity of the defect pairs with the strain being oriented in the direction of the pair. By observing the changes in emission from the defect pairs as the annealing temperature is increased, Reynolds *et al.*[35] investigated the relaxation of strain following an increase in the number of nearer pairs. Intrinsic energy bands were observed to adjust to these changes by moving to lower energies. For an annealing temperature of 800°C, shifts as much as 0.4 meV for the A-exciton transition and 3 meV for the B-exciton transition were quoted for σ polarization at 2 K.

Chichibu *et al.*[34] studied the photoreflectance spectra of high quality ZnO epitaxial layers grown on $ScAlMgO_4$ substrates by laser MBE and observed clear excitonic resonances due to three excitons. A larger broadening was observed for the B- and C-excitons compared to the A-exciton, which was interpreted to be due to a contribution by the exciton-polaritons in terms of large longitudinal-transverse splitting of respective excitons. Chichibu *et al.*[33] also used polarized photoreflectance spectroscopy to study excitonic polaritons in a ZnO single crystal. Following those observations of longitudinal-transverse splitting of ground-state exciton polaritons and resonances due to the first excited states of respective excitons, the authors confirmed the valence-band ordering to be A-Γ_{9v}, B-Γ_{7v}^{u}, and C-Γ_{7v}^{l}. The observed longitudinal-transverse splitting energies (1.5, 11.1, and 13.9 meV for the A-, B-, and C-excitons, respectively) were in good agreement with earlier results by Hümmer and Gebhardt[36] (2, 11, and 16 meV for A-, B-, and C-excitons, respectively).

By transmission measurements Muth *et al.*[31] determined the absorption coefficient, band gap, and exciton binding energies in epitaxial zinc oxide thin films grown by pulsed laser deposition on *c*-plane sapphire substrates. It was found that the excitonic absorption features were sharper in films annealed in oxygen. The absorption coefficient was consistent with the earlier report on thin ZnO platelets by Liang and Yoffe.[12] However, assignments by Muth *et al.* shown in Figure 5.3 for the B- and C-exciton energies are very different from those in other reports most probably due to misinterpretation of the C-exciton as the B-exciton and the exciton-LO phonon complex transitions[12,37] as the C-exciton. The absorption coefficient at the A-exciton transition energy is $\sim 2 \times 10^5\ \mathrm{cm}^{-1}$ at room temperature and increases to $\sim 2.9 \times 10^5\ \mathrm{cm}^{-1}$ at 77 K.

Another very powerful technique for studying exciton structure is PL. The low temperature PL spectrum of ZnO single crystal has been investigated in many aspects by many researchers.[21,27,28,30,32,35,38,39,40,41] Figure 5.4 shows a typical PL spectrum in the energy range of fundamental excitonic region taken at 10 K by Teke *et al.*[32] in the $E \perp c$ polarization geometry for a high quality ZnO crystal annealed under forming gas (5% H_2 and 95% N_2). The A-free excitons and their first excited state transitions are observed at $FX_A^{n=1} = 3.3771$ eV and $FX_A^{n=2} = 3.4220$ eV, respectively, for Γ_5 (3.3757 eV and 3.4202 eV for Γ_6) band

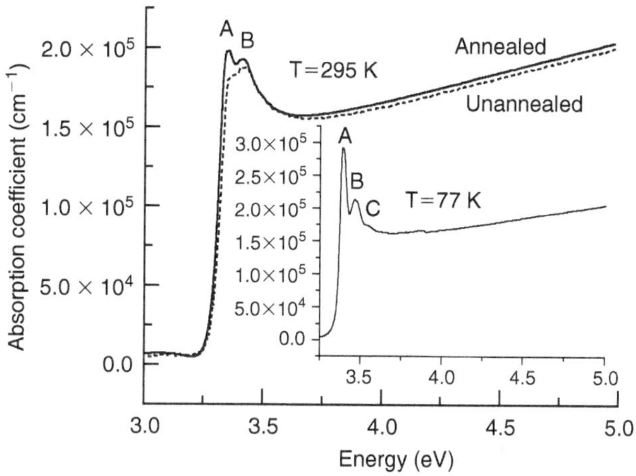

Figure 5.3: Absorption coefficient and excitonic structure for an annealed (solid lines) and an unannealed (dotted line) sample at room temperature. The inset shows the absorption coefficient for the annealed sample at 77 K. (After Muth *et al.*[31]) The C-exciton is misinterpreted as the B-exciton and the exciton-LO phonon complex transitions as the C exciton.

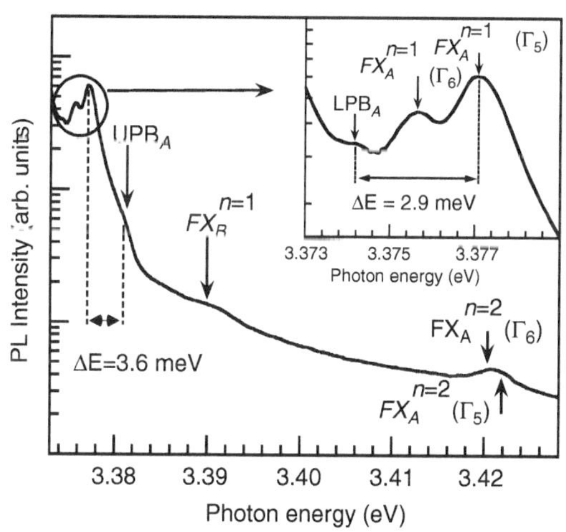

Figure 5.4: Free excitonic fine structure region of the 10 K PL spectrum for a forming gas annealed ZnO substrate. (After Teke *et al.*[32])

symmetry. Although, Γ_6 exciton is forbidden at $k = 0$ in the measurement mode of polarization, it is still observed evidently due to the fact that the photon has a finite momentum. Geometrical effects such as not having the sample oriented exactly perpendicular to electric field might also be responsible for the observation

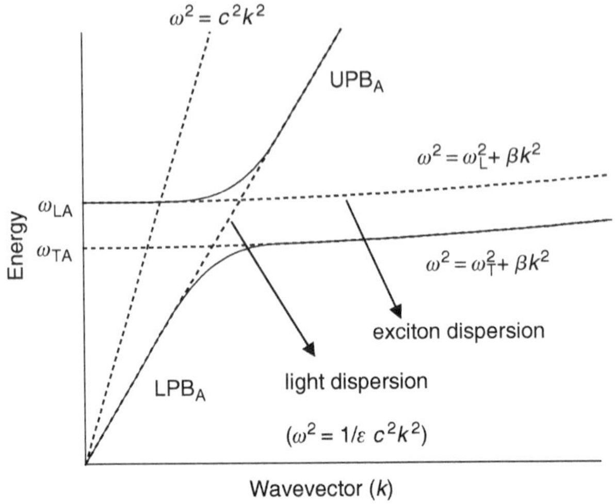

Figure 5.5: A simplified sketch of the exciton-polariton dispersion.

of Γ_6 transition. Using the energy separations of ground state and excited state peak positions, and assuming that exciton has a hydrogen-like set of energy levels, the exciton binding energy and bandgap energy can be predicted. The energy difference of about 45 meV between the ground and first excited states gives an A-free-exciton binding energy of 60 meV and a corresponding ZnO band gap energy of 3.4371 eV at 10 K. Based on the reported energy separation of the A- and B-free excitons (in the range of $E_{AB} = 9$–15 meV),[2,21,28] the weak emission centered at 3.3898 eV, which is about 12.7 meV apart from the A-exciton, is assigned to the B-exciton transition.

In strongly polar materials like ZnO, transverse Γ_5-excitons couple with photons to form polaritons. Γ_5-exciton couples to light polarized in the plane perpendicular to the *c*-axis. For wavevectors in the plane, these states are dipole coupled and split into a longitudinal state and a transverse state. In such a case, the usual transverse Γ_5-exciton dispersion $E_{ex} = E_{ex,0} + \hbar^2 k_{ex}^2/2m_{ex}$ will mix with the dispersion of the photon $E_{photon} = \hbar c k_{photon}$ to produce two new dispersion curves as seen in Figure 5.5, known as the upper polariton branch (UPB), and lower polariton branch (LPB). LPBs and UPBs can occur for excitons formed from all three valence bands, A, B, and C; however, in the PL spectra only LPB_A and UPB_A are observed.

In principle, although the polaritons can be formed anywhere along the dispersion curves, polariton lifetimes, which are higher at certain points, determine the observed peak positions. Therefore, as indicated in Figure 5.4 the $FX_A^{n=1}(\Gamma_5)$ exciton line has two components. The higher energy component at 3.3810 eV, which is

3.6 meV apart from the A-exciton, is assigned to the so-called longitudinal exciton, the energy of which is identical to that of the UPB_A at $k = 0$. The lower energy component at 3.3742 eV, which is about 2.9 meV apart from the A-exciton, most likely corresponds to the recombination from the "bottleneck" region of the lower polariton branch – LPB_A, where the photon and free-exciton dispersion curves cross. The emission line at 3.3771 eV (which is assigned as the $FX_A^{n=1}(\Gamma_5)$ transition energy) is believed to result from a long-lifetime region of the LPB_A, which of course is not at $k = 0$, but has nearly the same energy as that of the free A-exciton, at $k = 0$. These assignments are also consistent with the theory used to calculate the energy separation between the main exciton and polariton branches. The longitudinal-transverse splitting is given by $\Delta E = E(\Gamma_5)4\pi\alpha/2\varepsilon$, where $E(\Gamma_5)$ is the energy of the Γ_5-exciton with $4\pi\alpha$ denoting the polarizability and ε the optical dielectric constant. By using the reported values of 7.7×10^{-3} for polarizability[42] and 4.0–5.0 for optical dielectric constant[43,44] the calculated value will be in the range of 2.6–3.3 meV for $E(\Gamma_5) = 3.3771$ eV. The measured energy splitting (2.9 meV for LPB_A and 3.6 meV for UPB_A) is comparable to the predicted values and also to the experimental values of Chichibu *et al.*[33] (1.5 meV), Lagois[45] (1.8 meV), and Hümmer and Gebhardt[36] (2 meV), supporting the assignment of these two peaks. Since the Γ_6-excitons do not have a transverse character, they do not interact with light to form polaritons, and thus have only normal free-exciton dispersion curves as seen in the PL spectra. Therefore, the 3.3757 eV line in Figure 5.4 can be assigned to a low-k point of the Γ_6 dispersion curve. (Note that the transition cannot occur precisely at $k = 0$, because Γ_6-excitons are not allowed at $k = 0$.)

Low temperature reflectivity measurements performed by Teke *et al.*[32] validated the free excitonic features and corresponding peak assignments depending on PL measurements. Figure 5.6 shows reflectivity data obtained at 10 K for unpolarized and π polarized light. For unpolarized light, ground and first excited states of A- and B-excitons along with a weak C-exciton feature are observed. The reflection minima at $FX_A^{n=1} = 3.3772$ eV and $FX_B^{n=1} = 3.3901$ eV are in excellent agreement within the experimental resolution with the emission peaks for A- and B-free excitons in the PL spectra. The position of the first excited state ($FX_A^{n=2} = 3.421$ eV) and thereby the binding energy of the A-free exciton (~60 meV) were also confirmed by reflectivity measurements. Additionally, the reflection minima at 3.427 eV and 3.433 eV were assumed to be related to the second and the first excited states of the A- and B-free excitons, respectively. As will be discussed below, the temperature evolutions of the A- and B-exciton PL peaks also reveal characteristic features related to these exciton transitions supporting this premise.

Table 5.2 tabulates the observed excitonic peak energies reported for high quality ZnO single crystals from reflectance, photoreflectance, absorption, and PL measurements. The peak position of the A- and B-free excitons, and the first excited

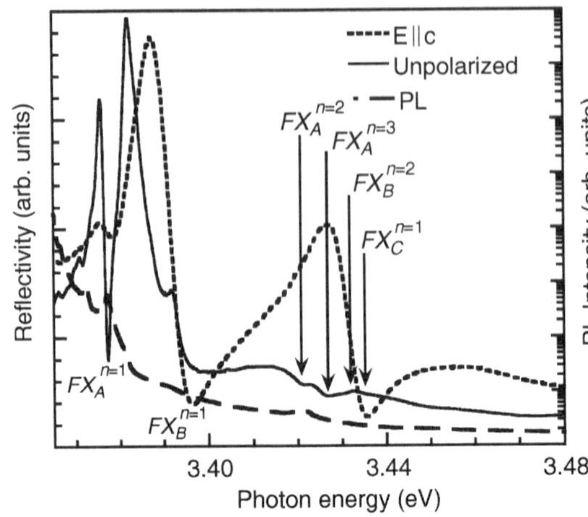

Figure 5.6: Reflection spectra of a forming gas annealed ZnO substrate measured at 10 K with unpolarized light and with E || c. The PL spectrum is also superimposed for comparison. (After Teke *et al.*[32])

states of the A-exciton from Teke *et al.*[32] are in very good agreement, within the experimental accuracy, with the results reported by Reynolds *et al.*[21] The observed polariton positions are also in reasonable agreement with the reported energies for bulk ZnO grown by Eagle-Picher using vapor-phase-transport. It should be noted that experimental resolution and the wavelength calibration of the particular setup used must be considered carefully to identify the exact peak positions of the very narrow excitonic lines.

5.2.2 Bound excitons

Bound excitons are extrinsic transitions and are related to dopants, native defects, or complexes, which usually create discrete electronic states in the bandgap, and therefore influence both optical absorption and emission processes. The electronic states of the bound excitons depend strongly on the semiconductor material, in particular the band structure. In theory, excitons could be bound to neutral or charged donors and acceptors. A basic assumption in the description of the bound exciton states for neutral donors and acceptors is a dominant coupling of the like particles in the bound exciton states. These two classes of bound excitons are by far the most important cases for direct bandgap materials. In high quality bulk ZnO substrates, the neutral shallow DBE often dominates because of the presence of donors due to unintentional (or doped) impurities and/or shallow donor-like defects. In samples

Table 5.2: Excitonic transition energies (eV) in ZnO single crystals.

	$FX_A(\Gamma_5)$			$FX_A(\Gamma_6)$			LPB_A UPB_A	FX_B			LPB_B UPB_B	FX_C			LPB_C UPB_C
	$n=1$	$n=2$	$n=3$	$n=1$	$n=2$	$n=3$		$n=1$	$n=2$	$n=3$		$n=1$	$n=2$	$n=3$	
Teke *et al.*[32] (melt grown)	3.3771	3.4220		3.3757	3.4206		3.3740 3.3810	3.3898							
Reynolds *et al.*[21] (vapor transport)	3.3773	3.4221	3.4303	3.3756	3.4209	3.4288		3.3895	3.4325						
Reynolds *et al.*[30] (vapor grown)	3.3793			3.3775			3.3743 3.3829								
Reynolds *et al.*[25] (melt grown)	3.3764	3.4217						3.3840	3.4271			3.4336	3.4657		
Reynolds *et al.*[25] (vapor grown)	3.3783	3.4229						3.3858	3.4283			3.4245	3.4667		
Hamby *et al.*[40] (vapor transport)	3.378						3.374 –	3.385							
Chichibu *et al.*[33] (chemical vapor deposition)		3.4231					3.3768 3.3783		3.4290		3.3830 3.3941		3.4679		3.4214 3.4353
Park *et al.*[4] (vapor grown)	3.3931	3.4243	3.4296					3.4331				3.4696			
Thomas[2] (needles)	3.3768	3.4225						3.3830	3.4275			3.4215	3.4650		
Liang and Yoffe[12] (vapor transport)	3.3781	3.4282	3.4375					3.3856	3.4324	3.4412		3.4264	3.4722	3.4808	
Lagois[45] (vapor grown)							3.3758 3.3776				3.3810 3.3912				3.4198 3.4317

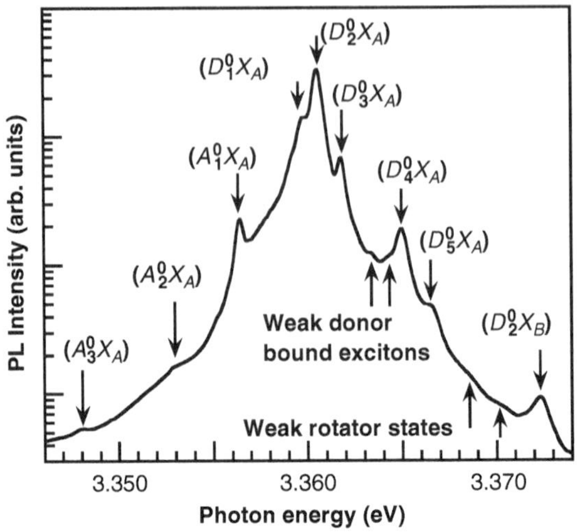

Figure 5.7: Bound excitonic region of the 10 K PL spectrum for a forming gas annealed ZnO substrate. (After Teke *et al.*[32])

containing acceptors, the ABE is observed. The recombination of bound excitons typically gives rise to sharp lines with a photon energy characteristic of each defect. Many sharp donor- and acceptor-bound exciton lines were observed in the low-temperature PL spectra of ZnO in a narrow energy range between 3.348 and 3.374 eV. In different reports, the total number of the observed emission lines vary depending on the sample and the measurement technique; but nevertheless, there are twelve transitions labeled from I_0 to I_{11}, following the general notation used in the earlier reports,[46] that form the basic set of bound exciton transitions characteristic of ZnO. However, the chemical origin and binding energy of the most underlying donor and acceptor atoms still remain unclear.

Figure 5.7 shows the bound exciton region of the 10 K PL spectrum for the melt grown bulk sample (from Cermet, Inc.) characterized by Teke *et al.*[32] The prominent lines are positioned at 3.3564 (I_9), 3.3598 (I_8), 3.3605 (I_6), 3.3618 (I_5), 3.3650 (I_4), 3.3664 (I_3), and 3.3724 (I_0) eV. The 3.3605 eV line is the most intense and has a full width at half maximum of about 0.7 meV, indicating the high quality of the sample. Several small peaks and shoulders can also be seen between these prominent lines. Different assignments for these emission peaks are present in the literature. Some of the recently reported values for bound exciton peak energies observed in low temperature PL are given in Table 5.3 for comparison. The emission peaks at 3.3598 (I_8), 3.3605 (I_6), 3.3618 (I_5), 3.3650 (I_4), and 3.3664 eV (I_3) have been assigned to the A excitons bound to neutral donors (D^0X_A).[32] The 3.3660–3.3665 eV (I_3) transition along with the 3.3674 eV (I_2) transition has also been attributed to the

Table 5.3: Bound exciton peak energies (eV) in ZnO single crystals.

Reference	A-exciton bound to neutral or ionized donors								Excited rotator states		B-exciton bound to neutral donor I_0
	I_{11}	I_{10}	I_9	I_8	I_7, I_6	I_5	I_4	I_3	I_2	I_1	
Teke *et al.*[32]	3.3481[a]	3.3530[a]	3.3564[a]	3.3598	3.3605	3.3618	3.3634 3.3643	3.3650 3.3664	3.3686	3.3702	3.3724
Reynolds *et al.*[38]			3.3562	3.3594	3.3602	3.3610	3.3624 3.3634	3.3652	3.3670 3.3662	3.3702 3.3714	
Alves *et al.*[41]			3.358[a]		3.361	3.362		3.364			
Thonke *et al.*[27]			3.3566[b]	3.3597	3.3606	3.3611	3.3628	3.364			
Boemare *et al.*[28]				3.3592			3.3622 3.3632	3.3653 3.3693	3.3693 3.3707	3.3741 3.3754	3.3707 3.3741 3.3754 3.3772
Meyer *et al.*[47]	3.3484	3.3531	3.3567	3.3598 3.3593	3.3608 3.3604 3.3600	3.3614	3.3628	3.3665 3.3660	3.3674[c]	3.3718[c] 3.3679[c]	3.3725[c]

[a] Assigned to neutral acceptor bound excitons.
[b] Not reported if acceptor- or donor-bound exciton.
[c] Assigned to donor or ionized donor bound excitons.

ionized donor bound exciton complexes,[26,47] based on the splitting in magnetic field.

In the lower energy region, the main peak at 3.3564 eV (I_9) was commonly observed and was previously attributed to Na or Li acceptors.[48] However, the binding energy of the deep Li_{Zn} acceptor (~500 meV) measured by optically detected magnetic resonance[49,50] rules out the possibility of observing such a near bandgap acceptor bound exciton. This 3.3564 eV (I_9) line which was initially assigned to an acceptor bound exciton by Teke *et al.* is in fact due to a neutral donor bound exciton, as verified by others based on its thermalization behavior in annealing/PL studies[38] and in temperature-dependent magnetotransmission and magnetophotoluminescence measurements.[26,47] Two other weak emissions at 3.3481 (I_{11}) and 3.3530 eV (I_{10}) are also visible in Figure 5.7, which may indicate the presence of deep acceptor states. However, their chemical origins have not yet been identified. Meyer *et al.*[47] could not observe any acceptor bound excitons and assigned the 3.3510 eV (I_{10}) line attributed to the neutral acceptor bound exciton (A^0X_A) by Teke *et al.* to a neutral donor bound exciton.

On the high-energy side of the neutral-DBE region, transitions between 3.3664 and 3.3724 eV have been attributed to the excited states or excited rotator states of the ground state neutral-donor-bound excitons. These excited states are analogous to the rotational states of the H_2 molecule. Several models have been proposed to explain the rotator states for different material systems. To identify these rotator states in bulk ZnO, Reynolds *et al.*[38] adopted the model, which is originally proposed by Rorison *et al.*[51] to explain their high magnetic field results in InP. In this model, DBEs are considered to be free excitons rotating around neutral donors, where one electron of the DBE state is strongly correlated with the hole and the other with the donor. The transitions observed at 3.3662 (Γ_6) and 3.3670 (Γ_5) eV were attributed to the rotator states associated with the ground state neutral donor bound exciton line at 3.3564 eV (I_9) based on the annealing studies and magnetophotoluminescence measurements. After annealing at 800°C, only these rotator states were observed with the only surviving donor bound exciton at 3.3564 eV. Using the energy difference of 10.6 meV Reynolds *et al.*[38] calculated an excitonic molecule radius of 1.5 times the excitonic radius of the free exciton (a_0) which agrees with the reported factors between 1.44 and 3.47.[52] Based on the same energy difference, the peaks at 3.702 (Γ_6) and 3.3714 eV (Γ_5) were assigned to the rotator state of the neutral donor bound exciton at 3.3594 eV (I_8). In measurements by Teke *et al.*,[32] very weak emissions at 3.3686 (Γ_6) and 3.3702 (Γ_5) eV with an energy separation of about 1.6 meV have been attributed to the rotator states associated with the main neutral donor bound exciton emission at 3.3605 eV. The splitting of these two peaks is consistent with the energy separation of the Γ_6 and Γ_5 band symmetries. The excited rotational states were also observed by Meyer *et al.*[47] for the I_4, I_6, and I_8 transitions with energy separations of 1.1, 1.2 and 1.5 eV, respectively. These separations are much smaller than those obtained by Reynolds *et al.*,[38] whose calculations would suggest an exciton radius of more than $4a_0$ for such small separations. However, one would expect the Bohr radius to be reduced in ZnO due to the relatively large binding energy.

Another relatively strong emission line at 3.3724 eV (I_0) was attributed to the transition due to the B-free exciton bound to the same main neutral donor (D^0X_B) by Teke *et al.*[32] The energy separation between this peak and the main peak at 3.3605 eV (I_6) is about 12 meV, which is consistent with the energy splitting of the A- and B-free exciton lines. Analysis of the temperature dependence of the PL spectrum also supports this assignment. Meyer *et al.*,[47] on the other hand, observed peaks around 3.364 eV which they assigned to neutral donor bound B-excitons. As suggested by Boemare *et al.*,[28] the emission peaks observed in this very narrow region on the high energy side of the well identified neutral donor bound exciton lines may be attributed to ionized donor bound excitons, excited rotator states, and/or neutral donor bound B-excitons. Identification of the fine structure in this energy range is not an easy task by just using the steady state photoluminescence.

Reynolds *et al.*[38] have investigated the bound-exciton region in detail by using low temperature PL measurements performed for different polarizations and applied magnetic fields. They resolved seven bound-exciton lines using the second-order grating configuration in the bound exciton spectral region. However, in this particular report there is almost a 2 meV shift between the first- and the second-order PL spectra that might be due to experimental issues associated with the setup used. Nevertheless, the peak energies of the neutral donor bound excitons reported by Teke *et al.*[32] are almost at the same positions, within the experimental resolution, as those reported by Reynolds *et al.*[38] Relative peak intensities of the particular donor-related exciton lines show some differences from sample to sample. For example, the most intense line was observed at 3.3628 eV by Thonke *et al.*,[27] at 3.3624 eV by Reynolds *et al.*,[38] at 3.3653 eV by Boemare *et al.*,[28] and at 3.364 eV by Alves *et al.*[41] and Hamby *et al.*,[40] whereas Teke *et al.*[32] observed the most intense neutral donor bound exciton line at 3.3605 eV. This is because the concentration of the particular donor as well as its capture cross-section could vary from sample to sample depending on the growth technique and post growth treatment used.

Even though the chemical origin of most of the donor/acceptor bound exciton lines is unknown, there is some experimental evidence relating a number of these lines to certain impurities. For example, the 3.3628 eV (I_4) line, which is typical for ZnO grown by hydrothermal and seeded vapor transport methods, and absent in samples grown from the vapor phase, is attributed to the Hydrogen donor.[47] Reduction of the Hydrogen-related electron nuclear double resonance signal with the decrease in the neutral donor resonance amplitude measured by electron paramagnetic resonance and IR absorption measurements on the vibrational properties of H in ZnO[53,54] strongly support this assignment.

Based on the Al-implantation studies of Schilling *et al.*,[55] who observed an increase in the intensity of the 3.3605 eV (I_6) emission line with increased Al concentration in ZnO, the I_6 transition is attributed to the Al impurity. Secondary ion mass spectroscopy (SIMS) revealed that Al was the dominant impurity with a concentration of more than one order of magnitude higher than that for other group III and group VII elements for samples where only the 3.3605 eV neutral donor bound exciton line was observed.[47] Al was suggested to be an omnipresent impurity in vapor grown ZnO.[49]

The 3.3598 eV (I_8) neutral donor bound exciton line was observed to be prominent in Ga-doped epitaxial ZnO films and in ZnO epitaxial films that were grown on GaN templates which resulted in Ga interdiffusion into ZnO, as verified by SIMS experiments.[47] The attribution of the I_8 line to the Ga impurity was also supported by the findings of Ko *et al.*[56] who also reported Ga-donor bound exciton recombination at 3.359 eV.

The 3.3564 (I_9) and 3.3530 eV (I_{10}) emission lines were initially thought to be related to Na and Li impurities, since they were typically observed in films doped with these elements. However, PL studies of Meyer *et al.*[47] on Na(Li)-diffused samples using different Na(Li) containing salts did not support this thesis. On the contrary, diffusion experiments using different concentrations of indium sulfate solutions led to the appearance of the I_9 line with In diffusion into bulk ZnO, which was confirmed by SIMS.[47] After the identification of In donor by optically detected magnetic resonance, it has also been stated that Indium donor in Li-doped ZnO is an unintentional, residual impurity.[49]

5.2.3 Two-electron satellites in PL

Another characteristic of the neutral-donor-bound exciton transition is the two-electron satellite (TES) transition in the spectral region of 3.32–3.34 eV. These transitions involve radiative recombination of an exciton bound to a neutral donor, leaving the donor in the excited state (2s, 2p states), thereby leading to a transition energy which is less than the DBE energy by an amount equal to the energy difference between the first excited (2s, 2p) and ground states (1s) of the donor. The usual observation of the (D^0X_A) line of course occurs when the donor final state is the ground state. In the effective-mass approximation, the energy difference between the ground-state neutral-donor bound excitons and their TES excited states can be used to determine the donor binding energies[27,41] and catalog the different species present in the material. When the central cell corrections are also neglected, the

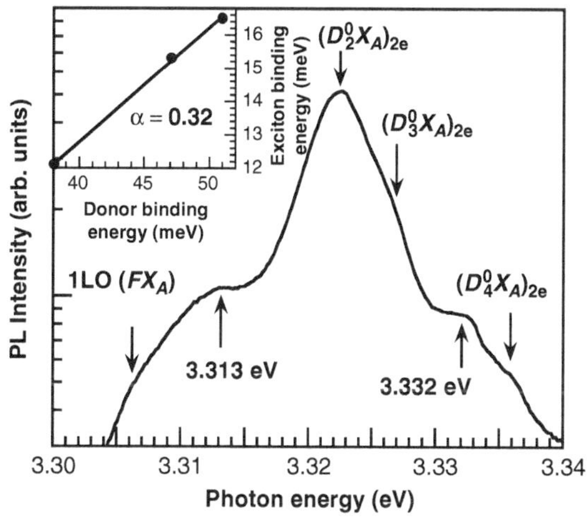

Figure 5.8: 10 K PL spectrum for a forming gas annealed ZnO substrate in the TES region of the main bound exciton lines. Inset shows the exciton binding energy versus donor binding energy. (After Teke *et al.*[32])

donor excitation energy from the ground state to the first excited state equals to 3/4 of the donor binding energy, E_D.

The spectral region for the expected two-electron satellite transitions is shown in Figure 5.8 for a forming gas annealed bulk ZnO sample. The main peak at 3.3224 eV is the TES $(D^0X_A)_{2e}$ associated with the most intense neutral donor bound exciton at 3.3605 eV (D^0X_A). The shoulder seen at about 3.3268 eV on the high-energy side of the main TES peak is related to the excited state of the donor whose ground state emission is at 3.3618 eV. A weak emission at 3.3364 eV is also attributed to the TES transition of the neutral donor whose ground state is at 3.3650 eV $(D_4^0X_A)$. From the separation of the ground state and the corresponding excited states, the donor binding energies are calculated as 51 meV for the donor at 3.3605 eV, 47 meV for the donor at 3.3618, and 38 meV for the donor at 3.3650 eV. Teke *et al.* could not identify two additional peaks in the TES region of the spectrum, at 3.332 and 3.312 eV which may be related to the excitons bound to structural defects, which is rather pervasive in GaN.

Meyer *et al.*[57] studied the TES spectral range of different bulk ZnO samples in detail to obtain the binding energies of various donors. They have observed the splitting of TES lines into 2s and 2p states, due to the effects of anisotropy and the polar interaction with optical phonons in polar hexagonal semiconductors. The effects of anisotropy and the polaron interactions were combined by employing the second order perturbation theory and the results of numerical calculations of the

Table 5.4: Binding energies and chemical identities of various donors.

PL line	Chemical identity[57]	Energy [eV] (donor binding energy [meV])			
		Meyer *et al.*[57]	Teke *et al.*[32]	Alves *et al.*[41]	Reynolds *et al.*[38]
I_3			3.3650 (38)	3.364 (43)	
I_4	H	3.3628 (46)			3.3636 (55.5)
I_5		3.3614 (51.4)	3.3618 (47)	3.362 (52)	3.3614 (56.7)
I_6	Al	3.3604 (53)	3.3805 (51)	3.361 (55)	
I_8	Ga	3.3598 (54.5)			
I_9	In	3.3567 (63.2)			
I_{10}		3.3531 (72.6)			
Proportionality constant (b), Haynes' constant (α)		$b = 0.365$	$b = 0.34$ $\alpha = 0.32$	$\alpha = 0.3$	$\alpha = 2$

ground and excited state energies of bound polarons in the isotropic approximation. More precise donor binding energies extracted from the data using this approach are listed in Table 5.4 along with the data mentioned above.

From the separation between the A-free exciton and the ground-state neutral-DBEs, Teke *et al.*[32] determined the localization energies as 16.6 meV (for 3.3605 eV), 15.3 meV (for 3.3618 eV), and 12.1 meV (for 3.3650 eV). According to the empirical Haynes rule,[58] the binding or localization energy of the DBEs is proportional to the binding energy of the corresponding donor, $E_{loc} = \alpha E_B$. However, a better fit was obtained using $E_{loc} = a + bE_B$, with $a = -1.1$ meV and $b = 0.34$. Indeed, the agreement with the linear fit is clearly seen in the inset of Figure 5.8. Meyer *et al.*[57] also observed a linear relationship between the donor binding energies and the exciton localization energies, with $a = -3.8$ meV and $b = 0.365$. For $a = 0$, data from Teke *et al.*[32] gives a Haynes' proportionality constant, $\alpha = 0.32$, which is close to the 0.3 reported by Alves *et al.*,[41] who calculated the binding energies of the donors as 43, 52, and 55 meV for the donors whose ground state bound exciton lines were at 3.364, 3.362, and 3.361 eV, respectively. Thonke *et al.*[27] also studied the TES transitions and found the binding energies of two shallow donors to be 39.9 and 55 meV. Reynolds *et al.*[38] reported binding energies of 55.5 and 56.7 meV for two donors at 3.3636 and 3.3614 eV, respectively. The Haynes proportionality constant obtained from these binding energies is about 2, much higher than the values given above.

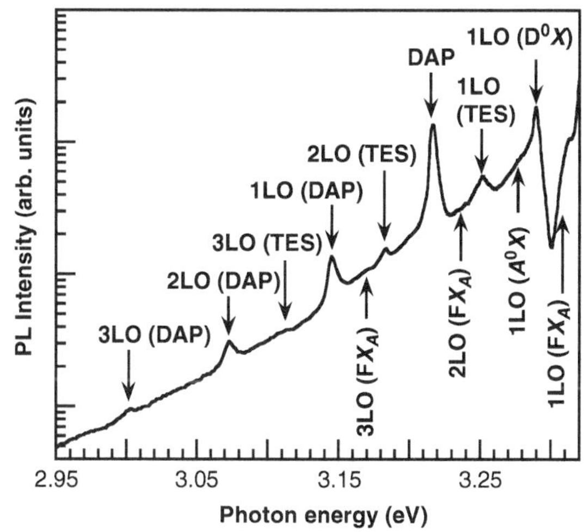

Figure 5.9: 10 K PL spectrum for a forming gas annealed ZnO substrate in the region where donor-acceptor-pair transition and LO-phonon replicas are expected to appear. (After Teke *et al.*[32])

5.2.4 DAP and shallow acceptor bound exciton transitions and LO-phonon replicas in PL

The spectral region containing the donor-acceptor-pair (DAP) transition and LO-phonon replicas of the main transitions has not been studied widely for single crystal ZnO. Figure 5.9 shows the corresponding spectrum taken by Teke *et al.*[32] at 10 K for a forming gas treated sample. It should be noted that LO-phonon replicas occur with a separation of 71–73 meV, the LO-phonon energy in ZnO.[59] Since some of the peaks are either very weak or mixed with other closely spaced peaks, temperature evolution of these peaks should be tracked carefully in order to make sure that the corresponding assignments are correct. As indicated in Figure 5.9, the bump at the higher energy side of the spectrum labeled as 1LO (FX_A) has a peak around 3.306 eV, which is at the expected position for the 1LO-phonon replica of the free exciton peak (about 71 meV apart from the $FX_A^{n=1}$ free-exciton peak). Although they are weak, second and third order LO-phonon replicas [labeled as 2LO (FX_A) and 3LO (FX_A)] are also observed in the PL spectrum.

The first order LO-phonon replicas of the main neutral donor bound excitons should fall between 3.290 and 3.295 eV in Figure 5.9. However, due to the line broadening, the peaks corresponding to each individual bound exciton could not be resolved very well. Indeed, the peak labeled as 1LO (D^0X) has a linewidth of about 6 meV, which prevents a definitive resolution. The LO-phonon replicas of the peak at 3.3650 eV can be separated from those of the two other closely spaced peaks at 3.3618 and 3.3605 eV. The peak at 3.2898 eV is the first LO-phonon replica of both the 3.3618 and 3.3605 eV lines, whereas the first LO-phonon replica of 3.3650 eV line is seen as a shoulder on the high-energy side of this intense peak. Resolving the second and higher order LO replicas is even harder because the energy position (3.218–3.223 eV) falls in the spectral region where the DAP transition and its LO-phonon replicas are expected to appear in the PL spectra. In fact, Teke *et al.*[32] observed a radiative recombination peak at 3.217 eV that is attributed to the donor-acceptor-pair (labeled as DAP in Figure 5.9) transition, along with its first, second, and third LO-phonon replicas at 3.145, 3.073, and 3.001 eV, respectively. Regarding the origin of the emission line at 3.2898 eV, Reynolds *et al.*[35] suggested that this line is associated with an acceptor-related transition based on their internal strain study by changing the post-annealing temperatures, which disagrees with the interpretation by Teke *et al.*[32] Although the latter investigation differs somewhat, it reports at least two closely spaced features at 3.2898 and 3.2920 eV, which are about 72 meV apart from the main neutral DBE lines at 3.3605 and 3.3650 eV. The temperature dependent measurements also show that relative intensities of these LO-phonon replicas follow those of the main bound excitons. Additionally, LO-phonon replicas are expected to be roughly two orders of magnitude less intense than the neutral donor bound exciton lines.[27] This is also similar to the case of GaN and other II-VI semiconductors, where donor-related bound exciton lines couple only weakly with the optical phonons.

The relatively broad peak around 3.280 eV in Figure 5.9 is the first LO-phonon replica associated with the bound exciton line at 3.3564 eV. Finally, the first, second, and third order LO-phonon replicas of the TES lines are also clearly observed in the PL spectra. These peaks are labeled as 1LO, 2LO, and 3LO (TES) and they are positioned at 3.252, 3.182, and 3.112 eV, respectively.

ZnO intentionally doped with N has also been investigated in detail, since nitrogen is believed to be the most promising acceptor in ZnO. In N-doped ZnO (with concentration up to 10^{19} cm^{-3}), a relatively broad line at 3.315 eV dominates the PL spectrum.[60] Look *et al.*[60] attributed this line to an acceptor-bound exciton (A^0X), associated with the N_O acceptor. A weaker and broader line (or shoulder) appearing at 3.238 eV has been attributed to the DAP emission, involving the N_O acceptor, superimposed with the LO-phonon replica of the dominant (A^0X) emission. The activation energy of the N_O acceptor has been estimated as 0.17–0.20 eV from the above PL experiments. In another study,[61] the PL spectrum of the *p*-type ZnO samples, doped with N by using NO gas, also contained a strong line at 3.309 eV followed by two LO-phonon replicas, that could be attributed to the shallow N_O acceptor. A broad peak at 3.245 eV followed by LO-phonon replicas has been observed also by Matsui *et al.*[62] in N-doped ZnO grown by pulse laser deposition on ZnO and glass substrates. This line has been attributed to DAP transition

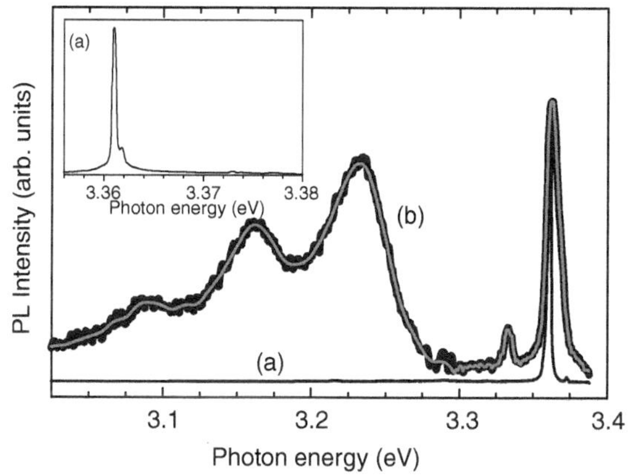

Figure 5.10: Low temperature PL spectra of (a) undoped and (b) nitrogen-doped ZnO films. (Zeuner *et al.*[63])

involving the N_O acceptor. The characteristic DAP emission with the zero-phonon line at 3.235 eV was observed in PL spectrum of N-doped ZnO (Figure 5.10) and studied in more detail by Zeuner *et al.*[63] Transient PL studies on this band confirmed its DAP nature. A non-exponential decay of PL intensity after pulse excitation was attributed to transitions between distant pairs. The acceptor binding energy has been estimated as 165 ± 10 meV from these experiments.

In ZnO doped with As, another potential acceptor in ZnO, a line at about 3.23–3.24 eV, followed by two LO-phonon replicas has been observed in the PL spectrum with an intensity exceeding the exciton emission intensity.[64] This emission has been attributed to the shallow As acceptor, the activation energy of which was estimated as 180 ± 10 meV from the temperature dependence of the PL spectrum. Concurrently with the DAP emission line, a sharp line at 3.325 eV appeared in these samples, tentatively attributed to the As acceptor bound exciton.[64] In yet another study,[65] the 3.322 and 3.273 eV lines greatly increased with As-doping, and the 3.219 and 3.172 eV lines dominated the PL spectrum at the highest chemical concentrations of As-dopant in the low 10^{20} cm^{-3} range. Ryu *et al.*[65] attributed the 3.359 eV line to the (A^0X) exciton, 3.322 and 3.273 eV lines to the As-related e-A transitions, and the 3.219 and 3.172 eV lines to the DAP transitions. The activation energies of the two As-related acceptors were estimated as 115 and 164 meV.

5.2.5 Temperature dependent PL measurements

In order to provide additional support for some of the peak assignments in the low temperature PL spectrum of the high quality ZnO substrate investigated, Teke *et al.*[32]

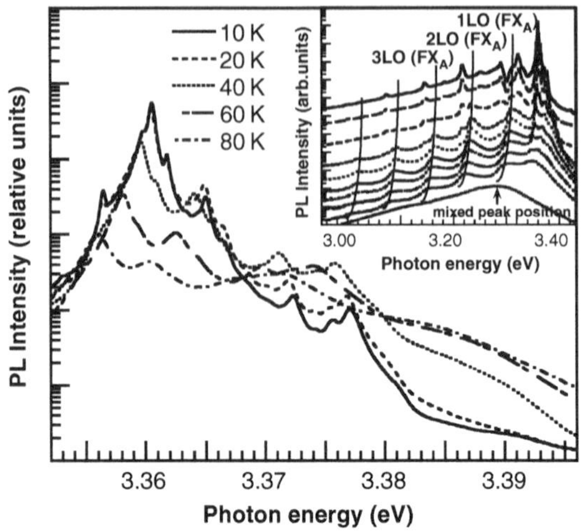

Figure 5.11: Temperature dependent PL spectrum for a forming gas annealed ZnO substrate. The inset shows the PL in the DAP and LO-phonon replicas region up to 160 K. The spectrum for each temperature is displaced vertically for clarity. The room temperature PL data is also included at the bottom. The lines drawn on some peaks are for guidance. (After Teke *et al.*[32])

studied the temperature evolution of these peaks. The temperature dependent measurements were performed between 10 and 300 K. Figure 5.11 shows only the PL spectra for temperatures up to 160 K (at 10 K, 20 K, and up to 160 K with 20 K steps) because most of the important changes occur in this range. The spectrum for each temperature is displaced vertically for clarity.

The variation of both A- and B-exciton peak positions with temperature is shown in Figure 5.11. The A- and B-exciton peaks can be tracked up to ~160 K, above which line broadening prevents a satisfactory distinction between the two. This is clearly seen for the room temperature PL spectrum (bottom curve in Figure 5.11 inset), where the peak position is at ~3.28 eV instead of the expected position of 3.31 eV (if the mostly accepted bandgap of 3.37 eV and the measured binding energy of 60 meV were used). As the temperature is increased, the convergence of the A- and B-excitons, and the 1LO-phonon replica coupled with the line broadening of each of these peaks hampers an accurate determination of the peak positions above 160 K. Therefore, the room temperature peak should be considered a combination of these multiple peaks.

The intensity of A- and B-free excitons increase with temperature up to 40 K and 80 K, respectively, and then decrease as the temperature is increased further. From this observation, Teke *et al.*[32] attributed the 3.3771 and 3.3898 eV emission lines at 10 K to the free excitons corresponding to the A and B bands, respectively. The lower energy bound exciton peaks at 3.3564, 3.3530, and 3.3481 eV quench gradually with increasing temperature, and disappear above 40 K. The temperature evolution of the main neutral-donor bound excitons was also tracked within the temperature range of 10–160 K (see also the Figure 5.11 inset). With increasing temperature, the main peak at 3.3605 eV and its TES along with their LO-phonon replicas quench. On the other hand, relative intensities of the bound exciton emissions lying between this main DBE peak and the A-free exciton peak increase initially, where the strength depends on the particular bound exciton, and then decrease at higher temperatures. The intensity of the bound exciton peak at 3.3724 eV, which was attributed to the B-exciton bound to the main donor rather than to a rotator state, follows the temperature behavior of the B-exciton up to 40 K supporting the assignment of Teke *et al.*[32] Although the B-free exciton emission continues to increase up to 80 K, further increase of the donor bound B-exciton intensity is prevented due to its partial dissociation above 40 K. The observed temperature characteristics of the free exciton and dominant bound exciton, where the relative intensity I_{FX}/I_{BX} is seen to increase with increasing temperature, can be interpreted using the approach of Viswanath *et al.*[66] for the study of GaN. Those authors also observed an increase consistent with the aforementioned ratio and inferred that with increasing temperature the dominant donor-bound exciton dissociates into a free exciton and a neutral donor. Based on this argument and the works of Reynolds *et al.*[38] and Hamby *et al.*,[40] Teke *et al.*[32] concluded that thermal dissociation of the 3.3605 eV bound exciton results in increase of the free excitons and the other shallow donor-bound excitons.

There is some controversy in the assignment of the ~3.23 eV peak, which is close to the main DAP transition at low temperatures. Hamby *et al.*[40] attributed this peak to the second LO-phonon replica of the A-free exciton. They attained a very good agreement in terms of the temperature dependent energy positions of this peak with that of the predicted values by taking into account the temperature broadening effect. Teke *et al.*[32] also assigned this peak to the 2LO-phonon replica of the A-free exciton depending on its temperature evolution. As seen in Figure 5.11, the main DAP line and its LO-phonon replicas quench with increasing temperature, while the adjacent line at ~3.23 eV on the high energy side increases and becomes more apparent at higher temperatures. The temperature dependence of this peak is similar to the A-free exciton; it increases with increasing temperature up to 60 K and then decreases at higher temperatures. Also, by following the approach of Hamby *et al.*,[40] the variation of A-free exciton and its 1LO and 2LO peak energies with temperature are plotted in Figure 5.12. As seen in this figure, the expected and the measured energy peak positions of the 1LO and 2LO replicas of the A-free exciton agree well, supporting the assignments on these peaks. It is also noted that the temperature variation of the A-free exciton peak energy follows the Varshni's empirical formula;[59] $E_g(T) = E_g(0) - \alpha T^2/(T + \beta)$, where $E_g(0)$ is the transition

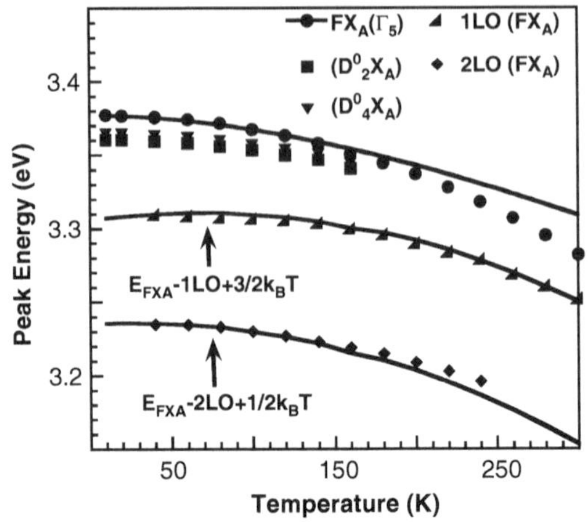

Figure 5.12: Temperature dependent peak positions of the A-free exciton, $FX_A^{n=1}(\Gamma_5)$ and its 1LO- and 2LO-phonon replicas. Also shown are the temperature evolutions of the peak positions of two major neutral-donor bound exciton transitions at 3.3606 and 3.3650 eV. The $FX_A^{n=1}(\Gamma_5)$ data were fit using Varshni equation and LO-phonon replicas were fit with the equations shown on the figure. (After Teke *et al.*[32])

energy at 0 K, T is the temperature, and α and β are the temperature coefficients. In Figure 5.12, a higher order LO-phonon replica of the A-free exciton also develops persistently with increasing temperature.

On the other hand, analyzing its temperature dependence, Thonke *et al.*[27] attributed the 3.236 eV peak to the free electron-acceptor (e, A^0) transition due to thermal ionization of the donor with increasing temperature. Figure 5.13 shows the appearance of the (e, A^0) peak with increasing temperature above 30 K as the DAP line at 3.220 eV quenches. Using the calculated donor binding energy of $E_D = 28$ meV (this is lower than the 39.9 meV deduced from the spacing of the TES lines) and the donor concentration $N_D = 1.7 \times 10^{17}$ cm^{-3} extracted from Hall measurements, they determined the acceptor ionization energy as $E_A = 195 \pm 10$ meV from the absolute spacing of the (D^0, A^0) line from the band gap:

$$E_A = [E_{gap} - E_D] - [E(D^0, A^0) - \alpha N_D^{1/3}] \qquad (5.5)$$

where α is a material dependent constant[27] and $E_{gap} = 3.438$ eV. The chemical species for the acceptor and the donor were proposed to be the substitutional nitrogen on O sites and hydrogen, respectively.

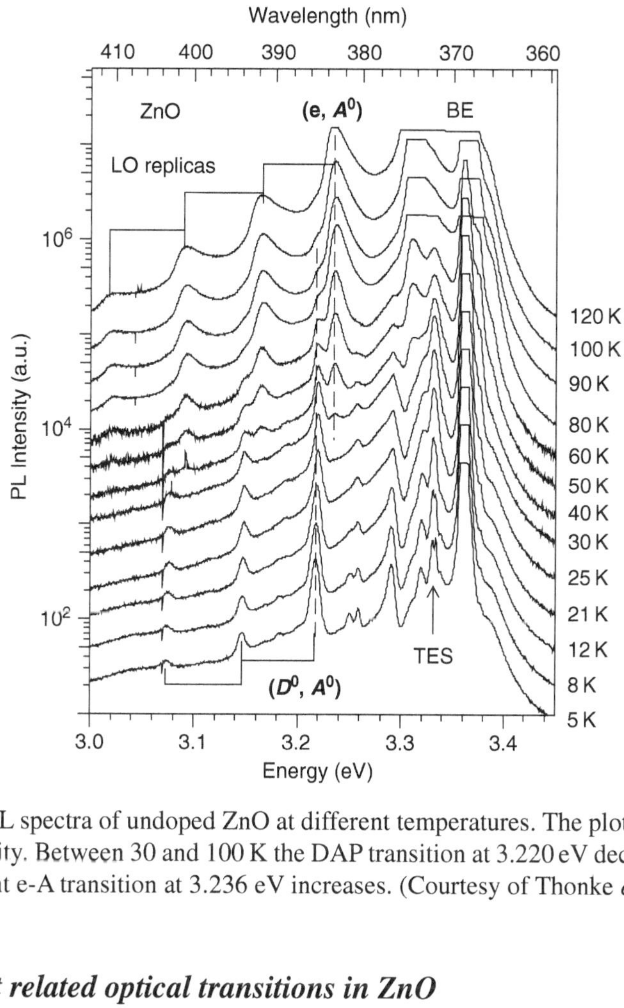

Figure 5.13: PL spectra of undoped ZnO at different temperatures. The plots are vertically shifted for clarity. Between 30 and 100 K the DAP transition at 3.220 eV decreases, and the adjacent e-A transition at 3.236 eV increases. (Courtesy of Thonke *et al.*[27])

5.2.6 Defect related optical transitions in ZnO

Most of the experimental results on point defects in ZnO have been obtained from the analysis of mainly the PL data. In undoped ZnO, the well-known green luminescence (GL) band peaking at about 2.5 eV usually dominates the defect-related part of the PL spectrum.[19] There are a few reports about other PL bands, in particular the yellow luminescence in Li doped ZnO.

5.2.6.1 Green luminescence band

The nature of the green luminescence (GL), appearing at about 2.5 eV in undoped ZnO, remained controversial for decades. Although in the early studies it was unambiguously attributed to copper impurities, strong evidence was later presented in

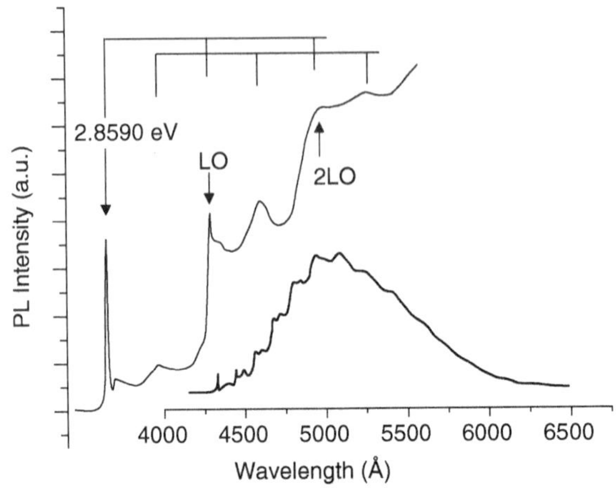

Figure 5.14: GL band in ZnO at 1.6 K. The enlarged portion shows fine structure at the high-energy side with the zero-phonon line at 2.859 eV. (After Dingle[67])

favor of the oxygen vacancy (V_O) as the defect responsible for the GL band. As it will be discussed below, the GL band with a characteristic fine structure is most likely related to copper impurities, whereas the structureless GL band with nearly the same position and width may be related to a native point defect such as V_O or V_{Zn}.

Dingle[67] studied wurtzite *n*-type ZnO crystals containing trace amounts of copper (4 ± 2 ppm), aluminum (<2 ppm), iron, magnesium, silicon, boron, and indium. The PL spectrum obtained was dominated by a broad GL band peaking at $\sim$2.45 eV and revealed the characteristic fine structure (Figure 5.14). The zero-phonon line was composed of two sharp lines (A and B components) separated by $\sim$0.1 meV (Figure 5.15). The ratio of the intensities of the two lines remained constant over the temperature range of 1.6–20.4 K and was independent of the crystal origin. Dingle[67] also studied the behavior of these lines and the band as a function of temperature, uniaxial pressure, magnetic field, and polarization of the excitation source. The defining information for the identification of the defect responsible for the GL band has been obtained from the splitting of the A and B lines in applied magnetic field. Each component of the zero-phonon line split into four symmetrically disposed components when the magnetic field was applied parallel or perpendicular to the crystal *c*-axis. The *g* factors obtained from the angle dependence of the splitting in magnetic field ($g_{\|} = 0.73 \pm 0.05$, $g_{\perp} = 1.48 \pm 0.05$) are nearly identical to those obtained from Electron Paramagnetic Resonance (EPR) studies of a divalent copper ion Cu^{2+} in a zinc lattice site of ZnO ($g_{\|} = 0.7383 \pm 0.0003$, $g_{\perp} = 1.5237 \pm 0.003$).[68] In addition to these observations, the absence of thermalization between

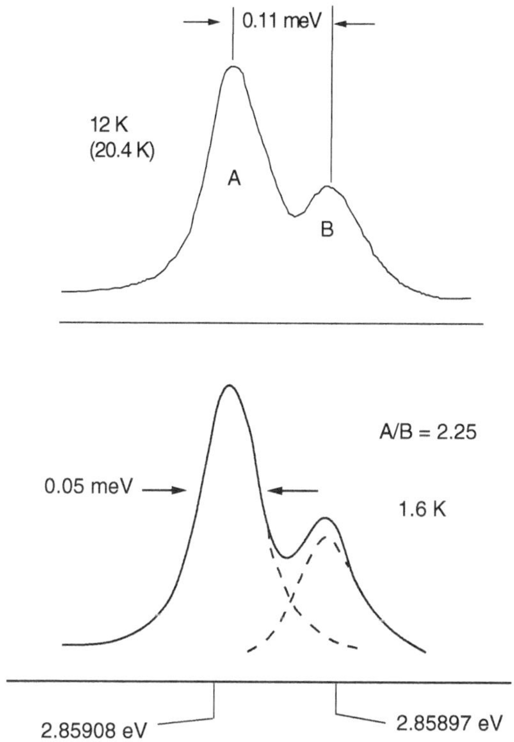

Figure 5.15: High-resolution spectra in the region of the zero-phonon line near 2.859 eV. The intensity ratio A/B $= 2.25 \pm 0.05$ is invariant in the temperature range of 1.6–20.4 K. (After Dingle[67])

the A and B lines in zero magnetic field, invariance of the A/B intensity ratio in different samples, and identical Zeeman patterns for the A and B lines provided the basis for the unequivocal identification of the GL band with its characteristic fine structure as being due to the Cu_{Zn} defect.[67]

The details of the ground state of the Cu_{Zn} acceptor in ZnO were established from the EPR and infrared absorption studies.[68] The low-temperature absorption spectrum of the Cu-doped ZnO contained two sharp lines at 717 and 722 meV. The details of the absorption spectra, the Zeeman splitting in magnetic field, and the EPR data allowed Dietz *et al.*[68] to construct the following model of Cu_{Zn} in ZnO. The free-ion term ^2D of the Cu^{2+} ion is split by the tetrahedral crystal field into the $^2E(D)$ and $^2T_2(D)$ states that in turn are split by a combination of weaker spin-orbit coupling and the trigonal crystal field. At low temperatures, intracenter transitions from the lowest sublevel of the $^2T_2$ state to two sublevels of the ^2E state are responsible for the absorption lines at 717 and 722 meV. Assignment of two components of the line at

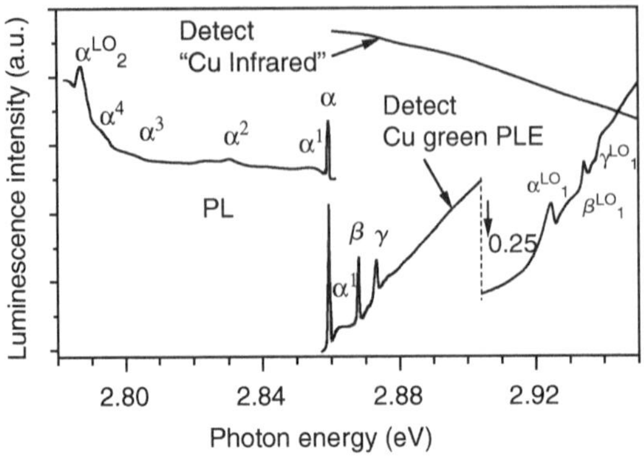

Figure 5.16: PL and PLE spectra near the zero-phonon transition of the GL associated with Cu in ZnO. (After Dean *et al.*[73])

717 meV (split by ~0.1 meV) as being due to two isotopes of Cu was confirmed by their transformation into a single line in the crystals doped with only ^{63}Cu isotope. Dietz *et al.*[68] concluded that the $Cu^{2+}t_2$ wave function is radially expanded (more than e wave function) relative to the d wave function of the free Cu ion, and the t_2 hole spends about 60% of its time on the Cu^{2+} ion, while it spends the rest of the time in the oxygen sp^3 orbitals. The inverse $^2E(D) \rightarrow {}^2T_2(D)$ transition of the Cu^{2+} ions in ZnO (a zero-phonon line at 717 meV) has been observed under electron-beam excitation at low temperatures.[69]

The excited states of the Cu_{Zn} acceptor in ZnO have been revealed in the absorption and PL excitation (PLE) spectra, namely a characteristic triplet α, β, γ at 2.8594, 2.8680, and 2.8733 eV, respectively (Figure 5.16).[70,71,72,73] The α line coincides with the zero-phonon line of the GL.[67] Remarkably, the isotope shift was also observed in the absorption and PLE spectra, at least for the α and β lines.[72] These lines have been attributed to transitions $Cu^{2+}(d^9) + h\nu \rightarrow (Cu^+(d^{10}), h)$ in which the excited state is split into three levels corresponding to three valence bands a, b, and c. Dahan *et al.*[71] have introduced the concept of intermediately bound excitons to explain the experimental data associated with Cu in ZnO. They suggested that the tenth electron, bound to the Cu d-shell, has a larger orbit (or swollen wave function) than the other d-shell electrons due to hybridization with the Bloch states of the host atoms (the covalent swelling). This tenth electron creates the potential capturing a hole, so that the tenth d-shell electron and a loosely bound hole can be considered as an exciton bound to Cu. Intermediately bound excitons are typically formed in crystals with hexagonal wurtzite-type lattice[74] such as ZnO. There are three close levels in the forbidden energy gap for such an exciton in ZnO:Cu according to three split levels of the top of the valence band. During the excitation process, all three excited states of the exciton can be observed, whereas only the lowest one is observed in emission at low temperature. The theory of the intermediately bound excitons accounts very well the isotope shift and g-factors of the α, β, γ lines in ZnO:Cu.[71,72]

Kuhnert and Helbig[75] studied the broad phonon sideband of the GL in ZnO (having maximum at 2.4 eV and FWHM 0.42 eV). The energy separation between the subsequent peaks is 72 meV, close to a LO-phonon mode of the ZnO crystal. The spectral distribution of the GL band can be described in terms of the one-dimensional configuration coordinate model. The Huang-Rhys factor $S = 6.5$ was obtained from the fitting. Later, Reynolds *et al.*[76] also analyzed the phonon-related fine structure of the GL in ZnO, but identified the peak at ~2.26 eV as the zero-phonon line, while attributing the high-energy peaks to hot luminescence. In another work, Reynolds *et al.*[77] suggested that transitions from the two shallow donor states to the same deep level are partially responsible for the fine structure of the GL band. However, this picture is in conflict with the fact that different authors observed the same shape of the GL fine structure in different ZnO samples.[67,70,73,75]

Garces *et al.*,[78] studying PL of a bulk ZnO sample before and after annealing in air at 900°C for one hour, observed a remarkable transformation of a broad structureless green band peaking at 500 nm into a structured emission peaking near 510 nm with a structure identical to the Cu-related GL. These authors suggested that there might be two distinct mechanisms giving rise to GL band, and that both involve Cu impurities but with different stable charge states (Cu^+ and Cu^{2+}) prior to optical excitation. Based on their EPR study of the samples they concluded that the structureless GL band of the as-grown ZnO sample is due to DAP recombination between an electron bound to a shallow donor and a hole loosely bound to the Cu^+ ion before annealing, while the structured green emission after the 900°C annealing was assigned to localized excitation of the isolated Cu^{2+} ion as originally suggested by Dingle.[67]

Structureless GL band in ZnO was attributed also to V_{Zn} acceptor,[79,80,81,82] a complex defect involving Zn_i,[83] O_{Zn},[84] and V_O.[85,86,87,88] Different authors suggested different types of electron transitions to explain the GL band, such as from the V_O donor level located near the conduction band to the valence band (*D-h* type recombination),[85] from V_O or another donor level to deep V_{Zn} acceptor level (DAP type),[81,82] from conduction band to the V_{Zn} acceptor (e-A type),[79] and between two states of V_O (intracenter transition).[87] The earlier notion of V_O in ZnO as being a donor with its 2+/+ and +/0 levels located near the conduction band has been recently reviewed by Van de Walle.[89] This author predicted that V_O in ZnO has only the level (2+/0) at about 2.7 eV above the valence band. Note also that the *D-h* type recombination is highly improbable in an n-type semiconductor.[90] Moreover, the DAP-type recombination observed in Ref. 82 may not be the same PL band as others detect because its maximum is at 2.3 eV (shifting to ~2.04 eV after time delay), i.e. in the yellow range.

Remarkable correlations attained between the intensity of the GL band, the concentration of free electrons, and the concentration of the defect identified as V_O at different annealing conditions or oxidation seemingly provided strong arguments in support of the model considering V_O as the defect responsible for the GL band.[85] However this assignment and interpretation of the experimental results should be regarded with caution because Vanheusden *et al.*[85] have also assigned the EPR signal with $g = 1.956$ to V_O, and the concentration of V_O was determined from the intensity of this signal. Later it was shown that isotropic EPR signal with $g = 1.956$ is related to a shallow donor, whereas EPR signal from V_O is characterized with $g_{\|} = 1.984$ and $g_{\perp} = 2.025$.[87] Optically detected magnetic resonance (ODMR) studies[87] have revealed that the structureless GL band peaking at about 2.45 eV anticorrelates with the isotropic signal at $g = 1.956$ and correlates with two resonances (with $g_{\|} = 1.984$ and $g_{\perp} = 2.025$) arising from a spin-triplet system ($S = 1$). Note that the g-values obtained on the structureless GL band[87] and on the GL with a distinct phonon structure[67,78] are incompatible, and thus, these two PL bands are related to different defects. The triplet resonances were detected only in the energy range of the structureless GL band, while the shallow donor signal (with $g = 1.956$) quenched the GL band and enhanced emission in the excitonic range. These findings strongly indicate that the shallow donor signal is transferred to the GL band by a shunt process or a spin-dependent energy transfer mechanism.[87] Leiter *et al.*[87,88] have identified V_O as the defect responsible for the structureless GL band in ZnO, and demonstrated striking similarities of this defect to the anion vacancy in other ionic host crystals: BaO, SrO, CaO, and MgO (F-centers). In the model of Leiter *et al.*,[87,88] the two-electron ground state of the neutral V_O is a diamagnetic singlet state. Absorption of a photon transfers the system into a singlet excited state. It relaxes nonradiatively into the emissive, paramagnetic state which can be detected by ODMR (Figure 5.17).

5.2.6.2 Yellow luminescence band

Doping ZnO with Li acceptor results in the yellow luminescence (YL) band with a peak at about 2.2 eV and a FWHM of ~0.5 eV.[15] In contrast to the GL band, the YL band decays very slowly after switching off the excitation source, and can be observed also in the thermoluminescence spectrum.[15] Li-doped ZnO is high-resistive due to deep location of the Li acceptor level (about 0.8 eV above the valence band).[15] The YL band is polarized at low temperatures, which was explained by two metastable orientations of the Li_{Zn} center in the ZnO lattice.[15] ODMR studies of the YL band in ZnO:Li suggest that the DAP-type transitions including a shallow donor and the Li acceptor dominate at low temperatures.[91]

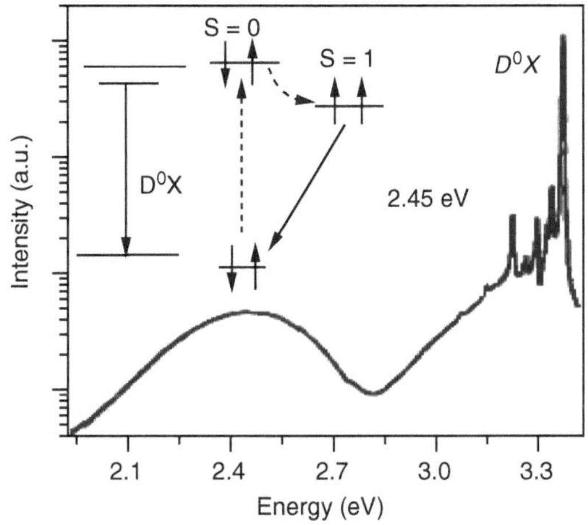

Figure 5.17: PL spectrum of undoped ZnO excited with the 325 nm line of an HeCd laser (5 K). The inset shows a recombination model for the GL band peaked at 2.45 eV. (After Leiter *et al.*[87])

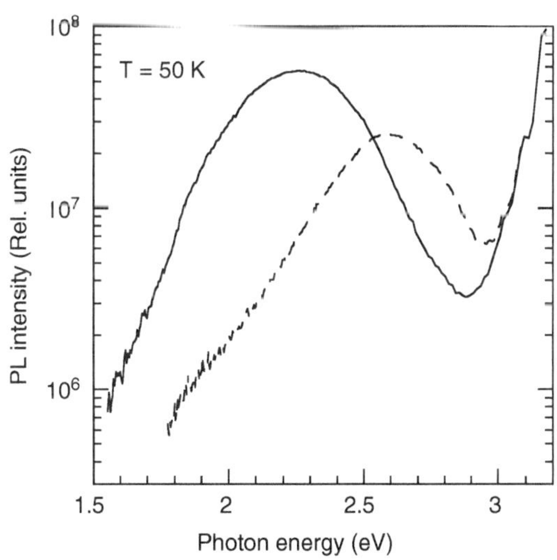

Figure 5.18: PL spectrum of undoped bulk ZnO annealed at 150°C in air for 1 hour. Solid (dashed) curve – after (before) HeCd irradiation for two hours with excitation density of 0.3 W/cm^2. Excitation density during the measurements was 10^{-3} W/cm^2.

YL with metastable behavior was observed also in undoped bulk ZnO.[92] Under irradiation with a HeCd laser, the GL band gradually bleached, while the YL band emerged and eventually replaced the GL band (Figure 5.18). The YL band saturated with excitation intensity above 10^{-3} W/cm^2, indicating low concentration of the

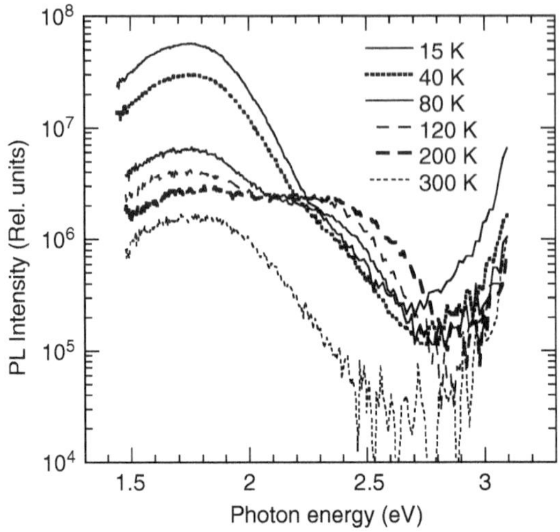

Figure 5.19: PL spectrum of undoped bulk ZnO at different temperatures. The sample was annealed at 700°C in air for 1 hour. Excitation density is 10^{-3} W/cm^2.

related defects. The decay of the YL was nonexponential at low temperature, allowing the possibility of the DAP transitions for this band. The YL band quenched at temperatures above 200 K with an activation energy of about 0.5 eV.

5.2.6.3 Red luminescence band

A red luminescence (RL) band emerged at about 1.75 eV in the PL spectrum of undoped bulk ZnO after it was annealed in air at 700°C (Figure 5.19).[92] The RL band is broad with a FWHM of about 0.5 eV, and its shape is Gaussian. With increasing temperature, the RL band quenches with an activation energy of about 15 meV in the range from 15 to 100 K. The quenching of the RL band apparently causes the emergence of the GL band (Figure 5.19). This may be a result of competition for holes between the acceptors responsible for the GL and RL bands. At $T > 200$ K, the GL band quenches rapidly, so that the RL band is observed alone at room temperature (Figure 5.19). Preliminarily results of transient PL demonstrate that the low-temperature PL decay of the RL band is nearly exponential with a characteristic time of about 1 μs.

5.3 Refractive index of ZnO and MgZnO

Knowledge of the dispersion of the refractive indices of semiconductor materials is necessary for accurate modeling and design of devices. The wurtzite ZnO lacks cubic symmetry and therefore has anisotropic optical properties. The anisotropy results in uniaxial birefringence, two different refractive indices for polarization parallel (n_e-extraordinary) and perpendicular (n_o-ordinary) to the *c*-axis. In this section a summary of reported results on ZnO dielectric constants and refractive index dispersion will be given.

In the early 1950s and 1960s, several workers reported the results of optical reflection measurements with light polarized parallel and perpendicular to the *c*-axis, and used Kramers-Kronig analysis to determine the dielectric functions. Refractive index dispersion of single crystal ZnO has been initially reported by Mollwo[10] and Bond[11] using the method of minimum deviation at room temperature for the visible and near infrared spectrum. Prisms fabricated from vapor-transport-grown ZnO were used. For light below the direct band edge, traditional angle-of-minimum deviation methods produced very accurate values of the refractive index for both $E \perp c$ and $E \parallel c$. Park and Schneider[18] then extended the measurements to the spectral region near the onset of the exciton absorption at temperatures down to liquid helium temperature using transmission interferometry. A ZnO platelet sample (with the *c*-axis lying in the plane of the platelet) grown by vapor phase transport was used for the measurements. By lowering the temperature, the dispersion curves blue shifted and were observed to get sharper. The refractive indices obtained at 4.2 K near the absorption edge are listed in Table 5.5. Freeouf[14] obtained the dielectric functions above and below the bandgap by measuring the reflectivity from 0.6 to 30 eV and applying the Kramers-Kronig analysis. Matz and Lütz[93] determined the optical functions of ZnO using nulling ellipsometry at several wavelengths defined by interference filters.

Hu *et al.*[94] used optical transmission to measure the optical functions of thin-film ZnO prepared by pulsed-laser deposition on α-SiO_2 substrates. The ordinary refractive indices deduced from transmittance oscillations were 0.02–0.03 lower than the bulk ZnO crystal data by Bond.[11] The existence of grain boundaries in the films was assumed to be the source for this difference. Additionally, the measurement technique introduced a considerably large error resulting in large variations in the data. The best fits for the ordinary index using the three-term Cauchy equation

$$n(\lambda) = A + \frac{B}{\lambda^2} + \frac{C}{\lambda^4}, \tag{5.6}$$

produced parameters of $A = 1.9281$, $B = -1.1157 \times 10^{-5}\ \mu m^2$, and $C = 5.9696 \times 10^{-3}\ \mu m^4$, whereas the bulk ZnO crystal[11] was best characterized by $A = 1.9436$, $B = -1.9714 \times 10^{-5}\ \mu m^2$, and $C = 7.0918 \times 10^{-3}\ \mu m^4$.

Table 5.5: Some values of the refractive indices of ZnO at 4.2 K near the absorption edge. (After Park and Schneider[18])

E ⊥ c		E ∥ c	
λ (nm)	*n*	λ (nm)	*n*
366.43	9.843	363.71	3.525
366.72	6.734	365.67	2.917
367.01	9.414	367.54	2.722
369.04	3.089	369.79	2.598
371.19	2.747	371.52	2.539
373.51	2.595	373.85	2.478
375.65	2.513	375.68	2.441
377.84	2.454	377.90	2.407
379.69	2.417	380.49	2.374
381.87	2.382	381.90	2.358
385.86	2.332	383.44	2.343
387.43	2.316	385.08	2.328
390.92	2.286	388.70	2.300
392.85	2.272	392.83	2.273
394.89	2.258	395.02	2.260
397.04	2.245	399.92	2.237
399.39	2.232	405.31	2.215
401.89	2.220	408.19	2.204
		411.27	2.194

The Kramers-Kronig technique is far less accurate than the ellipsometric technique, since it relies on less accurate data (reflectance) and requires extrapolations to perform the Kramers-Kronig integrals. More recent measurements of complex dielectric functions were reported by Yoshikawa and Adachi[95] using spectroscopic ellipsometry in the photon energy range of 1.5–5.0 eV at room temperature. Samples grown by vapor phase methods with surfaces oriented parallel to the *c*-axis were used and the data in the transparent region of ZnO have been analyzed. The data for polarization parallel (E ∥ c) and perpendicular (E ⊥ c) to the optical axis are shown in Figure 5.20 along with the data from Mollwo,[10] and Bond.[11] Also shown in the figures are fits to a first order Sellmeier equation:

$$n(\lambda)^2 = A + \frac{B\lambda^2}{\lambda^2 - C^2} \tag{5.7}$$

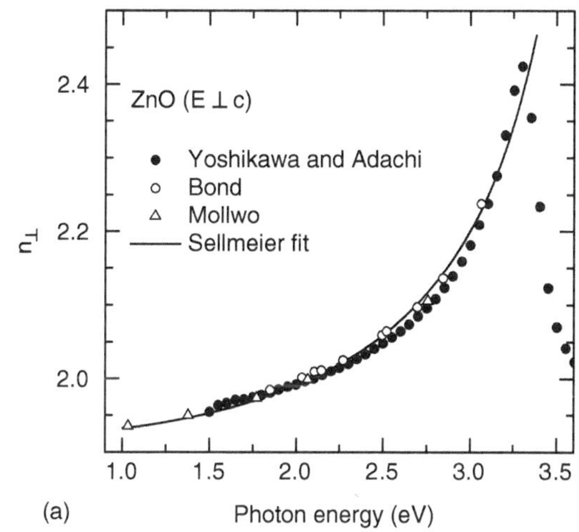

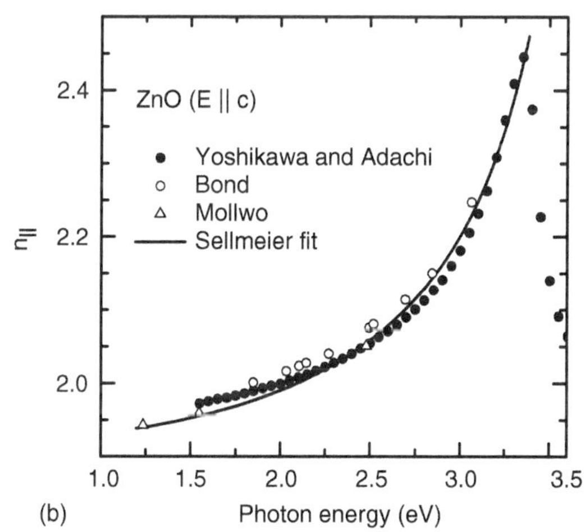

Figure 5.20: Refractive index dispersion of ZnO for (a) $E \perp c$ and (b) $E \parallel c$ below the fundamental absorption edge. The solid circles represent the spectroscopic ellipsometry data of Yoshikawa and Adachi.[95] The open circles and open triangles represent the data obtained by Bond[11] and Mollwo,[10] respectively. (Yoshikawa and Adachi[95])

where A, B, and C are the fitting parameters, and λ is the wavelength. Best fits were obtained for $A = 2.84$, $B = 0.84$, and $C = 0.319\,\mu m$ for $E \perp c$, and $A = 2.85$, $B = 0.87$, and $C = 0.310\,\mu m$ for $E \parallel c$.

Jellison and Boatner[96] employed a two-modulator generalized ellipsometry (2-MGE) technique to determine the anisotropic optical functions of ZnO, using samples grown by chemical vapor transport and hydrothermal methods. Near and above the band edge, the resulting dielectric functions from both samples

Table 5.6: Sellmeier coefficients for $Mg_xZn_{1-x}O$ films. (After Teng *et al.*[97])

	Parameter	$x=0$	$x=0.24$	$x=0.36$
Ordinary index	B	2.60 ± 0.02	2.37 ± 0.003	2.27 ± 0.004
E ⊥ c	C (μm)	0.2114 ± 0.0037	0.1793 ± 0.0012	0.1686 ± 0.0008
Extraordinary index	B	2.66 ± 0.02	2.43 ± 0.01	2.32 ± 0.01
E || c	C (μm)	0.2143 ± 0.0051	0.1815 ± 0.0016	0.1801 ± 0.0016

were identical within the experimental error, whereas the refractive indices of the hydrothermal-grown sample were ~0.012 ± 0.007 smaller than those of the vapor-transport-grown sample from 850 to 450 nm. 2-MGE is claimed to produce the most accurate results above the band edge, particularly more accurate than the results by Freeouf[14] in the measured spectral region of 3.3–5.0 eV, while the refractive indices determined below the direct band edge agree with the minimum-deviation methods[11] within an accuracy of ± 0.003.

Using the prism-coupled waveguide technique, Teng *et al.*[97] measured the indices of refraction for $Mg_xZn_{1-x}O$ epitaxial films grown on *c*-plane sapphire substrates by pulsed laser deposition. Measurements were performed for samples with x up to 0.36 in the wavelength range of 457–968 nm. The results are shown in Figure 5.21 for the ordinary and the extraordinary refractive indices. The data were fit by the least square method to the first order Sellmeier dispersion relationship with A = 1 in Equation 5.7. The data for the ZnO sample were best characterized with fitting parameters B = 2.60 ± 0.02 and C = 0.2114 ± 0.0037 μm for the ordinary index and B = 2.66 ± 0.02 and C = 0.2143 ± 0.0051 μm for the extraordinary index. The Sellmeier parameters for $Mg_xZn_{1-x}O$ films are summarized in Table 5.6.

One of the most recent reports of dielectric functions of ZnO is that by Ashkenov *et al.*[98] who characterized thin films grown by pulsed laser deposition on *c*-plane sapphire and a single crystalline sample grown by seeded chemical vapor transport method. The static dielectric constant was obtained from infrared spectroscopic ellipsometry measurements. The high frequency dielectric constant was calculated through the Lyddane–Sachs–Teller (LST) relation, $\varepsilon_0/\varepsilon_\infty = \omega_{LO}^2/\omega_{TO}^2$, using the static constant and the TO and LO phonon mode frequencies. The results are compared with the data from some of the previous reports in Table 5.7.

Schmidt *et al.*[99] also reported room temperature spectroscopic ellipsometry results on pulsed-laser-deposition grown wurtzite $Mg_xZn_{1-x}O$ ($0 < x < 0.29$) thin films. The refractive index data were fit to a three-term Cauchy approximation type formula (Equation 5.6) and the anisotropic Cauchy model parameters A, B, and C for ZnO were obtained as 1.916, 1.76×10^{-2} μm^2, and 3.9×10^{-3} μm^4 for E ⊥ c, and 1.844, 1.81×10^{-2} μm^2, and 3.6×10^{-3} μm^4 for E || c, respectively.

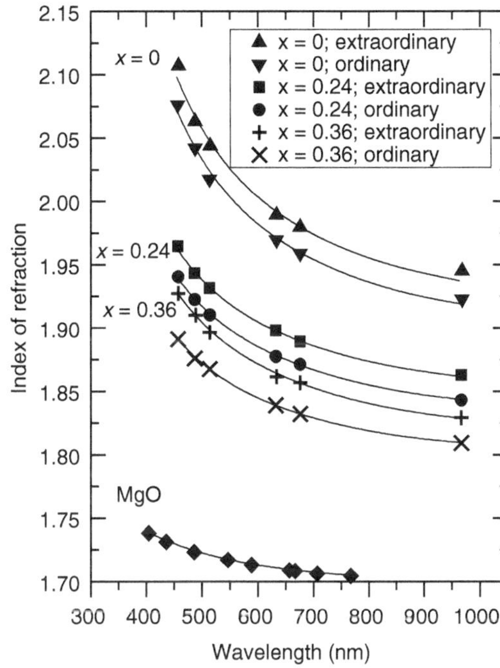

Figure 5.21: The ordinary and extraordinary refractive indices of $Mg_xZn_{1-x}O$ films ($x = 0$, 0.24, and 0.36). The solid curves are least-square fits to the first order Sellmeier dispersion. The index of refraction of cubic MgO crystal measured by Stephens and Malitson[17] is also plotted for comparison. (After Teng *et al.*[97])

Table 5.7: Static and high frequency dielectric constants of ZnO.

Sample		Ashkenov *et al.*[98]		Teng *et al.*[97]	Yoshikawa and Adachi[95]	Bond[11]
		Thin film	Bulk	Thin film	Bulk	Bulk
ε_0*	E ⊥ c	7.46	7.77	7.44	7.61	7.65
	E ∥ c	8.59	8.91	8.36	8.50	8.57
ε_∞	E ⊥ c	3.61	3.70	3.60	3.68	3.70
	E ∥ c	3.76	3.78	3.66	3.72	3.75

* Calculated via the LST relation with the phonon mode parameters obtained by Ashkenov *et al.*[98]

For the $Mg_xZn_{1-x}O$ alloy system, linear dependence of the Cauchy parameters on x was assumed:

$$n(\lambda) = \mathrm{A} + ax + \frac{\mathrm{B} + bx}{\lambda^2} + \frac{\mathrm{C} + cx}{\lambda^4}. \tag{5.8}$$

Table 5.8: Cauchy model parameters for the $Mg_xZn_{1-x}O$ alloy system. (Schmidt *et al.*[99])

	A	B (10^{-2} μm^2)	C (10^{-3} μm^4)	*a*	*b* (10^{-2} μm^2)	*c* (10^{-3} μm^4)
E ⊥ c	1.916	1.76	3.9	−0.574	−4.51	−4.90
E ∥ c	1.844	1.81	3.6	−0.782		

The coefficients *b* and *c* were found isotropic, and the parameters obtained from fits to Equation 5.8 are listed in Table 5.8. Surprisingly, for all the $Mg_xZn_{1-x}O$ thin films investigated the birefringence was found negative unlike ZnO, which showed positive birefringence (0.05) in the transparency region. This change of the sign of the birefringence going from ZnO to the $Mg_xZn_{1-x}O$ alloy system is in contrast to the results reported by Teng *et al.*,[97] where a positive birefringence of approximately 0.025 for ZnO and $Mg_xZn_{1-x}O$ for all Mg contents was found. When compared to the ZnO data by Teng *et al.*,[97] Jellison and Boatner,[96] Yoshikawa and Adachi,[95] and Hu *et al.*,[94] the ZnO ordinary and extraordinary refractive index values of Schmidt *et al.*[99] are about 0.02 lower and 0.03 higher, respectively, up to 700 nm, and the birefringence is larger.

5.4 Stimulated emission in ZnO

Even though *n*- and recently *p*-type doping have been reported in ZnO thin films,[100] there is no demonstration of electrically pumped lasing in ZnO-based structures. However, optically pumped stimulated emission (SE) has been observed by many researchers from ZnO epitaxial layers grown by a variety of methods.[101,102,103,104] There have been also earlier demonstrations, dating back to 1966, of SE and lasing in vapor-grown ZnO platelets which were cleaved parallel to the *c*-axis and pumped by an electron beam.[105,106,107]

Fabrication of low dimensional structures such as quantum wells and dots has been the focus of semiconductor laser research to decrease the threshold for lasing. Efficient stimulated emission may be obtained from these quantum structures since the transfer integral at the band edge is larger than that of the bulk semiconductor. Excitonic emission may also be used to obtain efficient lasing, which may be realized for ZnO due to its larger exciton binding energy compared to other wide band gap semiconductors. Exciton-exciton scattering-induced stimulated emission is very important for the realization of low-threshold lasers since it occurs at a threshold lower than that for the electron hole plasma (EHP) recombination. The demonstration of SE with excitonic origin paves the way for the realization of blue-violet laser diodes based on ZnO.

In the intermediate excitation density regime emissions due to biexcitonic, exciton-exciton, and exciton-carrier interactions may be observed. The inelastic collision between the excitons results in an exciton excited into a higher state and a photon with energy:[108]

$$E_n = E_{ex} - E_B^{ex}\left(1 - \frac{1}{n^2}\right) - \frac{3}{2}k_B T \quad (n = 2, 3, 4, ..., \infty), \tag{5.9}$$

where E_{ex} is the free exciton energy, E_B^{ex} is the binding energy of the exciton (60 meV for ZnO), n is the quantum number of the envelope function, and $k_B T$ is the thermal energy. At 300 K, Equation 5.9 gives $E_{ex} - E_\infty = 99$ meV, which is in good agreement with the experimental results which will be discussed below. At very high excitation intensities, an EHP forms beyond the "Mott density", which is calculated at 300 K as:[101]

$$n_M = \frac{k_B T}{2a_B^3 E_B^{ex}} \sim 3.7 \times 10^{19}\,\text{cm}^{-3}, \tag{5.10}$$

using an exciton Bohr radius of $a_B = 18$ Å for ZnO. Therefore, depending on the excitation intensity and the quality of the samples, SE originating from exciton interactions and/or EHP recombination can be observed.

Earliest experimental observations of SE and lasing in ZnO was reported under electron beam pumping for platelets prepared from vapor, with optical cavities formed by cleaving two surfaces parallel to the *c*-axis and perpendicular to the grown faces.[105,106,107,109] Pulsed operation was demonstrated with emission in the UV at temperatures near[105,106] and above[107] liquid nitrogen temperature. By investigating the spectral position of the SE peaks, Packard *et al.*[106] suggested that the laser emission is due to the annihilation of a bound exciton below 60 K and the annihilation of a free exciton with the simultaneous emission of one LO phonon at 77 K. The required exciton density for SE in ZnO at 77 K was calculated as 10^{17} cm^{-3}. Iwai *et al.*[107] could observe SE at higher temperatures with the SE peak shifting to longer wavelength side more than the A-exciton LO line, and argued that above 180 K the SE is due to the annihilation of a free exciton assisted by two LO phonons.

These reports were followed by demonstrations of optically pumped lasing in ZnO.[110,111] Using a Xe laser, Johnston[110] has reported SE at 275 K from ZnO platelets, but was unable to explain the data by exciton-LO phonon theories. Later on, Reynolds *et al.*[112] have shown optically pumped lasing in as grown ZnO platelets using a HeCd laser with very low pump power (4 W/cm^2) at 2 K. Extremely well

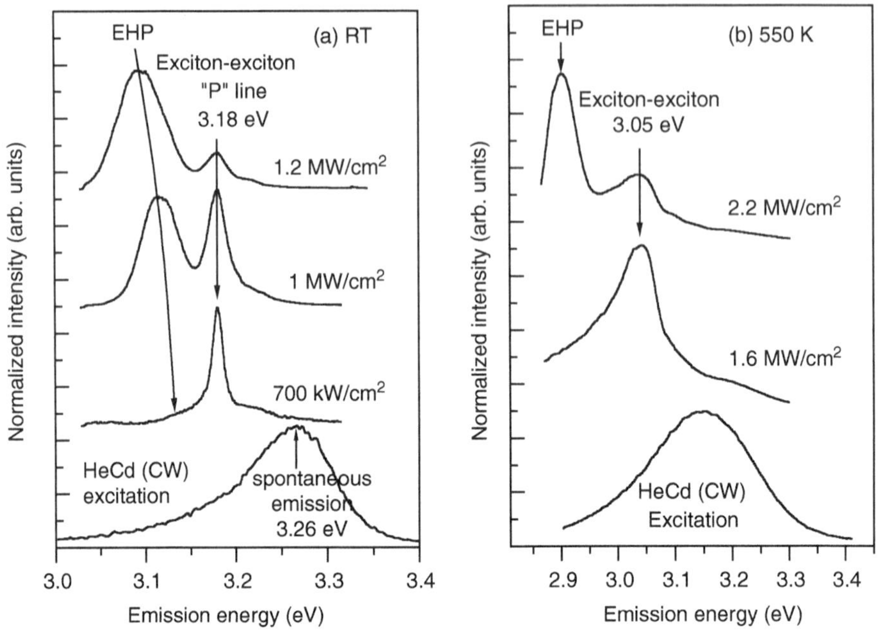

Figure 5.22: Normalized PL spectra for various excitation intensities at (a) room temperature and at (b) 550 K. (After Bagnall *et al.*[102])

formed lasing modes were observed from which a changeover from absorption to emission could be clearly detected.

As better quality ZnO layers and epitaxial cavities became available, more detailed investigations of optically pumped SE appeared in the literature. SE and lasing, which could survive even at temperatures as high as 550 K, have been observed by Bagnall *et al.*[102,113] in ZnO thin films grown by plasma enhanced MBE on *c*-plane sapphire. At room temperature, for excitation intensities exceeding 400 kW/cm^2 the exciton-exciton scattering related SE peak appeared at 3.18 eV and grew superlinearly. At higher excitation intensities (800 kW/cm^2) the electron hole plasma peak appeared at 3.14 eV and broadened and red shifted due to bandgap renormalization with further increase in the excitation intensity. The thresholds for the SE lines were 1.2 and 1.9 MW/cm^2 for the exciton-exciton scattering and EHP mechanisms, respectively. Figure 5.22 shows the SE observed at room temperature and at 550 K.

Bagnall *et al.*[114] reported optically pumped room temperature lasing from cleaved ZnO cavities even though they were unable to observe the longitudinal cavity modes directly. In a later study,[113] they could clearly see a very strong and stable mode

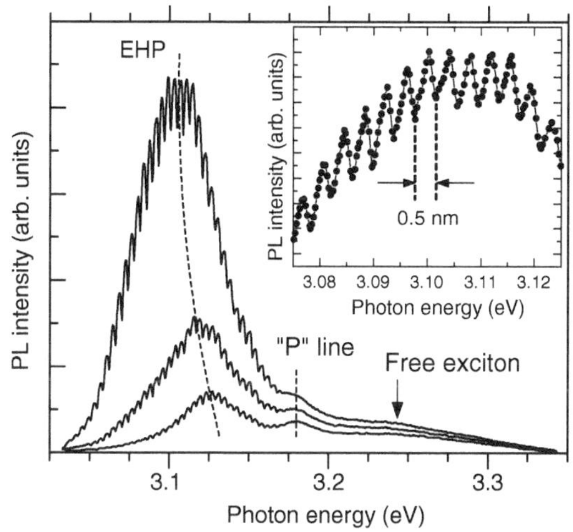

Figure 5.23: Room temperature laser emission spectra. Inset shows clearly resolved modes. (Bagnall *et al.*[113])

structure on the room temperature EHP emission peak of one of the ZnO epitaxial layers (see Figure 5.23). Although its origin was not explicitly identified, the lasing was claimed to be produced by a naturally formed or "accidental" ~70 μm cavity as calculated from the 0.5 nm mode spacing. In the same samples, pump-probe measurements revealed that the optical gain forms when photoexcited hot carriers cool down to an EHP state.[103] Yu and co-workers[104] also observed optically pumped lasing in 60 nm thick microcrystalline ZnO films grown on *c*-plane sapphire by laser MBE. However, the lasing mechanism was attributed to inelastic exciton-exciton scattering and a peak gain of 300 cm^{-1} was observed at a fluence of 3.8 $\mu J/cm^2$. Although different optical pumping conditions could be a reason for the observed discrepancy, the different crystal properties of the investigated epitaxial layers should also play a critical role in lasing mechanisms.

Özgür *et al.*[115] have shown that ZnO layers deposited directly on *c*-plane sapphire substrates by RF magnetron sputtering could have optical quality sufficient for excitonic laser action. Pulsed excitation, time-integrated PL (TI-PL) was performed using ~100 fs pulses at room temperature with average excitation energy densities varying between 5 $\mu J/cm^2$ and 700 $\mu J/cm^2$. To obtain the SE thresholds (I_{th}), the spectrally integrated PL intensities were plotted as a function of the pump energy density. SE features were observed for the samples annealed above 800°C. However, the as-grown samples didn't show any sign of SE for the maximum energy density used. The spectrally-resolved TI-PL for the sample annealed at 1000°C is shown in Figure 5.24 inset. For excitation densities above ~50 $\mu J/cm^2$ SE emerges at

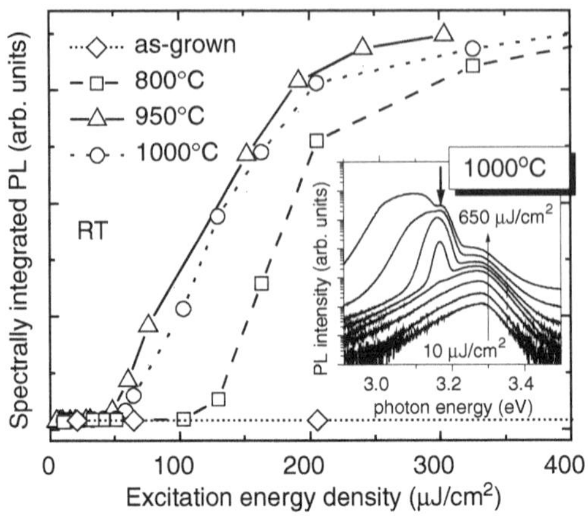

Figure 5.24: Room temperature (RT) spectrally integrated PL for the ZnO samples normalized to the spontaneous emission. The inset shows the spectrally-resolved PL for the 1000°C sample for different excitation densities. The downward-pointing arrow in the inset indicates the exciton-exciton scattering-induced SE peak. (After Özgür *et al.*[115])

3.167 eV as a sharp feature on the lower energy side of the spontaneous emission (SPE) peak and grows superlinearly. This SE peak has been attributed to exciton-exciton scattering and lies below the free exciton energy by an exciton binding energy plus the mean kinetic energy 3/2 k_BT,[101,102] where k_BT is the thermal energy. This peak then slightly redshifts to 3.160 eV since the inelastic exciton-exciton scattering leaves one exciton in an excited state which in turn reduces the emission energy of the recombining exciton.[101] As the excitation density is increased above 250 μJ/cm^2, a second peak emerges at 3.133 eV due to SE from the EHP. At these higher excitation densities, exciton wavefunctions start to overlap due to the increase in their density. Phase space filling and Coulomb interactions cause excitons to lose their individual character by ionization and eventually an EHP is formed. This EHP-induced SE peak shifts and broadens with increasing excitation as a result of bandgap renormalization. The coexistence of the exciton-exciton scattering and the EHP originates from the spatial nonuniformity of the sample as well as the laser beam profile, i.e. the EHP and the exciton-exciton scattering induced SE may come from different regions of the sample excited by the laser.

Özgür *et al.*[115] observed the SE peak attributed to exciton-exciton scattering also for another RF-sputtered sample annealed at 950°C, but not for the sample annealed at 800°C. Due to the existence of exciton-exciton scattering, I_{th} for the 950°C and 1000°C samples (49 and 58 μJ/cm^2, respectively) were significantly lower than that for the 800°C sample (130 μJ/cm^2). Figure 5.24 shows the PL data spectrally integrated between 3.1 and 3.4 eV for all the samples.

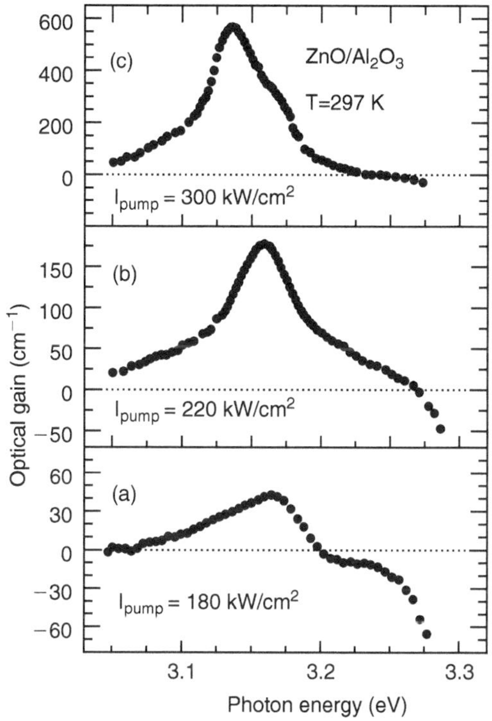

Figure 5.25: Optical gain spectrum of a ZnO epilayer at excitation density of (a) 180 kW/cm^2, (b) 220 kW/cm^2, and (c) 300 kW/cm^2 at RT. (After Chen *et al.*[101])

By using the variable stripe length method, Chen *et al.*[101] measured the optical gain spectrum of ZnO epilayers which were grown on *c*-plane sapphire by P-MBE employing a thin MgO buffer layer. The sharp increase of the emission intensity with excitation stripe length indicates the presence of optical gain. Based on the one-dimensional optical amplifier mode for an excited media, the spectra were analyzed by fitting to the equation

$$I(E) = \frac{I_0}{g}\left(e^{gL} - 1\right) \tag{5.11}$$

where g stands for the optical gain coefficient and L for the length of the excitation stripe. The evaluated optical gain as a function of photon energy is plotted in Figure 5.25. At an excitation density of 180 kW/cm^2, close to I_{th}, the peak gain is about 40 cm^{-1} and it increases to 177 cm^{-1} with a redshift of about 6 meV as the excitation density is increased to 220 kW/cm^2. The EHP peak appears at 3.135 eV with further increase of the excitation density to 300 kW/cm^2. Therefore, Chen *et al.*[101] argued that the observed optical gain resulted from the exciton-exciton scattering. For similar excitation densities, the peak gain seems smaller than that reported by Yu *et al.*,[104] which may result from different pumping conditions as well as the lack of light confinement in some of the samples.

Yu *et al.*[104] have also investigated the effect of film thickness on the optical gain. The film also forms the propagation medium in the waveguide structure, and its thickness determines the mode of propagation. They attributed the large gain to the modification of the spontaneous emission rate by the dielectric planar waveguide structure of thin film waveguides consisting of a thin ZnO layer ($n_2 = 2.45$) surrounded by air ($n_1 = 1$) on one side and sapphire ($n_3 = 1.8$) on the other. The film thickness values d, below which a guided mode of order m ceases to exist, can be calculated using the expression:

$$\left(\frac{d}{\lambda}\right)_{\mathrm{TE}} = \frac{1}{2\pi\sqrt{n_2^2 - n_3^2}}\left[m\pi + \tan^{-1}\left(\frac{n_3^2 - n_1^2}{n_2^2 - n_3^2}\right)^{1/2}\right], \tag{5.12}$$

where λ is the wavelength in ZnO. For a thickness of 60 nm, only the TE_0 mode is supported; therefore, lasing emission should be TE polarized. For sufficiently thin films where the lowest order TE guided mode exists, the spontaneous emission mostly goes into the TE_0 mode. This results in an increase in the SE and a reduction in the lasing threshold. There is an optimum thickness where the emission rate is maximum, and for thinner films this rate decreases rapidly and reaches zero at a cutoff value. Since all guided modes cease to exist below a finite layer thickness, no lasing occurs in very thin samples. According to Equation 5.12, the cutoff and the lowest threshold occur at thickness values of 29 nm and 55 nm, respectively, consistent with the observations of Yu *et al.*[104] Figure 5.26 shows the lasing threshold as a function of the film thickness. As observed in this figure, the discrepancy in the threshold values from different measurements may also be due to varying thickness.

Kawasaki *et al.*[116] and Zu *et al.*[117] investigated the effects of nanocrystal size on stimulated emission in ZnO thin films fabricated on *c*-plane sapphire substrates by laser MBE. The films having small nanocrystal size of about 50 nm showed excitonic SE as well as the SE from the EHP. The spontaneous emission generated by exciton-exciton collisions, which showed a quadratic intensity dependence, was followed by SE from the same process at higher excitation intensities exhibiting an eighth power dependence. With further increase of the excitation intensity the Coulomb interaction leads to ionization of excitons and the formation of a dense EHP. Therefore, the intensity of the excitonic SE peak is observed to reduce. This

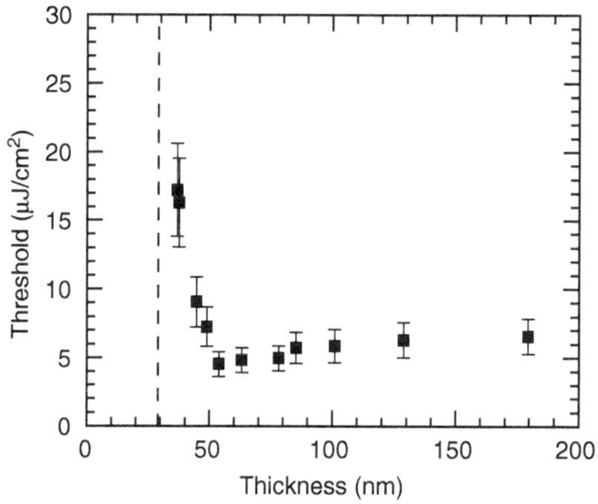

Figure 5.26: Lasing threshold as a function of film thickness. A 500 μm × 30 μm excitation stripe was used. Thickness values were determined from RHEED pattern intensity oscillations. (After Yu *et al.*[104])

was not observed in previously mentioned reports due to the sample imhomogeneity. Another major observation was that, unlike the EHP induced SE, the exciton-exciton scattering induced SE was not present in the films having larger nanocrystal size. Kawasaki *et al.*[116] concluded that there is an optimum hexagonally shaped microcrystallite size of $(50\,\text{nm})^3$ for observing excitonic SE. Analyzing similar results, Ohtomo and co-workers[118] speculated that the excitonic stimulated emission is due to the giant oscillator strength that can occur in high quality nanocrystals with a dimension larger than the exciton Bohr radius (~2 nm), but smaller than the optical wavelength (390 nm). For large grain samples, the exciton supperradiance effect becomes weak such that exciton-exciton collision process should not occur. Thus, the grain boundaries between nanocrystals are considered to serve as barriers resulting in the confinement of excitons in nanocrystals.

In high quality ZnO epitaxial films grown by MBE on GaN templates, biexciton formation was also reported and radiative recombination of the biexciton state was observed at ~3.35 eV for excitation densities lower than the threshold for exciton-exciton scattering-related SE.[119] The spectral shape of the biexciton PL, $I_{XX}(E)$, is generally expressed by an inverse Boltzmann distribution function,

$$I_M(E) \propto \sqrt{E_X - E_B^{XX} - E}\,\exp\left(-\frac{E_X - E_B^{XX} - E}{kT}\right) \tag{5.13}$$

where E_X, E_B^{XX}, k, and T are the energy of the exciton state, biexciton binding energy, Boltzmann factor, and effective temperature, respectively. From fitting to the experimental data taken at 88 K, the biexciton binding energy, E_B^{XX}, was obtained as 15 meV. In order to confirm the biexcitonic origin of the luminescence peak at ~3.55 eV Yamamoto *et al.*[119] measured the PL excitation (PLE) spectrum. They observed a peak at ~3.370 eV due to one-photon excitonic absorption (X) and a shoulder at around 3.363 eV. This shoulder was thought to be due to two-photon absorption of the biexciton state, since the giant oscillator strength of two-photon absorption is expected to occur at $E_X - E_B^{XX}/2 = 3.3625$ eV. Since the one-photon absorption band is broad, one and two-photon absorption presumably occur simultaneously in the energy region of the two-photon resonance. The broadening of the two-photon absorption band may result from local strain in the ZnO films. Thus, a shoulder rather than a clear PLE peak of two-photon absorption is seen. Therefore, the biexciton binding energy was verified to be 15 meV, which is close to the 14.7 meV value reported by Hvam *et al.*[120]

Stimulated emission has been observed also in quantum well structures utilizing the alloys of ZnO. It is expected that the observation of the excitonic SE phenomenon should be favored in quantum wells owing to the enhanced binding energy of excitons and hence by the larger stability of the exciton states due to reduced probability of dissociation by LO-phonons. Ohtomo *et al.*[121] reported the observation of SE in ZnO/ZnMgO multiple quantum wells (MQWs) up to 373 K, but did not elucidate the mechanism of SE. They investigated 10-period MQWs grown on $ScAlMgO_4$ (SCAM) substrates by laser MBE. The barriers were 5 nm thick, and the widths of the wells varied from 0.7 to 4.7 nm. The stimulated emission threshold was observed to increase with decreasing well width and also with increasing Mg composition in the barriers. Figure 5.27 shows the variation of the SE threshold with well width for two sets of samples having 12% and 26% Mg in the barriers.

Sun *et al.*[122] later investigated the mechanism of SE in the same MQW samples. They observed SE induced by both exciton-exciton scattering and EHP recombination. The exciton-originated recombination mechanism of SE was studied further by measuring the temperature dependence of the SE peak position. It was observed that the exciton-exciton band has the same temperature dependence characteristics as that of a ZnO single layer supporting the notion that this emission in ZnO/MgZnO MQWs is indeed due to exciton-exciton scattering. Additionally, the measurements at 5 K revealed that the exciton binding energy increases with decreasing QW width, reaching over 100 meV for QW widths smaller than 1.75 nm in ZnO/$Mg_{0.27}Zn_{0.73}O$ MQWs with 5 nm barriers.[123] These binding energies were deduced from the energy difference between the exciton-exciton scattering related and the free exciton emission lines.

As shown in Figure 5.28 luminescence due to the radiative recombination of localized biexcitons has also been observed for the same MQW samples at low temperatures (5 K).[124] The excitation threshold for biexciton emission (denoted by

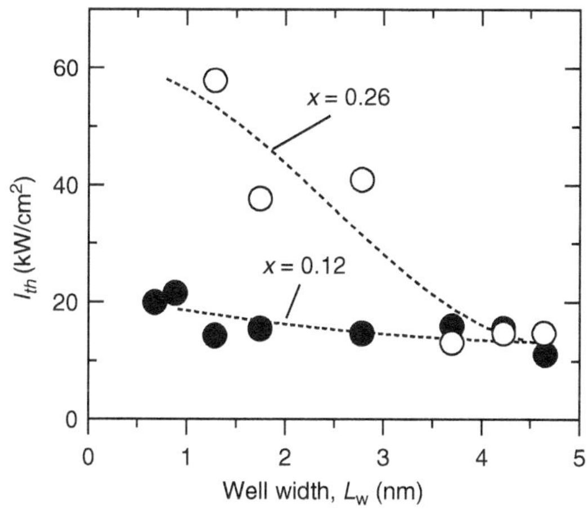

Figure 5.27: Well width (L_w) dependence of the stimulated emission threshold (I_{th}) in the superlattices with $x = 0.12$ (closed circles) and 0.26 (open circles). Stimulated emission did not take place for the $x = 0.26$ films with L_w below 1 nm since the excitation energy is lower than the absorption energy. (After Ohtomo *et al.*[121])

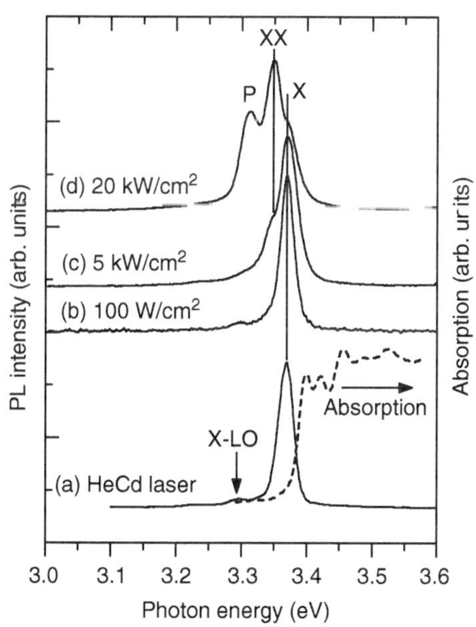

Figure 5.28: PL spectra at 5 K taken from a $ZnO/Zn_{0.74}Mg_{0.26}O$ MQW with a barrier/well widths of 5/3.7 nm. The lowest trace represent the low excitation PL and absorption spectra. (Sun *et al.*[124])

XX) was significantly lower than that for exciton-exciton scattering (denoted by P) and decreased significantly (to less than 50%) when the well width was reduced from 3.7 to 1.75 nm. The average biexciton binding energies were determined from the energy separation of the localized exciton (denoted by X) and XX peaks to be 19 and 28 meV for MQWs with well widths of 3.7 and 1.75 nm, respectively. Both of these values are larger than the bulk biexciton binding energy of 15 meV. In view of the thermal activation energy at $T = 300$ K (26 meV), it is expected that the biexcitonic effect in ZnO-based quantum structures can play an important role at higher temperatures or even at room temperature, which is desirable for the realization of ultra low threshold lasers.

5.5 Recombination dynamics in ZnO

Time-resolved PL (TRPL) is a nondestructive and powerful technique commonly used for the optical characterization of semiconductors. The free carrier or exciton lifetime, an important parameter related to material quality and device performance, can be measured by TRPL spectroscopy. The exciton lifetimes will vary with crystal quality, becoming longer as the quality improves. The efficiency of the radiative recombination is strongly related to decay time of the particular transition.

Reynolds *et al.*[125] measured the recombination lifetime of the allowed (Γ_5) and forbidden (Γ_6, allowed by induced strain) free excitons at 2 K in a strained single crystal ZnO grown by the hydrothermal method as 259 and 245 ps, respectively. The lifetime for the Γ_5 exciton was slightly higher, 322 ps, for an unstrained sample. They noted that free-exciton lifetimes are determined not only by the radiative decay, but also by the nonradiative decay and by capture processes leading to bound excitons.[48] Evidently, the measured single exponential decays reflect the effects from all three.

Teke *et al.*[32] measured the room temperature TRPL from bulk ZnO samples; one as received from Cermet, Inc. and the other annealed in forming gas (FG). Figure 5.29 shows the TRPL data for both samples at an excitation energy density of 540 μJ/cm^2. The instrument-limited rise implies that the relaxation processes to cool the carriers from 3.81 eV excitation energy-defined states to the zero momentum excitonic band edge states are very fast. For both samples, the decaying part of the TRPL data was well described by a bi-exponential decay function: $A_1 \exp(-t/\tau_1) + A_2 \exp(-t/\tau_2)$. Table 5.9 summarizes the decay constants and the amplitude ratios obtained from the fits.

The fast decay constant τ_1 is smaller for the as-received sample (170.4 ps), and most probably represents the effective non-radiative recombination at room temperature. The slow decaying component is attributed to the radiative lifetime of the

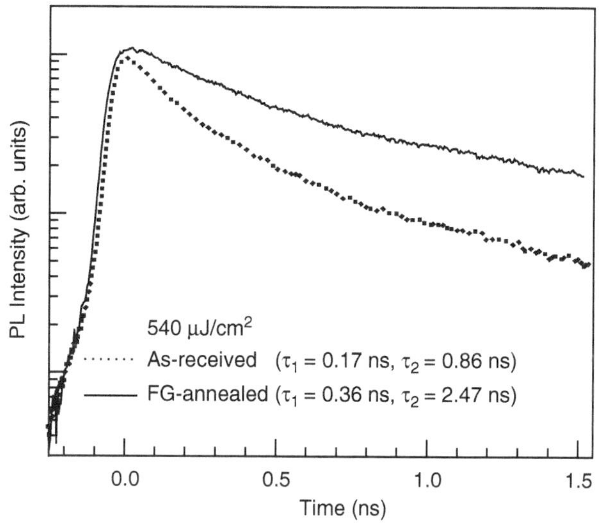

Figure 5.29: Room temperature time-resolved PL data for the as received and the forming gas treated samples. (After Teke *et al.*[32])

free exciton. The 0.86 ns value measured for the as-received sample is reasonably consistent with the 0.97 ns value measured by Koida *et al.*[126] for single crystal ZnO. The relative magnitude of the slow decaying component to the fast decaying component ($A_2/A_1 = 0.094$ for 540 μJ/cm^2) for the as-received sample suggests that the non-radiative processes are dominant. It has been proposed that the nonradiative recombination processes are governed by the defects introduced by the Zn vacancy complexes.[126] After forming gas annealing the decay constants increased remarkably ($\tau_1 = 0.86$ ns, $\tau_2 = 2.47$ ns), and the slow decaying component became dominant ($A_2/A_1 = 2.54$ for 540 μJ/cm^2), suggesting an increase in radiative recombination. This is also supported by the fact that the PL intensity increased by almost a factor of 4 in the forming gas annealed ZnO sample compared to the as-received one. The increase in the decay times is clearly observed in Figure 5.29.

When the excitation density was decreased from 540 to 54 μJ/cm^2, Teke *et al.*[32] observed a blueshift in the PL peak by 15 meV, most probably due to reducing strength of the bandgap renormalization. For the as-received sample the decay constants increase slightly with increasing excitation density, whereas the forming gas treated sample follows an opposite trend. In addition, compared to the forming gas treated sample, the as-received sample shows a more evident increase in the relative strength of the slow decaying component as the excitation density is increased, as the non-radiative centers begin to saturate.

Table 5.9: TRPL decay time constants and amplitude ratios for the ZnO samples at two different excitation energy densities. FX and DBE denote the free and donor bound excitons, respectively. (After Teke *et al.*[32])

	Samples	540 μJ/cm^2			54 μJ/cm^2		
		τ_1 (ps)	τ_2 (ps)	A_2/A_1	τ_1 (ps)	τ_2 (ps)	A_2/A_1
300 K	As-received	170.4 ± 1.8	863.9 ± 14.8	0.094	116.5 ± 1.5	585.0 ± 6.4	0.060
(FX)	FG-annealed	358.7 ± 8.8	2469 ± 256	2.536	428.3 ± 32.1	2969 ± 115	2.476
85 K	As-received	310.2 ± 2.5	1130 ± 6.6	0.653	286.8 ± 2.9	1000 ± 5.9	0.820
(DBE)	FG-annealed	474.0 ± 5.5	1285 ± 14.6	0.614	366.4 ± 4.1	1021 ± 7.3	0.869

The decay time constants of both as-received and forming gas annealed samples were also measured at 85 K.[32] At this temperature, the main DBE still dominates the overall continuous wave-PL spectrum even though A- and B-free excitons are clearly observed. However, a distinction between the bound and the free excitons could not be made in the TRPL measurements due to the resolution limitation of the experimental setup. Therefore, the TRPL data reflect mainly the decay due to the main DBE. In order to measure purely the free excitonic emission decay time, the measurements had to be performed at temperatures above 160 K where the bound exciton emission diminishes. The time constants measured at 85 K are also included in Table 5.9 for two different excitation densities. The main DBE emission lines in both samples are similar in terms of their intensities as observed from the time-integrated and the cw-PL. Compared to the as-received sample, the forming gas annealed sample showed slightly larger decay constants. Additionally, the decay times decreased with decreasing excitation energy density. In contrast to the significant improvement in free exciton lifetimes measured at room temperature, the post-growth treatment is not observed to have a strong effect on the DBE decay times.

Epitaxial ZnO layers exhibit shorter carrier lifetimes, since they suffer from higher defect densities compared to the bulk samples. Koida *et al.*[126] studied the correlation between the excitonic PL lifetime at room temperature and point defect density in bulk and epitaxial ZnO layers. The defect density, Zn vacancy being the most probable defect, was analyzed by positron annihilation. The single crystal sample showed biexponential behavior with decay constants of 0.97 and 14 ns, which were suggested to represent the free exciton lifetime, and the free carrier lifetime including trapping and emission processes, respectively. For ZnO epitaxial layers grown by laser MBE on $ScAlMgO_4$ (SCAM) substrates, the single exponential TRPL decay time, which is mainly governed by nonradiative processes, increased from 46 to 110 ps with increasing growth temperature from

570°C to 800°C. However, since a homoepitaxial film exhibited the shortest decay time (36 ps) in spite of the smallest number of point defects among the epilayers, i.e. no clear correlation was found between the PL decay time and the point defect density, the nonradiative process was considered to be governed by certain defect species introduced by the presence of Zn vacancies such as vacancy complexes.

Jung *et al.*[127] reported a biexponential decay for high quality ZnO epitaxial layers grown on sapphire by low pressure metalorganic vapor phase epitaxy. Room temperature TRPL measurements produced decay times of 180 ps and 1 ns, most probably representing the non-radiative and the radiative excitonic recombination times, respectively, consistent with the measurements of Teke *et al.*[32] and Koida *et al.*[126] on bulk ZnO samples.

In order to investigate the effects of annealing and SE on carrier dynamics in RF-sputtered ZnO thin films, TRPL spectroscopy was employed at room temperature and at 85 K by Özgür *et al.*[115] Figure 5.30 shows the TRPL data for three annealed samples at room temperature. The excitation densities were kept slightly below I_{th} ($\sim$30 μJ/cm^2) to measure the spontaneous emission (SPE) decay times, while high excitation densities ($\sim$200 μJ/cm^2) were used to observe the recombination dynamics under the influence of SE. Single exponential decay fits revealed the spontaneous recombination times as 74 ps, 59 ps, and 30 ps for the samples annealed at 1000°C, 950°C, and 800°C, respectively. The decay time for the as-deposited sample was below the system resolution. The increase of the decay times with annealing temperature suggests a reduction in the density of nonradiative recombination centers. As expected, the SE-induced recombination occurs very fast (<30 ps). The TRPL data for above I_{th} excitations also show a much weaker and slower decaying component visible after the SE is over ($\sim$55 ps) with the characteristic decay time of the spontaneous recombination.

The spontaneous recombination times observed by Özgür *et al.*[115] for the RF-sputtered ZnO thin films are comparable with other values reported in the literature. Guo *et al.*[81] reported 30 ps room temperature excitonic recombination times for ZnO thin films grown on Si by MOCVD. Koida *et al.*[126] measured recombination times of up to 110 ps for good quality ZnO thin films grown on $ScAlMgO_4$ substrates by MBE. These decay times, including the ones measured by Özgür *et al.*,[115] are much shorter than those reported for bulk ZnO[32,126] most probably due to effective nonradiative recombination in thin films at room temperature. Surprisingly, TRPL measurements performed on RF-sputtered ZnO thin films at 85 K didn't show any significant change in decay times. The inset in Figure 5.30 compares the room temperature and the 85 K TRPL data for the 950°C annealed sample for both below and above I_{th} (950°C). At 85 K the SPE decay time is 49 ps indicating that an effective nonradiative recombination mechanism is still present. However, the characteristic

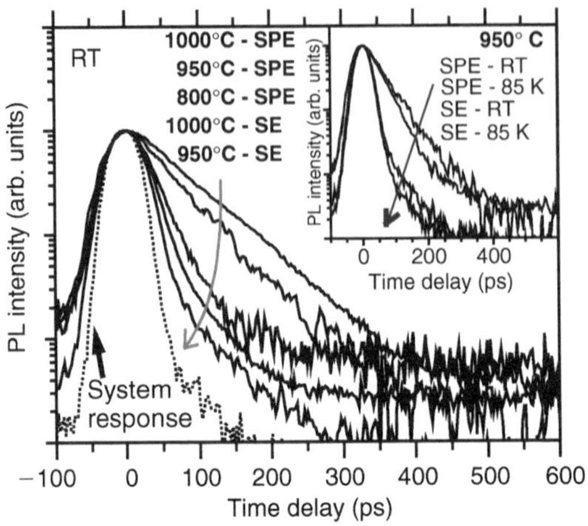

Figure 5.30: Room temperature (RT) TRPL for the spontaneous emission (SPE) and the stimulated emission (SE) from the annealed ZnO samples. The inset shows the 85 K and RT TRPL data for the sample annealed at 950°C. (After Özgür *et al.*[115])

single exponential decay along with the strong photon emission suggests that the radiative decay component is also fast. The slight decrease in the decay time at 85 K may be explained by increased absorption at low temperatures and the weak carrier density dependence of the recombination times.

5.6 Band gap engineering: a brief review

A crucial step in designing modern optoelectronic devices is the realization of band gap engineering to create barrier layers and quantum wells in device heterostructures. In order to realize such optoelectronic devices, modulation of the band gap is required. This has been demonstrated by the development of $Mg_xZn_{1-x}O$[121,128,129,130,131,132,133,134,135,136,137,138,139,140] and $Cd_yZn_{1-y}O$ alloys,[121,141,142,143,144,145,146,147] allowing band gap tuning in a wide range. The energy gap $E_g(x)$ of the ternary semiconductor $A_xZn_{1-x}O$ (where A = Mg or Cd) is determined by the following equation:[148]

$$E_g(x) = (1-x)E_{ZnO} + xE_{AO} - bx(1-x), \tag{5.14}$$

where b is the bowing parameter and E_{AO} and E_{ZnO} are the bandgap energies of compounds AO and ZnO, respectively. The bowing parameter b depends on the

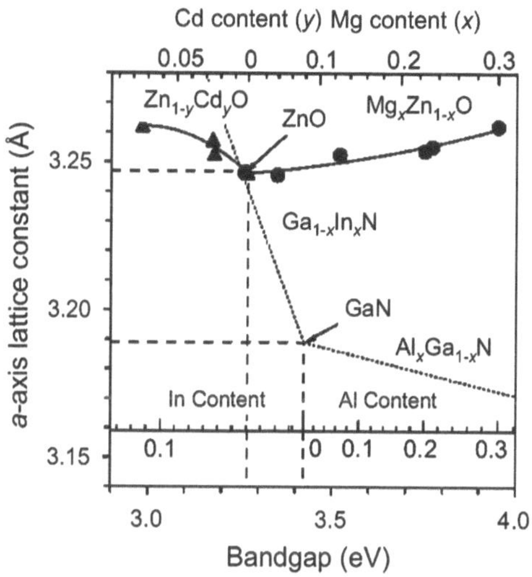

Figure 5.31: Optical and structural properties of $Cd_yZn_{1-y}O$ and $Mg_xZn_{1-x}O$ alloy films mapped out in a plane of *a*-axis length and room-temperature band gap energy. The same curves for (In, Ga)N and (Al, Ga)N alloy are also shown. Alloy compositions are shown on the top axis. (Courtesy of Makino *et al.*[140])

difference in electronegativities of the end binaries ZnO and AO. The band gap can be increased (decreased) by incorporating Mg[129,130,131] (Cd[141,142,145]) into ZnO.

Figure 5.31 shows the *a* lattice parameter as a function of room temperature E_g values and alloy compositions in $Cd_yZn_{1-y}O$ and $Mg_xZn_{1-x}O$. A (Cd, Zn)O/(Mg, Zn)O superlattice (SL), having a perfect lattice match between layers and a maximum barrier height of 0.9 eV, can be obtained by choosing an appropriate combination of Cd and Mg concentrations, because both *a*-parameters are monotonically increasing functions of alloy composition. This is a major advantage when compared to (In, Ga)N/(Al, Ga)N SLs since in the case of wurtzite structure, if the lattice constant of the well layer differs from that of the barrier layer, strain field exists inside the well layers which causes polarization charge. The same parametric plots for (In, Ga)N and (Al, Ga)N are shown by dashed curves in Figure 5.31.

5.6.1 $Mg_xZn_{1-x}O$ alloy

$Mg_xZn_{1-x}O$ alloy has been considered as a suitable material for the barrier layers in ZnO/(Mg, Zn)O superlattice structures[121] because alloying ZnO with MgO ($E_g \sim 7.7$ eV) enables widening of band gap of ZnO. According to the phase diagram of the ZnO-MgO binary system, the thermodynamic solid solubility of MgO in ZnO

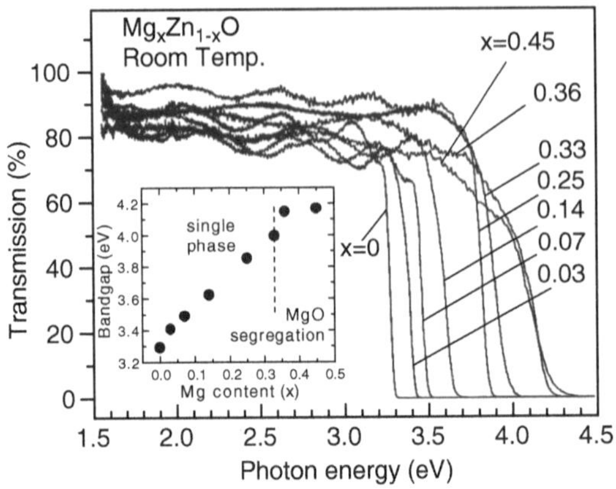

Figure 5.32: Transmittance spectra of $Mg_xZn_{1-x}O$ films measured at room temperature. The inset shows the band gap (E_g) determined from the spectra assuming an $\alpha^2 \propto (h\nu - E_g)$ dependence, where α and $h\nu$ are the absorption coefficient and the photon energy, respectively. (After Ohtomo *et al.*[128])

is less than 4 mol%.[149] In addition, ZnO has a wurtzite structure ($a = 3.24$ Å and $c = 5.20$ Å), while MgO has a cubic structure ($a = 4.24$ Å).

One of the first reports on the synthesis of $Mg_xZn_{1-x}O$ with a Mg content of up to 33% was by Ohtomo *et al.*[128] by PLD. Figure 5.32 shows the transmittance spectra measured at room temperature. The band gap E_g linearly increases with x up to 4.15 eV for $0 \leq x \leq 0.36$ and saturates at higher Mg concentrations due to MgO segregation (Figure 5.32, inset). This indicates that $Mg_xZn_{1-x}O$ is a suitable material for barrier layers in ZnO/(Mg, Zn)O heterostructures with a band gap offset up to 0.85 eV. Linear dependence of E_g on x up to $x = 0.36$ shows that the bowing parameter b is 0 eV, although a final declaration must wait for additional and more refined experiments. In other material systems, the accuracy of the bowing parameter determination was shown to depend on the quality of the measurement near the midpoint between the end binaries. This is, however, not quite feasible in the $Mg_xZn_{1-x}O$ as it begins to transform from the wurtzitic symmetry to a cubic one. Increasing the Mg composition further resulted in $Mg_xZn_{1-x}O$ films in metastable cubic-phase with band gaps above 5.0 eV.[130]

By transmission spectroscopy, Teng *et al.*[97] measured the room temperature absorption coefficients and exciton binding energies of $Mg_xZn_{1-x}O$ epitaxial films grown by PLD on sapphire substrates with x up to 0.36. The excitonic absorption features were clearly visible at room temperature despite alloy broadening. As expected, the exciton broadening parameters for these ternary alloys were found to be much larger than those of ZnO. However, the broadening is compensated by the extremely high binding energy of the exciton. The binding energies for A-, B-, and

C-excitons were independent of the Mg composition, x (60.4 meV for A, 51.5 meV for B, 50.5 meV for C-exciton).

Ohmoto *et al.*[129] studied the thermal stability of wurtzite-phase $Mg_xZn_{1-x}O$ alloy and ZnO/$Mg_xZn_{1-x}O$ bilayers grown by laser molecular-beam epitaxy. The bowing parameter of the as grown films calculated using the band gap values obtained from the absorption spectra is very high ($b = 3.11$).

Gruber *et al.*[131] grew MgZnO epilayers and ZnO/MgZnO quantum well structures by MOVPE and studied their properties. The Mg incorporation into the ZnO host material and its effect on the band gap was determined by PL measurements. An increase of 200 meV in the band gap at a Mg concentration of 10% has been achieved, and using MgZnO as the barrier material, ZnO/MgZnO quantum wells with different well widths were grown. The quantum well luminescence showed the quantization behavior and an enhancement in the exciton binding energy ($E_B > 96$ meV). The exciton binding energy is much higher than that in ZnO. The blueshift of the PL emission for the deposited $Mg_xZn_{1-x}O$ epilayers as a function of the Mg supply (composition) showed almost a linear dependence suggesting zero bandgap bowing.

5.6.2 $Cd_yZn_{1-y}O$ alloy

$Mg_xZn_{1-x}O$ alloy has been considered as a suitable material for barrier layers due to its wider band gap than that of ZnO. For narrower band gaps, which are desirable for wavelength tunability and attaining bandgaps corresponding to the visible spectrum, $Cd_yZn_{1-y}O$ alloy would be a good candidate because of the small direct band gap of CdO (2.3 eV).[150]

Makino *et al.*[140] have demonstrated the single-phase $Cd_yZn_{1-y}O$ alloy films grown by PLD on sapphire (0001) and $ScAlMgO_4$ (0001) substrates with Cd content of up to 7%. As shown in Figure 5.33, the band gap energy E_g decreases as Cd content increases and can be estimated as $E_g(y) = 3.28 - 4.40y + 5.93y^2$. The band gap observed to decrease from 3.28 eV down to 3.0 eV by introducing 7% of Cd. Figure 5.33 shows concentration (y) dependence of the room temperature absorption spectra in as-grown films. The spectrum of $Cd_{0.07}Zn_{0.93}O$ encompassed a broad shoulder on the lower energy side, indicating the formation of a cadmium-rich phase, the density of which is relatively low. Band gap energies (E_g) are plotted in the inset of Figure 5.33.

In 150-nm-thick $Cd_yZn_{1-y}O$ layers grown by MOVPE on c-plane sapphire substrates a decrease in the PL peak energy up to 300 meV was observed with increasing Cd composition, while introducing a lattice mismatch of only 0.5% with respect to binary ZnO. The redshift of the PL peaks of the samples correlated very well

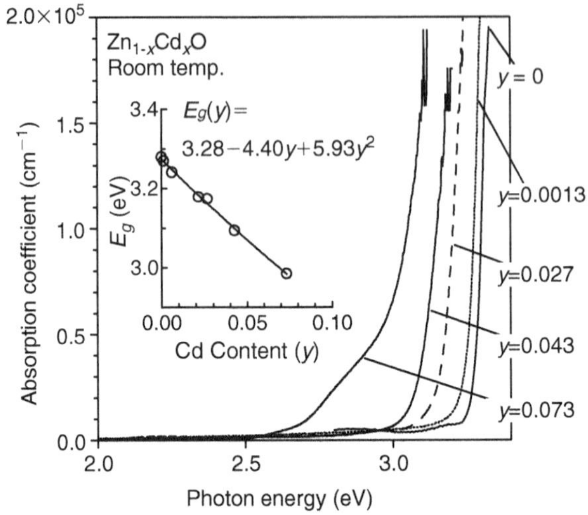

Figure 5.33: Concentration (y) dependence of absorption spectra of $Cd_yZn_{1-y}O$ epilayers obtained at room temperature. The curves, from right to left, correspond to those of the samples with $y = 0$, 0.0013, 0.027, 0.043, and 0.073. (Courtesy of Makino *et al.*[140])

with the Cd concentrations in the $Cd_yZn_{1-y}O$ layers determined by high resolution XRD and the resulting dependence strongly deviated from a linear behavior and was described by: $E_{PL} = (3.35 - 9.19y + 8.14y^2)$ eV. However, the samples in this work were shown to consist of laterally distinguishable regions with different Cd incorporation. Therefore, the corresponding very large bowing parameter ($b = 8.14$ eV) is not very reliable and is inconsistent with the results of Makino *et al.*[140]

Vigil *et al.*[146] used the spray pyrolysis method to grow $Cd_yZn_{1-y}O$ alloys. The band gap decreased from 3.28 to 2.51 eV when the composition y was changed from 0 to 1. The variation in the band gap as a function of Zn concentration was measured by optical transmission and a nonlinear dependence with a bowing parameter of approximately 0.54 was observed, which is in discord with the other results discussed above.

Ma *et al.*[142] succeeded in depositing good quality $Cd_xZn_{1-x}O$ thin films on glass and sapphire substrates using the dc reactive magnetron sputtering method. Samples exhibited good transmittance (over 85%) in the visible spectral range and a redshift was observed in the PL with increasing Cd composition. Films with $x \leq 0.53$ had structures of pure $Cd_xZn_{1-x}O$ with c-axis perpendicular to the substrate surface, while the films with $x = 0.77$ were a mixture of a hexagonal ZnO phase and a cubic

CdO phase. The band-gaps of the $Zn_{1-x}Cd_xO$ alloy films ($0 \leq x \leq 0.53$) could be nonlinearly tuned from 3.28 to 2.65 eV with the Cd contents from $x = 0$ to $x = 0.53$.

5.7 Conclusions

ZnO material has been a subject of varying degrees of research effort over the decades. Its optical properties have been studied extensively motivated by some of its unique properties and some advantages over other wide-gap materials. Among them is its large exciton binding energy of 60 meV which should pave the way for exciton-based lasing action with very low threshold currents. On the practical side, the availability of large size high quality bulk ZnO crystals is also a significant advantage.

In this chapter, a review of the optical properties of ZnO and its alloys is presented. Several principal intrinsic and extrinsic transitions involving free excitons, polaritons, bound excitons and their two electron satellites, donor acceptor pairs, and LO-phonon interactions, measured by absorption, reflection, photoreflection, and photoluminescence spectroscopy are analyzed. The refractive indices of ZnO and MgZnO, stimulated emission in ZnO thin films and quantum structures, and carrier recombination dynamics in ZnO are also discussed. Even though there is extensive data on ZnO some of its optical characteristics still remain unclear. There is no consensus on the valence band ordering in ZnO, and the chemical origin of some of the donor bound exciton transitions are yet to be determined.

Acknowledgments

The work of the authors is supported by grants from AFOSR under the leadership of Drs. G. L. Witt and G. Pomrenke. The authors would like to thank their colleagues at Cermet, J. Nause and B. Nemeth, and Air Force Research Laboratories (Wright Patterson Air Force Base), D. C. Reynolds and C. W. Litton for many useful discussions and inspiration. The use of time resolved photoluminescence facilities of Dr. H. O. Everitt and the efforts of Prof. A. Teke and Prof. M. Reshchikov in the investigation of optical transitions in ZnO are acknowledged.

References

[1] D. C. Reynolds and T. C. Collins, Phys. Rev. **185**, 1099 (1969).
[2] D. G. Thomas, J. Phys. Chem. Solids **15**, 86 (1960).

[3] J. J. Hopfield, J. Phys. Chem. Solids **15**, 97 (1960).
[4] Y. S. Park, C. W. Litton, T. C. Collins, and D. C. Reynolds, Phys. Rev. **143**, 512 (1965).
[5] B. Segall, Phys. Rev. **163**, 769 (1967).
[6] R. Dinges, D. Fröhlich, B. Staginnus, and W. Staude, Phys. Rev. Lett. **25**, 922 (1970).
[7] R. J. Collins and D. A. Kleinman, J. Phys. Chem. Solids **11**, 190 (1959).
[8] R. L. Weiher, Phys. Rev. **152**, 736 (1966).
[9] W. S. Bear, Phys. Rev. **154**, 785 (1967).
[10] E. Mollwo, Z. Angew. Phys. **6**, 257 (1954).
[11] W. L. Bond, J. Appl. Phys. **3**, 1674 (1965).
[12] W. Y. Liang and A. D. Yoffe, Phys. Rev. Lett. **20**, 59 (1968).
[13] A. R. Hutson, J. Appl. Phys. **32**, 2287 (1961).
[14] J. L. Freeouf, Phys. Rev. B **7**, 3810 (1973).
[15] O. F. Schirmer and D. Zwingel, Solid State Commun. **8**, 1559 (1970).
[16] J. J. Hopfield and D. G. Thomas, Phys. Rev. Lett. **15**, 22 (1965).
[17] R. E. Stephens and I. H. Malitson, J. Res. Natl. Bur. Stand. **49**, 249 (1952).
[18] Y. S. Park and J. R. Schneider, J. Appl. Phys. **39**, 3049 (1968).
[19] G. Heiland, E. Mollwo, and F. Stöckmann, Solid State Phys. **8**, 191 (1959).
[20] D. C. Look, Mat. Sci. Eng. B **80**, 381 (2001).
[21] D. C. Reynolds, D. C. Look, B. Jogai, C. W. Litton, G. Cantwell, and W. C. Harsch, Phys. Rev. B **60**, 2340 (1999).
[22] A. Mang, K. Reimann, and St. Rübenacke, Solid State Commun. **94**, 251 (1995).
[23] J. L. Birman, Phys. Rev. Lett. **2**, 157 (1959).
[24] W. R. L. Lambrecht, A. V. Rodina, S. Limpijumnong, B. Segall, and B. K. Meyer, Phys. Rev. B **65**, 075207 (2002).
[25] D. C. Reynolds, C. W. Litton, D. C. Look, J. E. Hoelscher, B. Claflin, T. C. Collins, J. Nause, and B. Nemeth, J. Appl. Phys. **95**, 4802 (2004).
[26] A. V. Rodina, M. Strassburg, M. Dvorzak, U. Haboeck, A. Hoffman, A. Zeuner, H. R. Alves, D. M. Hofmann, and B. K. Meyer, Phys. Rev. B **69**, 125206 (2004).
[27] K. Thonke, T. Gruber, N. Teofilov, R. Schönfelder, A. Waag, and R. Sauer, Physica B **308–310**, 945 (2001).
[28] C. Boemare, T. Monteiro, M. J. Soares, J. G. Guilherme, and E. Alves, Physica B **308–310**, 985 (2001).
[29] B. Gil, Phys. Rev. B **64**, 201310 (2001).
[30] D. C. Reynolds, D. C. Look, B. Jogai, and T. C. Collins, Appl. Phys. Lett. **79**, 3794 (2001).
[31] J. F. Muth, R. M. Kolbas, A. K. Sharma, S. Oktyabrsky, and J. Narayan, J. Appl. Phys. **85**, 7884 (1999).
[32] A. Teke, Ü. Özgür, S. Doğan, X. Gu, H. Morkoç, B. Nemeth, J. Nause, and H. O. Everitt, Phys. Rev. B **70**, 195207 (2004).
[33] S. F. Chichibu, T. Sota, G. Cantwell, D. B. Eason, and C. W. Litton, J. Appl. Phys. **93**, 756 (2003).
[34] S. F. Chichibu, A. Tsukazaki, M. Kawasaki, K. Tamura, Y. Segawa, T. Sota, and H. Koinuma, Appl. Phys. Lett. **80**, 2860 (2002).

[35] D. C. Reynolds, D. C. Look, B. Jogai, R. L. Jones, C. W. Litton, H. Harsch, and G. Cantwell, J. Luminescence **82**, 173 (1999).

[36] K. Hümmer and P. Gebhardt, Phys. Status Solidi B **85**, 271 (1978).

[37] T. Makino, C. H. Chia, N. T. Tuan, H. D. Sun, Y. Segawa, M. Kawasaki, A. Ohtomo, K. Tamura, and H. Koinuma, Appl. Phys. Lett. **77**, 975 (2000).

[38] D. C. Reynolds, D. C. Look, B. Jogai, C. W. Litton, T. C. Collins, W. Harsch, and G. Cantwell, Phys. Rev. B **57**, 12151 (1998).

[39] X. T. Zhang, Y. C. Liu, Z. Z. Zhi, J. Y. Zhang, Y. M. Lu, D. Z. Shen, W. Xu, X. W. Fan, and X. G. Kong, J. Luminescence **99**, 149 (2002).

[40] D. W. Hamby, D. A. Lucca, M. J. Klopfstein, and G. Cantwell, J. Appl. Phys. **93**, 3214 (2003).

[41] H. Alves, D. Pfisterer, A. Zeuner, T. Riemann, J. Christen, D. M. Hofmann, and B. K. Meyer, Optical Materials **23**, 33 (2003).

[42] J. J. Hopfield and D. G. Thomas, J. Phys. Chem. Solids **12**, 276 (1960).

[43] S. K. Suga, P. Cho, P. Heisinger, and T. Koda, J. Lumin. **12**, 109 (1967).

[44] C. Weisbuch and R. Ulbrich, Phys. Rev. Lett. **39**, 654 (1977).

[45] J. Lagois, Phys. Rev. B **23**, 5511 (1981).

[46] D. C. Reynolds, C. W. Litton, and T. C. Collins, Phys. Rev. **140**, A1726 (1965).

[47] B. K. Meyer, H. Alves, D. M. Hofmann, W. Kriegseis, D. Forster, F. Bertram, J. Christen, A. Hoffmann, M. Strassburg, M. Dworzak, U. Haboeck, and A. V. Rodina, Phys. Stat. Sol. (b) **241**, 231 (2004).

[48] A. Kobayashi, O. F. Sankey, and J. D. Dow, Phys. Rev. B **28**, 946 (1983).

[49] C. Gonzales, D. Block, R. T. Cox, and A. Hervé, J. Cryst. Growth **59**, 357 (1982).

[50] D. Block, A. Hervé, and R. T. Cox, Phys. Rev. B **25**, 6049 (1982).

[51] J. Rorison, D. C. Herbert, D. J. Dean, and M. S. Skolnick, J. Phys. C **17**, 6435 (1984).

[52] O. Akimoto and E. Hanamura, J. Phys. Soc. Jpn. **33**, 1537 (1972).

[53] E. V. Lavrov, J. Weber, F. Börrnert, C. G. van de Walle, and R. Helbig, Phys. Rev. B **66**, 165205 (2002).

[54] M. D. McCluskey, S. J. Jokela, K. K. Zhuravlev, P. J. Simpson, and K. G. Lynn, Appl. Phys. Lett **81**, 3807 (2002).

[55] M. Schilling, R. Helbig, and G. Pensl, J. Lumin. **33**, 201 (1985).

[56] H. J. Ko, Y. F. Chen, S. K. Hong, H. Wenish, and T. Yao, Appl. Phys. Lett. **77**, 3761 (2000).

[57] B. K. Meyer, J. Sann, D. M. Hofmann, C. Neumann, and A. Zeuner, Semicond. Sci. Technol. **20**, S62 (2005).

[58] J. R. Haynes, Phys. Rev. Lett. **4**, 361 (1960).

[59] Y. P. Varshni, Physica **34**, 149 (1967); L. Wang and N. C. Giles, J. Appl. Phys. **94**, 973 (2003).

[60] D. C. Look, D. C. Reynolds, C. W. Litton, R. L. Jones, D. B. Eason, and G. Cantwell, Appl. Phys. Lett. **81**, 1830 (2002).

[61] J. F. Rommeluère, L. Svob, F. Jomard, J. Mimila-Arroyo, A. Lusson, V. Sallet, and Y. Marfaing, Appl. Phys. Lett. **83**, 287 (2003).

[62] H. Matsui, H. Saeki, H. Tabata, and T. Kawai, J. Electrochem. Soc. **150**, G508 (2003).

[63] A. Zeuner, H. Alves, D. M. Hofmann, B. K. Meyer, A. Hoffmann, U. Haboeck, M. Strassburg, and M. Dworzak, Phys. Stat. Sol. (b) **234**, R7 (2002).
[64] C. Morhain, M. Teisseire, S. Vézian, F. Vigué, F. Raymond, P. Lorenzini, J. Guion, G. Neu, and J.-P. Faurie, Phys. Stat. Sol. (b) **229**, 881 (2002).
[65] Y. R. Ryu, T. S. Lee, and H. W. White, Appl. Phys. Lett. **83**, 87 (2003).
[66] A. K. Viswanath, J. I. Lee, S. Yu, D. Kim, Y. Choi, and C. H. Hong, J. Appl. Phys. **84**, 3848 (1998).
[67] R. Dingle, Phys. Rev. Lett. **23**, 579 (1969).
[68] R. E. Dietz, H. Kamimura, M. D. Sturge, and A. Yariv, Phys. Rev. **132**, 1559 (1963).
[69] B. M. Kimpel and H. J. Schulz, Phys. Rev. B **43**, 9938 (1991).
[70] I. J. Broser, R. K. F. Germer, H.-J. E. Schulz, and K. P. Wisznewski, Solid State Electronics **21**, 1597 (1978).
[71] P. Dahan, V. Fleurov, P. Thurian, R. Heitz, A. Hoffmann, and I. Broser, J. Phys.: Condens. Matter **10**, 2007 (1998).
[72] P. Dahan, V. Fleurov, P. Thurian, R. Heitz, A. Hoffmann, and I. Broser, Phys. Rev. B **57**, 9690 (1998).
[73] P. J. Dean, D. J. Robbins, S. G. Bishop, J. A. Savage, and P. Porteous, J. Phys. C: Solid State Phys. **14**, 2847 (1981).
[74] P. Dahan, V. Fleurov, and K. Kikoin, J. Phys.: Condens. Matter **9**, 5355 (1997).
[75] R. Kuhnert and R. Helbig, J. Luminescence **26**, 203 (1981).
[76] D. C. Reynolds, D. C. Look, B. Jogai, J. E. Van Nostrand, R. Jones, and J. Jenny, Solid State Commun. **106**, 701 (1998).
[77] D. C. Reynolds, D. C. Look, and B. Jogai, J. Appl. Phys. **89**, 6189 (2001).
[78] N. Y. Garces, L. Wang, L. Bai, N. C. Giles, L. E. Halliburton, and G. Cantwell, Appl. Phys. Lett. **81**, 622 (2002).
[79] A. F. Kohan, G. Ceder, D. Morgan, and C. G. Van de Walle, Phys. Rev. B **61**, 15019 (2000).
[80] X. Yang, G. Du, X. Wang, J. Wang, B. Liu, Y. Zhang, D. Liu, D. Liu, H. C. Ong, and S. Yang, J. Cryst. Growth **252**, 275 (2003).
[81] B. Guo, Z. R. Qiu, and K. S. Wong, Appl. Phys. Lett. **82**, 2290 (2003).
[82] H.-J. Egelhaaf and D. Oelkrug, J. Cryst. Growth **161**, 190 (1996).
[83] N. O. Korsunska, L. V. Borkovska, B. M. Bulakh, L. Yu. Khomenkova, V. I. Kushnirenko, and I. V. Markevich, J. Luminescence **102–103**, 733 (2003).
[84] B. Lin, Z. Fu, and Y. Jia, Appl. Phys. Lett. **79**, 943 (2001).
[85] K. Vanheusden, C. H. Seager, W. L. Warren, D. R. Tallant, and J. A. Voigt, Appl. Phys. Lett. **68**, 403 (1996); K. Vanheusden, W. L. Warren, C. H. Seager, D. R. Tallant, J. A. Voigt, and B. E. Gnade, J. Appl. Phys. **79**, 7983 (1996); K. Vanheusden, C. H. Seager, W. L. Warren, D. R. Tallant, J. Caruso, M. J. Hampden-Smith, and T. T. Kodas, J. Luminescence **75**, 11 (1997).
[86] S. A. Studenikin, N. Golego, and M. Cocivera, J. Appl. Phys. **84**, 2287 (1998).
[87] F. H. Leiter, H. R. Alves, A. Hofstaetter, D. M. Hoffmann, and B. K. Meyer, Phys. Stat. Sol. (b) **226**, R4 (2001).

[88] F. H. Leiter, H. R. Alves, N. G. Romanov, D. M. Hoffmann, and B. K. Meyer, Physica B **340–342**, 201 (2003).
[89] C. G. Van de Walle, Physica B **308–310**, 899 (2001).
[90] M. A. Reshchikov and R. Y. Korotkov, Phys. Rev. B **64**, 115205 (2001).
[91] R. T. Cox, D. Block, A. Hervé, R. Picard, C. Santier, and R. Helbig, Solid State Commun. **25**, 77 (1978).
[92] M. A. Reshchikov and H. Morkoç, unpublished.
[93] R. Matz and H. Lütz, Appl. Phys. **18**, 123 (1979).
[94] W. S. Hu, Z. G. Liu, J. Sun, S. N. Zhu, Q. Q. Xu, D. Feng, and Z. M. Ji, J. Phys. Chem. Solids **58**, 853 (1997).
[95] H. Yoshikawa and S. Adachi, Jpn. J. Appl. Phys. **36**, 6237 (1997).
[96] G. E. Jellison, Jr. and L. A. Boatner, Phys. Rev. B **58**, 3586 (1998).
[97] C. W. Teng, J. F. Muth, Ü. Özgür, M. J. Bergmann, H. O. Everitt, A. K. Sharma, C. Jin, and J. Narayan, Appl. Phys. Lett. **76**, 979 (2000).
[98] N. Ashkenov, B. N. Mbenkum, C. Bundesmann, V. Riede, M. Lorenz, D. Spemann, E. M. Kaidashev, A. Kasic, M. Schubert, M. Grundmann, G. Wagner, H. Neumann, V. Darakchieva, H. Arwin, and B. Monemar, J. Appl. Phys. **93**, 126 (2003).
[99] R. Schmidt, B. Rheinlander, M. Schubert, D. Spemann, T. Butz, J. Lenzner, E. M. Kaidashev, M. Lorenz, A. Rahm, H. C. Semmelhack, and M. Grundmann, Appl. Phys. Lett. **82**, 2260 (2003).
[100] K.-K. Kim, H.-S. Kim, D.-K. Hwang, J.-H. Lim, and S.-J. Park, Appl. Phys. Lett. **83**, 63 (2003).
[101] Y. Chen, N. T. Tuan, Y. Segawa, H.-J. Ko, S.-K. Hong, and T. Yao, Appl. Phys. Lett. **78**, 1469 (2001).
[102] D. M. Bagnall, Y. F. Chen, Z. Zhu, T. Yao, M. Y. Shen, and T. Goto, Appl. Phys. Lett. **73**, 1038 (1998).
[103] A. Yamamoto, T. Kido, T. Goto, Y. Chen, T. Yao, and A. Kasuya, Appl. Phys. Lett. **75**, 469 (1999).
[104] P. Yu, Z. K. Tang, G. K. L. Wong, M. Kawasaki, A. Ohtomo, H. Koinuma, and Y. Segawa, J. Cryst. Growth **184–185**, 601 (1998).
[105] F. H. Nicoll, Appl. Phys. Lett. **9**, 13 (1966).
[106] J. R. Packard, D. A. Campbell, and W. C. Tait, J. Appl. Phys. **38**, 5255 (1967).
[107] S. Iwai and A. Namba, Appl. Phys. Lett. **16**, 354 (1970).
[108] C. Klingshirn, Phys. Stat. Sol. (b) **71**, 547 (1975).
[109] J. Shewchun, B. K. Garside, B. S. Kawasaki, and T. Efthymiopoulos, J. Appl. Phys. **43**, 545 (1972).
[110] W. D. Johnston, Jr., J. Appl. Phys. **42**, 2731 (1971).
[111] T. Goto and D. W. Langer, J. Appl. Phys. **42**, 5066 (1971).
[112] D. C. Reynolds, D. C. Look, and B. Jogai, Solid State Commun. **99**, 873 (1996).
[113] D. M. Bagnall, Y. F. Chen, M. Y. Shen, Z. Zhu, T. Goto, and T. Yao, J. Cryst. Growth **184–185**, 605 (1998).
[114] D. M. Bagnall, Y. F. Chen, Z. Zhu, T. Yao, S. Koyama, M. Y. Shen, and T. Goto, Appl. Phys. Lett. **70**, 2230 (1997).

[115] Ü. Özgür, A. Teke, C. Liu, S.-J. Cho, H. Morkoç, and H. O. Everitt, Appl. Phys. Lett **84**, 3223 (2004).

[116] M. Kawasaki, A. Ohtomo, I. Ohkubo, H. Koinuma, Z. K. Tang, P. Yu, G. K. L. Wong, B. P. Zhang, and Y. Segawa, Mater. Sci. Eng. B **56**, 239 (1998).

[117] P. Zu, Z. K. Tang, G. K. L. Wong, M. Kawasaki, A. Ohtomo, H. Koinuma, and Y. Segawa, Solid State Commun. **103**, 459 (1997).

[118] A. Ohtomo, M. Kawasaki, Y. Sakurai, Y. Yoshida, H. Koinuma, P. Yu, Z. K. Tang, G. K. L. Wong, and Y. Segawa, Mater. Sci. Eng. B **54**, 24 (1998).

[119] A. Yamamoto, K. Miyajima, T. Goto, H. J. Ko, and T. Yao, J. Appl. Phys. **90**, 4973 (2001).

[120] J. M. Hvam, G. Blattner, M. Reuscher, and C. Klingshirn, Phys. Status Solidi B **118**, 179 (1983).

[121] A. Ohtomo, K. Tamura, M. Kawasaki, T. Makino, Y. Segawa, Z. K. Tang, G. K. Wong, Y. Matsumoto, and H. Koinuma, Appl. Phys. Lett. **77**, 2204 (2000).

[122] H. D. Sun, T. Makino, N. T. Tuan, Y. Segawa, Z. K. Tang, G. K. L. Wong, M. Kawasaki, A. Ohtomo, K. Tamura, and H. Koinuma, Appl. Phys. Lett. **77**, 4250 (2000).

[123] H. D. Sun, T. Makino, Y. Segawa, M. Kawasaki, A. Ohtomo, K. Tamura, and H. Koinuma, J. Appl. Phys. **91**, 1993 (2002).

[124] H. D. Sun, T. Makino, Y. Segawa, M. Kawasaki, A. Ohtomo, K. Tamura, and H. Koinuma, Appl. Phys. Lett. **78**, 3385 (2001).

[125] D. C. Reynolds, D. C. Look, B. Jogai, J. E. Hoelscher, R. E. Sherriff, M. T. Harris, and M. J. Callahan, J. Appl. Phys. **88**, 2152 (2000).

[126] T. Koida, S. F. Chichibu, A. Uedono, A. Tsukazaki, M. Kawasaki, T. Sota, Y. Segawa, and H. Koinuma, Appl. Phys. Lett. **82**, 532 (2003).

[127] S. W. Jung, W. I. Park, H. D. Cheong, G.-C. Yi, H. M. Jang, S. Hong, and T. Joo, Appl. Phys. Lett. **80**, 1924 (2002).

[128] A. Ohtomo, M. Kawasaki, T. Koida, K. Masubuchi, and H. Koinuma, Appl. Phys. Lett. **72**, 2466 (1998).

[129] A. Ohtomo, R. Shiroki, I. Ohkubo, H. Koinuma, and M. Kawasaki, Appl. Phys. Lett. **75**, 4088 (1999).

[130] S. Choopun, R. D. Vispute, W. Yang, R. P. Sharma, T. Venkatesan, and H. Shen, Appl. Phys. Lett. **80**, 1529 (2002).

[131] T. Gruber, C. Kirchner, R. Kling, F. Reuss, and A. Waag, Appl. Phys. Lett. **84**, 5359 (2004).

[132] K. Ogata, K. Koike, T. Tanite, T. Komuro, F. Yan, S. Sasa, M. Inoue, and M. Yano, J. Cryst. Growth **251**, 623 (2003).

[133] T. Minemmoto, T. Negami, S. Nishiwaki, H. Takakura, and Y. Hamakawa, Thin Solid Films **372**, 173 (2000).

[134] S. Krishnamoorthy, A. A. Iliadis, A. Inumpudi, S. Choopun, and R. D. Vispute, T. Venkatesan, Solid-State Electronics **46**, 1633 (2002).

[135] T. Makino, A. Ohtomo, C. H. Chia, Y. Segawa, H. Koinuma, and M. Kawasaki, Physica E **21**, 671 (2004).

[136] F. K. Shan, B. I. Kim, G. X. Liu, Z. F. Liu, J. Y. Sohn, W. J. Lee, B. C. Shin, and Y. S. Yu, J. Appl. Phys. **95**, 4772 (2004).
[137] D. J. Cohen, K. C. Ruthe, and S. A. Barnett, J. Appl. Phys. **96**, 459 (2004).
[138] A. Ohtomo, M. Kawasaki, I. Ohkubo, H. Koinuma, T. Yasuda, and Y. Segawa, Appl. Phys. Lett. **75**, 980 (1999).
[139] H. D. Sun, T. Makino, Y. Segawa, M. Kawasaki, A. Ohtomo, K. Tamura, and H. Koinuma, J. Appl. Phys. **91**, 1993 (2002).
[140] T. Makino, Y. Segawa, M. Kawasaki, A. Ohtomo, R. Shiroki, K. Tamura, T. Yasuda, and H. Koinuma, Appl. Phys. Lett. **78**, 1237 (2001).
[141] T. Makino, C. H. Chia, N. T. Tuan, Y. Segawa, M. Kawasaki, A. Ohtomo, K. Tamura, and H. Koinuma, Appl. Phys. Lett. **77**, 1632 (2000).
[142] D. W. Ma, Z. Z. Ye, and L. L. Chen, Phys. Stat. Sol. (a) **201**, 2929 (2004).
[143] K. Sakurai, T. Takagi, T. Kubo, D. Kajita, T. Tanabe, H. Takasu, S. Fujita, and S. Fujita, J. Cryst. Growth **237–239**, 514 (2002).
[144] T. Gruber, C. Kirchner, R. Kling, F. Reuss, A. Waag, F. Bertram, D. Forster, J. Christen, and M. Schreck, Appl. Phys. Lett. **83**, 3290 (2003).
[145] T. Tanabe, H. Takasu, S. Fujita, and S. Fujita, J. Cryst. Growth **237–239**, 514 (2002).
[146] O. Vigil, L. Vaillant, F. Cruz, G. Santana, A. Morales-Acevedo, and G. Contreras-Puente, Thin Solid Films **361–362**, 53 (2000).
[147] Y.-S. Choi, C.-C. Lee, and S. M. Cho, Thin Solid Films **289**, 153 (1996).
[148] J. A. Van Vechten and T. K. Bergstresser, Phys. Rev. B **1**, 3351 (1970).
[149] E. R. Segnit and A. E. Holland, J. Am. Ceram. Soc. **48**, 412 (1965).
[150] R. P. Koffyberg, Phys. Rev. B **13**, 4470 (1976).

Chapter 6

Minority Carrier Transport in ZnO and Related Materials

O. Lopatiuk[a], *A. Osinsky*[b] *and L. Chernyak*[a]

[a] Department of Physics, University of Central Florida, Orlando, FL 32816, USA
[b] SVT Associates, Eden Prairie, MN 55344, USA

Abstract: Transport properties of minority carriers in ZnO and related compounds are of critical importance for the functionality of bipolar devices. This review summarizes the available information on the subject of minority carrier transport in ZnO-based semiconductors, focusing on its temperature dependence and the dynamics of non-equilibrium carrier recombination. The influence of deep traps on minority carrier diffusion length and lifetime is discussed. A brief summary of the role of minority carrier transport in the performance of bipolar photovoltaic devices, as well as a discussion of techniques used for measurements of the minority carrier diffusion length and lifetime are also provided.

Keywords: ZnO, diffusion length, lifetime, electron injection.

6.1 Introduction

Recent developments in doping and growth of ZnO stimulated a renewal of interest in this material from the point of view of its applications in optoelectronic devices. As a direct wide bandgap semiconductor ($E_g \approx 3.35$ eV at room temperature) with high exciton binding energy (60 meV, compared to 25 meV in GaN), ZnO is a superior candidate for minority-carrier-based devices, such as light emitting diodes, laser diodes, and transparent *p-n* junctions. Moreover, it offers significant advantages over III-Nitride materials, which include availability of large area lattice-matched substrates, lower materials costs, and use of wet chemical etching, as opposed to reactive ion etching employed in GaN technology.

In the present state of the art, the development of the full potential of ZnO in bipolar device applications hinges in part on the availability of quality, highly conductive materials of both *n*- and *p*-type. Similar to GaN, achieving *n*-type conductivity in ZnO does not present a problem, since even nominally undoped material is generally *n*-type, due to the electrical activity of native defects, such as zinc interstitials,

Zinc Oxide Bulk, Thin Films and Nanostructures
C. Jagadish and S. Pearton (Editors)

zinc antisites, and oxygen vacancies [1,2], as well as hydrogen impurity [3]. Heterojunction devices employing n-ZnO have been demonstrated with p-AlGaN [4], p-SiC [5], p-diamond [6], and p-AlGaN/p-GaN heterostructure [7,8] as hole emitters.

On the other hand, *p*-type conductivity with sufficiently high carrier concentrations appears to be much more elusive, mainly due to high ionization energies of potential acceptors, such as nitrogen, phosphorus, and arsenic. This problem is compounded by high concentrations of native and unintentional donors, which act as compensating centers, thus further reducing the free carrier concentration. It has been shown that the background donor concentration can be lowered if Mg is incorporated into the ZnO lattice, since each percent of Mg increases the band gap of ZnO by 0.02 eV, suppressing the ionization of shallow donors [9]. The conversion to *p*-type can then be obtained by heavily doping the resultant material with phosphorus followed by annealing in O_2 atmosphere [10].

Recently, successful *p*-type doping of ZnO also has been attained using phosphorus [11,12], nitrogen [13], arsenic [14], or antimony [15] yielding net hole concentrations up to 10^{18} cm^{-3}. It should be noted that while ZnO homojunction devices have been reported, starting as early as the year 2000 [16–20], their quality is generally inferior to the heterojunctions employing n-ZnO.

Even with the advent of new technology enabling the production of viable *p*-type materials, the performance of bipolar devices is fundamentally limited by the transport properties of minority carriers. In direct bandgap semiconductors, including ZnO, minority carrier diffusion length is generally several orders of magnitude lower than in indirect gap materials such as silicon or germanium. In order to noticeably increase minority carrier diffusion length by reducing scattering by the dislocation walls, the edge threading dislocation density must be reduced by at least two orders of magnitude from a typical value of about 10^9 cm^{-2} in epitaxial ZnO layers [21–24].

Investigation of minority carrier diffusion lengths and lifetimes in both *n*- and *p*-type ZnO is an issue of practical importance, since it has direct implications on the performance of bipolar devices. Moreover, considering possible applications of these devices in high-temperature electronics, the insight into the temperature dependence of minority carrier properties is also of great value. Nonetheless, to the best of our knowledge, the reports on this subject are rather scarce. It is therefore the goal of this work to summarize the available information on the subject of minority carrier transport in ZnO and related compounds, focusing on its temperature dependence and the dynamics of non-equilibrium carrier recombination. This discussion will be preceded by a brief summary of the role of minority carrier transport in the performance of bipolar devices, as well as by the review of techniques of choice for measurement of the minority carrier diffusion length.

6.2 Role of minority carrier diffusion length in bipolar device performance

In general, when non-equilibrium carriers are generated in a material due to external excitation in the absence of electric field, they diffuse over a certain distance before undergoing recombination. This parameter, namely the average distance traveled in a particular direction between generation and recombination, is characterized by the diffusion length, L. The diffusion length is related to the carrier lifetime, τ (i.e., the time between generation and recombination of non-equilibrium carriers), through carrier diffusivity, D:

$$L = \sqrt{D\tau} \tag{6.1}$$

Diffusivity, or diffusion coefficient, is determined in turn by the mobility of the carriers, μ, according to the Einstein relation:

$$D = \frac{kT}{q}\mu \tag{6.2}$$

where k is the Boltzmann constant, T is absolute temperature, and q is fundamental charge.

The diffusion process is driven by concentration gradients; since external excitation has a much larger impact on the concentration of *minority* carriers than that of majority ones (because generation density is usually much lower than the majority carrier density), it is the minority carriers that are more susceptible to diffusion.

Diffusion of minority carriers is a process that is fundamental to the operation of bipolar, in particular photovoltaic, devices, with minority carrier diffusion length being the central parameter defining the device performance. In the presence of a *p-n* junction or a Schottky barrier, the non-equilibrium minority carriers generated by external excitation (e.g., light incident on a photodiode) within a few diffusion lengths of the space-charge region can be collected by the built-in field and thus contribute to the current flow across the device. The greater the diffusion length of the carriers, the more current can be collected, leading to the higher efficiency of the device. In photodiodes, it is usually only one side of the *p-n* junction that contributes to photocurrent. If the light is absorbed in the p-region of the junction, the quantum efficiency, η, can be represented as follows:

$$\eta = (1 - r)\left(1 - \frac{e^{-aW}}{1 + aL_n}\right) \tag{6.3}$$

where r and a are reflection and absorption coefficients, respectively, W is the width of the space-charge region, and L_n is the diffusion length of minority electrons.

Quantum efficiency is directly related to the spectral responsivity, R, of a photodiode:

$$R(E) = \frac{I_{ph}}{P_{op}} = \frac{q\eta}{E} \tag{6.4}$$

where I_{ph} is total photocurrent, P_{op} is optical power incident on the device, q is the fundamental charge, and E is the energy of the incident photons. The relationship between minority carrier diffusion length and the responsivity of Schottky photodiodes has been examined in great detail in ref. [25].

Schottky photodiodes are among the simplest photovoltaic devices, where the non-equilibrium minority carriers generated in the bulk of the semiconductor due to light absorption are collected by the built-in field of the Schottky barrier deposited on the surface of the semiconductor. In order for the carriers to contribute to device current, they have to be generated within a few diffusion lengths of the collector.

For incident energies greater than the band gap of the absorber material, non-equilibrium electron-hole pairs are generated only in the thin layer next to the surface of incidence, with the maximum depth of $1/a$. This value is on the order of 100 nm in ZnO [26,27] and is generally much smaller than the thickness of the absorber layer. Considering a front-illuminated configuration (in which the incident light passes through the semitransparent Schottky contact), if L is greater than the generation depth, most of the non-equilibrium minority carriers can diffuse a sufficient distance to be collected by the built-in field of the space-charge region. In this case, the internal quantum efficiency of the device approaches 100%, and the responsivity is independent of the diffusion length value. Below this threshold, the responsivity decreases with L, provided that the width of the space-charge region is smaller than the generation depth.

If, on the other hand, the energy of incident light is below the band gap, light penetration depth is large (on the order of several micrometers) [26,27], and a fair portion of the non-equilibrium carriers is generated in the neutral region of the semiconductor due to the ionization of the mid-gap levels. Since only the carriers within a few diffusion lengths of the space-charge region contribute to photocurrent, the responsivity at below-bandgap energies is limited by the diffusion length (unless the diffusion length exceeds the thickness of the absorber layer, in which case the latter is the limiting factor) [25].

6.3 Methods for determination of minority carrier lifetime and diffusion length

Although as of the date of this writing, minority carrier transport in ZnO remains, with a few exceptions, essentially unexplored, the measurement of minority carrier diffusion length is a well-established subject. This section reviews three of the most widely used techniques, namely Electron Beam Induced Current (EBIC), Surface Photovoltage (SPV), and Time-Resolved Photoluminescence (TRPL). While, to the best of the authors' knowledge, only the EBIC method has been applied to the studies of ZnO, SPV and TRPL techniques are also reviewed in an effort to provide a more complete discussion.

6.3.1 Electron Beam Induced Current technique

Due to a unique combination of convenience and reliability, EBIC method is among the most popular techniques for minority carrier diffusion length measurements. It requires comparatively simple sample preparation and is used *in situ* in a scanning electron microscope (SEM). Figure 6.1 shows a typical measurement configuration known as planar-collector geometry. The use of this configuration was pioneered by Ioannou, Davidson, and Dimitriadis [28,29]. As a charge collection technique, EBIC method employs a Schottky barrier or a *p-n* junction to collect the current resulting from the non-equilibrium minority carriers generated by the beam of the SEM. As the beam is moved away from the barrier/junction in a line-scan mode, the current decays as fewer and fewer minority carriers are able to diffuse to the space-charge region.

The mathematical model for EBIC was further developed by Boersma *et al.* [30], who showed that the decay of current, I_{EBIC}, can be described by the following

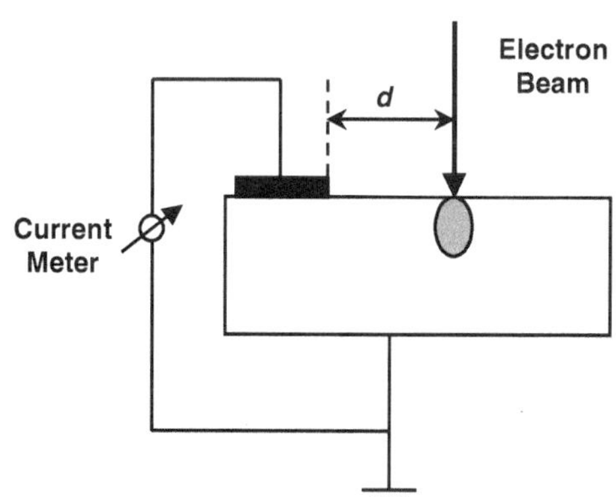

Figure 6.1: Experimental setup used for EBIC measurements in planar-collector configuration. The dashed ellipse represents the electron beam generation volume; d is the variable distance between electron beam and the Schottky contact, represented by a solid rectangle.

expression:

$$I_{EBIC} = Ad^{\alpha} \exp\left(-\frac{d}{L}\right) \tag{6.5}$$

where A is a scaling constant, d is beam-to-junction distance, and α is an exponent related to the surface recombination velocity, v_s. The diffusion length is usually extracted by rearranging the terms of Eqn. (6.5):

$$\ln(Id^{-\alpha}) = -\frac{d}{L} + \ln(A) \tag{6.6}$$

which yields a linear relationship between $\ln(Id^{-\alpha})$ and d with a slope equal to $-1/L$. This approach is taken to be accurate for $d > 2L$. In should be noted that in ref. [30], the authors analyzed only the two asymptotic cases, namely $v_s = 0$ and $v_s = \infty$, and found that $\alpha = -1/2$ for the former and $-3/2$ for the latter. Later, Chan *et al.* [31] demonstrated that this approach can be applied to materials with arbitrary surface recombination velocity by selecting α such that a linear relationship between $\ln(Id^{-\alpha})$ and d is obtained. However, even if *any* value of α is used (such that $-3/2 \leq \alpha \leq -1/2$), the result for the diffusion length changes by less than 20% [32]; this is due to the fact that the exponential term dominating the value of I_{EBIC} is independent of α.

6.3.2 Surface Photovoltage technique

SPV is an attractive technique because it is non-destructive and contactless. The sample preparation does not require deposition of contacts, junctions, or high-temperature processing. The use of SPV technique for minority carrier diffusion length measurements was first suggested by Moss [33] and is reviewed in great detail in ref. [34].

SPV method generally requires a special treatment to create an accumulation of charge on the surface of a semiconductor. The options available for surface charging include chemical treatment and deposition of corona charge. The nature of the chemical treatment depends on the composition of the semiconductor of interest, but usually involves immersion in a strongly oxidizing or reducing medium. Corona discharge method consists of the deposition of ions on the surface due to an electric field applied to a source of ions.

The deposited surface charge induces equal and opposite charge density just below the surface, thus forming a space-charge region. If the surface of the semiconductor is then illuminated with monochromatic, above-bandgap light of a wavelength λ, the minority carriers generated in the space-charge region as well as within a few diffusion lengths of the latter diffuse and/or drift toward the surface, establishing a surface potential which is proportional to the non-equilibrium minority carrier concentration at the edge of the space-charge region. Changing the wavelength of excitation allows to monitor the diffusion of the carriers generated at different distances from the surface, since the maximum light penetration depth is determined by the absorption coefficient $a(\lambda)$. Throughout this process, the intensity of the incident light, I, is adjusted in order to keep the surface potential constant. Under certain assumptions [34], the relationship between I and λ has the following form:

$$I = C\left(\frac{1}{a(\lambda)} + L\right) \tag{6.7}$$

where C is an arbitrary constant. The diffusion length can be determined by extrapolating the data to obtain the $1/a(\lambda)$ intercept. Although SPV is a relatively straightforward and reliable method [35–38], it is clear from Eqn. (6.7) that accurate analysis of the SPV data requires independent knowledge of $a(\lambda)$ in the measured specimen.

6.3.3 Time-Resolved Photoluminescence technique

TRPL technique is an indirect method that relies upon measuring the minority carrier lifetime in order to determine the diffusion length according to Eqns. (6.1) and (6.2). It provides a time-efficient way for characterizing the transport of non-equilibrium minority carriers and is most useful for materials with good luminescence properties. A significant advantage of this method is that it also provides valuable insight into the nature of the recombination mechanisms governing minority carrier lifetime. The detailed description of the TRPL measurements and analysis is provided in ref. [39].

TRPL measurements are based on recording the transient decay of near-bandgap (NBE) photoluminescence excited by a short pulse of light, such as from a pulsed laser source. If the concentration of non-equilibrium minority carriers decreases exponentially from its initial value ρ_o, the intensity of the photoluminescence, I_{PL}, also follows an exponential decay:

$$I_{PL} = \frac{\rho_o}{\tau_R}\exp\left(-\frac{t}{\tau}\right) \tag{6.8}$$

where τ_R is radiative recombination lifetime. In most cases, however, the photoluminescence intensity is affected not only by lifetime of minority carriers in the band, but also by the diffusion rate out of the absorber region, since the escape of the carriers from the active region is in this case equivalent to the reduction in their lifetime. Therefore, the lifetime obtained from the TRPL measurements is often less than the true minority carrier lifetime. In order to obtain an accurate measurement, it is often necessary to introduce a confinement structure to reduce the influence of diffusion. In such a confinement structure the semiconductor of interest is "sandwiched" between two layers with wider band gap and like doping type, so that the wide bandgap layers act as minority carrier mirrors.

6.4 Temperature dependence of minority carrier diffusion length and lifetime

Because of its intrinsic thermal stability, ZnO is a good candidate for high-temperature optoelectronic devices. However, the subject of the temperature dependence of minority carrier transport properties has not yet been adequately addressed in the literature. This section reviews the results obtained from variable-temperature diffusion length measurements performed on bulk ZnO [40].

The samples under investigation were weakly n-type ZnO substrates with electron concentration of $\sim 10^{14}\,\mathrm{cm}^{-3}$ and mobility of $\sim 150\,\mathrm{cm}^2/\mathrm{Vs}$. Secondary Ion Mass Spectroscopy (SIMS) measurements revealed the presence of Li in these substrates at the level of about $3 \times 10^{16}\,\mathrm{cm}^{-3}$ [41]. The Schottky contacts for EBIC measurements were deposited on the non-polar a-plane of ZnO crystals by electron beam evaporation of 80-nm-thick Au layer and subsequent lift off.

The studies of L as a function of temperature were carried out using EBIC method in a planar-collector configuration with a Schottky barrier (Section 6.3.1). At each temperature, several measurements were taken by scanning the beam of the SEM along a line perpendicular to the edge of the Schottky contact and recording the exponential decay of current. The recorded data were fitted with Eqn. (6.6) using $\alpha = -1/2$. This value corresponds to zero surface recombination velocity which, given the excellent luminescence properties of these samples and a good fit to the experimental results, is a reasonably good approximation. It should be noted that each EBIC line-scan was recorded on a previously unexposed area in order to avoid the influence of electron irradiation on the value of diffusion length (cf. Section 6.5 below).

Table 6.1 summarizes the results of EBIC measurements performed on one of the bulk ZnO samples and shows that the diffusion length of minority holes increases with increasing temperature, T. The increase of L with T is not unique to this semiconductor. Similar trends were previously observed in GaAs [42] and

Table 6.1: Temperature dependence of minority carrier diffusion length and cathodoluminescence intensity of the near-band-edge peak in n-ZnO [40]. Reprinted with permission from the American Institute of Physics.

Temperature (°C)	Diffusion length (μm)	CL intensity (10^3 counts)
25	0.438 ± 0.022	72.1 ± 3.7
50	0.472 ± 0.060	54.4 ± 3.8
75	0.493 ± 0.028	49.2 ± 2.4
100	0.520 ± 0.074	44.6 ± 4.7
125	0.547 ± 0.086	38.5 ± 6.8
E_A (eV)	0.045 ± 0.002	0.058 ± 0.007

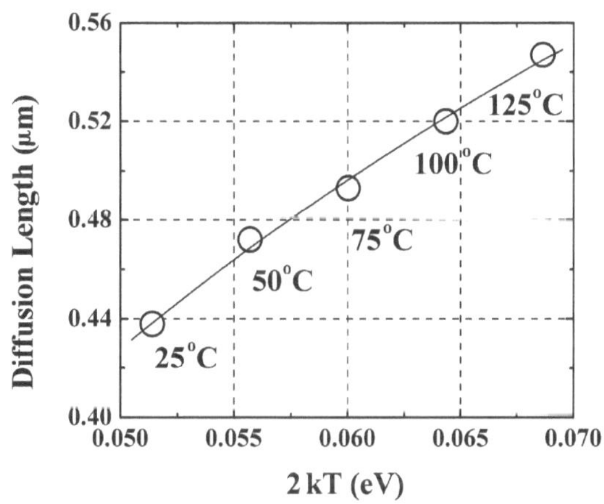

Figure 6.2: Experimental dependence of minority carrier diffusion length on temperature (open circles). The line shows the fit (Eqn. (6.9)) with activation energy of 45 ± 2 meV [40]. Reprinted with permission from the American Institute of Physics.

later in GaN epitaxial layers [43]. In all cases, this increase was exponential with temperature and was modeled with the following expression:

$$L = L_0 \exp\left(-\frac{\Delta E_A^T}{2kT}\right) \tag{6.9}$$

where L_0 is a scaling factor, ΔE_A^T is a thermal activation energy, and k is Boltzmann constant. Figure 6.2 shows the experimental results for n-ZnO and the fit using Eqn. (6.9), which yields activation energy of 45 ± 2 meV. This energy represents carrier de-localization energy, since it determines the increase of the diffusion

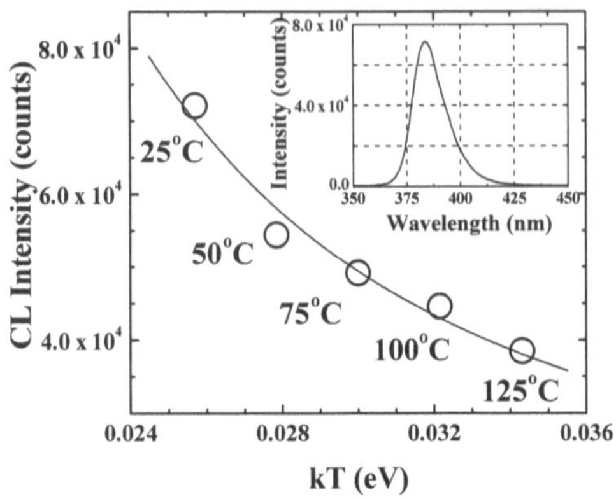

Figure 6.3: Experimentally obtained values for the peak NBE CL intensity in bulk ZnO as a function of temperature (open circles) and the fit (solid line; Eqn. (6.10)), yielding activation energy of 58 ± 7 meV. **Inset:** CL spectrum of n-ZnO showing the NBE transition at 25°C [40]. Reprinted with permission from the American Institute of Physics.

length due to reduction of recombination efficiency [42]. The smaller is the activation energy, the more efficient is the de-trapping of captured carriers at a fixed temperature (see also discussion below).

The role of increasing carrier lifetime is also supported by the results of cathodoluminescence (CL) measurements, which are presented in Fig. 6.3. The inset of Fig. 6.3 shows a cathodoluminescence spectrum in the vicinity of the NBE transition at 383 nm (3.24 eV). This feature in bulk ZnO has been attributed to the transition from the conduction band to a relatively deep acceptor level [44]. It was observed that the peak intensity, I_{CL}, of NBE luminescence decays systematically with increasing temperature, providing direct evidence that the number of recombination events decreases. Because the intensity of the NBE luminescence is inversely proportional to the lifetime of carriers in the band, the decay of I_{CL} indicates the increase of τ with temperature. The decay proceeds exponentially according to the equation below [45]:

$$I_{CL} = \frac{A}{\left(1 + B\exp\left(-\frac{\Delta E_A^T}{kT}\right)\right)} \tag{6.10}$$

where A and B are scaling factors, and ΔE_A^T is the activation energy, similar in nature to that in Eqn. (6.9). Based on the fit shown in Fig. 6.3, the activation energy was determined to be 58 ± 7 meV. This value is in excellent agreement with that obtained by photoluminescence measurements in Li-doped ZnO films [46]. It is also consistent with the results of the variable-temperature EBIC measurements, which suggests that the same underlying process is responsible for both the increase in the diffusion length and the CL intensity decay. This process is outlined below.

The increase in minority hole lifetime in the valence band is likely associated with a smaller recombination capture cross-section for this carrier at elevated temperatures. In GaAs, for example, detailed analysis for temperature dependence of capture cross-section indicates an order of magnitude decrease of recombination efficiency, measured in terms of an "effective capture radius", in the temperature range from 100 to 300 K [42]. Non-equilibrium electron-hole pairs are generated by the beam of the SEM and subsequently annihilate by recombining with each other. Since the hole capture cross-section is inversely proportional to temperature [42,47], the frequency of the recombination events (and, hence, the CL intensity) decreases as the temperature is raised. This means that non-equilibrium holes exist in the valence band for longer periods of time and, consequently, diffuse longer distances before undergoing recombination. Note that carrier diffusivity, D, is also a temperature-dependent quantity and, therefore, can affect the diffusion length (cf. Eqn. (6.1)). On the other hand, it has been demonstrated for n-ZnO that the mobility, μ, of majority carriers decreases in the temperature range of our experiments by about a factor of 2 [48]. Assuming that the mobility of the minority carriers exhibits the same behavior [43] and combining Eqns. (6.1) and (6.2), it is clear that the value of the diffusion length is dominated by the growing lifetime of minority holes. From the Einstein relation (Eqn. (6.2)), the above-referenced difference in mobility translates to about a 30% decrease in diffusivity at 125°C as compared to 25°C. Based on a 30% difference in diffusivity and using experimentally obtained values of diffusion length, we conclude that the lifetime of minority holes at 125°C is nearly 2.5 times greater than at room temperature.

Preliminary results indicate that temperature-induced increase in carrier lifetime also occurs in epitaxial ZnO. Nitrogen-doped ZnO samples grown using molecular beam epitaxy were provided by SVT Associates. Hall effect measurements revealed hole concentration of 4.5×10^{17} cm^{-3} and mobility of $\sim$1 cm^2/Vs. For one of these samples, Fig. 6.4 shows the decay of NBE CL intensity fitted with Eqn. (6.10). The measurements yielded an activation energy of 118 ± 12 meV. This value is comparable to the activation energy of the nitrogen acceptor in ZnO [49,50], which indicates possible non-equilibrium carrier trapping on nitrogen-related deep levels.

6.5 Studies of minority carrier recombination

Deep carrier traps have pronounced implications on minority carrier transport and often give rise to such undesirable phenomena as radiation-induced optical metastability, persistent photoconductivity, and optical quenching of photocurrent [51–53].

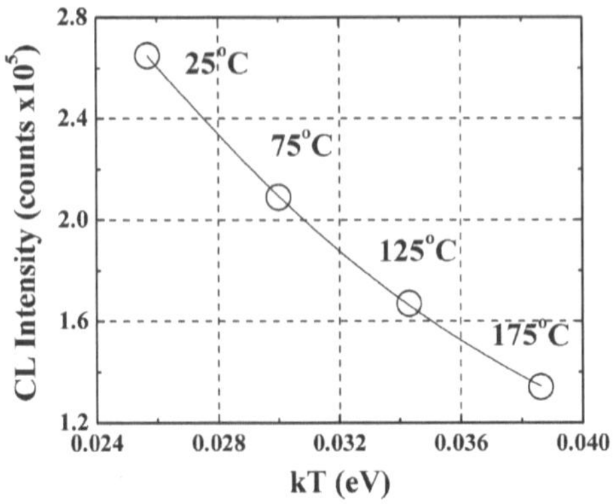

Figure 6.4: Maximum CL intensity of the NBE transition in ZnO:N as a function of temperature (open circles). Solid line shows the fit with Eqn. (6.10), resulting in activation energy of 118 ± 12 meV. After ref. [58].

On the other hand, it has been demonstrated that capture of non-equilibrium minority carriers by deep metastable traps is associated with the increase of minority carrier diffusion length and lifetime [32,40,54]. Moreover, in GaN this increase was shown to result in a significant (several-fold) improvement of photovoltaic detector response in agreement with Eqn. (6.3) [32,55].

Extensive studies aimed at achieving *p*-type conductivity in ZnO reveal that most potential acceptors, such as nitrogen, phosphorus, and arsenic, tend to form acceptor levels far from the valence band maximum [56]. Since the ionization fraction of such acceptors is low (due to their high activation energy), there is a large concentration of neutral states that may act as traps for non-equilibrium electrons. The effects of electron trapping on the diffusion length and lifetime of minority carriers can be probed by subjecting the material of interest to the excitation by the electron beam of the scanning electron microscope. The remainder of this review deals with the influence of electron irradiation on minority carrier diffusion length (Section 6.5.1) and lifetime (Section 6.5.2). The mechanism for the irradiation-induced effects and the role of deep carrier traps are discussed in detail in Section 6.5.3.

6.5.1 Influence of electron trapping on minority carrier diffusion length

The measurements of diffusion length as a function of electron beam irradiation duration were carried out on bulk n-ZnO [57], molecular beam epitaxy (MBE) p-ZnO doped with nitrogen [58], and phosphorus-doped ZnMgO grown by pulsed laser deposition (PLD) [54].

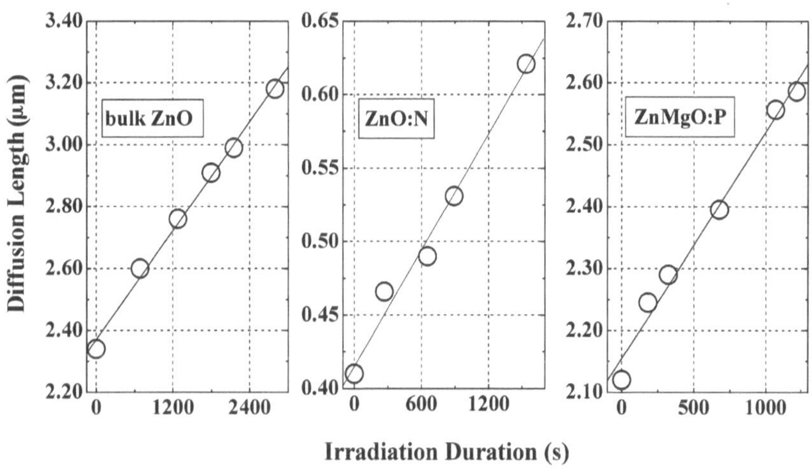

Figure 6.5: Experimental dependence of minority carrier diffusion length on duration of electron beam irradiation at room temperature (open circles) and the linear fit (solid line). After refs. [54,57,58].

As was already mentioned, bulk ZnO samples (Tokyo Denpa Co.) were weakly *n*-type, showing electron concentrations of $\sim 10^{14}\,cm^{-3}$ and mobility of $\sim 150\,cm^2/Vs$ at room temperature. SIMS measurements revealed the Li concentration of about $3 \times 10^{16}\,cm^{-3}$ [41] (Li is often added to ZnO to increase the resistivity of initially *n*-type samples). ZnO:N samples were grown using molecular beam epitaxy by SVT Associates. Hall effect measurements revealed hole concentration of $4.5 \times 10^{17}\,cm^{-3}$ and mobility of $\sim 1\,cm^2/Vs$. Phosphorus-doped $Zn_{0.9}Mg_{0.1}O$ layers were fabricated at the University of Florida using PLD. Capacitance-voltage profiling of similar films, grown using the same procedure, resulted in net acceptor concentration of $\sim 2 \times 10^{18}\,cm^{-3}$ after annealing.

EBIC experiments were conducted at room temperature *in situ* in a Philips XL30 SEM using a planar-collector configuration with a Schottky barrier (Section 6.3.1) to monitor the changes in minority carrier diffusion length as a function of time. After the initial line-scan, the electron beam was moved repeatedly along the same line in order to continuously irradiate the samples with electrons, with further EBIC measurements taken intermittently.

The results of EBIC experiments are presented in Fig. 6.5. Note that the results shown in the left panel of Fig. 6.5 and those discussed in Section 6.4 were obtained from different bulk ZnO samples, which may offer an explanation for the significant difference in the initial, room temperature values of diffusion length. Additionally, while the large diffusion length value of bulk ZnO can be attributed to the higher quality of the bulk material (as compared to epitaxial layers), the diffusion length of about 2 μm in PLD-grown ZnMgO is rather surprising because of the polycrystalline nature of the layers. It appears that in the latter sample, the transport of minority carriers is not significantly influenced by scattering from the grain boundaries.

The experiments revealed that diffusion length follows a general trend in all samples studied. Figure 6.5 shows that irradiation by the electron beam clearly results in a significant increase of the carrier diffusion length, and that this increase is linear with respect to the duration of electron irradiation, *t*. Up to 50% increase of diffusion length was achieved (cf. Fig. 6.5, center), with the rates of diffusion length increase ranging from ~13% to ~30% per 1000 s. This appears to be a common occurrence in wide bandgap semiconductors doped with species that create deep acceptor levels, as similar observations were made in (Al)GaN doped with Mg, Mn, Fe, and C [59–61]. It is also noteworthy that similar experiments conducted on bulk ZnO with no intentional dopants did not show any significant changes in minority carrier diffusion length [54].

The observed increase of *L* is ascribed to charging of the deep, neutral acceptor states by the electrons generated by the SEM beam, since trapping of non-equilibrium electrons prevents these levels from participating in recombination [54,57] (see also discussion in Section 6.5.3). Therefore, the difference in the rates of diffusion length increase is likely explained, at least in part, by the difference in the concentrations of these deep centers [62].

6.5.2 Optical studies of the effects of electron trapping on minority carrier lifetime

According to Eqn. (6.1), electron irradiation-induced increase in minority carrier diffusion length discussed above is associated with the increase of lifetime of non-equilibrium carriers. Experimental evidence for this dependence was obtained from the cathodoluminescence measurements performed on the same samples. Time-dependent CL measurements were conducted *in situ* in the Phillips XL30 SEM using an integrated Gatan MonoCL3 system. This setup allows combining periodic acquisition of CL spectra with continuous excitation of the sample by scanning the beam over the same location. For temperature-dependent CL measurements, the sample temperature was varied *in situ* using specially designed hot stage and an external temperature controller (Gatan). At each temperature, the electron beam irradiation and CL measurements were conducted at a different location.

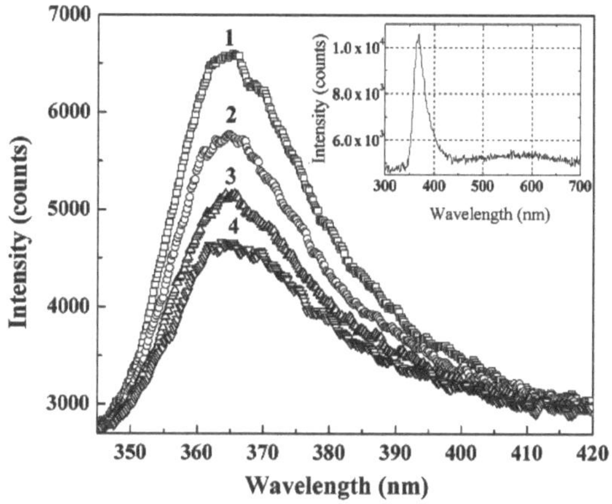

Figure 6.6: Room temperature CL spectra of $Zn_{0.9}Mg_{0.1}O$:P measured in the same location at different times of electron irradiation. **1** is a pre-irradiation spectrum; **2**, **3**, and **4** correspond to the duration of electron irradiation of 359, 793, and 1163 s, respectively. **Inset:** pre-irradiation broad-range CL spectrum taken in a different location than spectra in Fig. 6.6 [54]. Reprinted with permission from the American Institute of Physics.

6.5.2.1 Optical studies of $Zn_{0.9}Mg_{0.1}O$ *doped with phosphorus*

Figure 6.6 shows a series of NBE transitions in *p*-type $Zn_{0.9}Mg_{0.1}O$:P recorded under continuous electron irradiation and numbered in order of increasing irradiation duration. The edge of this spectrum at ~355 nm (see also inset of Fig. 6.6) is in agreement with 10% Mg content in ZnO lattice, since each atomic percent of Mg is known to increase the ZnO band gap by 0.02 eV [9]. The observed CL spectrum is attributed to the band-to-band as well as band-to-impurity (P-acceptor) optical transitions. The inset of Fig. 6.6 shows a wider range spectrum featuring a broad band, which is likely defect related [63].

While no changes were observed in the broad band CL (cf. inset of Fig. 6.6), the near-band-edge luminescence in Fig. 6.6 exhibits a continuous decay with increasing duration of electron beam irradiation. This demonstrates that exposure to the electron beam results in the increase of carrier lifetime (τ), since I_{CL} is proportional to $1/\tau$. Similar phenomena were also observed in GaN, where the decay of NBE CL intensity occurred concomitantly with increasing diffusion length [59–61].

To characterize the intensity decay, we relate it to the diffusion length, which is known to vary linearly with duration of irradiation (cf. Fig. 6.5). Since L is proportional to $\tau^{1/2}$ (Eqn. (6.1)), the inverse square root of normalized (with respect to the initial maximum value) intensity must also be proportional to L, and consequently, would be expected to change linearly with duration of electron irradiation. Figure 6.7 shows that this is indeed the case, indicating that the observed increase of the diffusion length is attributable to the growing lifetime of non-equilibrium carriers.

CL measurements were also conducted over a range of temperatures. The temperature of the samples was varied *in situ* from 25°C to 125°C. As is apparent from the inset of Fig. 6.7, the rate of the decrease in CL intensity (described by the slope, R, of the linear dependence of $I_{CL}^{-1/2}$ on t) diminishes with growing temperature. This suggests that while electron irradiation results in an increase of carrier lifetime, there exists another, thermally activated process that contributes to its decay.

Taking into account both irradiation-induced effect on R ($\propto \exp(\Delta E_A/kT)$) and its temperature dependence ($\propto \exp(\Delta E_A/-2kT)$) [42], and assuming that activation energies are similar in both cases, the temperature dependence of R can be described as follows:

$$R = R_0 \exp\left(\frac{\Delta E_A}{2kT}\right), \tag{6.11}$$

where R_0 is a scaling constant and ΔE_A is the activation energy for the overall process. Fitting the experimental results with this expression (inset of Fig. 6.7) yielded

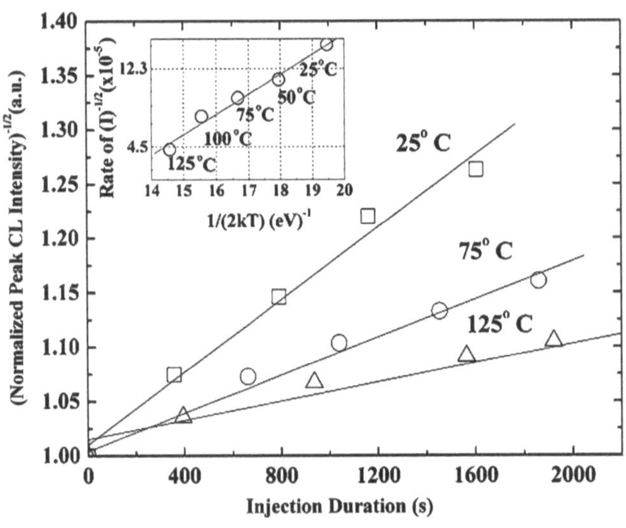

Figure 6.7: Variable-temperature dependence for the square root of inverse normalized intensity on duration of electron irradiation in $Zn_{0.9}Mg_{0.1}O$:P. The rate at every temperature is obtained from the slope of a linear fit. **Inset:** temperature dependence for the rate of the square root of inverse normalized intensity (open circles) and the fit. The slope of the graph yields $\Delta E_A = 256 \pm 20$ meV [54]. Reprinted with permission from the American Institute of Physics.

activation energy of 256 ± 20 meV. This activation energy is in good agreement with that for the phosphorus acceptor obtained based on the simple hydrogenic model. The model assumes phosphorus substitution on the oxygen site and predicts the activation energy of about 250–300 meV [64]. The experimentally obtained value of the activation energy, combined with the fact that no electron irradiation effects were observed in undoped ZnO [54], suggests that carrier trapping on phosphorus acceptor levels plays a crucial role in this phenomenon (see Section 6.5.3 for detailed mechanism).

6.5.2.2 Optical studies of bulk ZnO

Similarly to ZnMgO, the electron irradiation-induced increase of minority carrier diffusion length in bulk ZnO (Section 6.5.1) was found to correlate with the increase of minority carrier lifetime. Figure 6.8 shows a series of room temperature NBE spectra collected under continuous electron beam excitation, in which the intensity of the NBE transition can be seen to fall steadily with increasing t. The peak of this emission occurs at 383 nm (3.24 eV) and has been assigned to the transition from the conduction band to a deep acceptor level (e, A^0) [44]. The inverse square root of maximum CL intensity ($I_{CL}^{-1/2}$), a quantity proportional to $\sqrt{\tau}$ (and, therefore,

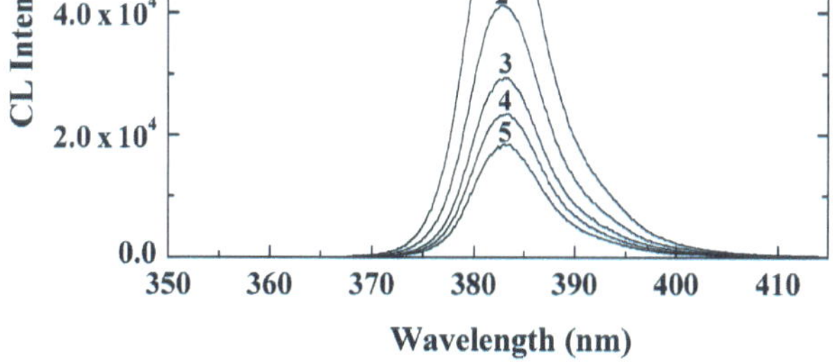

Figure 6.8: Room temperature cathodoluminescence spectra of bulk ZnO taken under continuous excitation by the electron beam. **1** is the pre-irradiation spectrum and **5** is the spectrum after 1450 s of electron irradiation. **Left Inset:** variable-temperature dependence of inverse square root of normalized intensity on duration of electron irradiation and the linear fit with the rate R. **Right Inset:** Arrhenius plot of R as a function of temperature yielding an activation energy ΔE_A^I of 283 ± 9 meV. After ref. [57].

to *L*), changes linearly with duration of irradiation (cf. left inset of Fig. 6.8), which is consistent with the results of EBIC measurements and indicates that the increase in *L* occurs due to the irradiation-induced growth of carrier lifetime.

CL measurements conducted at elevated temperatures confirmed the same trend for the irradiation-induced change of luminescence intensity. It can be seen from the left inset of Fig. 6.8 that the inverse square root of intensity increases linearly for all temperatures. The temperature dependence of rate, *R*, can be used to determine the activation energy of the irradiation-induced processes according to Eqn. (6.11). However, our earlier studies of the *temperature*-induced CL intensity decay (Section 6.4) yielded the activation energy of about 60 meV [54], thus allowing us to separate the two components as follows:

$$R = R_0 \exp\left(\frac{\Delta E_A^I}{kT}\right) \exp\left(-\frac{\Delta E_A^T}{2kT}\right) \tag{6.12}$$

where ΔE_A^I is the activation energy of electron irradiation effect and ΔE_A^T is the previously determined activation energy of thermally induced intensity decay. This treatment yielded the value for ΔE_A^I of 283 ± 9 meV. Incidentally, ref. [41] also reports high concentration of ~0.3 eV electron traps found in the same material by Deep Level Transient Spectroscopy (DLTS). The significant difference between ΔE_A^I and ΔE_A^T observed in bulk ZnO suggests that, unlike in other materials, temperature- and irradiation-induced changes of the minority carrier transport characteristics are two distinctly different processes.

Although several theoretical works have predicted a very shallow Li_{Zn} level [65,66], these predictions have not been substantiated experimentally, as most studies find a rather deep Li-acceptor with activation energies of several hundreds meV [49,67]. In fact, recent first-principles calculations by Wardle *et al.*, also suggest that the Li_{Zn} acceptor state lies at about 0.2 eV above the valence band maximum [67], which is in reasonable agreement with the ΔE_A^I of 283 ± 9 meV obtained in this work.

It should be clarified that the weak *n*-type character of the sample is not necessarily in contradiction with the dominant behavior of acceptor states observed in electron trapping phenomena. As was mentioned, the *n*-type conductivity in nominally undoped ZnO is due to the *shallow* donor states, whereas in the presence of deep electron traps the Fermi level may lie far below these states. Although shallow donors may capture non-equilibrium electrons under excitation, those are quickly released if the temperature is sufficiently high. Therefore, if the difference in the energetic position between the donor and trap states is large, the latter dominate the kinetics of electron trapping [68].

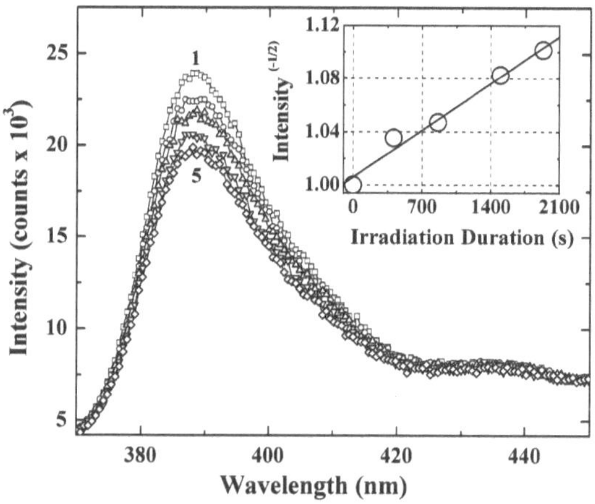

Figure 6.9: Room temperature CL spectra of ZnO:N taken under continuous excitation. Trace **1** corresponds to the pre-irradiation spectrum and trace **5** to the spectrum after 1940 s of electron irradiation. **Inset:** Linear dependence of the inverse square root of normalized peak intensity at room temperature. After ref. [58].

6.5.2.3 Optical studies of ZnO doped with nitrogen

CL measurements performed on MBE-grown, nitrogen-doped p-ZnO revealed behavior similar to that of bulk ZnO and PLD-grown ZnMgO [58]. The experiments were carried out at temperatures ranging from 25°C to 175°C. Room temperature ZnO:N spectra are shown in Fig. 6.9 and feature an NBE luminescence band with a maximum at about 388 nm (~3.20 eV). This band includes the conduction band – to a deep acceptor level (e, A^0) transition as well as the donor-acceptor pair (DAP) recombination, with nitrogen identified as the acceptor in both processes [69]. Additionally, a violet band centered around 435 nm has been attributed to the radiative recombination of the electrons trapped at grain boundaries with the holes in the valence band [70].

As expected, irradiation with electron beam resulted in decay of the intensity of NBE luminescence, indicating increasing lifetime. One can observe from the inset of Fig. 6.9 that, in agreement with the diffusion length measurements (cf. Fig. 6.5 in Section 6.5.1), the inverse square root of the peak normalized intensity of the NBE transition changes linearly with irradiation time, yielding the rate R. Note that the intensity of the violet band is not affected by electron irradiation, which suggests that electron trapping at the grain boundaries does not play a significant role in the irradiation-induced increase of carrier lifetime.

CL measurements at elevated temperatures (Fig. 6.10) confirmed that R decreases with temperature, indicating a thermally activated process that counteracts the effects of electron irradiation, similar to what occurs in bulk ZnO and ZnMnO:P.

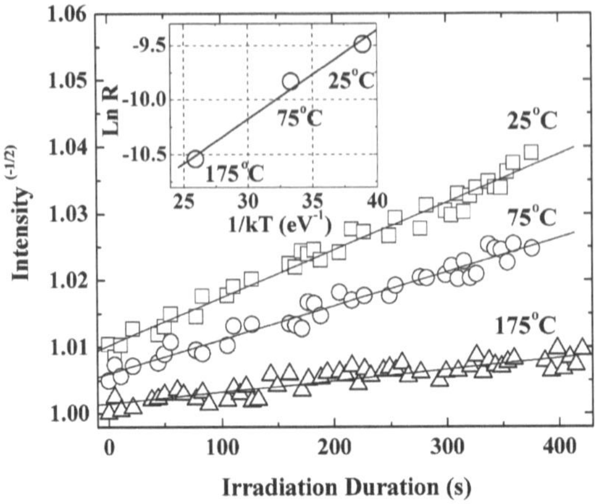

Figure 6.10: Variable-temperature dependence of the inverse square root of normalized intensity in ZnO:N on duration of electron irradiation and the linear fit with a rate *R*. The data are offset for clarity. **Inset:** Arrhenius plot of *R* as a function of temperature with a fit yielding $\Delta E_A^I = 134 \pm 10\,\text{meV}$. After ref. [58].

Note that while the values of *R* were obtained based on the intensity normalized with respect to its initial value, the data displayed in Fig. 6.10 are offset by shifting the normalized results along the y-axis to avoid the overlap of the data points. The activation energy (ΔE_A^I) of about $134 \pm 10\,\text{meV}$ was determined from the Arrhenius plot shown in the inset of Fig. 6.10, based on Eqn. (6.12) and using $\Delta E_A^T = 118\,\text{meV}$ obtained from the temperature-dependent CL measurements (Section 6.4). This value is in reasonable agreement with the ionization energy of the nitrogen acceptor in ZnO [49,50,69,71] and indicates that electron trapping by these levels plays an important role in the recombination dynamics of minority carriers. The details of the mechanism proposed to explain the influence of deep traps on minority carrier transport are presented in the following section.

6.5.3 Influence of deep electron traps on minority carrier transport

We have extensively applied the above-explained approach to investigate the influence of deep levels on minority carrier transport in acceptor-doped GaN [59–61] as well as ZnO compounds [54,57,58]. In all cases, the activation energy determined from the temperature dependence of irradiation-induced intensity decay was in agreement with the ionization energy of the dominant deep acceptor species.

When exciting radiation, such as that from the beam of the SEM, generates non-equilibrium electron-hole pairs in acceptor-doped wide bandgap semiconductors,

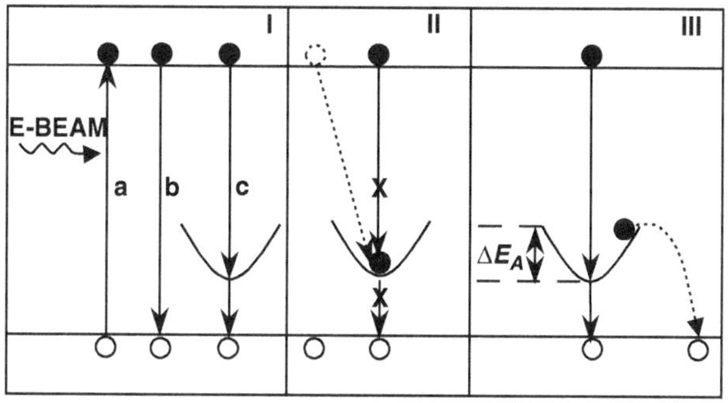

Figure 6.11: Electron beam generates non-equilibrium electron-hole pairs (**Ia**). Non-equilibrium carriers recombine either via the band-to-band transition (**Ib**) or through unoccupied (neutral) acceptor states (**Ic**). If a non-equilibrium electron is trapped by the acceptor level, recombination cannot proceed (**II**), leading to increased lifetime of non-equilibrium minority carriers in the band. Release of the trapped electron with activation energy ΔE_A^I restores the original recombination pathway (**III**), resulting in a slower rate of lifetime increase at elevated temperatures.

the dominant radiative recombination pathway generally involves the localized states of the acceptor atoms located far from the valence band edge (Fig. 6.11, I). The deep nature of these acceptors prevents their complete ionization at room temperature, and the large concentration of neutral (non-ionized) acceptor atoms acts as traps for non-equilibrium electrons generated by the SEM beam. As a consequence of electron trapping, these states are removed from participating in carrier recombination (Fig. 6.11, II). Since the transitions involving deep acceptor levels constitute a major part of radiative recombination events (cf. Section 6.5.2), electron trapping results in greater lifetime of non-equilibrium minority carriers in the band (due to a lower rate of recombination with non-equilibrium majority carriers) and, consequently, greater diffusion length.

The fact that the rate of the irradiation-induced lifetime increase, R, diminishes with increasing temperature indicates that there exists a competing, thermally activated process that counteracts the capture of non-equilibrium electrons by neutral acceptor levels. This process is the release of the trapped electrons, shown schematically in Fig. 6.11, III. As the temperature increases, the balance between capture and release shifts toward the latter, reducing the impact of electron irradiation on the carrier lifetime and diffusion length, and resulting in slower increase of these parameters at elevated temperatures.

It is interesting to note that in p-type GaN the increase of minority electron diffusion length due to electron trapping can also be induced by application of a forward bias to a p-n junction [55,72]. This increase was shown to lead to the multi-fold enhancement of the peak responsivity for a photovoltaic detector in agreement with Eqns. (6.3) and (6.4). Moreover, the increased responsivity persisted for at least several days at room temperature. Although conclusive evidence has not been obtained to date, the similarities in irradiation-induced phenomena, combined with the results of preliminary experiments, suggest that solid-state electron injection may produce the same results in ZnO.

6.6 Summary

Issues affecting minority carrier transport in ZnO have been discussed, with special attention given to the temperature dependence of minority carrier diffusion length and lifetime, as well as to the recombination dynamics of non-equilibrium minority carriers. The mechanisms governing temperature- and irradiation-induced effects have been presented. Additionally, device implications of minority carrier diffusion length control were considered.

Acknowledgements

This research was support in part by the National Science Foundation (ECS #0422604) and the American Chemical Society Petroleum Research Fund (40501-AC10).

References

[1] F. Oba, S. R. Nishitani, S. Isotani, H. Adachi and I. Tanaka, *J. Appl. Phys.*, **90,** 824 (2001).
[2] G. W. Tomlins, J. L. Routbort and T. O. Mason, *J. Appl. Phys.*, **87,** 117 (2000).
[3] C. G. Van de Walle, *Phys. Rev. Lett.*, **85,** 1012 (2000).
[4] Y. I. Alivov, E. V. Kalinina, A. E. Cherenkov, D. C. Look, B. M. Ataev, A. K. Omaev, M. V. Chukichev and D. M. Bagnall, *Appl. Phys. Lett.*, **83,** 4719 (2003).
[5] Y. I. Alivov, U. Ozgur, S. Dogan, D. Johnstone, V. Avrutin, N. Onojima, C. Liu, J. Xie, Q. Fan and H. Morkoc, *Appl. Phys. Lett.*, **86,** 241108 (2005).
[6] C. X. Wang, G. W. Yang, C. X. Gao, H. W. Liu, Y. H. Han, J. F. Luo and G. T. Zou, *Carbon,* **42,** 317 (2004).
[7] A. Osinsky, J. W. Dong, M. Z. Kauser, B. Hertog, A. M. Dabiran, P. P. Chow, S. J. Pearton, O. Lopatiuk and L. Chernyak, *Appl. Phys. Lett.*, **85,** 4272 (2004).
[8] J. W. Dong, A. Osinsky, B. Hertog, A. M. Dabiran, P. P. Chow, Y. W. Heo, D. P. Norton and S. J. Pearton, *J. Electron. Mater.*, **34,** 416 (2005).

[9] T. Gruber, C. Kirchner, R. Kling, F. Reuss and A. Waag, *Appl. Phys. Lett.*, **84,** 5359 (2004).
[10] K. Ip, Y. W. Heo, D. P. Norton, S. J. Pearton, J. R. LaRoche and F. Ren, *Appl. Phys. Lett.*, **85,** 1169 (2004).
[11] F. G. Chen, Z. Z. Ye, W. Z. Xu, B. H. Zhao, L. P. Zhu and J. G. Lv, *J. Cryst. Growth,* **281,** 458 (2005).
[12] V. Vaithianathan, B. T. Lee and S. S. Kim, *J. Appl. Phys.*, **98,** 043519 (2005).
[13] E. J. Egerton, A. K. Sood, R. Singh, Y. R. Puri, R. F. Davis, J. Pierce, D. C. Look and T. Steiner, *J. Electron. Mater.*, **34,** 949 (2005).
[14] V. Vaithianathan, B. T. Lee and S. S. Kim, *Appl. Phys. Lett.*, **86,** 062101 (2005).
[15] F. X. Xiu, Z. Yang, L. J. Mandalapu, D. T. Zhao, J. L. Liu and W. P. Beyermann, *Appl. Phys. Lett.*, **87,** 152101 (2005).
[16] Y. R. Ryu, W. J. Kim and H. W. White, *J. Cryst. Growth,* **219,** 419 (2000).
[17] J. M. Bian, X. M. Li, C. Y. Zhang, W. D. Yu and X. D. Gao, *Appl. Phys. Lett.*, **85,** 4070 (2004).
[18] F. Zhuge, L. P. Zhu, Z. Z. Ye, D. W. Ma, J. G. Lu, J. Y. Huang, F. Z. Wang, Z. G. Ji and S. B. Zhang, *Appl. Phys. Lett.*, **87,** 092103 (2005).
[19] S. K. Hazra and S. Basu, *Solid-State Electron.*, **49,** 1158 (2005).
[20] Y. W. Heo, Y. W. Kwon, Y. Li, S. J. Pearton and D. P. Norton, *J. Electron. Mater.*, **34,** 409 (2005).
[21] S. H. Lim, D. Shindo, H. B. Kang and K. Nakamura, *J. Vac. Sci. Technol. B,* **19,** 506 (2001).
[22] K. Miyamoto, M. Sano, H. Kato and T. Yao, *J. Cryst. Growth,* **265,** 34 (2004).
[23] Y. Wang, X. L. Du, Z. X. Mei, Z. Q. Zeng, Q. Y. Xu, Q. K. Xue and Z. Zhang, *J. Cryst. Growth,* **273,** 100 (2004).
[24] M. W. Cho, A. Setiawan, H. J. Ko, S. K. Hong and T. Yao, *Semicond. Sci. Technol.*, **20,** S13 (2005).
[25] E. Monroy, F. Calle, J. L. Pau, F. J. Sanchez, E. Munoz, F. Omnes, B. Beaumont and P. Gibart, *J. Appl. Phys.*, **88,** 2081 (2000).
[26] J. F. Muth, R. M. Kolbas, A. K. Sharma, S. Oktyabrsky and J. Narayan, *J. Appl. Phys.*, **85,** 7884 (1999).
[27] H. Yoshikawa and S. Adachi, *Jpn. J. Appl. Phys. Part 1 – Regul. Pap. Short Notes Rev. Pap.*, **36,** 6237 (1997).
[28] D. E. Ioannou and S. M. Davidson, *J. Phys. D – Appl. Phys.*, **12,** 1339 (1979).
[29] D. E. Ioannou and C. A. Dimitriadis, *IEEE Trans. Electron. Devices,* **29,** 445 (1982).
[30] J. Boersma, J. J. E. Iindenkleef and H. K. Kuiken, *J. Eng. Math,* **18,** 315 (1984).
[31] D. S. H. Chan, V. K. S. Ong and J. C. H. Phang, *IEEE Trans. Electron. Devices,* **42,** 963 (1995).
[32] L. Chernyak, A. Osinsky and A. Schulte, *Solid-State Electron.*, **45,** 1687 (2001).
[33] T. S. Moss, *J. Electron. Control,* **1,** 126 (1955).
[34] D. K. Schroder, *Meas. Sci. Technol.*, **12,** R16 (2001).
[35] R. Chakrabarti, J. Dutta, S. Bandyopadhyay, D. Bhattacharyya, S. Chaudhuri and A. K. Pal, *Solar Energy Mater. Solar Cells,* **61,** 113 (2000).

[36] J. Lagowski, A. Aleynikov, A. Savtchouk and P. Edelman, *Eur. Phys. J. – Appl. Phys.*, **27,** 503 (2004).
[37] H. O. Olafsson, J. T. Gudmundsson, H. G. Svavarsson and H. P. Gislason, *Physica B,* **274,** 689 (1999).
[38] R. Shikler, N. Fried, T. Meoded and Y. Rosenwaks, *Phys. Rev. B,* **61,** 11041 (2000).
[39] R. K. Ahrenkiel, *Solid-State Electron.,* **35,** 239 (1992).
[40] O. Lopatiuk, L. Chernyak, A. Osinsky, J. Q. Xie and P. P. Chow, *Appl. Phys. Lett.,* **87,** 162103 (2005).
[41] A. Y. Polyakov, N. B. Smirnov, A. V. Govorkov, E. A. Kozhukhova, S. J. Pearton, D. P. Norton, A. Osinsky and A. Dabiran, *J. Electron. Mater,* **35,** 663 (2006).
[42] M. Eckstein and H. U. Habermeier, *J. Phys. IV,* **1,** 23 (1991).
[43] L. Chernyak, A. Osinsky, H. Temkin, J. W. Yang, Q. Chen and M. A. Khan, *Appl. Phys. Lett.,* **69,** 2531 (1996).
[44] K. Thonke, T. Gruber, N. Teofilov, R. Schonfelder, A. Waag and R. Sauer, *Physica B,* **308,** 945 (2001).
[45] D. S. Jiang, H. Jung and K. Ploog, *J. Appl. Phys.,* **64,** 1371 (1988).
[46] A. Ortiz, C. Falcony, J. Hernandez, M. Garcia and J. C. Alonso, *Thin Solid Films,* **293,** 103 (1997).
[47] J. I. Pankove, *Optical Processes in Semiconductors* (Prentice-Hall, Englewood Cliffs, New Jersey, 1971).
[48] D. C. Look, D. C. Reynolds, J. R. Sizelove, R. L. Jones, C. W. Litton, G. Cantwell and W. C. Harsch, *Solid State Commun.,* **105,** 399 (1998).
[49] B. K. Meyer, H. Alves, D. M. Hofmann, W. Kriegseis, D. Forster, F. Bertram, J. Christen, A. Hoffmann, M. Strassburg, M. Dworzak, U. Haboeck and A. V. Rodina, *Phys. Status Solidi B – Basic Res.,* **241,** 231 (2004).
[50] S. Yamauchi, Y. Goto and T. Hariu, *J. Cryst. Growth,* **260,** 1 (2004).
[51] S. Dhar and S. Ghosh, *Appl. Phys. Lett.,* **80,** 4519 (2002).
[52] V. V. Ursaki, I. M. Tiginyanu, P. C. Ricci, A. Anedda, S. Hubbard and D. Pavlidis, *J. Appl. Phys.,* **94,** 3875 (2003).
[53] B. J. Ryan, D. P. Lowney, M. O. Henry, P. J. McNally, E. McGlynn, K. Jacobs and L. Considine, *Thin Solid Films,* **473,** 308 (2005).
[54] O. Lopatiuk, W. Burdett, L. Chernyak, K. P. Ip, Y. W. Heo, D. P. Norton, S. J. Pearton, B. Hertog, P. P. Chow and A. Osinsky, *Appl. Phys. Lett.,* **86,** 012105 (2005).
[55] L. Chernyak, G. Nootz and A. Osinsky, *Electron. Lett.,* **37,** 922 (2001).
[56] D. C. Look and B. Claftin, *Phys. Status Solidi B – Basic Res.,* **241,** 624 (2004).
[57] O. Lopatiuk, L. Chernyak, A. Osinsky and J. Q. Xie, *Appl. Phys. Lett.,* **87,** 162103 (2005).
[58] O. Lopatiuk, A. Osinsky and L. Chernyak, unpublished.
[59] L. Chernyak, W. Burdett, M. Klimov and A. Osinsky, *Appl. Phys. Lett.,* **82,** 3680 (2003).
[60] W. Burdett, O. Lopatiuk, L. Chernyak, M. Hermann, M. Stutzmann and M. Eickhoff, *J. Appl. Phys.,* **96,** 3556 (2004).

[61] O. Lopatiuk, A. Osinsky, A. Dabiran, K. Gartsman, I. Feldman and L. Chernyak, *Solid State Electron.*, **49,** 1662 (2005).
[62] L. Chernyak, A. Osinsky, V. Fuflyigin and E. F. Schubert, *Appl. Phys. Lett.*, **77,** 875 (2000).
[63] Y. W. Heo, K. Ip, S. J. Pearton and D. P. Norton, *Phys. Status Solidi A – Appl. Res.*, **201,** 1500 (2004).
[64] S. J. Pearton, D. P. Norton, K. Ip, Y. W. Heo and T. Steiner, *Prog. Mater. Sci.*, **50,** 293 (2005).
[65] C. H. Park, S. B. Zhang and S. H. Wei, *Phys. Rev. B*, **66,** 073202 (2002).
[66] E. C. Lee and K. J. Chang, *Phys. Rev. B*, **70,** 115210 (2004).
[67] M. G. Wardle, J. P. Goss and P. R. Briddon, *Phys. Rev. B*, **71,** 155205 (2005).
[68] M. Salis, A. Anedda, F. Quarati, A. J. Blue and W. Cunningham, *J. Appl. Phys.*, **97,** 033709 (2005).
[69] F. Reuss, C. Kirchner, T. Gruber, R. Kling, S. Maschek, W. Limmer, A. Waag and P. Ziemann, *J. Appl. Phys.*, **95,** 3385 (2004).
[70] R. Ghosh, B. Mallik, S. Fujihara and D. Basak, *Chem. Phys. Lett.*, **403,** 415 (2005).
[71] G. Xiong, K. B. Ucer, R. T. Williams, J. Lee, D. Bhattacharyya, J. Metson and P. Evans, *J. Appl. Phys.*, **97,** 043528 (2005).
[72] L. Chernyak, A. Schulte, A. Osinsky, J. Graff and E. F. Schubert, *Appl. Phys. Lett.*, **80,** 926 (2002).

Chapter 7

Contacts to ZnO

Jae-Hong Lim and Seong-Ju Park

Department of Materials Science and Engineering, Gwangju Institute of Technology,
1 Oryong-dong, Buk-gu, Gwangju, 500-712, Korea

7.1 Ohmic contacts

7.1.1 Introduction

An Ohmic contact is essential for realizing long-lifetime operation of optical and electrical devices. If the high contact resistance occurs between metal and semiconductors, device performance will be degraded severely through thermal stress and or contact failure. Thus, in order to attain high-performance ZnO-based optical and electrical devices, it is essential to achieve Ohmic contacts that have both low resistance and are thermally stable and reliable. This can be achieved either by per forming surface preparation to reduce the metal semiconductor barrier height which would allow an increase in carrier tunneling probability or by increasing the effective carrier concentration of the surface perhaps taking advantage of oxygen loss. Ohmic contact metallization should be one of the main goals in improving device performance which plays an important role in device technology. However, Ohmic contact technology in ZnO material has not been explored extensively until now.

7.1.2 Ohmic contact to n*-type ZnO*

The approaches are usually the oxygen desorption and the indiffusion of zinc in order to reduce the thickness of the Schottky barrier between the metal and the semiconductor and increase tunneling through the barrier. Another contact mechanism is surface cleaning technology to modify the surface and reduce barrier height.

Figure 7.1 shows that as-deposited and annealed Ti/Au contacts exhibit linear current-voltage (I-V) characteristics.[1,2] The Ti/Au scheme produces a specific

Zinc Oxide Bulk, Thin Films and Nanostructures
C. Jagadish and S. Pearton (Editors)

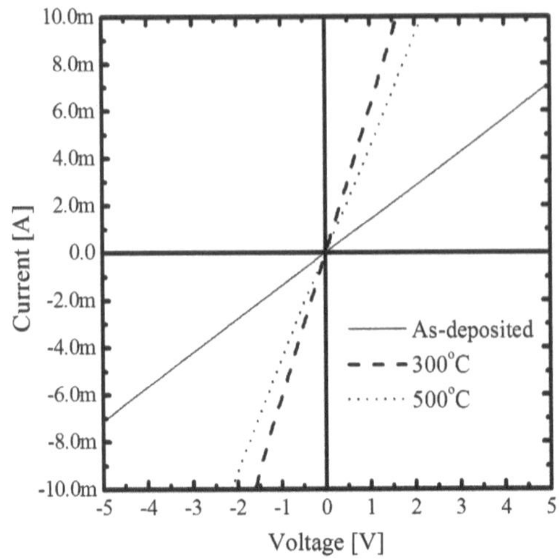

Figure 7.1: I-V characteristics for Ti/Au contacts on the annealed *n*-type ZnO layer. The as-deposited and annealed contacts exhibit linear I-V behavior, although the latter shows better characteristics. (After ref [1])

contact resistance of $2 \times 10^{-4}\,\Omega \cdot \mathrm{cm}^2$ when annealed at 300°C for 1 min in an N_2 atmosphere. This result is the reduction by two orders of magnitude in the specific contact resistance, compared to the as-deposited contact. However, thermal degradation occurred after annealing at temperatures in excess of 300°C. This degradation could be related to either the disruption of the interface, a reduction in the contact areas due to the formation of voids, as shown in Fig. 7.2. Contact to undoped ZnO, in addition, showed that annealing at 300°C for 5 min results in the degradation of its electrical property. The insertion of Al/Pt layer between Ti and Au layer reduce a specific contact resistance to $3.9 \times 10^{-7}\,\Omega \cdot \mathrm{cm}^2$ in phosphorus-doped *n*-type ZnO thin films samples with carrier concentrations of $6.0 \times 10^{19}\,\mathrm{cm}^{-2}$.[3,4] The thermal stability of this contact scheme for bulk *n*-type ZnO was also investigated. Higher annealing temperatures degraded the contact resistance, and Auger electron spectroscopy depth profiling revealed increasing intermixing of the metal layers.[5] The Al out-diffused to the surface at temperatures as low as 350°C, and the contact metallization was almost completely intermixed by 600°C. Zn/Au contact schemes became Ohmic with contact resistivity of $2.36 \times 10^{-5}\,\Omega \cdot \mathrm{cm}^2$ when annealed at 500°C due to the indiffusion of Zn atoms into ZnO and increase in the carrier concentration near surface region.[6] However, for the sample annealed at 600°C, the degraded electrical characteristics could be attributed to the formation of Au_3Zn phase, which may hinder the formation of Zn-rich (or oxygen deficient) oxide and so prevent the formation of an n+ layer.

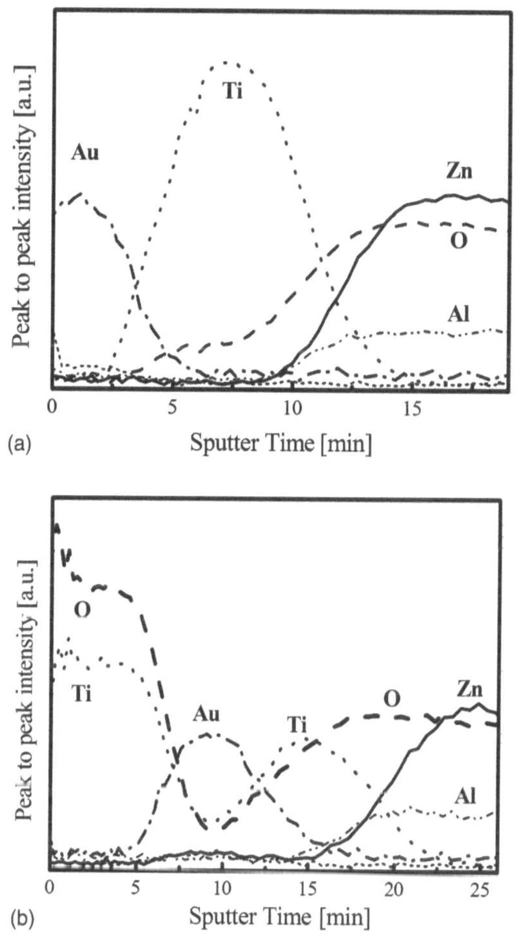

Figure 7.2: Auger depth profiles of the (a) as-deposited Ti/Au contact, (b) 500°C annealed contact on the ZnO layer. (After ref [2])

Nonalloyed Ohmic contact with low specific contact resistance is preferred, particularly for shallow junction and low voltage devices, as it provides smooth metal-semiconductor interfaces resulting from the limited interface reaction. A nonalloyed In based Ohmic contact has been formed on a hydrothermally grown *n*-type ZnO substrate by KrF excimer laser irradiation with 0.3 J/cm^2 and metal deposition without any impurity.[7] The In-deposited laser-irradiated sample exhibited Ohmic behavior with a resistivity of $7 \times 10^{-1}\ \Omega \cdot \text{cm}^2$. The same, however, was not the case for unirradiated samples which indicates that the laser irradiation enables current conduction into the *n*-type ZnO substrate. An increase in the contact resistance

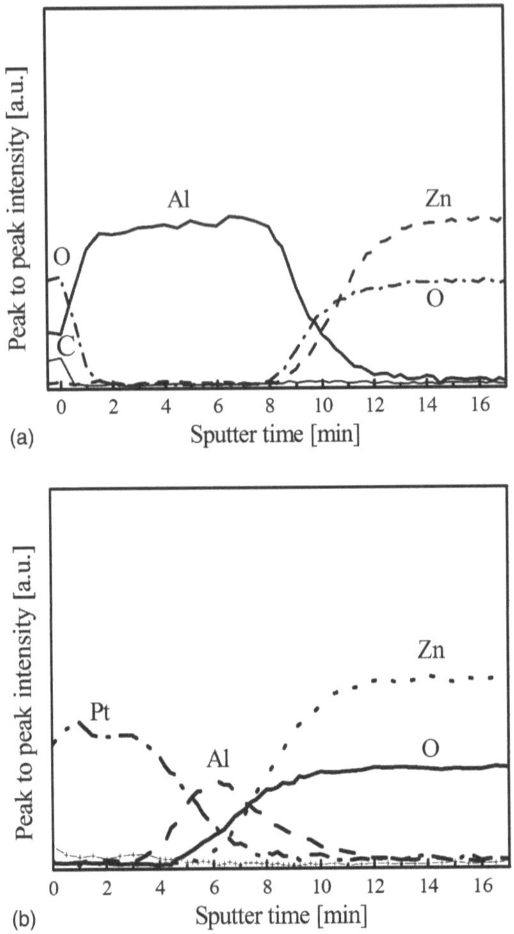

Figure 7.3: AES depth profiles obtained from (a) Al contact on ZnO and (b) Al/Pt contact on ZnO. (After ref [8])

was also observed after 5 min post annealing at 300°C, which resulted from the degradation of the sheet resistance of *n*-type ZnO substrate. The nonalloyed Al/Pt contacts produced a specific contact resistivity of $1.2 \times 10^{-5}\ \Omega \cdot cm^2$.[8] Pt overlayer on Al contact resulted in a large reduction in the specific contact resistivity on *n*-type ZnO, compared to the case without the overlayer, from the AES depth profile shown in Fig. 7.3. This reduction was attributed to the prevention of surface oxide layer (Al-O) by the Pt metal. In order to obtain high quality Ohmic contact on ZnO by using Al contact,[9] therefore, it is imperative to deposit the metal such as Pt and Au on Al layer for preventing the formation of surface oxide layer. Non-alloyed Al

Table 7.1: Specific contact resistance values obtained from TLM measurements for Al contacts to $Mg_xZn_{1-x}O$. (After ref [10])

Sample number	Mg composition percentage x (%)	Bandgap E_g (eV) at 300 K	Film resistivity ($\Omega \cdot cm$)	Specific contact resistance ρ_c ($\Omega \cdot cm^2$)
A	0	3.3	1.0	2.5×10^{-5}
B	11	3.5	7.7	2.1×10^{-3}
C	25	3.8	13.2	3.3×10^{-3}
D	34	4.0	34.1	9.1×10^{-3}

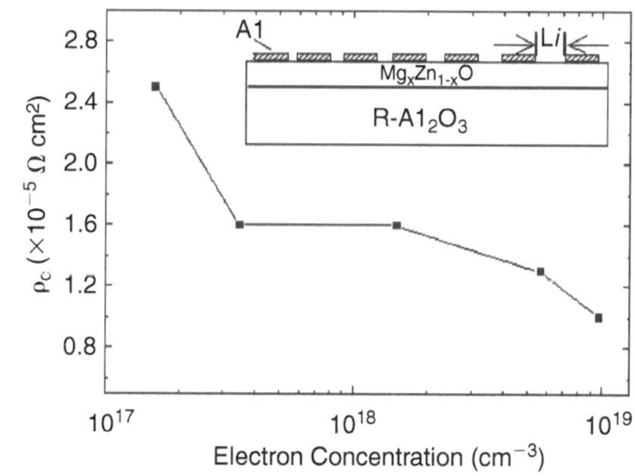

Figure 7.4: The plot of specific contact resistances as a function of electron concentrations in ZnO. The inset shows the TLM pattern. (After ref [10])

ohmic contacts was fabricated on epitaxial $Mg_xZn_{1-x}O$ ($0 \leq x \leq 0.34$).[10] Table 7.1 lists the specific contact resistance as a function of Mg composition in ZnO films and it has been shown that with increasing Mg composition (increasing bandgap) the specific contact resistance increased from 2.5×10^{-5} to 9.1×10^{-3} $\Omega \cdot cm^2$. For Al contact to $Mg_{0.34}Zn_{0.66}O$, specific contact resistance values were two orders of magnitude larger than that of Al ohmic contacts to ZnO with an electron concentration of 1.6×10^{17} cm^{-3}. The experimental specific contact resistance as a function of temperature for unintentionally and heavily Ga-doped ZnO is given in Fig. 7.4. It has been shown that if the electron concentrations increased (by Ga-doping during the growth), specific contact resistivity would decrease because of the increased tunneling. Low-resistance and non-alloyed ohmic contacts to epitaxially grown n-type ZnO has been achieved by exposing n-type ZnO to an inductively coupled argon and hydrogen plasma.[11] The I-V characteristics of as-deposited Ti/Au

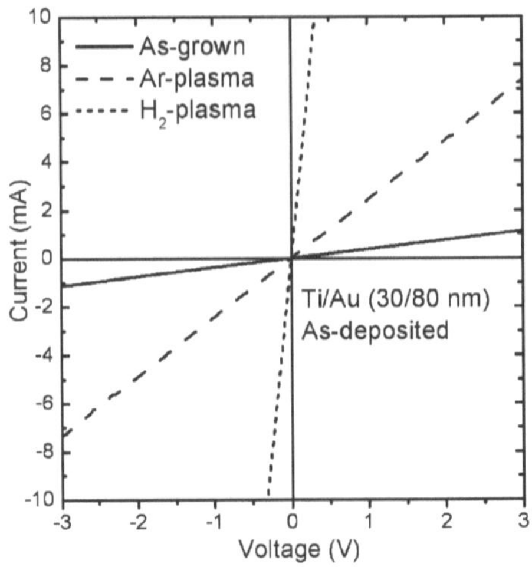

Figure 7.5: The I-V characteristics of as-deposited Ti/Au Ohmic contact on a plasma treated ZnO film. (After ref [11])

Ohmic contact on a plasma treated ZnO film are shown in Fig. 7.5. It has been shown that the specific contact resistivity of the Ar-plasma treated sample also decreased to $5.01 \times 10^{-4}\ \Omega \cdot cm^2$ which is lower by an order of the magnitude compared to that of the as-grown sample, and this decrease was attributed to the formation of oxygen vacancies on the ZnO surface by ion bombardment, which may act as donors. This result is comparable to the result reported by Kim *et al.*,[1,2] using the same metallization schemes after annealing at 300°C. The specific contact resistivity of the resulting Ohmic contacts drastically decreased from 7.1×10^{-3} to $4.31 \times 10^{-5}\ \Omega \cdot cm^2$ by hydrogen plasma treatment using Ti/Au metallization schemes. This was attributed that the hydrogen itself might act as carriers and hydrogen passivates the deep-level defects in the surface region of ZnO film.

The above-mentioned Ohmic schemes suffer from severe thermal degradation when annealed at high temperature. Therefore, contact schemes with thermal stability as well as low resistance for device applications were conducted as follows. Figure 7.6 shows the I-V characteristics of the Re/Ti/Au contacts on *n*-type ZnO as a function of the annealing temperature.[12] The as-deposited Re/Ti/Au contact was Ohmic with a contact resistivity of $2.1 \times 10^{-4}\ \Omega \cdot cm^2$. The electrical characteristics of the samples are further improved upon annealing, namely, the sample produces a specific contact resistance of $1.7 \times 10^{-7}\ \Omega \cdot cm^2$ when annealed at 700°C for 1 min in a nitrogen ambient. The insertion of Re layer was fairly effective for improving the thermal stability due to the formation of Ti-O and Re-O phases at

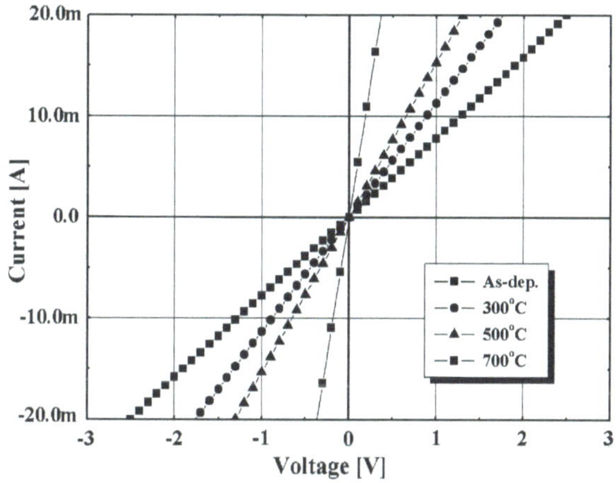

Figure 7.6: The I-V characteristics of the Re/Ti/Au contacts on *n*-type ZnO as a function of the annealing temperature. (After ref [12])

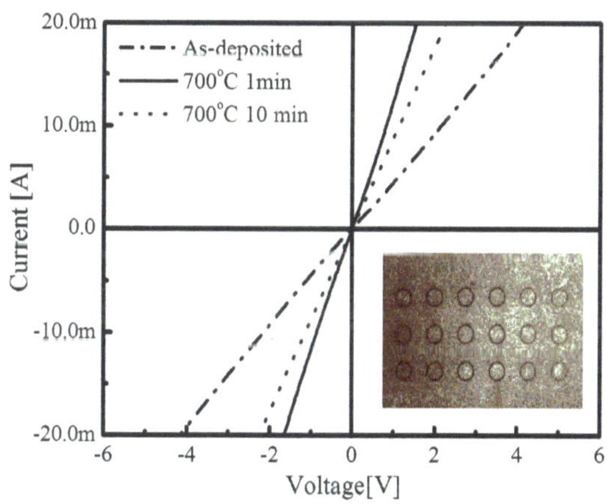

Figure 7.7: The I-V characteristics of the Ru contacts on the annealed ZnO. It is shown that both the as-deposited and annealed contacts exhibit linear I-V characteristics, although the latter shows better behavior. The c-TLM test pattern is shown in the inset. (After ref [13])

the interface and the suppression of Zn out-diffusion from the ZnO layer. In addition, the as-deposited Ru contact scheme yields a specific contact resistance of $2.1 \times 10^{-3}\ \Omega \cdot cm^2$.[13] However, annealing of the contact at 700°C for 1 min results in a resistance of $3.2 \times 10^{-5}\ \Omega \cdot cm^2$, as shown in Fig. 7.7. The annealing process results in a reduction in the specific contact resistance (by about two orders of magnitude), compared with the as-deposited sample. Oxygen was outdiffused from

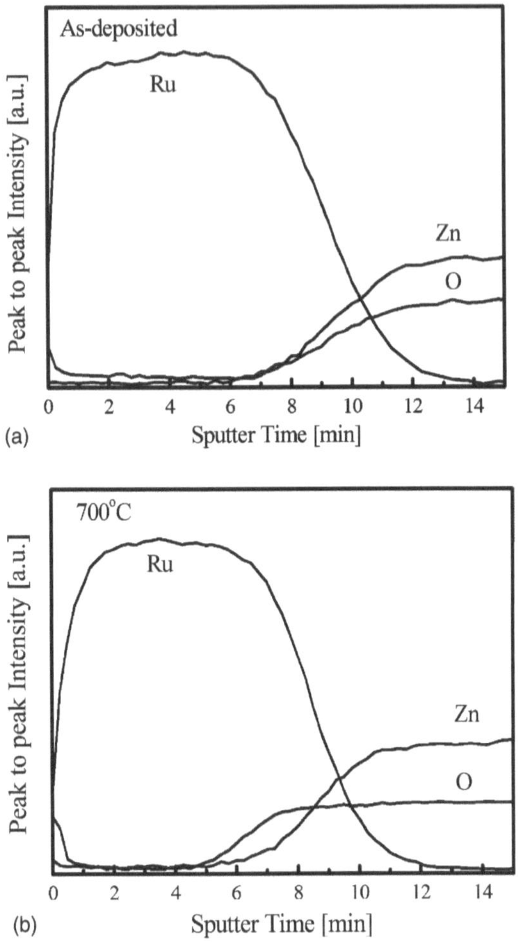

Figure 7.8: AES depth profiles of (a) the as-deposited and (b) 700°C-annealed Ru contacts on ZnO. (After ref [13])

the ZnO layer and participated in the formation of the RuO_2 interfacial product, resulting in the accumulation of oxygen vacancies near the ZnO surface. The prolonged annealing treatment caused negligible degradation of electrical and thermal properties. From Fig. 7.8, the Ru-O interfacial layer may prevent the outdiffusion of Zn (and hence the formation of zinc vacancies), acting as a diffusion barrier after the annealing process.

Surface modification using focused ion beams (FIBs) consisting of Ga ions and direct write Pt contact showed the contact resistance values for surface-modification

Table 7.2: Various Ohmic contact schemes for *n*-type ZnO.

Contact scheme	ZnO type	Contact resistance ($\Omega \cdot cm^2$)	Annealing condition (°C)	Thermal stability (°C)	Ref
Ti/Au	n (Al doped)	2×10^{-4}	300, N_2	300	1,2
Ti/Al/Pt/Au	n (P doped)	3.9×10^{-7}	200, N_2	<600	3,4
Zn/Au	n (Al doped)	2.36×10^{-5}	500, N_2	500	6
In	n	7×10^{-1}		<300	7
Al	n (Al doped)	8×10^{-4}			9
Al/Pt	n (Al doped)	1.5×10^{-5}			8
Al	n (MgZnO)	$2.1 \times 10^{-3} \sim 9.1 \times 10^{-3}$		<600	10
Re/Ti/Au	n (Al doped)	1.7×10^{-7}	700, N_2	700	12
Ru	n	3.2×10^{-5}	700, N_2	700	13
Pt-Ga	n (Ga doped)	3.1×10^{-4}			14

doses $D_{s1} = 1 \times 10^{17}$ cm^{-2}, $D_{s2} = 3 \times 10^{17}$ cm^{-2}, and $D_{s3} = 3 \times 10^{16}$ cm^{-2}, were 3.1, 3.3, and $3.7 \times 10^{-4}\ \Omega \cdot cm^2$, respectively.[14] When the contacts are direct-write deposited on the ZnO surface without the surface-modification step, the contact resistance value obtained is $4.1 \times 10^{-3}\ \Omega \cdot cm^2$, which is over one order of magnitude higher than the lowest value for contact resistance of the surface modified and direct-write Pt contacts.

The various Ohmic contact metallization schemes for *n*-type ZnO are summarized in Table 7.2 together with ZnO type, specific contact resistance, and annealing condition.

7.1.3 Ohmic contact to p*-type ZnO*

Because *p*-type ZnO does not exist in nature, considerable efforts have been made to grow *p*-type ZnO by impurity doping. *p*-type doping in ZnO may be possible by substituting either group-I elements (Li, Na, and K) for Zn sites or group-V elements (N, P, and As) for O sites. Recently ZnO-based homojunction and heterojunction devices have been reported with poor performance on the *p*-type ohmic contact. Therefore, it is necessary to investigate the electrical behaviors and ohmic contact mechanisms for *p*-type ZnO to realize the ZnO-based light-emitting diodes. The approaches to reduce the contact resistance to *p*-type ZnO are usually the oxygen indiffusion (formation of oxygen interstitials) and the outdiffusion of zinc

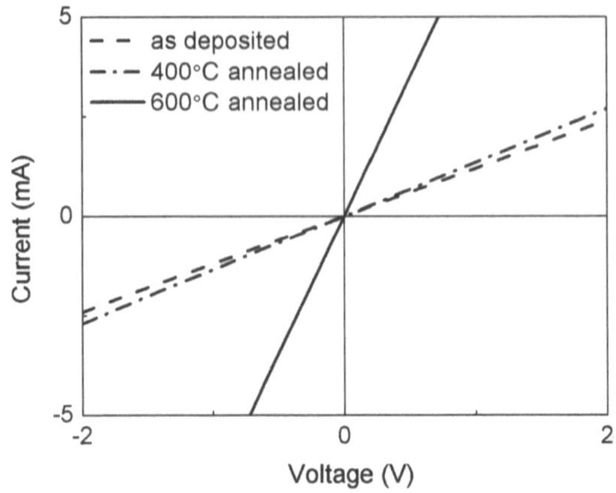

Figure 7.9: I-V characteristics of the Ni (30 nm)/Au (80 nm) contacts to *p*-type ZnO. All contacts show linear Ohmic behavior. (After ref [15])

(formation of zinc vacancies) in order to increase the carrier concentration, reduce the thickness of the Schottky barrier between the metal and the semiconductor and increase tunneling through the barrier.

A Ni/Au metallization scheme was reported for low resistance Ohmic contacts to the phosphorus doped *p*-type ZnO with a hole concentration of 1.0×10^{18} cm^{-3}.[15] I-V characteristics of the Ni/Au contacts to *p*-type ZnO is shown in Fig. 7.9. As-deposited Ni/Au contacts to *p*-type ZnO showed a specific contact resistance of $7.67 \times 10^{-3}\ \Omega \cdot cm^2$. When the Ni/Au contact was annealed at 600°C for 30 s in an air ambient the specific contact resistance was greatly decreased to $1.72 \times 10^{-4}\ \Omega \cdot cm^2$. The improved Ohmic property was attributed to an increase in the hole concentration by the formation of Ni-Zn and Au-Zn phases due to the out-diffusion of Zn during the thermal annealing process. The outdiffusion of Zn results in the accumulation of Zn vacancies at the surface region of the ZnO layer, which can act as acceptors in ZnO.[15] Thus, it was believed that the increase in the hole concentration at the surface region of ZnO could be responsible for the improved Ohmic property of the annealed Ni/Au contacts on the *p*-type ZnO. The oxygen in NiO was originated from the air atmosphere during the thermal annealing process. This was also consistent with the fact that the ZnO formation is more favorable than the NiO formation because the formation enthalpy of NiO ($\Delta Hf = -239.9$ kJ/mole) is larger than that of ZnO ($\Delta Hf = -350.9$ kJ/mole). Although the as-deposited Ni/ITO contact was highly resistant and opaque, contact resistance and transparency were greatly improved by a thermal annealing process.[16] A specific contact resistance as low as $6.2 \times 10^{-5}\ \Omega \cdot cm^2$ was obtained after thermal annealing at 400°C for 1 min

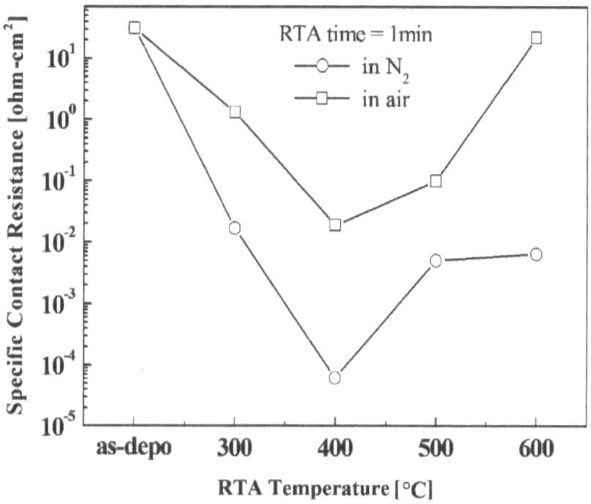

Figure 7.10: The effect of annealing temperature on the specific contact resistance of Ni/ITO contacts to *p*-type ZnO under air and nitrogen ambients. (After ref [16])

under an N_2 ambient. Figure 7.10 shows the effect of RTA temperature on the specific contact resistance of Ohmic contacts annealed under different gas ambients. The sheet resistance of Ni/ITO contacts was not changed before and after thermal annealing under N_2, while the sheet resistance of Ni/ITO annealed under an air ambient was increased from 170 to 1100 Ω/cm^2. The measured light transmittance of the Ni/ITO bilayer on *p* type ZnO was determined to be above 80% at a wavelength of 400~600 nm. The Ni/Au contacts to ZnO:As films show linear I-V characteristics, whereas In/Au and Ti/Au contacts show back-to-back Schottky barriers.[17] The Ni/Au metal bilayer forms an ohmic contact to the ZnO:As film, however, the other two metal bilayers do not. In addition, Pt/indium tin oxide (ITO) scheme on the ZnO:As film produce contact resistivities of $8.0 \times 10^{-4} \sim 3.5 \times 10^{-3}\ \Omega \cdot cm^2$ upon annealing at 300~600°C as shown in Fig. 7.11.[18] It could be associated with the improved sheet resistance of the ITO and the removal of the native oxide or contamination layers. Another possibility may be related to the breaking up of the Pt layer. The formation of the discontinuous Pt layer at the interface implies that inhomogeneous Schottky barriers might form. It is noted that Pt and ITO have work functions of 5.65 and 4.7 eV, respectively. The difference between the Schottky barriers heights of the Pt and the ITO can lead to an increase in the electric field at the metal/semiconductor interface. The increase of the electric field was shown to cause the lowering of a barrier height.

For large band gap *p*-type ZnMgO, Ti/Au and Pt/Au contacts to polycrystalline phosphorus-doped $Zn_{0.9}Mg_{0.1}O$ films show Ohmic characteristics after annealing at >500°C.[19] A specific contact resistivity of $3 \times 10^{-3}\ \Omega \cdot cm^2$ was achieved for

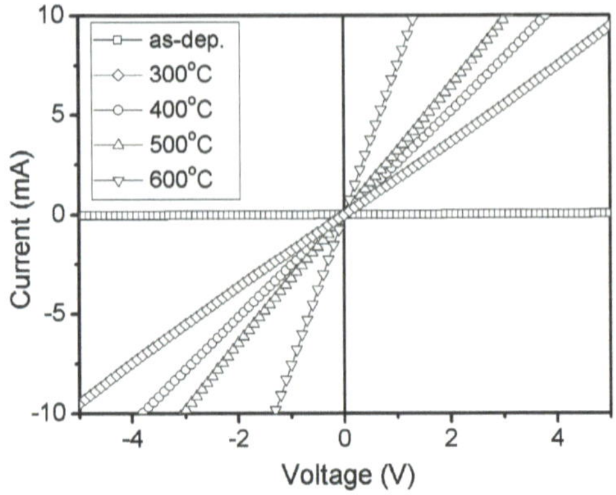

Figure 7.11: The I-V characteristics of the Pt/ITO contacts on the as-grown *p*-type ZnO layer as a function of the annealing temperature, measured on the 3 μm spaced pads. (After ref [18])

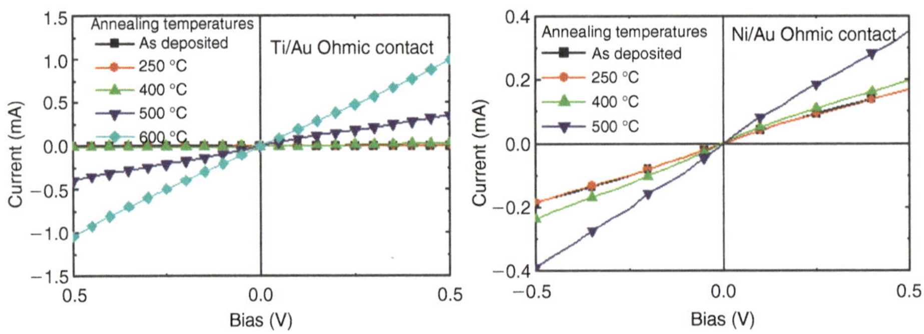

Figure 7.12: I-V characteristics as a function of annealing temperature for TLM patterns with Ti/Au (top) or Ni/Au (bottom) Ohmic contacts. (After ref [19])

Ti/Au annealed at 600°C. However, the Ni/Au contacts were less thermally stable than the Ti/Au and could only be annealed to 500°C without reaction and severe degradation of the contact morphology, as shown in Fig. 7.12.

7.2 Schottky contact to *n*-type ZnO

In spite of the fact that high quality Schottky contacts are critical for ZnO device applications, there is little information about the Schottky contacts on ZnO to date. To create Schottky barrier with undoped ZnO, a high work function can be applied to the surface of a ZnO crystal. Taking into account their metal work function,

Table 7.3: Schottky barrier heights (Φ_{SH}) and ideality factors (n) for Au Schottky contacts on ZnO:N layers, and resistivities (ρ) and incorporated N concentrations (N) of ZnO:N layers. (After ref [23])

Sample number	T_g (°C)	Polar dir.	Φ_{SH}	n (−)	ρ ($\Omega \cdot$cm)	N (cm^{-3})
A	400	O-polar	...	...	8.75×10^{-3}	4×10^{18}
B	400	Zn-polar	0.55	3.2	1.05×10^{-2}	8×10^{19}
C	300	O-polar	0.37	3.5	1.03×10^{-2}	7×10^{19}
D	300	Zn-polar	0.66	1.8	1.41×10^{-2}	2×10^{20}

one expects Pt (5.64 eV), Au (5.47 eV), Ni (5.22 eV), Pd (5.22 eV), Ag (4.64 eV), and Ti (4.33 eV) to be good candidates to form Schottky contacts with different barrier heights to ZnO.[20] Ultraviolet photoelectron spectroscopy measurements have shown that the electron affinity of a clean ZnO(000-1) surface is 4.1 eV. Hence, estimations of the barrier height on n-type ZnO result in values of 1.55, 1.37, 1.12, 1.12, 0.54, and 0.23 eV for Pt, Au, Ni, Pd, Ag, and Ti, respectively. However, the trend in barrier heights did not correlate with the metal work functions, suggesting that either intrinsic surface states or residual surface contamination is playing a role in determining the electrical properties of the contacts.

The Schottky contact to ZnO was first reported by Mead *et al.*[21] The energy barriers for Au and Pd were determined to be 0.71 and 0.65 eV, respectively. However, their results did not include its detailed properties. Recently, there have been several reports on Schottky contacts to ZnO. Ag Schottky contact was fabricated on the (11-20) epitaxial ZnO film grown by MOCVD.[22] The Schottky barrier heights, at room temperature were determined to be 0.89 and 0.92 eV using I-V-T, and C-V methods. The ideality factor decreased from 1.37 to 1.29 as the temperature was increased from 265 K to 340 K. The reason for the high ideality factor could most probably be due to the existence of an interfacial layer and surface states. The characteristics of Au Schottky contacts were investigated on ZnO:N layers grown on (0001) GaN/Al_2O_3 substrates by plasma-assisted molecular-beam epitaxy.[23] It is found that the Schottky characteristics are dependent on the growth temperature and polar direction of ZnO:N layers. The Schottky barrier height for the Au contact to a ZnO:N layer (300°C, Zn-polar) is estimated to be 0.66 and 0.69 eV by I-V and C-V measurements, respectively. It is found that the Schottky barrier height is proportional to the resistivity and incorporated N concentration of ZnO:N layers, as shown in Table 7.3. Consequently, the low growth temperature and Zn-polar direction are favored for N incorporation in the growth of ZnO:N layers, which contributes to the increased resistivity in ZnO:N layers and results in good Schottky characteristics. Schottky barrier contacts have been formed by electron beam evaporation of Pd on the (000-1) face of hydrothermally grown n-type single-crystalline ZnO.[24] The contacts can be operated under reverse bias voltages up to -5 V and in the temperature range between 130 and 350 K. A barrier height of 0.83 eV is deducted at room temperature. The ideality factor for the current-voltage characteristics varies between 1.01 and 1.03 in the voltage range of -5.0 to $+0.5$ V, indicating that thermionic emission is a dominant mechanism for charge-carrier transport.

To improve Schottky behaviors on n-type ZnO, surface treatments using solutions, oxygen plasma, and ozone have been performed.[25–30] Neville and Mead reported ideality factors very close to unity ($n = 1.05$) after depositing Au and Pd on the n-type ZnO surface etched for 15 min in concentrated phosphoric acid, followed by 5 min etch in concentrated HCl and rinsed in organic solvents.[25] However, the thermal stability of Schottky contacts was not been extensively studied. Au and Ag Schottky contacts have been investigated on Zn surface of undoped n-type ZnO samples comparing their barrier heights and thermal stabilities.[26] Schottky barrier heights of 0.65~0.70 eV and diode ideality factors of 1.6~1.8 have been obtained by capacitance-voltage and current-voltage measurements, respectively. Saturation current densities ranged from 10^{-5} A cm^{-2} on surfaces etched in HCl to 8×10^{-7} A cm^{-2} on solvent cleaned samples. The diode ideality factors were in the range 1.6~1.8 under all conditions. Table 7.4 shows the characteristics of Schottky contacts depending on contact materials and surface treatments. The properties of both the Au and the Ag Schottky diodes were degraded by heating in vacuum to temperatures even as low as 365 K. The degradation mechanisms during annealing were the reaction with the ZnO surface and the localized delamination for Au and Ag contacts, respectively. The properties of Pt Schottky contacts on the Zn-terminated

Table 7.4: Ideality factor n, reverse current at 20.5 V I_r, saturation current density J_s, the activation energy for the saturation current E_a and the cutoff voltage V_c in C-V for various Au/ZnO and Ag/ZnO Schottky diodes measured at room temperature. (After ref [26])

Metal	Surface treatment	I_r (A)	n	J_s (A cm^{-2})	E_a (eV)	V_c (V)
Au	Organic solvents	8.6×10^{-8}	1.8	8.4×10^{-7}	0.35	0.65
Au	HCl	2.1×10^{-6}	1.6	8×10^{-6}	0.4	0.64
Au	HNO_3	1.6×10^{-6}	1.8	4.8×10^{-6}	0.29	0.65
Ag	HCl	2.5×10^{-6}	1.6	1×10^{-5}	0.3	0.69
Ag	HNO_3	2×10^{-6}	1.8	6×10^{-6}	0.35	0.68

n-type ZnO surfaces was improved using surface treatment with a hydrogen peroxide solution and sulfide solution.[27,28] The Pt contacts on organic solvent-cleaned ZnO show leaky behavior with a high leakage current of 0.05 A under −5 V reverse bias voltage, whereas the hydrogen peroxide-treated and sulfide-treated contacts show Schottky behavior with very low leakage current of 6.5×10^{-8} and 3.75×10^{-10} A under −5 V reverse bias, respectively. Schottky barrier heights and ideality factors are 0.89 eV and 1.15 for the hydrogen peroxide-treated contact and 0.79 eV and 1.51 for sulfide-treated contacts respectively. Au Schottky contacts was improved exposing n-type ZnO(000-1) to 20% O_2/80% He plasma at 525 ± 20°C for 30 min.[29] Plasma cleaning resulted in highly ordered, stoichiometric, and smooth surfaces. Contacts on as-received material showed mA leakage currents and ideality factors >2. Contacts on plasma-cleaned wafers cooled in vacuum showed 36 ± 1 nA leakage current to 24 V, a barrier height of 0.67 ± 0.05 eV, and an ideality factor of 1.866 ± 0.05. UV-ozone cleaning prior to metal deposition of either e-beam Pt contacts or sputtered W contacts on n-type single-crystal ZnO is found to significantly improve their rectifying characteristics.[30] Pt contacts deposited directly on the as-received ZnO surface are Ohmic but show rectifying behavior with ozone cleaning. The Schottky barrier height of these Pt contacts was 0.70 eV, with ideality factor of 1.5 and a saturation current density of 6.2×10^{-6} A cm^{-2}. In contrast, the as-deposited W contacts are Ohmic, independent of the use of ozone cleaning. The postdeposition annealing did not cause any detectable intermixing up to temperatures of 500°C but after 700°C anneal, the AES depth profiles showed that Zn diffused out to the Au-Pt interface, as shown in Fig. 7.13. Postdeposition annealing at 700°C produces rectifying behavior with

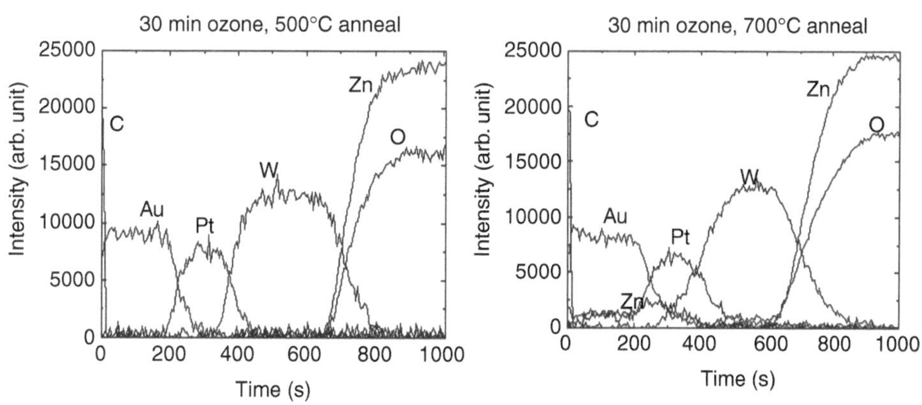

Figure 7.13: AES depth profiles from W/Pt/Au contacts on ZnO after annealing at either 500 or 700°C. (After ref [30])

Schottky barrier heights of 0.45 eV for control samples and 0.49 eV for those cleaned with ozone exposure. The improvement in rectifying properties of both the Pt and W contacts is related to removal of surface carbon contamination from the ZnO.

References

[1] H. K. Kim, S. H. Han, T. Y. Seong, and W. K. Choi, Appl. Phys. Lett. 77, 1647 (2000).
[2] H. K. Kim, S. H. Han, T. Y. Seong, and W. K. Choi, J. Electrochem. Soc. 148, G114 (2001).
[3] K. Ip, Y. W. Heo, K. H. Baik, D. P. Norton, S. J. Pearton, and F. Ren, Appl. Phys. Lett. 84, 544 (2004).
[4] K. Ip, Y. W. Heo, K. H. Baik, D. P. Norton, S. J. Pearton, and F. Ren, J. Vac. Sci. Technol. B 22, 171 (2004).
[5] K. Ip, K. H. Baik, Y. W. Heo, D. P. Norton, S. J. Pearton, J. R. LaRoche, B. Luo, F. Ren, and J. M. Zavada, J. Vac. Sci. Technol. B 21, 2378 (2003).
[6] S. H. Kim, S. W. Jeong, D. K. Hwang, S. J. Park, and T. Y. Seong, Electrochem. Solid-State Lett. 8, G198 (2005).
[7] T. Akane, K. Sugioka, and K. Midorikawa, J. Vac. Sci. Technol. B 18, 1406 (2000).
[8] H. K. Kim, K. K. Kim, S. J. Park, T. Y. Seong, and I. Adesida, J. Appl. Phys. 94, 4225 (2003).
[9] H. K. Kim, T. Y. Seong, K. K. Kim, S. J. Park, Y. S. Yoon, and I. Adesida, Jpn. J. Appl. Phys. Part 1 43, 976 (2004).
[10] H. Sheng, N. W. Emanetoglu, S. Muthukumar, S. Feng, and Y. Lu, J. Electron. Mater. 31, 811 (2002).
[11] J. M. Lee, K. K. Kim, S. J. Park, and W. K. Choi, Appl. Phys. Lett. 78, 3842 (2001).
[12] S. H. Kim, K. K. Kim, S. J. Park, and T. Y. Seong, J. Electrochem. Soc. 152, G169 (2005).
[13] H. K. Kim, K. K. Kim, S. J. Park, T. Y. Seong, and Y. S. Yoon, Jpn. J. Appl. Phys. 41, Part 2 (5B), L546 (2002).
[14] A. Inumpudi, A. A. Iliadis, S. Krishnamoorthy, S. Choopun, R. D. Vispute, and T. Venkatesan, Solid State Electron. 46, 1665 (2002).
[15] J. H. Lim, K. K. Kim, D. K. Hwang, H. S. Kim, J. Y. Oh, and S. J. Park, J. Electrochem. Soc. 152, G179 (2005).
[16] S. H. Kang, D. K. Hwang, and S. J. Park, Appl. Phys. Lett. 86, 211902 (2005).
[17] Y. R. Ryu, T. S. Lee, J. H. Leem, and H. W. White, Appl. Phys. Lett. 83, 4032 (2003).
[18] S. H. Kim, J. T. Maeng, C. J. Choi, J. H. Leem, M. S. Han, and T. Y. Seong, Electrochem. Solid-State Lett. 8, G167 (2005).
[19] S. Kim, B. S. Kang, F. Ren, Y. W. Heo, K. Ip, D. P. Norton, and S. J. Pearton, Appl. Phys. Lett. 84, 1904 (2004).
[20] *83rd CRC Handbook of Chemistry and Physics*, D. R. Lide and Editors, CRC, Boca Raton (2002).

[21] C. A. Mead, Phys. Lett. 18, 218 (1965).
[22] H. Sheng, S. Muthukumar, N. W. Emanetoglu, and Y. Lu, Appl. Phys. Lett. 80, 2132 (2002).
[23] D. C. Oh, J. J. Kim, H. Makino, T. Hanada, M. W. Cho, T. Yao, and H. J. Ko, Appl. Phys. Lett. 86, 042110 (2005).
[24] U. Grossner, S. Gabrielsen, T. M. Børseth, J. Grillenberger, A. Y. Kuznetsov, and B. G. Svensson, Appl. Phys. Lett. 85, 2259 (2004).
[25] R. C. Neville, C. A. Mead, J. Appl. Phys. 41, 3795 (1970).
[26] A. Y. Polyakov, N. B. Smirnov, E. A. Kozhukhova, V. I. Vdovin, K. Ip, Y. W. Heo, D. P. Norton, and S. J. Pearton, Appl. Phys. Lett. 83, 1575 (2003).
[27] S. H. Kim, H. K. Kim, and T. Y. Seong, Appl. Phys. Lett. 86, 112101 (2005).
[28] S. H. Kim, H. K. Kim, and T. Y. Seong, Appl. Phys. Lett. 86, 022101 (2005).
[29] B. J. Coppa, R. F. Davis, and R. J. Nemanich, Appl. Phys. Lett. 82, 400 (2003).
[30] K. Ip, B. P. Gila, A. H. Onstine, E. S. Lambers, Y. W. Heo, K. H. Baik, D. P. Norton, S. J. Pearton, S. Kim, J. R. LaRoche, and F. Ren, Appl. Phys. Lett. 84, 5133 (2004).

Chapter 8

Ion Implantation into ZnO

S. O. Kucheyev

Lawrence Livermore National Laboratory, Livermore, CA 94551, USA

C. Jagadish

Department of Electronic Materials Engineering, Research School of Physical Sciences and Engineering, The Australian National University, Canberra, ACT 0200, Australia

Abstract: The current status of ion-beam processing of ZnO is reviewed. In particular, we discuss the following aspects of ion implantation into ZnO: (i) damage accumulation and amorphization, (ii) thermal stability of radiation defects, (iii) electrical doping, and (iv) electrical isolation. Emphasis is given to the current understanding of the physical processes responsible for the formation and evolution of radiation damage.

Keywords: Ion implantation, ZnO, amorphization, annealing, doping, defects, electrical isolation.

8.1 Introduction

Zinc oxide is a II-VI semiconductor with a room temperature direct band gap of ~3.4 eV. At 20°C and atmospheric pressure, it has the wurtzite lattice structure. Current technological applications of polycrystalline ZnO ceramics and powders are numerous, as has been extensively reviewed elsewhere [1,2]. In addition, over the past several decades, single-crystal ZnO has been considered as a potential candidate for a range of optoelectronic devices as well as devices for high-temperature/high-power electronics [1–8]. Practical advantages of ZnO over another wide band-gap semiconductor – GaN, which currently dominates the market of short-wavelength optoelectronics – include a bulk growth capability, amenability to conventional chemical wet etching, and convenient cleavage planes. However, at present, there are still serious challenges to overcome with processing single-crystal ZnO.

In the fabrication of ZnO-based devices, ion bombardment represents a very attractive processing tool for several technological steps, including selective-area

Zinc Oxide Bulk, Thin Films and Nanostructures
C. Jagadish and S. Pearton (Editors)

doping, electrical isolation, dry etching, quantum-well intermixing, and ion slicing. However, ion-beam-produced lattice disorder and its often undesirable consequences can limit technological applications of ion implantation. Hence, numerous ion implantation studies have been dealing with ion-beam-defect processes in ZnO.

Most studies of ion-beam-defect processes in solids, in fact, focus on understanding the following fundamental aspects: (i) the degree of dynamic annealing, (ii) the nucleation of an amorphous phase, (iii) the nature of ion-beam-produced defects, (iv) the effect of electronic energy loss, (v) the influence of the density of collision cascades, (vi) chemical effects of implanted species, (vii) material decomposition during bombardment, (viii) mobility and lifetime of defects, and (ix) thermal stability of radiation damage. In addition, more applied studies often focus on the effects of implantation damage on technologically important material properties such as various electrical and optical characteristics.

The present review focuses on the current progress made in the understanding of ion-beam processes in ZnO. Section 8.2 introduces the concept of dynamic annealing and the difference between atomic displacements ballistically generated in collision cascades and stable lattice disorder. Section 8.3 treats the structural characteristics of ZnO exposed to ion bombardment. The current progress in the annealing of implantation damage is discussed in Section 8.4. Studies of the lattice sites of dopants implanted into ZnO are reviewed in Section 8.5. Electrical characteristics of ion-irradiated ZnO are reviewed in Sections 8.6 and 8.7. Throughout this chapter, we also emphasize the current problems of ion-beam processing of ZnO and outline some directions for future studies in this field.

A comment should also be made on what is not included in this review due to space limitations. We omit the discussion of specific defect levels studied by deep level transient spectroscopy (DLTS), lattice disorder produced by dry etching, the effects of hydrogen implanted into ZnO, defects produced by irradiation with energetic electrons, and magnetic properties of ion-implanted ZnO. For the same reason, we only briefly discuss studies of luminescence from ZnO implanted with rare-earth ions. Most of these studies have recently been reviewed in Refs 6 to 8 and in Chapters 2, 9 and 16.

8.2 Ballistic displacements and dynamic annealing processes

An energetic ion penetrating through a ZnO crystal generates a collision cascade which consists of vacancies (in both Zn and O sub lattices), interstitials (of different symmetries), and, presumably, amorphous zones in dense regions of the collision cascade. The characteristic time of generation and thermalization of collision cascades is ~1–10 ps [9]. Figure 8.1 shows a schematic of a collision cascade and

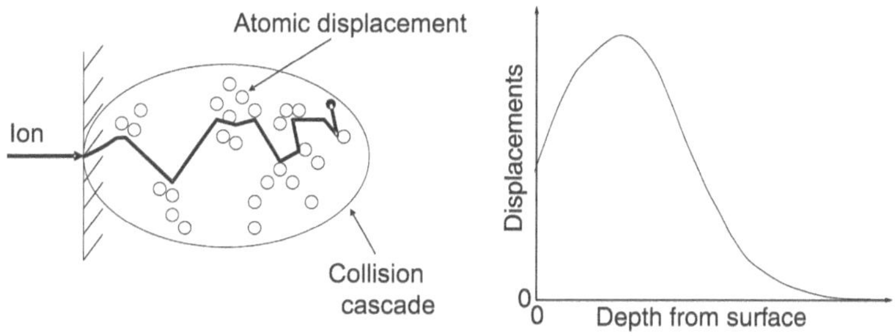

Figure 8.1: Schematic of a collision cascade and a typical depth profile of atomic displacements created by keV ions.

a typical depth profile of atomic displacements (averaged over many cascades). It is seen from Fig. 8.1 that ion implantation produces Gaussian-like (uni-modal) depth profiles of atomic displacements. The thickness of the layer modified by ion bombardment, the total number of atomic displacements generated, and the shape (i.e., effective width and skewness) of the depth profiles of atomic displacements and implanted atoms depend on ion mass and energy.

Ion ranges and atomic displacements can be readily calculated by several methods. The currently most common approach in the ion-beam community to obtain such information is calculations with the TRIM code [10], a Monte Carlo computer simulation program. Such theoretical calculations of atomic displacements, however, take into account only ballistic processes and completely neglect *dynamic annealing* (i.e., migration and interaction of ion-beam-generated defects *during* ion irradiation). Ion generated simple point defects, which survive after the thermalization of collision cascades, may migrate through the lattice and experience annihilation and cluster formation. As a result, the total amount and the depth distribution of lattice disorder in solids experimentally observed after ion bombardment in most cases depart from predictions based only on collisional processes.

Complex dynamic annealing processes in solids under ion bombardment are also highly dependent on implant conditions such as ion mass, energy, dose,[1] substrate temperature, and beam flux. In contrast to our ability to calculate atomic displacements in solids, stable lattice disorder produced by ion bombardment is usually very difficult to predict. In most cases, experimental studies are essential to understand the effects of implant conditions on ion-beam-produced stable lattice disorder.

[1] In this chapter, we are using the term "ion dose" (in ions/cm^2), which is well accepted in the literature on ion implantation into semiconductors. However, it should be noted that, in the literature on irradiation effects in ceramics, ion dose is often referred to as "integrated ion flux" or "ion fluence".

8.3 Accumulation of structural disorder and lattice amorphization

Lattice disorder is always a concomitant effect of bombardment with energetic ions. Therefore, studies of implantation disorder and its consequences are necessary for a successful application of ion implantation in device fabrication. The structural characteristics of semiconductors exposed to ion bombardment are usually studied by Rutherford backscattering/channeling (RBS/C) spectrometry, cross-sectional transmission electron microscopy (XTEM), and x-ray diffraction (XRD). RBS/C is a very powerful, depth-sensitive (quantitative) technique often used to monitor the gross level of lattice disorder produced by ion bombardment. XTEM is currently one of the most powerful techniques used to identify the nature of implantation-produced defects. Lattice distortions (stress) produced by ion bombardment are often studied by XRD. In addition, ion-beam-induced stoichiometric imbalance is commonly monitored by x-ray photoelectron spectroscopy (XPS).

The structural characteristics of ion-bombarded ZnO have been studied by several groups. A summary of structural (as well as electrical and optical) studies of ion-implanted ZnO is given in Table 8.1. It is seen from this table that damage buildup (i.e., lattice disorder as a function of ion dose) has been studied using RBS/C, XTEM, and/or XRD by several groups for ions of different masses (such as Ne, Si, Ar, Cu, Er, and Au) and for different ion energies. However, there has been only one *systematic* study [20] of the effects of ion dose, mass, and implantation temperature (for the temperature range of –196 to 20°C). Most studies have been limited to irradiation to a single dose (or to a few doses) of a potentially electrically or optically active species (see Table 8.1). Moreover, we are not aware of any reports on the flux effect (i.e., implantation to the same dose with different values of beam flux [dose rate]). Hence, additional systematic studies are needed to fill the gaps in our understanding of the effects of ion irradiation conditions on ion-beam processes in ZnO. The current understanding of dynamic annealing processes in ZnO and the main features of damage accumulation in this material is discussed below.

8.3.1 Amorphization

It has been firmly established by several research groups that ZnO exhibits very strong dynamic annealing and remains crystalline (though heavily damaged) even after bombardment to very high doses of keV heavy ions [11,13,16,17,20,23,28,46]. For example, it has been found that, under bombardment at –196°C with 300 keV ^{197}Au or 360 keV ^{208}Pb ions (which generate dense collision cascades), the ZnO lattice does not exhibit an ion-beam-induced crystalline-to-amorphous transition

Table 8.1: A summary of ion-implantation studies of ZnO. Abbreviations of the experimental techniques which are not discussed in the text are spelled out in table footnotes.

Ion	Energy (keV)	Doses (cm^{-2})	Beam flux ($cm^{-2}\,s^{-1}$)	T_{impl}[a] (°C)	T_{ann}[b] (°C)	Ambient	Analysis	Ref.
N	50–380[1]	$<10^{14}$ total	1.2×10^{12}	20	Oven, ≤1200	N_2	PAS, CL,[c] Hall	[37]
Ne	100	1×10^{15}–1×10^{17}	6×10^{12}	20	–	–	XRD, OA,[d] *I-V*[e]	[25]
Ne, Ar	100	1×10^{15}–1×10^{17}	6×10^{12}–3×10^{13}	20	200–700	Vacuum	XRD, Hall, AFM	[44]
Al	50–380[1]	$<10^{15}$ total	1.2×10^{12}	20	Oven, ≤900	N_2	PAS, CL, Hall	[36]
Si	60	2×10^{15}–8×10^{16}	1.2×10^{13}	−196	–	–	RBS/C, XTEM	[20]
P	50–380[1]	$<4.2 \times 10^{15}$ total	1.2×10^{12}	20	Oven, ≤1100	N_2	PAS, CL, Hall	[43]
Ar	1	–	–	20	–	–	RHEED[f]	[11]
Ar	200	1×10^{14}–1×10^{16}	NR[g]	20	950	NR	RBS/C, PL[h]	[13]
Ar	6	–	–	20	–	–	XTEM	[18]
Cr	93	7×10^{15}	NR	20	500–1100	Air	RBS/C	[17]
Mn	96	7×10^{15}	NR	20	500–1100	Air	RBS/C	[17]
Fe	100	1×10^{16}	NR	20	Oven, 300–1050	Air, vacuum	RBS/C, PL	[28]
Cu	60	1×10^{15}–1×10^{17}	1.2×10^{13}	20	200–600	Vacuum	RBS/C, XRD	[23]
Ga	150	2×10^{16}	NR	20	RTA,[a] 900	Air	AFM, SIMS[b]	[45]
As	300	1.4×10^{17}	2.5×10^{12}	20	Oven, 1000, 1100, 1200	Ar	RBS/C, XTEM, AFM	[46]

[a] T_{impl} = implantation temperature.
[b] T_{ann} = annealing temperature.
[c] CL = cathodoluminescence.
[d] OA = optical absorption.
[e] *I-V* = current–voltage measurements.
[f] RHEED = reflection high energy electron diffraction.
[g] NR = not reported.
[h] PL = photoluminescence.

(*Continued*)

Table 8.1: (*Continued*)

Ion	Energy (keV)	Doses (cm^{-2})	Beam flux ($cm^{-2}\,s^{-1}$)	T_{impl}[a] (°C)	T_{ann}[b] (°C)	Ambient	Analysis	Ref.
As	70	1×10^{15}	NR	20	Oven, 700–900	O_2	PL, *I-V*, XRD, AFM	[38,42]
In	170	1×10^{16}	NR	20	750–850	Air	SIMS, PL	[24]
Er	150	5×10^{14}, 5×10^{15}	NR	20	Oven, 800, 1050	Air	RBS/C, PL	[27]
Tm	150	5×10^{15}, 5×10^{16}	NR	20	Oven, 800–1050	Air	RBS/C, PL	[32,39]
Au	300	8×10^{13}–4×10^{16}	3.0×10^{12}	−196, 20	–	–	RBS/C, XTEM, XPS	[20]
Pb	360	3×10^{16}, 1×10^{17}	NR	−196	–	–	RBS/C	[16]
Bi	120	7×10^{15}	NR	20	500–1100	Air	RBS/C	[17]
C	90000	1.3×10^{14}	NR	20	400	NR	CL	[15]
N	350	3×10^{14}, 1×10^{15}, 3×10^{15}	NR	20	Oven, 400–800	O_1	PL, SIMS	[22]
O	110,000	1.3×10^{14}	NR	20	400	NR	CL	[15]
Al	40–260[1]	2×10^{11}–2×10^{14} total	NR	20	Oven, 200–1000	O_2	PL	[14]
S	50	1×10^{16}	NR	20	Oven, 800	Ar	PL, SIMS	[40]
Ar	60–350[1]	2×10^{13} total	NR	20	Oven, 200–1000	O_2	PL	[14]
Co, Mn, Fe, Cr	250	3×10^{16}	NR	20	RTA, 700	N_2	CL	[29,34]
Cu	125	10^{12}–10^{14}	NR	20	Oven, 800	Air	PL	[41]
Er	180	6×10^{13}–6×10^{15}	NR	20	RTA, 600–1100	NR	PL	[26]

Fe	60	2.0×10^{13}	NR	20	Oven, 200–1000	Vacuum	ECS	[33]
Cu	60	2.3×10^{13}	NR	20	Oven, 200–800	Vacuum	ECS	[35]
Ag	60	2.0×10^{13}	NR	20	Oven, 200–800	Vacuum	ECS	[31]
Er	60	1.3×10^{13}, 2.8×10^{13}	NR	20	Oven, 100–900	Vacuum	ECS	[30]
P	200	5×10^{15}	NR	20	Oven, 100–500	Vacuum	Hall	[12]
Ar	250	5×10^{15}	NR	20	Oven, 100–500	Vacuum	Hall	[12]
V	300	5×10^{15}	NR	20	Oven, 100–500	Vacuum	Hall	[12]
H	600	1.0×10^{12}–7.8×10^{16}	4×10^{9}–2×10^{11}	20	RTA, 100–800	Ar	*I-V*	[19]
Li	3500	5.0×10^{11}–1.2×10^{15}	1.9×10^{11}	20	RTA, 100–800	Ar	*I-V*	[19]
O	2000	5.0×10^{11}–1.0×10^{16}	1.9×10^{11}, 1.9×10^{12}	20, 350	RTA, 100–800	Ar	*I-V*	[19,21]
Si	3500	3.0×10^{11}–1.0×10^{15}	1.0×10^{11}	20	RTA, 100–800	Ar	*I-V*	[19]
Cr	100–1000[1]	1.5×10^{15} total	$\sim 10^{11}$	20	RTA, 100–1000	Ar	*I-V*	[21]
Fe	100–1000[1]	1.5×10^{15} total	$\sim 10^{10}$	20	RTA, 100–1000	Ar	*I-V*	[21]
Ni	100–1000[1]	1.5×10^{15} total	$\sim 10^{12}$	20	RTA, 100–1000	Ar	*I-V*	[21]

[a] RTA = rapid thermal annealing.
[b] SIMS = secondary ion mass spectrometry.
[1] Multiple energy implants.

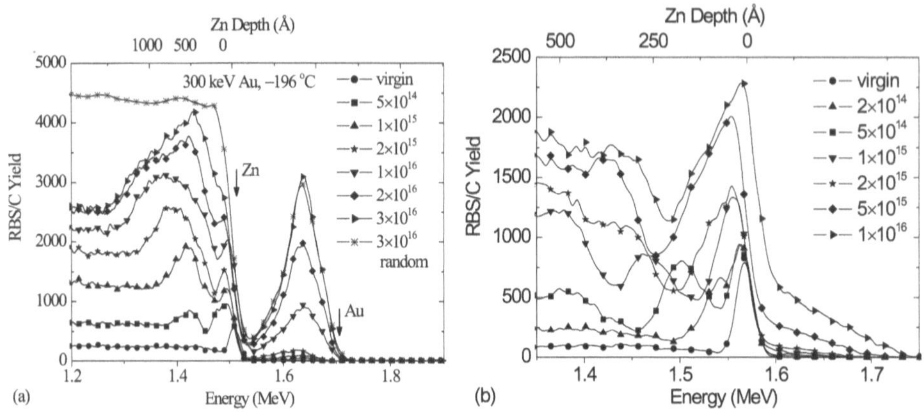

Figure 8.2: Selected RBS/C spectra [acquired with 25° (a) and 8° (b) glancing angle detector geometries] showing the damage buildup in ZnO bombarded at −196°C with 300 keV Au ions with a beam flux of $\sim 3 \times 10^{12}\,cm^{-2}\,s^{-1}$. Implantation doses (in cm^{-2}) are indicated in the legend. The positions of the surface peaks of Au and Zn are denoted by arrows in (a). Note that the near-surface random yield in (b) corresponds to ~2800 counts. After Ref. 20.

even for extremely high ion doses (up to $\sim 10^{17}\,cm^{-2}$) when each lattice atom has been ballistically displaced hundreds of times [16,20]. This result indicates that most Frenkel pairs (i.e., lattice vacancies and interstitials), ballistically generated in ZnO by the ion beam, experience annihilation, and only a very small portion of such migrating point defects avoids annihilation to agglomerate into energetically stable defect complexes. Such strong dynamic annealing processes and the absence of lattice amorphization, also found in all the other studies [11,17,23,24,27,28,32,46], are consistent with a large ionicity of Zn–O bonds (0.616, based on calculations of Phillips [47]), according to the well known ionicity criterion for amorphization [48,49].

A detailed study of the evolution of lattice defects in single-crystal ZnO bombarded with relatively light 60 keV ^{28}Si and heavy 300 keV ^{197}Au ions at −196 and 20°C has been reported in [20]. Figure 8.2(a) shows RBS/C spectra illustrating the damage buildup in ZnO bombarded at −196°C with 300 keV ^{197}Au ions. It is seen from Fig. 8.2(a) that the depth profiles of ion-beam-produced lattice defects are bimodal. Indeed, Fig. 8.2(a) shows that, with increasing ion dose, lattice disorder in the Zn sub lattice accumulates both in the crystal bulk and near the sample surface. Such bimodal damage-depth profiles are not uncommon for ion-implanted semiconductors under irradiation conditions with pronounced dynamic annealing. In this case, disorder typically accumulates in the crystal bulk (where the nuclear energy loss profile is maximum, see Fig. 8.1) and at the sample surface (which is often a sink for ion-beam-generated point defects) [50].

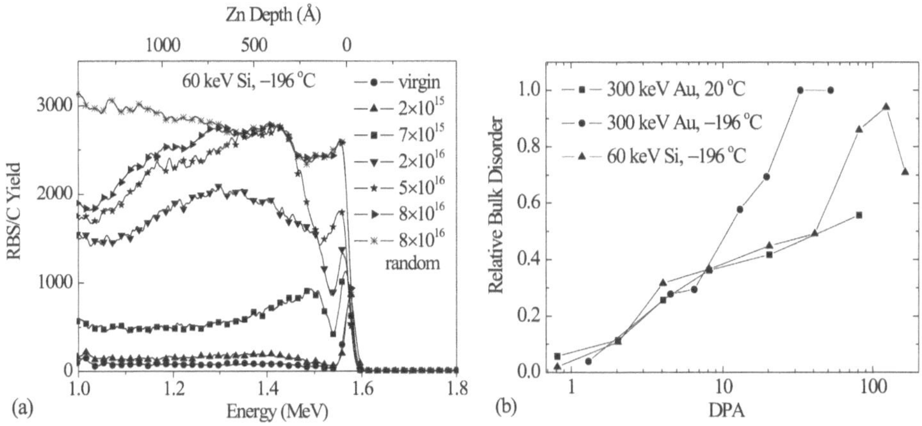

Figure 8.3: (a) Selected RBS/C spectra showing the damage buildup in ZnO bombarded at $-196°C$ with 60 keV Si ions with a beam flux of $\sim 1.2 \times 10^{13}\ cm^{-2}\ s^{-1}$. Implantation doses (in cm^{-2}) are indicated in the legend. (b) Dependence of the maximum relative disorder in the bulk defect peak on the DPA produced in ZnO samples by bombardment at $-196°C$ with 60 keV Si ions as well as at $-196°C$ and 20°C with 300 keV Au ions. After Ref. 20.

Although ZnO is not amorphized by keV heavy-ion bombardment [Fig. 8.2(a)], a crystalline-to-amorphous phase transition in ZnO can be induced by irradiation with relatively light ^{28}Si ions. To illustrate this, Fig. 8.3(a) shows selected RBS/C spectra, illustrating the evolution of lattice disorder in ZnO bombarded at $-196°C$ with 60 keV Si ions. These RBS/C spectra reveal that, similar to the case of heavy-ion bombardment shown in Fig. 8.2(a), lattice disorder accumulates both at the surface and in the crystal bulk. Figure 8.3(a) also shows that, for a dose of $5 \times 10^{16}\ cm^{-2}$, a buried amorphous layer nucleates in the crystal bulk, as suggested by the RBS/C yield reaching the random level and confirmed by transmission electron microscopy (see Fig. 8.4 in Section 8.3.5). Finally, Fig. 8.3(a) shows that irradiation with 60 keV Si ions to a dose of $8 \times 10^{16}\ cm^{-2}$ amorphizes the entire ~700-Å-thick near-surface layer.

The damage buildup behavior in ZnO bombarded with heavy (Au) and light (Si) ions is further illustrated in Fig. 8.3(b), showing the dependence of the maximum relative disorder in the bulk defect peak (extracted from RBS/C spectra) on the number of displacements per atom (DPA). Such DPA values are defined here as concentrations of ion-beam-generated lattice vacancies normalized to the atomic concentration of ZnO.[2] It is seen from Fig. 8.3(b) that, for ion doses resulting in ≤ 10 DPA, the

[2] Ballistic calculations were performed with the TRIM code (version SRIM-2003.17) [10] with threshold displacement energies of 57 eV for Zn and O sublattices (chosen based on experimental data by Locker and Meese [51,52]).

level of bulk disorder gradually increases with increasing ion dose for both Au and Si irradiation regimes. For the case of bombardment with 300 keV Au ions at −196°C to doses resulting in ≥50 DPA, Fig. 8.3(b) reveals a strong increase in the amount of relative disorder, as measured by RBS/C. Such an increase in the level of ion-beam-produced lattice damage for very large doses of Au ions has been attributed to ion-beam-induced material decomposition revealed by XPS [20]. For DPA ≥ 10, Fig. 8.3(b) also reveals a fast buildup of lattice damage up to complete disordering, as measured by the ion channeling technique, for irradiation with Si ions. Such a rapid disorder accumulation in the very high-dose regime has been attributed to chemical effects of implanted Si ions [20], which we discuss in more detail below.

8.3.2 Chemical effects of implanted species

Amorphization of ZnO with high doses of 60 keV Si ions (see Fig. 8.3) has been attributed to strong chemical effects of Si atoms implanted into the ZnO lattice, resulting in the stabilization of an amorphous phase [20]. The scale of such chemical effects depends on the particular ion species used. A chemical enhancement of damage buildup, as observed by RBS/C, can be due to (i) the trapping of ion-beam-generated mobile point defects by implanted impurity atoms, (ii) second phase formation and associated lattice distortion, and/or (iii) enhanced stability of irradiation-induced defects in ZnO containing a high concentration of impurities [50]. Indeed, the trapping of mobile defects at impurity atoms may result in stable defect clusters which are observed as enhanced scattering centers in RBS/C or directly by XTEM. At high concentrations (above a few atomic percent), precipitation of a second phase involving the implanted atoms is increasingly likely, thus leading to disordering of the surrounding crystalline matrix. Alternatively, large concentrations of implanted atoms may enhance the stability of an amorphous phase by defect trapping and/or influencing energy barriers for defect migration and interactions. Additional studies are currently needed to ascertain the contributions to the chemical effect in ZnO from each of the above three processes.

8.3.3 Effect of irradiation temperature

Figure 8.3(b) also shows that an increase in irradiation temperature from −196 up to 20°C has a negligible effect on the disorder buildup in the crystal bulk for ion doses resulting in <50 DPA. This experimental fact suggests that, for −196 to 20°C, sample temperature has a minor effect on the stability of ion-beam-produced defect complexes and the efficiency of defect formation and annihilation processes. Such a negligible temperature effect may suggest an athermic mechanism for defect agglomeration and annihilation processes driven by electrostatic interaction between charged point defects whose formation is possible in ionic crystals such as ZnO [20]. In this case, even very large dose heavy-ion bombardment at $-196°C$ and above does not result in lattice amorphization, suggesting an instability of the amorphous phase in this temperature interval. Additional damage buildup studies at temperatures below $-196°C$ would be desirable to better understand the influence of sample temperature on dynamic annealing processes and the stability of the amorphous phase in ZnO.

8.3.4 Role of collision cascade density

The fact that damage–DPA curves in Fig. 8.3(b) for ^{28}Si and ^{197}Au ion bombardment regimes essentially overlap for DPA values ≤ 10 suggests that the density of collision cascades has a negligible effect on the damage buildup behavior in ZnO (at least, at -196 to 20°C). This is in contrast to the case of many other semiconductors for which, under certain ion irradiation conditions, collision cascade density can have a dramatic effect on the amount of stable post-irradiation disorder (see, for example, [50]). For example, it has previously been shown that, for GaN, both the level of ion-beam-produced stable lattice disorder (as measured by RBS/C) and the main features of the damage buildup behavior (such as surface vs bulk amorphization, saturation of the damage level in the crystal bulk, the sigmoidality of damage–dose curves, etc.) strongly depend on the density of collision cascades [50].

The fact that the buildup of structural defects in ZnO under keV ion bombardment is essentially independent of the density of collision cascades [20] is consistent with studies of the electrical properties of ZnO irradiated with MeV light ions (which generate dilute collision cascades) [19,21]. Indeed, it has been shown in [19] (and discussed in more detail in Section 8.7) that the ion doses necessary for electrical isolation of ZnO inversely depend on the number of lattice displacements produced by the ion beam, suggesting a negligible effect of collision cascade density on changes in electrical properties in the case of MeV light-ion bombardment.

8.3.5 Defect microstructure

The microstructure of ion-beam-produced defects in ZnO has been studied by XTEM in [20,46] and by positron annihilation spectroscopy (PAS) in [36,37,43]. It has been found in [20] that, for both light (^{28}Si) and heavy (^{197}Au) ion bombardment regimes, ion irradiation produces energetically favorable *planar defects*

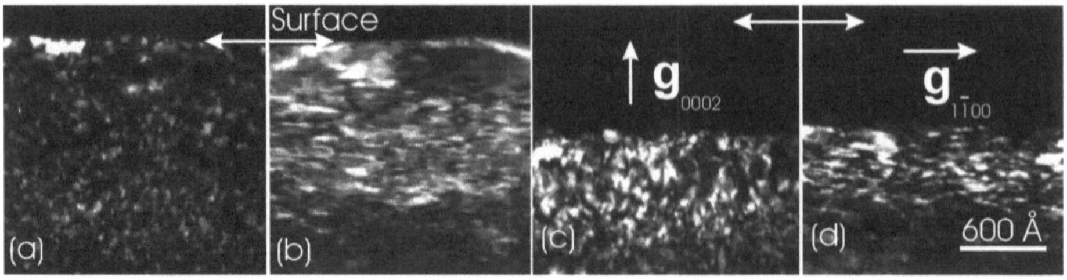

Figure 8.4: Dark-field XTEM images [(a),(c) **g** = 0002* and (b),(d) **g** = $1\bar{1}00$*] of ZnO bombarded at −196°C with 60 keV Si ions with a beam flux of $\sim 1.2 \times 10^{13}$ cm^{-2} s^{-1} to doses of 7×10^{15} [(a) and (b)] and 8×10^{16} cm^{-2} [(c) and (d)]. All four images are of the same magnification. After Ref. 20.

which are parallel to the basal plane of the wurtzite structure of ZnO. As an example, Fig. 8.4 shows XTEM images of ZnO bombarded at −196°C with 60 keV Si ions to doses of 7×10^{15} [Figs. 8.4(a) and 8.4(b)] and 8×10^{16} cm^{-2} [Figs. 8.4(c) and 8.4(d)]. Images from Figs. 8.4(a) and 8.4(b) show that irradiation to a dose of 7×10^{15} cm^{-2} creates a dense band of point defect clusters and planar defects similar to those observed in the case of heavy-ion bombardment [20]. Figure 8.4(b) also shows that, in the case of bombardment with 60 keV Si ions to a dose of 7×10^{15} cm^{-2}, the ∼300-Å-thick near-surface layer is essentially free from planar defects, which are concentrated in the deeper layer from ∼300 up to ∼700 Å from the sample surface.[3] This result is consistent with RBS/C data from Fig. 8.3(a), reflecting the depth profile of ion-beam-produced defects in this sample.

In the case of a sample implanted with 60 keV Si ions to a higher dose of 8×10^{16} cm^{-2}, Figs. 8.4(c) and 8.4(d) (as well as selected-area diffraction analysis reported in [20]) confirm the amorphization of the ∼725-Å-thick near-surface layer, as suggested by RBS/C data from Fig. 8.3(a). However, a close examination of Figs. 8.4(c) and 8.4(d) and electron diffraction patterns has revealed that amorphization is not complete, with some crystallites imbedded into the amorphous

[3] It should be noted that irregular contrast observed in both implanted and unimplanted regions in **g** = 0002* images shown in Fig. 8.4 can be attributed to lattice defects introduced into the near-surface region of a XTEM sample during the ion-beam thinning process. The absence of such an ion-beam-milling-induced contrast in other semiconductors (such as Si, Ge, GaAs, and GaN) can be attributed to the fact that, in these latter materials, bombardment with low keV Ar ions, used for ion-beam milling, results in the amorphization of the near-surface layer. In contrast, the ZnO lattice remains crystalline (though heavily damaged) after bombardment with keV Ar ions, resulting in a strong defect-induced diffraction contrast experimentally observed. A high-resolution TEM study of the lattice defects produced in ZnO by ion-beam milling (with 6 keV Ar ions) can be found in [18].

matrix. Such imperfect amorphization has been attributed to redistribution and segregation of Si atoms in the ZnO lattice during ion bombardment with the formation of localized regions enriched or depleted with Si [20].

Figure 8.4, discussed above, illustrates that energetically favorable defect structures in ion-implanted ZnO consist of point defect clusters and planar defects which are parallel to the basal plane of the ZnO wurtzite structure. It should be noted that a detailed XTEM study of the atomic structure of ion-beam-produced defects has not been possible in [20] due to a high density of implantation-produced defects. However, radiation defects have also been studied by Yoshiie et al. [53,54] in ZnO irradiated with a (less damaging) electron beam in the transmission electron microscope. These XTEM studies [53,54] have shown that high-energy (>700 keV) electron bombardment of ZnO at temperatures of 20–500°C results in the formation of dislocation loops which are interstitial in nature and have Burgers vectors of 1/2 [0001] and 1/3 $[11\bar{2}0]$. It is plausible that similar dislocation loops form in ZnO as a result of dynamic annealing processes during keV light or heavy-ion bombardment. Verification of this possibility must await more detailed XTEM studies.

A very recent XTEM study [46] has revealed a band of cavities ~15 nm in size in ZnO single crystals bombarded at 20°C with 300 keV As ions to a dose of $1.4 \times 10^{17}\,cm^{-2}$. The presence of large vacancy clusters in N-, Al-, and P-implanted ZnO has also been suggested by PAS studies of Chen et al. [36,37,43]. However, further studies are currently needed to identify whether these cavities (or large vacancy clusters) are voids or gas bubbles and to understand the physical mechanism of their formation.

8.3.6 Anomalous defect peak

The RBS/C study [20] has also revealed the formation of an *intermediate defect peak* (IDP) between the surface and bulk peaks of disorder in 300-keV-Au-implanted ZnO. As an example, Fig. 8.2(b) illustrates this effect, showing that the position of the IDP changes with increasing ion dose. As ion bombardment proceeds, the IDP originates at the sample surface and moves toward the bulk defect peak. For the sample temperatures studied in [20] (−196 and 20°C), the behavior of the IDP is independent of irradiation temperature, within experimental error.

The formation of this IDP is rather unexpected and, to our knowledge, has not been observed in any other material. Indeed, typical depth profiles of ion-beam-produced lattice disorder in crystalline solids have a Gaussian shape with one maximum (in addition to the surface defect peak), reflecting the depth profile of atomic displacements ballistically generated by the ion beam.

Results from [20] have essentially ruled out the possibilities that the IDP in RBS/C spectra is related to measurement artifacts, such as segregation of Au, large surface nonuniformities, or gross ion-beam-induced stoichiometric changes. This suggests that the mechanism for IDP formation is disorder-related. It has been suggested that the IDP can be caused by the formation of a localized band of lattice defects resulting from imperfect defect annihilation processes during ion bombardment [20]. A possible nucleation site for such a defect cluster formation, giving rise to a direct scattering peak in RBS/C spectra, can be an excess of ion-beam-generated lattice vacancies near the sample surface. Indeed, it is well known that vacancies and interstitials are spatially separated in a collision cascade, with an interstitial excess at the ion end of range and a vacancy excess closer to the surface. This effect is expected to be pronounced for bombardment regimes with strong dynamic annealing, which is the case for ZnO at least at −196°C and above. The importance of such a ballistic separation of interstitials and vacancies in the damage buildup behavior in ZnO bombarded with ^{197}Au ions is also supported by the fact that the bulk defect peak is shifted, relative to the maximum of the nuclear energy loss profile, closer to the ion end of range, the region with an interstitial excess (see Fig. 8.2) [20].

This scenario for a defect-related formation of the IDP appears to be consistent with the experimental observations from [20]. The fact that the IDP is observed only in ZnO samples bombarded with (heavy) ^{197}Au ions (not in samples irradiated with lighter ^{28}Si ions) [20] may suggest that the nucleation of the defect band giving rise to the IDP requires a large vacancy excess and the formation of dense collision cascades generated by heavy ions. The fact that the IDP moves deeper into the crystal bulk, with increasing ion dose, can be associated with the onset of ion-beam-induced loss of O from the sample surface as ion irradiation proceeds. Loss of O can stimulate defect annihilation processes in the region between the sample surface and the defect band giving rise to a separation of the IDP from the surface peak. Indeed, such O-loss has been measured by XPS for higher ion doses when ion-beam-induced stoichiometric changes become significant [20]. However, this intriguing dose dependence of the IDP position can also be attributed to a possible increased defect annihilation at the sample surface. Additional work is presently needed to clarify the role of the sample surface and material stoichiometry in the formation and evolution of the IDP.

8.4 Thermal annealing of implantation damage

Although we are not aware of any studies of thermal annealing of amorphous ZnO (stabilized by impurities), a number of studies of the annealing of *preamorphous disorder* in ion-implanted ZnO have been reported in the literature. These reports are summarized in Table 8.1. Experimental results reported by different research groups are in overall agreement. However, the interpretation of experimental results has been somewhat controversial, and we discuss it in this section.

It has been found in many studies that lattice defects produced in ZnO by intermediate- and high-dose bombardment with keV ions can be removed by thermal annealing at ~800–1100°C (see Table 8.1). This finding is somewhat surprising since, typically, extended defects in ion-implanted semiconductors are completely removed by annealing at temperatures which are ~2/3 of the melting temperature (in K) [55]. Hence, one can expect that relatively large thermal budgets (with temperatures $\gtrsim$1225°C) will be necessary to anneal irradiation-produced defects in ZnO, given its melting point of ~1975°C [56].

To better understand why such relatively low thermal budgets (~800–1100°C) result in an apparent recovery of even heavily-damaged ZnO, the thermal stability of ion-implanted ZnO has been studied in [46], where ZnO implanted at 20°C with high-dose (1.4×10^{17} cm^{-2}) 300 keV As ions was annealed at 1000–1200°C in the proximity geometry (when samples are placed face-to-face). Damage recovery was studied by a combination of RBS/C, XTEM, and atomic force microscopy (AFM). Interestingly, results from [46] have shown that such a thermal treatment leads to the *decomposition and evaporation* of the heavily-damaged layer instead of defect recovery and recrystallization that could be inferred from RBS/C and XTEM data alone. Such a relatively poor thermal stability of heavily-damaged ZnO has significant implications for understanding results on thermal annealing of ion-implanted ZnO, and we discuss it in more detail below.

Figures 8.5 and 8.6 illustrate the importance of such thermally-induced decomposition of heavily-damaged ZnO on the interpretation of experimental data. Figure 8.5(e) shows RBS/C spectra of as-implanted and annealed samples (for samples implanted at 20°C with 300 keV As ions to a dose of 1.4×10^{17} cm^{-2}). It can be seen from Fig. 8.5(e) that, following implantation, the channeled and random spectra of the as-implanted sample match for a depth up to ~200 nm, indicating the formation of a heavily-damaged layer. Figure 8.5(a) shows a XTEM image of the as-implanted sample and reveals that the implanted region consists of a heavily-damaged layer extending ~200 nm into the bulk, which is consistent with RBS/C data from Fig. 8.5(e).

The RBS/C spectra in Fig. 8.5(e) also show that thermal annealing in the range of 1000–1200°C appears to effectively recover the crystallinity of the implanted layer. In particular, the level of lattice disorder progressively decreases with increasing annealing temperature, and the spectrum from the sample annealed at 1200°C is almost identical to that of the virgin sample, suggesting a complete recovery of implantation damage. This is consistent with other reports [17,23,27,28,32], where

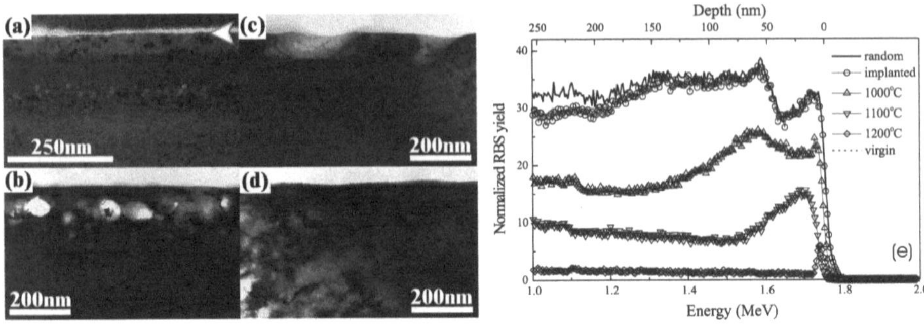

Figure 8.5: (a)–(d): Bright-field XTEM images of ZnO implanted at 20°C with 300 keV As ions to a dose of 1.4×10^{17} cm^{-2}. Image (a) is the as-implanted sample. Images (b)–(d) are implanted samples annealed at 1000, 1100, and 1200°C, respectively. The arrow in (a) indicates the sample surface. (e) RBS/C spectra showing effective damage recovery upon annealing of ZnO bombarded at 20°C with 300 keV As ions to a dose of 1.4×10^{17} cm^{-2}. Annealing temperatures are indicated in the legend. Note that the random spectrum was taken from the as-implanted sample. After Ref. 46.

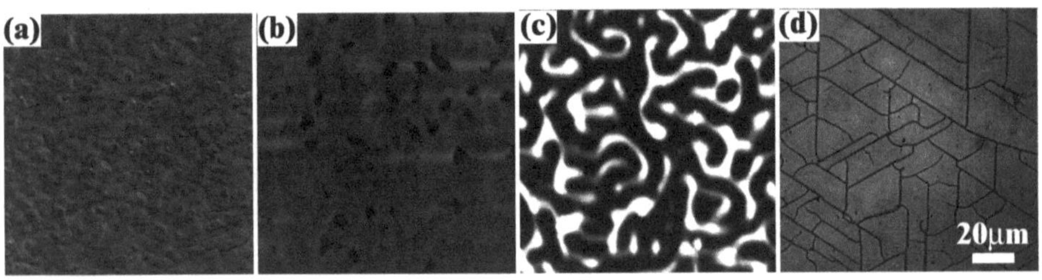

Figure 8.6: AFM [(a)–(c)] and optical (d) images of ZnO samples implanted at 20°C with 300 keV As ions to a dose of 1.4×10^{17} cm^{-2} and annealed at 1000°C [(a),(d)], 1100°C (b), and 1200°C (c). All AFM images were taken with the same height scale and are for a field of view of $5 \times 5\,\mu$m^2. After Ref. 46.

thermal annealing of lattice disorder produced by high-dose ion implantation has been studied by RBS/C.

Figures 8.5(b)–8.5(d) show XTEM images taken from annealed samples. Figure 8.5(b) reveals large cavities of ~60 nm in size in the sample annealed at 1000°C. After annealing at 1100°C [Fig. 8.5(c)], further cavity growth is evident, and these cavities are now open to the surface. Although the material surrounding these cavities is highly crystalline, some dislocations in the surface layer can also be seen from Fig. 8.5(c). Finally, Fig. 8.5(d) shows that most implantation-induced damage has effectively been removed by annealing at 1200°C. However, a significant number of dislocations can still be seen in Fig. 8.5(d).

The thermally-induced damage recovery revealed by RBS/C [Fig. 8.5(e)] and XTEM [Figs. 8.5(a)–8.5(d)] data can be due to either lattice recrystallization or a decomposition and evaporation of the implanted layer. To clarify this, the surface morphology of as-implanted and annealed samples has been studied by AFM and is illustrated in Fig. 8.6. It is seen from Fig. 8.6(a) that, after annealing at 1000°C, the surface is relatively smooth, having a root-mean square (rms) roughness of ~3.5 nm. This is, however, rougher than both the unimplanted ZnO that has been annealed at 1000°C and the as-implanted sample, which have rms roughness values of ~0.6 and ~2.5 nm, respectively. Note that the rms roughness of the annealed, unimplanted ZnO is the same as that of the as-received ZnO (i.e., ~0.6 nm). In addition, there is a negligible height difference between the implanted and unimplanted regions of the sample annealed at 1000°C.

Interestingly, after annealing at 1000°C, the surface of the implanted region is covered in cracks, as clearly illustrated in the optical image in Fig. 8.6(d). The pattern of surface cracks in Fig. 8.6(d) reflects the low index crystallographic directions of the wurtzite structure of ZnO. Such cracking suggests that tensile stress develops in the implanted layer during annealing, which could be related to local changes in stoichiometry in the near-surface region, discussed above. Moreover, the presence of stress also explains the observation of (stress-induced) dislocations in XTEM images of annealed samples [Figs. 8.5(c) and 8.5(d)] [57,58].

Figure 8.6(b) shows that the surface of the sample annealed at 1100°C exhibits cavities, which is also consistent with XTEM data from Fig. 8.5(c). For this sample, the interface between the implanted and unimplanted regions has a step height of ~30 nm, indicating some material loss from the implanted region during annealing. At 1200°C, these effects are further exacerbated. Indeed, Fig. 8.6(c) shows that the surface is covered by ridges and troughs. For this sample, the step height between the implanted and unimplanted regions is ~50–100 nm. Note that the roughness of the unimplanted region remains essentially unchanged after annealing at 1200°C. Thus, instead of the apparent recovery of heavily-damaged ZnO, decomposition and evaporation of the implanted layer occurs during the high-temperature heat treatment, rather than defect recovery and recrystallization. After such a thermally-induced decomposition (into Zn-rich ZnO_x and O_2), material evaporation is not unexpected since Zn metal has relatively low melting and boiling points of ~420 and ~910°C, respectively [56].

The decomposition behavior discussed above for the case of heavily damaged ZnO is also supported by a recent study by Khanna et al. [59] of the thermal stability of as-grown ZnO. They [59] have shown that a thermally-induced degradation of the ZnO surface starts at relatively low temperatures of 200–300°C, depending on the annealing ambient. The degree of thermally-induced decomposition should depend on the quality of the as-grown material and on the level of ion-beam-produced lattice disorder. Additional annealing studies with a better control of the ZnO surface to prevent it from decomposition are currently needed. A discussion of the approaches known to prevent the surface decomposition of ion-implanted semiconductors during thermal annealing can be found, for example, in [55] and will not be reproduced here.

8.5 Lattice sites of dopants

The position of dopants in the lattice is important since it determines their electrical and optical properties. Lattice sites of several dopants (or common impurities), including Fe [33], Cu [35], Ag [31], Er [27,30], and Tm [32] have been studied by RBS/C and emission channeling spectrometry (ECS), as summarized in Table 8.1. Because ZnO exhibits very strong dynamic annealing (see Section 8.3 for details), defect migration and annihilation processes are expected to promote the migration of implanted species into energetically favorable lattice sites [55]. Hence, a high fraction of implanted dopants in substitutional positions even directly after implantation at room temperature is not unexpected and has been found experimentally for all the dopants studied (i.e., Fe, Cu, Ag, Er, and Tm) in [27,30–33,35]. Interestingly, it has also been found that the fraction of substitutional impurities decreases with increasing annealing temperature above ~600°C [27,30–33,35], which could be attributed to the thermally-induced material decomposition described in detail in the previous section.

8.6 Electrical doping

Electrical doping of ZnO by ion implantation has been studied in [12,36–38,43] (see Table 8.1 for additional details of these studies). Since *n*-type doping does not seem to be a major problem for the future ZnO-based (opto)electronic devices, these studies [12,36–38,43] have focused on implantation of potential *p*-type dopants such as N, P, V, and As. Unfortunately, all these attempts to dope ZnO with *p*-type dopants by ion implantation have been unsuccessful [12,36–38,43]. The problem with doping by ion implantation is likely to be related to dopant activation efficiency and undesirable effects of ion-beam-produced lattice defects on the electrical properties of the material. In order for *p*-type doping of ZnO by ion implantation to be successful and reproducible, a thorough understanding of damage accumulation and thermal stability of implantation-produced defects in ZnO is highly desirable.

8.7 Electrical isolation

Ion irradiation under appropriate conditions can render semiconductors highly resistive and, hence, can be used for selective-area electrical isolation (see, for example, a review [60]). It is generally believed that irradiation-induced degradation of carrier mobility as well as the trapping of carriers at deep centers associated with irradiation-produced damage (*defect isolation*) or with implanted species (*chemical isolation*) is the mechanism responsible for electrical isolation of semiconductors [60].

8.7.1 Defect isolation

Defect isolation of *n*-type ZnO epilayers (grown on sapphire substrates) by MeV light-ion irradiation has been studied in [19,21]. For MeV light ions, projected ranges are considerably greater than the thickness of the ZnO epilayers used in [19,21]. In this case, the profiles of ion-generated atomic displacements are essentially uniform throughout the conductive ZnO film. A single MeV implant is, hence, sufficient to isolate a relatively thick ZnO film. Moreover, irradiation with MeV ions allows one to separate the effects of defect isolation from chemical isolation and to avoid the formation of a layer with substantial defect-induced (hopping) conduction, which inevitably forms at the ion end-of-range region in the case of keV implants and usually complicates the interpretation of data [60].

Figure 8.7(a) shows the evolution of sheet resistance (R_s) of ZnO films irradiated with 0.6 MeV H, 0.7 MeV Li, 2 MeV O, and 3.5 MeV Si ions at 20°C. It is seen from Fig. 8.7(a) that, similar to the situation for other semiconductors [60–65], the ion dose dependence of R_s, for each particular ion, has three distinct regions: (i) a low dose region where R_s increases only slightly with increasing ion dose, (ii) a relatively narrow intermediate dose region characterized by a very rapid increase in R_s (by ~7 orders of magnitude), and (iii) the third region for large doses where R_s reaches its maximum value, remaining approximately constant with further increasing ion dose. In all the samples studied (except for the case of H ion irradiation when even larger ion doses are needed for complete isolation), the values of R_s reach the maximum after some characteristic dose (the so called threshold isolation dose [61,62] $-\Phi_{th}$). The levels of R_s at the plateaus shown in Fig. 8.7(a) for doses above Φ_{th} are $\sim 4 \times 10^{10}$ Ω/sq. However, it should be noted that the real maximum values of R_s can be even larger because the R_s values measured have a contribution from the parasitic resistances of the experimental setup, which are of the same order of magnitude.

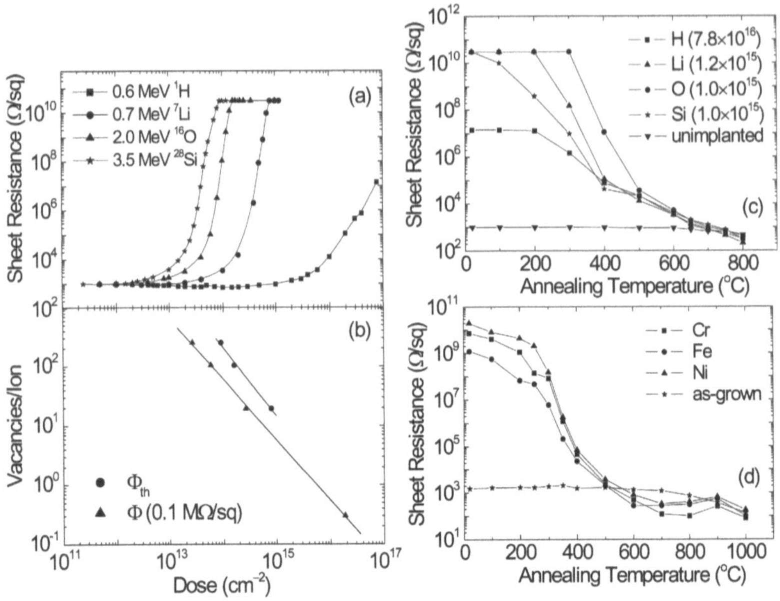

Figure 8.7: (a) Dose dependence of sheet resistance of ZnO epilayers bombarded with different MeV ions, as indicated in the legend. (b) The average number of ion-beam-generated lattice vacancies within the 0.6-μm-thick ZnO epilayer is plotted vs threshold ion doses Φ_{th} and doses $\Phi(0.1\,\mathrm{M\Omega/sq})$ required to produce a sheet resistance of 0.1 MΩ/sq, as indicated in the legend. (c) Evolution of sheet resistance as a function of temperature of isochronal annealing (for 60 s in an Ar ambient) of ZnO epilayers bombarded with different ions to different doses (in cm^{-2}), as indicated in the legend. (d) Evolution of sheet resistance as a function of temperature of isochronal annealing (for 60 s in an Ar ambient) of ZnO epilayers bombarded with Cr, Fe, or Ni ions, as indicated in the legend. After Refs 19(a)–(c) and 21(d).

Figure 8.7(a) also shows that the isolation curves progressively shift toward lower doses with increasing ion mass. Such a shift is caused by an increase in the number of ion-beam-generated atomic displacements with increasing ion mass. This ion mass effect is better illustrated in Fig. 8.7(b), showing the average number of lattice vacancies produced by different ions within ZnO epilayers vs both Φ_{th} and doses required to produce a R_s of 0.1 MΩ/sq [denoted as $\Phi(0.1\,\mathrm{M\Omega/sq})$]. Straight line fits with slopes of -1.14 ± 0.10 and -1.02 ± 0.02 represent the best fits to Φ_{th} and $\Phi(0.1\,\mathrm{M\Omega/sq})$ dependencies, respectively. Thus, results from Fig. 8.7(b) indicate that the efficiency of the carrier removal process close-to-linearly depends on the number of ion-beam-generated atomic displacements.

8.7.2 Thermal stability of defect isolation

The thermal stability of ion-irradiation-induced electrical isolation of n-type single-crystal ZnO epilayers has been studied in [19,21]. Figure 8.7(c) shows the evolution of R_s as a function of temperature of isochronal annealing of ZnO epilayers bombarded with 0.6 MeV H ions to a dose of $7.8 \times 10^{16}\,cm^{-2}$, 0.7 MeV Li ions to a dose of $1.2 \times 10^{15}\,cm^{-2}$, and 2 MeV O and 3.5 MeV Si ions to a dose of $10^{15}\,cm^{-2}$. Note that all these doses (except for the case of H ions) are above Φ_{th}, as can be seen from Fig. 8.7(a). It is clearly seen from Fig. 8.7(c) that, in all the cases studied, defect-induced electrical isolation of ZnO is unstable to annealing at temperatures above ~300°C. Figure 8.7(c) also shows that, in all the cases studied, R_s is recovered to its original value after annealing at ~650°C. For comparison, in Fig. 8.7(c), it is shown that annealing at temperatures below ~700°C has a negligible effect on R_s of as-grown samples. In addition, Fig. 8.7(c) shows that the thermal stability of electrical isolation is different for different ion species and doses, consistent with previous studies of electrical isolation in other compound semiconductors [60,63].

Previous studies of electrical isolation in other compound semiconductors have shown that the thermal stability of isolation can be improved by several approaches [60,64]. These include (i) increasing ion dose, (ii) carrying out ion irradiation at elevated temperatures to promote the formation of more stable defect complexes, and (iii) implantation of impurities which form thermally stable deep levels in the band gap and trap free carriers (so-called chemical isolation, which is discussed in detail in Section 8.7.3).

Results from [21] have shown that an increase in implantation temperature (up to 350°C) and ion dose (up to ~2 orders of magnitude above Φ_{th}) has only a minor effect on the thermal stability of ion-beam-produced defects responsible for electrical isolation of ZnO. Hence, *defect-induced* isolation of ZnO by ion irradiation currently faces difficulties if desirable device processing and/or operating temperatures are above ~400°C.

8.7.3 Chemical isolation

Previous isolation studies of compound semiconductors [60] have shown that thermally-stable electrical isolation can be achieved by implantation of impurities which form deep levels in the band gap and trap free carriers (so-called chemical isolation). Some previous studies have also suggested that transition metals in ZnO can form relatively deep levels in the band gap [3]. Hence, the effect of Cr, Fe, or Ni impurities, introduced by ion implantation, on electrical properties of single-crystal ZnO has been studied in [21].

In [21], relatively uniform depth profiles of Cr, Fe, or Ni impurities (with a bulk concentration of $\sim 2 \times 10^{19}$–$5 \times 10^{19}\,\mathrm{cm}^{-3}$) were achieved by a multiple-energy implantation scheme. Shown in Fig. 8.7(d) is the evolution of R_s as a function of temperature of isochronal annealing of ZnO resistors bombarded with Cr, Fe, or Ni ions. It is seen from Fig. 8.7(d) that, directly after ion irradiation (i.e., before post-irradiation annealing), R_s of all three samples reaches the values of $\sim 10^9$–$10^{10}\,\Omega/\mathrm{sq}$. Such a dramatic increase in the value of R_s is due to ion-beam-produced lattice defects, which is similar to the case of bombardment with light MeV ions described above. Figure 8.7(d) also shows that, similar to the case of irradiation with the other ion species discussed above, annealing at temperatures above ~300°C effectively recovers implantation-induced defects, resulting in a dramatic decrease in R_s.

Interestingly, Fig. 8.7(d) also shows that, for annealing temperatures from ~600 to 900°C, R_s is lower than that of as-grown ZnO, indicating the formation of shallow (rather than deep) levels as a result of implantation of Cr, Fe, and Ni impurities (as well as ion-beam-produced lattice defects) and subsequent thermal annealing at temperatures up to 1000°C. Hence, introduction of Cr, Fe, or Ni into single-crystal ZnO by ion implantation and subsequent thermal annealing at temperatures up to ~1000°C does not result in an improvement in the thermal stability of chemical isolation. At present, more work is needed to study whether thermal budgets larger than 1000°C can electrically activate Cr, Fe, or Ni impurities implanted into ZnO, resulting in the formation of deep levels in the band gap. Additional work is also highly desirable to study the effect of other impurities which can potentially form thermally-stable deep acceptor levels in ZnO if desirable device processing and/or operating temperatures are above ~400°C.

8.7.4 Mechanism of isolation

In this section, we discuss the mechanism for electrical isolation in ZnO. We also discuss possible physical reasons for a relatively poor thermal stability of isolation in this material.

A better understanding of defect processes responsible for electrical isolation can often be provided by studying the influence of ion-beam flux. Shown in Fig. 8.8(a) is the evolution of R_s of ZnO resistors bombarded at 20°C with different fluxes of 2 MeV O ions. Figure 8.8(a) reveals a negligible beam flux effect with the variation in the flux value over an order of magnitude. This is consistent with the fact that ion-beam-generated point defects are highly mobile in ZnO at least at −196°C and above (see Section 8.3). Indeed, a high defect diffusivity in ZnO results in a very fast stabilization of collision cascades produced by energetic light ions. In this case

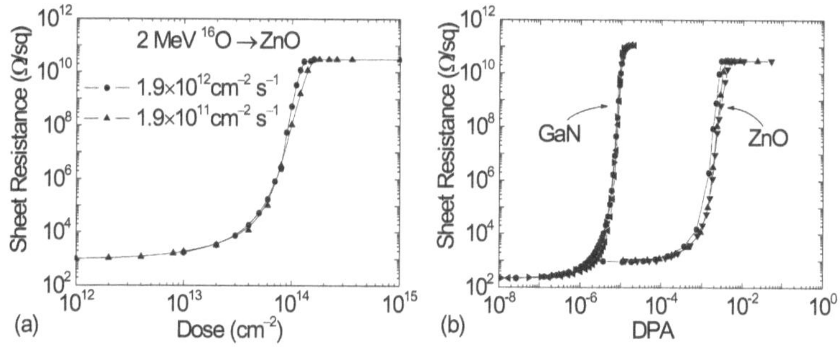

Figure 8.8: (a) Dose dependence of sheet resistance of ZnO samples irradiated with 2 MeV O ions at 20°C with different values of beam flux, as indicated in the legend. (b) Sheet resistance of GaN and ZnO epilayers (with original free carrier concentrations of $\sim 3 \times 10^{17}$ and $\sim 1 \times 10^{17}$ cm^{-3}, respectively) as a function of displacements per atom (DPA). Samples were bombarded at 20°C with MeV Li, C, O, and Si ions. See Refs 19 and 61 for the details of implant conditions. When plotted as a function of DPA, isolation curves for these ions essentially overlap (within experimental error). After Refs 21 and 66.

of very efficient dynamic annealing, a negligible flux effect is typically observed for common values of beam flux. This has recently been discussed in [62], where the influence of beam flux on the process of electrical isolation in another wide band gap semiconductor – GaN – has been considered in detail. Hence, the negligible flux effect in ZnO, coupled with efficient dynamic annealing processes, strongly suggests that centers responsible for isolation are not simple Frenkel pairs but the product of defect migration and interaction processes. In addition, an analysis of the temperature dependence of sheet resistance has suggested that effective levels associated with irradiation-produced defects are rather shallow (<50 meV).

The fact that defect-induced isolation is thermally unstable indicates that (i) annealing recovers defect complexes associated with deep acceptor levels responsible for irradiation-induced dramatic changes in R_S and/or (ii) annealing results in the transformation of some ion-beam-produced defects, with deep levels in the band gap before annealing, into new defect structures which have shallow donor levels and, hence, significantly reduce the value of R_S. Results available in the literature cannot distinguish between these two scenarios, and, at present, more work is needed to better understand the electrical properties of ion-beam-produced defects in ZnO.

A quantitative model for electrical isolation of ZnO by light-ion bombardment has recently been developed in [67,68]. In this model, a decrease in the concentration of free carriers responsible for isolation has been assumed to be due to the formation of complexes of ion-beam-generated point defects with shallow donor or acceptor dopants. The defect interaction processes have been described in terms of quasi-chemical reactions. Results have shown that the model developed in [67,68] can adequately describe experimental data for electrical isolation in the case of MeV light-ion irradiation of *n*-ZnO. In particular, this model [67,68] describes well the dependence of sheet resistance on ion dose and the number of ion-beam-generated atomic displacements as well as the dependence of the threshold isolation dose on the original free electron concentration and on the number of atomic displacements. Future studies should be able to extend the theoretical approach developed in [67,68] to include ion beam flux and temperature dependencies as well as to describe electrical isolation in other semiconductors.

8.7.5 Comparison with other materials

It is instructive to compare the isolation behavior of ZnO with that of other semiconductors. For example, previous studies [55,60,65] have shown that ion irradiation of *n*-type GaAs or GaN creates a highly resistive material. The isolation process in these materials (i.e., GaAs and GaN) has been attributed to the formation of complexes of simple point defects such as lattice vacancies, interstitials, as well as antisite defects which are carrier traps with electronic levels close to the midgap. The thermal stability of ion-irradiation-induced electrical isolation in GaAs and GaN matches the expected annealing behavior where annealing temperatures scale with the melting point T_{melt} of the material [55,60]. For example, electrical isolation in GaAs ($T_{GaAs}^{melt} = 1240°C$) and GaN ($T_{GaN}^{melt} \approx 2500°C$) can be stable to thermal annealing at ~600 and >900°C, respectively [55,60]. In addition, activation energies extracted from $R_s(1/T)$ dependencies for GaAs and GaN suggest the formation of deep carrier traps with levels close to the midgap [60,69].

Recent studies [19,21], discussed above, have revealed a quite different behavior of electrical isolation in ZnO. In particular, electrical isolation produced in ZnO by ion bombardment under a rather wide range of implant conditions is unstable to annealing at temperatures above ~300–400°C, whereas one can expect that significantly larger thermal budgets would be necessary to anneal irradiation-produced defects in ZnO, given its melting point of ~1975°C.

Finally, Fig. 8.8(b) shows R_s of GaN and ZnO epilayers as a function of displacements per atom (DPA) for irradiation with Li, C, O, and Si ions. Figure 8.8(b) clearly illustrates that the isolation process for both ZnO and GaN is essentially independent of ion mass (within experimental error), provided that ion doses are converted to DPA. It is also seen from Fig. 8.8(b) that the efficiency of the formation of lattice defects, responsible for electrical isolation, is rather different in GaN and ZnO, with ZnO being significantly more resistant to defect-induced electrical isolation. Indeed, Fig. 8.8(b) shows that two orders of magnitude more atomic displacements are needed to isolate ZnO as compared to the case of GaN with a similar original free electron concentration. This result is consistent with extremely efficient dynamic annealing processes in ZnO, which have been extensively discussed throughout this chapter.

8.8 Concluding remarks

In this chapter, we have reviewed the current status of the field of ion implantation into ZnO. Due to very efficient dynamic annealing and thermally-induced decomposition, this material exhibits a range of intriguing and complex defect-related processes. Although our understanding of ion-beam damage and defect annealing behavior in ZnO has matured considerably in the past several years, much more work is obviously needed before ion-damage effects in this material are fully understood. Such understanding is highly desirable for the development of ion bombardment as a powerful processing tool in the fabrication of future ZnO-based (opto)electronic devices.

Acknowledgments

We are grateful to a number of people involved in this work through collaboration and discussion, including J. S. Williams, J. Zou, V. A. Coleman, H. H. Tan, P. N. K. Deenapanray, S. J. Pearton, C. Evans, A. J. Nelson, A. V. Hamza, M. Yano, K. Koike, S. Sasa, M. Inoue, and K. Ogata. C.J. gratefully acknowledges the financial support of the Australian Research Council. Work at LLNL was performed under the auspices of the U.S. DOE by the University of California, LLNL under Contract No. W-7405-Eng-48.

References

[1] W. Hirschwald, B. Bonasewicz, L. Ernst, M. Grade, D. Hofmann, S. Krebs, R. Littbarski, G. Neumann, M. Grunze, D. Kolb, and H. J. Schulz, "Zinc Oxide," in: Current Topics in Materials Science **7**, 143 (1981).
[2] D. R. Clarke, J. Am. Ceram. Soc. **82**, 485 (1999).
[3] V. A. Nikitenko, Zhurnal Prikladnoi Spectroskopii **57**, 367 (1992).
[4] J. E. Nause, III-V Rev. **12**, 28 (1999).
[5] D. C. Look, Mater. Sci. Eng. B **80**, 383 (2001).

[6] S. J. Pearton, D. P. Norton, K. Ip, Y. W. Heo, and T. Steiner, Superlattices and Microstructures **34**, 3 (2003).
[7] S. J. Pearton, D. P. Norton, K. Ip, Y. W. Heo, and T. Steiner, J. Vac. Sci. Technol. B **22**, 932 (2004).
[8] S. J. Pearton, D. P. Norton, K. Ip, Y. W. Heo, and T. Steiner, Prog. Mater. Sci. **50**, 293 (2005).
[9] D. A. Thompson, Rad. Effects **56**, 105 (1981).
[10] J. F. Ziegler, J. P. Biersack, and U. Littmark, *The Stopping and Range of Ions in Solids* (Pergamon, New York, 1985).
[11] A. Ladage, Z. Phys. **144**, 354 (1956).
[12] B. W. Thomas and D. Walsh, J. Phys. D: Appl. Phys. **6**, 612 (1973).
[13] D. Walsh, Solid-State Electronics **20**, 813 (1977).
[14] M. Schilling, R. Helbig, and G. Pensl, J. Lumin. **33**, 201 (1985).
[15] V. P. Regel, V. A. Nikitenko, I. P. Kuzmina, V. G. Galstyan, T. P. Dolukhanyan, S. F. Nikul'shin, N. L. Sizova, and V. A. Skuratov, Sov. Phys. Tech. Phys. **32**, 183 (1987).
[16] C. W. White, L. A. Boatner, P. S. Sklad, C. J. McHargue, S. J. Pennycook, M. J. Aziz, G. C. Farlow, and J. Rankin, Mat. Res. Soc. Symp. Proc. **74**, 357 (1987).
[17] E. Sonder, R. A. Zuhr, and R. E. Valiga, J. Appl. Phys. **64**, 1140 (1988).
[18] J.-J. Couderc, J.-J. Demai, G. Vanderschaeve, and A. Peigney, Microsc. Microanal. Microstruct. **6**, 229 (1995).
[19] S. O. Kucheyev, P. N. K. Deenapanray, C. Jagadish, J. S. Williams, M. Yano, K. Koike, S. Sasa, M. Inoue, and K. Ogata, Appl. Phys. Lett. **81**, 3350 (2002).
[20] S. O. Kucheyev, J. S. Williams, C. Jagadish, J. Zou, C. Evans, A. J. Nelson, and A. V. Hamza, Phys. Rev. B **67**, 094115 (2003).
[21] S. O. Kucheyev, P. N. K. Deenapanray, C. Jagadish, J. S. Williams, M. Yano, K. Koike, S. Sasa, M. Inoue, and K. Ogata, J. Appl. Phys. **93**, 2972 (2003).
[22] A. N. Georgobiani, A. N. Gruzintsev, V. T. Volkov, M. O. Vorobiev, V. I. Demin, and V. A. Dravin, Nucl. Instrum. Meth. A **514**, 117 (2003).
[23] K. Kono, S. K. Arora, and N. Kishimoto, Nucl. Instrum. Meth. B **206**, 291 (2003).
[24] I. Sakaguchi, D. Park, Y. Takata, S. Hishita, N. Ohashi, H. Haneda, and T. Mitsuhashi, Nucl. Instrum. Meth. B **206**, 153 (2003).
[25] N. Matsunami, M. Itoh, Y. Takai, M. Tazawa, and M. Sataka, Nucl. Instrum. Meth. B **206**, 282 (2003).
[26] T. Fukudome, A. Kaminaka, H. Isshiki, R. Saito, S. Yugo, and T. Kimura, Nucl. Instrum. Meth. B **206**, 287 (2003).
[27] E. Alves, E. Rita, U. Wahl, J. G. Correia, T. Monteiro, J. Soares, and C. Boemare, Nucl. Instrum. Meth. B **206**, 1047 (2003).
[28] T. Monteiro, C. Boemare, M. J. Soares, E. Rita, and E. Alves, J. Appl. Phys. **93**, 8995 (2003).
[29] A. Y. Polyakov, A. V. Govorkov, N. B. Smirnov, N. V. Pashkova, S. J. Pearton, M. E. Overberg, C. R. Abernathy, D. P. Norton, J. M. Zavada, and R. G. Wilson, Solid-State Electronics **47**, 1523 (2003).

[30] U. Wahl, E. Rita, J. G. Correia, E. Alves, and J. P. Araujo, Appl. Phys. Lett. **82**, 1173 (2003).
[31] E. Rita, U. Wahl, A. M. L. Lopes, J. P. Araujo, J. G. Correia, E. Alves, and J. C. Soares, Physica B **340**, 240 (2003).
[32] E. Rita, E. Alves, U. Wahl, J. G. Correia, A. J. Neves, M. J. Soares, and T. Monteiro, Physica B **340**, 235 (2003).
[33] E. Rita, U. Wahl, J. G. Correia, E. Alves, and J. C. Soares, Appl. Phys. Lett. **85**, 4899 (2004).
[34] A. Y. Polyakov, A. V. Govorkov, N. B. Smirnov, N. V. Pashkova, S. J. Pearton, K. Ip, R. M. Frazier, C. R. Abernathy, D. P. Norton, J. M. Zavada, and R. G. Wilson, Mater. Sci. Semicond. Processing **7**, 77 (2004).
[35] U. Wahl, E. Rita, J. G. Correia, E. Alves, and J. G. Soares, Phys. Rev. B **69**, 012102 (2004).
[36] Z. Q. Chen, M. Maekawa, S. Yamamoto, A. Kawasuso, X. L. Yuan, T. Sekiguchi, R. Suzuki, and T. Ohdaira, Phys. Rev. B **69**, 035210 (2004).
[37] Z. Q. Chen, T. Sekiguchi, X. L. Yuan, M. Maekawa, and A. Kawasuso, J. Phys.: Condens. Matter **16**, S293 (2004).
[38] T. S. Jeong, M. S. Han, and C. J. Youn, J. Appl. Phys. **96**, 175 (2004).
[39] M. Peres, J. Wang, M. J. Soares, A. J. Neves, T. Monteiro, E. Rita, U. Wahl, J. G. Correia, and E. Alves, Superlattices and Microstructures **36**, 747 (2004).
[40] I. Sakaguchi, Y. Sato, D. Park, N. Ohashi, H. Haneda, and S. Hishita, Nucl. Instrum. Meth. B **217**, 417 (2004).
[41] I. Sakaguchi, S. Hishita, and H. Haneda, Appl. Surf. Sci. **237**, 358 (2004).
[42] T. S. Jeong, M. S. Han, J. H. Kim, C. J. Youn, R. Y. Ryu, and H. W. White, J. Cryst. Growth **275**, 541 (2005).
[43] Z. Q. Chen, A. Kawasuso, Y. Xu, H. Naramoto, X. L. Yuan, T. Sekiguchi, R. Suzuki, and T. Ohdaira, J. Appl. Phys. **97**, 013528 (2005).
[44] N. Matsunami, O. Fukuoka, M. Tazawa, and M. Sataka, Surface and Coatings Technology **196**, 50 (2005).
[45] I. Sakaguchi, S. Hishita, H. Ryoken, Y. Sato, and H. Haneda, Surf. Interface Analysis **37**, 332 (2005).
[46] V. A. Coleman, H. H. Tan, C. Jagadish, S. O. Kucheyev, and J. Zou, Appl. Phys. Lett. **87**, 231912 (2005).
[47] J. C. Phillips, *Bonds and Bands in Semiconductors* (Academic, New York, 1968).
[48] P. V. Pavlov, E. I. Zorin, D. I. Tetelbaum, V. P. Lesnikov, G. M. Ryzhkov, and A. V. Pavlov, Phys. Status Solidi A **19**, 373 (1973).
[49] H. M. Naguib and R. Kelly, Radiat. Eff. **25**, 1 (1975).
[50] S. O. Kucheyev, J. S. Williams, C. Jagadish, J. Zou, G. Li, and A. I. Titov, Phys. Rev. B **64**, 035202 (2001).
[51] J. M. Meese and D. R. Locker, Solid State Commun. **11**, 1547 (1972).
[52] D. R. Locker and J. M. Meese, IEEE Trans. Nucl. Sci. **19**, 237 (1972).
[53] T. Yoshiie, H. Iwanaga, N. Shibata, and S. Takeuchi, Phil. Mag. A **41**, 935 (1980).

[54] T. Yoshiie, H. Iwanaga, N. Shibata, K. Suzuki, M. Ichihara, and S. Takeuchi, Phil. Mag. A **47**, 315 (1983).
[55] S. O. Kucheyev, J. S. Williams, and S. J. Pearton, Mater. Sci. Eng. R **33**, 51 (2001).
[56] *CRC Handbook of Chemistry and Physics*, 84th ed., edited by D. R. Lide (CRC Press, New York, 2003).
[57] S. O. Kucheyev, J. E. Bradby, J. S. Williams, C. Jagadish, and M. V. Swain, Appl. Phys. Lett. **80**, 956 (2002).
[58] J. E. Bradby, S. O. Kucheyev, J. S. Williams, C. Jagadish, M. V. Swain, P. Munroe, and M. R. Phillips, Appl. Phys. Lett. **80**, 4537 (2002).
[59] R. Khanna, K. Ip, Y. W. Heo, D. P. Norton, S. J. Pearton, and F. Ren, Appl. Phys. Lett. **85**, 3468 (2004).
[60] S. J. Pearton, Mater. Sci. Rep. **4**, 313 (1990).
[61] H. Boudinov, S. O. Kucheyev, J. S. Williams, C. Jagadish, and G. Li, Appl. Phys. Lett. **78**, 943 (2001).
[62] S. O. Kucheyev, H. Boudinov, J. S. Williams, C. Jagadish, and G. Li, J. Appl. Phys. **91**, 4117 (2002).
[63] T. v. Lippen, H. Boudinov, H. H. Tan, and C. Jagadish, J. Appl. Phys. **80**, 264 (2002).
[64] J. P. de Souza, I. Danilov, and H. Boudinov, Appl. Phys. Lett. **84**, 4757 (1998).
[65] J. P. de Souza, I. Danilov, and H. Boudinov, Appl. Phys. Lett. **68**, 535 (1996).
[66] S. O. Kucheyev, J. S. Williams, and C. Jagadish, Vacuum **73**, 93 (2004).
[67] A. I. Titov and S. O. Kucheyev, J. Appl. Phys. **92**, 5740 (2002).
[68] A. I. Titov, P. A. Karaseov, and S. O. Kucheyev, Nucl. Instrum. Meth. B **243**, 79 (2006).
[69] X. A. Cao, S. J. Pearton, G. T. Dang, A. P. Zhang, F. Ren, R. G. Wilson, and J. M. Van Hove, J. Appl. Phys. **87**, 1091 (2000).

Chapter 9

Advances in Processing of ZnO

K. Ip, S. J. Pearton and D. P. Norton

Department of Materials Science and Engineering, University of Florida, Gainesville, FL 32611, USA

F. Ren

Department of Chemical Engineering, University of Florida, Gainesville, FL 32611, USA

Abstract: Recent advances in the control of background *n*-type conductivity, synthesis of high quality bulk substrates and in device processing techniques for ZnO have renewed interest in this material for possible applications to UV light emitters, spin functional devices, chemical and gas sensors and transparent electronics. We summarize recent progress in device fabrication techniques such as ion implantation doping and isolation, wet and dry etching and Ohmic and Schottky contact formation and new understanding of the role of hydrogen.

9.1 Introduction

ZnO has a long history as a material for gas sensing, varistors, piezoelectric transducers, as a light transmitting electrode in optoelectronic devices, electro-optic modulators and as a sunscreen.[1] In the last few years, major advances have been made in the areas of conductivity control and availability of high quality bulk ZnO substrates. This work has refocused attention on ZnO for UV light-emitters and transparent electronics. ZnO can be grown at relatively low temperatures on cheap substrates such as glass and has a larger exciton binding energy (~60 meV) than GaN (~25 meV).[2–22] In addition, the advances in growth of ZnO nanorods and nanowires have suggested that these may have applications to bio-detectio,[22] because of their large surface areas. Finally, incorporation of percent levels of transition metal impurities such as Mn lead to ferromagnetism in ZnO with practical Curie temperatures. This suggests that ZnO may be an ideal host as a spintronic material.

The purpose of this chapter is to provide a summary of recent advances in developing improved control of doping and unintentional impurities during growth of ZnO and also advances in device processing methods.

Zinc Oxide Bulk, Thin Films and Nanostructures
C. Jagadish and S. Pearton (Editors)

9.2 Ion implantation

Ion implantation is a precise method of introducing dopants into a semiconductor and can also be used to create high resistance regions for inter-device isolation. Implant doping is in its infancy in ZnO and there has been no clear demonstration of activation of an implanted donor or acceptor. The residual implant damage remaining after annealing appears to have donor-like character. To minimize this damage, it may be necessary to adopt techniques used for other compound semiconductors, such as elevated temperatures during the implant step to take advantage of so-called dynamic annealing in which vacancies and interstitials created by the nuclear stopping process are annihilated before they can form stable complexes. It may also be necessary to co-implant O in order to maintain the local stoichiometry of the implanted region since the O recoil is much more significant than the Zn recoils.

Kucheyev et al.[23,24] have recently reported a systematic study of implant isolation in *n*-type ZnO epi layers grown on sapphire. At 350°C, the maximum sheet resistance only reached values of around 10^5 ohms per square while at 25°C the maximum sheet resistance was as high as 10^{11} ohms per square for samples also implanted at 25°C. This type of defect isolation is stable only to ~300–400°C and the sheet resistance is returned to its pre-implantation value by ~600°C. There was no significant chemical effect noted for Cr, Fe or Ni implantation relative to O, indicating that these elements do not introduce a large concentration of deep acceptors into ZnO. Arrhenius plots of sheet resistance of the O-implanted material yielded activation energies of 15–47 meV. The thermal stability of the defect isolation is much lower than is the case in GaN, where the high resistance persists to anneal temperatures >900°C.

9.3 Etching of ZnO

ZnO is readily etched in many acid solutions, including HNO_3/HCl and HF.[25–28] In most cases the etching is reaction-limited, with activation energies of $>6\,\text{kCal}\cdot\text{mol}^{-1}$. We have found that the etching of the ZnO is strongly dependent on material quality. For example, Table 9.1 shows some wet etch rates for bulk and thin film ZnO in different solutions. If the ZnO is very thin, the wet etch rates are high in all acid solutions. A particular problem encountered with the wet etching was the presence of very significant undercut (as much as around 10 μm), which occurred mainly at the end of the selective removal of the ZnO from the underlying AlGaN. Initial results have also appeared on plasma etching of sputter-deposited thin films,[28–31] while plasma-induced damage from high ion density Ar or

Table 9.1: Wet etch rates of thin and bulk ZnO in different acid solutions. In some cases only a lower bound for etch rates could be measured because the films were completely etched away during a short exposure to the acids.

Etch solution	Etch rate for 1000 Å film (Å/sec)	Etch rate for 5000 Å film (Å/sec)	Etch rate for bulk ZnO (Å/sec)
HF/H_2O 1/20	45	2.1	2.1
HCl/H_2O 1/100	>10	2.2	2.2
HCl/H_2O 1/50	>65	2.4	2.4
H_3PO_4/H_2O 1/10	>10	18.5	8

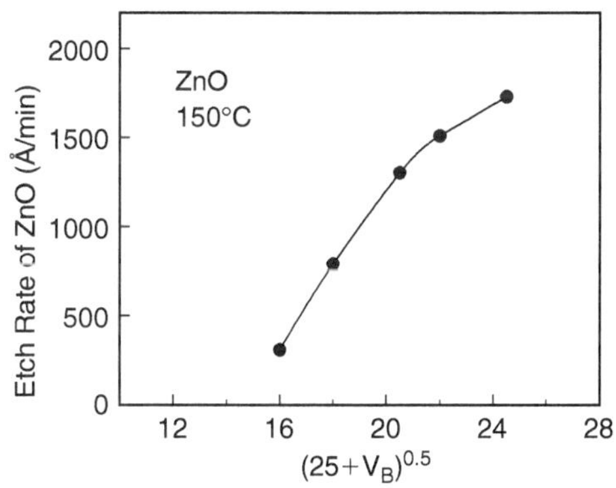

Figure 9.1: Etch rate of ZnO in Cl_2/Ar at 150°C as a function of the average ion kinetic energy (plasma potential of +25 V minus the measured dc bias voltage).

H_2 discharges was found to increase the conductivity of the near surface of similar samples and lead to improved *n*-type ohmic contact resistivities.[32]

Figure 9.1 shows the ZnO etch rate at 150°C in Inductively Coupled Plasmas of Cl_2/Ar as a function of the substrate bias, V_b. The *x*-axis is plotted as the square root of the average ion energy, which is the plasma potential of ~25 V minus the dc self-bias.[33] A commonly accepted model for an etching process occurring by ion-enhanced sputtering in a collision-cascade process predicts the etch rate will be proportional to $E^{0.5}$-$E_{TH}{}^{0.5}$, where E is the ion energy and E_{TH} is the threshold energy.[34] Therefore, a plot of etch rate versus $E^{0.5}$ should be a straight line with an *x*-intercept equal to E_{TH}.[35–39] For Cl_2/Ar this threshold is ~170 eV. In the case of $CH_4/H_2/Ar$, the value of E_{TH} is ~96 eV. The fact that both Cl_2/Ar and $CH_4/H_2/Ar$ exhibit an ion-assisted etch mechanism is consistent with the moderate vapor pressure for the expected group II etch product, namely $(CH_3)_2Zn$, with a vapor pressure of 301 mTorr at 20°C[32–39] and the high bond strength of ZnO. To form the etch product, the Zn-O bonds must first be broken by ion bombardment. The $ZnCl_2$ etch product has a lower vapor pressure (1 mTorr at 428°C), consistent with the slower etch rates for this chemistry. Given the higher etch rates for $CH_4/H_2/Ar$, we then focused on results with that plasma chemistry.

300 K PL spectra from the samples etched in $CH_4/H_2/Ar$ at different rf chuck powers showed an overall PL intensity decrease from both the bandedge (3.2 eV) and the deep-level emission bands (2.3–2.6 eV). The magnitude of this decrease scales with ion energy up to approximately −250 eV, and saturates for larger energies. This may be due to more efficient dynamic annealing of ion-induced point defects at higher defect production rates, producing essentially a saturation damage level as is observed in III-V compound semiconductor etching.[40] We did not observe the increase in bandedge intensity and suppression of the deep-level emission reported for H_2 plasma exposure of ZnO, suggesting that the Ar ion bombardment component dominates during $CH_4/H_2/Ar$ etching at room temperature.

The near-surface stoichiometry of the ZnO was unaffected by $CH_4/H_2/Ar$ etching, as measured by AES. This indicates that the $CH_4/H_2/Ar$ plasma chemistry is capable of equi-rate removal of the Zn and O etch products during ICP etching. Given this result and the fact that the etching occurs through an ion-assisted mechanism, we would expect to observe smooth, anisotropic pattern transfer. Figure 9.2 shows an SEM micrograph of features etched in ZnO using a Ni/Cr mask and $BCl_3/Cl_2/Ar$ discharge. The vertical sidewalls are an indication of the fact that the etch products are volatile only with additional ion-assistance. In addition, the etched field shows only a slight degree of roughening, consistent with the fact that the surface retains

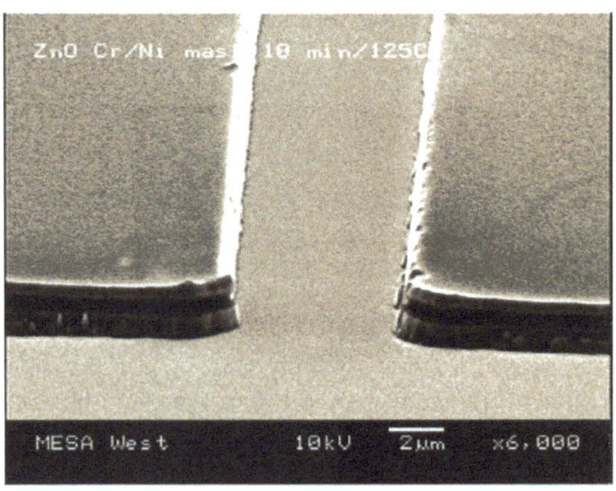

Figure 9.2: SEM micrograph of features etched into ZnO using a $BCl_3/Cl_2/Ar$ plasma. The Cr/Ni mask is still in place.

its stoichiometry. The etch mechanism for ZnO in plasma chemistries of $CH_4/H_2/Ar$ and Cl_2/Ar is ion-assisted.[40–42] For both chemistries the etch rate increases with ion energy as predicted from an ion-assisted chemical sputtering process. The near-surface stoichiometry is unaffected by $CH_4/H_2/Ar$ etching, but the PL intensity decreases, indicating the creation of deep level recombination centers.

9.4 Ohmic contacts

9.4.1 N-type Ohmic contacts

The achievement of acceptable ZnO device characteristics relies heavily on developing low specific contact resistance ohmic metallization schemes.[42–51] The usual approaches involve surface cleaning to reduce barrier height or increase of the effective carrier concentration of the surface through preferential loss of oxygen.[46,47] Specific contact resistances of $\sim 3 \times 10^{-4}\ \Omega\text{-cm}^{-3}$ were reported for Pt-Ga contacts on *n*-ZnO epitaxial layers,[46,48] $2 \times 10^{-4}\ \Omega\text{-cm}^{-3}$ for Ti/Au on Al-doped epitaxial layers,[52,53] $0.7\ \Omega\text{-cm}^{-3}$ for non-alloyed In on laser-processed *n*-ZnO substrates,[54] $2.5 \times 10^{-5}\ \Omega\text{-cm}^{-3}$ for non-alloyed Al on epitaxial *n*-type ZnO,[55] $7.3 \times 10^{-3}\ \Omega\text{-cm}^{-3}$ to $4.3 \times 10^{-5}\ \Omega\text{-cm}^{-3}$ for Ti/Au on plasma exposed, Al-doped *n*-type epitaxial ZnO[56,57] and $9 \times 10^{-7}\ \Omega\text{-cm}^{-3}$ for Ti/Al on n^+-epitaxial ZnO.[58,59] Several points are clear from the past works, namely that the minimum contact resistance generally occurs for post-deposition annealing temperatures of 200°C to 300°C on doped samples which must be treated so as to further increase the near-surface carrier concentration. Table 9.1 summarizes results of previous work.[42–95]

As an example of the typical results obtained from contacts on epi *n*-type ZnO, Figure 9.3 shows results from Ti/Al/Pt/Au Ohmic contacts on phosphorus-doped ZnO epitaxial films grown by pulsed-laser deposition (PLD) on single crystal (0001) Al_2O_3. The films were grown at 400°C in an oxygen pressure of 20 mTorr. The samples were annealed in the PLD chamber at temperatures ranging from 425 to 600°C in O_2 ambient (100 mTorr) for 60 min. The resulting film thickness ranged from 350 nm to 500 nm. The carrier concentration in the ZnO films and the specific contact resistance of as-deposited contacts on these films, measured at room temperature, are shown in Figure 9.3 as a function of post-growth anneal temperatures. The carrier concentration ranged from $7.5 \times 10^{15}\ \text{cm}^{-3}$ after 600°C post-growth anneal to $1.5 \times 10^{20}\ \text{cm}^{-3}$ for the unannealed sample. The as-deposited ohmic contacts with the lowest specific contact resistance ($8.7 \times 10^{-7}\ \Omega\text{-cm}^2$) were obtained for the sample with the highest carrier concentration, as expected.

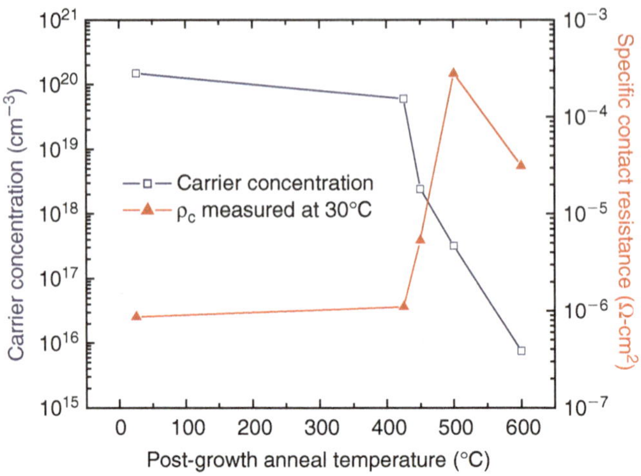

Figure 9.3: Carrier concentration of epi-ZnO and specific contact resistance of as-deposited Ohmic contacts measured at 30°C versus post-growth anneal temperature.

AES depth profiles of the Ti/Al/Pt/Au contact performed after annealing at 250°C showed the initially sharp interfaces between the different metals were degraded by reactions occurring, especially between the Ti and the ZnO to form Ti-O phases and between the Pt and Al. The O appears to diffuse outward while the Pt diffuses inward. The as-deposited contacts showed very sharp interfaces between the different metals and between the Ti and the ZnO. Anneals at 600°C almost completely intermixed the contact metallurgy, as shown in Figure 9.4. Low thermal stability of both Ohmic and Schottky contacts on ZnO appears to be a significant problem in this materials system and clearly there is a need to investigate refractory metals with better thermal properties if applications such as high temperature electronics or lasers operating at high current densities are to be realized. The relatively low thermal stability of these metals on ZnO is in sharp contrast to the stability of the same metal system on GaN.

In summary, specific contact resistances in the range 10^{-7}–10^{-8} Ω-cm^2 were obtained for Ti/Al/Pt/Au contacts on heavily *n*-type ZnO thin films, even in the as-deposited state. However, the contacts show significant changes in morphology even for low temperature (200°C) anneals and this suggests that more thermally stable contacts schemes should be investigated.

9.4.2 P-type Ohmic contacts

In addition to achieving stable and high hole concentrations, work is also needed to develop low resistance *p*-Ohmic contacts. Previous reports have shown that Au,

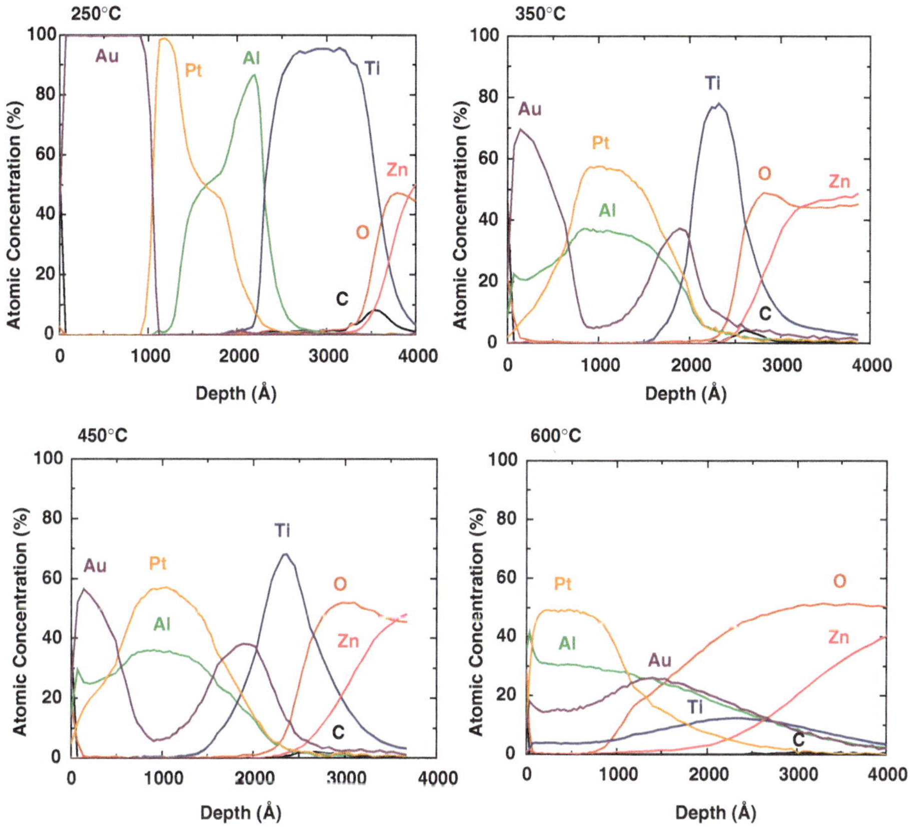

Figure 9.4: AES depth profiles of Ti/Al/Pt/Au contacts on ZnO after annealing at 250°C (top left), 350°C (top right), 450°C (bottom left) or 600°C (bottom right).

Ni, Pt and Ni/Au metallurgy can be used as Ohmic metallization on thin films of *p*-type ZnO or ZnMgO.[52,94] A specific contact resistance of 1.7×10^{-4} Ω-cm^2 was reported after annealing at 600°C in air was achieved for Ni/Au on P-doped ZnO produced by sputtering. Both Au and Au/Ni/Au are found to provide low specific contact resistance on lightly doped *p*-ZnMgO after annealing at 600°C. In both cases, the as-deposited contacts are rectifying and the transition to Ohmic behavior is associated with out-diffusion of Zn from the ZnMgO, as shown in Figure 9.5. A minimum specific contact resistance of 7.6×10^{-6} Ω-cm^2 was obtained with Au/Ni/Au, which is about a factor of 3 lower than for pure Au contacts annealed at the same temperature. Table 9.2 shows a summary of reported *p*-Ohmic contacts on ZnO.

Ohmic and Schottky contacts to *p*-type $Zn_{0.9}Mg_{0.1}O$ have recently been studied in our lab, on material that is P-doped during pulsed laser deposition and

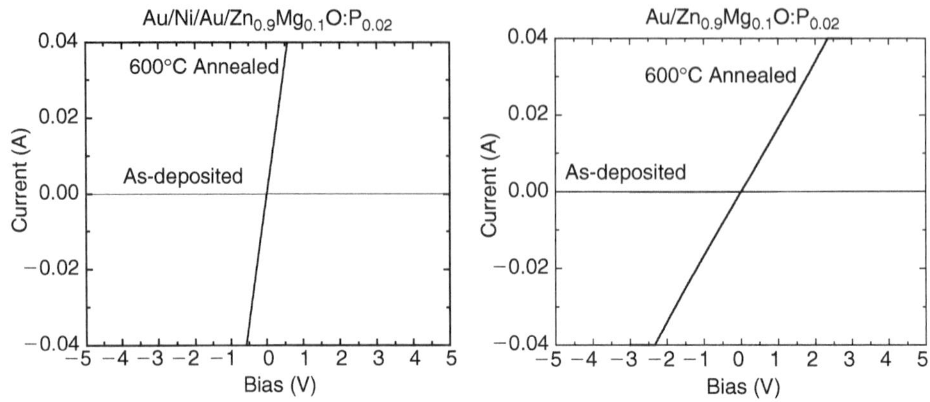

Figure 9.5: I-V characteristics from Au/Ni/Au (top) or Au (bottom) contacts on *p*-type ZnMgO before and after annealing at 600°C.

Table 9.2: Ohmic contacts ZnO and their respective specific contact resistance from published works.

Description	Metal	Lowest ρ_c (Ω-cm^2)	Ref
(a) *n*-type			
n-ZnO epi on Al_2O_3	Ti/Au	1.5×10^{-5}	59
n-ZnO epi on Al_2O_3	FIB direct write Ga-Pt	3.1×10^{-4}	43,45
n-ZnO epi on Al_2O_3	E-beam Ti/Au	2×10^{-4}	15,52
n-ZnO epi on Al_2O_3	E-beam Ti/Au	4.3×10^{-5}	20
n-ZnO epi on Al_2O_3	E-beam Re/Ti/Au	1.7×10^{-7}	58
n-ZnO epi on Al_2O_3	E-beam Ti/Al	9.0×10^{-7}	50
n-ZnO epi on Al_2O_3	E-beam Al/Pt	2×10^{-6}	54,55
n-ZnO epi on Al_2O_3	E-beam Ru	2.1×10^{-5}	55
n-ZnO epi on Al_2O_3	E-beam Al	8×10^{-4}	62
n-ZnO bulk	E-beam Ti/Au	5×10^{-5}	60
n-ZnO bulk	E-beam In	7×10^{-1}	67
(b) *p*-type			
p-ZnO epi on Al_2O_3	E-beam Ti/Au	1.72×10^{-4}	52
p-ZnO epi on Al_2O_3	E-beam Pt/ITO	7.7×10^{-4}	54
p-ZnMgO epi on glass	E-beam Ti/Au or Ni/Au	3×10^{-3}	63
p-ZnMgO epi on glass	E-beam Ni/Au	4×10^{-5}	64
p-ZnO on SiC	E-beam In/Au, Ti/Au, or Ni/Au	—	71

Table 9.3: Sheet resistance, transfer length and specific contact resistance of annealed Au or Au/Ni/Au contacts on *p*-ZnMgO.

	R_s (Ω/sample)	L_t (μm)	R_c (Ω-cm^2)
Au (600°C annealed)	7.5×10^3	0.57	2.5×10^{-5}
Au/Ni/Au (600°C annealed)	1.7×10^3	0.68	7.6×10^{-6}

subsequently annealed to reduce the background *n*-type conduction. At anneal temperatures around 500–600°C, there is a conversion to *p*-type conduction with carrier density of order 10^{16} cm^{-3}. For Ohmic contacts on this material, the lowest specific contact resistivity of 3×10^{-3} Ω-cm^2 was obtained for Ti/Au annealed at 600°C for 30 sec. Ni/Au was less thermally stable and showed severe degradation of contact morphology at this annealing temperature. Table 9.3 shows the electrical properties of our contacts to *p*-ZnMgO before and after 600°C annealing. Both Pt and Ti with Au overlayers showed rectifying characteristics on *p*-ZnMgO, with barrier heights of ~0.55–0.56 eV and ideality factors of ~1.9. Comparison of these results with the same metals on *n*-type ZnO indicates that high surface state densities play a significant role in determining the effective barrier height. Most of these studies are in their infancy due to the lack of readily available *p*-type material and much more work is needed to improve both the barrier height and thermal stability.

To further delineate the character of the phosphorus dopant in the (Zn,Mg)O matrix, simple metal-insulator-semiconductor structures and junctions with *n*-type ZnO were fabricated. Using the Pt as a bottom electrode, vertically-oriented device structures were realized. The presence of this bottom electrode enabled the applied voltage to appear across the diode active region. In contrast, the epitaxial (Zn,Mg)O:P films on sapphire required a lateral device geometry in which significant series resistance between the electrode and junction made the C-V and I-V measurements indirect in measuring junction properties.

The as-deposited phosphorus doped ZnO films were heavily *n*-type due to a compensating donor defect. The moderate temperature anneal suppressed the *n*-type behavior considerably. The samples were annealed in situ at a temperature of 500°C to 600°C in a 100 Torr O_2 ambient for 60 min. In this case, the gate oxide selected was (Ce,Tb)$MgAl_{11}O_{19}$, which is a wide bandgap insulator that can be deposited as an amorphous dielectric using pulsed-laser deposition. The top gate metal was Al. To delineate the carrier type in the (Zn,Mg)O:P film, capacitance-voltage characteristics of metal-insulator-semiconductor diode structures were measured in which the (Zn,Mg)O:P film served as the semiconductor. Capacitance-voltage behavior for the metal-insulator-semiconductor diodes provides an indirect means of determining

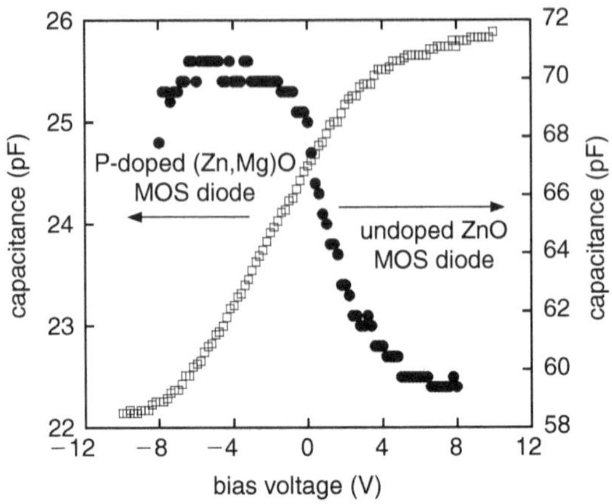

Figure 9.6: Capacitance-voltage characteristics for MOS devices employing ZnO and P-doped (Zn,Mg)O thin films.

the majority carrier type in a semiconductor material. In particular, the symmetry of the capacitance-voltage behavior in MIS diodes is directly dependent on the carrier type underneath the gate. For *n*-type material, a positive voltage relative to the semiconductor will maximize capacitance. For *p*-type, a negative voltage is needed to accumulate majority carrier holes. MIS structures were fabricated with both *n*-type ZnO and phosphorus-doped (Zn,Mg)O thin films. Figure 9.6 shows the capacitance-voltage characteristics for both structures. The polarity of the capacitance-voltage characteristic for the device employing nominally undoped ZnO is clearly *n*-type, with capacitance decreasing with an applied negative voltage. For the (Zn,Mg)O:P device, the symmetry is *p*-type. For the structures considered, the dopant density of the *n*-type and *p*-type structures is on the order of 10^{18}–10^{19} cm^{-3} for both cases. With a phosphorus content of 2 at.%, one can estimate the magnitude of the acceptor level using $n = [2]^{-1/2}[N_V N_A]^{1/2} \exp[-E_A/2kT]$. Assuming that all of the P is on the O site, the activation energy is estimated to be on the order of 180–200 meV.

9.5 Schottky contacts

9.5.1 N-type Schottky contacts

The formation of high quality Schottky diodes on *n*-type material is necessary for some ZnO device applications. Metals such as Au, Ag, Pd form relatively Schottky barriers of 0.6–0.8 eV with *n*-ZnO[77–93] but the barrier heights do not seem to follow the difference in the work function values, indicating the non-negligible impact of the interface defects states. The thermal stability of the Schottky diodes on *n*-ZnO has not been extensively studied but many authors indicate that for Au serious problems arise for temperatures higher than 330 K.[81,82,88–91] At least one study has shown the thermal stability of the Ag/*n*-ZnO Schottky diodes to be higher than that of Au Schottky diodes.[88] The effect of various surface treatments on the values of reverse currents and of the ideality factor of the forward current characteristics has also not been systematically studied. Neville and Mead[77] reported ideality factors very close to unity (n = 1.05) after depositing Au and Pd on the *n*-ZnO surface etched for 15 minutes in concentrated phosphoric acid, followed by 5 minutes etch in concentrated HCl and rinsed in organic solvents. However, similar surface treatments used in other reports led to unacceptably high leakage for Au diodes. Low ideality factors of 1.19 and excellent reverse currents of 1 nA at −1 V were obtained for the Au/n-ZnO Schottky diodes prepared on the O surface of undoped bulk crystals boiled in organic solvents and given a short etch in concentrated HCl.[51] Some authors have reported good electrical performance of the Schottky diodes after etching in concentrated nitric acid. However, in the majority of papers the ideality factors of the ZnO Schottky diodes are considerably higher than unity which has been explained by the prevalence of tunneling, the impact of interface states or the influence of deep recombination centers. Table 9.4 summarizes the results reported in the literature for barrier heights on both *n*- and *p*-type ZnO.

We have recently investigated the electrical properties of Au and Ag Schottky contacts on bulk, *n*-type ZnO whose background carrier density was of the order of 10^{17} cm^{-3}. Capacitance-voltage measurements yielded cutoff voltages of 0.65 V in each case, in good agreement with the reported 0.65 eV values of the Schottky barrier heights of Au on the (0001) ZnO surface. The I-V characteristics of all studied samples showed ideality factors *n* close to 2. This is commonly observed for *n*-ZnO Schottky diodes and is often ascribed to the prevalence of tunneling. The temperature dependence of the saturation current showed activation energies much lower than expected from the barrier height obtained from the C-V plots. DLTS spectra measured on all these diodes showed similar spectra, with the presence of two electron traps with a activation energies close to 0.2 eV and 0.3 eV and low concentration of about 10^{14} cm^{-3} often observed in undoped *n*-ZnO. Diodes prepared on epiready (0001) Zn surface of undoped bulk *n*-ZnO crystals without etching showed a considerably lower reverse current while having virtually the same ideality factor, C-V characteristics and DLTS spectra as the diodes with etched surfaces. Anneal temperatures near 370 K were sufficient to produce surface reactions involving Au, leading to the formation of the near surface layer with reduced electron concentration and a high density of deep level defects.

Table 9.4: Schottky contacts to ZnO and their respective barrier height and ideality factor from published works.

Description	Metal	ϕ_B (eV)	η	Ref
(a) *n*-type				
n-ZnO epi on Si	E-beam Au (Schottky) E-beam Al (Ohmic)	0.59	1.5	72
n-ZnO epi on Al_2O_3	E-beam Ag (Schottky) E-beam Al (Ohmic)	0.69–0.83	1.33	81
n-ZnO epi on GaN/ Al_2O_3	Evaporated Au (Schottky) In (Ohmic)	0.37–0.66	1.8–3.2	82
n-ZnO epi on Al_2O_3 or *n*-ZnO bulk	Evaporated Pd and Ag (Schottky)	0.59–0.68	1.4–1.95	73
n-ZnO epix on Al_2O_3	E-beam Ag or Au (Schottky)	0.84	1.5	74
n-ZnO bulk	E-beam Au (Schottky) E-beam Ti (Ohmic)	0.60–0.71	1.03–1.17	68,85,86
n-ZnO bulk	Vacuum deposition Au or Ag (Schottky) In (Ohmic)	0.64–0.69	1.6–1.8	88
n-ZnO bulk	Pt (Schottky) Ti/Au (Ohmic)	0.89–0.93	1.15	60
n-ZnO bulk	Pt (Schottky) Ti/Au (Ohmic)	0.79	1.51	62
n-ZnO bulk	Pt	—	—	65
n-ZnO bulk	Vacuum deposition Ag or Au (Schottky) In (Ohmic)	0.5–0.6	1.3–1.6	75
n-ZnO bulk	Resistively evaporated Au (Schottky) InGa (Ohmic)	—	—	76
n-ZnO bulk	Resistively evaporated Au or Pd (Schottky)	0.59–0.67	—	77
(b) *p*-type				
p-ZnMgO on glass	E-beam Ti/Au or Pt/Au (Schottky) E-beam Ti/Au (Ohmic)	0.55–0.56	1.9	63
p-ZnO on glass	Evaporated Au (Schottky) Mn (Ohmic)	—	2.7–3.5	64

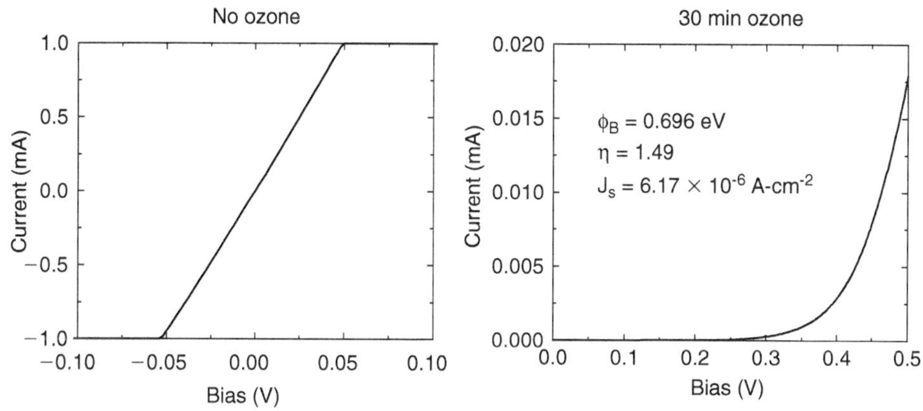

Figure 9.7: I-V characteristic from Pt/Au contacts on ZnO without any ozone cleaning prior to metal deposition (top) and forward (top) I-V characteristics from Pt/Au contacts on ozone cleaned ZnO.

Ag Schottky diodes prepared with similar surface treatments showed no meaningful differences with the Au contacts in terms of I-V characteristics, C-V characteristics, DLTS spectra or thermal stability. Once again, solvent cleaning produced the lowest reverse currents. The low thermal stability of Au and Ag Schottky contacts on ZnO limits their use in device applications.

UV-ozone cleaning prior to metal deposition of either e-beam Pt contacts or sputtered W contacts on n-type single-crystal ZnO is found to significantly improve their rectifying characteristics. Pt contacts deposited directly on the as-received ZnO surface are Ohmic but show rectifying behavior with ozone cleaning (Figure 9.7). The Schottky barrier height of these Pt contacts was 0.70 eV, with ideality factor of 1.5 and a saturation current density of 6.2×10^{-6} A $\cdot$ cm^{-2}. In contrast, the as-deposited W contacts are Ohmic, independent of the use of ozone cleaning. Post-deposition annealing at 700°C produces rectifying behavior with Schottky barrier heights of 0.45 eV for control samples and 0.49 eV for those cleaned with ozone exposure. The improvement in rectifying properties of both the Pt and W contacts is related to removal of surface carbon contamination from the ZnO.

9.5.2 P-type Schottky contacts

Table 9.4 summarizes reported barrier heights on both n- and p-type ZnO. The p-contacts also show relatively low barrier heights for metals such as Au. The fact that the barrier height is not the bandgap minus the barrier height of this metal on *n*-type ZnO also points to the strong influence of surface states on ZnO. There is clearly much scope for work on improved rectifying contacts to ZnO, both in terms of increasing the barrier height and finding more thermally stable contact metallurgy for device applications requiring good thermal stability. In particular there is a need to improve or standardize the cleaning procedures prior to contact deposition.

9.6 Properties of hydrogen in ZnO

There is particular interest in the properties of hydrogen in ZnO, because of the predictions from density functional theory and total energy calculations that it should be a shallow donor.[96,98] The generally observed *n*-type conductivity, therefore, may at least in fact be explained by the presence of residual hydrogen from the growth ambient, rather than to native defects such as Zn interstitials or O vacancies. There is some experimental support for the predicted observations of its muonium counterpart[99,100] and from electron paramagnetic resonance of single-crystal samples.[112] There have been many other studies on the effects of hydrogen on the electrical and optical properties of ZnO.[95–112]

9.6.1 Proton implantation

An investigation into the retention of implanted hydrogen in single-crystal, bulk ZnO as a function of annealing temperature and the effects of the implantation on both the crystal quality and optical properties of the material showed that the temperatures at which implanted hydrogen is evolved from the ZnO are considerably lower then for GaN which is a more developed materials system for visible and UV light-emitters.

Figure 9.8 shows the SIMS profiles of implanted ^{2}H in bulk ZnO as a function of subsequent annealing temperature. Note the effect of the annealing is an evolution of ^{2}H out of the ZnO crystal, with the remaining deuterium atoms of each temperature decorating the residual implant damage. The peak in the as-implanted profile occurred at 0.96 μm, in good agreement with the projected range from Transfer-of-Ion-in-Matter (TRIM) simulations. The thermal stability of the implanted ^{2}H is considerably lower in ZnO than in GaN,[107,108] where temperatures of ~900°C are needed to remove deuterium to below the detection limit ($\sim 3 \times 10^{15}$ cm^{-3}) of SIMS and this suggests that slow-diffusing H_2 molecules or larger clusters do not form during the anneal. Since we did not observe conventional out-diffusion profiles we were unable to estimate a diffusion coefficient for the ^{2}H in ZnO. Our results are consistent with an implant-damaged trap-controlled release of ^{2}H from the ZnO lattice for temperatures >500°C.

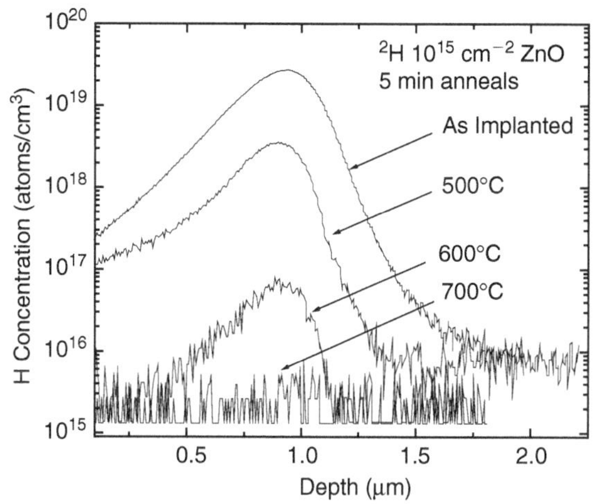

Figure 9.8: SIMS profiles of ^{2}H implanted into ZnO (100 keV, 10^{15} cm^{-2}) before and after annealing at different temperatures (5 min anneals).

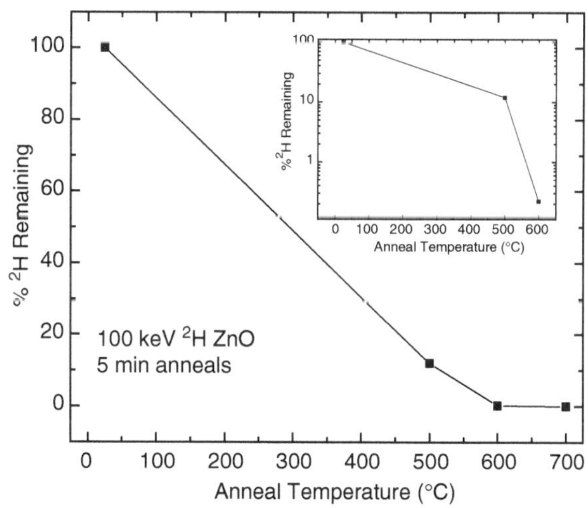

Figure 9.9: Percentage of retained ^{2}H implanted into ZnO (100 keV, 10^{15} cm^{-2}) as a function of annealing temperature (5 min anneals). The inset shows the data on a log scale.

Figure 9.9 shows the percentage of ^{2}H remaining in the ZnO as a function of anneal temperature. The ^{2}H concentrations were obtained by integrating the area under the curves in the SIMS data. It is evident that the thermal stability of implanted deuterium is not that high with 12% of the initial dose retained after 500°C anneals and ~0.2% after 600°C anneals.

RBS/C showed that implantation of ^1H, even to much higher doses (10^{16} cm^{-2}), did not affect the backscattering yield near the ZnO surface. However, there was a small (but detectable) increase in scattering peak deeper in the sample, in the region where the nuclear energy loss profile of 100 keV H^+ is a maximum. The RBS/C yield at this depth was ~6.5% of the random level before H^+ implantation and ~7.8% after implantation to the dose of 10^{16} cm^{-2}.

While the structural properties of the ZnO were minimally affected by the hydrogen (or deuterium) implantation, the optical properties were severely degraded. Room temperature cathodoluminescence showed that even for a dose of 10^{15} cm^{-2} $^1H^+$ ions, the intensity of near-gap emission has reduced by more than 3 orders of magnitude as compared to control values. This is due to the formation of effective non-radiative recombination centers associated with ion-beam produced defects. Similar results were obtained from PL measurements. The band-edge luminescence was still severely degraded even after 700°C anneals where the ^2H has been completely evolved from the crystal. This indicates that point defect recombination centers are still controlling the optical quality under these conditions. Kucheyev et al.[23,24] have found that resistance of ZnO can be increased by about 7 orders of magnitude as a result of trap introduction by ion irradiation.

The results of deep level transient spectroscopy (DLTS) and capacitance-voltage (C-V) profiling studies of undoped *n*-ZnO[113–116] showed that the carrier removal rate in this material is considerably lower than in other wide-bandgap semiconductors such as GaN. The dominant electron traps introduced by such irradiation were the E_c–0.55 eV and the E_c–0.78 eV centers. Implant isolation studies have shown low thermal stability (350°C) for the electrical effects of implantation.[23,24]

Proton implantation was performed with energy of 50 keV with doses of 5×10^{13}–5×10^{15} cm^{-2}. Implantation with 5×10^{13} cm^{-2} 50 keV protons led to quite measurable decrease of the carrier concentration. For the dose of 5×10^{15} cm^{-2}, the capacitance dependence on voltage was very slight and the apparent concentration deduced from the slope of $1/C^2$ versus voltage curve was quite high, on the order of 10^{18} cm^{-3}. The depth of this region of high electron concentration was close to the estimated range of 50 keV protons in ZnO. The apparent room temperature electron concentrations are close to 2×10^{16} cm^{-3} for the dose of 5×10^{14} cm^{-2} and close to 8×10^{15} cm^{-3} for the dose of 10^{15} cm^{-2}. At lower measurement temperatures (85 K), the profiles looked similar, with slightly lower absolute concentrations due to the freeze-out of some of the deeper shallow donors (admittance spectroscopy data suggests that these frozen-out donors have the activation energy of about 70 meV). The 5×10^{15} cm^{-2} dose sample still showed a high electron concentration wall near the end of the range of implanted protons, exceeding the initial shallow donor concentration. This suggests a strong freeze-out of the major centers in the implanted region occurred at low temperature leaving the sample depleted of carriers down to the high-electron concentration region near the end of the protons range. Illumination of the proton implanted samples with UV light generated by a deuterium lamp led to persistent increase of capacitance of these samples and to persistent changes in the measured profiles as though some donors were activated within the part of the space charge region (SCR) that had been depleted in the dark. The changes in capacitances persisted for many hours at 85 K and could still be observed even at room temperature. Two electron traps with apparent activation energy of 0.2 eV and 0.3 eV and very low concentration of respectively 7×10^{13} cm^{-3} and 1.8×10^{14} cm^{-3} were observed. Implantation of the sample with 50 keV protons to the dose of 5×10^{13} cm^{-2} had little effect on the density of these two traps but introduced two additional traps with activation energy 0.55 eV and 0.75 eV with concentrations of 3.7×10^{14} cm^{-3} and 3.75×10^{14} cm^{-3}. At higher temperatures three traps were observed, namely the 0.55 eV traps detected in the 5×10^{13} cm^{-2} dose implanted sample (however, the signal from this trap was very weak, most likely because the freeze-out of carriers was still a factor at the temperatures of the peak), the 0.75 eV traps with the peak near 280 K and 0.9 eV traps with the peak near 370 K. The carrier removal rate deduced from room temperature C-V profiling was significantly higher than the removal rate reported for 1.8 MeV protons. The range of the 50 keV protons is about 50 times shorter than the range of the 1.8 MeV protons and consequently, the actual density of radiation defects causing the carriers to be removed should be much higher in the former case.

9.6.2 Hydrogen plasma exposure

Bulk ZnO samples were exposed to ^{2}H plasmas at temperatures of 100–300°C at 900 mTorr with 50 W of 13.56 MHz power. Some of these samples were subsequently annealed at temperatures up to 600°C under flowing N_2 ambients for 5 min.

SIMS profiles of ^{2}H in plasma exposed ZnO, for different sample temperatures during the plasma treatment showed the profiles follow those expected for diffusion from a constant or semi-infinite source, i.e.

$$C(x,t) = C_0\, erfc\left(\frac{X}{\sqrt{4Dt}}\right)$$

where $C(x,t)$ is the concentration at a distance x for diffusion time t, C_0 is the solid solubility and D is the diffusivity of ^{2}H in ZnO.[114] The incorporation depths of ^{2}H are very large compared to those in GaN or GaAs under similar conditions, where depths of 1–2 μm are observed. It is clear that hydrogen must diffuse as an interstitial, with little trapping by the lattice elements or by defects or impurities. The position of H in the lattice after immobilization has not yet been determined experimentally, but from theory the lowest energy states for H^+ is at a bond-centered position forming an O-H bond, while for H_2 the anti-bonding Zn site is most stable.[97]

Using a simple estimate of the diffusivity D, from $D = X^2/4t$, and where X is taken to be the distance at which ^2H concentration has fallen to $5 \times 10^{15}\,cm^{-3}$ in Figure 9.10, we can estimate the activation energy for diffusion. The extracted activation energy, E_a, is $0.17 \pm 12\,eV$ for ^2H in ZnO. Note that the absolute diffusivities of ^1H would be ~40% larger because of the relationship for diffusivities of isotopes, i.e.[114]

$$\frac{D^1H}{D^2H} = \left(\frac{M^2H}{M^1H}\right)^{1/2}$$

The small activation energy is consistent with the notion that the atomic hydrogen diffuses in interstitial form.

Figure 9.10 shows SIMS profiles of a ZnO sample exposed to a ^2H plasma of 0.5 h at 200°C, then annealed for 5 min under N_2 at different temperatures. There is significant loss of ^2H even after a short anneal at 400°C, with virtually all of it evolved out of the crystal by 500°C. This is in sharp contrast to ^2H in GaN, where much higher temperatures (≥800°C) are needed to evolve the deuterium out of the sample.

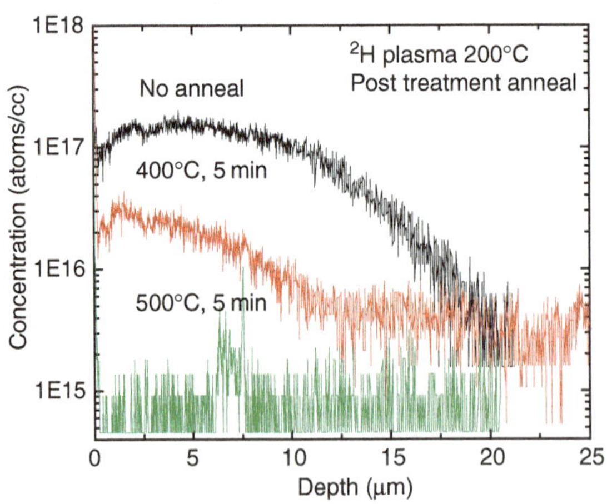

Figure 9.10: SIMS profiles of ^2H in ZnO exposed to deuterium plasmas for 0.5 h at different temperatures.

Lavrov et al.[115] have identified two hydrogen-related defects in ZnO, by using local vibrational mode spectroscopy. The H-I center consists of a hydrogen atom at the bond centered site, while the H-II center contains two inequivalent hydrogen atoms bound primarily to two oxygen atoms.

The ^{2}H plasma treatment causes an increase in donor concentration consistent with past reports.[95–117] In that case, the effect was attributed to hydrogen passivation of compensating acceptor impurities present in the as-grown ZnO epitaxial layers. An alternative explanation is that the hydrogen induces a donor state and thereby increases the free electron concentration. Subsequent annealing reduces the carrier density to slightly below the initial value in the as-received ZnO, which may indicate that it contained hydrogen as a result of the growth process. We emphasize that the *n*-type conductivity probably arises from multiple impurity sources and we cannot unambiguously assign all of the changes to hydrogen.

Hydrogen is found to exhibit a very rapid diffusion in ZnO when incorporated by plasma exposure, with D of $8.7 \times 10^{-10}\,\text{cm}^2/\text{V}_\text{S}$ at 300°C. The activation energy for diffusion is indicative of interstitial motion. All of the plasma-incorporated hydrogen is removed from the ZnO by annealing at $\geq$500°C. When the hydrogen is incorporated by direct implantation, the thermal stability is somewhat higher, due to trapping at residual damage. The free electron concentration increases after plasma hydrogenation, consistent with the small ionization energy predicted for H in ZnO and the experimentally measured energy of 60 ± 10 meV for muonium in ZnO. The electrical activity and rapid diffusivity of H or ZnO must be taken into account when designing device fabrication processes such as deposition of dielectrics using SiH_4 as a precursor or dry etching involving use of $CH_4/H_2/Ar$ plasmas since these could lead to significant changes in near-surface conductivity.

For many other semiconductor materials it has been shown that the hydrogen introduced, e.g. from a plasma at relatively low temperatures, readily forms complexes with various donor and acceptor species and often renders them electrically neutral (the phenomenon widely known as hydrogen passivation. There seems to be a general agreement that the hydrogen can be relatively easily introduced from a hydrogen plasma at moderate temperatures of 200–300°C. At least two papers report considerable increase of the electron concentration in the near-surface regions of the hydrogen plasma treated *n*-ZnO bulk crystals and thin films. After the plasma exposure the concentration of donors became quite high, close to 4×10^{17} cm^{-3}, near the surface and then showed a plateau at the level of 1.8×10^{17} cm^{-3} down to the depth of about 0.15 μm available to C-V measurements without causing an electrical breakdown. One can observe a striking similarity in the SIMS profile of hydrogen and the profile of the uncompensated shallow donors measured by C-V. The DLTS spectrum was dominated by two peaks corresponding to the electron traps with apparent activation energies of 0.2 eV and 0.3 eV. The concentrations of these traps were 8×10^{13} cm^{-3} and 1.4×10^{14} cm^{-3}. The second trap at E_c–0.3 eV seems to be the same trap as the E2 center and is often dominant in DLTS spectra measured by several groups.

One feature of the hydrogen plasma treatment results for our ZnO samples is the lack of hydrogen passivation effect for deep electron and hole traps which is the hallmark of hydrogen passivation phenomena in many other semiconductors. We have shown that hydrogen plasma treatment of good-quality *n*-ZnO crystals leads to about a factor of 2 increase of the electron concentration in the surface region of the samples which is in almost one-to-one correspondence with the density of introduced hydrogen atoms according to SIMS measurements. This increase correlates with the similar increase of the intensity of the bound exciton 3.35 eV line suggesting that the type of shallow donors introduced by hydrogen plasma treatment is the same as in the virgin sample. From admittance spectroscopy measurements and from comparison with the results of shallow donors studies in undoped ZnO we associate these donors with the 37 meV donors and attribute them to either the hydrogen donors per se or to hydrogen donor complexes with native defects.

9.7 Conclusions and summary

Major advances have been made by many groups around the world recently in demonstrating *p*-type ZnO, fabricating improved Ohmic contacts, developing high resolution dry etch pattern transfer processes and implant isolation processes and in realizing room temperature ferromagnetism in ZnO. There are still many areas that need additional work, including

1. Higher *p*-type doping levels in epitaxial films. This will require even better control of the background *n*-type conductivity in the material that arises from native defects and impurities such as hydrogen.
2. Realization of high quality *p-n* junctions by epitaxial growth, with good breakdown characteristics. These are the building blocks for devices such as UV LEDs or lasers.
3. Improved Schottky contacts with higher barrier height than reported to date. This will need a better understanding of surface cleaning processes and metal schemes that do not disrupt the ZnO surface during their deposition. In particular it is necessary to understand that nature of the defects present on ZnO surfaces that may be pinning the Fermi level.
4. *p*-type Ohmic contacts with lower specific contact resistance. A desirable goal is to achieve a contact resistance of 10^{-5} Ohm $\cdot$ cm^2 or less for ZnO-based

light emitters so that self-heating in the contact region does not degrade their reliability.

5. Improved plasma etching processes that utilize simple photoresist masking and are able to achieve practical etch rates for ZnO. It is also necessary to develop selective etch processes for the ZnMgO/ZnO heterostructure.
6. Development of implant doping processes, to realize both n and *p*-type layers. This will require a better understanding of the defects created by the implant step and their thermal stability and also a better understanding of the stability of the ZnO surface during activation annealing.

It is likely that many of these areas will see significant progress in the near future as the promise of ZnO for low cost transparent electronics and UV light emission becomes more widely known.

Acknowledgments

The work at UF was supported in part by the Army Research Office under grant no. DAAD19–01–1–0603, the Army Research Laboratory, the National Science Foundation (DMR 0400416, Dr. L. Hess), the Air Force Office of Scientific Research under grant F49620–03–1–0370 and by the U.S. Department of Energy under contact DE-AC05–000R22725.

References

[1] See for example, Semiconducting Transparent Thin Films, H. L. Hartnagel, A. L. Dawar, A. K. Jain and C. Jagadish (IOP Publishing, Bristol, 1995).
[2] D. C. Look, Mater. Sci. Eng. *B80* 383 (2001).
[3] P. Zu, Z. K. Tang, G. K. L. Wong, M. Kawasaki, A. Ohtomo, K. Koinuma and Y. Sagawa, Solid-State Commun. *103* 459 (1997).
[4] D. M. Bagnall, Y. R. Chen, Z. Zhu, T. Yao, S. Koyama, M. Y. Shen and T. Goto, Appl. Phys. Lett. *70* 2230 (1997).
[5] M. Wraback, H. Shen, S. Liang, C. R. Gorla and Y. Lu, Appl. Phys. Lett. *74* 507 (1999).
[6] J.-M. Lee, K.-K. Kim, S.-J. Park and W.-K. Choi, Appl. Phys. Lett. *78* 2842 (2001).
[7] J. E. Nause, III-V's Review *12* 28 (1999).
[8] Y. Chen, D. Bagnell and T. Yao, Mat. Sci. Eng. *B75* 190 (2000).
[9] D. C. Look, D. C. Reynolds, J. W. Hemsky, R. L. Jones and J. R. Sizelove, Appl. Phys. Lett. *75* 811 (1999).
[10] D. C. Look, J. W. Hemsky and J. R. Sizelove, Phys. Rev. Lett. *82* 2552 (1999).

[11] F. D. Auret, S. A. Goodman, M. Hayes, M. J. Legodi, H. A. van Laarhoven and D. C. Look, Appl. Phys. Lett. *80* 956 (2002).
[12] S. O. Kucheyev, J. E. Bradley, J. S. Williams, C. Jagadish and M. V. Swain, Appl. Phys. Lett. *80* 956 (2002).
[13] D. C. Reynolds, D. C. Look and B. Jogai, Solid-State Commun. *99* 873 (1996).
[14] M. Wraback, H. Shen, S. Liang, C. R. Gorla and Y. Lu, Appl. Phys. Lett. *76* 507 (1999).
[15] T. Aoki, D. C. Look and Y. Hatanaka, Appl. Phys. Lett. *76* 3257 (2000).
[16] C. C. Chang and Y. E. Chen, IEEE Trans. Ultrasonics, Ferroelectrics and Frequency Control *44* 624 (1997).
[17] P. M. Verghese and D. R. Clarke, J. Appl. Phys. *87* 4430 (2000).
[18] C. R. Gorla, N. W. Emanetoglu, S. Liang, W. E. Mayo, Y. Lu, M. Wraback and H. Shen, J. Appl. Phys. *85* 2595 (1999).
[19] H. Ohta, K. Kawamura, M. Orita, M. Hirano, N. Sarukura and H. Hosono, Appl. Phys. Lett. *77* 475 (2000).
[20] M. Joseph, H. Tabata and T. Kawai, Jap. J. Appl. Phys. *38* L1205 (1999).
[21] S. Krishnamoorthy, A. A. Iliadis, A. Inumpudi, S. Choopun, R. D. Vispute and T. Venkatesan, Solid-State Electron. *46* 1631 (2002).
[22] Y. Li, G. S. Tompa, S. Liang, C. Gorla, C. Lu and J. Doyle, J. Vac. Sci. Technol. *A15* 1663 (1997).
[23] S. O. Kucheyev, C. Jagadish, J. S. Williams, P. N. K. Deenapanray, M. Yano, K. Koike, S. Sasa, M. Inoue and K. Ogata, J. Appl. Phys. *93* 2972 (2003).
[24] S. O. Kucheyev, P. N. L. Deenapanray, C. Jagadish, J. S. Williams, M. Yano, K. Kioke, S. Sasa, M. Inoue and K. Ogata, Appl. Phys. Lett. *83* 3350 (2002).
[25] Y. Li, G. S. Tompa, S. Liang, C. Gorla, C. Lu and J. Doyle, J. Vac. Sci. Technol. *A15* 1663 (1997).
[26] J. G. E. Gardeniers, Z. M. Rittersma and G. J. Burger, J. Appl. Phys. *83* 7844 (1998).
[27] H. Maki, T. Ikoma, I. Sakaguchi, N. Ohashi, H. Haneda, J. Tanaka and N. Ichinose, Thin Solid Films *378* 232 (2004).
[28] S.-C. Chang, D. B. Hicks and R. C. O. Laugal, Technical Digest. IEEE Solid-State Sensor and Actuator Workshop *41* (1992).
[29] J. S. Wang, Y. Y. Chen and K. M. Larkin, Proc. 1982 IEEE Ultrasonic Symp, pp. 345–348 (1982).
[30] G. D. Swanson, T. Tanagawa and D. L. Polla, Electrochem. Soc. Ext. Abstracts 90–2 1082 (1990).
[31] J.-M. Lee, K.-M. Chang, K.-K. Kim, W.-K. Choi and S. J. Park, J. Electrochem. Soc. *148* G1 (2001).
[32] J.-M. Lee, K.-K. Kim, S.-J. Park and W.-K. Park, Appl. Phys. Lett. *78* 3842 (2001).
[33] R. J. Shul, M. C. Lovejoy, A. G. Baen, J. C. Zolper, D. J. Rieger, M. J. Hafich, R. F. Corless and C. B. Vartuli, J. Vac. Sci. Technol. *A13* 912 (1995).
[34] C. Steinbruchel, Appl. Phys. Lett. *55* 1960 (1989).

[35] K. Pelhos, V. M. Donnelly, A. Kornblit, M. L. Green, R. B. Van Dover, L. Manchanda and E. Bower, J. Vac. Sci. Technol. *A19* 1361 (2001).
[36] C. C. Chang, K. V. Guinn, V. M. Donnelly and I. P. Herman, J. Vac. Sci. Technol. *A13* 1970 (1995).
[37] H. Kuzami, R. Hamasaki and K. Tago, Jap. J. Appl. Phys. *36* 4829 (1997).
[38] A. C. Jones and P. O'Brien, CVD of Compound Semiconductors (VCH, Weinheim, Germany, 1997).
[39] S. W. Pang, in Handbook of Advanced Plasma Processing Techniques, ed. R. J. Shul (Springer-Verlag, Berlin, 2000).
[40] K. Ip, K. Baik, M. E. Overberg, E. S. Lambers, Y. W. Heo, D. P. Norton, S. J. Pearton, F. Ren and J. M. Zavada, Appl. Phys Lett. *81* 3546 (2002).
[41] H. K. Kim, J. W. Bae, T. K. Kim, K. K. Kim, T. Y. Seong and I. Adesida, J. Vac. Sci. Technol. *B21* 1273 (2003).
[42] S. Krishnamoorthy, A. A. Iliadis, A. Inumpudi, S. Choopun, R. D. Vispute and T. Venkatesan, Solid-State Electron. *46* 1631 (2002).
[43] A. A. Iliadis, R. D. Vispute, T. Venkatesan and K. A. Jones, Thin Solid Films *420–421* 478 (2002).
[44] V. Hoppe, D. Stachel and D. Beyer, Physica Scripta *T57* 122 (1994).
[45] A. Inumpudi, A. A. Iliadis, S. Krishnamoorthy, S. Choopun, R. D. Vispute and T. Venkatesan, Solid-State Electron. *46* 1665 (2002).
[46] H.-K. Kim, S.-H. Han, T.-Y. Seong and W.-K. Choi, Appl. Phys. Lett. *77* 1647 (2000).
[47] H.-K. Kim, S.-H. Han, T.-Y. Seong and W.-K. Choi, J. Electrochem. Soc. *148* G114 (2001).
[48] T. Akane, K. Sugioka and K. Midorikawa, J. Vac. Sci. Technol. *B18* 1406 (2001).
[49] H. Sheng, N. W. Emanctoglu, S. Muthukumar, S. Feng and Y. Lu, J. Electron. Mater. *31* 811 (2002).
[50] S. Y. Kim, H. W. Jang, J. K. Kim, C. M. Jeon, W. I. Park, G. C. Yi and J.-L. Lee, J. Electron. Mater. *31* 868 (2002).
[51] J.-M. Lee, K.-K. Kim. S.-J. Park and W.-K. Choi, Appl. Phys. Lett. *78* 3842 (2001).
[52] S.-H. Kim, K.-K. Kim, S.-J. Park and T.-Y. Seong, J. Electrochem. Soc. *152* G169 (2005).
[53] S. H. Kim, H. W. Jang, J. K. Kim, C. M. Jeon, W. I. Park, G.-C. Yi and J.-L. Lee, J. Electron. Mat. *31* 868 (2002).
[54] H.-K. Kim, I. Adesida, K.-K. Kim and T.-Y. Seong, J. Electrochem. Soc. *151* G223 (2004).
[55] H.-K. Kim, K.-K. Kim, S.-J. Park, T.-Y. Seong and I. Adesida, J. Appl. Phys. *94* 4225 (2003).
[56] J.-H. Lim, K.-K. Kim, D.-K. Hwang, H.-S. Kim, J.-Y. Oh and S.-J. Park, J. Electrochem. Soc. *152* G179 (2005).
[57] U. Grossner, S. Gabrielsen, T. M. Børseth, J. Grillenberger, A. Y. Kuznetsov and B. G. Svensson, Appl. Phys. Lett. *85* 2259 (2004).
[58] S.-H. Kim, J.-T. Maeng, C.-J. Choi, J. H. Leem, M. S. Han and T.-Y. Seong, J. Electrochem. Soc. *8* G167 (2005).

[59] H.-K. Kim, J. W. Bae, K.-K. Kim, S.-J. Park, T.-Y. Seong and I. Adesida, Thin Solid Films *447* 90 (2004).
[60] S.-H. Kim, H.-K. Kim and T.-Y. Seong, Appl. Phys. Lett. *86* 112101 (2005).
[61] S.-H. Kim, H.-K. Kim and T.-Y. Seong, Appl. Phys. Lett. *86* 022101 (2005).
[62] U. Grossner, S. Gabrielsen, T. M. Børseth, J. Grillenberger, A. Y. Kuznetsov and B. G. Svensson, Appl. Phys. Lett. *85* 2259 (2004).
[63] S. Kim, B. S. Kang, F. Ren, Y. W. Heo, K. Ip, D. P. Norton and S. J. Pearton, Appl. Phys. Lett. *84* 1904 (2004).
[64] H. S. Yang, Y. Li, D. P. Norton, K. Ip, S. J. Pearton, S. Jang and F. Ren, Appl. Phys. Lett. *86* 192103 (2005).
[65] W. T. Petrie and J. M. Vohs, J. Chem. Phys. *101* 8098 (1994).
[66] H.-K. Kim, K.-K. Kim, S.-J. Park, T.-Y. Seong and Y. S. Yoon, Jpn. J. Appl. Phys. *41* L546 (2002).
[67] H.-K. Kim, T.-Y. Seong, K.-K. Kim, S.-J. Park, Y. S. Yoon and I. Adesida, Jpn. J. Appl. Phys. *43* 976 (2004).
[68] B. J. Coppa, C. C. Fulton, P. J. Hartlieb, R. F. Davis, B. J. Rodriguez, B. J. Shields and R. J. Nemanich, J. Appl. Phys. *95* 5856 (2004).
[69] T. Akane, K. Sugioka and K. Midorikawa, J. Vac. Sci. Technol. *B18* 1406 (2000).
[70] H. Fabricius, T. Skeetrup and P. Bisgaard, Appl. Optics *25* 2764 (1986).
[71] Y. R. Ryu, T. S. Lee, J. H. Leem and H. W. White, Appl. Phys. Lett. *83* 4032 (2003).
[72] G. Yuan, Z. Ye, L. Zhu, J. Huang, Q. Qian and B. Zhao, J. Cryst. Growth *268* 169 (2004).
[73] H. von Wenckstern, E. M. Kaidashev, M. Lorenz, H. Hockmuth, G. Biehne, J. Lenzner, V. Gottschalch, R. Pickenhain and M. Grundmann, Appl. Phys. Lett. *84* 79 (2004).
[74] S. Liang, H. Sheng, Y. Liu, Z. Huo, Y. Lu and H. Shen, J. Cryst. Growth *225* 110 (2003).
[75] N. Ohashi, J. Tanaka, T. Ohgaki, H. Haneda, M. Ozawa and T. Tsurumi, J. Mater. Res. *17* 1529 (2002).
[76] F. D. Auret, S. A. Goodman, M. J. Legodi, W. E. Meyer and D. C. Look, Appl. Phys. Lett. *80* 1340 (2002).
[77] R. C. Neville and C. A. Mead, J. Appl. Phys. *41* 3795 (1970).
[78] H. S. Yang, Y. Li, D. P. Norton, K. Ip, S. J. Pearton, S. Jang and F. Ren, Appl. Phys. Lett. *86* 192103 (2005).
[79] G. S. Marlow and M. B. Das, Solid-State Electron. *25* 91 (1982).
[80] C. A. Mead, Phys. Lett. *18* 218 (1965).
[81] J. C. Simpson and F. Cordaro, J. Appl. Phys. *63* 1781 (1988).
[82] H. Sheng, S. Muthukumar, N. W. Emanetoglu and Y. Lu, Appl. Phys. Lett. *80* 2132 (2002).
[83] D. C. Oh, J. J. Kim, H. Makino, T. Hanada, M. W. Cho and T. Yao, Appl. Phys. Lett. *86* 042110 (2005).
[84] F. D. Auret, S. A. Goodman, M. Hayes, M. J. Legodi and H. A. van Laarhoven, Appl. Phys. Lett. *79* 3074 (2001) and F. D. Auret, private communication.
[85] B. J. Coppa, R. F. Davis and R. J. Nemanich, Appl. Phys. Lett. *82* 400 (2003).

[86] B. J. Coppa, C. C. Fulton, S. M. Kiesel, R. F. Davis, C. Pandarinath, J. E. Burnette, R. J. Nemanich and D. J. Smith, J. Appl. Phys. *97* 103517 (2005).

[87] A. Y. Polyakov, N. Smirnov, E. Kozhukhova, V. Vdovin, K. Ip, D. P. Norton and S. J. Pearton, J. Vac. Sci. Technol. *A21* 1603 (2003).

[88] A. Y. Polyakov, N. Smirnov, E. Kozhukhova, V. Vdovin, K. Ip, Y. W. Heo, D. P. Norton and S. J. Pearton, Appl. Phys. Lett. *83* 1575 (2003).

[89] Y. W. Heo, L. C. Tien, Y. Kwon, S. J. Pearton, B. S. Kang, F. Ren and J. R. LaRoche, Appl. Phys. Lett. *85* 3107 (2004).

[90] K. Ip, Y. W. Heo, R. Mehandru, B. P. Gila, C. R. Abernathy, F. Ren and S. J. Pearton, Appl. Surf. Sci. 236/1–4 (2004) pp. 387–393.

[91] K. Ip, B. P. Gila, A. H. Onstine, E. S. Lambers, Y. W. Heo, K. H. Baik, D. P. Norton, S. J. Pearton, S. Kim, J. R. LaRoche and F. Ren, Appl. Phys. Lett. *84* 5133 (2004).

[92] S. J. Pearton, D. P. Norton, K. Ip, Y. Heo and T. Steiner, J. Vac. Sci. Technol. *B22* 932 (2004).

[93] K. Ip, Y. Heo, K. Baik, D. Norton, S. J. Pearton, S. Kim, J. LaRoche and F. Ren, Appl. Phys. Lett. *84* 2835 (2004).

[94] S. Kim, B. Kang, F. Ren, K. Ip, Y. Heo, D. Norton and S. J. Pearton, Appl. Phys. Lett. *84* 1698 (2004).

[95] S. J. Pearton, D. P. Norton, K. Ip, Y. W. Heo and T. Steiner, Prog. Mater. Sci. *50* 293 (2005).

[96] C. Kilic and Z. Zunger, Appl. Phys. Lett. *81* 73 (2002).

[97] C. van de Walle, Physica *B308–310* 899 (2001).

[98] S. J. F. Cox, E. A. Davis, S. P. Cottrell, P. J. C. King, J. S. Lord, J. M. Gil, H. V. Alberto, R. C. Vilao, J. P. Duarte, N. A. de Campos, A. Weidinger, R. L. Lichti and S. J. C. Irving, Phys. Rev. Lett. *86* 2601 (2001).

[99] D. M. Hofmann, A. Hofstaetter, F. Leiter, H. Zhou, F. Henecker, B. K. Meyer, S. B. Schmidt and P. G. Baranov, Phys. Rev. Lett. *88* 045504 (2002).

[100] S. J. Baik, J. H. Jang, C. H. Lee, W. Y. Cho and K. S. Lim, Appl. Phys. Lett. *70* 3516 (1997).

[101] N. Ohashi, T. Ishigaki, N. Okada, T. Sekiguchi, I. Sakaguchi and H. Haneda, Appl. Phys. Lett. *80* 2869 (2002).

[102] V. Bogatu, A. Goldenbaum, A. Many and Y. Goldstein, Phys. Stat. Solidi *B212* 89 (1999).

[103] T. Sekiguchi, N. Ohashi and Y. Terada, Jpn. J. Appl. Phys. *36* L289 (1997).

[104] C. S. Han, J. Jun and H. Kim, Appl. Surf. Sci. *175/176* 567 (2001).

[105] Y. Natsume and H. Sakata, J. Mater. Sci. Materials in Electronics *12* 87 (2001).

[106] Y.-S. Kang, H. Y. Kim and J. Y. Lee, J. Electrochem. Sci. *147* 4625 (2000).

[107] R. G. Wilson, S. J. Pearton, C. R. Abernathy and J. M. Zavada, J. Vac. Sci. Technol. *A13* 719 (1995).

[108] S. J. Pearton, J. C. Zolper, R. J. Shul and F. Ren, J. Appl. Phys. *86* 1 (1999).

[109] F. D. Auret, S. A. Goodman, M. Hayes, M. J. Legodi, H. A. Van Laarhoven and D. C. Look, Appl. Phys. Lett. *79* 3074 (2001).

[110] F. D. Auret, S. A. Goodman, M. Hayes, M. J. Legodi, H. A. van Laarhoven and D. C. Look, J. Phys.: Condens. Matter. *13* 8989 (2001).

[111] K. Ip, M. E. Overberg, Y. W. Heo, D. P. Norton, S. J. Pearton, C. E. Stutz, B. Luo, F. Ren, D. C. Look and J. M. Zavada, Appl. Phys. Lett. *82* 385 (2003).

[112] I. P. Kuz'mina and V. A. Nikitenko, Zinc Oxide: Growth and Optical Properties (Moscow, Nauka, 1984) (in Russian).

[113] H. Sheng, S. Muthukumar, N. W. Emanetoglu and Y. Lu, Appl. Phys. Lett. *80* 2132 (2002).

[114] R. J. Borg and C. J. Dienes, An Introduction to Solid State Diffusion (Academic Press, Boston, 1988).

[115] E. V. Lavrov, J. Weber, F. Borrnet, C. G. Van de Walle and R. Helbig, Phys. Rev. B (in press).

[116] D. C. Reynolds, D. C. Look, B. Jogai, C. W. Litton, T. C. Collins, W. Harsch and G. Cantwell, Phys. Rev. *B57* 12151 (1998).

[117] K. Thonke, Th. Gruber, N. Teofilov, R. Schönfelder, A. Waag and R. Sauer, Physica *B308–310* 945 (2001).

Chapter 10

Novel Nanostructures and Nanodevices of ZnO

Zhong Lin Wang

School of Materials Science and Engineering, Georgia Institute of Technology, Atlanta, GA 30332-0245, USA

Abstract: Zinc oxide is a unique material that exhibits semiconducting and piezoelectric dual properties. Using a solid–vapor phase thermal sublimation technique, nanowires, nanobelts, nanocages, nanocombs, nanosprings, seamless nanorings and nanohelixes of ZnO have been synthesized under specific growth conditions. These unique nanostructures unambiguously demonstrate that ZnO probably has the richest family of nanostructures among all materials, both in structures and in properties. The nanostructures could have novel applications in optoelectronics, sensors, transducers and biomedical sciences.

Keywords: ZnO, nanobelt, nanowire, nanocages, nanocomb, nanoring, nanohelix, nanospring, field effect transistor, nanocantilever.

10.1 Introduction

Nanostructured ZnO materials have received broad attention due to their distinguished performance in electronics, optics and photonics. From the 1960s, synthesis of ZnO thin films has been an active field because of their applications as sensors, transducers and catalysts. In the last few decades, especially since the nanotechnology initiative led by the US, study of one-dimensional (1D) materials has become a leading edge in nanoscience and nanotechnology. With reduction in size, novel electrical, mechanical, chemical and optical properties are introduced, which are largely believed to be the result of surface and quantum confinement effects. Nanowire-like structures are the ideal system for studying the transport process in 1D confined objects, which are of benefit not only for understanding the fundamental phenomena in low-dimensional systems, but also for developing new generation nanodevices with high performance. ZnO is a key technological material. The lack of a centre of symmetry in wurtzite, combined with large electromechanical coupling, results in

Zinc Oxide Bulk, Thin Films and Nanostructures
C. Jagadish and S. Pearton (Editors)

strong piezoelectric and pyroelectric properties and the consequent use of ZnO in mechanical actuators and piezoelectric sensors. In addition, ZnO is a wide band-gap (3.37 eV) compound semiconductor that is suitable for short wavelength optoelectronic applications. The high exciton binding energy (60 meV) in ZnO crystal can ensure efficient excitonic emission at room temperature and room temperature ultraviolet (UV) luminescence has been reported in disordered nanoparticles and thin films. ZnO is transparent to visible light and can be made highly conductive by doping. ZnO is a versatile functional material that has a diverse group of growth morphologies, such as nanobelts [1], nanowires [2], nanocages [3], nanocombs [4], nanosprings [5], nanorings [6] and nanohelixes [7,8]. The objective of this chapter is to review the unique nanostructures that have been grown for ZnO and their corresponding growth mechanisms. The potential applications and novel nanodevices demonstrated for ZnO nanostructures will be reviewed.

10.2 Synthesis technique

The oxide nanostructures to be reviewed in this chapter were synthesized by a solid–vapor process. In principle, the thermal evaporation technique is a simple process in which condensed or powder source material(s) is/are vaporized at elevating temperature and then the resultant vapor phase(s) condense(s) under certain conditions (temperature, pressure, atmosphere, substrate, etc.) to form the desired product(s). The processes are usually carried out in a horizontal tube furnace, as shown in Figure 10.1, which is composed of a horizontal tube furnace, an alumina tube, a rotary pump system and a gas supply and control system. A viewing window

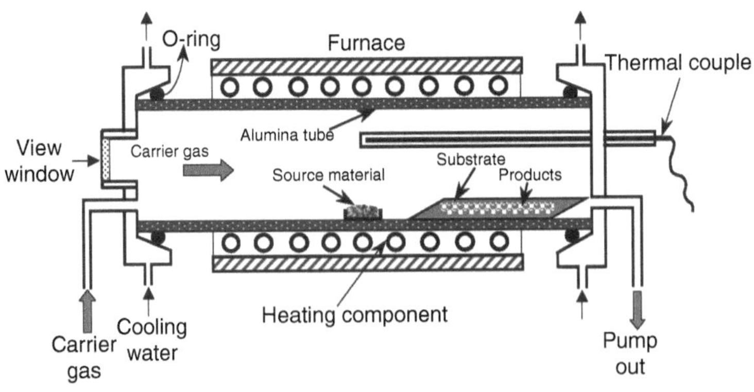

Figure 10.1: A schematic diagram of the experimental apparatus for growth of oxides nanostructures by the solid–vapor phase process.

is set up at the left end of the alumina tube, which is used to monitor the growth process. The right-hand end of the alumina tube is connected to the rotary pump. Both ends are sealed by rubber O-rings. The ultimate vacuum for this configuration is $\sim 2 \times 10^{-3}$ Torr. The carrying gas comes in from the left end of the alumina tube and is pumped out at the right end. The source material(s) is/are loaded on an alumina boat and positioned at the centre of the alumina tube, where the temperature is the highest. Alumina substrates were placed downstream for collecting growth products. This simple set-up can achieve high control of the final product.

There are several processing parameters such as temperature, pressure, carrier gas (including gas species and its flow rate), substrate and evaporation period, which can be controlled and need to be selected properly before and/or during the thermal vaporization [9]. The source temperature selection mainly depends on the volatility of the source material(s). Usually, it is slightly lower than the melting point of the source material. The pressure is determined according to the evaporation rate or vapor pressure of the source material(s). The substrate temperature usually drops with the distance of its location from the position of the source material(s). The local temperature determines the type of product that will be obtained. It is also noted that the thermal evaporation process is very sensitive to the concentration of oxygen in the growth system. Oxygen influences not only the volatility of the source material(s) and the stoichiometry of the vapor phase, but also the formation of the product(s). In the present study, after evacuating the alumina tube to $\sim 2 \times 10^{-3}$ Torr, thermal vaporization was conducted at a designed heating rate and at a pressure of 200–600 Torr and Ar carrier gas at 50 sccm (standard cubic centimetres per minute). The on and off times as well as the duration of inlet of the carrier gas could greatly affect the growth kinetics.

10.3 Crystal structure of ZnO and its polar surfaces

Zinc oxide has a hexagonal structure (space group $P6_3mc$) with lattice parameters $a = 0.3296$ and $c = 0.52065$ nm. The structure of ZnO can be simply described as a number of alternating planes composed of tetrahedrally coordinated O^{2-} and Zn^{2+} ions, stacked alternatively along the c-axis (Figure 10.2a). The tetrahedral coordination in ZnO results in non-central symmetric structure and piezoelectricity. Another important characteristic of ZnO is the polar surfaces. The most common polar surface is the basal plane. The oppositely charged ions produce positively charged Zn-(0001) and negatively charged O-(000$\bar{1}$) polar surfaces (Figure 10.2b), resulting in a normal dipole moment and spontaneous polarization along the c-axis. Another

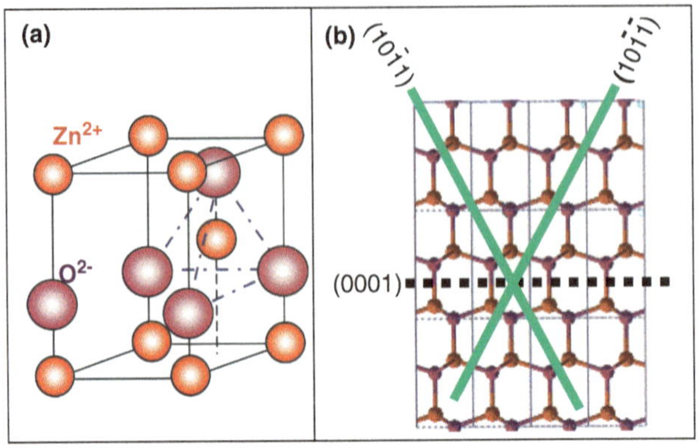

Figure 10.2: (a) Wurtzite structure model of ZnO. The tetrahedral coordination of ZnO is shown. (b) The structure model of ZnO projected along [2$\bar{1}\bar{1}$0], displaying the ±(0001), ±(10$\bar{1}\bar{1}$) and ±(10$\bar{1}$1) polar surfaces.

polar surface is the {01$\bar{1}$1}. By projecting the structure along [1$\bar{2}$10], as shown in Figure 10.2b, beside the most typical ±(0001) polar surfaces that are terminated with Zn and oxygen, respectively, ±(10$\bar{1}$1) and ±(10$\bar{1}\bar{1}$) are also polar surfaces. The {10$\bar{1}$1} type surfaces are not common for ZnO, but they have been observed in a nanohelical structure found recently [7]. The charges on the polar surfaces are ionic charges, which are non-transferable and non-flowable. Because the interaction energy among the charges depends on the distribution of the charges, the structure is arranged in such a configuration to minimize the electrostatic energy. This is the main driving force for growing the polar surface dominated nanostructures.

10.4 Nanostructures and growth processes

The different surface structures of ZnO could induce anisotropic growth. Under thermodynamic equilibrium conditions, the facet with higher surface energy is usually small in area, while the lower energy facets are larger. Specifically, in the ZnO growth, the highest growth rate is along the *c*-axis and the large facets are usually {01$\bar{1}$0} and {2$\bar{1}\bar{1}$0}. By controlling the growth kinetics, it is possible to change the growth behavior of ZnO nanobelts. Here, we received various ZnO nanostructures by means of the controllable synthesized parameters, such as deposition temperatures and pressures as well as carrier gas flux. The ZnO nanostructures presented here include nanowires, nanobelts, nanocages, nanocombs, nanosprings, seamless nanorings and nanohelixes.

10.4.1 Nanorods/nanowires

Growth of 1D nanostructure usually follows the vapor–liquid–solid (VLS) approach, in which a liquid alloy droplet composed of a metal catalyst component (such as Au, Fe) and a nanowire component (such as Si, III–V compound, II–V compound, oxide) is first formed under the reaction conditions. The metal catalyst can be rationally chosen from the phase diagram by identifying metals in which the nanowire component elements are soluble in the liquid phase but do not form solid compounds more stable than the desired nanowire phase. For the 1D ZnO nanowires grown via a VLS process, the commonly used catalyst for ZnO is Au [10]. The liquid droplet serves as a preferential site for absorption of gas phase reactant and, when supersaturated, the nucleation site for crystallization. Nanowire growth begins after the liquid becomes supersaturated in reactant materials and continues as long as the catalyst alloy remains in a liquid state and the reactant is available. During growth, the catalyst droplet directs the nanowire's growth direction and defines the diameter of the nanowire. Ultimately, the growth terminates when the temperature is below the eutectic temperature of the catalyst alloy or the reactant is no longer available. As a result, a nanowire obtained from the VLS process typically has a solid catalyst nanoparticle at its tip with a diameter comparable to that of the connected nanowires. Thus, one can usually determine whether the nanowire growth was governed by a VLS process from whether there is a catalyst particle present at one end of the nanowire.

The VLS crystal growth mechanism was first proposed by Wagner and Ellis [11] in 1964 for Si whisker growth, in which Si whiskers with diameters of up to the micrometre scale were grown by hydrogen reduction of $SiCl_4$ with the presence of Au, Pt, Ag, Pd, Cu and Ni as the catalysts. Westwater et al. [12] and Lieber et al. [13] then developed this mechanism and successfully prepared nanometer scale Si wires by pyrolysis of SiH_4 with Au as the catalyst and laser ablation of $Si_{0.95}Fe_{0.05}$ targets, respectively.

Figure 10.3 shows an SEM image of uniform ZnO nanorods/nanowires grown using Au catalyst on a polycrystalline alumina substrate. The nanorods show no alignment and are dispersively distributed, but their diameters and lengths are rather uniform, as defined by the size of the Au catalyst. The catalyst particles locate at the tips of the nanowires, as can be seen in the enlarged image. The nanorods grow along [0001] and their side surfaces are enclosed by $\{2\bar{1}\bar{1}0\}$.

Recently, we have found that Sn can also be an effective catalyst for growing ZnO nanorods [14]. Using the epitaxial growth of ZnO nanorods on ZnO crystal, aligned nanorods have been grown (Figure 10.4) [2]. In this case, the growth direction of the nanorods is led by the Sn catalyst and the epitaxial orientation is defined by the substrate that determines the aligned growth. The nanorods are also aligned in

Figure 10.3: (a) ZnO nanorods grown using gold as a catalyst. (b) An enlarged image of the nanorods, showing gold particles at the tips.

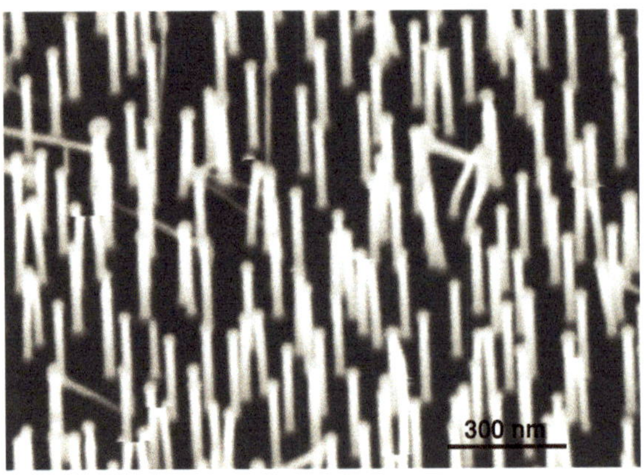

Figure 10.4: Aligned ZnO nanorods/nanowires epitaxially grown on ZnO substrate using Sn as a catalyst [2].

orientation because they tend to take the least mismatch orientation on the substrate, to reduce the interface mismatch energy. The choice of substrate is important for the epitaxial growth. One must consider the crystallographic structure as well as the surfaces to be used for the growth. We have found recently that the atomic termination and charge status of the substrate can strongly affect the morphology of the grown nanorods [14]. The nanorods growing out of Zn-terminated (0001) ZnO substrates are very different from those grown out of the oxygen-terminated $(000\bar{1})$ ZnO substrates although the same type of Sn catalyst was used.

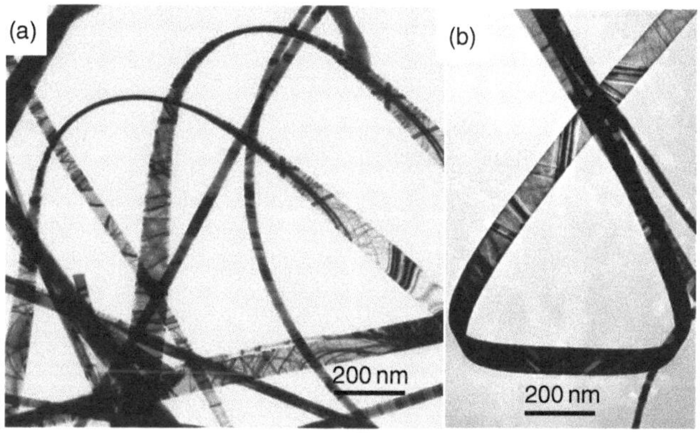

Figure 10.5: TEM images of the as-synthesized ZnO nanobelts, showing uniform morphology [1].

10.4.2 Nanobelts

Nanobelts are nanowires that have a well-defined geometrical shape and side surfaces. Nanobelts of ZnO are usually grown [1] by sublimation of ZnO powder without introducing a catalyst. Figure 10.5 shows a TEM image of ZnO nanobelts, displaying different morphologies from the nanowires and nanorods. Each nanobelt has a uniform width along its entire length and the typical widths of the nanobelts are in the range of 50–300 nm and thicknesses are 10–30 nm. No particle was observed at the ends of the nanobelts. A ripple-like contrast that appeared in the TEM image is due to strain resulting from the bending of the belt. The nanobelt grows along $[01\bar{1}0]$, with top and bottom flat surfaces $\pm(2\bar{1}\bar{1}0)$ and side surfaces $\pm(0001)$. The cross-section of the nanobelt is rectangular, with typical thickness and width-to-thickness ratios of ~5–10.

10.4.3 Nanocages

By changing the composition of the source material, nanocages of ZnO shells have been grown [15]. A mixture of commercial ZnO, SnO_2 and graphite powders in an atomic ratio of 2:1:1 was used as the source material. Thermal evaporation was conducted at 1150°C for 15 min under pressure of 200–300 Torr and with an Ar flow rate 25 sccm; then the sample was cooled down to 1000°C and kept isothermal for 30 min without supplying any Ar gas. In the temperature zone of 300–500°C that we are interested in, balls and polyhedra are observed. Some of them are dispersed on the substrate and there appears to be no correlation among them (Figure 10.6a), while some of them line up along a ZnO nanobelt (Figure 10.6b) that was formed in

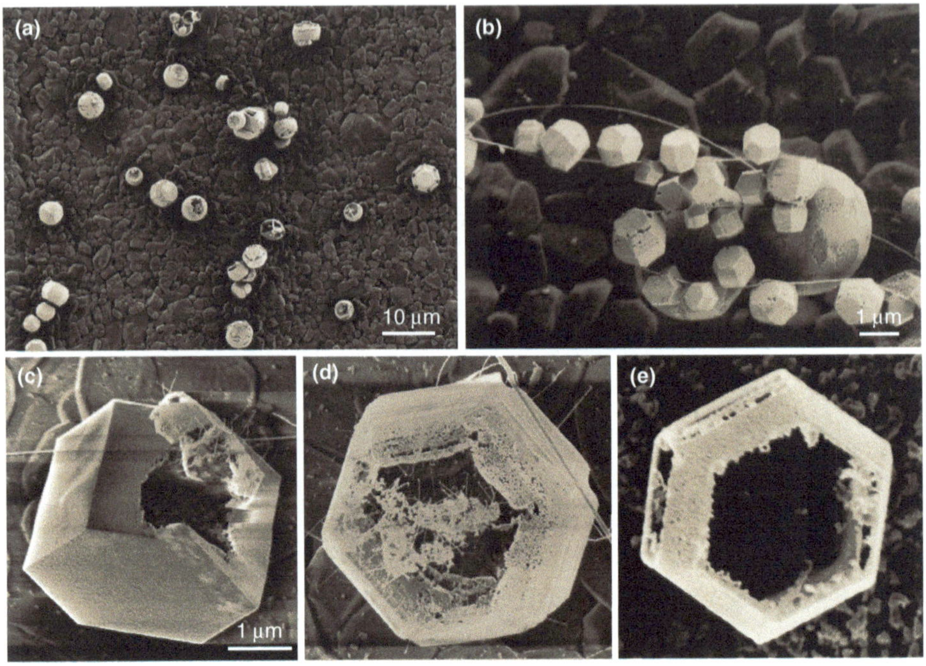

Figure 10.6: Low magnification SEM images of ZnO polyhedral cages and shells (a) distributed dispersively on the substrate and (b) aligned along a nanobelt, respectively. (c–e) Typical drum-shaped cages of truncated hexagon-based shells [15].

a higher temperature zone of typically 700–900°C. Hexagon-based rods are often seen (Figure 10.6c) but they have shell structure. The polyhedron is enclosed by (0001) (top and bottom surface), $\{10\bar{1}0\}$ (side surfaces), stepped $\{10\bar{1}1\}$ (inclined surfaces) and high index planes with rough surfaces. Some truncated hexagon-based drums show open corners, such as the one displayed in Figure 10.6d. In some cases, one side of the shell collapses possibly due to the high growth temperature as well as the small shell thickness (Figure 10.6e). A common feature is that the shell exhibits mesoporous structure. Although the polyhedral shell structure appears to be composed of nanocrystals, transmission electron diffraction and imaging indicates that they have a "single-crystalline" textured orientation and the side surfaces are $\{10\bar{1}0\}$. These types of shell structures could be candidates for drug delivery use.

10.4.4 Nanocoms and nanosaws

"Comb-like" structures of ZnO have been reported by others [16], but the mechanism that drives the growth was not elaborated until recently. The comb structures

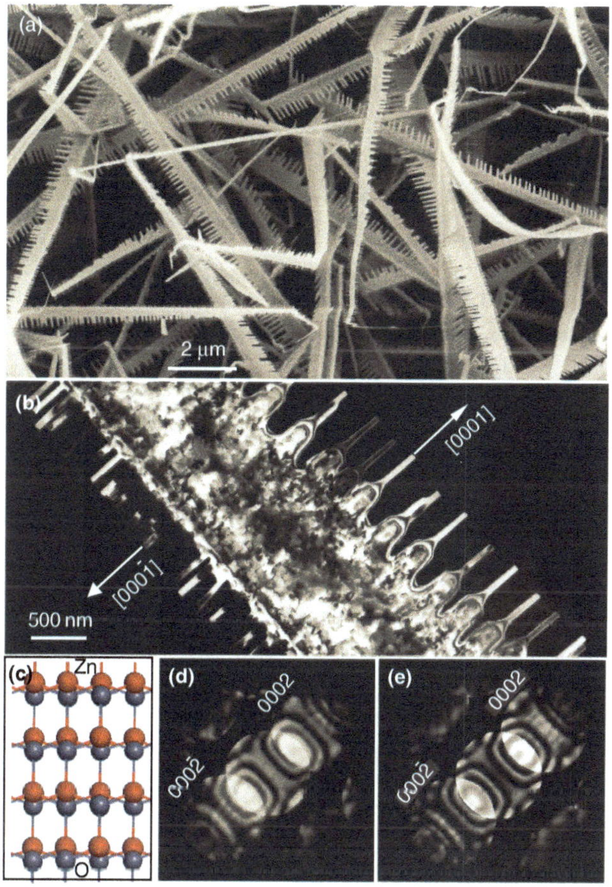

Figure 10.7: (a) SEM image of "comb-like" cantilever arrays of ZnO, which is the result of surface polarization induced growth due to the chemically active (0001)-Zn. (b) High-resolution TEM (HRTEM) recorded from the tip of the comb-teeth, showing possible evidence of Zn segregation at the growth front, which is likely to be effective for driving self-catalyzed growth. (c) Structure model of ZnO projected along [01$\bar{1}$0], showing the termination effect of the crystal. (d,e) Experimentally observed and theoretically simulated convergent beam electron diffraction (CBED) patterns for determining the polarity of the nanocombs, respectively [4].

we have received (Figure 10.7a,b) with the comb-teeth growing along [0001], the top/bottom surfaces being $\pm(01\bar{1}0)$ and side surfaces $\pm(2\bar{1}\bar{1}0)$. Using CBED [17], which relies on dynamic scattering effect and is an effective technique for determining the polarity of wurtzite structure, we have found that the comb structure is an asymmetric growth along Zn-[0001] (Figure 10.7b) [4]. This conclusion is received by comparing the experimentally observed CBED pattern and the theoretically calculated pattern by matching the fine detailed structure features in the (0002) and $(000\bar{2})$ diffraction disks (Figure 10.7d,e). The positively charged Zn-(0001) surface is chemically active and the negatively charged O-$(000\bar{1})$ surface is relatively inert, resulting in a growth of long fingers along [0001]. Using HRTEM, we found that the Zn-terminated (0001) surface has tiny Zn clusters, which could lead to self-catalyzed growth without the presence of foreign catalyst [4]. The chemically inactive $(000\bar{1})$ surface typically does not grow nanobelt structure.

The anisotropic growth appears to be a common characteristic for the wurtzite family. Similar saw-teeth growth has been observed for ZnS [18] and CdSe [19], and it is suggested to be induced by the Zn- and Cd-terminated (0001) surfaces, respectively.

10.4.5 Nanospiral or nanosprings

Due to differences in surface energies among (0001), $\{01\bar{1}0\}$ and $\{2\bar{1}\bar{1}0\}$, freestanding nanobelts and nanowires of ZnO are usually dominated by the lower energy, non-polar surfaces, such as $\{01\bar{1}0\}$ and $\{2\bar{1}\bar{1}0\}$. Recently, by introducing doping, such as In and/or Li, ZnO nanobelts dominated by the (0001) polar surfaces have been grown [5]. The nanobelt grows along $[2\bar{1}\bar{1}0]$ (the *a*-axis), with its top/bottom large surface ±(0001) and the side surfaces $\pm(01\bar{1}0)$. Due to the small thickness of 5–20 nm and large aspect ratio of ~1:4, the flexibility and toughness of the nanobelts are extremely high. A polar-surface-dominated nanobelt can be approximated to be a capacitor with two parallel charged plates (Figure 10.8a). The polar nanobelt tends to roll over into an enclosed ring to reduce the electrostatic energy (Figure 10.8b). A spiral shape is also possible for reducing the electrostatic energy (Figure 10.8c) [20]. The formation of the nanorings and nanohelixes can be understood from the nature of the polar surfaces [5]. If the surface charges are uncompensated during the growth, the spontaneous polarization induces electrostatic energy due to the dipole moment, but rolling up to form a circular ring would minimize or neutralize the overall dipole moment, reducing the electrostatic energy. On the other hand, bending of the nanobelt produces elastic energy. The stable shape of the nanobelt is determined by the minimization of the total energy contributed by spontaneous polarization and elasticity. If the nanobelt is rolled loop-by-loop, the repulsive force between the charged surfaces stretches the nanospring, while the elastic deformation force pulls the loops together; the balance between the two forms the nanospring that has elasticity (Figure 10.8d). The nanospring has a uniform shape with radius of ~500–800 nm and evenly distributed pitches. Each is made of a uniformly deformed single-crystal ZnO nanobelt.

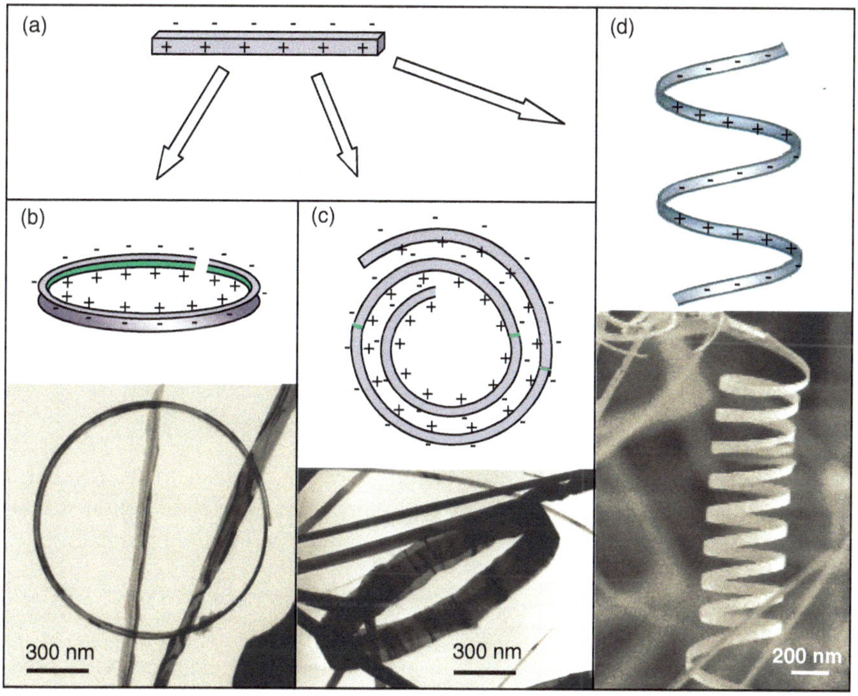

Figure 10.8: (a) Model of a polar nanobelt. Polar surface induced formation of (b) nanoring, (c) nanospiral and (d) nanohelixes of ZnO and their formation processes.

10.4.6 Seamless nanorings

By adjusting the raw materials with the introduction of impurities, such as indium, we have synthesized a nanoring structure of ZnO (Figure 10.9) [6]. TEM image indicates that the nanoring is a single-crystal entity with circular shape. The single-crystal structure referred here means a complete nanoring that is made of a single crystalline ribbon bent evenly at the curvature of the nanoring. Although the radius of the ring is large, its thickness is only ~10 nm. The nanoring is made of a loop-by-loop co-axial, uni-radius, epitaxial-coiling of a nanobelt [6].

The growth of the nanoring structures can be understood from the polar surfaces of the ZnO nanobelt. The polar nanobelt, which is the building block of the nanoring, grows along $[10\bar{1}0]$, with side surfaces $\pm(1\bar{2}10)$ and top/bottom surfaces $\pm(0001)$, and have a typical width of ~15 nm and thickness ~10 nm. The nanobelt has polar charges on its top and bottom surfaces (Figure 10.10a). If the surface charges are uncompensated during growth, the nanobelt may tend to fold itself as

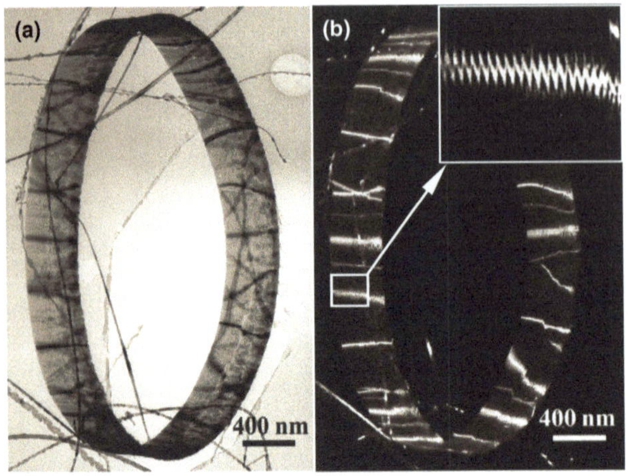

Figure 10.9: Seamless single-crystal nanorings of ZnO. (a) Bright-field and (b) dark-field TEM image of the nanoring. The inset in (b) is an enlargement of a local region, displaying the contrast produced by the uni-radius, loop-by-loop self-coiling nanobelt [6].

its length getting longer to minimize the area of the polar surface. One possible way is to interface the positively charged Zn-(0001) plane (top surface) with the negatively charged O-(0001) plane (bottom surface), resulting in neutralization of the local polar charges and the reduced surface area, thus, forming a loop with an overlapped end (Figure 10.10b). This type of folding is 90° with respect to the folding direction for forming the nanospring or nanospiral (see Figure 10.8), possibly due to the difference in aspect ratio of the nanobelts and relative size of the polar surfaces. The radius of the loop may be determined by the initial folding of the nanobelt at the initial growth, but the size of the loop cannot be too small to reduce the elastic deformation energy. The total energy involved in the process coming from polar charges, surface area and elastic deformation. The long-range electrostatic interaction is likely to be the initial driving force for folding the nanobelt to form the first loop for the subsequent growth. This is the nucleation of the nanoring.

As the growth continues, the nanobelt may be naturally attracted onto the rim of the nanoring due to electrostatic interaction and extends parallel to the rim of the nanoring to neutralize the local polar charge and reduce the surface area, resulting in the formation of a self-coiled, co-axial, uni-radius, multi-looped nanoring structure (Figure 10.10c). The self-assembly is spontaneous, which means that the self-coiling along the rim proceeds as the nanobelt grows. The reduced surface area and the formation of chemical bonds (short-range force) between the

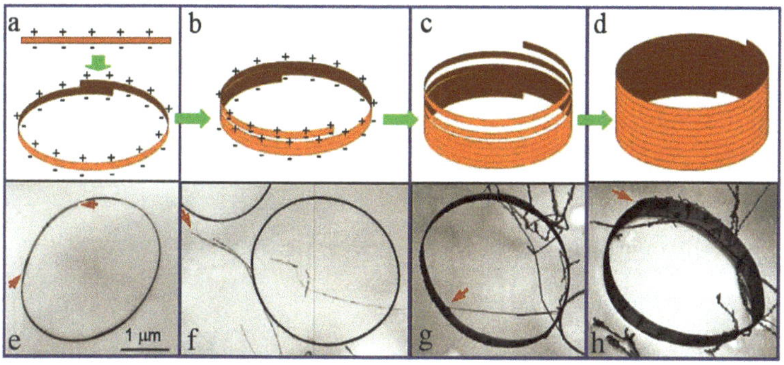

Figure 10.10: (a) Structure model of ZnO, showing the ±(0001) polar surfaces. (b–d) Proposed growth process and (e–h) corresponding experimental results showing the initiation and formation of the single-crystal nanoring via self-coiling of a polar nanobelt. The nanoring is initiated by folding a nanobelt into a loop with overlapped ends due to long-range electrostatic interaction among the polar charges; the short-range chemical bonding stabilizes the coiled ring structure; and the spontaneous self-coiling of the nanobelt is driven by minimizing the energy contributed by polar charges, surface area and elastic deformation.

loops stabilize the coiled structure. The width of the nanoring increases as more loops winding along the nanoring axis, and all of them remain in the same crystal orientation (Figure 10.10d). Since the growth was carried out in a temperature region of 200–400°C, "epitaxial sintering" of the adjacent loops forms a single-crystal cylindrical nanoring structure, and the loops of the nanobelt are joined by chemical bonds as a single entity. A uni-radius and perfectly aligned coiling is energetically favorable because of the complete neutralization of the local polar charges inside the nanoring and the reduced surface area. This is the "slinky" growth model of the nanoring. The charge model of the nanoring is analogous to that of a single DNA helix.

In the literature, almost all of the reported nanowires and nanobelts of ZnO are dominated by the non-polar surfaces such as $\{2\bar{1}\bar{1}0\}$ and $\{01\bar{1}0\}$. The question here is why we have received the polar surface dominated nanobelts? An important fact that we have noticed is that planar defects are present for stabilizing the nanobelts dominated by polar surfaces [21]. Figure 10.11a is a TEM image of a broken piece of a nanoring, showing periodically distributed planar defects with sharp contrast. The periodic spacing between the defects in the nanoring represents the width of the self-coiling nanobelt. The electron diffraction pattern recorded from the region, as shown in the insert, presents that the radial direction of the nanoring is $[2\bar{1}\bar{1}0]$ (e.g., the direction pointing to the center of the ring), tangential direction of the nanoring is $[01\bar{1}0]$, and [0001] is parallel to central symmetric nanoring axis.

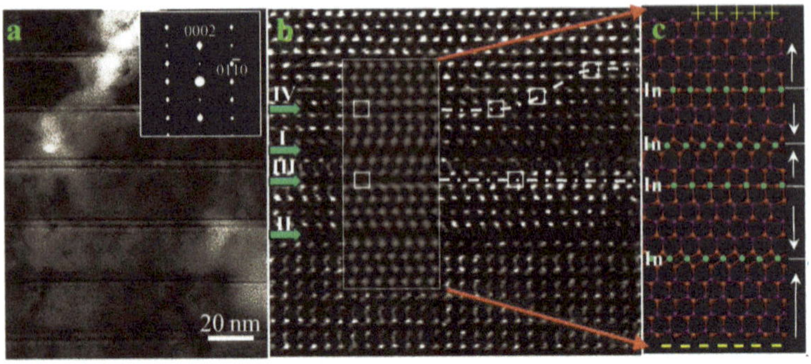

Figure 10.11: Planar defect in the formation of nanorings and nanosprings. (a) Low-magnification TEM image recorded from a broken nanoring; the inset is its electron diffraction pattern. (b) HRTEM image recorded from a broken nanoring, the inset is a simulated image based on the model depicted in (c). The sample thickness is 2.924 nm and defocus is −28.67 nm. An excellent match between the simulated image and the experimental image supports our model. The four green arrowheads point to four In ion layers. Four inversion domain boundaries, labeled I, II, III and IV, have been identified in the image [21].

Combined the chemical information received from energy dispersive X-ray spectroscopy (EDS) and the known results on defect structure in indium doped ZnO ceramics, the planar defects in our nanobelts and nanoring may be related to a local segregation of In ions [21]. We now construct a model and carry out some detailed image simulations in reference to the experimental data. Figure 10.11b presents a nanobelt that has two planar defects with sharp contrast. If we consider the two dark layers (labeled **I** and **II** in the image) as two In-O octahedral layers, which was originally proposed by Yan et al. [22] and Li et al. [23], we can build a model as displayed in Figure 10.11c. If we define the (0001) surface is Zn-terminated and the $(000\bar{1})$ surface is oxygen-terminated, so that the polarization is along *c*-axis, the two slabs of ZnO on both sides of the In-O octahedral layer must have opposite polarization, which means that the In-O layer effectively induces a "head-to-head" polarization domain, so-called inversion domain boundary (IDB). To configure the two sharply contrasted planar defects (**I** and **II**), there must exist another type of defect (labeled as **III**) between the **I** and **II** layers, and it should correspond to a "tail-to-tail" IDB. The **III** layer does exit in the image, and the bright spots at its very adjacent forms a rectangle pattern, as indicated by a rectangle between the **I** and **II** layers. On the other hand, based on the structural model of In_2O_3, the two slabs of ZnO on either sides of In-O octahedral layer can take not only head-to-head polarity, but also tail-to-tail polarity. In the first case, the four-fold symmetry axis of the In-O octahedra lie in the ZnO *c*-plane to form a head-to-head IDB. In the second case, the four-fold symmetry axis of the In-O octahedron is parallel to the

c-axis to form a tail-to-tail IDB. Such tail-to-tail layer also exists above the **I** layer, as labeled to be the **IV** layer. In order to confirm our model, image simulation was carried out using dynamic electron diffraction theory [24]. A simulated image based on the model in Figure 10.11c is inserted in Figure 10.11b. A key fact presented by Figure 10.11 is that, the top and bottom surfaces of the nanobelt still preserve the negative and positive ionic charges, respectively, although with presence of the two inversion domain boundaries. Therefore, the polarity of the nanobelt building block is preserved!

Our data consistently show that the presence of planar defects is the key for the growth of polar nanobelts, and doping is a key factor for inducing planar defects [5,6,21]. The planar defects that are parallel to (0001) and are introduced by doping may lead to the fastest growth along directions parallel to the (0001) plane, and the most logic choices would be $[01\bar{1}0]$ or $[2\bar{1}\bar{1}0]$. Such a growth must lead to the formation of larger size polar surfaces and subsequently the nanorings and nanosprings. If the energy lowed by the planar defects is sufficient to overcome the energy increased by the polar surface, the nanobelt structure is likely to be stable.

10.4.7 Deformation-free nanohelix

As noticed in Figure 10.8, the nanosprings usually have a radius of 0.5–1.5 μm, which cannot be too small as constrained by the elastic deformation energy. We have recently found a nanohelical structure that has radius as small as ~50 nm (Figure 10.12a), much smaller than the nanospring as presented in Figure 10.8 [7]. In order to understand this structure, the intrinsic crystal structure of the nanohelixes has been investigated. TEM image presents the uniform shape and contrast of the nanohelix (Figure 10.12b). HRTEM imaging reveals that the nanohelix has an axial direction of [0001], although the growth direction of the nanowire changes periodically along the length. Detailed HRTEM images from the regions labeled c and d in Figure 10.12b are displayed in Figure 10.12c and d, respectively, which show that the nanowire that constructs the nanohelix grows along $[01\bar{1}1]$. Because the incident electron beam is parallel to $[2\bar{1}\bar{1}0]$, the two side surfaces of the nanowire are $\pm(01\bar{1}\bar{2})$. No dislocations were found in the nanohelixes. It is important to note that the image recorded from the "twist" point of the nanohelix shows no change in crystal lattice (Figure 10.12b), and the traces of the two sides are visible, indicating the non-twisted single-crystal structure of the entire nanohelix.

The nature of the $\pm\{10\bar{1}\bar{1}\}$ and $\pm(01\bar{1}\bar{2})$ planes can be understood from the atomic model of ZnO. By projecting the structure along $[1\bar{2}10]$, beside the most typical $\pm(0001)$ polar surfaces that are terminated with Zn and oxygen, respectively,

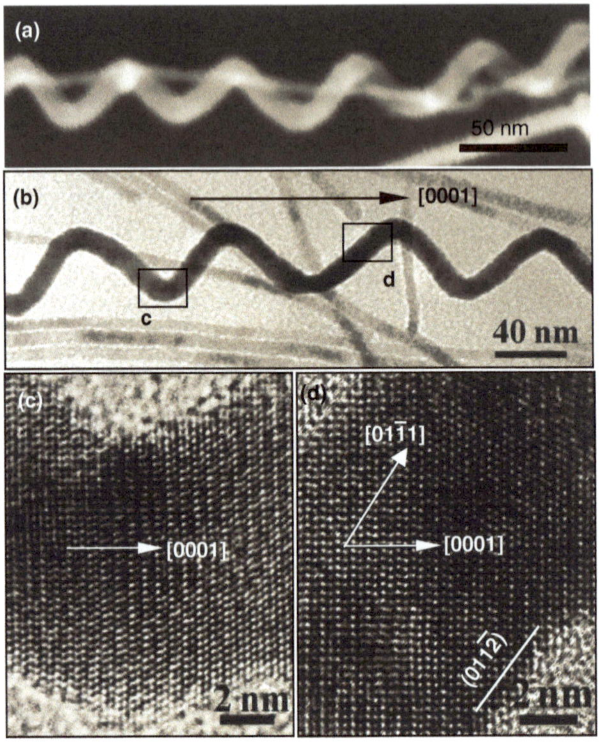

Figure 10.12: (a) SEM image of a left-handed nanohelix. (b) A bright-field TEM image of a nanohelix. No significant strain contrast is found (apart from the overlap effect between the nanohelix and nanowires). (c,d) HRTEM images recorded from the c and d areas labeled in (a), respectively, showing the growth direction, side surfaces and dislocation-free volume [7].

$\{10\bar{1}\bar{1}\}$ are also polar surfaces (see Figure 10.2b). From the structure information provided by Figure 10.13, the structure of the nanowire that self-coils to form the nanohelix can be constructed (Figure 10.13b). The nanowire grows along $[01\bar{1}1]$, the two end surfaces being $\pm(0001)$, side surfaces being non-polar $\pm(01\bar{1}\bar{2})$ (represented by yellow), Zn^{2+}-terminated $(1\bar{1}01)$ and $(\bar{1}011)$ (represented by red) and O^{2-}-terminated $(\bar{1}10\bar{1})$ and $(10\bar{1}\bar{1})$ (represented by green) surfaces (see Figure 10.13b,c).

The structure model presented in Figure 10.13a is the basic building block/segment for constructing the nanohelix via a self-coiling process during the growth. Because there are a totally of six crystallographically equivalent $<0\bar{1}11>$ directions: $[01\bar{1}1]$, $[\bar{1}101]$, $[\bar{1}011]$, $[0\bar{1}11]$, $[1\bar{1}01]$ and $[10\bar{1}1]$, and there is a 60°

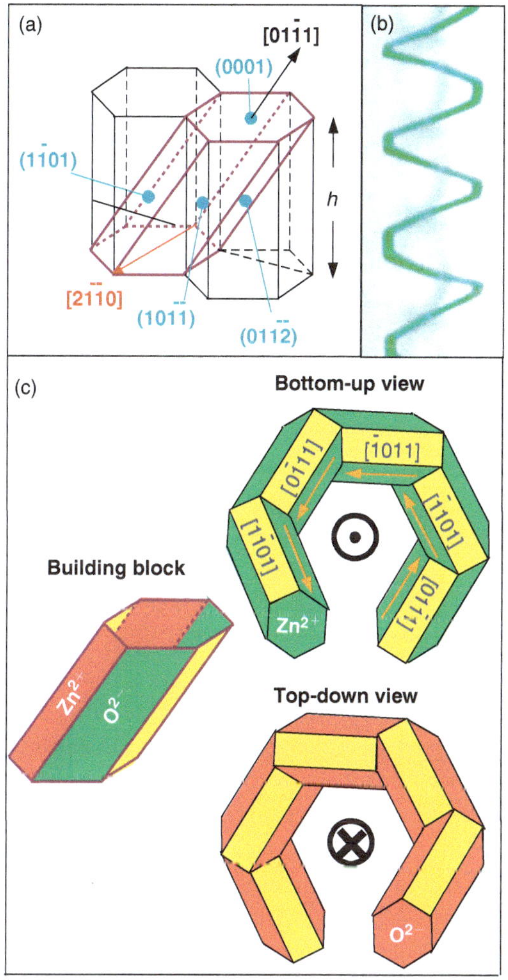

Figure 10.13: Structure of the deformation-free nanohelix. (a) The fundamental building block of the nanowire (in purple) for constructing the nanohelixes, and its growth direction and surfaces. (b) A model of the nanohelical structure. (c) The fundamental building block with Zn^{2+}-terminated polar surfaces (red) and O^{2-}-terminated surfaces (green) and the non-polar surfaces (yellow), bottom-up and top-down views of the nanohelical model. The distribution of charges on the surfaces of the nanohelical is analogous to the charge model of a DNA helix.

rotation between the two adjacent directions, thus, there are six equivalent orientations to stack the building block along the [0001] axial direction without introducing deformation or twist. A realistic three-dimensional model of the nanohelix is presented in Figure 10.11b, which is a stacking of the building blocks around the [0001] axis following the sequences of the six directions described above. The interface between the two building blocks is perfectly coherent and the same piece of crystal, without mismatch, translation or twist.

The distribution of the polar charges on the surfaces of the polar nanowire is analogous to the charge model of an alpha-helix model, and it is best seen through the top and bottom views of the model (Figure 10.13c). If viewing the nanohelix from bottom-up, the non-polar $(01\bar{1}2)$, Zn^{2+}-terminated $(1\bar{1}01)$ and $(\bar{1}011)$ are seen. The six growth directions of the building blocks are indicated. It is important to point out that there is no deformation introduced in the hexagonal screw-coiling stacking process, thus, no dislocations are needed to accommodate deformation. If viewing the nanohelix from top-down, the non-polar $(0\bar{1}12)$, and the O^{2-}-terminated $(\bar{1}10\bar{1})$ and $(10\bar{1}\bar{1})$ surfaces are seen. The Zn-terminated (0001) leads the nanowire growth due to self-catalysis [4].

The structure model presented in Figure 10.13 is supported by the energy calculation for the surface charge [7]. By folding a polar nanowire into a helix without introducing deformation, the change in total energy drops more than 15%.

10.4.8 Nanohelix of superlatticed nanobelt

The ZnO nanohelixes were grown with high reproducibility by the solid-vapor process [1] through a few steps in controlling growth kinetics. First, two grams of commercial ZnO powder was compacted and loaded into an alumina boat and positioned at the center of the alumina tube as the source material. Then the system was pre-pumped to $\sim 2 \times 10^{-2}$ mbar. The ramp rate was controlled at 20–25°C/min while the temperature was raised from room temperature to 800°C. The furnace was then held at 800°C for 20 minutes. The ramp rate was then maintained at 20°C/min from 800°C to 1400°C. Once the temperature reached 1000°C, argon was introduced as a carrier gas to raise the pressure from $\sim 2 \times 10^{-2}$ mbar to the desired synthesis pressure of 200–250 mbar within ~2.5 minutes. The solid-vapor deposition was carried out at 140°C for ~2 hours under a pressure of 200–250 mbar. The argon carrier gas was kept at a flow rate of 50 sccm. The as-grown nanohelixes of ZnO were deposited onto a polycrystalline Al_2O_3 substrate at a local temperature of 700–800°C.

The nanohelix is a self-coiled result of a superlattice-structured ZnO nanobelt [8]. Low magnification SEM images of the nanohelixes (Figure 10.14a) clearly present their crystalline structure and the winding shape of the nanobelt. A dark-field TEM image recorded from a segment of a nanobelt (Figure 10.14b) shows that the nanobelt has a periodic superlattice-structure across its entire width, with a uniform periodicity of ~3.5 nm. With a 5° offset, the stripes are nearly parallel to the nanobelt growth direction and run along the entire length of the nanohelix.

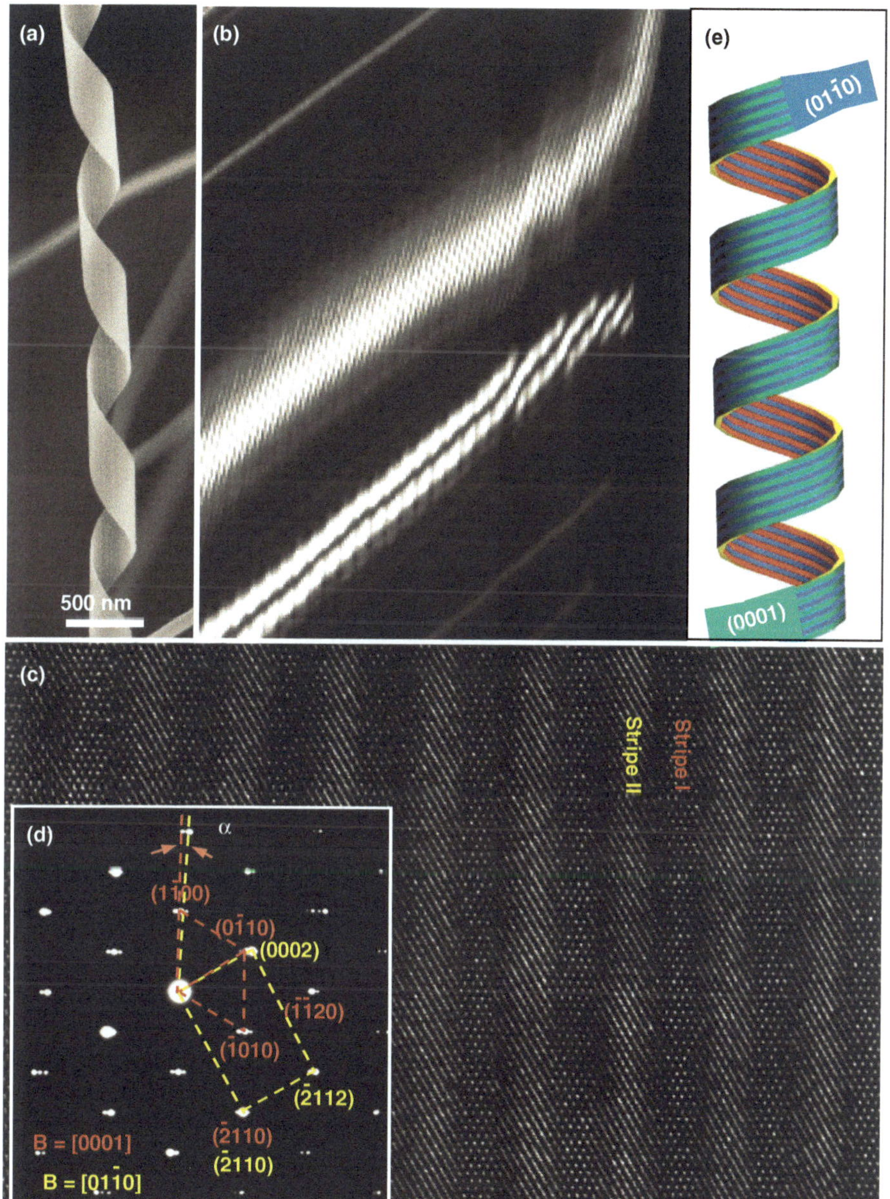

Figure 10.14: (a) Typical low-magnification SEM image of a ZnO nanohelix. (b) Dark-field TEM image from a segment of a nanohelix. (c) High-magnification TEM image and (d) the corresponding electron diffraction pattern of a ZnO nanohelix with the incident beam perpendicular to the surface of the nanobelt, respectively. (e) Initiation and formation process of a nanohelix [8].

HRTEM clearly displays the superlattice structure of the nanobelt with a periodic alternating arrangement of two types of stripes (Figure 10.14c), which are oriented epitaxially in the following orientation relationship: Stripe I: growth direction $[1\bar{1}00]$, top and bottom surfaces $\pm(0001)$ (polar surfaces); Stripe II: growth direction $[2\bar{1}\bar{1}2]$, top and bottom surfaces $\pm(01\bar{1}0)$ (non-polar surfaces). The corresponding electron diffraction pattern of the nanobelt is given in Figure 10.14d.

The initiation of the nanohelix is a direct result of superlattice formation and is driven by minimizing the surface area of the polar surfaces (Figure 10.14e). At the starting point, a stiff-nanoribbon is a single crystal object dominated by large c-plane surfaces. The superlattice is formed spontaneously and abruptly across the entire width of the nanoribbon. An important fact is that the width of the nanobelt increases after the structural transformation from a single-crystal, c-plane-dominated stiff-nanoribbon (in green) into a superlattice-structured, flexible nanobelt. The stripes may not be exactly parallel to the growth direction of the nanobelt, and a small in-plane rotation for an angle of α exists between the growth directions of the two adjacent stripes, which may induce the geometrical curvature required for rigidly twisting its shape. The polar charges remaining on the surfaces of Stripe I also help to bend the nanobelt into a ring structure especially with the decrease of the nanobelt thickness. Thus, the accumulation of rigid structural rotation/twist across the width of the superlattice nanobelt and a continuous growth of the stripes along the length of the nanobelt may result in the formation of a helical structure, which could be left-handed or right-handed. The winding of the nanobelt is terminated after fully transforming the partial-polar-surface-dominated into a non-polar-surface-dominated nanoribbon (in blue).

10.5 Properties, potential applications and novel devices

10.5.1 Luminescent property

ZnO exhibits a direct band-gap of 3.37 eV at room temperature with a large exciton energy of 60 meV. The strong exciton binding energy, which is much larger than that of GaN (25 meV), and the thermal energy at room temperature (26 meV) can ensure an efficient exciton emission at room temperature under low excitation energy. As a consequence, ZnO is recognized as a promising photonic material in the blue–UV region.

Single-crystalline ZnO nanowires have been synthesized using high-temperature VLS growth methods discussed above. Room temperature UV lasing in ZnO nanowires has been demonstrated [10]. PL spectra of the ZnO nanorods were measured with a fluorescence spectrophotometer using a Xe lamp with an excitation

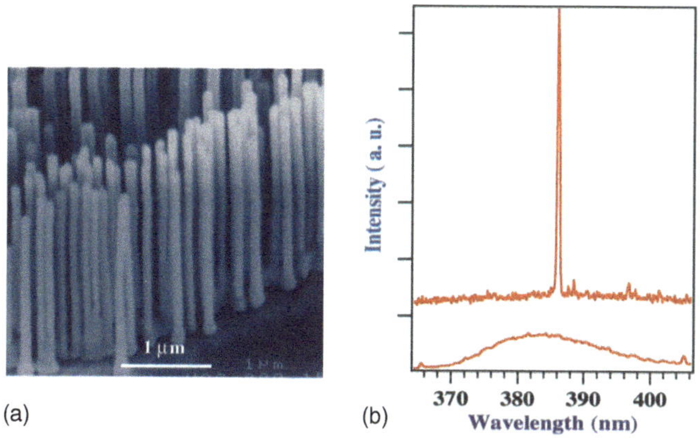

Figure 10.15: (a) An SEM image of aligned ZnO nanowire arrays grown on sapphire substrate coated with a 3-nm thick layer of Au. (b) Emission spectra from nanowire arrays at pump powers of 20 and 100 kW/cm^2. The spectra are offset for easy comparison. Stimulated emission from the nanowires was collected in the direction along the nanowire's end-plane normal (the symmetric axis) with a monochromator. All experiments were carried out at room temperature (Courtesy of Prof. P. D. Yang, UC Berkeley) [10].

wavelength of 325 nm at room temperature. Figure 10.15 shows the PL spectrum of the ZnO nanorods with a diameter in the range 60 ± 80 nm. Three emitting bands, including a strong UV emission at around 386 nm and a very weak blue band (440 ± 480 nm) as well as an almost negligible green band ($510 \perp 580$ nm), were observed. The UV emission must be contributing to the near band edge emission of the wide band-gap ZnO. It has been suggested that the green band emission corresponds to the singly ionized oxygen vacancy in ZnO and results from the recombination of a photogenerated hole with the singly ionized charge state of this defect. The stronger the intensity of the green luminescence, the more singly ionized oxygen vacancies there are. Thus, the almost negligible green band in Figure 10.15 shows that there is a very low concentration of oxygen vacancies in the well-aligned ZnO nanorods. The observation of blue band emission (440 ± 480 nm) of ZnO film has also been reported, using cathodoluminescence.

10.5.2 Field effect transistor

Field effect transistors (FETs) have been fabricated using individual nanobelts [25]. Large bundles of ZnO nanobelts were dispersed in ethanol by ultrasonication until mostly individual nanobelts were isolated. These ethanol dispersions were dried onto an SiO_2/Si substrate for imaging by non-contact mode atomic force microscopy

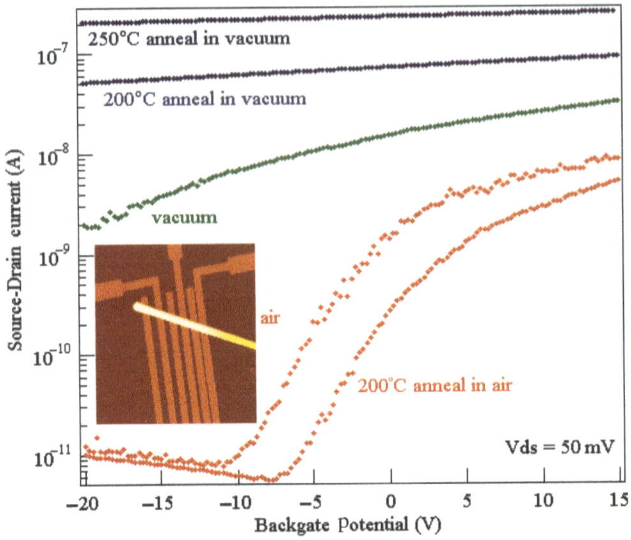

Figure 10.16: ZnO nanobelt FET. Source–drain current versus gate bias for an SnO_2 FET after various treatments measured in this order: air, vacuum, 200°C vacuum anneal, 250°C vacuum anneal, 200°C air anneal. The inset shows an FET device made using a single nanobelt [25].

(AFM). FETs were fabricated by depositing dispersed ZnO nanobelts on predefined gold electrode arrays. The SiO_2 gate dielectric thickness was 120 nm and the back gate electrode was fabricated by evaporation of gold on the Si (p^+) side of the substrate. In addition, in both fabrication schemes, the electrode arrays were variably spaced. They included electrode gaps as small as 100 nm and as large as 6 μm. An AFM image of the FET and a schematic diagram are shown in Figure 10.16. The principle of this device is that controlling the gate voltage would control the current flowing from the source to the drain.

By forming metal electrode/nanostructure electrical contacts and capacitively coupling the nanostructure to a nearby gate electrode, an FET is produced using a nanobelt that allows the exploration of new aspects of the physical and chemical properties of the nanostructures [26–28].

The alternative way of contacting the nanobelts by simply depositing them on top of pre-fabricated gold electrodes led to very resistive contacts. A typical ZnO FET showed a gate threshold voltage of -15 V, a switching ratio of nearly 100 and a peak conductivity of 1.25×10^{-3} Ω/cm. A completely analogous behavior has been observed in the case of carbon nanotubes deposited on top of Au electrodes or covered by Ti electrodes [41].

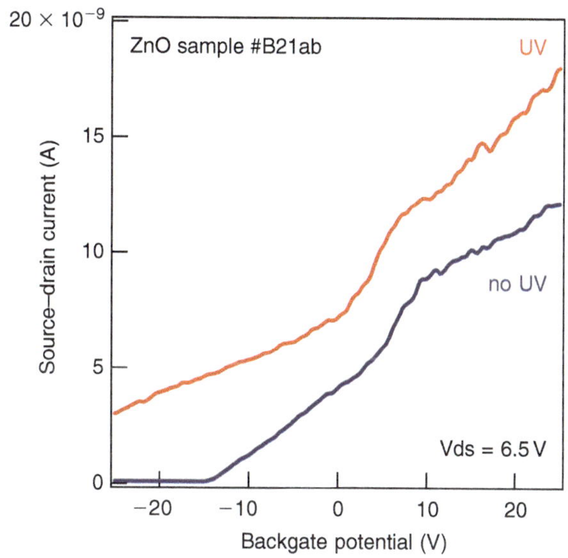

Figure 10.17: Source–drain current measured from a ZnO nanobelt FET, showing a strong dependence on UV illumination.

10.5.3 Photoconductivity

UV light irradiation of the nanobelt diode of ZnO in air is observed to result in a significant increase of the conductivity (Figure 10.17). Light with a wavelength of 350 nm ($E_\lambda = 3.54$ eV) was used, exceeding the direct band-gap of ZnO. The increase in the conductivity results from both photogeneration of electron hole pairs as well as doping by UV-light-induced surface desorption [29–31]. These processes could be observed by introducing a shutter between the light source and the ZnO nanobelt so that the flux of UV photons could be turned "ON" and "OFF".

The unique geometrical shape of nanobelts is ideal for field emission. MoO_3 nanobelts have been shown to exhibit superior performance. The work function at the tips of individual ZnO nanobelts has been measured by a novel technique [32].

10.5.4 Nanocantilever

The cantilever-based scanning probe microscopy (SPM) technique is one of the most powerful approaches in imaging, manipulating and measuring nanoscale properties and phenomena. The most conventional cantilever used for SPM is based on silicon, Si_3N_4 or SiC, which is fabricated by an e-beam or optical lithography technique and has, typically, dimensions of thickness ~100 nm, width ~5 μm and length

~50 μm. Utilization of nanowire- and nanotube-based cantilevers can have several advantages for SPM. Carbon nanotubes can be grown on the tip of a conventional cantilever and used for imaging surfaces with a large degree of abrupt variation in the surface morphology [33]. We have demonstrated the manipulation of nanobelts by AFM and its potential for use in nanocantilevers [34].

Combining microelectromechanical system (MEMS) technology with self-assembled nanobelts we are able to produce cost-effective cantilevers with much heightened sensitivity for a range of devices and applications. Force, pressure, mass, thermal, biological and chemical sensors are all prospective devices. Semiconducting nanobelts are ideal candidates for cantilever applications. Structurally they are defect free single crystals, providing excellent mechanical properties. The reduced dimensions of nanobelt cantilevers offer a significant increase in cantilever sensitivity. Combining the aforementioned techniques with micromanipulation has led to the horizontal alignment of individual ZnO nanobelts onto silicon chips. The aligned ZnO cantilevers, shown in Figure 10.18, were manipulated to have a range of lengths. This exemplifies our ability to tune the resonance frequency of each cantilever and thus modify cantilevers for different applications such as contact, non-contact and tapping mode AFM. Periodic contrast of the ZnO cantilevers is observed as a result of electronic charge induced vibrations during SEM operation. Such contrast is absent in regions where the nanobelts are in direct contact with the silicon substrate, suggesting adequate adhesion forces between the cantilevers and the silicon chip.

10.5.5 Piezoelectricity of the polar nanobelts

Piezoelectricity is due to the atomic scale polarization. To illustrate the piezoelectricity, one considers an atom with a positive charge that is surrounded tetrahedrally by anions (Figure 10.19a). The center of gravity of the negative charges is at the

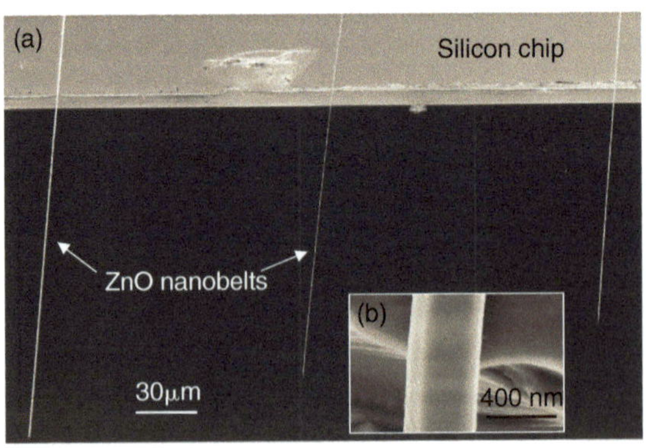

Figure 10.18: (a) Nanobelts as ultrasmall nanocantilever arrays aligned on a silicon chip. (b) An enlarged SEM image recorded from the nanobelt cantilever [34].

center of the tetrahedron. By exerting a pressure on the crystal along the cornering direction of the tetrahedron, the tetrahedron will experience a distortion and the center of gravity of the negative charges will no longer coincide with the position of the positive central atom, an electric dipole is generated. If all of the tetrahedra in the crystal have the same orientation or some other mutual orientation that does not allow for a cancellation among the dipoles, the crystal will have a macroscopic dipole. The two opposite faces of the crystal have opposite electric charges.

The piezoelectricity refers to a reverse process in which a contraction or elongation is created to the crystal once it is positioned in an electric field. Crystals can only be piezoelectric if they are non-central symmetric to ensure the non-compensation among the dipoles created by the tetrahedra. Piezoelectric effect can convert a mechanical vibration into an electric signal or vice versa. It is widely used in resonators, controlling tip movement in scanning probe microscopy, sensors for vibration waves in air and under sea, etc.

Piezoelectricity is difference from ferroelectricity or ferromagnetism. Ferroelectricity is originated from electric dipole moment induced by the spontaneous polarization of the crystal, and ferromagnetism is based on atomic magnetic moment induced by electron spin. Ferroelectricity and ferromagnetism have many common characteristics, such as domains and hysteresis loops. If the dipoles can be canceled each other by the randomly oriented ferroelectric domains, the material does not exhibit a macroscopic dipole. It is called paraelectric. If these dipole domains cannot be canceled out with each other, the residual dipoles will add up, forming a macroscopic dipole, which is the ferroelectricity. Piezoelectricity, however, is originated from the polarization of the tetrahedral coordinated unit and it is a crystal structure determined effect. In nanometer scale, ferroelectricity and ferromagnetism may

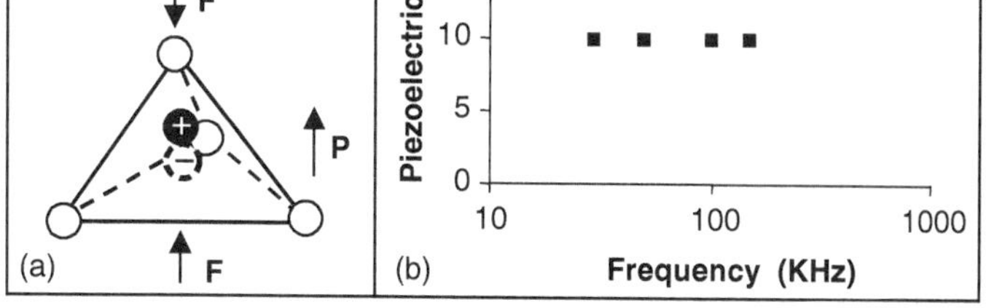

Figure 10.19: (a) Schematics showing piezoelectric effect in tetrahedrally coordinated cation-anion unit. (b) Experimentally measured piezoelectric coefficient d_{33} for ZnO and its comparison to that of the bulk (Courtesy of Dr. Scott Mao and Minhua Zhao) [35].

be greatly reduced or vanished, but piezoelectricity preserves with a possibility of enhanced performance due to the free boundary for volume expansion/contraction.

The piezoelectric coefficient of ZnO nanobelt has been measured by AFM using a conductive tip [35]. After coating (100) Si wafer with a 100-nm Pd, ZnO nanobelts were dispersed on the conductive surface. Then the whole surface was coated with another 5-nm Pd coating, which served as an electrode on ZnO nanobelt to get uniform electric field and avoid electrostatic effect. Extra care was taken to ensure that top and bottom surface of the nanobelt was not short circuited after Pd deposition. The ZnO nanobelt was located by a commercially available AFM under tapping mode. Piezoresponse force microscopy (PFM) is used to measure the effective piezoelectric coefficient (d_{33}) of individual (0001) surface dominated zinc oxide nanobelt lying on conductive surface. Based on references of bulk (0001) ZnO and x-cut quartz, effective piezoelectric coefficient d_{33} of ZnO nanobelt is found to be frequency dependent and varies from 14.3 to 26.7 pm/V (Figure 10.19b), which is much larger than that of the bulk (0001) ZnO of 9.93 pm/V. The results project the applications of ZnO nanobelts as nanosensors and nanoactuators.

10.5.6 Quantum confinement

For investigation of quantum confine effects, nanobelts of small sizes are required. We have recently grown ultrasmall nanobelts with the VLS growth using a novel catalyst [36]. Instead of using dispersive nanoparticles as the catalyst for seeded growth, a uniform thin film (~10 nm) of tin was coated on the silicon substrate. The ZnO nanobelts produced by the tin film catalyst are rather narrow, thin and uniform (Figure 10.20a). Electron diffraction pattern and HRTEM image show that the nanobelts grow along [0001], its top surfaces are $(2\bar{1}\bar{1}0)$ and the side surfaces are $(0\bar{1}10)$. The average diameter of the nanobelts is 5.5 nm with a standard deviation of ±1.5 nm, indicating a very good size uniformity.

To examine the size-induced quantum effect of the ultra thin ZnO nanobelts, photoluminescence (PL) measurements were performed at room temperature using a Xe lamp with an excitation wavelength of 330 nm (Figure 10.20b). In comparison to the PL from nanobelts of an average width of ~200 nm, the 6 nm nanobelts have a 14 nm shift in the emission peak, which possibly indicate quantum confinement due to the reduced size of the nanobelts.

10.5.7 Piezoelectric nanogenerators

Developing novel technologies for wireless nanodevices and nanosystems is of critical importance for *in-situ*, real-time and implantable biosensing, biomedical

(a)

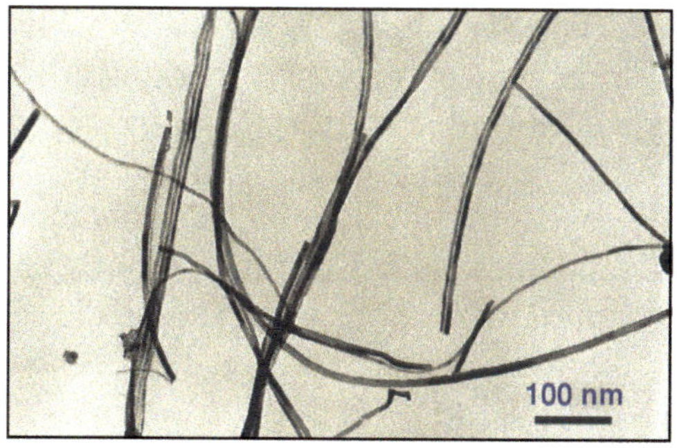

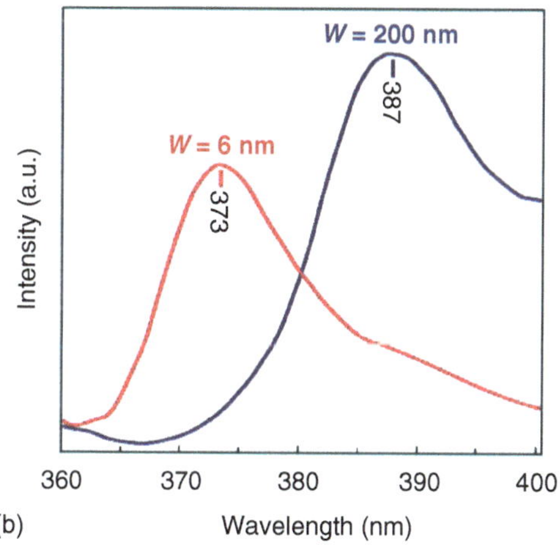

(b)

Figure 10.20: Ultra-narrow ZnO nanobelts. (a) TEM image of a ZnO nanobelts grown using tin thin film as the catalyst. (b) PL spectra recorded from the wide nanobelts ($W = 200$ nm) and the narrow nanobelts ($W = 6$ nm), showing drastic blue shift in the emission peak due to size effect.

monitoring and biodetection. An implanted wireless biosensor requires a power source, which may be provided directly or indirectly by charging of a battery. It is highly desired for wireless devices and even required for implanted biomedical devices to be self-powered without using a battery. Therefore, it is essential to explore innovative nanotechnologies for converting mechanical energy (such as body movement, muscle stretching), vibration energy (such as an acoustic/ ultrasonic wave), and hydraulic energy (such as body fluid and blood flow) into electric energy that will be used to power nanodevices without using batteries. It also has a huge impact to miniaturizing the size of the integrated nanosystems by reducing the size of the power generator and improving its efficiency and power density.

We have demonstrated an innovative approach for converting nano-scale mechanical energy into electric energy by piezoelectric zinc oxide nanowire (NW) arrays [38]. By deflecting the aligned NWs using a conductive atomic force microscopy (AFM) tip in contact mode, the energy that was first created by the deflection force and later converted into electricity by piezoelectric effect has been measured for demonstrating a nano-scale power generator. The operation mechanism of the electric generator relies on the unique coupling of piezoelectric and semiconducting dual properties of ZnO as well as the elegant rectifying function of the Schottky barrier formed between the metal tip and the NW. Each nanowire can output 10–20 mV of electricity, and the efficiency of the NW-based piezoelectric power generator is $\sim$17–30% [38].

10.6 Conclusion remark

Among the functional oxides with peroskite, rutile, CaF_2, spinel and wurtzite structures [37], ZnO is unique because it exhibits the semiconducting and piezoelectric dual properties. Structurally, ZnO is a material that has diverse structures, whose configurations are much more rich than any of the known nanomaterials including carbon nanotubes. Using a solid-state thermal sublimation process, and by controlling the growth kinetics, local growth temperature and the chemical composition of the source materials, a wide range of nanostructures of ZnO have been synthesized in our laboratory (Figures 10.21 and 10.22). The striking new structures reviewed in this article are those induced by the spontaneous polarization. These are ideal objects for understanding piezoelectricity and polarization-induced phenomena at nano-scale, building nanodevices, actuators and sensors. The piezoelectricity offers the potential of fabricating electro-mechanical coupled devices, pumps and switches for biofluid, and piezoelectric-based resonators. It can be predicted that these novel structures will open many new applications in nanotechnology. Future research will focus on the following issues:

Fundamental understanding about the growth mechanisms. Although we have shown some success in synthesis and control over the received nanostructure, understanding the initial nucleation and growth of the nanostructures remains a key challenge.

Fundamental understanding on growth kinetics. We have elaborated the key role played by growth kinetics in the growth of 1D nanostructures. Growth of 1D nanostructure is a thermodynamically non-equilibrium process and it is controlled by kinetics, but growth kinetics is a rather complex process. A solid understanding on growth kinetics is essential for controlling the growth process.

Structurally, morphologically and dimensionally controlled synthesis. For large-scale integration, controlling the surface structure of nanostructures would have equivalent importance as controlling the helical angle of carbon nanotubes, which

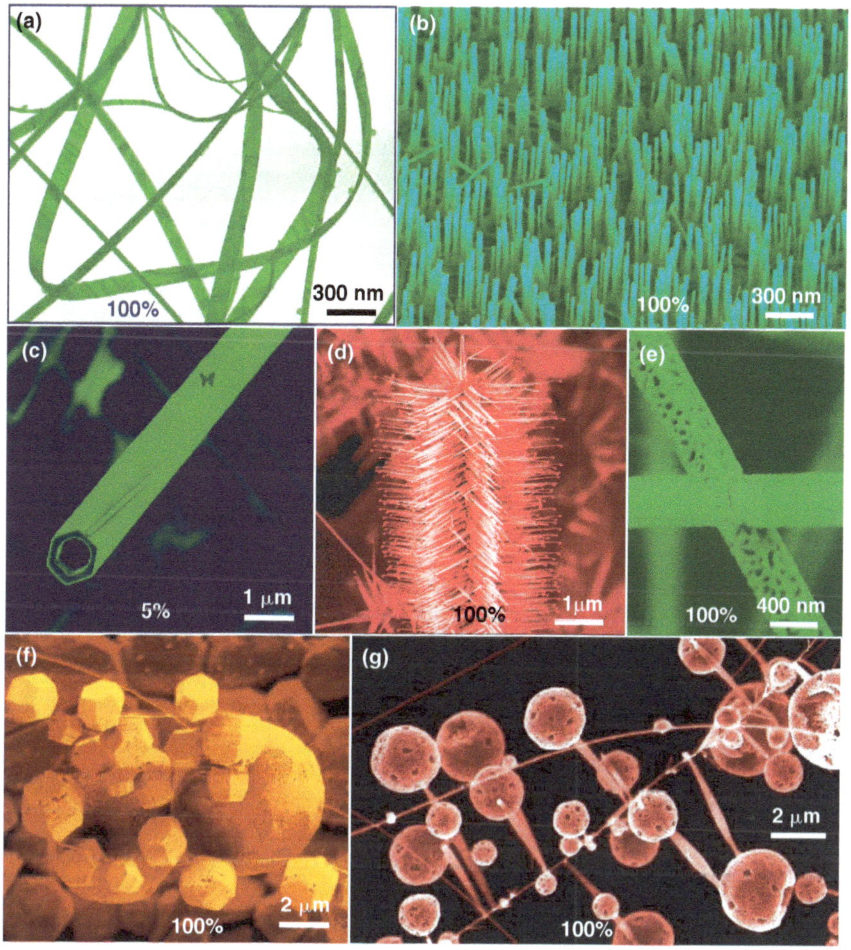

Figure 10.21: A collection of novel nanostructures of ZnO dominated by non-polar surfaces, synthesized under controlled conditions by thermal evaporation of solid powders by controlling source materials, growth temperature, temperature gradient and substrate. (a) Nanobelt; (b) aligned nanowire arrays; (c) nanotubes; (d) array of propellers; (e) mesoporous nanowires; (f) cages and shell structures and (g) hierarchical shell and propeller structure. The percentage in each Figure indicates the purity of the as-synthesized sample for the specific nanostructure in a specific local temperature region.

determines the semiconductor or metallic behavior of the nanotube. Property control is possible only if structural control is achieved. Techniques are required to grow the designed structure with superior controllability in size, size distribution, shape, crystal structure, defect distribution and even surface structure.

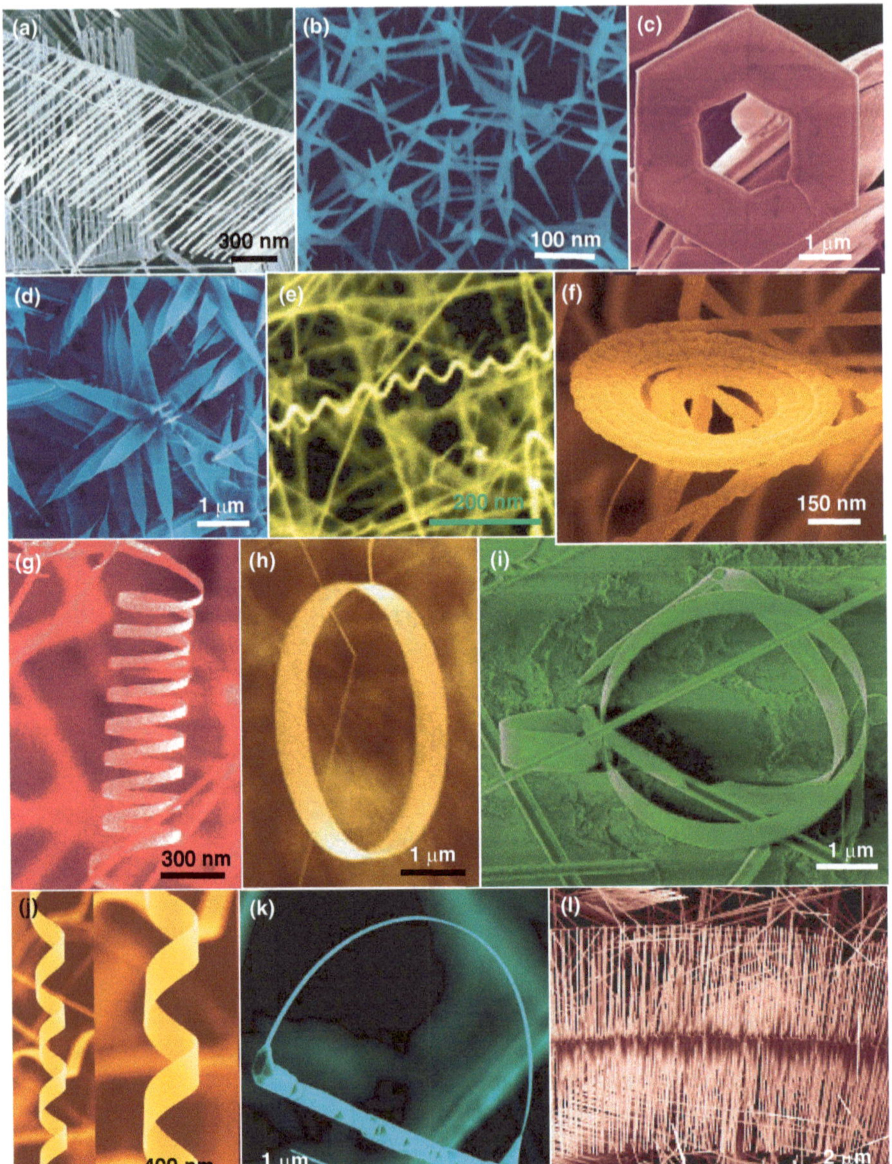

Figure 10.22: A collection of polar-surface-induced/dominated nanostructures of ZnO, synthesized under controlled conditions by thermal evaporation of solid powders unless notify otherwise: (a) nanocombs induced by asymmetric growth on the Zn-(0001) surface; (b) tetraleg structure due to catalytically active Zn-(0001) surfaces; (c) hexagonal disks/rings synthesized by solution-based chemical synthesis; (d) nanopropellers created by fast growth along <01̄10> and *c*-axis; (e) deformation-free nanohelixes as a result of

(Continued)

Large-scale patterned and designed/targeted growth as well as self-organization. Future applications and nanomanufacturing will strongly rely on designed growth and self-assembly technology. We would like to control the location, the number and orientation of the grown nanostructure. This is an essential step in integrating nanostructures with existing technologies.

Modeling the growth process. Theoretical simulation and modeling about surface process and growth kinetics in 1D nanostructure growth are important for understanding the growth process.

Acknowledgement

The results reviewed in this paper were partially contributed from our group members and collaborators: Puxian Gao, Yong Ding, Xiang Yang Kong, Xuedong Bai, Carsten Ronning, Zhengwei Pan, William Hughes, Daniel Moore, Chris Ma, Xudong Wang, Jinhui Song, Wenjie Mai, Changshi Lao, Yolande Berta, C. Yu, Li Shi, M.S. Arnold, Ph. Avouris, E. Comini, G. Faglia, G. Sberveglieri, Enge Wang, Ying Dai, Yue Zhang, Minhua Zhao and Scott X. Mao, to whom we are very grateful. Research supported by NSF, NASA and DARPA.

References

[1] Z. W. Pan, Z. R. Dai and Z. L. Wang, *Science* **291**, 1947 (2001).
[2] P. X. Gao, Y. Ding and Z. L. Wang, *Nano Lett.* **3**, 1315 (2003).
[3] P. X. Gao and Z. L. Wang, *J. Am. Chem. Soc.* **125**, 11299 (2003).
[4] Z. L. Wang, X. Y. Kong and J. M. Zuo, *Phys. Rev. Lett.* **91**, 185502 (2003).
[5] X. Y. Kong and Z. L. Wang, *Nano Lett.* **3**, 1625 (2003).
[6] X. Y. Kong, Y. Ding, R. S. Yang and Z. L. Wang, *Science* **303**, 1348 (2004).

block-by-block self-assembly; (f) spiral of a nanobelt with increased thickness along the length; (g) nanosprings; (h) single-crystal seamless nanoring formed by loop-by-loop coiling of a polar nanobelt; (i) a nanoarchitecture composed of a nanorod, nanobow and nanoring. The percentage in each figure indicates the purity of the as-synthesized sample for the specific nanostructure in a specific local temperature region; (j) double-sided nanocombs formed with the presence of an inversion domain boundary parallel to the center stem, and the symmetric teeth are the result of self-catalysis of Zn-termined (0001) and (k) nanobow structure produced by a change in growth direction from *c*-axis to *a*-axis. (l) Rigid helix of ZnO formed by superlattice-structured nanobelt for reducing the surface energy of the polar surfaces.

[7] R. S. Yang, Y. Ding and Z. L. Wang, *Nano Lett.* **4**, 1309 (2004).
[8] P. X. Gao, Y. Ding, W. Mai, W. L. Hughes, C. S. Lao and Z. L. Wang, *Science* **309**, 1700 (2005).
[9] P. X. Gao and Z. L. Wang, *Small* **1**, 945–949 (2005).
[10] M. H. Huang, Y. Y. Wu, H. Feick, N. Tran, E. Weber and P. D. Yang, *Adv. Mater.* **13**, 113 (2001).
[11] R. S. Wagner and W. C. Ellis, *Appl. Phys. Lett.* **4**, 89 (1964).
[12] J. Westwater, D. P. Gosain, S. Tomiya, S. Usui and H. Ruda, *J. Vac. Sci. Technol. B* **15**, 554 (1997).
[13] A. M. Morales and C. M. Lieber, *Science* **279**, 208 (1998).
[14] P. X. Gao and Z. L. Wang, *J. Phys. Chem. B* **106**, 12653 (2002).
[15] P. X. Gao and Z. L. Wang, *J. Am. Chem. Soc.* **125**, 11299 (2003).
[16] J. Y. Lao, J. Y. Huang, D. Z. Wang and Z. F. Ren, *Nano Lett.* **3**, 235 (2003).
[17] F. Vigue, P. Vennegues, S. Vezian, M. Laugt and J.-P. Faurie, *Appl. Phys. Lett.* **79**, 194 (2001).
[18] D. Moore, C. Ronning, C. Ma and Z. L. Wang, *Chem. Phys. Lett.* **385**, 8 (2004).
[19] C. Ma, Y. Ding, D. Moore, X. D. Wang and Z. L. Wang, *J. Am. Chem. Soc.* **126**, 708 (2004).
[20] X. Y. Kong and Z. L. Wang, *Appl. Phys. Lett.* **84**, 975 (2004).
[21] Y. Ding, X. Y. Kong and Z. L. Wang, *Phys. Rev. B* **70**, 235408 (2004).
[22] Y. Yan, S. J. Pennycook, J. Dai, R. P. H. Chang, A. Wang and T. J. Marks, *Appl. Phys. Lett.* **73**, 2585 (1998).
[23] C. Li, Y. Bando, M. Nakamura and N. Kimizuka, *Micron* **31**, 543 (2000).
[24] J. M. Cowley and A. Moodie, *Acta Cryst.* **10**, 609 (1957).
[25] M. S. Arnold, Ph. Avouris, Z. W. Pan and Z. L. Wang, *J. Phys. Chem. B* **107**, 659 (2003).
[26] Y. Cui and C. M. Lieber, *Science* **291**, 851 (2001).
[27] P. G. Collins, M. S. Arnold and Ph. Avouris, *Science* **292**, 706 (2001).
[28] J. Kong, N. Franklin, C. Wu, S. Pan, K. J. Cho and H. Dai, *Science* **287**, 622 (2000).
[29] D. H. Zhang, *Mater. Chem. Phys.* **45**, 248 (1996).
[30] P. Bonasewicz, W. Hirschwald and G. Neumann, *J. Electrochem. Soc.* **133**, 2270 (1986).
[31] Y. Shapira, S. M. Cox and D. Lichtma, *Surf. Sci.* **54**, 43 (1976).
[32] X. Bai, E. G. Wang, P. Gao and Z. L. Wang, *Nano. Lett.* **3**, 1147 (2003).
[33] H. J. Dai, J. H. Hafner, A. G. Rinzler, D. T. Colbert and R. E. Smalley, *Nature* **384**, 147 (1996).
[34] W. Hughes and Z. L. Wang, *Appl. Phys. Lett.* **82**, 2886 (2003).
[35] M. H. Zhao, Z. L. Wang and S. X. Mao, *Nano Lett.* **4**, 587 (2004).
[36] X. D. Wang, Y. Ding, C. J. Summers and Z. L. Wang, *J. Phys. Chem. B* **108**, 8773 (2004).
[37] Z. L. Wang and Z. C. Kang, *Functional and Smart Materials – Structural Evolution and Structure Analysis*, Plenum Publishing Co., New York (1998).
[38] Z. L. Wang and J. H. Song, *Science* **312**, 242 (2006).

Chapter 11

ZnO/ZnMgO Heterojunction FETs

Mitsuaki Yano, Kazuto Koike, Shigehiko Sasa and Masataka Inoue

New Materials Research Center, Osaka Institute of Technology,
Asahi-ku Ohmiya, Osaka 535-8585, Japan

Abstract: Molecular beam epitaxial (MBE) growth techniques of ZnO, $Zn_{1-x}Mg_xO$, and $Zn_{1-x}Mg_xO$/ZnO on sapphire substrates toward the application to heterojunction field-effect transistors (HFETs) are discussed by focusing on the control of alloy composition of the $Zn_{1-x}Mg_xO$ at high Mg contents, the properties of layered structures, the internal electric field by polarization, and the performance of a prototype HFET. The MBE growth was conducted using metal sources of Zn, and Mg in Knudsen cells, and a gas source of O* (radical oxygen) from a plasma cell. Each epitaxial film was grown on *a*-plane sapphire substrates via a ZnO initiation layer to enhance uniform nucleation. The grown films were *c*-axis oriented O-polarity single-crystals and free from in-plane rotational domains.

The conduction of undoped ZnO films was *n*-type with a typical electron mobility of $\sim 100\,cm^2/Vs$ and a density of $\sim 3 \times 10^{17}\,cm^{-3}$ at room temperature, although there was a low quality degenerated region nearby the substrate. The doping of Mg atoms into ZnO films was performed to obtain a heterojunction, the Mg content of which was successfully controlled by using a semi-empirical calculation based on a simplified growth mechanism. In spite of the large miscibility gap between ZnO and MgO, homogeneous wurtzite $Zn_{1-x}Mg_xO$ films at high Mg contents ($x \leq 0.45$) were obtained using the growth at low temperatures on the ZnO initiation layer. The band gap energy of these $Zn_{1-x}Mg_xO$ films at room temperature was found to increase in accordance with $3.37 + 2.51x$ eV as a function of Mg content x.

By using the increased band-gap energy, several types of $Zn_{1-x}Mg_xO$/ZnO heterostructures were fabricated to study carrier confinement effect. This study on heterostructures revealed that the carrier confinement was dependent on the polarization-induced internal electric field in addition to the depth and width of the potential well. The polarization in $Zn_{1-x}Mg_xO$/ZnO heterostructure was analyzed, and a probable model to understand the experimental data was proposed.

Consequently, the growth technique was applied to fabricate a prototype HFET based on a single-quantum well (SQW) as the conduction channel where a two-dimensional electron gas was formed at the interface due to the internal electric field. A large field-effect mobility of about $140\,cm^2/Vs$ was obtained for the HFET at room temperature in consistent with the expectation from the data obtained by Hall measurement on SQW samples. Although the device performance stays at a preliminary stage yet, the fabrication

Zinc Oxide Bulk, Thin Films and Nanostructures
C. Jagadish and S. Pearton (Editors)

techniques and material parameters presented in this study will be useful to leap up ZnO-based devices toward the next generation.

Keywords: ZnO, ZnMgO, molecular beam epitaxy, control of alloy composition, miscibility gap, heterojunction, pseudomorphic growth, quantum confined stark shift, piezoelectric polarization, spontaneous polarization, two-dimensional electron gas, heterojunction field effect transistor.

11.1 Introduction

The complementary semiconductors to Si in the next generation, the next of GaAs, are requested to be ecological and economical. Zinc Oxide (ZnO) is one of the important candidates because: (1) it has a long history in the use of cosmetics and antiphlogistics, (2) both oxygen and zinc are ubiquitous elements on earth as shown by Fig. 11.1 [1], (3) zinc is a fifteenth rich element in human body, ~33 mg/kg [2], and zinc has an established position as a mass-produced nonferrous metal in the world. As a semiconductor material, the wide and direct band-gap of ZnO is of great advantage to short wavelength optoelectronic devices [3–5]. Its high electron saturation-velocity [6] shown in Fig. 11.2 is also promising for electronic devices. In addition, ZnO has superiority in high temperature endurance in air as a nature of oxides. Hence we can expect it to be advantageous to the transistors for high-temperature and high-power operation.

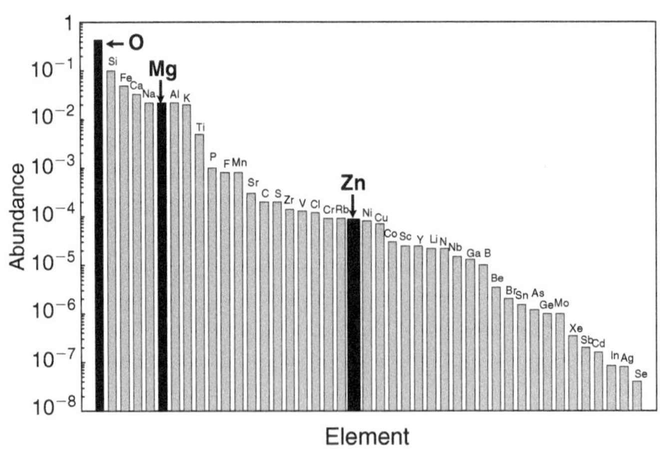

Figure 11.1: Abundance of major elements in the earth.

Although some of these strong points overlap with another wide band-gap semiconductor GaN, particular advantages of ZnO include bulk growth capability [7,8,9], amenability to conventional chemical etching [10,11], low growth temperature [12], high radiation hardness [13,14], and large exciton binding energy [15]. Accordingly, much effort has been devoted first to develop such a transparent electronic circuitry on glass substrates as composed of performance ZnO-based thin film transistors (TFTs) wired by conductive polycrystalline ZnO-based films for instance [16].

$Zn_{1-x}Mg_xO$/ZnO heterojunction field-effect transistors (HFETs), however, are a subject of unexplored application in spite of the capability to achieve two-dimensional electron gas (2DEG); we demonstrated previously [17] the formation of 2DEG in a sheet carrier density of $\sim 10^{13}$ cm^{-2} at $Zn_{0.6}Mg_{0.4}O$/ZnO heterointerface with no intentional doping. In this chapter, we review the state of the art of $Zn_{1-x}Mg_xO$/ZnO heterosystem and discuss on the preliminary experimental results of a $Zn_{0.7}Mg_{0.3}O$/ZnO HFET to study potential applications to the future development. The contents in this chapter will be rolled out as follows:

(1) Introduction (this section).
(2) Molecular beam epitaxial (MBE) growth and characterization of ZnO films on *a*-plane sapphire substrates.
(3) Extension of the MBE growth technique from ZnO to $Zn_{1-x}Mg_xO$.
(4) Structural, optical, and electrical properties of $Zn_{1-x}Mg_xO$/ZnO heterostructures.
(5) Performance of a prototype $Zn_{1-x}Mg_xO$/ZnO HFET.

Our data used in this chapter are the latest ones as of the end of August, 2005, and contain many unpublished original data especially in the Sections 11.3 and 11.4.

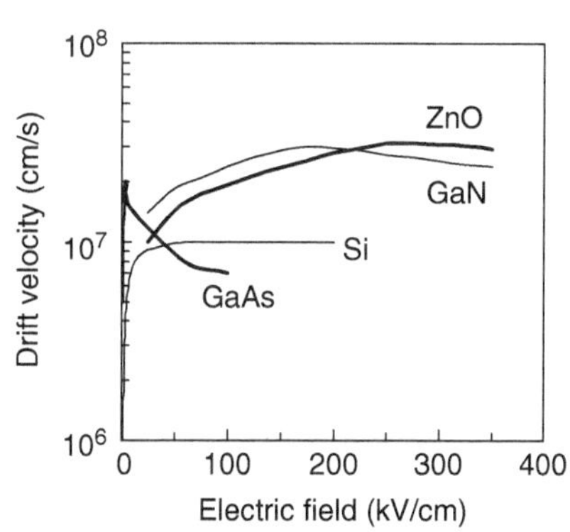

Figure 11.2: Electron saturation-velocities of major semiconductors and ZnO as a function of electric field.

Therefore, we did our best for the interpretation of these data but apologize in advance for any misunderstanding of confusion this article might contain.

11.2 MBE growth and characterization of ZnO films on *a*-plane sapphire substrates

Although such lattice matched bulk crystals as ZnO and $ScAlMgO_4$ [5] are available to the substrates for ZnO epitaxial growth, we used in this experiment sapphire plates by considering the wide distribution of mass-produced high-quality crystals. Sapphire as substrates has advantages of the stability at high temperatures suitable for epitaxial growth and the electrical insulation suitable for measuring the electrical properties of epitaxial films. The orientation of sapphire substrates used here was *a*-plane $(11\bar{2}0)$ although *c*-plane (0001) was another candidate for ZnO growth. Indeed there are many reports on the ZnO growth on *c*-plane sapphire substrates [4,18,19] and we also confirmed that both *a*- and *c*-plane substrates were available to grow single-crystalline *c*-axis oriented ZnO films, but the use of *a*-plane substrates was found to have an advantage in obtaining single-crystalline ZnO films in a wide range of growth condition; the ZnO growth on *c*-plane substrates sometimes resulted in the inclusion of 30° in-plane rotational domains in the *c*-axis oriented films.

The easy inclusion of rotational domains is related to the large mismatch of about ~18% between the *a*-axis lattice constants of sapphire and ZnO, although both materials have the same six-fold symmetry in *c*-axis orientation as shown by the material properties in Table 11.1 [20–27]. The *c*-axis lattice constant of sapphire substrate, on the other hand, closely coincides with four times of the *a*-axis lattice constant of ZnO, suggesting a successful epitaxial relationship in the $ZnO[11\bar{2}0]//Al_2O_3[0001]$ configuration. The smaller thermal mismatch in this direction will also be favorable for the epitaxial growth. This superiority in *c*-axis oriented ZnO growth on *a*-plane substrate to that on *c*-plane substrate has been pointed out and demonstrated by Fons et al. [28] for the first time, which was supported by the recent work by Coleman et al. [29] on the difference in mechanical properties of the ZnO films grown on these two kinds of substrates. We followed these works and used *a*-plane sapphire substrates for the ZnO growth in this study.

The ZnO growth was performed as shown in Fig. 11.3 by using an EpiQuest-made plasma-assisted MBE system equipped with Knudsen cells (K-cells) for Zn, Mg, and a plasma cell for O* (radical oxygen) operated at 350 W. A set of planar electrodes was installed in the plasma cell to trap ions electrostatically, to which electrodes we applied 200 V during the growth. The purity of these sources was 99.99999% (7*N*) for Zn, 4*N* for Mg, and 6*N* for O_2. The growth chamber was evacuated by

Table 11.1: Material properties of ZnO, MgO and sapphire.

	ZnO	MgO	α-Al_2O_3
Crystal structure	Wurtzite	Rocksalt	Rhombohedral
Band-gap energy (eV)	3.44@1.6 K[20]	7.83@85 K[23]	–
Lattice constant (nm)	c: 0.5204[21]		c: 1.299[26]
	a: 0.3250[21]	0.4216[24]	a: 0.476[26]
Thermal expansion	$\alpha_{\parallel c}$: 3.0×10^{-6} [22]		$\alpha_{\parallel c}$: 6.6×10^{-6} [27]
coefficient (K^{-1})	$\alpha_{\perp c}$: 6.5×10^{-6} [22]	10.5×10^{-6} [25]	$\alpha_{\perp c}$: 5.6×10^{-6} [27]

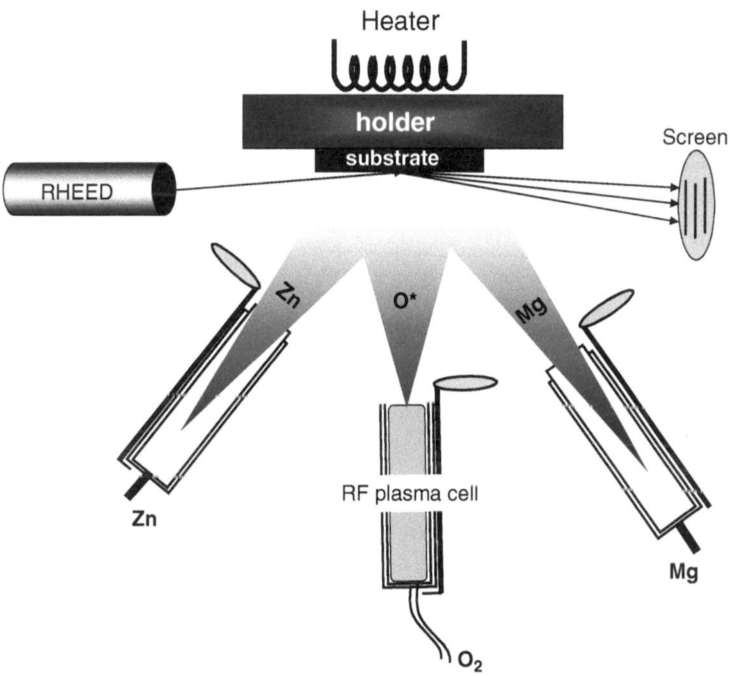

Figure 11.3: Schematic of the plasma-assisted MBE system used in this study.

the combination of a cryogenic pump and a turbo molecular pump to a background pressure of less than 1×10^{-7} Pa. The ZnO growth was conducted under a typical flow rate of 0.40 sccm oxygen gas with a 2.0×10^{-4} Pa beam equivalent pressure (BEP) of Zn, resulting in a working pressure of less than 1.0×10^{-4} Pa. The growth rate under the condition was determined by the oxygen flow rate, in other words, the ZnO growth in this study was conducted under a Zn-excess condition. The growth

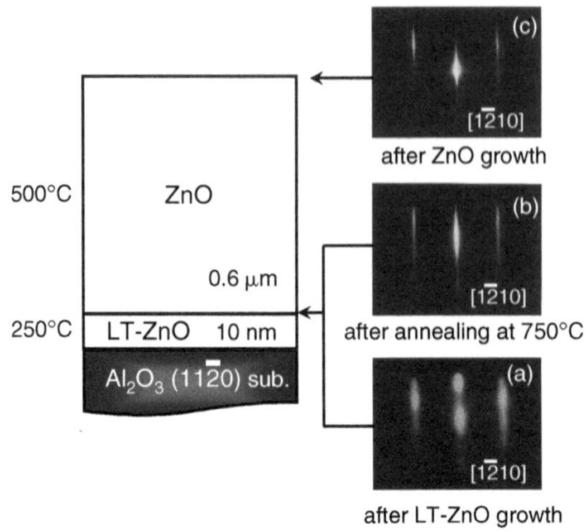

Figure 11.4: Schematic cross-section of a ZnO film grown on *a*-plane sapphire substrate, and the corresponding RHEED patterns observed at respective growth stages.

temperature was measured using a thermocouple and a pyrometer that were calibrated by the melting points of InSb and Al blocks (2 mm cube) pasted on sapphire substrates.

We used double-side polished *a*-plane sapphire substrates which were chemically etched by a solution of $H_2SO_4 : H_3PO_4 = 3 : 1$ at 160°C and then rinsed in de-ionized water. This wet etching was effective to improve surface morphology; the measurement by atomic force microscopy (AFM) revealed that the root-mean-square (rms) roughness of the substrate surface in $3 \times 3\ \mu m^2$ area was reduced from 0.54 nm to 0.13 nm. After loading into the MBE growth chamber, the substrate was heated at 750°C for 10 min and then irradiated by oxygen radicals at 500°C for 10 min to terminate the sapphire surface by O atoms. Then, we grew a 10-nm-thick ZnO initiation layer at a low growth temperature of 250°C with a low growth rate of 200 nm/h. This low-temperature ZnO (LT-ZnO) growth was followed by an *in-situ* annealing at 750°C under a Zn-beam irradiation to improve the crystalline quality and to flatten the surface on an atomic scale. Successively, a 600-nm-thick ZnO film was grown at 500°C with a rate of 400 nm/h on the annealed LT-ZnO initiation layer. The growth rates mentioned here were determined by measuring the film thickness after 1 h growth using a stylus method (α-STEP, TENCOR Instruments).

Figure 11.4 shows the schematic sample structure and the corresponding reflection high-energy electron diffraction (RHEED) patterns to the respective layers during growth. Just after starting the LT-ZnO growth, the streaky but faint and

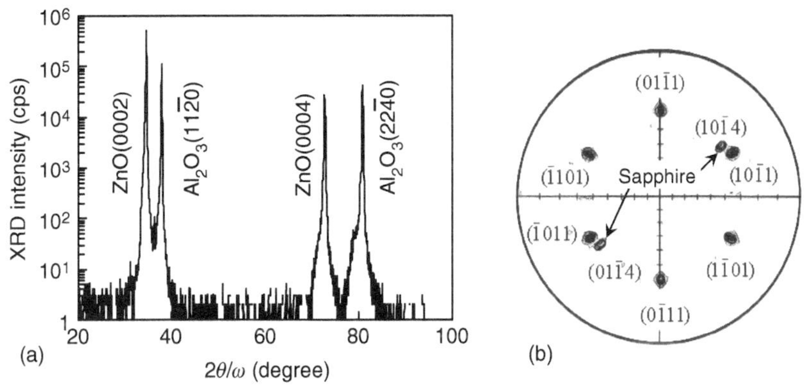

Figure 11.5: XRD patterns in $2\theta/\omega$ profile (a) and pole-figure (b) configurations from a ZnO film grown on *a*-plane sapphire substrate.

broad RHEED pattern from sapphire substrate immediately changed to a spotty and diffused one with a corresponding rod-spacing to relaxed ZnO. This RHEED pattern continued to the end of the LT-ZnO growth as shown in Fig. 11.4a which had six-fold symmetry on substrate rotation, suggesting a single-crystalline growth in the *c*-axis orientation. After the 750°C annealing, this spotty pattern changed to a streaky one as shown in Fig. 11.4b due to a progression of solid phase epitaxy. This streaky pattern was succeeded to the main ZnO growth with increasing the reflection intensity. The surface reconstruction was (1×1) as shown in Fig. 11.4c, indicating the characteristic of O-polarity $(-c)$ ZnO growth with O-termination [30]. This appearance of O-terminated surface under Zn-excess growth condition is not clearly understood yet, but it changed to (3×3) by closing the oxygen radicals during growth. This (3×3) reconstruction is reported to be a characteristic of O-polarity ZnO growth with Zn-termination [30,31]. In addition, we also confirmed that our ZnO films are easily etched by diluted acid solutions such as 0.1% HNO_3 and 0.1% HCl. This high resolvability to acid solutions is another characteristic of O-polarity ZnO surface [11]. After the growth, the rms roughness of the ZnO surface in $3 \times 3\ \mu m^2$ area was measured to be less than 0.5 nm by AFM.

In Fig. 11.5, we show typical x-ray diffraction (XRD) patterns in the $2\theta/\omega$ equator scan (a) and the polar scan (b). In the $2\theta/\omega$ XRD profile of Fig. 11.5a, only the (0002) and (0004) reflections from ZnO are observed in addition to the reflections from the *a*-plane sapphire substrate, indicating that the ZnO film is grown in the *c*-axis orientation. In the pole-figure of Fig. 11.5b, only the $\{10\bar{1}1\}$ spots with six-fold symmetry are observed in addition to the reflection spots with two-fold symmetry from the *a*-plane sapphire substrate [28,32]. Therefore, we can conclude that the

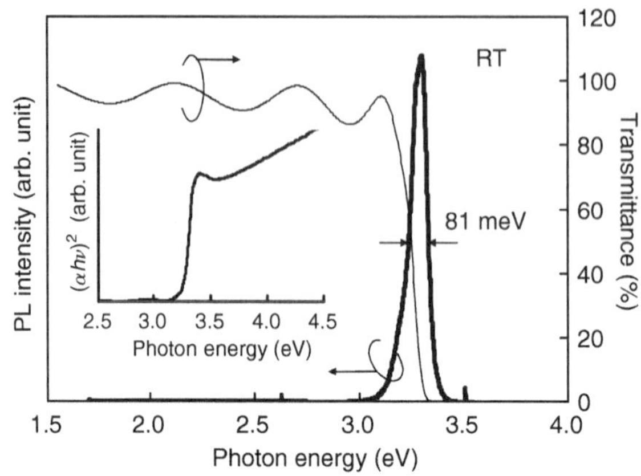

Figure 11.6: Transmittance and PL spectra at room temperature from a ZnO film grown on *a*-plane sapphire substrate. Inset is the $(\alpha h\nu)^2$ vs. $h\nu$ plot of the transmittance curve.

ZnO film is a single-crystal grown in the *c*-axis orientation and does not include in-plane rotational domains. The epitaxial relationship expected from these XRD data is $[11\bar{2}0]$ZnO(0001)//[0001]$Al_2O_3(11\bar{2}0)$, in agreement with the RHEED patterns during growth.

Figure 11.6 shows the optical transmittance and photoluminescence (PL) spectra of the ZnO film at room temperature. Note that the absorption by sapphire substrate has been subtracted from the transmittance spectrum. The PL measurement was conducted using the 325 nm line of a He-Cd laser with a power of 10 mW as an excitation source. Inset is the $(\alpha h\nu)^2$ vs. $h\nu$ plot of the transmittance spectrum, where α is the absorption coefficient and $h\nu$ the photon energy. The oscillations in the lower energy side of transmittance spectrum are due to interference, and the transmittance without interference is close to 100%. There exists an apparent peak in $(\alpha h\nu)^2$ vs. $h\nu$ curve close to the absorption edge, which is due to the absorption by free excitons [33]. The optical absorption edge estimated from the $(\alpha h\nu)^2$ vs. $h\nu$ curve is ~3.31 eV in agreement with the reported band-gap energy of 3.37 eV at room temperature [34] if added the exciton binding energy in literature of ~60 meV [15]. This appearance of exciton peak at room temperature is consistent with the larger exciton binding energy compared to the thermal energy at 300 K (26 meV), and the domination of excitonic transition at room temperature agrees with the PL spectrum of which major peak at ~3.30 eV is due to the recombination of free excitons.

We show in Fig. 11.7a the PL spectrum at 5 K, and in Fig. 11.7b the shift of respective peaks in the spectrum with increasing temperature. In the PL spectrum, only three major peaks with a shoulder at the low energy side are observed. Referring to literature [35], the major peak at the highest energy of 3.372 eV, which is denoted

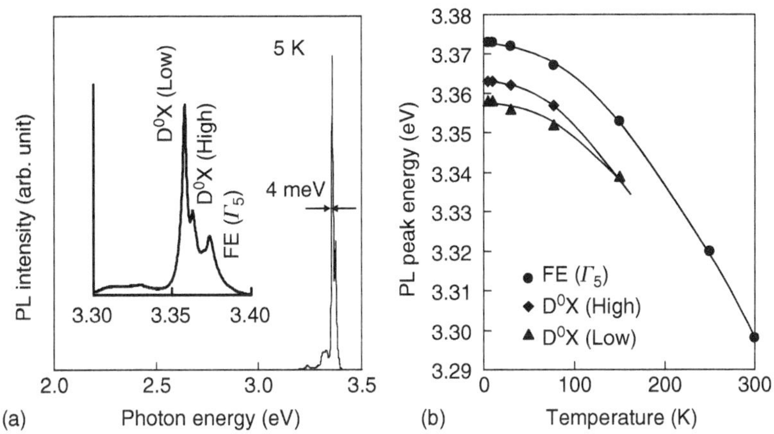

Figure 11.7: A typical PL spectrum at 5 K from a ZnO film grown on *a*-plane sapphire substrate (a), and the shift of major peaks with temperature (b). Inset of (a) shows the major peaks in a magnified horizontal scale.

by FE(Γ_5), is attributed to the higher-energy excited rotational transitions of free excitons, and the other two at 3.362 eV denoted by D^0X(High) and 3.358 eV by D^0X(Low) are to the highest-energy and the lowest-energy donor-bound-excitonic transitions, respectively. The broad and weak emission band at the lower energy side (3.30–3.35 eV) is composed of phonon replicas of these peaks. Since the emission in visible region is less than one-thousandth of the highest major peak in intensity, we can expect the ZnO film to be high quality with low defect and impurity concentrations.

In the next step, we measured electrical properties of the ZnO film by forming a Hall bridge with In ohmic electrodes. As shown by the solid circles in Fig. 11.8a and b, the conduction at 300 K was *n*-type with a nominal electron density of 3.1×10^{17} cm^{-3} and a nominal electron mobility of 105 cm^2/Vs. While a mobility of ~150 cm^2/Vs is the highest record in our group, these values in Fig. 11.8a and b are the typical for our MBE grown films, which is compatible with the reports by others on high quality undoped ZnO films grown on sapphire substrates [36–38]. As shown by the solid circles in Fig. 11.8a, this nominal electron density decreased slowly with temperature till ~150 K, however, it increased with the further decrease in temperature. A similar anomalous behavior has been reported for GaN films grown on sapphire substrates [39], and explained using a two-layer conduction model; a degenerated region with a low mobility and a temperature-independent electron density is formed near the substrate, and its contribution to electrical conduction becomes dominant at very low temperatures.

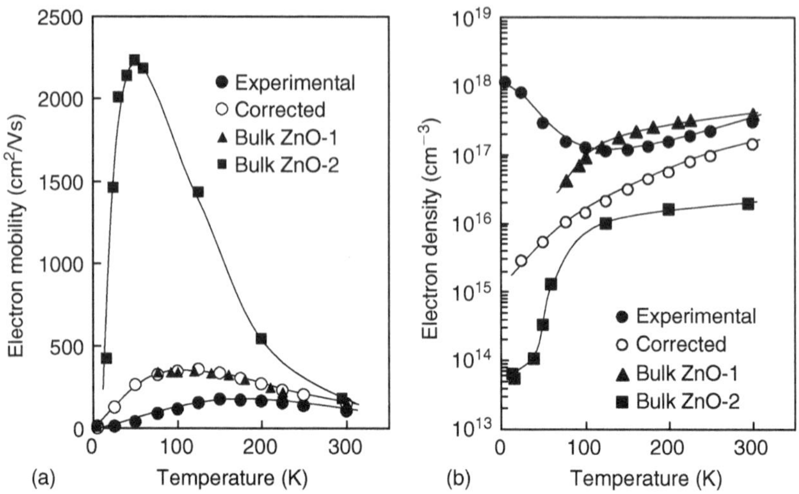

Figure 11.8: Temperature dependent mobility (a) and density (b) of the electrons in a ZnO film grown on a-plane sapphire substrate and two different bulk crystals in literature [41,42]. Solid circles are shown for the experimental data on the ZnO film before the correction, and the open circles for those after the correction by using a two-layer conduction model. Solid triangles and squares are the data for bulk crystals.

We employed the two-layer model to explain the electrical properties of our ZnO film, assuming the presence of a degenerated region near the substrate with a sheet electron density of $\sim 7 \times 10^{13}\,\mathrm{cm}^{-2}$ and a mobility of $\sim 9\,\mathrm{cm}^2/\mathrm{Vs}$. These values are selected to accord with the experimental data at 5 K. This sheet electron density is in agreement with a reported one of $\sim 6 \times 10^{13}\,\mathrm{cm}^{-2}$ for the degenerated region of a ZnO film (40 nm in thickness with an electron density of $\sim 1.5 \times 10^{19}\,\mathrm{cm}^{-3}$) grown on a-plane sapphire substrate [40]. As shown by the open circles in Fig. 11.8a and b, this calculation revealed a reasonable temperature dependence of the rest ZnO film. The electron density after correction increases with temperature as the typical of semiconductor with an activation energy of ~43 meV. This magnitude of activation energy is close to a reported one for the donors related to a Zn-excess defect [41]. For comparison, we also depict in Fig. 11.8a and b the data by others [41,42] for two bulk crystals with different electron densities.

In Fig. 11.8a, we can see a good agreement between the corrected mobility by using the two-layer model and the solid triangles reported for a ZnO bulk (ZnO-1 [41]). On the other hand, the electron density of the corresponding ZnO bulk, which is shown in Fig. 11.8b, is several times higher than that of the MBE grown sample after the correction. This difference is presumably caused by the presence of ionized defects such as dislocation lines in the latter sample. It is easily expected that dense dislocations will be generated in the ZnO film due to the lattice mismatched epitaxial growth, although most of them are confined in the degenerated region nearby substrate [38,39]. In fact, presence of 10^8 cm^{-2} order threading dislocations was confirmed for the surface layer of the ZnO film by counting dislocation-related dark spots in the in-plane monochromatic mapping image of cathodoluminescence (CL) intensity. It is note worthy that the electron mobility in high quality bulk crystal (ZnO-2 [42]) also decreases to a small value close to zero at very low temperatures as shown by the solid squares in Fig. 11.8a, in accordance with the modeling of the two-layer conduction used in this study.

In conclusion, we have shown the single-crystalline ZnO growth on a-plane sapphire substrates by MBE, and analyzed their optical and electrical properties. It is revealed that the ZnO films are grown in the O-polarity of c-axis orientation with a very high degree of optical transparency. Due to a large lattice mismatch, however, a low quality degenerated region is inevitably formed nearby the substrates. The electrical properties of the ZnO film except the degenerated region, however, are nearly compatible to those of a bulk crystal. In other words, it will be requested to isolate the degenerated region from the device-active layer by inserting highly resistive buffer layers to apply the ZnO films to electronic devices. $Zn_{1-x}Mg_xO$ alloys and $Zn_{1-x}Mg_xO$/ZnO superlattices might be the candidate buffer layers for the isolation since their band-gap energies are expected to increase with the Mg content in $Zn_{1-x}Mg_xO$.

11.3 Extension of the MBE growth technique from ZnO to $Zn_{1-x}Mg_xO$

In this section, we discuss the growth of $Zn_{1-x}Mg_xO$ alloy films on a-plane sapphire substrates by extending the growth technique in previous section. Firstly, how to control the Mg content in $Zn_{1-x}Mg_xO$ alloy films was examined [43,44]. We grew a series of $Zn_{1-x}Mg_xO$ films with different Mg contents as a function of oxygen flow rate by maintaining other growth conditions at constant; the growth was performed for 1 h at 500°C with fixed BEPs of 2.0×10^{-4} Pa for Zn, and 3.7×10^{-6} Pa for Mg. We show in Fig. 11.9a the film thickness and the Mg content x in the $Zn_{1-x}Mg_xO$ films grown at various oxygen flow rates, which are converted in Fig. 11.9b to the "net" thicknesses of *ZnO* and *MgO* components in alloy films by distributing the $Zn_{1-x}Mg_xO$ thickness in proportion to the values of $1 - x$ and x. In the calculation, we ignored the change of lattice constants with x, which will be discussed in the next section, due to the small contribution within experimental

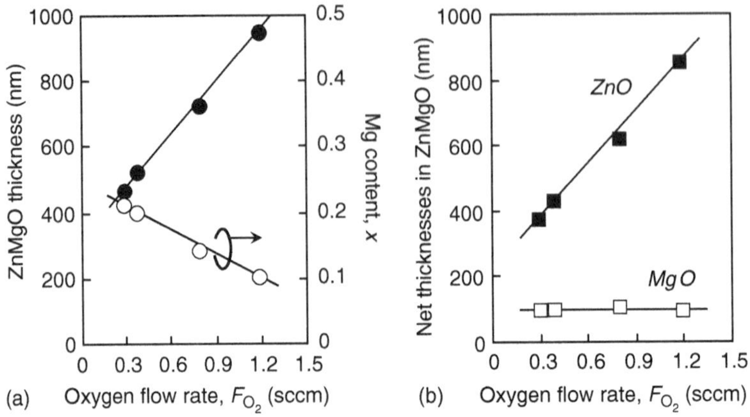

Figure 11.9: Change of the thickness and Mg content of $Zn_{1-x}Mg_xO$ films as a function of the oxygen flow rate during growth (a), the vertical axis of which was converted into the "net" thicknesses of *ZnO* and *MgO* components in the alloy films (b).

accuracy [33,43]. The Mg content x was determined using an electron probe microanalyzer calibrated by inductively coupled plasma mass spectroscopy [43,44]. As well, we also grew a series of $Zn_{1-x}Mg_xO$ films with different Mg contents as a function of growth temperature by maintaining other growth conditions at constant; BEPs of 2.0×10^{-4} Pa for Zn, and 3.7×10^{-6} Pa for Mg, and an oxygen flow rate of 0.35 sccm. Figure 11.10a depicts the experimental data on the film thickness and the Mg content, which are converted into the "net" thicknesses of *ZnO* and *MgO* components as shown in Fig. 11.10b.

It is seen from Fig. 11.9b that the "net" *ZnO* thickness increases in proportion to the oxygen flow rate whereas that of *MgO* is constant, indicating that the growth was proceeded under a Zn-excess condition. We might be able to approximate the growth mechanism under the Zn-excess condition by simply taking the incorporation rate of O* from the plasma cell to be unity. In the simplified model proposed here, we assume that the O* react preferentially with Mg flux Γ_{Mg} first, and then the rest of the O* reacts with Zn flux Γ_{Zn}. This assumption might be acceptable by taking account of the much larger enthalpy of MgO (602 KJ/mol) [45] than that of ZnO (348 KJ/mol) [45]. The ratio of *MgO/ZnO* in "net" amount of components under the assumption is expressed as follows using a parameter S_{Mg} of Mg incorporation rate;

$$\frac{MgO}{ZnO} = \frac{x}{1-x} = \frac{\Gamma_{Mg} \times S_{Mg}}{\Gamma_{O*} - \Gamma_{Mg} \times S_{Mg}}, \tag{11.1}$$

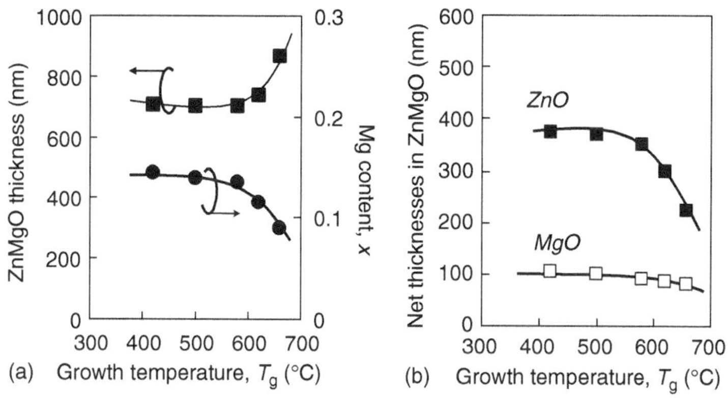

Figure 11.10: Change of the thickness and Mg content of $Zn_{1-x}Mg_xO$ films as a function of the growth temperature (a), the vertical axis of which was converted into the "net" thicknesses of *ZnO* and *MgO* components in the alloy films (b).

$$\Gamma_{Mg} = \frac{A \times P_{Mg}}{\sqrt{2\pi m_{Mg} k (T_{Mg} + 273)}}, \tag{11.2}$$

where m_{Mg} is the atomic mass of Mg, k the Boltzman constant, T_{Mg} the temperature of the K-cell for Mg, A a constant, and P_{Mg} the Mg equilibrium vapor pressure at T_{Mg}. The Mg incorporation rate S_{Mg} at 400°C was taken to be unity by considering the data on "net" *MgO* thickness in Fig. 11.10b which became independent of the growth temperature below 500°C, and the S_{Mg} at higher growth temperatures T_g was also determined using the data in Fig. 11.10b as $S_{Mg} = MgO(T_g)/MgO(400°C)$ where $MgO(T_g)$ and $MgO(400°C)$ were the "net" *MgO* thicknesses at T_g and 400°C, respectively.

By substituting a set of experimental data (x, T_{Mg}) at a selected growth temperature at a constant oxygen flow rate F_{O_2}, we can fix the values of A and Γ_{O*} in Eqns. (11.1) and (11.2) using the P_{Mg} data in literature. Once the magnitude of A and Γ_{O*} is fixed, the Mg content x is easily calculated as a function of T_{Mg} [43,44]. We show in Fig. 11.11 three typical calculated results at $F_{O_2} = 0.35$ sccm which are compared to the corresponding experimental data. In spite of the simple and assertive modeling used here, we can see that the calculation can simulate the experimental data in a wide range of growth condition. Although not shown here, we also found that the consistency between the calculation and experimental data became worse at $T_g > 600°C$; the Mg content in experimental data was larger than that by calculation. This discrepancy at high growth temperatures is caused by the change in growth condition from Zn-excess to Zn-shortage due to a rapid reduction

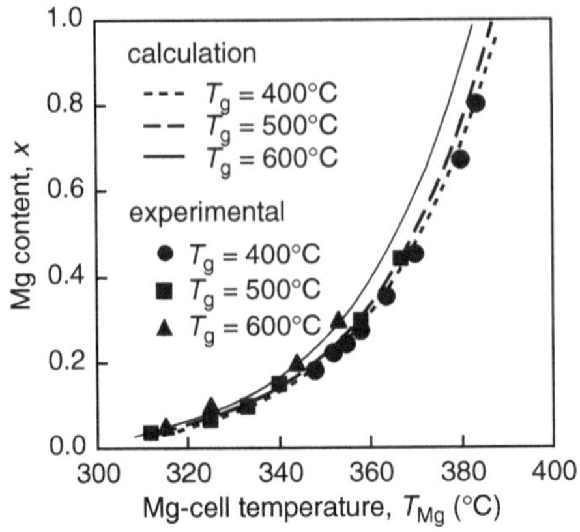

Figure 11.11: Comparison between the calculated and experimental results on the Mg content x in $Zn_{1-x}Mg_xO$ films as a function of Mg-cell temperature.

of Zn incorporation rate S_{Zn} as shown in Fig. 11.10b; under the condition, the premise of Zn-excess growth condition, $\Gamma_{Zn} \times S_{Zn} > \Gamma_{O^*} - \Gamma_{Mg} \times S_{Mg}$, is no more satisfied and Eqn. (11.1) becomes invalid [44].

The Mg content x discussed above was a macroscopic one, and was determined by assuming a homogenous composition in the whole films. Due to the difference in crystal structure between ZnO (wurtzite) and MgO (rocksalt), however, existence of a wide miscibility gap is predicted based on thermodynamics [46]. In fact, phase separation by the miscibility gap was observed in our experiment [43,47,48] as well as the reports by others [33,49–53]. The size of phase-separated rocksalt domains, however, was so small that the macroscopic discussion mentioned above on the Mg content x was found to be approved independent of the phase separation. As will be discussed later, we also found that the range of miscibility gap in Mg content was able to be decreased by changing the growth condition [48] since the growth by MBE is apart from thermal equilibrium condition.

Figure 11.12a shows typical $2\theta/\omega$ XRD profiles on logarithmic scale of 400-nm-thick $Zn_{1-x}Mg_xO$ films with different Mg contents. These alloy films were grown at 400°C on a-plane sapphire substrates via annealed 10-nm-thick LT-ZnO buffer layers. In addition to the intense reflection from the sapphire substrates, two peaks from different crystal structures are contained in these spectra; one at $\sim$35° is the (0002) reflection from wurtzite $Zn_{1-x}Mg_xO$ and the other at $\sim$43° is the (200) reflection from rocksalt $Zn_{1-x}Mg_xO$. As shown by Fig. 11.12a, we obtained single-phase c-axis oriented $Zn_{1-x}Mg_xO$ films in wurtzite structure up to $x \approx 0.4$ (the maximum x value we obtained was 0.43 except reproducibility). Over

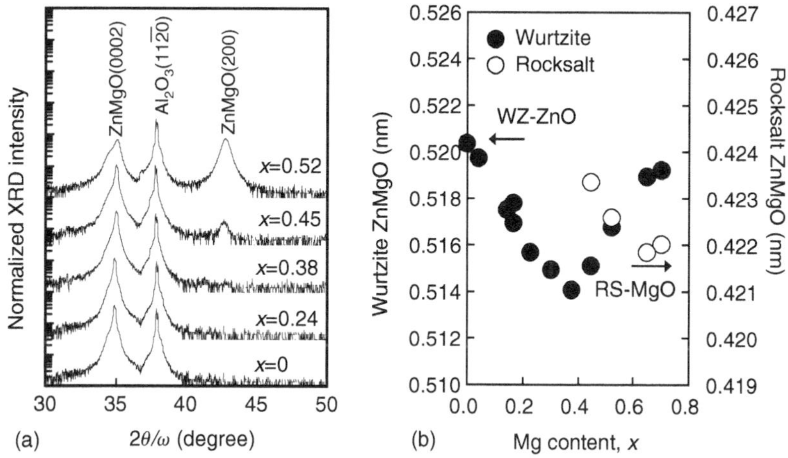

Figure 11.12: $2\theta/\omega$ XRD profiles from $Zn_{1-x}Mg_xO$ films with different Mg contents (a), and the change of lattice constant as a function of the Mg content (b).

the solubility limit at $x \approx 0.4$, however, inclusion of rocksalt domains due to phase separation became clear with increasing Mg content.

The change of lattice constant in these alloys is calculated from their XRD diffraction angles as shown in Fig. 11.12b. In agreement with the reports by others [33,49–52], we can see a nearly linear decrease of the c-axis lattice constant as $0.5204 - 0.017x$ nm with increasing the Mg content x in wurtzite $Zn_{1-x}Mg_xO$. After the phase separation, on the contrary, the c-axis lattice constant of wurtzite $Zn_{1-x}Mg_xO$ domains increased again with Mg content and went back to that of ZnO bulk crystals. At the same time, the a-axis lattice constant of newly appeared rocksalt $Zn_{1-x}Mg_xO$ domains decreased to close to that of rocksalt MgO bulk crystals ($a = 0.4216$ nm) [24].

In Fig. 11.13a and b, we show optical transmittance spectra of these $Zn_{1-x}Mg_xO$ films at room temperature and their absorption edge energies, respectively. It is seen that the transmittance spectra in Fig. 11.13a have a clear absorption edge till the limit of the homogeneous wurztite growth at $x \approx 0.4$ ($x = 0.43$ in this case). As shown in Fig. 11.13b, the linear increase of the absorption edge energy in this region is from ~3.31 eV at $x = 0$ to ~4.39 eV at $x = 0.43$, yielding a relationship for optical band-gap energy $E_g \approx 3.37 + 2.51x$ eV. In this evaluation of band-gap energy, we simply added an exciton binding energy of 60 meV [15] to the experimentally determined absorption edge energies although its realistic value might increase with x. This relationship is roughly consistent with the Vegard's low (E_g of pseudo-wurtzite MgO at 0 K has been estimated to be 6.06 eV theoretically [54]) as an evidence

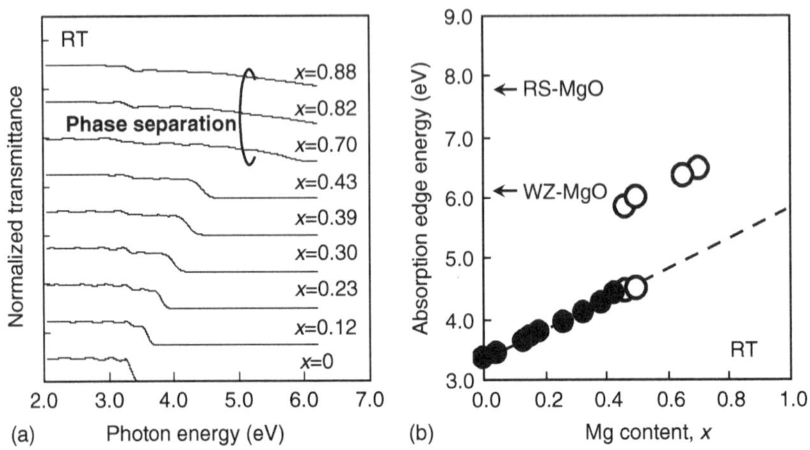

Figure 11.13: Transmittance spectra at room temperature from the $Zn_{1-x}Mg_xO$ films with different Mg contents x (a), and the change of their absorption edge energies with x (b). The open circles in (b) denote the presence of phase separation in the alloy films.

of single-phase wurtzite growth. When the Mg content exceeded the solubility limit, on the other hand, an additional absorption edge appeared at the much higher energy side than the expectation by the Vegard's law, and then it dominated the transmittance spectra as shown in Fig. 11.13a. This behavior should be resulted from the development of Mg-rich rocksalt domains in agreement with the XRD data shown in Fig. 11.12b.

It should be noted that this type of phase separation also appeared even at $x = 0.22$ when the $Zn_{1-x}Mg_xO$ films were grown without using the LT-ZnO initiation layer. Similar result has been reported by Fujita et al. [52] and hence we conclude that the use of c-axis oriented wurtzite ZnO initiation layer is a key issue to force the $Zn_{1-x}Mg_xO$ films to crystallize in the same structure. Another important point is the effect of growth temperature; we grew a series of $Zn_{1-x}Mg_xO$ films at different temperatures and found a considerable increase of the solubility limit in wurtzite structure with the decrease of growth temperature [48], which is shown in Fig. 11.14. It increased from $x \approx 0.3$ at 600°C to $x \approx 0.45$ at 350°C although the increase was obtained as a trade-off of crystalline quality. With decreasing the growth temperature, the crystalline quality evaluated by XRD intensity and PL intensity became worse and single-crystalline growth itself was difficult at lower temperatures than 250°C. Since a very small value of $x = 0.04$ is reported for the solubility limit under thermal equilibrium condition [46], our results indicate a considerable alleviation of the limit by utilizing the metastable MBE growth condition at low temperatures.

Indeed the metastable growth at low temperatures is effective to obtain the wurtzite $Zn_{1-x}Mg_xO$ films at high Mg contents, but it is not clear whether the

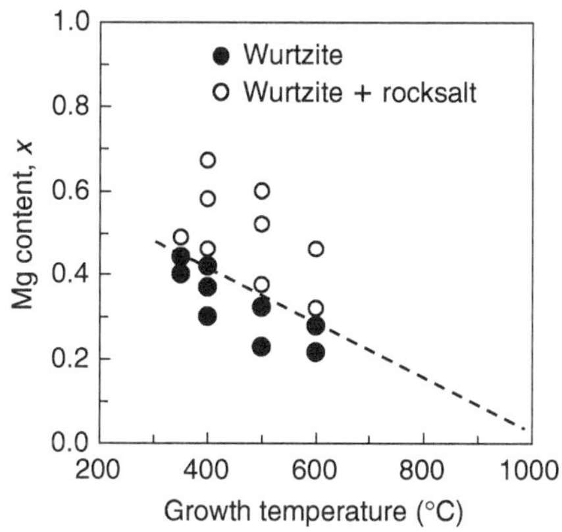

Figure 11.14: Change of the Mg solubility limit in wurtzite $Zn_{1-x}Mg_xO$ films with growth temperature. Solid circles indicate the growth in homogeneous wurtzite structure, and open circles the growth in phase separated structure.

grown structure is stable at elevated temperatures. In order to study the thermal stability, we performed a post growth annealing in O_2 ambient for a 400-nm-thick $Zn_{0.64}Mg_{0.34}O$ film which was grown at 350°C on a-plane sapphire substrate via an annealed LT-ZnO initiation layer. We measured its $2\theta/\omega$ XRD profile after each 1 h annealing treatment that was repeatedly performed for the same sample by increasing the temperature by 100°C in every step. As shown in Fig. 11.15, we scarcely observed any structural change at lower annealing temperatures than 700°C.

At 800°C, however, a small new peak from rocksalt ZnMgO(200) reflection appeared, indicating the onset of phase separation. With increasing the post growth annealing temperature up to 1100°C, this peak intensity became high and the wurtzite $Zn_{1-x}Mg_xO(0002)$ reflection moved toward the lower angle side (toward the ZnO side) due to the enhancement of phase separation as indicated by Fig. 11.12b. Similar enhancement of the phase separation with increasing annealing temperature has also been reported by others [55]. Note that the Mg content x of these wurtzite domains at 1100°C was estimated to be ~0.08 by referring the c-axis lattice constant to the data shown in Fig. 11.15. In addition, at over 900°C, other new peaks from spinel $Zn(Mg)Al_2O_3(311)$ and (222) appeared at around 37° and 38.5°, respectively, due to the thermal reaction between the $Zn_{1-x}Mg_xO$/LT-ZnO layer and the sapphire substrate [34,56].

It is important to note that this onset temperature of the phase separation by post growth annealing, 800°C in this case, is much higher than that by the growth temperature; we can expect the latter one at $x = 0.34$ to be 500°C from the growth-temperature dependent solubility limit shown in Fig. 11.14. Therefore, we can conclude that the metastable MBE growth at low temperatures actually provides

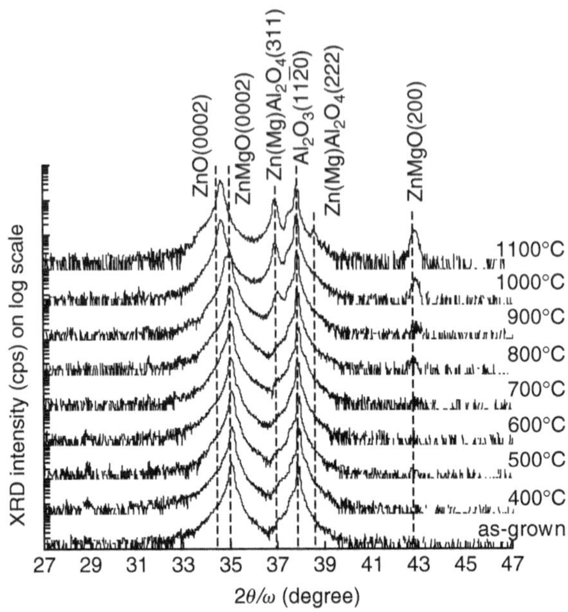

Figure 11.15: $2\theta/\omega$ XRD profiles from a $Zn_{0.64}Mg_{0.34}O$ alloy film after the post growth annealing treatments at various temperatures in O_2 ambient for 1 h. The alloy film was grown on a-plane sapphire substrate at 350°C.

us an effective tool to obtain wurtzite $Zn_{1-x}Mg_xO$ films at high Mg contents. Although the detail will be reported elsewhere, we should add a comment that the onset temperature of the phase separation by post growth annealing was dependent on the Mg content x; it increased with decreasing x, and reached at around 1000°C for a low Mg content film of $Zn_{0.85}Mg_{0.15}O$ for instance.

11.4 Structural, optical and electrical properties of $Zn_{1-x}Mg_xO$/ZnO heterostructures

In order to understand the strain in $Zn_{1-x}Mg_xO$/ZnO heterostructures by the lattice mismatch between wurtzite $Zn_{1-x}Mg_xO$ and ZnO, we measured XRD profiles for two different model samples of Fig. 11.16a and b; (a) is consisted of a 50-nm-thick ZnO capping layer on a 500-nm-thick $Zn_{0.7}Mg_{0.3}O$ buffer layer (ZnO/$Zn_{0.7}Mg_{0.3}O$) and (b) of a 50-nm-thick $Zn_{0.7}Mg_{0.3}O$ capping layer on a 500-nm-thick ZnO buffer layer ($Zn_{0.7}Mg_{0.3}O$/ZnO). Respective layers of both samples were grown at 400°C successively on a-plane sapphire substrates via an

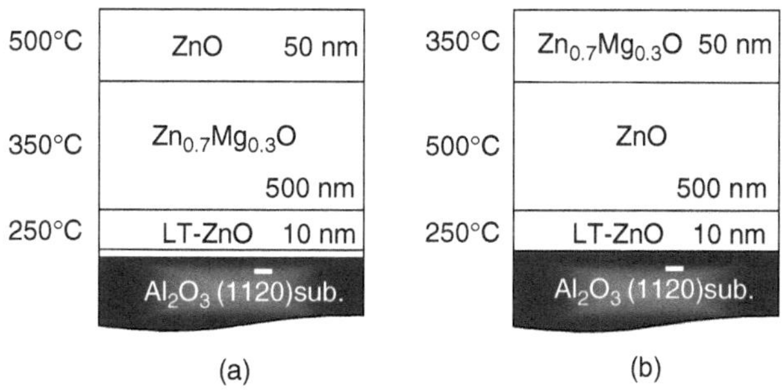

Figure 11.16: Schematic cross-sections of model heterostructures of ZnO/$Zn_{0.7}Mg_{0.3}O$ (a), and $Zn_{0.7}Mg_{0.3}O$/ZnO (b). Both of them were grown on *a*-plane sapphire substrates.

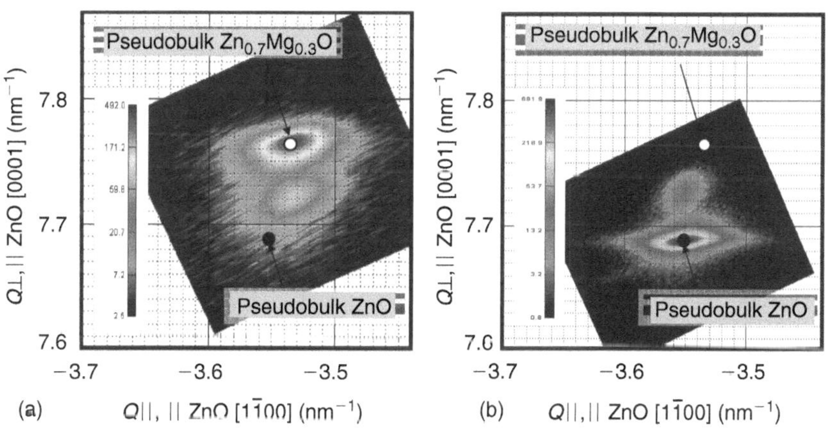

Figure 11.17: XRD reciprocal space contour mappings around the (1$\bar{1}$04) diffraction from a ZnO/$Zn_{0.7}Mg_{0.3}O$ sample (a), and a $Zn_{0.7}Mg_{0.3}O$/ZnO sample (b). (a) and (b) in this figure correspond to those in Fig. 16, respectively.

annealed 10-nm-thick LT-ZnO initiation layer. In Fig. 11.17a and b, we show the XRD reciprocal space contour mappings around the (1$\bar{1}$04) diffraction spots measured by ATX-G, Advanced Thin Film X-ray System, Rigaku Co., for the samples (a) and (b) in Fig. 11.16, respectively. Each mapping in Fig. 11.17a and b is clearly composed of two peaks from the capping and buffer layers whereas the signal from the underlying thin LT-ZnO initiation layer is not seen in this intensity scale. It has been know by referring the experimental data by others [33] that the *a*-axis lattice constant in $Zn_{1-x}Mg_xO$ expands nearly as $0.3250 + 0.005x$ nm with increasing the Mg content x, while the *c*-axis one in Fig. 11.12b reduces as $0.5204 - 0.017x$ nm.

Therefore, an in-plane (a-axis) lattice mismatch of ~0.45% is expected for both model samples.

In Fig. 11.17a, both the ZnO capping and the $Zn_{0.7}Mg_{0.3}O$ buffer layers have an identical $Q_{\|}$ value within experimental accuracy. The a-axis lattice constant corresponding to this $Q_{\|}$ value is 0.3265 nm, which is well accorded with the expectation from the relationship $0.3250 + 0.005x$ nm and $x = 0.30$ for strain-free $Zn_{0.7}Mg_{0.3}O$. The $Q_{\|}$ value, on the other hand, has a difference in these two peaks, and the corresponding c-axis lattice constants for the ZnO capping and the $Zn_{0.7}Mg_{0.3}O$ buffer layers are 0.5180 nm and 0.5151 nm, respectively. Although the latter value for the buffer layer nearly agrees with the estimation of 0.5153 nm for strain-free $Zn_{0.7}Mg_{0.3}O$ (using the relationship $0.5204 - 0.017x$ nm and $x = 0.30$), the former value for capping layer is ~0.46% smaller than that of strain-free ZnO. Note that if the ZnO layer were free from strain, its peak should have the coordinates marked by a solid circle. By considering the Poisson ratio being close to unity [57], the result in Fig. 11.17a indicates that most of the $Zn_{0.7}Mg_{0.3}O$ buffer layer is grown free from strain, and the ZnO capping layer is tensile strained in the lateral direction to form a pseudomorphic structure.

In the case of Fig. 11.17b where the composed materials are reversed, the coordinates marked by an open circle are the expected position for strain-free $Zn_{0.7}Mg_{0.3}O$. By a similar analysis to that for Fig. 11.17a, accordingly, it is revealed that only the $Zn_{0.7}Mg_{0.3}O$ capping layer is compressive strained in the lateral direction when grown on the strain-free ZnO buffer layer.

$2\theta/\omega$ XRD measurement for the LT-ZnO initiation layer, although the data are not shown here, revealed that the (0002) reflection peak was broad but with a diffraction angle close to that of bulk crystals. This result indicates that a considerable part of the strain by the ~18% lattice mismatch to sapphire substrate has been accommodated at the initial growth stage of the ZnO initiation layer. Since the $Zn_{0.7}Mg_{0.3}O$ buffer layer on LT-ZnO initiation layer is completely relaxed after the 500-nm-thick growth, we can conclude from the data in Fig. 11.17a and b that the critical thickness of the pseudomorphic $Zn_{0.7}Mg_{0.3}O$/ZnO growth at 400°C is larger than 50 nm and smaller than 500 nm.

In the next, we applied the growth technique to prepare $Zn_{0.7}Mg_{0.3}O$/ZnO multiple quantum wells (MQWs) to check the interface control in atomic scale. After growing a thick (1 μm) ZnO buffer layer on a LT-ZnO initiation layer at 500°C, an MQW with 40 periods was fabricated at 400°C followed by a capping with a 50-nm-thick $Zn_{0.7}Mg_{0.3}O$ layer at the same temperature. The ZnO well layer thickness (L_W) of the MQW was varied from 1 to 8 nm whereas the $Zn_{0.7}Mg_{0.3}O$ barrier layer thickness (L_B) was maintained at a constant value of 5 nm. During the MQW growth, we took a growth interruption of 30 s at each heterointerface to make the interface abrupt.

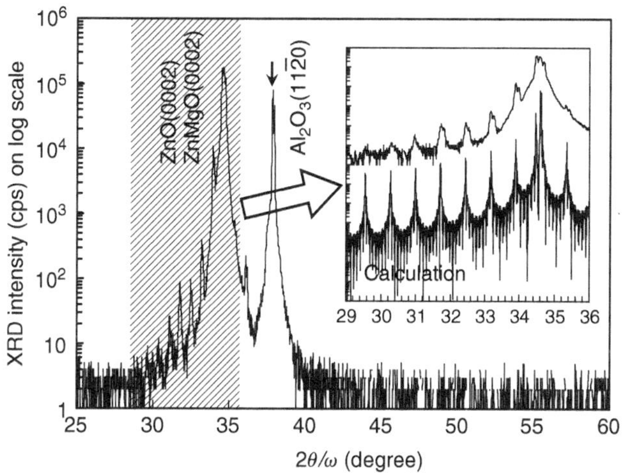

Figure 11.18: A $2\theta/\omega$ XRD profile from a $Zn_{0.7}Mg_{0.3}O/ZnO$ MQW with $L_W = 8$ nm and $L_B = 5$ nm. Inset is the horizontally magnified satellite peaks that are compared to a theoretical calculation.

Figure 11.18 shows a $2\theta/\omega$ XRD profile on logarithmic scale measured by a conventional diffractometer in our laboratory from an MQW with $L_W = 8$ nm. The major spectrum consists of only (0002) reflection peaks from the ZnO buffer/wells and the wurtzite $Zn_{0.7}Mg_{0.3}O$ barrier layers, and neither peak from phase-separated nor misoriented domains is detected, indicating a *c*-axis oriented MQW growth keeping the single-phase wurtzite structure. We can see clear MQW satellite peaks up to 8th order although each of them consists of $CuK\alpha_1$ and $CuK\alpha_2$ reflections. For the $CuK\alpha_1$ component, we fitted a simple calculation to these satellite peaks of –7th to +1st by assuming abrupt heterointerface as shown by the inset of Fig. 11.18. The parameters used for the calculation were a 7.83-nm-thick L_W with 0.520 nm *c*-axis lattice constant and a 4.85-nm-thick L_B with 0.518 nm *c*-axis lattice constant. The calculated MQW period of 12.7 nm agrees with the designed one within experimental accuracy.

In more detail, the *c*-axis lattice constant of ZnO wells agrees with that of strain-free one ($c = 0.5204$ nm). On the other hand, the *c*-axis lattice constant of $Zn_{0.7}Mg_{0.3}O$ barriers is ~0.5% larger than that of strain-free wurtzite $Zn_{0.7}Mg_{0.3}O$ ($c = 0.5153$ nm). This result is consistent with the data shown in Fig. 11.17b where the *c*-axis lattice constant of the $Zn_{0.7}Mg_{0.3}O$ layer on relaxed ZnO is ~0.5% expanded due to the pseudomorphic growth. Consequently, we can expect that the whole MQW is grown in coherence to the thick and relaxed ZnO buffer layer, and

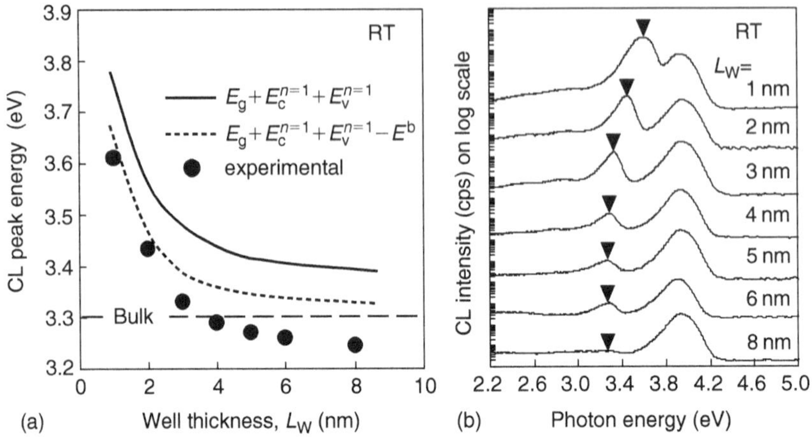

Figure 11.19: Shift of CL peak energy with the thickness L_W of the ZnO wells in $Zn_{0.7}Mg_{0.3}O$/ZnO MQWs ($L_B = 5$ nm) (a), and the corresponding CL spectra (b). The solid circles in (a) are experimental data and the solid and dashed curves are theoretical calculations.

only the $Zn_{0.7}Mg_{0.3}O$ barrier and capping layers in MQW are under a compressive strain of ~0.5% in the lateral direction to result in the increase of *c*-axis lattice constant by elastic deformation.

The solid circles in Fig. 11.19a show the CL peak energies at room temperature of confined excitons in the MQWs with different L_W values. The acceleration voltage of the incident electron beam for CL measurement was 4 kV, and the penetration depth from the surface was estimated to be ~100 nm by using a Monte-Carlo simulation, in agreement with a report by others [58]. Each corresponding CL spectrum shown in Fig. 11.19b consists of two major peaks; one is due to the bound excitons in $Zn_{0.7}Mg_{0.3}O$ barrier/capping layers, and the other at the lower energy side should be due to the confined excitons in ZnO wells.

The energy of the former peaks was independent of the well thickness L_W, and located at around 3.95 eV in agreement with the CL peak energy for thick single-layered (pseudobulk) $Zn_{0.7}Mg_{0.3}O$ films [48,59]. Although this peak energy is a little smaller than their optical band-gap energy of 4.12 eV even considering the large exciton binding energy in $Zn_{0.7}Mg_{0.3}O$, such a difference can be explained by the spatial inhomogeneity of cation sublattice because localized excitons in $Zn_{0.7}Mg_{0.3}O$ films easily experience different Coulomb potentials due to their small Bohr radii [33,43]. We should mention that the defect-related emission band at around 2.2 eV [43,60–62] is negligibly weak in these spectra indicating a high crystalline quality, and that the signal from the ZnO buffer layer is not included in the CL spectra in Fig. 11.19b due to the short penetration depth of the incident electron beam.

As shown by the solid circles in Fig. 11.19a, the emission energy of confined excitons increases with decreasing L_W by the quantum size effect in ZnO well. This emission energy is compared to the calculation based on a Krönig-Penny type finite-square potential well model using the following parameters; $E_g(\mathrm{ZnO}) = 3.37$ eV, $E_g(\mathrm{Zn_{0.7}Mg_{0.3}O}) = 4.12$ eV, $m_e^*(\mathrm{ZnO}) = 0.28m_0$ [63,64], and $m_{hh}^*(\mathrm{ZnO}) = 0.78m_0$ [64], where E_g is the band-gap energy at 300 K, and m_e^* and m_{hh}^* are the electron and heavy-hole masses, respectively. We also used band offsets in conduction and valence bands, ΔE_c and ΔE_v, assuming a relationship of $\Delta E_c/\Delta E_v = 9$ in literature [65]. The solid line in Fig. 11.19a shows the calculated result for the transition energy from the first electron subband $E_c^{n=1}$ to the first heavy-hole subband $E_v^{n=1}$. Since excitons in ZnO wells are known to be responsible for the emission at 300 K, we subtracted the exciton binding energy E^b in literature [64,66] from the calculated energy to obtain the excitonic transition energy E_{ex} as shown by the dashed line in Fig. 11.19a ($E_{ex} = E_g + E_c^{n=1} + E_v^{n=1} - E^b$).

We can see that the experimental data at $L_W \leq 2$ nm nearly agree with the dashed line. However, those at $L_W > 3$ nm fall short of the dashed line in proportion to the increase of well thickness, and have gone down at last less than 3.30 eV which is the excitonic emission energy in pseudobulk ZnO as has been shown by the PL data in Fig. 11.6. We attribute this excess decrease proportional to L_W to the quantum-confined Stark (QCS) effect by internal electric field [17,64,67]. Similar to the case of $\mathrm{Ga_{1-x}Al_xN/GaN}$ MQWs [68–70], the presence of internal electric field along the c-axis direction can be expected for these coherent $\mathrm{Zn_{0.7}Mg_{0.3}O/ZnO}$ MQWs by the difference in spontaneous polarizations between adjacent materials and by the piezoelectric polarization in compressed barriers. Although the degree of the redshift by QCS effect in Fig. 11.19a is more than several times smaller compared to the reported values for $\mathrm{Ga_{1-x}Al_xN/GaN}$ MQWs [68–70], similar weak redshift on $\mathrm{Zn_{1-x}Mg_xO/ZnO}$ MQWs has also been observed by others [17,64,67]. Apart from the degree of redshift, the rapid decrease in intensity of the QCS emission with the increase of L_W, which is seen in Fig. 11.19b, is consistent with the increase of spatial separation of electron and hole wave-functions by the internal electric field in ZnO wells.

In order to understand the different QCS shifts between $\mathrm{Zn_{1-x}Mg_xO/ZnO}$ and $\mathrm{Ga_{1-x}Al_xN/GaN}$, we compared the material constants related to the polarization in these two heterosystems as shown in Table 11.2 [21,24,68,71–79]. According to the strain in MQW, the a-axis lattice constant in $\mathrm{Zn_{1-x}Mg_xO}$ is known to have a ~35% smaller dependence on x compared to that in $\mathrm{Ga_{1-x}Al_xN}$ [71], which is summarized in Table 11.3 with the lattice constants and polarizations in related materials [21,68,71,78]. In Table 11.3, we assumed the Vegard's law to estimate the

Table 11.2: Comparison of strain-related material constants.

	WZ-GaN	WZ-AlN	WZ-ZnO	RS-MgO
a [nm]	0.3189[71]	0.3112[71]	0.3250[21]	0.4216[24]
c [nm]	0.5185[71]	0.4982[71]	0.5204[21]	–
C_{11} [GPa]	367[72]	396[72]	207[76]	297.1[79]
C_{12} [GPa]	135[72]	137[72]	117.7[76]	96.5[79]
C_{13} [GPa]	103[72]	108[72]	106.1[76]	–
C_{33} [GPa]	405[72]	373[72]	209.5[76]	–
C_{44} [GPa]	95[72]	116[72]	44.8[77]	155.7[79]
e_{33} [C/m^2]	0.73[74]	1.46[74]	0.96[76]	–
e_{31} [C/m^2]	−0.49[74]	−0.60[74]	−0.62[76]	–
e_{15} [C/m^2]	−0.30[75]	−0.48[73]	−0.37[76]	–
P_{SP} [C/m^2]	−0.029[68]	−0.081[68]	−0.057[78]	–

Table 11.3: Comparison of a-axis lattice constants and polarizations as a function of alloy composition/strain.

Wurtzite	GaN	$Ga_{1-x}Al_xN$	AlN	ZnO	$Zn_{1-x}Mg_xO$	MgO
a [nm]	0.3189[71]	$0.3189 - 0.0077x$	0.3112[71]	0.3250[21]	$0.3250 + 0.005x$	–
P_{SP} [C/m^2]	−0.029[68]	$-(0.029 + 0.052x)$	−0.081[68]	−0.057[78]	$\mathbf{-(0.057 + 0.064x)}$*	–
P_{PE} [C/m^2]	$-1.35\varepsilon_{x,y}$	$-(1.35 + 0.70x)\varepsilon_{x,y}$	$-2.05\varepsilon_{x,y}$	$-2.21\varepsilon_{x,y}$	–	–

* Estimated in this study.

parameters for alloy films, and the expressions for piezoelectric polarizations $\boldsymbol{P}_{PE}$ using the strain $\varepsilon_{x,y}$ will be elucidated in the latter part of this section. It is clearly seen that the magnitude of the piezoelectric polarization in ZnO well is not smaller than that in GaN well if these wells are strained by the respective barrier layers of $Zn_{1-x}Mg_xO$ and $Ga_{1-x}Al_xN$ having the same x value without strain. However, it is difficult to evaluate the piezoelectric polarization in our $Zn_{0.7}Mg_{0.3}O$/ZnO MQW since, different from the case mentioned above, piezoelectric constants are not known for wurtzite $Zn_{0.7}Mg_{0.3}O$ which is the strained layer in our case. As well, the contribution from the spontaneous polarization is not clear yet for $Zn_{1-x}Mg_xO$/ZnO due to the absence of the data on wurtzite MgO. Therefore, unfortunately, it is difficult at present to estimate the magnitude of internal electric field in our $Zn_{1-x}Mg_xO$/ZnO MQWs by using a simple calculation. According to the spontaneous polarization in $Zn_{1-x}Mg_xO$, we will discuss later together with the expressions for $\boldsymbol{P}_{PE}$.

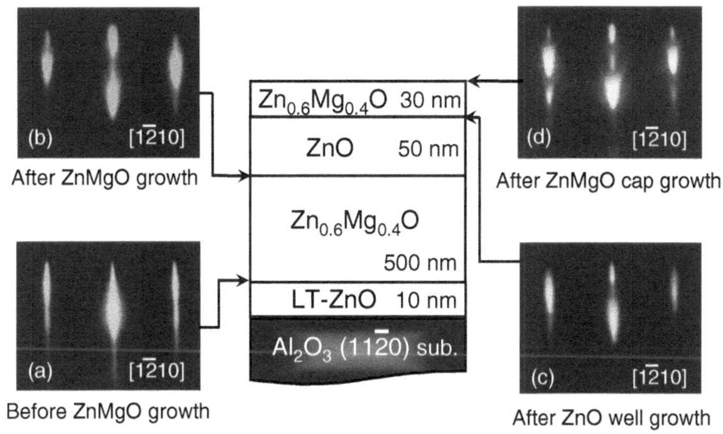

Figure 11.20: Schematic cross-section of a $Zn_{0.6}Mg_{0.4}O/ZnO/Zn_{0.6}Mg_{0.4}O$ SQW sample, and the corresponding RHEED patterns observed at respective growth stages.

In order to analyze the internal electric field in $Zn_{1-x}Mg_xO/ZnO$ potential wells, we further studied the optical and electrical properties of a single quantum well (SQW) as the simplest model of heterostructures. In Fig. 11.20, we show a schematic of the $Zn_{0.6}Mg_{0.4}O/ZnO/Zn_{0.6}Mg_{0.4}O$ SQW structure studied in this experiment. The SQW was grown at a low temperature of 350°C to suppress the phase separation. We also show the RHEED patterns in the $[1\bar{2}10]$ direction from the surfaces before the bottom $Zn_{0.6}Mg_{0.4}O$ barrier growth (a), after the bottom $Zn_{0.6}Mg_{0.4}O$ barrier growth (b), after the ZnO well growth (c), and after the top $Zn_{0.6}Mg_{0.4}O$ barrier growth (d). During the bottom $Zn_{0.6}Mg_{0.4}O$ barrier growth, a sharp and streak pattern (a) from the annealed LT-ZnO initiation layer changed into a chevronlike pattern (b) by maintaining the same six-fold symmetry on substrate rotation. A similar chevronlike pattern (d) was observed after the top $Zn_{0.6}Mg_{0.4}O$ barrier growth although the pattern after the ZnO well growth (c) was observed to be streaky. These results suggest that the SQW grows in the *c*-axis orientation of wurtzite structure while three-dimensional facets may exist at the bottom heterointerface [17,59,80]. This suggestion on surface morphology was supported by the results of AFM measurement that the rms roughness of top $Zn_{0.6}Mg_{0.4}O$ surface (d) was 0.24 nm in $3 \times 3\,\mu m^2$ area whereas that of the ZnO surface before bottom $Zn_{0.6}Mg_{0.4}O$ barrier growth (a) was less than 0.1 nm.

We confirmed this single-crystalline wurtzite growth by the XRD measurement of $2\theta/\omega$ profile and $\{10\bar{1}1\}$ pole-figure. Although not shown here, the $2\theta/\omega$ XRD profile contained only the (0002) and (0004) reflections from $Zn_{0.6}Mg_{0.4}O$ and ZnO in addition to the reflections from the *a*-plane sapphire substrate, indicating

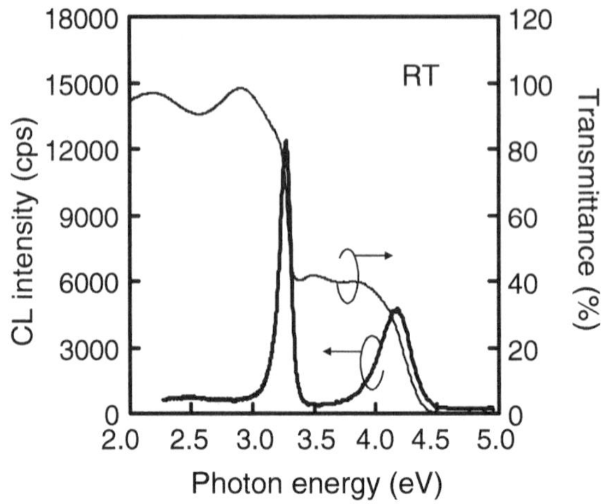

Figure 11.21: Optical transmittance and CL spectra from an SQW at room temperature.

that the growth was performed in the strict *c*-axis orientation without any inclusion of phase separated rocksalt domains. The pole-figure measurement revealed that the film was homogeneous in the lateral direction and did not include rotational domains.

Figure 11.21 shows the optical transmittance and CL spectra from the SQW at room temperature. The transmittance spectrum indicates a high degree of optical transparency close to 100% in the visible region with oscillations due to interference and steps due to the band-edge absorptions by ZnO and $Zn_{0.6}Mg_{0.4}O$. We applied the $(\alpha h\nu)^2$ vs. $h\nu$ plot for the transmittance curve in the staircase regions, and obtained absorption edge energies of ~3.3 eV and ~4.35 eV for the ZnO well and the $Zn_{0.6}Mg_{0.4}O$ barrier layers, respectively. By taking the exciton binding energy into account, these values agree with the band-gap energies in Fig. 11.13b measured for pseudobulk films. The band-gap discontinuity at the heterointerface is estimated to be $\Delta E_c \approx 0.9$ eV and $\Delta E_v \approx 0.1$ eV using the reported ratio of $\Delta E_c/\Delta E_v = 9$ [65], and hence, we can expect the SQW to form a deep potential well for electrons.

The CL spectrum in Fig. 11.21 consists of two major peaks at ~3.29 eV and ~4.13 eV. These peaks can be related to the near-band-edge emissions from the ZnO well and the $Zn_{0.6}Mg_{0.4}O$ barrier layers, respectively. The former peak energy nearly agrees with that of free excitonic emission in Fig. 11.6 which was measured for a pseudobulk ZnO film at room temperature. The latter peak energy agrees with that of a localized excitonic emission in pseudobulk $Zn_{0.6}Mg_{0.4}O$ film although the value for both is ~220 meV smaller than their band-gap energy due to the exciton binding energy and the spatial inhomogeneity of cation sublattice [17,48].

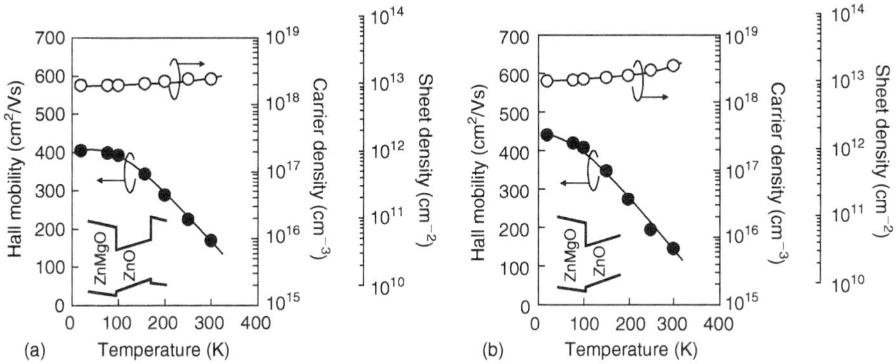

Figure 11.22: Temperature dependent mobility and density of the electrons in 50-nm-thick ZnO well of SQW structure (a), and those in 50-nm-thick ZnO capping layer of single-heterojunction structure (b). Insets are the band diagrams expected for respective heterostructures.

In the next step, we measured temperature-dependent Hall mobility and electron density in the SQW. The conduction was n-type without intentional doping as shown by the experimental data in Fig. 11.22a. Since our $Zn_{0.6}Mg_{0.4}O$ has been confirmed to have roughly three orders of magnitude higher resistivity than that of ZnO [48], carriers in ZnO well should be responsible for the measured data. The electrical properties in Fig. 11.22a, however, are very different from the data in Fig. 11.8a and b of a pseudobulk ZnO film even compared with those after correction; the mobility is much higher and does not decrease at very low temperatures, and the electron density is nearly independent of temperature. This temperature-independent high mobility at very low temperatures is also different from the characteristic of high quality ZnO bulk crystals [42] which has been shown in Fig. 11.8a. The electron density in SQW is as high as $\sim 2 \times 10^{18}\,cm^{-3}$ nearly independent of temperature, the value of which is calculated assuming a uniform distribution in the 50-nm-thick ZnO well for the electrons with a sheet density of $\sim 1 \times 10^{13}\,cm^{-2}$. Such a high electron mobility with a high density at very low temperatures strongly suggests a formation of 2DEG in the SQW [17,59,80,81].

Considering the well thickness of 50 nm, however, it seems difficult to expect the 2DEG to distribute uniformly in the whole width of the potential well, and hence, presence of internal electric field along the c-axis direction should be expected to yield the maldistribution of electrons, in agreement with the discussion on the CL data of MQWs. As has been pointed out on the CL data, we ascribe the internal electric field to the polarization along the c-axis direction. In the case of SQW studied here, we can expect from the data in Fig. 11.17a that the thick bottom $Zn_{0.6}Mg_{0.4}O$

barrier is completely relaxed, and only the ZnO well is under a ~0.62% tensile strain in the lateral direction, by considering the a-axis lattice constant in $Zn_{1-x}Mg_xO$ as $0.3250+0.005x$ nm. This tensile strain should induce piezoelectric polarization, and therefore, both piezoelectric and spontaneous polarizations should be responsible for the internal electric field. It is important to note that this expectation suggests the presence of internal electric field even in the absence of top $Zn_{0.6}Mg_{0.4}O$ barrier.

In fact, nearly the same electrical properties were obtained from single-heterojunction structures like the sample (a) in Fig. 11.16; to the bare ZnO films grown on 400-nm-thick $Zn_{0.6}Mg_{0.4}O$ barrier without $Zn_{0.6}Mg_{0.4}O$ capping, the electron mobility and density at very low temperatures were found to be 500–450 cm^2/Vs and 1.4×10^{13}–1.7×10^{13} cm^{-3}, respectively, for the relevant ZnO thicknesses of 50–100 nm. Typical data on a 50-nm-thick ZnO/400-nm-thick $Zn_{0.6}Mg_{0.4}O$ single-heterojunction sample is shown in Fig. 11.22b. These experimental data strongly suggest that both 2DEGs in SQW and single-heterojunction samples are localized at the bottom heterointerface by the internal electric field. In other words, only the SQW structure has localized holes at the top heterointerface since the single-heterojunction sample lacks the barrier to accumulate holes as shown by the corresponding band diagrams of the insets in Fig. 11.22a and b. Therefore, difference in PL spectra is expected between these two heterostructures despite of the similarity in electrical properties.

We hence measured the PL spectra at 4.5 K both from the SQW and the single-heterojunction samples under various excitation powers, the traces of which are shown in Fig. 11.23a and b, respectively. The spectra in Fig. 11.23a from the SQW consist of a broad and weak peak which shifts toward the higher energy side with increasing excitation power, in addition to an intense near-band-edge emission at 3.36 eV. Such a blueshift with increasing excitation power is a characteristic of the QCS effect as has been reported by others on GaN/$Al_{1-x}Ga_xN$ MQWs [82]. In the PL spectra from the single-heterojunction sample of Fig. 11.23b, on the other hand, only the near-band-edge emission at 3.36 eV exists and the signal corresponds to the broad and weak peak in Fig. 11.23a is not observed. Since no signal corresponds to that was also observed from our pseudobulk ZnO films, we consider that the characteristic of the interested peak in Fig. 11.23a should be resulted from the QCS effect by the internal electric field in SQW.

Compared to the CL peak shift shown in Fig. 11.19a and b for MQWs, the well thickness dependence of which by the QCS effect is ~8 meV/nm, the corresponding shift in Fig. 11.23a of roughly 0.4 eV for 50-nm-thick well seems reasonable even though there are differences in excitation method, Mg content in barrier layers, and strain conditions. According to the dependence on the Mg content x in $Zn_{1-x}Mg_xO$ barrier layers, both spontaneous and piezoelectric polarizations will change with

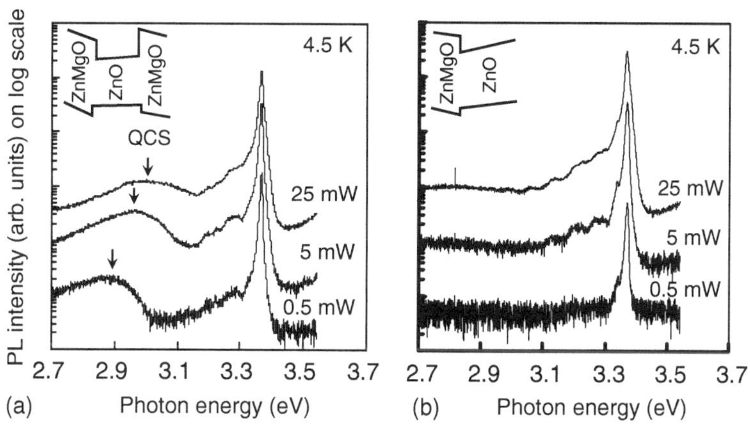

Figure 11.23: PL spectra at 4.5 K from a 50-nm-thick ZnO well of SQW structure (a), and those from a 50-nm-thick ZnO capping layer of single-heterojunction structure (b).

x. By assuming the Vegard's law, the magnitude of the spontaneous polarization in $Zn_{1-x}Mg_xO$ decrease/increase in proportion to x, from which the discontinuity of spontaneous polarization $\Delta \boldsymbol{P}_{SP}$ at heterointerface is brought about. As well, the magnitude of the piezoelectric polarization $\boldsymbol{P}_{PE}$ in strained ZnO well is proportional to the x in barrier layers. Therefore, both $|\Delta \boldsymbol{P}_{SP}|$ and $|\boldsymbol{P}_{PE}|$ are expected to increase with the Mg content x in $Zn_{1-x}Mg_xO$ barrier layers.

In the case of noncentrosymmetric compound crystals like ZnO, the polarity of heteroepitaxial films cannot be predicted in a straightforward way. However, our ZnO films grown on O-terminated a-plane sapphire substrates by MBE are known to be O-polarity ($-c$) as has been mentioned in the previous section on ZnO growth. We depict in Fig. 11.24 a schematic drawing of the crystal structure in the O-polarity growth. Since ZnO has negative spontaneous polarization as shown in Table 11.1, we can expect the direction to face the surface side as indicated by the arrows for $\boldsymbol{P}_{SP}$ in Fig. 11.25a although the magnitude is not clear for the $Zn_{0.6}Mg_{0.4}O$ buffer layer. To the piezoelectric polarization $\boldsymbol{P}_{PE}$ in strained ZnO well, on the other hand, we can evaluate the magnitude by adopting the ~0.62% tensile strain, to the value of which we calculated the a-axis lattice constant in $Zn_{1-x}Mg_xO$ as $0.3250 + 0.005x$ nm, into the following formula;

$$\begin{aligned}\boldsymbol{P}_{PE} &= e_{33}\varepsilon_z + e_{31}(\varepsilon_x + \varepsilon_y) \\ &= 2\frac{a - a_0}{a_0}\left(e_{31} - e_{33}\frac{C_{13}}{C_{33}}\right),\end{aligned} \tag{11.3}$$

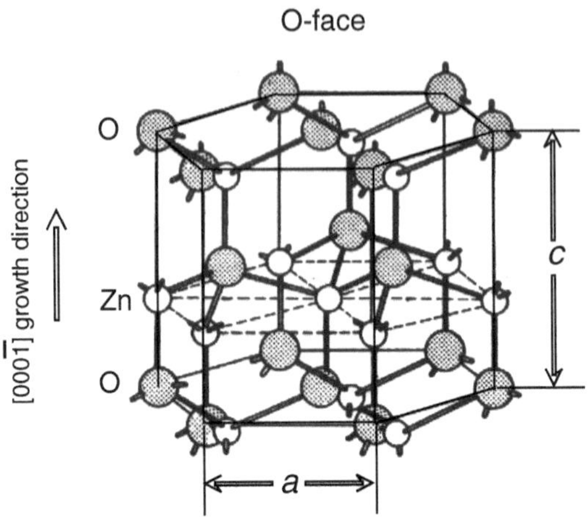

Figure 11.24: Schematic drawing of the crystal growth of wurtzite ZnO in O-polarity.

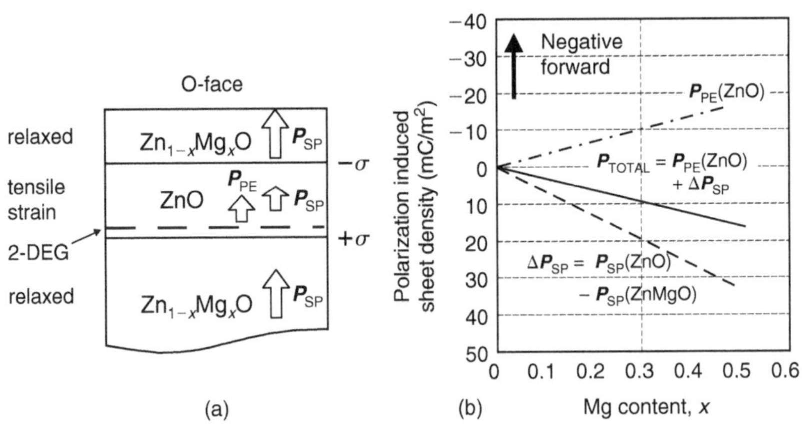

Figure 11.25: Polarization induced sheet charge density σ and directions of the spontaneous and piezoelectric polarizations in O-polarity ZnO/$Zn_{1-x}Mg_xO$ SQW (a), and the discontinuity of spontaneous polarization at heterointerface, piezoelectric and total polarizations in strained ZnO well, as a function of the Mg content x in $Zn_{1-x}Mg_xO$ barriers (b).

where ε_x and ε_y are the strain in the in-plane direction, ε_z is the strain along the c-axis direction, e_{ij} the piezoelectric constant [76], and C_{ij} the elastic constant [68,76]. Since $[e_{31} - e_{33}(C_{13}/C_{33})]$ is negative for the numerical data in Table 11.2, the piezoelectric polarization becomes negative for tensile and positive for compressive strained wells.

We calculated the $\boldsymbol{P}_{\mathrm{PE}}$ as a function of in-plane strain $\varepsilon_{\mathrm{x,y}} = (a - a_0)/a_0$, and show the expressions in Table 11.3 by comparing those for GaN, AlN (and $Ga_{1-x}Al_xN$ assuming the Vegard's law). Since $\varepsilon_{\mathrm{x,y}}$ is positive for the tensile strained ZnO on $Zn_{1-x}Mg_xO$, which is opposite to the compressive strained GaN on $Ga_{1-x}Al_xN$, the $\boldsymbol{P}_{\mathrm{PE}}$ should be negative as shown by the arrow in Fig. 11.25a. By using the $\boldsymbol{P}_{\mathrm{SP}}$ and $\boldsymbol{P}_{\mathrm{PE}}$ of respective layers, the amount of the polarization-induced sheet charge density σ at heterointerface is calculated as [68];

$$\begin{aligned}\sigma &= \boldsymbol{P}(\mathrm{ZnO}) - \boldsymbol{P}(\mathrm{ZnMgO}) \\ &= \{\boldsymbol{P}_{\mathrm{SP}}(\mathrm{ZnO}) + \boldsymbol{P}_{\mathrm{PE}}(\mathrm{ZnO})\} - \{\boldsymbol{P}_{\mathrm{SP}}(\mathrm{ZnMgO}) + \boldsymbol{P}_{\mathrm{PE}}(\mathrm{ZnMgO})\} \\ &= \boldsymbol{P}_{\mathrm{PE}}(\mathrm{ZnO}) + \{\boldsymbol{P}_{\mathrm{SP}}(\mathrm{ZnO}) - \boldsymbol{P}_{\mathrm{SP}}(\mathrm{ZnMgO})\} \\ &= \boldsymbol{P}_{\mathrm{PE}}(\mathrm{ZnO}) + \Delta\boldsymbol{P}_{\mathrm{SP}}. \end{aligned} \tag{11.4}$$

where we regarded the $\boldsymbol{P}_{\mathrm{PE}}(\mathrm{ZnMgO})$ as zero in accordance with the experimental data in Fig. 11.17a.

Our experimental results in Figs. 11.22a, b and 11.23a, b suggest the presence of 2DEG at the bottom heterointerface of SQW while this 2DEG is accumulated by the polarization-induced sheet charge density σ. Therefore, the σ at bottom heterointerface must be positive, which requires positive $\Delta\boldsymbol{P}_{\mathrm{SP}}$ with a larger value than that of $|\boldsymbol{P}_{\mathrm{PE}}(\mathrm{ZnO}) < 0|$ to satisfy Eqn. (11.4). Since both $\boldsymbol{P}_{\mathrm{SP}}(\mathrm{ZnO})$ and $\boldsymbol{P}_{\mathrm{SP}}(\mathrm{ZnMgO})$ are negative, $|\boldsymbol{P}_{\mathrm{SP}}(\mathrm{ZnO})| < |\boldsymbol{P}_{\mathrm{SP}}(\mathrm{ZnMgO})|$ is requested to obtain positive $\Delta\boldsymbol{P}_{\mathrm{SP}}$. These relationships are schematically shown in Fig. 11.25b as a probable model for strained $ZnO/Zn_{1-x}Mg_xO$ heterosystem. Although the model in Fig. 11.25b is only a qualitative one, it makes possible to explain the much smaller QCS shift in $Zn_{0.7}Mg_{0.3}O/ZnO$ MQW than that in $Ga_{0.73}Al_{0.27}N/GaN$ MQW [69]; in the former case, the internal electric field is determined by $|\Delta\boldsymbol{P}_{\mathrm{SP}}| - |\boldsymbol{P}_{\mathrm{PE}}|$, while that in the latter case by $|\Delta\boldsymbol{P}_{\mathrm{SP}}| + |\boldsymbol{P}_{\mathrm{PE}}|$ due to the difference in strain direction between the ZnO and GaN wells.

As shown by Fig. 11.25b, it is suggested that the magnitude of σ, i.e., the density of the 2DEG in ZnO well, will increase with the Mg content x in $Zn_{1-x}Mg_xO$ barrier layer. We examined the dependence on x by preparing SQWs with the same structure as that in Fig. 11.20 but have different Mg contents x of the $Zn_{1-x}Mg_xO$ barriers. The experimental data are shown in Fig. 11.26a, indicating that the 2DEG density in ZnO well linearly increases with x. This linear increase with x is consistent with the expectation from the $\boldsymbol{P}_{\mathrm{TOTAL}}$ line in Fig. 11.25b. On the other hand, when the ZnO well thickness d is varied by maintaining the Mg content at $x = 0.4$, we also observed a slight increase of 2DEG density with d as shown in Fig. 11.26b although the σ in Fig. 11.25b is independent of d. We notice that the slope of this increase, $\sim 5 \times 10^{12}\,\mathrm{cm}^{-2}$ per $d = 100\,\mathrm{nm}$, i.e., $\sim 5 \times 10^{17}\,\mathrm{cm}^{-3}$, agrees with the

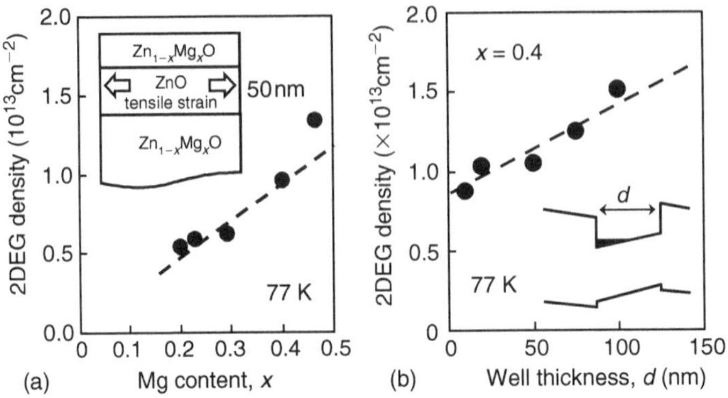

Figure 11.26: 2DEG density in SQWs at 77 K as a function of the Mg content x in $Zn_{1-x}Mg_xO$ barriers (a), and that as a function of the well thickness d (b).

typical carrier density in our pseudobulk ZnO films grown at 350°C, and hence attribute the slight increase to the contribution from the native donors in ZnO well.

Although the polarization-induced sheet charge density σ accumulates the 2DEG in quantum well, the 2DEG density also depends on the Fermi level in SQWs [83]. By the native donors in ZnO and $Zn_{1-x}Mg_xO$, accordingly, our 2DEG density in Fig. 11.26a and b might be larger than the "intrinsic" 2DEG density n_S that corresponds to σ. In awareness of this uncertainty, we tentatively estimated a value of $-83\,\mathrm{mC/m^2}$ as the maximum magnitude of the spontaneous polarization in $Zn_{0.6}Mg_{0.4}O$ using Eqn. (11.4). The following values were used in the estimation; $n_S \approx 7.5\times10^{12}\,\mathrm{cm^{-2}}$ ($\approx 12\,\mathrm{mC/m^2}$) by subtracting the conduction electrons in 50-nm-thick ZnO well of $\sim 2.5\times10^{12}\,\mathrm{cm^{-2}}$ (assuming the presence of $\sim 5\times10^{17}\,\mathrm{cm^{-3}}$ native donors) from the observed 2DEG density of $\sim 1\times10^{13}\,\mathrm{cm^{-2}}$ at $x=0.4$, $\boldsymbol{P}_{SP}(ZnO) = -57\,\mathrm{mC/m^2}$ from the numerical data in Table 11.2, and $\boldsymbol{P}_{PE}(ZnO) = -14\,\mathrm{mC/m^2}$ from the relationship in Table 11.3 of $\boldsymbol{P}_{PE}(ZnO) = -2.21\varepsilon_{x,y}$ where $\varepsilon_{x,y} = 0.62\%$ at $x = 0.4$. The calculated result might overestimate the n_S value, and hence overestimate the $|\boldsymbol{P}_{SP}(ZnMgO)|$ value since contribution from the donors in $Zn_{0.6}Mg_{0.4}O$ barrier layer was not taken into account. By assuming the Vegard's law, this amount of spontaneous polarization as the maximum value at $x = 0.4$ leads a relationship of $|\boldsymbol{P}_{SP}(ZnMgO)| \leq |-(0.057 + 0.064x)|\,\mathrm{C/m^2}$ for $Zn_{1-x}Mg_xO$ alloys with the Mg content x, which is given in Table 11.3 as ***only a rough estimation*** to compare with that of $Ga_{1-x}Al_xN$.

Although the stage of these experimental data are more than one decade behind of those on $Ga_{1-x}Al_xN$/GaN, it is strongly suggested that $Zn_{1-x}Mg_xO$/ZnO heterosystem is a realistic candidate for device applications. For instance, although the 2DEG mobility in this study is more than ten times smaller than that of $Ga_{0.75}Al_{0.25}N$/GaN

[84], we believe that a part of such an weak point is due to technical lag and will be overcome in the near future by improving the crystalline quality using a commercially available bulk ZnO substrates. To make the most of ZnO strong points, therefore, trial fabrication of realistic devices is a matter of great importance to keep the interest in device applications as a counterpart of GaN. We hence demonstrate in the next section characteristics of a prototype $Zn_{1-x}Mg_xO$/ZnO HFET to exploit the application field.

11.5 Performance of a prototype $Zn_{1-x}Mg_xO$/ZnO HFET

The prototype HFET studied here was based on a $Zn_{0.7}Mg_{0.3}O$/ZnO/$Zn_{0.7}Mg_{0.3}O$ SQW that was grown without doping on *a*-plane sapphire substrate. On a 10-nm-thick LT-ZnO initiation layer annealed at 750°C, a 400-nm-thick bottom $Zn_{0.7}Mg_{0.3}O$ barrier layer, a 20-nm-thick ZnO well layer, a 30-nm-thick top $Zn_{0.7}Mg_{0.3}O$ barrier layer, and a 20-nm-thick MgO insulation layer were successively grown at 350°C. Their growth rates were ~300 nm/h and ~350 nm/h for the ZnO well layer and the $Zn_{0.7}Mg_{0.3}O$ barrier layers, respectively, under an O_2 flow rate of 0.35 sccm, and ~150 nm/h for the MgO insulation layer under an O_2 flow rate of 0.20 sccm. From the optical transmittance data shown in Fig. 11.13b, we estimate a band-gap energy of ~4.1 eV for the $Zn_{0.7}Mg_{0.3}O$ barrier layer at room temperature. By referring a reported ratio of $\Delta E_c/\Delta E_v \cong 9$ [65], the band-edge discontinuities at heterointerface are roughly estimated to be $\Delta E_c \sim 0.7$ eV for conduction band and $\Delta E_v \sim 0.08$ eV for valence band, and hence the SQW forms a deep potential well for electrons.

The grown film was confirmed to be *c*-axis oriented single-crystal using XRD analysis, and was fabricated into the HFET structure using a conventional photolithography technique combined with electron-cyclotron-resonance (ECR) plasma etching. The ECR plasma etching was performed using an Elionix-made EIS-200ER system operated at 2.45 GHz with a microwave power of 80 W and a dc biasing of 850 V in a 0.01 Pa mixture gas of CF_4 and CH_4 (2 : 3).

Figure 11.27 depicts the etching rate of ZnO as a function of the mixing ratio x of CF_4 and CH_4 gasses [$x = F_{CF_4}/(F_{CH_4} + F_{CF_4})$, where F_{CF_4} and F_{CH_4} are the flow rates of CF_4 and CH_4 gasses, respectively]. These data were taken under a microwave power of 80 W and a dc biasing of 1000 V. We found that the etching rate at $x = 0$ (pure CH_4) was larger than that at $x = 1$ (pure CF_4). Interestingly, a dramatic increase in the etching rate was found for the mixture of two gasses, and the maximum etching rate of 40 nm/min was obtained at $x = 0.6$, maintaining the total flow rate at 1.0 sccm. This tendency can be interpreted as follows by comparing the data with the mechanism of ZnSe etching using CF_4 and CH_4 gasses [85–87]. Namely,

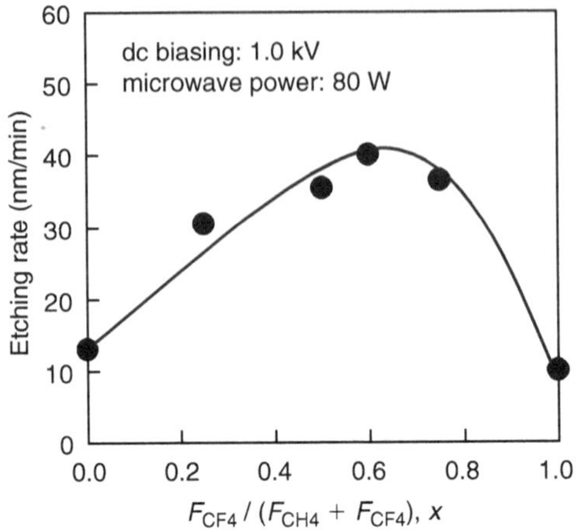

Figure 11.27: ZnO etching rate as a function of the mixing ratio of CF_4 and CH_4 gasses. The total flow rate of the mixture gas was 1.0 sccm independent of the mixing ratio.

$Zn(CH_3)_2$ and H_2O are the by-products of the ZnO etching using CH_4, whereas fluoride compounds such as ZnF_2 are those of the ZnO etching using CF_4. Due to the low vapor pressures of fluoride compounds compared to that of $Zn(CH_3)_2$, the etching using CF_4 is proceeded by sputtering-off the fluoride compounds that are formed on the ZnO surface. Similar etching mechanism has also been reported by others for the etching process using inductively-coupled plasma [88]. By mixing these two gasses, both CF_4 and CH_4 might play complementary roles in the etching process presumably due to the simultaneous occurrence of both chemical and physical etchings. Such an enhancement of etching rate with the gas mixing will be more important in the case of $Zn_{1-x}Mg_xO$ processing since it contains Mg-O bonds that are much strongly connected each other compared to Zn-O bonds [45].

Schematic of the prototype HFET is shown by Fig. 11.28 a with a top view photograph of Fig. 11.28b where the dimension of the gate electrode is 50-μm-long and 50-μm-wide. We used a Au/In (200 nm/20 nm in thickness) bilayer metal that was evaporated on the ZnO recesses formed by the ECR plasma etching and then alloyed at 400°C for 5 min to obtain the ohmic contacts for source and drain electrodes, and a Au/Ti (70 nm/20 nm in thickness) bilayer metal that was evaporated on the MgO surface without alloying to obtain the Schottky contact for gate electrode.

Figure 11.29 shows typical temperature dependences of the electron mobility and sheet density in the SQW channel. These data were measured in the dark by forming a Hall bridge from the same piece of the film as used for the HFET. We can see that the conduction electrons in SQW are distinguished by a nearly temperature independent sheet density of $\sim 6 \times 10^{12}\,cm^{-2}$ and a high Hall mobility (μ_{Hall}) of $130\,cm^2/Vs$

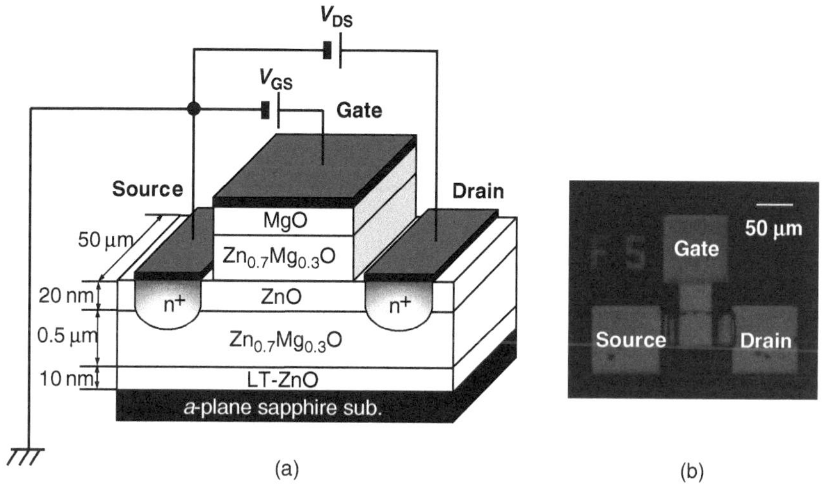

Figure 11.28: Schematic of a prototype HFET (a), and its top view photograph (b).

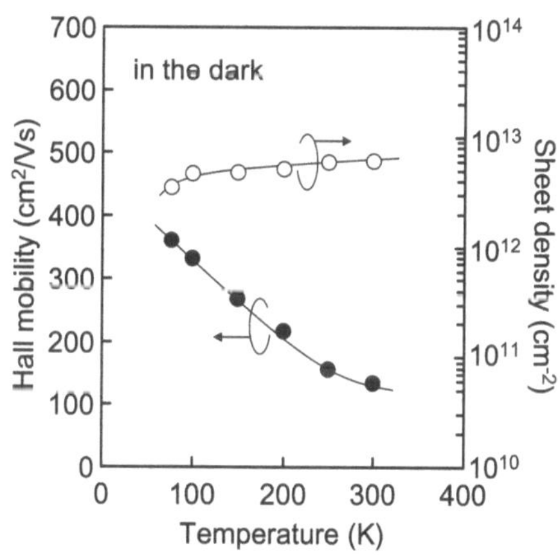

Figure 11.29: Temperature dependent Hall mobility and density of the conduction electrons in 20-nm-thick ZnO well of SQW structure.

at room temperature with a continuous increase even at the lower temperatures than 100 K. Although both values are a little smaller than those in Fig. 11.22a for a SQW with a 50-nm-thick ZnO well in $Zn_{0.6}Mg_{0.4}O$ barriers, their temperature dependences are in agreement with the characteristic of the 2DEG that is generated at the bottom heterointerface by the internal electric field in c-axis oriented SQWs.

In Fig. 11.30a, we show a typical drain-source current dependence on drain-source voltage (I_{DS}-V_{DS}) for the prototype HFET at room temperature in the dark, where the gate-source voltage (V_{GS}) was changed from $+1.0$ V to -8.0 V by 1.0 V

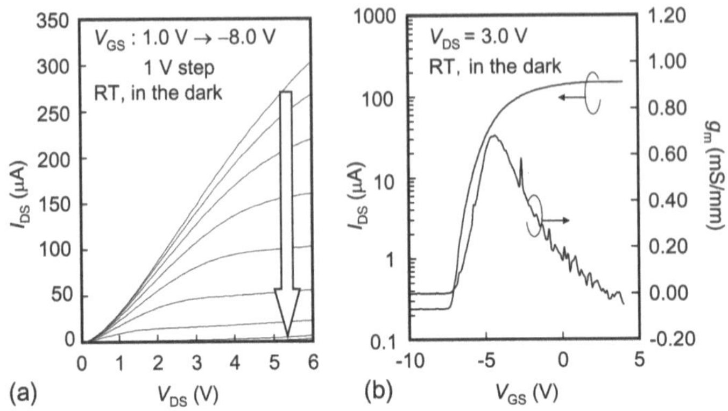

Figure 11.30: I_{DS}-V_{DS} characteristics (a), and I_{DS}-V_{GS} and g_m-V_{GS} characteristics at $V_{DS} = 3.0\,V$ (b), of a prototype HFET.

steps. It is obvious that the operation of the device is an n-channel depletion mode in agreement with the expectation from the 2DEG formation in SQW. Although the initial rise up in I_{DS} current is not sharp at small V_{DS} voltages, which might be due to an incompleteness of ohmic contact as source and drain electrodes, a clear pinch-off and a current saturation is observed, indicating that this operation is in accordance with the standard theory of FET. We should mention that the incompleteness in ohmic contact formation is presumably resulted from the remaining of top $Zn_{0.7}Mg_{0.3}O$ barrier layer on ZnO recesses; precise control in etching depth was not enough in this experiment due to the difficulty in selective etching. We found a much higher etching rate of ZnO layer compared to those of $Zn_{0.7}Mg_{0.3}O$ and MgO layers.

Figure 11.30b shows the I_{DS}-V_{GS} relationship at $V_{DS} = 3.0\,V$. The saturation is about 150 μA at around $V_{GS} = 2.0\,V$. The threshold voltage (V_{TH}) derived from a straight line fitted to the plots of the square root of I_{DS} vs. V_{GS} at the saturation region was $\sim -7.0\,V$, according to the following formulae;

$$I_{DS} = \frac{W \mu_{FE} C_i}{2L}(V_{GS} - V_{TH})^2 \quad \text{for} \quad V_{DS} > V_{GS} - V_{TH}, \tag{11.5}$$

$$1/C_i = 1/C_{MgO} + 1/C_{ZnMgO} + 1/C_{ZnO}, \tag{11.6}$$

$$C_{MgO} = \frac{\varepsilon_0 \varepsilon_{MgO}}{d_1}, \quad C_{ZnMgO} = \frac{\varepsilon_0 \varepsilon_{ZnMgO}}{d_2}, \quad C_{ZnO} = \frac{\varepsilon_0 \varepsilon_{ZnO}}{d_3}, \tag{11.7}$$

where μ_{FE} is the field-effect mobility, C_i the capacitance per unit area of the gate insulator, ε_0 the permittivity of vacuum, ε_{MgO}, ε_{ZnMgO} and ε_{ZnO} the relative permittivities of MgO, $Zn_{0.7}Mg_{0.3}O$ and ZnO, respectively, and d_1, d_2 and d_3 the thicknesses of top MgO layer, top $Zn_{0.7}Mg_{0.3}O$ barrier layer and ZnO well layer, respectively. The relative dielectric constant in $Zn_{0.7}Mg_{0.3}O$ was taken as 9.02 by assuming a linear increase from $\varepsilon_{ZnO} = 8.75$ [89] to $\varepsilon_{MgO} = 9.64$ [90] with the Mg content in $Zn_{1-x}Mg_xO$. Here, we assume that the 2DEG is localized at the bottom interface in the ZnO well according to the experimental results discussed in the previous section. The evaluated value for μ_{FE} is ~140 cm^2/Vs, and the maximum transconductance g_m is estimated to be ~0.035 mS (~0.70 mS/mm) at $V_{GS} = -4.6$ V as shown by the g_m-V_{GS} curve in Fig. 11.30b.

This μ_{FE} value evaluated is well accorded with that of μ_{Hall} measured by Hall effect, suggesting a sharp interface at the ZnO well/bottom $Zn_{0.7}Mg_{0.3}O$ barrier with no degradation of crystalline quality despite of the ion bombardment during ECR etching. As has been reported elsewhere [85], we believe that the annealing process at 400°C to obtain the ohmic contacts was also effective to recover the damage of the ion bombardment. This large μ_{FE} as high as 140 cm^2/Vs is 10^2–10^3 times larger in magnitude compared to those reported as typical values for conventional TFTs with polycrystalline ZnO channels grown on glass substrates [91,92]. We also prepared a top-gate type TFT with the gate electrode dimensions of 40-μm-long and 50-μm-wide but using a 70-nm-thick single-crystalline ZnO film on *a*-plane sapphire substrate as the channel and a 70-nm-thick evaporated Al_2O_3 film on the channel as the gate insulation layer [93]. As shown by the I_{DS}-V_{DS} characteristics in Fig. 11.31, however, its typical device performance was $g_m \cong 0.10$ mS/mm and $\mu_{FE} \cong 10$ cm^2/Vs. These values are one order of magnitude smaller than those obtained for the present HFET. The μ_{FE} value of 140 cm^2/Vs is even higher than the reported one for a back-gate type TFT using a high quality single-crystalline ZnO film grown on a lattice matched $ScAlMgO_4$ substrate by pulsed laser deposition ($\mu_{FE} = 70$ cm^2/Vs and $g_m \cong 0.018$ μS at room temperature) [94,95], demonstrating much advantageous feature of our HFET for practical electronic operations.

We also examined the I_{DS}-V_{GS} relationship at $V_{DS} = 3.0$ V by illuminating the prototype HFET from the substrate-side. Although the data are not shown here, the saturation current in I_{DS} was found to increase in proportion to the illumination intensity when a fluorescence lamp was used, whereas it scarcely changed when an incandescent lamp was used. These characteristics are consistent with the optical transmittance spectrum shown in Fig. 11.6 for a pseudobulk ZnO film, and are pronounced for the applications to transparent transistors insensitive to visible light illumination.

However, the on/off ratio of I_{DS} shown in Fig. 11.30b is only ~800, which might be not enough for realistic applications. We found that the I_{DS} in subthreshold region

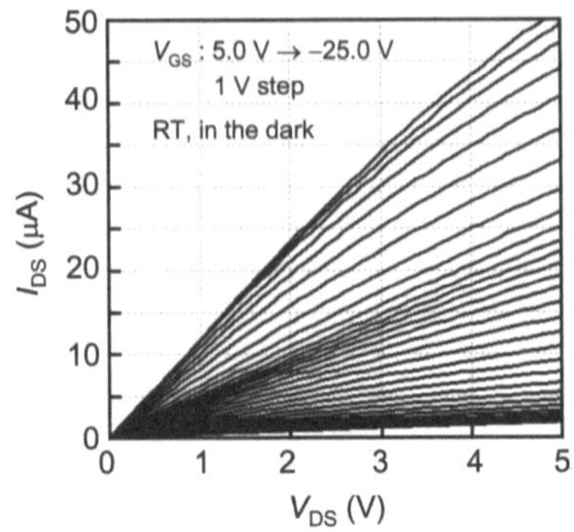

Figure 11.31: I_{DS}-V_{GS} characteristics of a top-gate type TFT with a single-crystalline ZnO channel.

($V_{GS} < -7$ V) increased in proportion to the V_{DS} used for the I_{DS}-V_{GS} measurement, suggesting the presence of a leakage current through the bottom $Zn_{0.7}Mg_{0.3}O$ barrier. In fact, the resistance of the bottom $Zn_{0.7}Mg_{0.3}O$ barrier expected from the geometric dimensions of HFET and the resistivity of a pseudobulk $Zn_{0.7}Mg_{0.3}O$ film (~500 Ωcm) is close to 10 MΩ which is consistent with the observed resistance at $V_{DS} = 3$ V. Therefore, the low on/off ratio, we believe, will be improved by increasing the resistivity of the bottom $Zn_{0.7}Mg_{0.3}O$ barrier using Li doping [96,97] or N doping [98,99] techniques for instance.

In conclusion, we demonstrated the fabrication process and characteristics of a prototype $Zn_{0.7}Mg_{0.3}O$/ZnO HFET in relation to the optical and electrical properties of the heterosystem. Although there remain several points to be improved, this HFET showed a high g_m of 0.70 mS/mm and a high μ_{FE} of 140 cm^2/Vs as a potential device performance for the future development in such a new application field as transparent central processing units. Needless to say, the presented data are based on the ZnO films grown on sapphire substrates, and therefore, much higher device performance can be expected when lattice-matched substrates are used. Another important point is that the single-crystalline ZnO growth in *c*-axis orientation has been performed on (111)-oriented Si substrates using a low-temperature growth by MBE [43,61,62,100,101], therefore, it will also be able to integrate the ZnO-based HFETs with Si circuitry in the near future to make the best use of the complementary characteristics of both materials.

11.6 Conclusion

In this chapter, we reviewed recent activities on $Zn_{1-x}Mg_xO$/ZnO heterosystem based on the properties of respective constituent materials of ZnO and

$Zn_{1-x}Mg_xO$ epitaxial films. These films were grown on *a*-plane sapphire substrates using plasma-assisted MBE. The films were grown in *c*-axis oriented O-polarity single-crystals without the inclusion of in-plane rotational domains. The ZnO films were *n*-type conduction with no intentional doping, and had a typical mobility of $\sim 100\,cm^2/Vs$ and a density of $\sim 3 \times 10^{17}\,cm^{-3}$ at room temperature. Although the major part of the ZnO films has compatible electrical properties to that of bulk crystals, due to the large lattice mismatch to the substrates, there exists a low quality degenerated region nearby the substrates. According to the Mg doping into ZnO, a semi-empirical relationship between the Mg content x in $Zn_{1-x}Mg_xO$ and the growth parameters is proposed successfully based on a simplified growth mechanism. The existence of miscibility gap was confirmed in accordance with a prediction based on thermodynamics, however, the solubility limit of the Mg content x in wurtzite $Zn_{1-x}Mg_xO$ was found to be extended to $x \approx 0.4$ by using the metastable growth condition of MBE at low temperatures. Although the metastable growth was performed at low growth temperatures, the grown alloy films at high Mg contents were found to be stable against the post growth annealing over 700°C.

The strained growth of $Zn_{1-x}Mg_xO$/ZnO heterosystem due to lattice mismatch was studied, and it was found that the pseudomorphic growth was stable at 400°C for both 50-nm-thick ZnO/400-nm-thick $Zn_{0.7}Mg_{0.3}O$ and 50-nm-thick $Zn_{0.7}Mg_{0.3}O$/400-nm-thick ZnO heterostructures. This strained growth was applied to prepare $Zn_{0.7}Mg_{0.3}O$/ZnO MQWs with different ZnO well thicknesses. Their controlled multilayered structures in atomic scale were confirmed by measuring the satellite peaks in XRD profiles. Presence of internal electric field in the ZnO well was found by measuring the QCS effect in the CL spectra from these MQWs. The degree of the redshift in CL peak by the QCS effect, however, was more than several times smaller than the reported values for $Ga_{1-x}Al_xN$/GaN MQWs.

To study the internal electric filed in $Zn_{1-x}Mg_xO$/ZnO, we further studied optical and electrical properties of SQWs, and revealed the presence of 2DEG of which density increased with the Mg content x in $Zn_{1-x}Mg_xO$ barrier layers. It was also strongly suggested that the 2DEGs were localized at the bottom heterointerface in SQWs. Although the presence of the QCS effect in CL and the 2DEG formation at heterointerface are similar to the case of $Ga_{1-x}Al_xN$/GaN system, there exists a difference in strain direction; the piezoelectric polarization $\boldsymbol{P}_{PE}$ is negative in the tensile strained ZnO well while positive in the compressive strained GaN well. To understand the difference in polarization, we take the total polarization $\boldsymbol{P}_{TOTAL} = \boldsymbol{P}_{PE} + \Delta\boldsymbol{P}_{SP}$ into account, where $\Delta\boldsymbol{P}_{SP} = \boldsymbol{P}_{SP}(\text{well}) - \boldsymbol{P}_{SP}(\text{barrier})$ is the discontinuity of spontaneous polarization at heterointerface. Assuming a positive $\Delta\boldsymbol{P}_{SP}$ similar to that in $Ga_{1-x}Al_xN$/GaN and a small $\boldsymbol{P}_{PE}$ value of $|\boldsymbol{P}_{PE}| < \Delta\boldsymbol{P}_{SP}$, we obtain a positive $\boldsymbol{P}_{TOTAL}$ for $Zn_{1-x}Mg_xO$/ZnO with a much smaller magnitude than that of $Ga_{1-x}Al_xN$/GaN. By using the values of elastic and piezoelectric constants and spontaneous polarization of ZnO in literature, and the values of experimentally determined strain and 2DEG density, we tentatively estimated the spontaneous polarization in $Zn_{1-x}Mg_xO$ to be $-(0.057 + 0.064x)\,C/m^2$ as a function of the Mg content x.

A prototype HFET with a 2DEG in SQW was fabricated using a photolithography technique combined with ECR plasma etching. The characteristic of HFET was n-channel depletion mode, and a current saturation in I_{DS}-V_{DS} relationship was observed, indicating that this operation was in accordance with the standard theory of FET. A large μ_{FE} of about $140\,cm^2/Vs$ was obtained for the HFET at room temperature, the value of which was close to that of μ_{Hall}. The resulted g_m of 0.70 mS/mm was much higher than those reported for ZnO based TFTs as expected from the electrical properties of the 2DEGs in SQWs. Although there remain several points to be improved, this HFET has a high performance as a potential device in the near future.

Acknowledgements

A part of this work was performed as a project of Bio Venture Center in OIT. One of the authors (K.K.) was supported for this work by a Grant-in-Aid for Young Scientists (B) from the Ministry of Education, Culture, Sports, Science and Technology of Japan (#15760248). We express our thanks to Dr. Ken-ichi Ogata, a former member of our project, for his contribution to the optical analysis and valuable discussions, and also to Dr. Katsuhiko Inaba of X-ray Research Laboratory at Rigaku Corporation for his help on the XRD analysis of strained structure using a contour mapping technique in reciprocal space.

References

[1] Y. Makita, H. Tanoue, Mater. Sci. Technol. 37 (2000) 1 [in Japanese].
[2] CRC Handbook of Chemistry and Physics, ed. by D. R. Lide, 77th ed. (1996) 14–14, 7–14.
[3] D. C. Reynolds, D. C. Look, B. Jogai, J. Solid State Commun. 99 (1996) 873.
[4] Y. Chen, N. T. Tuan, Y. Segawa, H. Ko, S. Hong, T. Yao, Appl. Phys. Lett. 78 (2001) 1469.
[5] A. Tsukazaki, A. Ohtomo, T. Onuma, M. Ohtani, T. Makino, M. Sumiya, K. Ohtani, S. F. Chichibu, S. Fuke, Y. Segawa, H. Ohno, H. Koinuma, M. Kawasaki, Nature Mater. 4 (2005) 42.
[6] J. D. Albrecht, P. P. Ruden, S. Limpijumnong, W. R. L. Lambrecht, K. F. Brennan, J. Appl. Phys. 86 (1999) 6864.

[7] N. Sakagami, M. Yamashita, T. Sekiguchi, S. Miyashita, K. Obara, T. Shishido, J. Cryst. Growth 229 (2001) 98.
[8] E. Ohshima, H. Ogino, I. Niikura, K. Maeda, M. Sato, M. Ito, T. Fukuda, J. Cryst. Growth 260 (2004) 166.
[9] K. Maeda, M. Sato, I. Niikura, T. Fukuda, Semicond. Sci. Technol. 20, S49 (2005).
[10] Y. Li, G. S. Tompa, S. Liang, C. Gorla, Y. Lu, J. Doyle, J. Vac. Sci. Technol. A 15 (1997) 1063.
[11] A. N. Mariano, R. E. Hanneman, J. Appl. Phys. 34 (1963) 384.
[12] W. Yang, R. D. Vispute, S. Choopun, R. P. Sharma, T. Venkatesan, H. Shen, Appl. Phys. Lett. 78 (2001) 2787.
[13] S. O. Kucheyev, C. Jagadish, J. S. Williams, P. N. K. Deenapanray, M. Yano, K. Koike, S. Sasa, M. Inoue, K. Ogata, J. Appl. Phys. 93 (2003) 2972.
[14] C. Coskun, D. C. Look, G. C. Farlow, J. R. Sizelove, Semicond. Sci. Technol. 19 (2004) 752.
[15] D. G. Thomas, J. Phys. Chem. Solids 15 (1960) 86.
[16] E. M. C. Fortunato, P. M. C. Barquinha, A. C. M. B. G. Pimentel, A. M. F. Gonçalves, A. J. S. Marques, R. F. P. Martins, L. M. N. Pereira, Appl. Phys. Lett. 85 (2004) 2541.
[17] K. Koike, K. Hama, I. Nakashima, G. Takada, M. Ozaki, K. Ogata, S. Sasa, M. Inoue, M. Yano, Jpn. J. Appl. Phys. 43 (2004) L1372.
[18] Y. Chen, D. M. Bagnall, H. Ko, K. Park, K. Hiraga, Z. Zhu, T. Yao, J. Appl. Phys. 84 (1998) 3912.
[19] E. M. Kaidashev, M. Lorenz, H. Wenckstern, A. Rahm, H. C. Semmelhack, K. H. Han, G. Benndorf, C. Bundesmann, H. Hochmuth, M. Grundmann, Appl. Phys. Lett. 82 (2003) 3901.
[20] K. Hümmer, Phys. Stat. Sol. 56 (1973) 249.
[21] H. Karzel, W. Potzel, M. Köfferlein, W. Schiessl, M. Steiner, U. Hiller, G. M. Kalvius, D. W. Mitchell, T. P. Das, P. Blaha, K. Schwarz, M. P. Pasternak, Phys. Rev. B 53 (1996) 11425.
[22] S. J. Pearton, D. P. Norton, K. Ip, W. Heo, T. Steiner, J. Vac. Sci. Technol. B 22 (2004) 932.
[23] R. C. Whited, C. J. Flaten, W. C. Walker, Solid State Commun. 13 (1973) 1903.
[24] K. Hirata, K. Moriya, Y. Waseda, J. Mater. Sci. 12 (1977) 838.
[25] G. K. White, O. L. Anderson, J. Appl. Phys. 37 (1966) 430.
[26] W. D. Kingery, J. Nucl. Mater. 24 (1967) 21.
[27] J. Zussman, G. W. Brindley, Am. Mineralogist 42 (1957) 666.
[28] P. Fons, K. Iwata, A. Yamada, K. Matsubara, S. Niki, K. Nakahara, T. Tanabe, H. Takasu, Appl. Phys. Lett. 77 (2000) 1801.
[29] V. A. Coleman, J. E. Bradby, C. Jagadish, P. Munroe, Y. W. Heo, S. J. Pearton, D. P. Norton, M. Inoue, M. Yano, Appl. Phys. Lett. 86 (2005) 203105.
[30] K. Iwata, P. Fons, S. Niki, A. Yamada, K. Matsubara, K. Nakahara, H. Takasu, Phys. Stat. Sol. (a) 180 (2000) 287.
[31] Y. Chen, H. Ko, S. Hong, T. Yao, Y. Segawa, J. Cryst. Growth 214/215 (2000) 87.

[32] P. Fons, K. Iwata, S. Niki, A. Yamada, K. Matsubara, M. Watanabe, J. Cryst. Growth 209 (2000) 532.
[33] A. Ohtomo, M. Kawasaki, T. Koida, K. Masubuchi, H. Koinuma, Y. Sakurai, Y. Yoshida, T. Yasuda, Y. Segawa, Appl. Phys. Lett. 72 (1998) 2466.
[34] H. Kato, K. Miyamoto, M. Sano, T. Yao, Appl. Phys. Lett. 84 (2004) 4562.
[35] D. C. Reynolds, D. C. Look, B. Jogai, C. W. Litton, T. C. Collins, W. Harsch, G. Cantwell, Phys. Rev. B 57 (1998) 12151.
[36] K. Nakahara, T. Tanabe, H. Takasu, P. Fons, K. Iwata, A. Yamada, K. Matsubara, R. Hunger, S. Niki, Jpn. J. Appl. Phys. 40 (2001) 250.
[37] K. Miyamoto, M. Sano, H. Kato, T. Yao, Jpn. J. Appl. Phys. 41 (2002) L1203.
[38] A. Ohtomo, H. Kimura, K. Sato, T. Makino, Y. Segawa, H. Koinuma, M. Kawasaki, J. Cryst. Growth 214–215 (2000) 284.
[39] D. C. Look, R. J. Molnar, Appl. Phys. Lett. 70 (1997) 3377.
[40] H. Tampo, A. Yamada, P. Fons, H. Shibata, K. Matsubara, K. Iwata, K. Nakahara, S. Niki, Phys. Stat. Sol. (c) 1 (2004) 888.
[41] K. J. Hagemark, L. C. Chacka, J. Solid State Chem. 15 (1975) 261.
[42] R. Helbig, P. Wagner, J. Phys. Chem. Solids 35 (1974) 327.
[43] M. Yano, K. Ogata, F. P. Yan, K. Koike, S. Sasa, M. Inoue, Mater. Res. Soc. Symp. Proc. 744 (2003) 75.
[44] Y. Fengping, J. Shuisheng, K. Ogata, K. Koike, S. Sasa, M. Inoue, M. Yano, Science in China Series E-Engineering & Mater. Sci. 47 (2004) 166.
[45] D. D. Wagman, W. H. Evans, V. B. Parker, R. H. Schumm, I. Halow, S. M. Bailey, K. I. Churney, R. L. Nuttall, "The NBS tables of chemical thermodynamic properties", [J. Phys. Chem. Ref. Data 11 (1982) Suppl.2], ACS and AIP for NBS (1982).
[46] J. F. Sarver, F. L. Katnack, F. A. Hummel, J. Electrochem. Soc. 106 (1959) 960.
[47] K. Ogata, K. Koike, T. Tanite, T. Komuro, F. Yan, S. Sasa, M. Inoue, M. Yano, J. Cryst. Growth 251 (2003) 623.
[48] K. Koike, K. Hama, I. Nakashima, G. Takada, K. Ogata, S. Sasa, M. Inoue, M. Yano, J. Cryst. Growth 278 (2005) 288.
[49] Y. Matsumoto, M. Murakami, Z. Jin, A. Ohtomo, M. Lippmaa, M. Kawasaki, H. Koinuma, Jpn. J. Appl. Phys. 38 (1999) L603.
[50] W. I. Park, G. C. Yi, H. M. Jang, Appl. Phys. Lett. 79 (2001) 2022.
[51] S. Choopun, R. D. Vispute, W. Yang, R. P. Sharma, T. Venkatesan, H. Shen, Appl. Phys. Lett. 80 (2002) 1529.
[52] Sz. Fujita, T. Takagi, H. Tanaka, Sg. Fujita, Phys. Stat. Sol. (b) 241 (2004) 599.
[53] S. Sadofev, S. Blumstengel, J. Cui, J. Puls, S. Rogaschewski, P. Schäfer, Y. G. Sadofyev, F. Henneberger, Appl. Phys. Lett. 87 (2005) 091903.
[54] S. Limpijumnong, W. R. L. Lambrecht, Phys. Rev. B 63 (2001) 104103.
[55] A. Ohtomo, R. Shiroki, I. Ohkubo, H. Koinuma, M. Kawasaki, Appl. Phys. Lett. 75 (1999) 4088.
[56] C. R. Gorla, W. E. Mayo, S. Liang, Y. Lu, J. Appl. Phys. 87 (2000) 3736.
[57] T. Makino, T. Yasuda, Y. Segawa, A. Ohtomo, K. Tamura, M. Kawasaki, H. Koinuma, Appl. Phys. Lett. 79 (2001) 1282.

[58] H. C. Ong, A. S. K. Li, G. T. Du, Appl. Phys. Lett. 78 (2001) 2667.
[59] K. Koike, I. Nakashima, K. Hashimoto, S. Sasa, M. Inoue, M. Yano, Appl. Phys. Lett. 87 (2005) 112106.
[60] X. L. Wu, G. G. Siu, C. L. Fu, H. C. Ong, Appl. Phys. Lett. 78, 2285 (2001).
[61] K. Koike, T. Tanite, S. Sasa, M. Inoue, M. Yano, Mater. Res. Soc. Symp. Proc. 692 (2002) 655.
[62] K. Koike, T. Komuro, K. Ogata, S. Sasa, M. Inoue, M. Yano, Physica E 21 (2004) 679.
[63] E. Mollwo, R. Till, Z. Phys. 216 (1968) 315.
[64] B. P. Zhang, N. T. Binh, K. Wakatsuki, C. Y. Liu, Y. Segawa, N. Usami, Appl. Phys. Lett. 86 (2005) 032105.
[65] T. Makino, C. H. Chia, N. T. Tuan, H. D. Sun, Y. Segawa, M. Kawasaki, A. Ohtomo, K. Tamura, H. Koinuma, Appl. Phys. Lett. 77 (2000) 975.
[66] H. D. Sun, T. Makino, Y. Segawa, M. Kawasaki, A. Ohtomo, K. Tamura, H. Koinuma, J. Appl. Phys. 91 (2002) 1993.
[67] T. Makino, A. Ohtomo, C. H. Chia, Y. Segawa, H. Koinuma, M. Kawasaki, Physica E 21 (2004) 671.
[68] O. Ambacher, J. Smart, J. R. Shealy, N. G. Weimann, K. Chu, M. Murphy, W. J. Schaff, L. F. Eastman, R. Dimitrov, L. Wittmer, M. Stutzmann, W. Rieger, J. Hilsenbeck, J. Appl. Phys. 85 (1999) 3222.
[69] N. Grandjean, B. Damilano, S. Dalmasso, M. Leroux, M. Laügt, J. Massies, J. Appl. Phys. 86 (1999) 3714.
[70] J. S. Im, H. Kollmer, J. Off, A. Sohmer, F. Scholz, A. Hangleiter, Phys. Rev. B 57 (1998) R9435.
[71] *Properties of Group III Nitrides*, ed. by J. H. Edgar (INSPEC, London, 1994).
[72] A. F. Wright, J. Appl. Phys. 82 (1997) 2833.
[73] K. Tsubouchi, N. Miskoshiba, IEEE Trans. Sonics Ultrason. SU-32 (1985) 634.
[74] F. Bernardini, V. Fiorentini, D. Vanderbilt, Phys. Rev. B 56 (1997) 10024.
[75] G. D. O'Clock, M. T. Duffy, Appl. Phys. Lett. 23 (1973) 55.
[76] I. B. Kobiakov, Solid State Commun. 35 (1980) 305.
[77] R. M. Martin, Phys. Rev. B 5 (1972) 1607.
[78] Y. Noel, C. M. Zicovich-Wilson, B. Civalleri, Ph. D'Arco, R. Dovesi, Phys. Rev. B 65 (2001) 014111-1.
[79] Y. Sumino, I. Ohno, T. Goto, J. Phys. Earth 24 (1976) 263.
[80] K. Koike, K. Hama, I. Nakashima, S. Sasa, M. Inoue, M. Yano, Jpn. J. Appl. Phys. 44 (2005) 3822.
[81] T. Edahiro, N. Fujimura, T. Ito, J. Appl. Phys. 93 (2003) 7673.
[82] E. L. Waldron, E. F. Schubert, J. W. Graff, A. Osinsky, M. J. Murphy, W. F. Schaff, Mater. Res. Soc. Symp. Proc. 639 (2000) G1.11.
[83] E. T. Yu, G. J. Sollivan, P. M. Asbeck, C. D. Wang, D. Qiao, S. S. Lau, Appl. Phys. Lett. 71 (1997) 2794.
[84] M. J. Manfra, N. G. Weimann, J. W. P. Hsu, L. N. Pfeiffer, K. W. West, S. Syed, H. L. Stormer, W. Pan, D. V. Lang, S. N. G. Chu, G. Kowach, A. M. Sergent,

J. Caissie, K. M. Molvar, L. J. Mahoney, R. J. Molnar, J. Appl. Phys. 92 (2002) 338.
[85] K. Ogata, T. Honden, T. Tanite, T. Komuro, K. Koike, S. Sasa, M. Inoue, M. Yano, J. Vac. Sci. Technol. A 22 (2004) 531.
[86] C. R. Eddy, Jr., D. Leonhardt, V. A. Shamamian, J. E. Butler, J. Electron. Mater. 30 (2001) 538.
[87] J. M. Lee, K. M. Chang, K. K. Kim, W. K. Choi, S. J. Park, J. Electrochem. Soc. 148 (2001) G1.
[88] S. R. Droes, T. T. Kodas, M. J. Hampden-Smith, Adv. Mater. (Weinheim, Ger.) 10 (1998) 1129.
[89] E. C. Heltemes, H. L. Swinney, J. Appl. Phys. 38 (1967) 2387.
[90] J. Jasperse, A. Kahan, J. N. Plendl, S. S. Mitra, Phys. Rev. 146 (1966) 526.
[91] S. Masuda, K. Kitamura, Y. Okumura, S. Miyatake, H. Tabata, T. Kawai, J. Appl. Phys. 93 (2003) 1624.
[92] R. L. Hoffman, B. J. Norris, J. F. Wager, Appl. Phys. Lett. 82 (2003) 733.
[93] A. Nakashima, T. Tanite, T. Komuro, K. Koike, K. Ogata, S. Sasa, M. Yano, M. Inoue, Extended Abstracts, No.1 (The 50th Spring Meeting, 2003); The Japan Society of Applied Physics and Related Societies, 2003, p.344 [in Japanese].
[94] T. I. Suzuki, A. Ohtomo, A. Tsukazaki, F. Sato, J. Nishii, H. Ohno, M. Kawasaki, Adv. Mater. 16 (2004) 1887.
[95] A. Ohtomo, M. Kawasaki, H. Ohno, *Bulletin of Solid State Physics and Applications* (Japan Society of Applied Physics, 11, 2005), p.26 [in Japanese].
[96] Y. J. Zeng, Z. Z. Ye, W. Z. Xu, L. L. Chen, D. Y. Li, Li. P. Zhu, B. H. Zhao, Y. L. Hu, J. Cryst. Growth 283 (2005) 180.
[97] D. Zwingel, J. Lumin. 5 (1972) 385.
[98] D. C. Look, D. C. Reynolds, C. W. Litton, R. L. Jones, D. B. Eason, G. Cantwell, Appl. Phys. Lett. 81 (2002) 1830.
[99] N. Y. Garces, N. C. Giles, L. E. Halliburton, G. Cantwell, D. B. Eason, D. C. Reynolds, D. C. Look, Appl. Phys. Lett. 80 (2002) 1334.
[100] M. Fujita, N. Kawamoto, T. Tatsumi, K. Yamagishi, Y. Horikosi, Jpn. J. Appl. Phys. 42 (2003) 67.
[101] C. Jin, R. Narayan, A. Tiwari, H. Zhou, A. Kvit, J. Narayan, Mater. Sci. & Engineering B 117 (2005) 348.

Chapter 12

ZnO Thin-Film Transistors

R. L. Hoffman

Hewlett-Packard Company, Corvallis, Oregon, USA

12.1 Introduction

Although the thin-film transistor (TFT) was first conceptualized in the 1930s [1,2], the rise to prominence of TFT technology has occurred only relatively recently, in conjunction with the emergence of the active-matrix liquid crystal display (AMLCD) toward its current status as the cornerstone of a rapidly growing, multi-billion dollar industry [3,4]. In contrast to wafer-based Si CMOS technology, in which both performance improvements and cost reductions are linked closely to continued miniaturization and increasing circuit density, TFT technology is nicely suited to applications in which distributed, low-density circuitry is applied across a large area, as exemplified in the TFT matrix of an AMLCD backplane [3].

The supremacy of Si in the microelectronics industry holds also for TFT technology, in which amorphous-Si (a-Si) and polycrystalline-Si (poly-Si) TFTs effectively dominate current TFT manufacturing processes [3]. However, as one might expect, numerous alternate technologies compete for entry into the mainstream, often attempting to leverage unique characteristics so as to differentiate from conventional Si-based technologies, and targeting appropriate niche applications for effective market entry. A prime example is that of TFTs based on organic semiconductors, which have gained prominence over the past decade; these devices offer numerous potential advantages, including low-temperature processing, mechanical flexibility, and (in certain cases) compatibility with solution-based deposition methods [3].

More recently, interest has arisen in the possibility of fabricating active electronic devices exclusively from optically transparent materials, thus enabling the manufacture of optically transparent circuitry, so-called "transparent electronics" or "invisible electronics" [5,6]. This interest was sparked in the late 1990s by

Zinc Oxide Bulk, Thin Films and Nanostructures
C. Jagadish and S. Pearton (Editors)

reports of p-type conductivity in several wide-bandgap oxide semiconductors [7,8] which, in combination with known n-type transparent semiconductors such as ZnO, SnO_2, and In_2O_3, opened the door to the fabrication of optically transparent oxide-based pn junction diodes, uv-emitters, and the like [9–12]. With these motivators in place it was, perhaps, inevitable that the pursuit of TFTs based on wide-bandgap, transparent semiconductor materials would follow.

Reference to the use of ZnO to fabricate optically transparent TFTs is seen as early as 2000 [13,14]; however, detailed reports of such devices and their electrical characteristics did not appear until early 2003 when several publications issued, nearly simultaneously, from independent research groups [15–17]. These and subsequent reports, although still rather limited in number, comprise an emerging framework of evidence for the viability of ZnO as a candidate TFT channel material [18–37]. It should be noted here that field-effect transistors (FETs) using single-crystal ZnO as a channel material were in fact explored, at least briefly, in the late 1960s; this work, however, appears to have gained little traction at the time [38].

Perhaps the most striking trait of TFTs built around ZnO is their optical transparency. In view of the predominate use of TFTs as pixel switches in AMLCDs and other display technologies, one is lead to speculate as to the potential value of an optically transparent TFT in this application. In addition to the possibility of increased pixel aspect ratio (the fraction of the pixel area through which light can pass to the viewer), ZnO TFTs offer the (potential) intrinsic advantage of insensitivity to visible light (since the wide bandgap of ZnO precludes absorption of visible photons), thus allowing potential process and architecture simplifications (with associated reductions in manufacturing cost) as compared to Si-based TFTs, for which shielding from display light may be necessary to avoid impacting electrical characteristics [3].

12.2 Device structure and fabrication

From a generalized perspective, ZnO TFTs share many of the basic structural and process-related features common to thin-film transistors, including a-Si and poly-Si devices. As for issues of specific relevance to ZnO TFTs, structural considerations (for the bulk of the work reported to date) have been primarily concerned with the appropriate use of simplified test structures with which key elements of material and device functionality may be assessed. Using such structures, various aspects of device fabrication (deposition and processing; material selection; and the like) have been explored, with the cumulative set of results providing an emerging framework around the fabrication of ZnO TFTs.

12.2.1 TFT structure and operation

Figure 12.1 provides a schematic representation of the basic TFT structure. The channel layer – a semiconductor such as a-Si, poly-Si, or, of course, ZnO – is provided with source and drain electrodes that are capable of efficient carrier injection into (source) and extraction from (drain) the channel region; the gate electrode is separated from the channel, source, and drain by a thin gate dielectric film. The key dimensions of this structure are the channel width (W) and length (L), and the gate dielectric thickness (t_{gd}); the active channel region comprises the rectangular area between the source and drain electrodes, as shown in Fig. 12.1.

Figure 12.2 illustrates the central elements of TFT operation, with voltage and charge polarities shown for an n-channel device (i.e., one in which the channel current is carried by electrons). A voltage applied to the gate electrode (with the source electrode being the usual voltage reference) modulates the charge density in the channel through the capacitance of the gate dielectric – the gate-to-channel

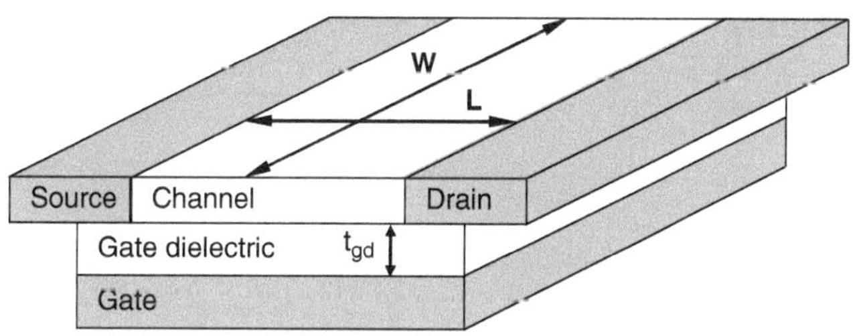

Figure 12.1: Simplified schematic depiction of the basic TFT structure (after Ref. 39).

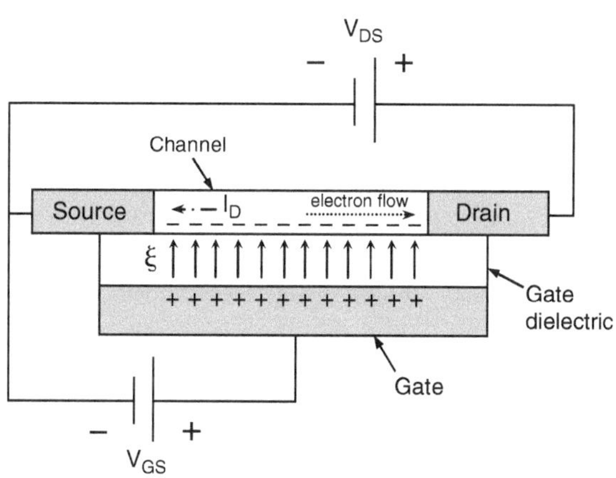

Figure 12.2: Schematic illustration of the central elements of TFT operation.

voltage differential is accompanied by a corresponding gate-to-channel electric field (ξ), mediated by the formation of space charge layers of equal magnitude and opposite sign, located on opposite sides of the gate dielectric. The magnitude of the induced charge density is proportional to the applied gate voltage; in the channel, this variation in charge density (comprised, at least partially, of free carriers) produces a corresponding variation in source-to-drain conductance. For an n-channel device, as depicted in Fig. 12.2, a positive gate voltage draws additional electrons into the channel, while a negative gate voltage reduces the channel electron density. This modulation of source-to-drain conductance (and, consequently, drain current [I_D]) via an applied gate voltage comprises the fundamental component of field-effect transistor (FET) functionality.

TFT operation is typically quantified in the form of drain current (I_D) as a function of gate-to-source voltage (V_{GS}) and drain-to-source voltage (V_{DS}). With the appropriate selection of simplifying assumptions, the widely recognized ideal metal-insulator-semiconductor field-effect transistor (MISFET) equation,

$$I_D = \begin{cases} 0 & \ldots V_{GS} \leq V_T \\ \mu\, C_{ins} \dfrac{W}{L} \left[(V_{GS} - V_T) V_{DS} - \dfrac{V_{DS}^2}{2} \right] & \ldots V_{DS} \leq V_{GS} - V_T, \\ \dfrac{1}{2} \mu\, C_{ins} \dfrac{W}{L} [V_{GS} - V_T]^2 & \ldots V_{DS} > V_{GS} - V_T \end{cases} \qquad (12.1)$$

is obtained (μ, C_{ins}, and V_T are the channel mobility, gate dielectric capacitance per unit area, and threshold voltage, respectively) [3,40,41]. While this expression may not provide quantitative agreement with devices that deviate substantially from one or more of the relevant model assumptions, it nonetheless comprises a useful qualitative basis for understanding the key elements of TFT operation.

12.2.2 ZnO TFT test structures

Practically useful, fully integrated TFTs are, of course, more complex than the simplified, schematic structure shown in Fig. 12.1 [3,4,42]. However, although ZnO TFTs have in fact been produced using a conventional, fully integrated TFT structure and process [19], to date a preponderance of reported ZnO TFTs have employed simplified test structures in which the performance of the ZnO channel and its interaction with the selected gate dielectric material can be most effectively observed, analyzed, and optimized; Fig. 12.3 shows such a simplified TFT structure. A substrate is coated with a conductor and a dielectric, comprising the gate electrode and gate dielectric, respectively; these layers are typically left unpatterned,

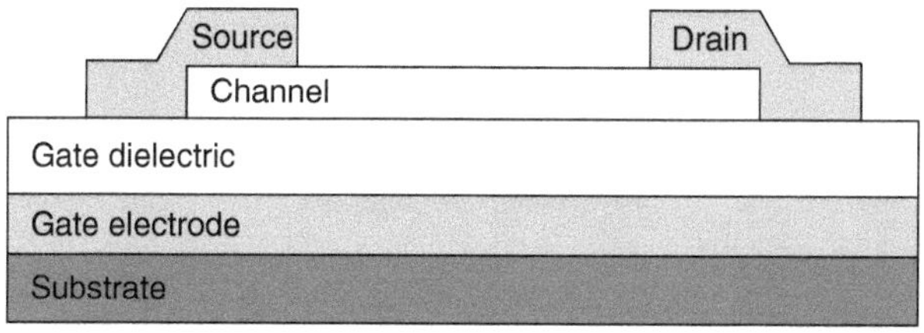

Figure 12.3: Cross-sectional view of a commonly-used test structure for the development of ZnO TFTs.

such that multiple TFTs on a single substrate share a common gate electrode. This blanket-coated gate electrode/gate dielectric structure is advantageous in several respects, including ease of fabrication and lack of topography. The remaining device layers – the channel (ZnO) and source/drain electrodes – are deposited and patterned, either via standard photolithography or, in many cases, by deposition through shadow masks. Although limited in achievable minimum dimensions, the use of shadow mask-based patterning follows in the vein of structure and process simplification, circumventing the potential for unexpected interactions with photolithographic patterning processes.

A commonly-used variant of the simplified structure of Fig. 12.3 employs a heavily-doped Si wafer to serve as both substrate and gate electrode; the gate dielectric is then formed via thermal oxidation of the Si wafer, thus yielding a high-quality, thermally-grown silicon dioxide (SiO_2) gate dielectric layer [17,22,24]. In instances where a gate dielectric other than SiO_2 is desired, a second dielectric layer may be deposited over the thermal SiO_2 layer to provide the appropriate channel/gate dielectric interface, while retaining the low defect density, high breakdown field, and low current leakage of the thermal SiO_2 layer.

The structure shown in Fig. 12.3 may be further simplified by leaving the channel layer unpatterned; patterning is then necessary only for the source/drain electrode layer, with the gate electrode, gate dielectric, and channel layers comprising a blanket-coated, multi-layer film stack. However, while this structure has clear advantages in terms of maximizing process and structural simplicity, it is not without its drawbacks. Of primary concern is the potential for yielding an overstated quantification of drain current (with respect to channel width-to-length ratio), from which is then extracted a similarly inflated value for channel mobility (mobility extraction will be discussed in detail in Section 12.3.2); this may be a significant issue in assessing the validity of published results, since channel mobility is generally the principal metric used in quantifying TFT performance.

The nature of this issue is illustrated in Fig. 12.4a, which shows the layout of a TFT structure with a channel layer patterned so as to provide a well-defined rectangular

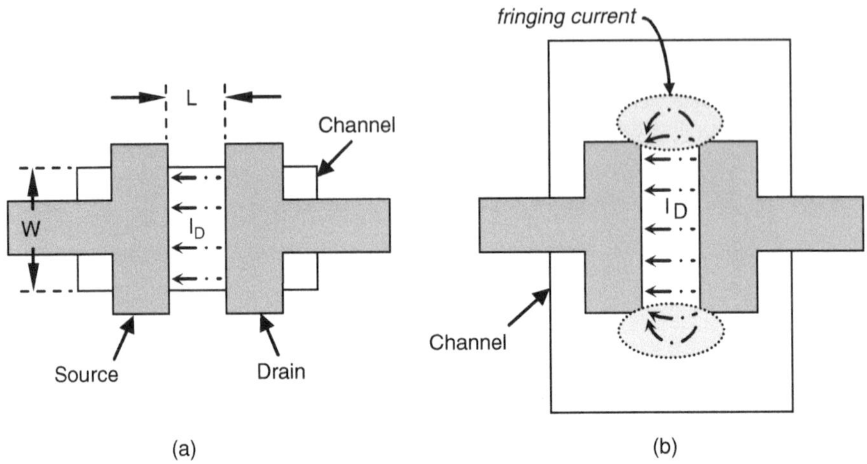

Figure 12.4: Schematic layout of a TFT structure corresponding to the cross-section of Fig. 12.3 (unpatterned gate electrode and gate dielectric layers are not shown explicitly here), in which (a) the channel layer is patterned so as to yield a precisely-defined channel width-to-length ratio (W/L), and (b) the channel layer is unpatterned, resulting in an ill-defined channel width (and thus width-to-length ratio) due to drain current fringing beyond the ends of the source and drain electrodes, as indicated.

active region whose width (W) and length (L) are defined by the width of the patterned channel region and the separation between source and drain electrodes, respectively; in this case, drain-to-source current flows only within this well-defined channel region, and the channel width-to-length ratio (W/L) is clearly defined. Conversely, in Fig. 12.4b, the channel layer is not patterned (or, more precisely, the channel region extends beyond the ends of the source and drain electrodes, and may or may not be patterned over the substrate in general); while the channel length is still precisely defined by the separation between source and drain electrodes, the channel width is no longer equivalent to any drawn dimension of the TFT layout. As shown in Fig. 12.4b, drain-to-source current is no longer restricted to the rectangular region directly between the source and drain electrodes, but rather includes an additional fringing component that extends into the channel region beyond the electrode ends. This is an inherent trait of such a simplified structure (i.e., one in which both gate electrode and channel layers are unpatterned), since the applied gate voltage exerts a uniform field over the entire channel layer area, leaving no physical restriction to confine drain-to-source current flow to the area directly between the source and drain electrodes. The practical consequence of this fringing current is an effective increase in the TFT width-to-length ratio as compared to that expected based on electrode separation and electrode width, and can result in a substantial error in calculated vs. actual mobility values, particularly for small width-to-length ratios where the additive correction factor becomes comparable in magnitude to the W/L value obtained from drawn electrode dimensions. Thus, if W/L values are not corrected to account for this fringing current component, extracted mobilities will show an approximately inverse dependence on W/L for small W/L values, approaching the true mobility only for sufficiently large W/L.

12.2.3 ZnO TFT fabrication

Within the context of device structure and fabrication, the primary source of differentiation between ZnO TFTs and conventional Si-based TFTs is found in the identity of the channel material; analogous structural and process considerations apply, at least from a high level, except where directly influenced by channel material selection. Issues relating to ZnO deposition and processing are, of course, unique to ZnO-channel devices. Second-order interactions include selection, deposition, and processing of gate dielectric and source/drain electrode materials; the channel/gate dielectric interface is critical in establishing both mobility (Section 12.3.2) and turn-on voltage (Section 12.3.1), while the source and drain electrodes must be capable of efficient (i.e., low-resistance) carrier injection into and extraction from the ZnO channel layer.

ZnO TFTs have been fabricated using a variety of deposition methods for the ZnO channel layer, including radio frequency (RF) sputtering [17,22,24,31,36], ion beam sputtering [16], pulsed-laser deposition (PLD) [15,19,23], and solution processing from zinc acetate-based (dip-coating) [18,33], zinc nitrate-based (spin-coating) [21], and ZnO nanoparticle-based [30] precursors; devices with ZnO nanowires as the TFT channel layer have also been reported [27]. Of these methods, RF sputtering from a ceramic ZnO target has been the leading choice for channel deposition, although perhaps largely by reason of convenience.

Although the impact of source/drain electrode material selection (and associated processing) on ZnO TFT electrical performance has not been explicitly reported (for example, via the quantification of contact resistance), a survey of the literature shows that a rather broad range of materials have been successfully employed in this context. These include various metals (Al [19,22,36,38], Au [17,20,30,33], In [15], Ti [34], and Cu [20]), doped ZnO (ZnO:In [15], ZnO:Al [24] and ZnO:Ga [31]), and other conductive oxides (ITO [16,21,23] and NiO_x [32]). While it is interesting to note that functional devices have been obtained with such a diverse group of source/drain electrode materials, it is important to recognize that, because in many cases the associated ZnO TFT structures are relatively large (due to the common use of shadow mask-based patterning, as discussed in the previous section), the effect of an otherwise poor (i.e., high-resistivity) source/drain contact may be at least partially obscured by the large physical dimensions of the contact (since contact resistance is inversely proportional to contact dimensions).

Several gate dielectric materials have also been shown to provide functional ZnO TFTs. As previously noted, thermally-grown SiO_2 affords a convenient gate dielectric for TFT test structures (the oxide is grown over a heavily-doped Si wafer which serves as both substrate and gate dielectric) [17,24]. Although this approach is useful only for early-stage material and process development (as it does not allow the integration of multiple devices), its widespread use has generated strong evidence for the viability of SiO_2 as a gate dielectric for ZnO TFTs. SiO_2 offers also the advantages arising from its status as a widely-used and well-understood material in the microelectronics field, as sophisticated deposition tools and methods are readily available.

In TFT structures using non-Si substrates such as glass, a deposited dielectric layer is, of course, required; the fabrication of ZnO TFTs with deposited gate dielectric (as opposed to the use of a thermally-grown SiO_2 layer as discussed above) comprises both a step toward practically useful ZnO TFTs and an opportunity to demonstrate the optical transparency attainable for devices with a transparent ZnO channel. Functional ZnO TFTs have been obtained using a number of deposited dielectric materials, including silicon nitride (Si_3N_4) deposited by plasma-enhanced chemical vapor deposition (PECVD) [15], silicon oxynitride (SiO_xN_y) by RF sputtering [26], aluminum oxide (Al_2O_3) by electron-beam evaporation [36] and atomic layer deposition (ALD) [16,31,36], hafnium oxide (HfO_2) by RF sputtering [23], calcium hafnium oxide ($CaHfO_3$) by PLD [19], (Ce,Tb)$MgAl_{11}O_{19}$ [27,34], $Bi_{1.5}Zn_{1.0}Nb_{1.5}O_7$ [37], and a laminated organic fluoropolymer [20].

12.3 Electrical characterization, performance, and modeling

Although the electrical characteristics of ZnO TFTs do show the essential qualitative traits predicted by the ideal TFT model (Equation 12.1), there are also significant areas of deviation, such that Equation 12.1 is inadequate for a quantitative reproduction of ZnO TFT current-voltage (I-V) characteristics. Figures 12.5 and 12.6 plot exemplary drain current vs. drain-to-source voltage (I_D-V_{DS}) and drain current vs. gate-to-source voltage (I_D-V_{GS}) data, for a ZnO TFT structure similar to that depicted in Fig. 12.3; these data will be used throughout this section, as convenient, to illustrate important characterization methods, key aspects of ZnO TFT functionality, and the like.

A minimal set of parameters used to quantify the electrical performance of ZnO TFTs (and TFTs in general) typically includes an assessment of mobility (channel electron transport efficiency) and turn-on voltage or threshold voltage (the

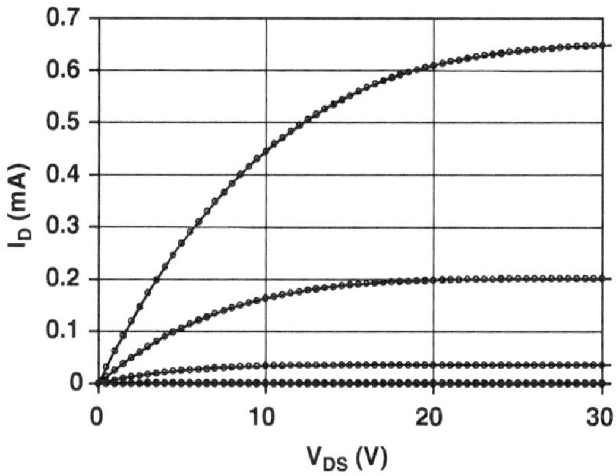

Figure 12.5: Drain current vs. drain-to-source voltage (I_D-V_{DS}) curves for a ZnO TFT with Si substrate, 100 nm thermal SiO_2 gate dielectric, ZnO:Al/ITO source and drain electrodes, $L = 200\ \mu m$, $W/L = 9$; $V_{GS} = 0$ to 40 V in 10 V increments, I_D increases with increasing V_{GS} (data from Ref. 24).

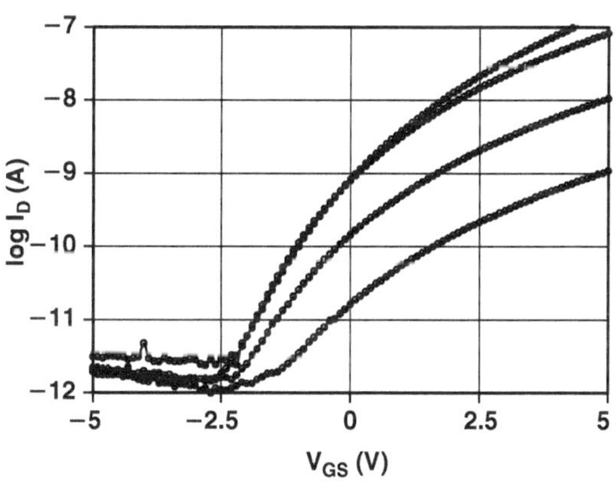

Figure 12.6: Drain current vs. gate-to-source voltage (I_D-V_{GS}) curves for the ZnO TFT with I_D-V_{DS} data shown in Fig. 12.5; $V_{DS} = 10$ mV to 10 V in 1 decade increments, I_D increases with increasing V_{DS} (after Ref. 24).

gate-to-source voltage at which the channel begins to turn "on", i.e., channel conductance begins to increase from its minimum "off" level); the composition of this minimal parameter set should not be a surprise, in view of the fact that mobility (μ) and threshold voltage (V_T) are the sole intrinsic device parameters in the ideal TFT model (Equation 12.1).

Of course, TFT characterization in general extends far beyond this foundation; further TFT characterization may include the evaluation of source/drain contact resistance, sub-threshold slope, dynamic response, and stability of measured electrical characteristics over time; however, as little data is yet available for these aspects of ZnO TFT performance, they will not be explicitly discussed here. The interested reader is encouraged to consult one of the many available references for a detailed treatment of TFT technology [3,4,40–42].

12.3.1 Turn-on

The simplest use of a TFT is as a voltage-controlled switch, in which a voltage applied to the gate electrode controls the resistance of the channel such that the drain-to-source resistance can be modulated between a high-resistance "off" state and a low-resistance "on" state. This is the basic mode of operation, for example, of TFTs in AMLCD displays, where each pixel includes a TFT which controls the voltage across the associated LCD pixel element [3]. A key metric in this switching mode of operation is the gate voltage at which the channel conductance begins to increase from its minimum "off" level. This parameter establishes, among other things, the identity of a TFT as "enhancement-mode" or "depletion-mode"; an enhancement-mode TFT is "off" with an applied gate voltage of 0 V, while a depletion-mode device requires the application of a non-zero gate voltage to reach the off-state (enhancement-mode devices are preferable in many applications).

12.3.1.1 Parameterization of TFT turn-on

The TFT parameter most frequently used to quantify turn-on is the threshold voltage (V_T), as per the ideal TFT model (Equation 12.1); for devices that do not deviate substantially from the assumptions of the ideal model, V_T can be a useful parameter. However, in the case of ZnO TFTs, Equation 12.1 does not provide a satisfactory quantitative representation of measured electrical characteristics. While the primary source of this deviation (namely, a strong dependence of channel mobility on gate voltage) will be discussed in the following section (Section 12.3.2), suffice it to say here that such quantitative deviation from the ideal model equation clearly calls into question the validity of extraction methods based on direct manipulation of this equation.

Figure 12.7 illustrates the futility of applying conventional threshold voltage extraction methods to ZnO TFT electrical characteristics. In the linear regions of TFT operation (i.e., for small V_{DS}, specifically such that $V_{DS} \ll V_{GS} - V_T$), Equation 12.1 simplifies to

$$I_D = \mu C_{ins} \frac{W}{L} V_{DS} \left[V_{GS} - V_T \right]. \qquad (12.2)$$

Thus, if I_D is measured as a function of V_{GS} (for a small, fixed value of V_{DS}), Equation 12.2 suggests that a linear extrapolation of the I_D-V_{GS} curve should intercept

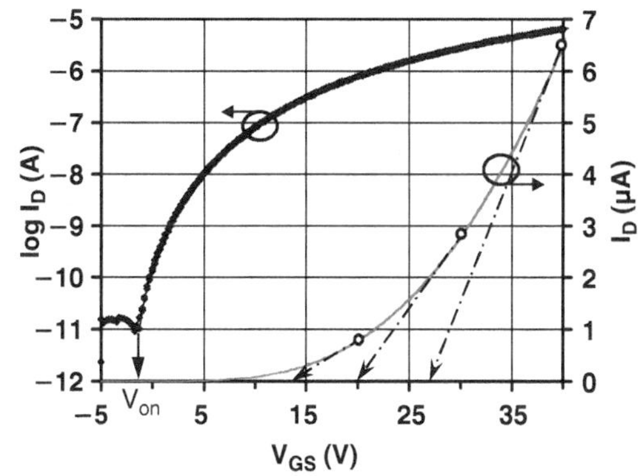

Figure 12.7: Drain current vs. gate-to-source voltage (I_D-V_{GS}) data, plotted on both a log scale (left axis) and a linear scale (right axis), for the ZnO TFT with I_D-V_{DS} data shown in Fig. 12.5; $V_{DS} = 100\,mV$. Dashed arrows show extrapolation to the V_{GS} axis from three selected points ($V_{GS} = 20$, 30, and 40 V) on the linear I_D-V_{GS} curve, as per conventional threshold voltage extraction procedure (data from Ref. 24).

the V_{GS} axis at $V_{GS} = V_T$ [43]. Figure 12.7 plots such an I_D-V_{GS} data set, for a ZnO TFT, on both a log scale (left axis) and a linear scale (right axis). As per Equation 12.2, the linear I_D-V_{GS} curve is extrapolated – by way of example, from three selected points, $V_{GS} = 20$, 30, and 40 V – to the V_{GS} axis in an attempt to identify the threshold voltage of this device. Two problems are immediately apparent. First, the threshold voltage thus obtained varies dramatically with the selected point on the I_D-V_{GS} curve from which extrapolation begins; starting V_{GS} values of 20, 30, and 40 V yield approximate V_T values of 13.5, 20, and 27 V, respectively. Second, while V_T is intended to provide a quantification of TFT turn-on (as discussed above), the true V_{GS} at which device turn-on begins, as clearly seen in the log scale I_D-V_{GS} curve (left axis), is approximately −2 V, far from any V_T value that one might obtain via extrapolation from the linear I_D-V_{GS} curve.

As illustrated in the above example, conventional threshold voltage extraction is problematic for ZnO TFTs in two significant respects: (1) a single, uniquely-defined V_T value cannot be obtained, and (2) resulting V_T values do not provide a physically-meaningful quantification of device turn-on behavior.

In order to avoid this ambiguity associated with the quantification and interpretation of threshold voltage, TFT turn-on may be quantified via direct examination of log(I_D)-V_{GS} data (e.g., Fig. 12.7, left axis), from which an alternate parameter, the device turn-on voltage (V_{on}), is readily identified as the value of V_{GS} at which the drain current begins its initial increase from the minimum "off" level [24]; for example, from the log(I_D)-V_{GS} plot shown in Fig. 12.7, V_{on} is seen to occur at approximately −2 V. A more precise assessment is enabled by examination of Fig. 12.6, which includes I_D-V_{GS} curves for a range of V_{DS} values. Note that the apparent V_{on} tends to shift slightly toward more positive voltages as V_{DS} decreases, most noticeably between the lowermost two curves in Fig. 12.6 ($V_{DS} = 10$ and 100 mV), thus illustrating that V_{on} is best assessed using a reasonably large V_{DS} value for which the measured drain current signal is maximized. In any case, it is evident that this minimal uncertainty in the V_{on} value thus obtained is negligible as compared to that seen for estimated V_T values

Thus, for ZnO TFTs, V_{on} comprises a physically meaningful and uniquely-defined parameter with which to quantify device turn-on; V_T, conversely, is a largely ill-defined and ambiguous metric (assuming conventional V_T extraction methods are used), as illustrated in Fig. 12.7 and the associated discussion. Unfortunately, reports of ZnO TFTs found in the literature predominately include V_T, as per typical TFT parameterization techniques; without the corresponding I_D-V_{GS} data, this parameter should be interpreted with caution in view of its inherent ambiguity, as discussed above.

12.3.1.2 Turn-on characteristics of ZnO TFTs

It is well known that ZnO tends easily toward n-type conductivity, as a result of extrinsic donor doping (intentional or unintentional) and/or intrinsic defects accompanying an oxygen-deficient stoichiometry; the associated equilibrium population of free electrons plays a primary role in establishing the turn-on voltage (V_{on}) of a ZnO TFT. Simply stated (in the case that the ZnO channel has a non-zero equilibrium free electron concentration), V_{on} is equivalent to the gate voltage required to fully deplete the channel of free electrons. Thus, although in general a number of material and device parameters (including also gate dielectric charge, gate-to-channel work function difference, traps or other defect states in the channel bulk and/or at the channel/gate dielectric interface, and the like [41,44]) must be considered in fully evaluating and controlling the turn-on characteristics of a TFT, the primary consideration in the case of ZnO TFTs is that the equilibrium (i.e., with zero applied gate bias) free electron concentration in the ZnO channel be controllably minimized such that a reasonable turn-on voltage can be achieved.

Such control of the equilibrium free electron concentration can be achieved in a number of ways. ZnO deposition and post-deposition annealing conditions are, of course, key factors in determining the properties of the resulting film. Bae *et al.* report significant effects of both deposition (substrate temperature) and post-deposition annealing (temperature, duration, and ambient) on turn-on voltage for RF sputtered ZnO TFTs (see Ref. 25, Figs. 5b and 6b). In addition, the selection of a suitable oxygen partial pressure during ZnO channel deposition has been generally recognized as a critical factor in optimizing ZnO TFT performance [15,17,36]. Suppression of the equilibrium electron concentration may also be achieved through doping with extrinsic species. Kwon *et al.* employ acceptor (phosphorus) doping to reduce the electron concentration of PLD deposited $(Zn_{0.9}Mg_{0.1})O$; the substitution of Mg into the ZnO lattice acts to increase the material bandgap, thus further reducing the electron concentration [23].

In general, with respect to ZnO TFT turn-on, a primary observation to be drawn from the existing literature is that enhancement-mode operation (i.e., $V_{on} > 0\,V$), having been demonstrated by multiple groups (for example, see Ref. 15, Fig. 8b; Ref. 19, Fig. 2a; Ref. 20, Fig. 8b; Ref. 21, Fig. 3; Ref. 31, Fig. 5a), is in fact achievable for these devices. Thus, while substantial work remains (for example, with respect to the stability of turn-on voltage, and of electrical characteristics in general), it is apparent that the properties of the ZnO channel can be effectively controlled to yield acceptable turn-on characteristics for practically useful TFT operation.

12.3.2 Channel mobility

In combination with turn-on voltage, mobility (μ) completes the minimal set of parameters typically used to quantify basic TFT electrical performance. Mobility provides a measure of the efficiency with which charge carriers (electrons, for ZnO TFTs) move from source to drain, through the TFT channel. More specifically, mobility is defined as the constant of proportionality between the average electron drift velocity (v_d) and the strength of the applied electric field (ξ), i.e., $v_d = \mu\xi$ [44]. Mobility comprises a key TFT performance metric; both current drive capability and maximum operating frequency, two of the primary TFT performance criteria, are directly proportional to mobility. As such, an accurate and precise assessment of mobility is critical to providing a useful characterization of TFT performance.

12.3.2.1 Mobility extraction methodology

The usual approach used to evaluating TFT mobility relies on fitting measured current-voltage data to one of several expressions obtained via manipulation of the ideal transistor equation (Equation 12.1), thus yielding one of the three typically-reported field-effect transistor (FET) mobility types: field-effect (μ_{FE}), saturation (μ_{sat}), and effective mobility (μ_{eff}) [43]. Unfortunately, the use of these typical mobility extraction methods for the characterization of ZnO TFTs is somewhat problematic, particularly when they are applied without a clear understanding of their limitations. In order to further elucidate these issues, a brief description of the conventional TFT mobility extraction methods is included below; see Ref. 43 for further detail.

Field effect mobility (μ_{FE}) is obtained from measured I_D-V_{GS} data with small, fixed V_{DS} such that the device operates in the linear region (thus satisfying the assumptions of Equation 12.2), by way of either transconductance [$g_m(V_{GS}) \equiv dI_D/dV_{GS}$] or drain conductance [$g_d(V_{GS}) \equiv dI_D/dV_{DS}$]. In the case of transconductance, differentiating Equation 12.2 with respect to V_{GS} (for the moment, assuming μ to be independent of V_{GS}) and rearranging the resulting expression yields

$$\mu_{FE}(V_{GS}) = \frac{g_m(V_{GS})}{C_{ins}\frac{W}{L}V_{DS}}, \tag{12.3}$$

where $\mu(V_{GS})$ and $g_m(V_{GS})$ indicate the functional dependence of μ and g_m on V_{GS}. Alternatively, Equation 12.2 may be differentiated with respect to V_{DS}, yielding the drain conductance (g_d); further differentiating g_d with respect to V_{GS} (again, assuming μ to be independent of V_{GS}) and rearranging yields

$$\mu_{FE}(V_{GS}) = \frac{g'_d(V_{GS})}{C_{ins}\frac{W}{L}}, \tag{12.4}$$

where g'_d indicates the derivative of g_d with respect to V_{GS}. Thus μ_{FE} may be obtained by calculating either g_m or g'_d from a measured I_D-V_{GS} data set (again, with small V_{DS} so as to satisfy the assumptions of Equation 12.2) and substituting this data into Equation 12.3 (g_m) or 12.4 (g'_d). Equations 12.3 and 12.4 are functionally equivalent, i.e., they will produce identical μ_{FE} data from a given I_D-V_{GS} data set.

Saturation mobility (μ_{sat}) is obtained also from I_D-V_{GS} data, however in this case V_{DS} must be sufficiently large that the device operates in the saturation region (i.e., $V_{DS} > V_{GS} - V_T$, as per Equation 12.1). In this case, the ideal model (Equation 12.1) suggests that the square root of the drain current in saturation ($I_{D,sat}$),

$$\sqrt{I_{D,sat}} = \sqrt{\frac{1}{2}\mu C_{ins}\frac{W}{L}}[V_{GS} - V_T], \tag{12.5}$$

should be linearly dependent on V_{GS}. Taking the derivative of Equation 12.5 with respect to V_{GS} (once again assuming μ to be independent of V_{GS}) and rearranging yields the saturation mobility,

$$\mu_{sat}(V_{GS}) = \frac{\left(\frac{d\sqrt{I_{D,sat}}}{dV_{GS}}\right)^2}{\frac{1}{2}C_{ins}\frac{W}{L}}. \tag{12.6}$$

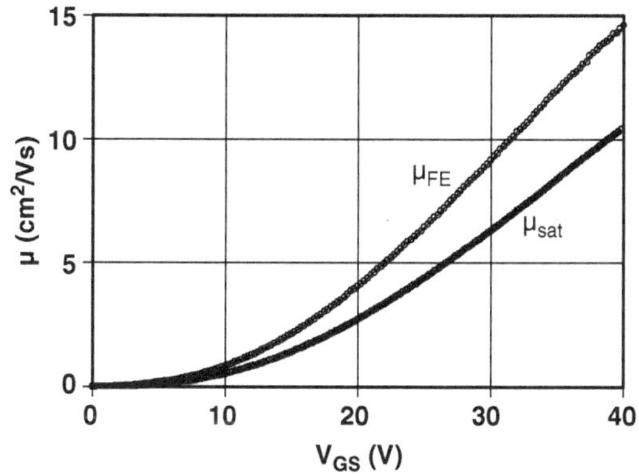

Figure 12.8: Field-effect (μ_{FE}) and saturation mobility (μ_{sat}) vs. gate-to-source voltage (V_{GS}), obtained using conventional TFT mobility extraction methods, for the ZnO TFT with I_D-V_{DS} data shown in Fig. 12.5.

Substitution of measured $I_{D,sat}$-V_{GS} data into Equation 12.6, then, gives μ_{sat} as a function of V_{GS}.

Figure 12.8 plots μ_{FE} and μ_{sat}, obtained using conventional mobility extraction methods as described above, for the ZnO TFT with current-voltage characteristics shown in Figs. 12.5 and 12.6.

The issues associated with μ_{FE} and μ_{sat} should at this point be apparent. First, the derivation of the associated expressions (Equations 12.3, 12.4 [μ_{FE}] and 12.6 [μ_{sat}]) begins from the ideal TFT model (Equation 12.1), which (as noted in the previous section) does not provide quantitative agreement with measured ZnO TFT characteristics; this fact alone calls into question the validity of the resulting data. Second, for both μ_{FE} and μ_{sat}, the derivation explicitly assumes that mobility is *independent* of V_{GS} (in differentiating with respect to V_{GS}) while, as seen in Fig. 12.8, the resulting mobility values may be *strongly* V_{GS}-dependent for ZnO TFTs, thus directly invalidating the entire analysis. In fact, this strong dependence of mobility on gate voltage comprises also the primary source of deviation between the ideal model and measured ZnO TFT current-voltage data [39].

Of the three commonly used mobility metrics, the effective mobility (μ_{eff}), when properly assessed, does not directly require the assumption of V_{GS}-independent mobility. The derivation of the expression for effective mobility begins from a modified version of Equation 12.2; recognizing that the term $C_{ins}(V_{GS} - V_T)$ is an approximation for the channel charge (Q_{CH}), Equation 12.2 becomes

$$I_D = \mu \frac{W}{L} V_{DS} Q_{CH}(V_{GS}), \tag{12.7}$$

where $Q_{CH}(V_{GS})$ indicates the functional dependence of Q_{CH} on V_{GS}. Now, differentiating Equation 12.7 with respect to V_{DS} (yielding g_d) and rearranging, μ_{eff} is given by

$$\mu_{eff}(V_{GS}) = \frac{g_d(V_{GS})}{\frac{W}{L}Q_{CH}(V_{GS})}. \tag{12.8}$$

Note that the derivation of Equation 12.8 does not include differentiation with respect to V_{GS}, thus avoiding the complications thus encountered in the derivation of the corresponding expressions for μ_{FE} and μ_{sat}. However, in order to employ Equation 12.8 to assess TFT mobility, Q_{CH} must be quantified as a function of V_{GS}. While in some instances this may be accomplished through measurement of the gate-to-channel capacitance [43], it may be difficult to implement for ZnO TFTs, in which contributions to measured capacitance from bulk and interface traps or other defects act to complicate measured capacitance data [24]. The usual approach to quantification of Q_{CH} is to substitute the approximation $C_{ins}(V_{GS} - V_T)$ back into Equation 12.8; however, as discussed in the previous section, the identification of a unique and well-defined value for V_T is essentially futile in the case of ZnO TFTs. Thus, as for μ_{FE} and μ_{sat}, μ_{eff} (at least, as typically employed) does not provide a suitable means for accurately assessing the mobility of ZnO TFTs.

Since the fundamental issue in the application of conventional mobility extraction techniques to ZnO TFTs, as seen above, is found in the strong dependence of mobility on gate voltage, a suitable mobility extraction technique must avoid restricting this dependence. This can be accomplished in a rather straightforward manner if, rather than beginning from the ideal TFT model, expressions are generated from basic charge transport equations within the context of the basic TFT structure. Such an analysis yields expressions which, although similar to those derived above for μ_{FE} and μ_{eff}, are able to self-consistently include gate voltage-dependent mobility. In order to avoid the ambiguity associated with the use of the conventional mobility metrics discussed above, two additional mobility metrics are introduced: the average (μ_{avg}) and incremental (μ_{inc}) mobilities (see Ref. 24 for an extended discussion and derivation of these mobility metrics).

The derivation of μ_{avg} begins with an expression for channel conductance (G_{CH}; equivalent to g_d, as used above, with the restriction of small V_{DS}) derived from simple charge transport theory and assuming drift-dominated current flow:

$$G_{CH}(V_{GS}) \equiv \lim_{V_{DS}\to 0}\left(\left.\frac{\partial I_D}{\partial V_{DS}}\right|_{V_{GS}}\right) = \mu_{avg}(V_{GS})\frac{W}{L}Q_{ind}(V_{GS}), \tag{12.9}$$

where, again, $G_{CH}(V_{GS})$, $\mu_{avg}(V_{GS})$, and $Q_{ind}(V_{GS})$ indicate the functional dependence of each parameter on V_{GS}; Q_{ind} is defined as the cumulative charge (per unit area) induced in the channel by the applied gate voltage in excess of V_{on}. Note here a key aspect of the derivation of μ_{avg} and μ_{inc}: V_{on} is used exclusively to quantify device turn-on, rather than V_T as employed for conventional mobility metrics. As discussed in the previous section, V_T is a rather ill-defined parameter for ZnO TFTs, whereas V_{on} provides a uniquely-defined, unambiguous value for parameterization of ZnO TFT functionality.

As discussed in the context of μ_{eff} (specifically, Equation 12.7), while Q_{ind} might in theory be quantified through measurement of gate-to-channel capacitance, it is practically preferable in many cases to make the approximation

$$Q_{ind}(V_{GS}) \cong C_{ins}[V_{GS} - V_{on}]. \quad (12.10)$$

While this approximation introduces some degree of error in the resulting mobility values, its worst-case magnitude can be quantified based on channel and gate dielectric properties (thickness, dielectric constant) [24], and is generally in the vicinity of ten to twenty percent for reasonable device parameters; furthermore, this error results always in an underestimate of the true channel mobility. Note again the use of V_{on} rather than V_T, which here provides a solid, physically meaningful estimate of Q_{ind}, without the ambiguity introduced at a similar point in the derivation of μ_{eff} (Equation 12.8).

Now, combining Equations 12.9 and 12.10, μ_{avg} is given by

$$\mu_{avg}(V_{GS}) = \frac{G_{CH}(V_{GS})}{\frac{W}{L}Q_{ind}(V_{GS})} \cong \frac{G_{CH}(V_{GS})}{\frac{W}{L}C_{ins}[V_{GS} - V_{on}]}. \quad (12.11)$$

Note the similarity between the expressions for μ_{avg} (Equation 12.11) and μ_{eff} (Equation 12.8) (assuming, in Equation 12.8, that $Q_{ind}(V_{GS})$ is approximated as $C_{ins}(V_{GS} - V_T)$) – the sole substantive difference is the use of V_{on} (Equation 12.11) in the place of V_T (Equation 12.8). However, due to the troublesome nature of V_T for parameterization of ZnO TFT behavior, this seemingly minor difference is in fact crucial to the nature of μ_{avg} as a well-defined, unambiguous mobility metric. Furthermore, μ_{avg} provides a physically meaningful and practically useful quantification of mobility: μ_{avg} represents the average mobility of the cumulative charge induced in the channel by the gate voltage in excess of V_{on} and, as such, comprises a direct measure of attainable TFT current drive as a function of maximum gate voltage [24].

Whereas μ_{avg} characterizes the cumulative induced channel charge, the incremental mobility (μ_{inc}) quantifies the mobility of a differential quantity of charge as it is added to or removed from the total channel charge by a differential change in gate voltage [24]. Taking this definition for μ_{inc}, an expression analogous to

Equation 12.9 can be written to quantify the incremental change in channel conductance (ΔG_{CH}) corresponding to an incremental change in induced charge density (ΔQ_{ind}),

$$\Delta G_{CH}(V_{GS}) = \mu_{inc}(V_{GS}) \frac{W}{L} \Delta Q_{ind}(V_{GS}). \tag{12.12}$$

Based on this expression, and again estimating Q_{ind} as per Equation 12.10 (i.e., $\Delta Q_{ind} \sim C_{ins} \Delta V_{GS}$), μ_{inc} is given by

$$\mu_{inc}(V_{GS}) = \frac{\Delta G_{CH}(V_{GS})}{\frac{W}{L} \Delta Q_{ind}(V_{GS})} \cong \frac{G'_{CH}(V_{GS})}{\frac{W}{L} C_{ins}}, \tag{12.13}$$

where G'_{CH} ($=\Delta G_{CH}/\Delta V_{GS}$) is the differential channel conductance (i.e., the derivative of G_{CH} with respect to V_{GS}). As the expression for μ_{avg} was similar to that for μ_{eff}, note here that the expression for μ_{inc} (Equation 12.13) is substantively identical to that for μ_{FE} (Equation 12.4). While this may lead one to question the value of introducing μ_{inc} as an alternate mobility metric, it is important to recall that the interpretation of μ_{inc} is closely tied to physical device operation, and that the derivation of Equation 12.13 at no point requires the assumption that μ_{inc} be independent of V_{GS}; conversely, μ_{FE} is the result of manipulation of the ideal TFT equation, and violates its own assumption of V_{GS}-independence. Thus, while μ_{inc} is not a quantitatively unique mobility metric, it provides important insight into device operation that is not found in the usual understanding of μ_{FE}.

12.3.2.2 Mobility characteristics of ZnO TFTs

Figure 12.9 plots μ_{avg}, μ_{inc}, μ_{FE}, and μ_{sat} for the ZnO TFT with current-voltage characteristics shown in Figs. 12.5 and 12.6 (μ_{FE} and μ_{sat}, previously shown in Fig. 12.8, are included here for comparison). Figure 12.9 illustrates two important points with regard to the quantification of ZnO TFT mobility. First, regardless of the selected mobility metric, the dependence of mobility on gate voltage is substantial. Second, although each of the mobility types shown here exhibits a qualitatively similar dependence on gate voltage, quantitative differences are quite large; furthermore, μ_{avg} – the mobility metric that correlates directly to TFT current drive – provides the smallest values of the four. In combination, these observations show clearly the difficulty in interpreting and making comparisons among reported mobility values for ZnO TFTs: if one or both of the chosen mobility extraction method and the gate voltage dependence are not explicitly given (for example, if mobility is reported as a single value, rather than as a function of gate voltage), there remains a substantial degree of uncertainty (both quantitative and qualitative) as to the "true" mobility of the device.

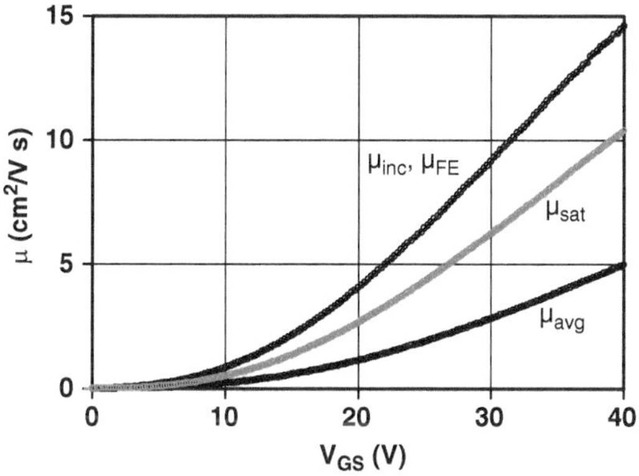

Figure 12.9: Average (μ_{avg}), incremental (μ_{inc}), field-effect (μ_{FE}) and saturation mobility (μ_{sat}) vs. gate-to-source voltage (V_{GS}) for the ZnO TFT with I_D-V_{DS} data shown in Fig. 12.5 (note: μ_{inc} and μ_{FE} data are identical to one another; μ_{FE} and μ_{sat} were previously shown in Fig. 12.8) (data from Ref. 24).

As noted previously, the strong dependence of mobility on gate voltage (as seen in Fig. 12.9) is primarily responsible for the deviation of ZnO TFT current-voltage characteristics from those predicted by the ideal TFT model (Equation 12.1), which assumes that all carriers in the channel are characterized by a single, gate voltage-independent mobility. This non-ideality appears to be a generally-observed aspect of ZnO TFT functionality, consistently seen in reports where mobility behavior is explicitly examined [17,19,24]. While the underlying mechanism(s) responsible for these mobility characteristics have not yet been extensively studied, a likely explanation derives from the poly-crystalline nature of the ZnO channel film, in which negatively-charged grain boundary defects create a potential energy barrier at each grain boundary, thus inhibiting the effective inter-grain transport of free electrons in the channel. As the applied gate voltage increases (i.e., becomes more positive), the ratio of free carrier charge to fixed grain boundary charge increases along with the growing charge density of the induced free electron population, resulting in a concomitant reduction in the impact of the grain boundary charge on free electron transport. This mechanism is analogous to that seen in poly-Si TFTs [41,42], and comprises the growing consensus in the ZnO TFT literature as the probable explanation for the highly non-ideal mobility characteristics of ZnO TFTs [17,19,20,24,36,45]. Within this context, Carcia *et al.* have employed the model described by Levinson for charge transport in polycrystalline TFTs [46] to extract grain boundary scattering parameters from measured mobility characteristics [17,20,36]; similarly, Hossain *et al.* have undertaken numerical modeling and simulation to further explore the effects of grain boundary-limited charge transport in ZnO TFTs [45].

At this point it is worthwhile to note that, while TFT mobilities found in the literature are generally obtained using an extraction methodology associated with one of μ_{FE}, μ_{sat}, and μ_{eff} (as defined by Schroder) [43], and while one of these three mobility terms is generally used to describe the resulting mobility data, the use of terminology is quite inconsistent. Thus, for example, the use of the term "field-effect mobility" cannot in general be taken as an unambiguous identification of a particular extraction methodology, without some accompanying clarification of the author's choice of nomenclature; for this reason, appropriate caution should be exercised in quantitatively evaluating and comparing reported mobility values.

In view of the preceding discussion, it should be apparent that a quantitative review of reported ZnO TFT mobilities is not a simple matter, if useful and valid conclusions are to be obtained; mobility values reported by disparate groups are not, in general, directly comparable, due to potential differences in mobility extraction methodology, use of terminology, choice of bias conditions (e.g., gate voltage for which mobility is reported), and the like. Further, mobility is of course dependent on processing conditions (e.g., ZnO deposition method and post-deposition treatment); device structure (e.g., top-gate vs. bottom-gate configuration, electrode configuration, and the like); and material selection and processing for other device layers (e.g., gate dielectric and source/drain electrodes).

Despite these limitations, however, an assessment of the relevant literature does provide an emerging picture of a range of achievable mobilities for ZnO TFTs. Mobility values as high as $\mu_{inc} \sim 25\,cm^2/V\,s$ have been reported [24], with a 600°C post-deposition anneal and high gate voltage (V_{GS}) bias (70 V across a 100 nm thermal SiO_2 gate dielectric). While this result comprises an important validation of potential ZnO TFT performance, significant reductions in both processing temperature and maximum gate voltage are critical to demonstrating the viability of ZnO in a manufacturable TFT process; in this endeavor, Nishii *et al.* have reported mobilities of $\mu_{sat} \sim 7\,cm^2/V\,s$ with a maximum processing temperature of 300°C and $V_{GS} = 30\,V$ [19]. Several groups have shown further reduction in processing temperature to at or near room temperature, while retaining reasonable mobility values. Carcia *et al.* obtained mobilities of $\mu_{FE} \sim 1$–$2\,cm^2/V\,s$, for ZnO channel films RF sputtered at "near room temperature" with no post-deposition anneal; this process was used also to fabricate ZnO TFTs on flexible organic (polyimide) substrates, yielding a mobility of $\mu_{sat} \sim 0.4\,cm^2/V\,s$ [17,20]. With further material and process development (including the use of an electron-beam evaporated Al_2O_3 gate dielectric), mobilities of $\sim$5–50 $cm^2/V\,s$ were obtained, although the change in gate dielectric introduced a substantial degree of hysteresis in current-voltage characteristics [36]. Fortunato *et al.* have also reported well-performing, room-temperature processed devices [26,28,31], although quantitative mobility values are unclear, due to the use of structures with unpatterned channel layer and small width-to-length ratio, as per Fig. 12.4 and related discussion (Section 12.2.2).

12.3.3 Incorporation of gate voltage-dependent mobility into the ideal TFT model

In addition to a reduced degree of ambiguity in quantifying ZnO TFT electrical performance (as discussed in the preceding sections), the use of μ_{avg} and V_{on} is further advantageous in that these parameters provide the basis for an analytical model that explicitly includes the gate voltage-dependence of channel mobility, thus allowing the precise reproduction of the DC current-voltage characteristics of ZnO TFTs [39].

The derivation process and associated assumptions (as described in Ref. 39) follow those of the ideal TFT model, with the exception of an unrestricted dependence of mobility on gate voltage, to produce a set of expressions analogous to those of the ideal model (Equation 12.1),

$$I_D = \begin{cases} 0 & \ldots V_{GS} \leq V_{on} \\ C_{ins}\frac{W}{L}\int_{V_{GS}-V_{on}-V_{DS}}^{V_{GS}-V_{on}} \mu_{avg}(V_{eff})V_{eff}\,dV_{eff} & \ldots V_{DS} \leq V_{GS}-V_{on}. \\ C_{ins}\frac{W}{L}\int_{0}^{V_{GS}-V_{on}} \mu_{avg}(V_{eff})V_{eff}\,dV_{eff} & \ldots V_{DS} > V_{GS}-V_{on} \end{cases} \tag{12.14}$$

With the exception of V_{eff}, the variables in Equation 12.14 are as previously defined. V_{eff} quantifies the portion of the applied gate voltage that acts to induce mobile channel charge, and is given by

$$V_{eff}(y) \equiv V_{GS} - V_{on} - V(y), \tag{12.15}$$

in which V(y) is the electron quasi-Fermi potential (referenced to the source electrode) and y is the position along the channel length, measured from the source.

In order to employ Equation 12.14, the measured $\mu_{avg}(V_{GS})$ data set must be manipulated to yield $\mu_{avg}(V_{eff})$. Since the I_D-V_{GS} measurement used to generate $\mu_{avg}(V_{GS})$ data, if properly carried out, employs a small drain bias V_{DS} (ideally, approaching zero), and since $V(y) < V_{DS}$ throughout the channel, V(y) can be reasonably neglected in the context of this measurement, yielding $\mu_{avg}(V_{eff}) \approx \mu_{avg}(V_{GS}\text{-}V_{on})$. Thus, $\mu_{avg}(V_{eff})$ is obtained through a rigid shift (by $-V_{on}$) of the associated $\mu_{avg}(V_{GS})$ data set [39].

Now, to obtain a closed-form expression (based on Equation 12.14) suitable for ZnO TFTs, $\mu_{avg}(V_{eff})$ must be parameterized so as to allow the integration of

Equation 12.14 to be carried out. Recognizing that the $\mu_{avg}(V_{eff})$ data set is shifted so as to be non-zero only for positive V_{eff}, an *n*th-order polynomial curve fit,

$$\mu_{avg(V_{eff})} \cong \begin{cases} 0 & \ldots V_{eff} \leq 0 \\ \sum_{i=0}^{n} c_i [V_{eff}]^i & \ldots V_{eff} > 0 \end{cases}, \tag{12.16}$$

should be suitable to provide a reasonable representation of a $\mu_{avg}(V_{eff})$ data set. Substituting this expression for $\mu_{avg}(V_{eff})$ into Equation 12.14 and carrying out the integration yields

$$I_D \cong \begin{cases} 0 & \ldots V_{GS} \leq V_{on} \\ C_{ins}\frac{W}{L}\sum_{i=0}^{n}\left(\frac{c_i}{i+2}\left[(V_{GS}-V_{on})^{i+2}-(V_{GS}-V_{on}-V_{DS})^{i+2}\right]\right) & \ldots V_{DS} \leq V_{GS}-V_{on}, \\ C_{ins}\frac{W}{L}\sum_{i=0}^{n}\left(\frac{c_i}{i+2}(V_{GS}-V_{on})^{i+2}\right) & \ldots V_{DS} > V_{GS}-V_{on} \end{cases} \tag{12.17}$$

in which $\{c_i\}$ are the polynomial coefficients of the *n*th-order polynomial fit (Equation 12.16). Note that, for gate voltage-independent mobility (i.e., $n = 0$, $c_0 = \mu$), Equation 12.17 collapses directly to the ideal TFT model (Equation 12.1).

Figure 12.10 shows the I_D-V_{DS} data previously plotted in Fig. 12.5, along with data obtained using the model of Equation 12.17 and a sixth-order polynomial fit for $\mu_{avg}(V_{eff})$ (from $\mu_{avg}(V_{GS})$, Fig. 12.9); nearly perfect agreement between measured and calculated drain current values is obtained. While the sixth-order polynomial fit employed here may be somewhat excessive from the standpoint of computational efficiency, it serves here to verify model validity and to demonstrate achievable precision; in practice, model precision and complexity may be traded off against one another, as appropriate to the relevant circumstance.

12.3.4 Drain current on/off ratio

An additional performance metric often included in TFT characterization is the on/off ratio, the ratio of maximum to minimum drain current (I_D) for a fixed (and, typically, relatively large) drain-to-source voltage (V_{DS}). Assessment of on/off ratio is straightforward, typically obtained from an I_D-V_{GS} transfer curve with V_{GS} swept through an appropriate range (either defined by the intended application of the TFT, or arbitrarily selected in the device characterization process); for example, for the $V_{DS} = 10\,V$ curve of Fig. 12.6, the maximum and minimum I_D values ($\sim 2 \times 10^{-7}$

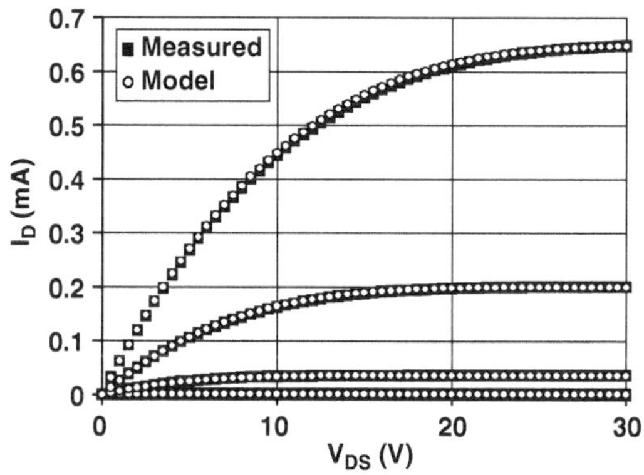

Figure 12.10: Drain current vs. drain-to-source voltage (I_D-V_{DS}) curves for the ZnO TFT with I_D-V_{DS} data shown in Fig. 12.5, $\mu_{avg}(V_{GS})$ in Fig. 12.9; $V_{GS} = 0$ to 40 V in 10 V increments, I_D increases with increasing V_{GS}; black squares show measured data, white circles show data calculated using Equation 12.17 and a sixth-order polynomial fit for $\mu_{avg}(V_{eff})$. Model parameters: $C_{ins} = 34.5\,nF/cm^2$; W/L = 9; n = 6; $\{c_0, c_1, c_2, c_3, c_4, c_5, c_6\} = \{0, 5.0834 \times 10^{-3}, -1.4634 \times 10^{-3}, 2.7185 \times 10^{-4}, -6.3302 \times 10^{-6}, 6.2682 \times 10^{-8}, -2.4285 \times 10^{-10}\}$; $V_{on} = -2.1$ V (data from Ref. 39).

and 2×10^{-12} A, respectively) yield an on/off ratio of ~10^5 for the selected V_{GS} range (−5 to 5 V).

It is important, however, to understand the practical significance of reported on/off ratios for ZnO TFTs, particularly in the case where simplified test structures (as discussed in Section 12.2.2), often with rather large channel lengths on the order of ~100 μm, are employed. TFT on/off ratio is determined by two measured values: the maximum ("on") and minimum ("off") drain currents. While on-current is determined primarily by ideal, expected aspects of device operation (i.e., mobility and turn-on voltage), off-current (i.e., drain current for gate voltages below the device turn-on voltage) is a consequence of one or more current leakage mechanisms not accounted for in the ideal transistor model (Equation 12.1), which predicts zero drain current in the off state. Off-state current leakage mechanisms can be grouped into two categories: source-to-drain leakage (i.e., lateral charge flow within the channel layer), and gate-to-channel leakage (i.e., charge flow through the gate dielectric). For a given voltage bias condition (assuming an equivalent gate dielectric), source-to-drain leakage increases with *decreasing* channel length (source-to-drain leakage is related to source-to-drain field strength, which is inversely proportional to channel length), while gate-to-channel leakage increases with *increasing* channel length (leakage through the gate dielectric scales with gate area, which is directly proportional to channel length). Thus, the off-current component of the on/off ratio reported for a ZnO TFT test structure with relatively large channel length is likely to be influenced primarily by gate dielectric leakage, whereas source-to-drain leakage mechanisms may become dominant in establishing measured off-current (and thus on/off ratio) as channel length decreases.

In view of the above, since on/off ratio may be strongly influenced by TFT design (e.g., structure, layout), it is best understood as a device parameter (i.e., relating to a particular device instantiation); as such, an on/off ratio reported for a large TFT test structure does not necessarily provide a good indication of the performance to be expected for practically useful devices with channel lengths on the order of $\sim$5–10 μm or smaller, and should be interpreted accordingly. However, while in most cases ZnO TFTs reported to date have employed structures with channel lengths of $\sim$50–100 μm, Nishii *et al.* have reported the fabrication of well-behaved ZnO TFTs with 5 μm channels, using a device structure similar to that employed in the fabrication of a-Si TFTs used in AMLCD technology, and achieving low off-currents of $\sim$1 pA and on/off ratios of $\sim 10^7$ [19]; likewise, Volkman *et al.* explored channel lengths down to 3 μm for their TFTs with solution-processed, nanoparticle-based ZnO channels, observing generally well-behaved current-voltage characteristics for channel lengths as small as 5 μm [30]. These results provide evidence that off-state source-to-drain current leakage can in fact be effectively controlled in ZnO TFTs with useful channel dimensions.

12.4 Optical properties

Interest in ZnO TFTs is in large part related to the fact that, as a wide-bandgap semiconductor, ZnO is optically transparent through the visible region of the electromagnetic spectrum. Since typical gate dielectric materials are optically transparent, the fabrication of a fully-transparent ZnO TFT requires only the use of a transparent conductor such as ITO (indium tin oxide) or doped ZnO for source, drain, and gate electrodes. Thus, many reports of ZnO TFTs have included the fabrication of all-transparent devices, along with basic characterization of the optical properties (e.g., percent transmittance) of the device stack. Figure 12.11 plots an exemplary set of optical transmittance vs. wavelength data, through the channel and source/drain electrode regions of an all-transparent ZnO TFT; in both regions, the average transmittance through the visible portion of the electromagnetic spectrum ($\sim$400–700 nm) is about 75% [16]. A survey of the literature indicates a typical range of transparency (averaged over the visible spectrum) of $\sim$75–85% for all-transparent ZnO TFTs [15,16,21,23,31,33].

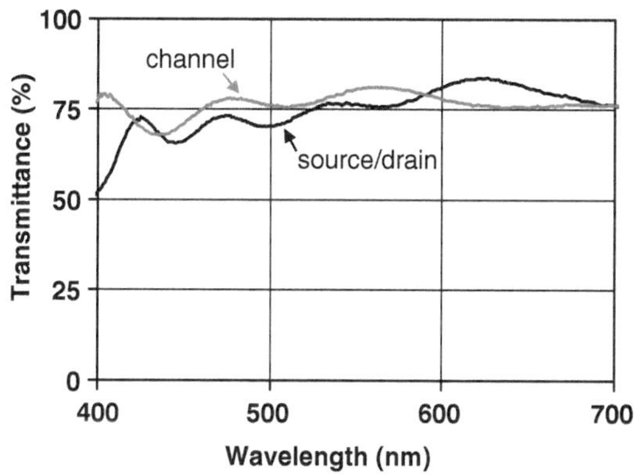

Figure 12.11: Optical transmittance vs. wavelength through the channel and source/drain electrode regions of a bottom-gate, all-transparent ZnO TFT; gate electrode 200 nm ITO (indium tin oxide), gate dielectric 200 nm ATO (aluminum-titanium oxide), channel 100 nm ZnO, source/drain 300 nm ITO (after Ref. 16).

In addition to enabling the fabrication of fully-transparent TFTs, the associated lack of interaction with visible photons is potentially advantageous in displays and other applications where the device may be exposed to light from one or more sources (e.g., ambient light, display backlight, and the like). While several reports have noted that the current-voltage characteristics of ZnO TFTs show little if any sensitivity to "ambient light" [16,31], a more systematic evaluation of the optical response of ZnO TFTs by Bae *et al.* found, in addition to the expected strong interaction with UV (340 nm) photons, a significant increase in measured drain current upon exposure to both blue (450 nm) and green (540 nm) light [22,32]. While it is likely that such visible light sensitivity may be substantially reduced or eliminated through optimization of deposition and processing methods targeted toward improved material quality (specifically, to reduce the population of optically-active mid-gap states), these results point to the need for further examination of the electro-optic properties of ZnO TFTs, if their unique properties are to be effectively leveraged.

12.5 Conclusion

This chapter has provided an overview of the current state of ZnO thin-film transistor (TFT) technology, with particular attention given to electrical characterization techniques and the interpretation of resulting data. The highly non-ideal mobility characteristics of these devices necessitate particular care in extracting even the most basic electrical performance parameters, and in making a useful assessment of ZnO TFT performance reported in the literature.

Despite their immaturity, as compared to established amorphous- and polycrystalline-Si TFTs (and even to the relatively less-developed organic semiconductor-based devices), ZnO TFTs are attracting mounting interest, particularly within the context of envisioned "transparent electronics" technology, for which optically transparent TFTs comprise a key enabling component. While substantial work remains to fully elucidate electrical and optical properties, to develop and optimize suitable fabrication processes, and to better understand their potential for use in displays and other applications, the growing body of evidence suggests that ZnO TFTs are likely to become a viable contender within the field of established and emerging TFT technologies.

References

[1] J. E. Lilienfield, "Device for Controlling Electric Current", US Pat. 1,900,018 (1933).
[2] O. Heil, "Improvements in or Relating to Electrical Amplifiers and Other Control Arrangements and Devices", Brit. Pat. 439,457 (1935).
[3] Cherie R. Kagan and Paul Andry, eds. "Thin-Film Transistors" (Marcel Dekker, New York, 2003).
[4] Yue Kuo, ed. "Thin Film Transistors Materials and Processes, Volume 1: Amorphous Silicon Thin Film Transistors" (Kluwer Academic Publishers, Boston, 2004).
[5] John F. Wager, Science **300**, 1245 (2003).
[6] Hiromichi Ohta and Hideo Hosono, Materials Today **7**, 42 (2004).
[7] Hiroshi Kawazoe, Masahiro Yasukawa, Hiroyuki Hyodo, Masaaki Kurita, Hiroshi Yanagi, and Hideo Hosono, Nature **389**, 939 (1997).
[8] Atsushi Kudo, Hiroshi Yanagi, Hideo Hosono, and Hiroshi Kawazoe, Appl. Phys. Lett. **73**, 220 (1998).
[9] Hiromichi Ohta, Ken-ichi Kawamura, Masahiro Orita, Masahiro Hirano, Nobuhiko Sarukura, and Hideo Hosono, Appl. Phys. Lett. **77**, 475 (2000).
[10] Hiroshi Kawazoe, Hiroshi Yanagi, Kazushige Ueda, and Hideo Hosono, MRS Bull. **25**, 28 (2000).
[11] Hiroshi Yanagi, Kazushige Ueda, Hiromichi Ohta, Masahiro Orita, Masahiro Hirano, and Hideo Hosono, Solid-State Commun. **121**, 15 (2002).
[12] D. C. Look, B. Claflin, Y. I. Alivov, and S. J. Park, Phys. Stat. Sol. (a) **201**, 2203 (2004).
[13] M. Kawasaki, K. Tamura, K. Saikusa, T. Aita, A. Tsukazaki, A. Ohtomo, Z. G. Jin, Y. Matsumoto, T. Fukumura, H. Koinuma, Y. Ohmaki, S. Kishimoto, Y. Ohno, F. Matsukura, H. Ohno, T. Makino, N. T. Tuan, H. D. Sun, C. H. Chia, Y. Segawa, Z. K. Tang, and G. K. L. Wang, Ext. Abstr. of the 2000 Int. Conf. on Solid-State Dev. and Mat., 128 (2000).

[14] A. Ohtomo and M. Kawasaki, IEICE Trans. Electron. **E83-C**, 1614 (2000).
[15] S. Masuda, K. Kitamura, Y. Okumura, S. Miyatake, H. Tabata, and T. Kawai, J. Appl. Phys. **93**, 1624 (2003).
[16] R. L. Hoffman, B. J. Norris, and J. F. Wager, Appl. Phys. Lett. **82**, 733 (2003).
[17] P. F. Carcia, R. S. McLean, M. H. Reilly, and G. Nunes, Jr., Appl. Phys. Lett. **82**, 1117 (2003).
[18] Y. Ohya, T. Niwa, T. Ban, and Y. Takahashi, Jpn. J. Appl. Phys. Part 1 **40**, 297 (2001).
[19] J. Nishii, F. M. Hossain, S. Takagi, T. Aita, K. Saikusa, Y. Ohmaki, I. Ohkubo, S. Kishimoto, A. Ohtomo, T. Fukumura, F. Matsukura, Y. Ohno, H. Koinuma, H. Ohno, and M. Kawasaki, Jpn. J. Appl. Phys. Part 2 **42**, L347 (2003).
[20] P. F. Carcia, R. S. McLean, M. H. Reilly, I. Malajovich, K. G. Sharp, S. Agrawal, and G. Nunes, Jr., MRS Symp. Proc. **769**, H7.2.1 (2003).
[21] B. J. Norris, J. Anderson, J. F. Wager, and D. A. Keszler, J. Phys. D: Appl. Phys. **36**, L105 (2003).
[22] H. S. Bae, M. H. Yoon, J. H. Kim, and S. Im, Appl. Phys. Lett. **83**, 5313 (2003).
[23] Y. Kwon, Y. Li, Y. W. Heo, M. Jones, P. H. Holloway, D. P. Norton, Z. V. Park, and S. Li, Appl. Phys. Lett. **84**, 2685 (2004).
[24] R. L. Hoffman, J. Appl. Phys. **95**, 5813 (2004).
[25] H. S. Bae and Seongil Im, J. Vac. Sci. Technol. B **22**, 1191 (2004).
[26] E. Fortunato, A. Pimentel, L. Pereira, A. Goncalves, G. Lavareda, H. Aguas, I. Ferreira, C. N. Carvalho, and R. Martins, J. Non-Cryst. Sol. **338–340**, 806 (2004).
[27] Y. W. Heo, L. C. Tien, Y. Kwon, D. P. Norton, S. J. Pearton, B. S. Kang, and F. Ren, Appl. Phys. Lett. **85**, 2274 (2004).
[28] E. M. C. Fortunato, P. M. C. Barquinha, A. C. M. B. G. Pimentel, A. M. F. Goncalves, A. J. S. Marques, R. F. P. Martins, and L. M. N. Pereira, Appl. Phys. Lett. **85**, 2541 (2004).
[29] H. S. Bae, J. H. Kim, and S. Im, Electrochem. Solid-State Lett. **7**, G279 (2004).
[30] Steven K. Volkman, Brian A. Mattis, Steven E. Molesa, Josephine B. Lee, Alejandro de la Fuente Vornbrock, Teymur Bakhishev, and Vivek Subramanian, 2004 IEEE International Electron Devices Meeting (IEDM), IEDM Technical Digest, 769 (2004).
[31] E. M. C. Fortunato, P. M. C. Barquinha, A. C. M. B. G. Pimentel, A. M. F. Goncalves, A. J. S. Marques, L. M. N. Pereira, and R. F. P. Martins, Adv. Mat. **17**, 590 (2005).
[32] H. S. Bae, C. M. Choi, J. H. Kim, and S. Im, J. Appl. Phys. **97**, 076104 (2005).
[33] Yutaka Ohya, Takashi Kume, and Takayuki Ban, Jpn. J. Appl. Phys. Part 1 **44**, 1919 (2005).
[34] C. J. Kao, Yong Wook Kwon, Y. W. Heo, D. P. Norton, S. J. Pearton, F. Ren, and G. C. Chi, J. Vac. Sci. Technol. B **23**, 1024 (2005).
[35] Y. J. Li, Y. W. Kwon, M. Jones, Y. W. Heo, J. Zhou, S. C. Luo, P. H. Holloway, E. Douglas, D. P. Norton, Z. Park, and S. Li, Semicond. Sci. Technol. **20**, 720 (2005).
[36] P. F. Carcia, R. S. McLean, and M. H. Reilly, J. SID **13**, 547 (2005).
[37] Il-Doo Kim, YongWoo Choi, and Harry L. Tuller, Appl. Phys. Lett. **87**, 043509 (2005).
[38] G. F. Boesen and J. E. Jacobs, Proc. IEEE **56**, 2094 (1968).
[39] R. L. Hoffman, Solid-State Electron. **49**, 648 (2005).

[40] J. Torkel Wallmark and Harwick Johnson, eds. "Field-Effect Transistors: Physics, Technology, and Applications" (Prentice-Hall, Englewood Cliffs, NJ, 1966).
[41] David W. Greve, "Field Effect Devices and Applications: Devices for Portable, Low-Power, and Imaging Systems" (Prentice-Hall, Englewood Cliffs, NJ, 1998).
[42] Yue Kuo, ed. "Thin Film Transistors Materials and Processes, Volume 2: Polycrystalline Silicon Thin Film Transistors" (Kluwer Academic Publishers, Boston, 2004).
[43] Dieter K. Schroder, "Semiconductor Material and Device Characterization", 2nd ed. (Wiley, New York, 1998).
[44] S. M. Sze, "Physics of Semiconductor Devices", 2nd ed. (Wiley, New York, 1981).
[45] Faruque M. Hossain, J. Nishii, S. Takagi, A. Ohtomo, T. Fukumura, H. Fujioka, H. Ohno, H. Koinuma, and M. Kawasaki, J. Appl. Phys. **94**, 7768 (2003).
[46] J. Levinson, F. R. Shepherd, P. J. Scanlon, W. D. Westwood, G. Este, and M. Rider, J. Appl. Phys. **53**, 1193 (1982).

Chapter 13

ZnO Piezoelectric Devices

Yicheng Lu, Nuri W. Emanetoglu and Ying Chen

Department of Electrical and Computer Engineering, Rutgers, The State University of New Jersey, 94 Brett Road, Piscataway, NJ 08854-8058, USA

Abstract: ZnO is an important piezoelectric material, which finds broad applications in signal processing, telecommunications, micro- and nano-electromechanical systems (MEMS and NEMS), and sensors. The current progress of piezoelectric ZnO is reviewed. We focus on the ZnO piezoelectric properties, growth and compensation doping, as well as various piezoelectric ZnO-based surface acoustic wave (SAW) and bulk acoustic wave (BAW) devices. Recent advances, including tailoring of piezoelectric properties using $Mg_xZn_{1-x}O$ and its heterostructures, integration of piezoelectric ZnO with GaN, SiC, etc., are reviewed. The integration of semiconducting and piezoelectric ZnO to implement novel sensors is also discussed.

13.1 Introduction

ZnO is a multifunctional material possessing unique electrical, optical, acoustic, and mechanical properties. It belongs to the 6mm wurtzite crystal class; therefore is piezoelectric. ZnO has been used for thin film piezoelectric devices since the early 1970s, due to its moderate coupling coefficients and ease of deposition. Its applications include bulk acoustic wave (BAW) and surface acoustic wave (SAW) resonators, filters and sensors, and micro- and nano-electromechanical systems (MEMS and NEMS). Piezoelectric ZnO thin films have been used for implementing acoustic wave devices on semiconductor substrates (e.g. Si [1], GaAs [2] and InP [3]); non-piezoelectric materials that have high acoustic velocity and low propagation loss (e.g. diamond [4–6] and sapphire [7–12]); weakly piezoelectric materials (e.g. quartz [13]); and low-cost amorphous materials, such as glass [14,15]. The most ubiquitous application of ZnO thin films to date has been SAW filters. The advantages of using piezoelectric ZnO thin film based multilayered structures depend on the particular substrates and applications, and

Zinc Oxide Bulk, Thin Films and Nanostructures
C. Jagadish and S. Pearton (Editors)

include: (i) direct integration of piezoelectric devices with semiconductor ICs (e.g. GaAs, Si); (ii) higher frequency capabilities for a given electrode periodicity (non-piezoelectric, high acoustic velocity substrates); (iii) the low propagation loss characteristics of certain substrates (e.g. diamond, sapphire); (iv) the existence of high coupling factor, higher order generalized SAW modes and high velocity pseudo surface acoustic wave (HVPSAW) modes; and (v) a higher power handling capability. Possible disadvantages of such structures are their dispersive velocity characteristic, and losses that may be introduced by multilayer structures and their interfaces.

This review focuses on the recent progress in piezoelectric ZnO and its device applications. Section 13.2 provides the most important piezoelectric properties of ZnO material. Section 13.3 describes the major growth technologies to obtain piezoelectric films. The compensation doping process is also discussed. Section 13.4 presents the characteristics of surface acoustic waves in ZnO and ZnO-based multilayer structures. The integrated SAW devices, including sensors, are also covered. Section 13.5 presents ZnO thin film bulk acoustic devices. The conclusion and outline of future research work can be found in Section 13.6.

13.2 Piezoelectric properties of ZnO

Piezoelectricity is the mechanism of converting mechanical energy into electrical energy, and vice versa. The generation of electrical polarization when a piezoelectric material is strained is known as the direct piezoelectric effect. The converse piezoelectric effect refers to the generation of strain in a piezoelectric material when an electric field is applied. Piezoelectricity occurs only in twenty of the thirty-two crystal classes where the unit cells are hemimorphic, devoid of a center of symmetry.

Zinc oxide (ZnO) belongs to the hexagonal wurtzite (6mm) crystal class, in which the z-reference coordinate is chosen along the Z-axis. Several important parameters associated with the piezoelectricity of ZnO are listed below:

a) ZnO dielectric coefficients under the constant strain:

$$\begin{bmatrix} \varepsilon_{11} & 0 & 0 \\ 0 & \varepsilon_{11} & 0 \\ 0 & 0 & \varepsilon_{33} \end{bmatrix} = \begin{bmatrix} 7.57 & 0 & 0 \\ 0 & 7.57 & 0 \\ 0 & 0 & 9.03 \end{bmatrix} \qquad (13.1)$$

b) ZnO stiffness coefficients (10^{10} N/m²):

$$\begin{bmatrix} c_{11} & c_{12} & c_{13} & 0 & 0 & 0 \\ c_{12} & c_{11} & c_{13} & 0 & 0 & 0 \\ c_{13} & c_{13} & c_{33} & 0 & 0 & 0 \\ 0 & 0 & 0 & c_{44} & 0 & 0 \\ 0 & 0 & 0 & 0 & c_{44} & 0 \\ 0 & 0 & 0 & 0 & 0 & c_{66} \end{bmatrix} = \begin{bmatrix} 20.97 & 12.11 & 10.51 & 0 & 0 & 0 \\ 12.11 & 20.97 & 10.51 & 0 & 0 & 0 \\ 10.51 & 10.51 & 21.09 & 0 & 0 & 0 \\ 0 & 0 & 0 & 4.247 & 0 & 0 \\ 0 & 0 & 0 & 0 & 4.247 & 0 \\ 0 & 0 & 0 & 0 & 0 & 4.43 \end{bmatrix} \tag{13.2}$$

where $c_{66} = (c_{11} - c_{12})/2$

c) ZnO compliance coefficients (10^{-12} m²/N):

$$\begin{bmatrix} s_{11} & s_{12} & s_{13} & 0 & 0 & 0 \\ s_{12} & s_{11} & s_{13} & 0 & 0 & 0 \\ s_{13} & s_{13} & s_{33} & 0 & 0 & 0 \\ 0 & 0 & 0 & s_{44} & 0 & 0 \\ 0 & 0 & 0 & 0 & s_{44} & 0 \\ 0 & 0 & 0 & 0 & 0 & s_{66} \end{bmatrix} = \begin{bmatrix} 11.12 & -4.56 & -1.4 & 0 & 0 & 0 \\ -4.56 & 11.12 & -1.4 & 0 & 0 & 0 \\ -1.4 & -1.4 & 8.47 & 0 & 0 & 0 \\ 0 & 0 & 0 & 34.4 & 0 & 0 \\ 0 & 0 & 0 & 0 & 34.4 & 0 \\ 0 & 0 & 0 & 0 & 0 & 31.36 \end{bmatrix} \tag{13.3}$$

where $s_{66} = 2(s_{11} - s_{12})$

d) ZnO piezoelectric stress constants (C/m²):

$$\begin{bmatrix} 0 & 0 & 0 & 0 & e_{15} & 0 \\ 0 & 0 & 0 & e_{15} & 0 & 0 \\ e_{31} & e_{31} & e_{33} & 0 & 0 & 0 \end{bmatrix} = \begin{bmatrix} 0 & 0 & 0 & 0 & -0.48 & 0 \\ 0 & 0 & 0 & -0.48 & 0 & 0 \\ -0.573 & -0.573 & 1.32 & 0 & 0 & 0 \end{bmatrix} \tag{13.4}$$

e) ZnO piezoelectric strain coefficients (10^{-12} C/N):

$$\begin{bmatrix} 0 & 0 & 0 & 0 & d_{15} & 0 \\ 0 & 0 & 0 & d_{15} & 0 & 0 \\ d_{31} & d_{31} & d_{33} & 0 & 0 & 0 \end{bmatrix} = \begin{bmatrix} 0 & 0 & 0 & 0 & -11.34 & 0 \\ 0 & 0 & 0 & -11.34 & 0 & 0 \\ -5.43 & -5.43 & 11.67 & 0 & 0 & 0 \end{bmatrix} \tag{13.5}$$

13.3 Material growth and compensation doping

The most common techniques used for depositing ZnO films are laser ablation, metalorganic chemical vapor deposition (MOCVD), molecular beam evaporation

(MBE) and sputtering. MBE and MOCVD growth technologies result in the excellent epitaxial quality and in-situ modulation doping capability. However, the as-grown ZnO films generally show the *n*-type semiconducting properties, due to the oxygen vacancies and Zn interstitials, as well as other defects and impurities. In order to effectively stimulate the piezoelectric effect, this *n*-type conductivity has to be compensated. This requires very heavy doping, which can be difficult for MBE and MOCVD technologies. Therefore, a post-growth diffusion step is frequently used to achieve piezoelectric ZnO films grown by MBE and MOCVD. Furthermore, most piezoelectric thin film applications need thick layers, 1 μm or above. Thus MBE is generally not used for growing piezoelectric ZnO thin films. Sputtering is the common production scale deposition method, due to its relative ease of use and low cost. Sputtering targets can be made to contain the compensated doping species, such as Li, Ni, Cu, etc. Therefore, piezoelectricity excitation can be obtained in sputtered ZnO films without post-deposition diffusion.

13.3.1 Pulsed laser deposition

Recent developments in pulsed laser deposition (PLD) have demonstrated its flexibility with respect to the realization of novel, complex materials and device structures. PLD has been used to deposit ZnO films with good optical properties [16,17]. The classical experimental conditions for the deposition of high quality ZnO films by PLD are the following: use of an excimer (ArF or KrF) or frequency quadrupled laser, with a fluence around 2 J/cm^2, a substrate temperature in the 600∼800°C range under an oxygen pressure lying between 10^{-6} and 10^{-1} mbar, and a target to substrate distance in the 5∼10 cm range [18].

Piezoelectric ZnO thin films can be achieved using PLD in high purity oxygen atmosphere without compensation doping. F. Craciun *et al.* [19] deposited *c*-axis oriented piezoelectric ZnO films on silicon, sapphire and Corning glass substrates. The deposition temperature was varied in the range of 200∼320°C. G. H. Kim *et al.* [20] reported ZnO films deposited by PLD for film bulk acoustic resonators (FBAR). Piezoelectric ZnO films with 1 μm thickness were deposited on the Al bottom electrode by the PLD technique using a Nd:YAG laser with a wavelength of 355 nm, and laser energy density of 2.5 J cm^{-2}. The base vacuum in the chamber was down to 10^{-5} Torr, and high purity oxygen was injected into the chamber by a mass flow controller. A 99.999% pure zinc oxide target was used to preclude pit formation and to ensure uniform ablation of the target. Three main processing factors, deposition temperature, oxygen gas pressure, and laser frequencies were investigated. It was found that the optimized parameters for FBARs were 100°C, 6.7 Pa, and 1 Hz. Under these conditions, the ZnO sample had x-ray diffraction

(XRD) full width of half maximum (FWHM) of 0.58°, and electrical resistivity of 2.77×10^7 Ω cm, which is suitable for film bulk acoustic wave resonators.

13.3.2 Sputtering

Sputtering is a popular deposition technique, as it is able to obtain a large size, well-oriented and uniform ZnO films even on amorphous substrates at high deposition rates [10,21]. M. Katoda *et al.* reported that ZnO films deposited by DC-mode electron cyclotron resonance (ECR) and RF-mode ECR sputtering systems showed excellent piezoelectric properties and *c*-axis orientation [15]. The ECR sputtering system added magnets to the outside of a cylindrical zinc metal (Zn) target. There was no evidence of column and fiber grains in the cross section of the ZnO films. A ZnO on interdigital transducer (IDT)/glass structure was able to drive a 1.1 GHz fundamental Rayleigh SAW. J. Lee *et al.* deposited ZnO thin films on SiO_2/Si(100) substrates by RF planar magnetron sputtering using two-step fabrication methods for SAW applications: the first deposition with argon only for 30 min at 100 W and the second deposition with oxygen in the range $O_2/(Ar + O_2)$ up to 50% [22]. In comparison with films deposited by the one-step fabrication method, these films exhibited a lower insertion loss and a closer agreement between experimental coupling coefficients and the theoretical values. Post-deposition annealing plays an important role in improving the structural and electrical properties of piezoelectric ZnO films. S. Y. Chu *et al.* reported the post-deposition annealing effects on ZnO film quality [23]. When the deposited ZnO films were annealed for 1 hour at 400°C in vacuum, the stress was relieved, resistivity increased by one order of magnitude, and the loss factor was kept at minimum. When annealing at 600°C and above, the loss factor and stress were found to increase. The interface between piezoelectric ZnO films and the substrates, and interface between ZnO film and metal electrodes, are critical important for SAW and BAW device performances. Y. Yoshino *et al.* studied the microstructures of ZnO/substrate interfaces [24]. It was found that an amorphous layer exists at the ZnO/glass interface, while a thicker amorphous layer exists at the ZnO/Al interface. No amorphous layer was observed at the ZnO/Au and ZnO/sapphire interfaces, and direct orientation of ZnO thin films begin at the interface of both these substrates.

Unintentionally doped ZnO films generally are conductive, *n*-type semiconducting material. This electrical conductivity will short out the piezoelectric effect. Therefore, a compensation dopant, such as Li or Ni, must be added to the sputtering target. T. Mitsuyu *et al.* added lithium carbonate (Li_2CO_3) to ZnO targets (purity, 99.9%) before the sintering process, in order to obtain high resistivity films [7]. H. Ieki *et al.* obtained a Ni-doped ZnO target by mixing the appropriate quantities of 99.995% pure ZnO powder with 1~2% wt NiO power before the sintering process [10].

The electron cyclotron resonance (ECR) sputtering process is used to deposit $(01\bar{1}2)$ plane epitaxial ZnO films on *r*-plane sapphire substrate at 250~550°C [25]. The ZnO films deposited on glass or sapphire substrates had a high electrical resistivity ($\rho > 10^{10}\ \Omega$ cm) and strongly generated higher-frequency SAWs with a large electromechanical coupling factor for the Sezawa SAW mode.

SAW acoustic-optic (AO) devices which use ZnO piezoelectric thin films deposited on InGaAs/InP structure can be applied to optical fiber communications. S. J. Chang *et al.* reported ZnO thin films which were deposited on (100)-oriented InP substrates by RF-magnetron sputtering [26]. An undoped and high purity ZnO target was used for the deposition. The best sputtering condition to achieve a small XRD FWHM of 0.183°C, was at room temperature with 200 W RF power, and mTorr gas pressure with a gas ratio of $Ar{:}O_2 = 2{:}1$. The resistivity of the deposited films reached $10^{10}\ \Omega$ cm and their refractive index was close to that of bulk ZnO. ZnO/InP based SAW devices were then successfully fabricated [3].

13.3.3 Metalorganic chemical vapor deposition

Chemical vapor deposition (CVD) is a thin film growth process in which a compound condenses onto a substrate from gas phase to produce a solid deposition, involving either thermal decomposition or a chemical reaction on the substrate surface. Metalorganic CVD (MOCVD) uses metalorganic, also known as organometallic, species as the precursors for the deposition. The advantages of MOCVD include: large area deposition, *in-situ* doping, excellent control of composition and thickness uniformity, good reproducibility, flexibility of low or high operating pressure, high deposition rate, ability to coat complex shapes, fast gas switching to produce sharp interfaces for heterostructures and multilayers, and availability of a wide variety of source materials.

Significant progress has been made in ZnO growth by MOCVD in recently years. The improvement of the material quality is related to improved reactor design and/or the use of less-reactive precursors, allowing one to minimize parasitic prereactions in the gas phase. The gas phase reaction can be eliminated in a modified MOCVD system by the separation of metalorganics and oxygen flow, as well as by process control of flow patterns and gas residence time. The stable metal-organic source zinc acetylacetonate was successfully used in combination with oxygen for the growth of high-quality ZnO films on *c*- and *a*-plane [27] as well as on *r*-plane [28,29] sapphire substrates by atmospheric pressure MOCVD. The *a*-plane $(11\bar{2}0)$ZnO on *r*-plane $(01\bar{1}2)$sapphire substrate provides in-plane

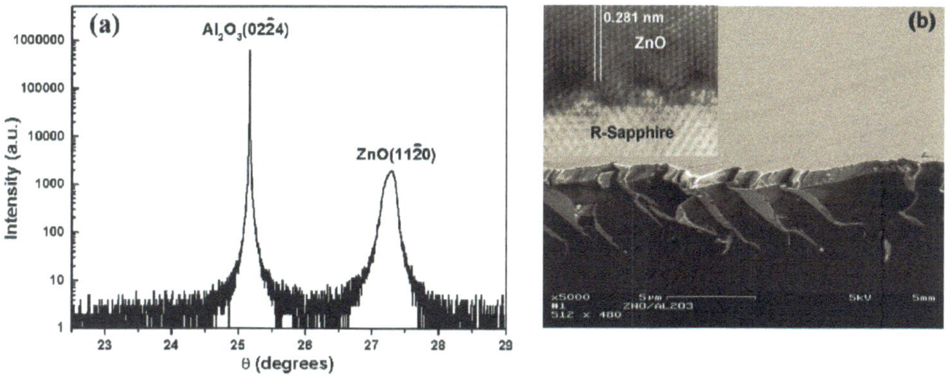

Figure 13.1: (a) Semi-logarithmic plot of X-ray ω-2θ scans of 1.3 μm thick ZnO thin film grown on r-Al_2O_3, (b) A SEM picture of ZnO/r-Al_2O_3 interface. The insert shows a TEM picture of the interface.

anisotropy, which is particularly useful for multi-mode SAW devices and sensors. C. R. Gorla *et al.* reported the optimum growth conditions for growing high epitaxial quality ZnO on r-sapphire substrates at 350°C to 450°C [29]. As shown in Figure 13.1(a), the epitaxial relationships of ZnO on r-sapphire are determined to be $(11\bar{2}0)$ZnO//$(01\bar{1}2)$r-Al_2O_3 and [0001]ZnO//$[0\bar{1}11]$r-Al_2O_3. Figure 13.1(b) shows a SEM picture of the ZnO/r-Al_2O_3; the inset shows a high resolution TEM image of the interface. As can be seen in the Figure 13.1, the ZnO film is dense with a smooth surface, and the interface between r-sapphire and the deposited ZnO is atomically sharp. A series of misfit dislocations exist within 5 nm of the ZnO/r-Al_2O_3 interface, resulting in nearly stress-free ZnO film. The quality of this interface is crucial for efficient acoustic energy transfer from the ZnO film to the r-Al_2O_3 substrate. As-grown ZnO is *n*-type conductive. Compensation doping using Li [9,30] or Cu [31] has been used to achieve the piezoelectric films though the post-growth diffusion. Lithium carbonate (Li_2CO_3) was used as the Li diffusion source for compensatory doping. After Li diffusion, resistivities above 10^7 Ω cm were achieved.

13.3.4 Growth of piezoelectric $Mg_xZn_{1-x}O$ films

Magnesium zinc oxide ($Mg_xZn_{1-x}O$) is a new piezoelectric material formed by alloying ZnO and MgO. MgO is a non-piezoelectric material with a rock-salt cubic structure, while piezoelectric ZnO has a wurtzite crystal structure. For Mg composition less than 35% ($x < 35\%$), the $Mg_xZn_{1-x}O$ crystal has a wurtzite structure [32,33]. Compared with ZnO, MgO has higher bulk acoustic wave velocities.

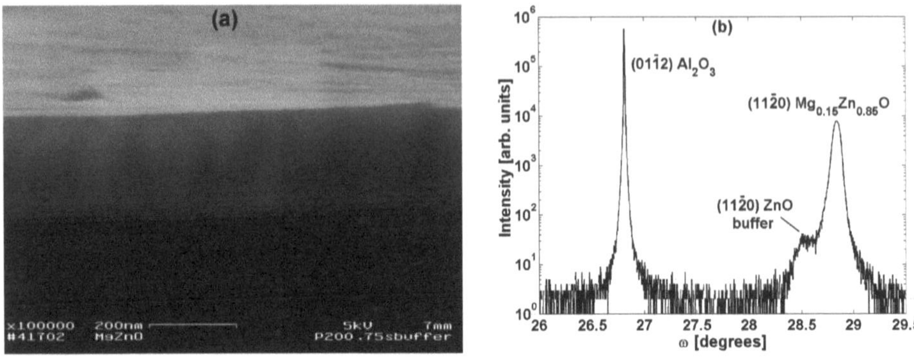

Figure 13.2: (a) FE-SEM picture of $Mg_{0.3}Zn_{0.7}O$/r-Al_2O_3. (b) Semi-logarithmic plot of X-ray ω-2θ scans scans of 1.3 μm thick $Mg_{0.15}Zn_{0.85}O$ thin film grown on r-Al_2O_3 [34].

Therefore, $Mg_xZn_{1-x}O$'s acoustic velocity increases and its piezoelectric coupling coefficient decreases as the Mg content increases, within the piezoelectric range. This allows for piezoelectric device design flexibility, as acoustic wave properties can be tailored by controlling Mg content as well as using ZnO/$Mg_xZn_{1-x}O$ multilayer structures.

$Mg_xZn_{1-x}O$ ($0 \le x \le 0.3$) films can be grown on *r*-plane (01$\bar{1}$2) sapphire substrate [34] and Si substrate [35] using MOCVD technology. The metalorganic sources for Zn and Mg were Diethyl Zinc (DEZn) and Bis(methylcyclopentadienyl) Magnesium ($(CH_3C_5H_4)_2Mg$, i.e. CP_2Mg), respectively. The vapors of Zn and Mg were introduced into the chamber, onto the rotating substrate, using a carrier gas flowing through the bubblers. The oxidizer was oxygen gas, which was introduced through a separate line into the chamber. Gas phase reaction is one of the main barriers for high quality $Mg_xZn_{1-x}O$ film growth. The reactions between DEZn, CP_2Mg, and oxygen can occur even at room temperature. Such gas phase reactions generate particles, which get incorporated into the deposited films, degrading film crystallinity and morphology. N. W. Emanetoglu *et al.* found a ZnO buffer layer to be critical for the growth of epitaxial wurtzite-type $Mg_xZn_{1-x}O$ films on r-sapphire substrates [34]. The minimum buffer thickness needed to obtain wurtzite-type $Mg_xZn_{1-x}O$ ($x \le 0.33$) films of single crystal quality was determined to be of the order of 50 Å. Figure 13.2(a) shows a FESEM picture of the $Mg_{0.15}Zn_{0.85}O$/ r-Al_2O_3. Figure 13.2(b) shows semi-logarithmic plots of X-ray θ-2θ scans obtained from $Mg_{0.15}Zn_{0.85}O$ thin films grown on r-sapphire substrates. The $Mg_{0.15}Zn_{0.85}O$ film peak has shifted 0.035° towards the substrate compared to the ZnO film peak. After growth, a Li-diffusion process was performed to compensate the weak *n*-type conductivity and achieve piezoelectric $Mg_xZn_{1-x}O$.

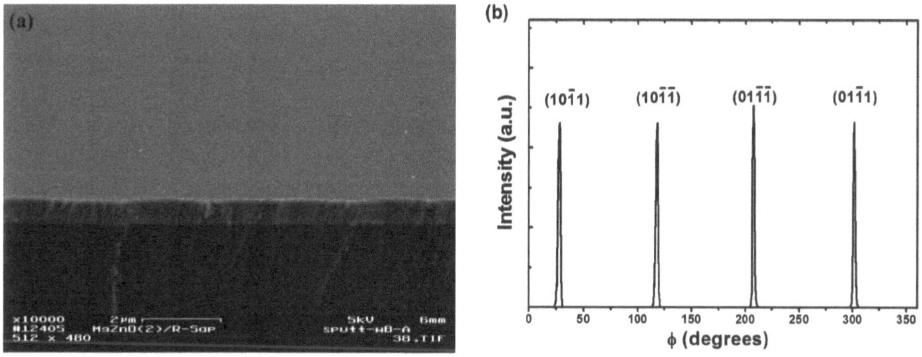

Figure 13.3: (a) FESEM image and (b) phi scan of {10$\bar{1}$1} family of planes of hybrid deposited $Mg_{0.2}Zn_{0.8}O$ films on r-sapphire substrates, followed by anneal at 700°C for 2 hours [37].

Sputtering technology has also been used to obtain piezoelectric $Mg_xZn_{1-x}O$ ($0 \leq x \leq 0.3$) films. The sputtering targets are made up of appropriate quantities of pure ZnO and MgO powder with 2% wt NiO. Highly resistive and well *c*-axis oriented $Mg_xZn_{1-x}O$ piezoelectric films were deposited on SiO_2/Si substrates for BAW devices using such targets [36]. G. Saraf *et al.* deposited piezoelectric $Mg_xZn_{1-x}O$ ($0 \leq x \leq 0.3$) films on *r*-plane (01$\bar{1}$2) sapphire substrates using a hybrid deposition technique: MOCVD followed by RF sputtering [37]. A vertical-flow rotating disc MOCVD was used to deposit the highly crystalline ZnO buffer on the r-sapphire substrate. Diethylzinc (DEZn) and oxygen were used as the Zn metal-organic source and oxidizer, respectively. Argon was used as the carrier gas. Typical growth temperatures of the ZnO buffer ranged from 400°C to 500°C. The growth rate was reduced to 60–100 nm/hr for thickness control and to obtain better crystallinity of the ZnO buffer. Thick piezoelectric $Mg_xZn_{1-x}O$ films were then deposited by RF sputtering on MOCVD grown ZnO buffer on r-sapphire substrates. Following the sputter deposition, the $Mg_xZn_{1-x}O$ films were annealed in an O_2 atmosphere for 2 hours at temperatures ranging from 600°C to 700°C, to improve film quality.

Figure 13.3(a) shows FESEM images of the piezoelectric $Mg_{0.20}Zn_{0.80}O$ films deposited by this hybrid deposition technique, followed by anneal at 700°C for 2 hours. It is observed that the deposited films are dense and uniform. The in-plane registry between the r-sapphire substrate and the piezoelectric $Mg_xZn_{1-x}O$ film was investigated using an x-ray φ-scan of the {10$\bar{1}$1} family of planes, as shown in Figure 13.3(b). It was observed that (11$\bar{2}$0) $Mg_xZn_{1-x}O$||(01$\bar{1}$2) Al_2O_3 and [0001] $Mg_xZn_{1-x}O$||[0$\bar{1}$11] Al_2O_3, which is identical to the epitaxial relationship observed for MOCVD grown $Mg_xZn_{1-x}O$ film on *r*-plane sapphire. Figure 13.4(a) to (c),

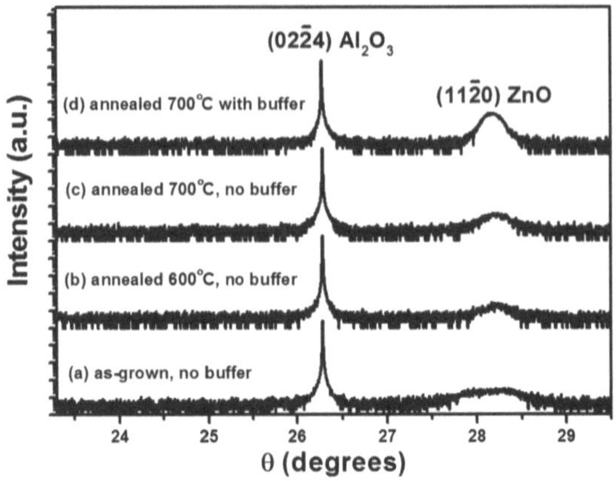

Figure 13.4: X-ray diffraction of ZnO samples on R-sapphire substrates: (a) as-deposited without a buffer, (b) post-deposition annealed at 600°C without a buffer, (c) post-annealed at 700°C without a buffer and (d) with a buffer (hybrid deposition) and post-deposition annealed at 700°C [37].

shows the XRD scans of piezoelectric ZnO films directly sputtered on r-sapphire substrates without the MOCVD grown ZnO buffer, (a) as-deposited, (b) annealed at 600°C and (c) annealed at 700°C. It is observed that the as-deposited sputtered films have the expected a-axis (11$\bar{2}$0) orientation, however, the crystallinity is poor as indicated by the large FWHM of the (11$\bar{2}$0) ZnO reflection. A post-deposition anneal improves the crystallinity, with the highest temperatures giving the best results. The annealing temperature was limited to 700°C to prevent the possible formation of a $ZnAl_2O_4$ spinel at the (11$\bar{2}$0) ZnO/r-sapphire interface. Figure 13.4(d), shows the XRD scan of a piezoelectric ZnO film deposited on r-sapphire substrate using the hybrid deposition method, which was annealed at 700°C for 2 hours. It is observed that the MOCVD grown ZnO buffer greatly improves the FWHM and the intensity of the (11$\bar{2}$0) ZnO diffraction peak of the following sputter deposited films.

13.4 ZnO-based surface acoustic wave devices

13.4.1 Background

SAW devices have been widely used for signal processing since 1964, when the interdigital transducer (IDT) was invented. The basic principle of SAW devices is to use a piezoelectric material to convert an electrical signal into a mechanical wave

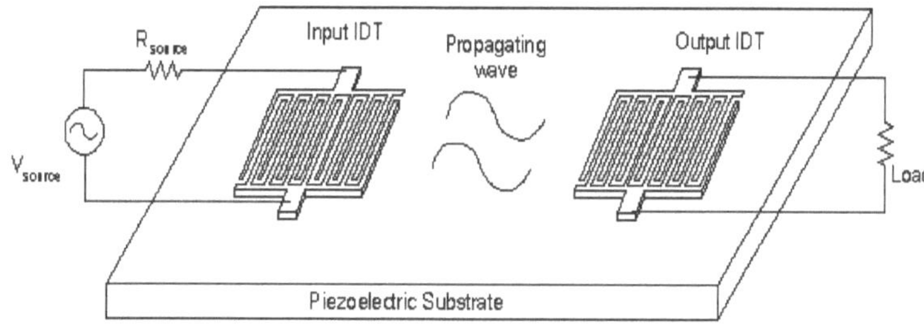

Figure 13.5: Schematic of a SAW delay line device.

in the launching transducer, and conversely, to convert the mechanical wave into an electrical signal at the receiving transducer. The IDT electrodes represent a sampled data structure in space, which is related to time via the acoustic velocity. Thus, arbitrary finite impulse response (FIR) filters can be implemented in a single device, by proper design of the electrode structure. SAW devices have been demonstrated for operating frequencies from tens of MHz up to 10 GHz. These devices are rugged, reliable, cheap, and have no DC power consumption, thus are widely used in consumer and military electronics, especially for portable applications.

A schematic of a SAW delay line is shown in Figure 13.5. An ac voltage applied to the input IDT will cause strain, inducing movement of crystal cells, thus causing the lattice to vibrate. This acoustic wave will propagate along the surface, polarizing the lattice cells, generating an electric field. When this wave reaches the output IDT the accompanying electric field will induce a potential difference between neighboring electrodes. Only alternating current signals can be transmitted across a SAW device. Therefore, low pass filters can not be implemented with SAW devices. High pass filters cannot be realized as well, due to the quadratic dependence on frequency of the propagation loss. Thus, SAW devices are always bandpass, or bandstop filters.

The two important surface acoustic wave parameters of a piezoelectric material are the acoustic wave velocity, v_{SAW}, and the electromechanical coupling coefficient, K^2. Both of these parameters are dependent on the material properties, and the specific crystal orientation of the surface and the propagation direction. The electromechanical coupling coefficient is a measure of the conversion efficiency between electrical energy and mechanical energy. Piezoelectric stiffening increases the effective elastic stiffness of a medium according to the following relationship,

$$c_{\text{stiffened}} = c(1 + K^2) \tag{13.6}$$

where $K^2 = e^2/c\varepsilon$ and c, e and ε are the elastic, piezoelectric and dielectric constants along the propagation direction of the acoustic wave in the solid. K is the electromechanical coupling coefficient, and is a useful criterion for determining the piezoelectric coupling strength of a piezoelectric material. For many practical applications, circuit modeling and device design, K^2, instead of K, is used as the coupling coefficient metric. For SAW devices, the coupling coefficient is related to the difference of the metallized surface (i.e. short-circuited or infinite conductivity) and the free surface (i.e. open circuit) velocities of the piezoelectric material.

The most important specifications for SAW device design are center frequency and bandwidth. For delay line applications, the delay length is also specified. In many important applications, including sensors, the temperature coefficient of delay (TCD), or temperature coefficient of frequency (TCF), is also specified. The periodicity of the IDT and the SAW velocity of the substrate determine the center frequency, f_c, of the device:

$$v_{\mathrm{SAW}} = \lambda_o f_c \tag{13.7}$$

where λ_o is the wavelength at f_c, which is determined by the periodicity of the IDT. The bandwidth of a SAW device is inversely proportional to the number of electrodes in the IDTs:

$$BW_{3\,\mathrm{dB}} = \frac{90}{N} \tag{13.8}$$

where N is the number of electrode pairs in the IDT.

SAW resonators, delay lines and filters have found wide applications in communication systems, as they are easy to fabricate, rugged, cheap, and can implement complex filtering functions in a single device. Although digital signal processing chips are competitive in signal processing functionality up to several tens of MHz, SAW devices have the advantage of being passive devices, hence with low power consumption. This is critical for most of the modern SAW applications which involve portable wireless communications devices. An important application of SAW devices is in the field of chemical and biochemical sensing. Surface acoustic waves are very sensitive to changes in surface properties, including mass loading, conductivity, temperature, stress, or viscosity in a liquid. For chemical/biochemical sensing applications, the surface of the delay path is coated with a chemically selective coating which bonds with the target chemical. This delay line is used in the feedback path of an oscillator circuit. As the mass loading increases, the phase velocity decreases, changing the phase shift across the SAW device, and therefore the oscillation frequency of the overall circuit. The most common SAW sensor architecture is the dual delay line oscillator, which uses a reference device along with the sensor device, in order to eliminate the effect of environmental factors such as temperature and stress on the sensor device.

13.4.2 SAW in ZnO-based multilayer structures

Piezoelectric ZnO films have been deposited on various films and substrates, such as sapphire (Al_2O_3), quartz, Si or SiO_2/Si, GaAs, diamond thin films and SiC to fabricate SAW devices. These layered structures have been used to design SAW filters for various applications. A multilayer material system offers additional design flexibility as acoustic velocity and piezoelectric coupling coefficient will be functions of the film-thickness to wavelength ratio h/λ, where h stands for the film thickness and λ stands for SAW wavelength. In general, there is velocity dispersion in a material system composed of a thin film layer on a substrate, when the film has a different acoustic velocity compared to the substrate. The energy of a surface acoustic wave decays exponentially with depth, and is thus mostly confined within one wavelength of the surface. Therefore, in a system composed of a high SAW velocity substrate and a thin film with a slower acoustic velocity, when h/λ is small, most of the energy is transported in the substrate. As a result, the acoustic velocity will be slower than, but close to, the substrate velocity. When h/λ is approximately equal to 1, most of the energy is trapped within the film itself. And the velocity will be close to, but somewhat larger than, the film velocity. When h/λ is much greater than 1, almost all of the energy is trapped within the film, and the thin film can practically be viewed as the sole transport layer, with the acoustic velocity being equal to that of the thin film.

(i) ZnO on Si or SiO_2/Si

While the direct integration of SAW filters with Si ICs appears attractive, SAW filters consume relatively large chip area of the circuits, making this integration an expensive proposition. Further, though silicon, which is non-piezoelectric, has relatively high acoustic velocities, it suffers high propagation losses due to its electrical conductivity. ZnO/SiO_2/Si SAW-devices with high quality polycrystalline ZnO films have been studied since the early 1970s. In 1991, Weber *et al.* studied the propagation loss and the coupling coefficient of three different types of devices: Rayleigh wave devices with ZnO-layers $h/\lambda = 0.4$ and $h/\lambda = 0.04$, and Sezawa wave devices [38]. At 550 MHz, devices with thin ZnO-layers exhibited a propagation loss of only 7 dB/cm, compared to 24 dB/cm for Rayleigh wave devices with a thick ZnO-layer, and 14 dB/cm for the Sezawa wave devices. The phase velocity and coupling coefficient were examined as functions of the film thickness, and compared with their theoretical values. The insertion loss behavior also confirmed the better coupling efficiency of the Sezawa wave, compared to that of the fundamental Rayleigh wave. The measured average values for the coupling coefficient K^2 were 1.9% for the Sezawa mode, and 0.4% and 1.7% for the fundamental Rayleigh mode. P. Wu *et al.* simulated the temperature compensation of the second general SAW modes for different thickness ratios of ZnO to SiO_2 [39]. The simulation results showed good temperature stability at 2.4 GHz when the ZnO:SiO_2 thickness ratio was 2:3 and total thickness was 3 μm.

(ii) ZnO on GaAs

ZnO deposited on GaAs substrate can monolithically integrate SAW devices with GaAs electronics. Furthermore, such a structure has low dispersion due to the similarity of the SAW velocities of both materials. The acoustic properties of ZnO films sputtered on {001}-cut (110)-propagating GaAs substrates were investigated by Kim *et al.* [2], including SAW velocity, effective piezoelectric coupling constant, propagation loss, diffraction, velocity surface, and reflectivity of shorted and open metallic gratings. The measurements of the velocity surface, which directly affects the SAW diffraction, on the bare and metallized ZnO on SiO_2 or Si_3N_4 on {001}-cut GaAs samples were reported using two different techniques: (1) knife-edge laser probe, (2) line-focus-beam scanning acoustic microscope. It was found that near the (110) propagation direction, the focusing SAW property of the bare GaAs changes into a nonfocusing one for the layered structure, but a reversed phenomenon exists near the (100) direction. Furthermore, to some extent the diffraction of the substrate can be controlled with the film thickness. The reflectivity of shorted and open gratings were also analyzed and measured. Zero reflectivity is observed for a shorted grating. There was good agreement between the measured data and theoretical values. A monolithic integration of these filters on GaAs is highly desirable to miniaturize the outer dimensions of the cellular phones. A. Haddou *et al.* presented an analysis of a ladder SAW filter built up on a two-layered structure made up of a ZnO film on a GaAs substrate in the 900 MHz frequency range [40]. V. Y. Zhang *et al.* reported the lowest order SAW mode in ZnO/GaAs structure [41]. The velocity and coupling coefficient were simulated as a function of film thickness, film orientation and IDT location. The velocity varied with frequency and thickness product (hf) and the K^2 nearly monotonically increased with hf when IDTs were placed on the upper-most surface with a free interface. For the metallized interface, the maximum K^2 is 1% for $hf = 230$ m/s and the minimum K^2 is 0.5% for $hf = 750$ m/s. When IDTs are located at the interface, a higher K^2 (1.5%) is obtained for $hf = 1500$ m/s.

(iii) ZnO on quartz

ZnO films have been deposited on quartz substrates for the development of SAW devices where temperature stability is of primary importance. Such SAW devices can be used as gas or liquid sensors in critical environments. Tomar *et al.* fabricated ZnO thin film SAW filters on fused quartz substrates operating at the frequency range of 38 to 64 MHz [42]. Temperature stability was achieved using a ZnO over-layer. A 5.4 μm thick ZnO over-layer was found to reduce the negative TCD of

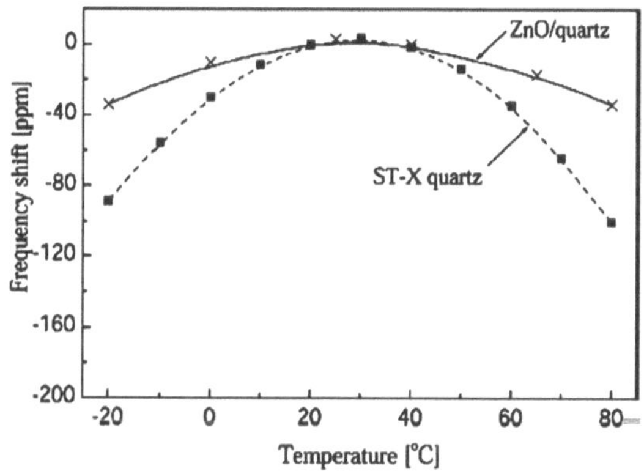

Figure 13.6: Measured frequency shifts of N-CDMA IF filter consisting of ZnO/IDT/26° Y-X (solid line) and ST-X quartz (broken line) relative to temperature [43].

an IDT/ZnO/fused quartz layered structure to almost zero (−3 ppm/°C) while a coupling coefficient, $K^2 = 1.1\%$ and insertion loss, $IL = 27$ dB were measured. M. Kadota *et al.* studied the Rayleigh wave propagating in ZnO/ST-cut X propagation quartz substrates [13,43,44]. The Rayleigh wave had a large coupling factor and an excellent TCF by combining a quartz substrate with a specific cut angle and propagation direction with a positive TCF (+10–25 ppm/°C) and a zinc oxide (ZnO) film with a negative TCF. A 29°45′ and a 42°45′ rotated Y plates were selected as the ST-cut plate, and the Rayleigh wave propagating in the direction of 35° from X-axis on their plates (ST-35°X) was used as it had a positive TCF. Figure 13.6 shows the frequency shift ($\Delta f/f$) per 1°C in the range from −20°C to 80°C. The ZnO/IDT/quartz structure has the smallest (best) TCF, the slowest velocity and the largest coupling factor. The insertion losses of their IF filters on these layered structure were 3–5 dB better and their TCF was superior ($\Delta f/f = 0.37$ ppm/°C). The combination of this specific-cut-angle quartz and ZnO film has been applied to IF filters for wideband code division multiple access (W-CDMA) and narrow-band CDMA (N-CDMA) systems.

(iv) ZnO on sapphire

Sapphire is a readily available non-piezoelectric substrate that has high acoustic velocities and very low propagation losses. Piezoelectric ZnO thin films have been deposited on *c*-, *m*- and *r*-plane sapphire substrates to yield high piezoelectric coupling coefficients. Piezoelectric ZnO thin films on c-sapphire substrates have been used for SAW devices. Surface acoustic wave properties of IDT/ZnO/sapphire

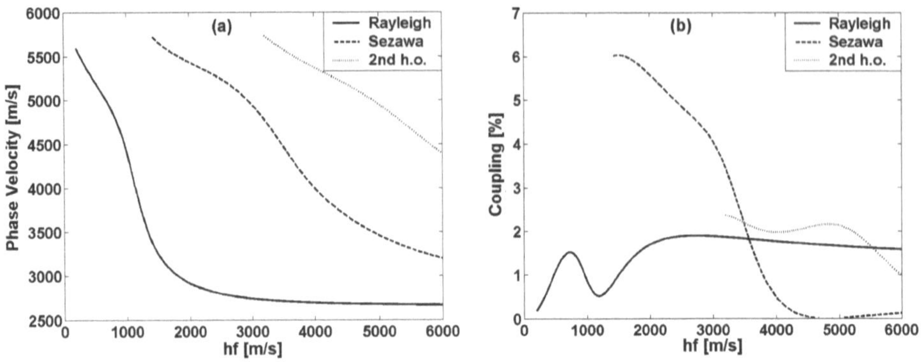

Figure 13.7: (a) SAW phase velocities of the Rayleigh wave and the first two higher modes propagating in ZnO/r-sapphire, (b) piezoelectric coupling coefficient of these modes.

layered structure including velocity, electromechanical coupling coefficient, and propagation loss were recently analyzed and simulated by T. Wu *et al.* [45].

In the a-ZnO/r-Al_2O_3 system, the ZnO film's c-axis lies in the surface plane, resulting in in-plane anisotropy. Three types of acoustic wave modes can be excited in this system, the generalized SAW, the Rayleigh-type and the Love-type wave modes. The Rayleigh wave has two components, a longitudinal component and a sheer vertical (SV) component. The Love wave differs from the Rayleigh wave in that it has a sheer horizontal (SH) component instead of the SV component. The Rayleigh-type waves have maximum electromechanical coupling for propagation parallel to the c-axis of the film, while Love-type waves have maximum electromechanical coupling for propagation perpendicular to the c-axis. It should be noted that when one wave type has maximum coupling, the other wave type has zero coupling. This anisotropic property allows for unambiguous identification of wave modes for device design and fabrication. In all other propagation directions, the generalized SAW mode propagates. Henceforth in this text, Rayleigh-type waves will refer only to Rayleigh-type waves propagating parallel to the c-axis of the piezoelectric film, while Love-type waves will refer only to Love-type waves propagating perpendicular to the c-axis.

The base SAW modes exist at all acoustic wavelengths (frequencies). As the frequency is increased, or wavelength decreased, other wave modes get trapped in the layered structure due to the dispersive characteristics, analogous to electromagnetic waveguides. The first Rayleigh-like SAW mode to get trapped is the Sezawa wave, which is especially important in the a-plane ZnO/r-plane Al_2O_3 system. As shown in Figure 13.7, the Sezawa wave mode appears around an hf value of 1335 m/s, and at its peak has a piezoelectric coupling coefficient which is slightly larger than 6% at $hf = 1500$ m/s, with an acoustic velocity of 5658.43 m/s. Thus, it has a significantly

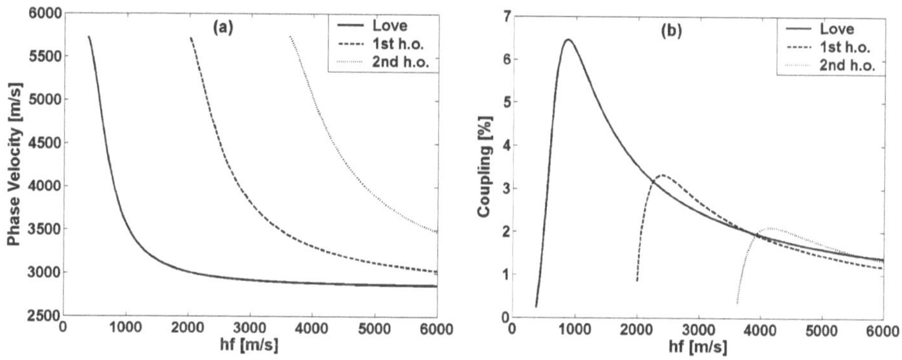

Figure 13.8: (a) SAW phase velocities of the Love wave and the first two higher modes propagating in ZnO/r-sapphire, (b) piezoelectric coupling coefficient of these modes.

larger maximum coupling coefficient at a larger SAW velocity compared to the base Rayleigh wave mode, and is therefore attractive for high frequency SAW devices. For example, for a 2 μm thick ZnO film, the Sezawa wave mode would appear at 667.5 MHz and have maximum coupling at 750 MHz.

Figure 13.8 shows the SAW phase velocities and piezoelectric coupling coefficients of Love wave calculated for ZnO/r-Al_2O_3 with propagation perpendicular to the *c*-axis of ZnO, plotted versus the hf product. The Love wave mode maximum coupling is 6.46%, at an hf value of 870, with 3813.12 m/s velocity. Compared to the Sezawa wave mode, the maximum coupling is slightly higher, but with a lower SAW velocity at maximum coupling. Thus, the Sezawa wave mode is more attractive for higher frequency applications, with less demanding line width requirements.

(v) ZnO on SiC

Silicon carbide has been developed for high temperature, high voltage and high power semiconductor applications. SiC is one of the materials with the highest acoustic velocities, however current SiC substrates suffer from high cost and high electrical conductivity for SAW applications. The advantages for ZnO on high velocity, low loss substrates, such as SiC, include higher frequency capabilities for a given photolithography technology, high coupling factor and high order wave modes, and a higher power handling capability.

F. S. Hickernell *et al.* characterized the SAW propagation properties of ZnO films on SiC both theoretically and experimentally [46]. Experimental SAW measurements were made with 400 nm sputter-deposited piezoelectric ZnO films on commercially available SiC substrates to obtain an initial assessment of the mode structure, velocity characteristics, propagation loss and transducer conversion efficiency. The measurements characterized the velocity and propagation loss

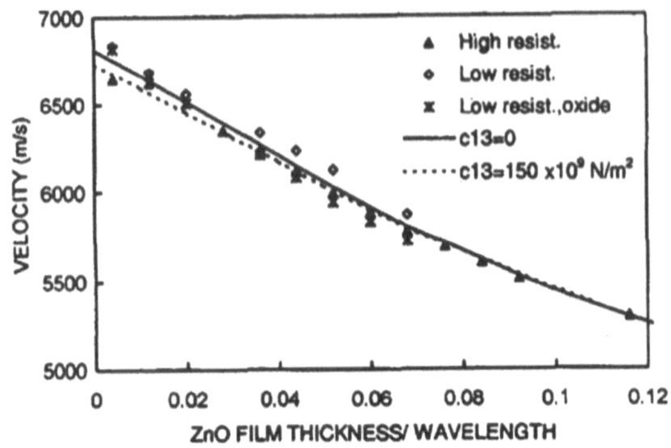

Figure 13.9: Measured and calculated velocity dispersion for the 1st GSAW mode of IDT/ZnO/SiO_2/SiC structure [46].

of two wave modes, a generalized SAW mode with velocities between 6000 to 7000 m/s, and a high velocity pseudo SAW mode with velocities in the range of 8500 to 12500 m/s. For one of the low resistivity substrates, a 60 nm SiO_2 layer was deposited on the SiC substrate before ZnO deposition to isolate the IDT electrodes from the semiconducting SiC surface (IDT/ZnO/SiO_2/SiC). The SAW was propagating along the (100) direction. The highest piezoelectric coupling of 5.2% was theoretically predicted for the first GSAW mode and for the electrode configuration with the IDT at the film-substrate interface and metallized upper surface. An increase of the unknown elastic modulus c_{13} resulted in a decrease of GSAW velocities. The measured velocity dispersion characteristics of the GSAW (the first-order mode) are shown in Figure 13.9. The IDT array is at the upper ZnO film surface. In Figure 13.9, two calculated velocity dispersion functions of the GSAW are presented that correspond to $c_{13} = 0$ and $c_{13} = 150 \times 10^9$ N/m^2. The HVPSAW characteristics including velocities and propagation loss were also simulated.

(vi) ZnO on diamond

Diamond has the highest Young's modulus among all materials, and thus yields the highest acoustic velocities. Diamond and diamond-like carbon thin films on Si have been investigated for SAW device applications. The high quality synthetic diamond substrates may be attractive for future ultra high frequency SAW devices. When a diamond wafer or a diamond film deposited on Si is used as substrate for ZnO deposition, high velocity SAW is expected. In 1992, a 1.07 GHz SAW filter of ZnO/IDT/diamond/Si structure was reported for the first time by Nakahata *et al.* using 8 μm wavelength IDT with a SAW velocity of 8600 m/s [5]. The dispersive SAW properties of ZnO/diamond/Si layered structure were analyzed by Wu *et al.* [6].

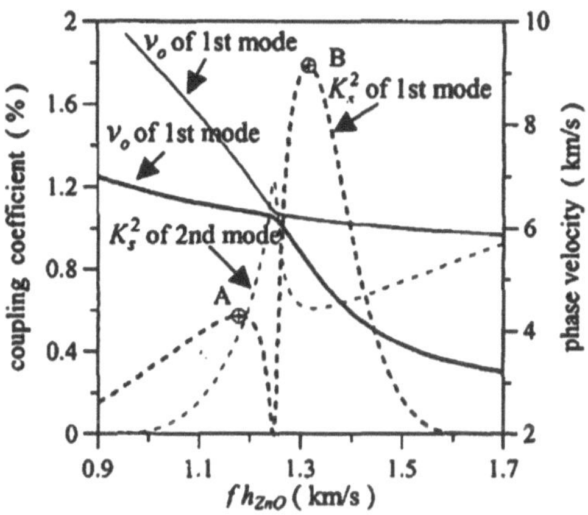

Figure 13.10: Electromechanical coupling coefficients and dispersion curves of the first two SAW modes in ZnO/diamond/Si structure [6].

Table 13.1: Structures and characteristics of diamond SAW filters [48].

Structure	IDT/ZnO/diamond/Si	SiO_2/ZnO/IDT/diamond/Si
Velocity (m/s)	11600	9000
Coupling coefficient (%)	1.1	0.4
Temperature coefficient of frequency (ppm/°C)	−28	−0
Application	Power durable RF	Narrow band (retiming, resonator)

It can be seen in Figure 13.10 that this structure has high SAW velocities up to 9000 m/s, and electro-mechanical coupling coefficient up to 1.8%. Dreifus *et al.* analyzed the SAW velocity dispersion of the Rayleigh and Love modes [47].

K. Higaki *et al.* demonstrated a 2.5 GHz low loss narrow-band filter and a 1.5 GHz IF filter using the SiO_2/ZnO/diamond structure and the ZnO/diamond structure, respectively [48]. The structures of diamond SAW filters, including their schematic diagrams and their SAW characteristics, are summarized in Table 13.1. The IDT/ZnO/diamond structure had a velocity larger than 10000 m/s and a coupling coefficient (K^2) value of more than 1%. Utilizing this high SAW velocity, a 2.5 GHz filter was fabricated with almost 1 μm line-and-space (L&S) IDTs. SAW filters with center frequencies 3.4 GHz and 4.7 GHz were also successfully fabricated with almost 0.8 μm and 0.6 μm L&S IDTs, respectively. These results suggest

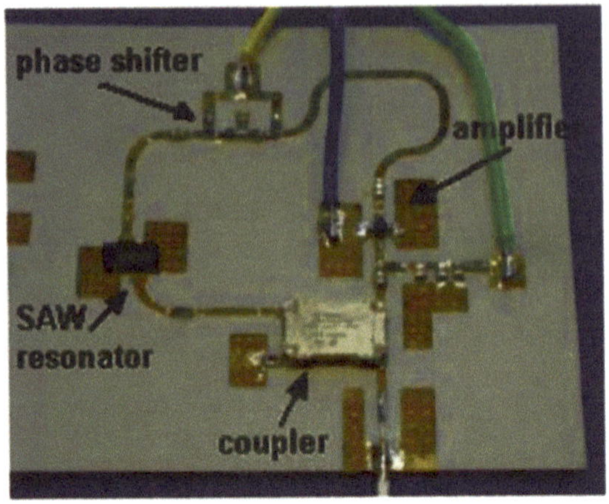

Figure 13.11: Layout of the oscillator circuit at 1 GHz using the first Sezawa mode of ZnO/diamond/Si SAW resonator [50].

the great advantages of the diamond SAW filters in manufacturing, which is a loosening of the minimum feature requirement, and the increased reliability of these high frequency devices compared with those made with conventional materials. The SiO_2/ZnO/IDT/diamond structure can provide zero TCF as well as a high velocity of 9000 m/s and relatively large K^2, which indicates that this structure is suitable for narrow band filters in high frequencies, such as resonators and retiming filters in optical communication systems. Nakahata *et al.* fabricated two kinds of narrow band SAW filters on the SiO_2/IDT/ZnO/diamond structure [49]. A 10 μm thick polycrystalline diamond film was deposited on a Si substrate by the hot filament chemical vapor deposition (CVD) method. Devices with 0.5 μm wide electrodes had center frequency of 5 GHz, with a minimum insertion loss of 13 dB, Q value of 650 and zero TCF. Thus, this material system can be used for narrow band filters and resonators operating at frequencies up to 5 GHz, without introducing special photolithography processes, such as electron-beam photolithography. Devices with 0.65 μm wide electrodes had a center frequency of 10 GHz, a minimum insertion loss of 22 dB, Q values of 730, a zero TCF and a relatively large K^2 of 1.5%, using the fifth harmonic wave mode of the second Sezawa mode. The kh_{ZnO} and kh_{SiO_2} values were 2.8 and 4.5, respectively. Recently, V. U. Prechtel *et al.* fabricated diamond based SAW oscillators operating at 1 GHz using the first Sezawa mode [50]. The ZnO film thickness and IDT period were 900 nm and 9.43 μm, respectively. The oscillator loop shown in Figure 13.11 consisted of a phase shifter, a SiGe amplifier, and a 10 dB coupler. The oscillator frequency was 1.008 GHz, the tuning range 2 MHz, the power consumption 0.48 W, and the noise floor −171 dBc

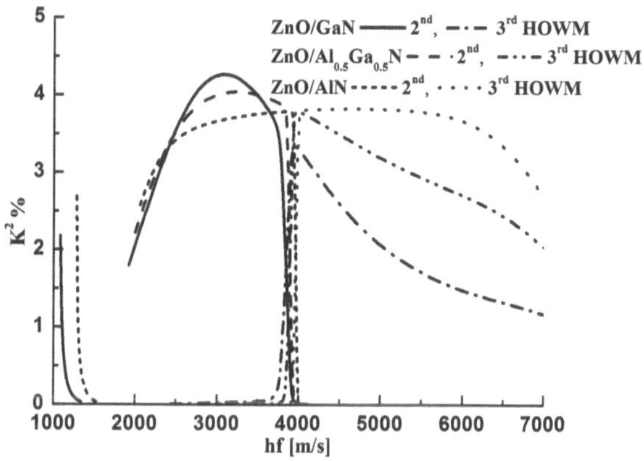

Figure 13.12: Piezoelectric coupling coefficients of the second and third higher order wave modes propagating in the ZnO (h_1)/$Al_xGa_{1-x}N$ (h_2)/c-Al_2O_3 ($h_1 = h_2, h = h_1 + h_2$) system, with $x = 0$, 0.5, and 1 [51].

at 1 MHz offset from the carrier. The loading power for the SAW resonator was 24 dBm.

(vii) ZnO on GaN/c-Al_2O_3 and $Al_xGa_{1-x}N$/c-Al_2O_3

Recently, Y. Chen *et al.* have deposited piezoelectric ZnO films on $Al_xGa_{1-x}N$/c-Al_2O_3 ($0 \leq x \leq 1$) substrates to study the SAW characteristics of these layered structures [51]. In comparison with a single $Al_xGa_{1-x}N$ layer deposited on c-Al_2O_3 with the same total film thickness, a ZnO/$Al_xGa_{1-x}N$/c-Al_2O_3 multilayer structure provides several advantages, including higher order wave modes with higher velocity and larger electro-mechanical coupling coefficient. SAW velocities and coupling coefficients of the ZnO/$Al_xGa_{1-x}N$/c-Al_2O_3 structure were tailored as a function of the Al mole percentage in $Al_xGa_{1-x}N$ films, and as a function of the ZnO (h_1) to $Al_xGa_{1-x}N$ (h_2) thickness ratio. It was found that a wide thickness-frequency product (hf) region where coupling is close to its maximum value, K^2_{max}, can be obtained. As shown in Figure 13.12, the K^2_{max} of the second order wave mode ($h_1 = h_2$) was estimated to be 4.3% for ZnO/GaN/c-Al_2O_3, and 3.8% for ZnO/AlN/c-Al_2O_3. The bandwidth of second and third order wave modes, in which the coupling coefficient is within ±0.3% of K^2_{max}, was calculated to be 820 *hf* for ZnO/GaN/c-Al_2O_3, and 3620 *hf* for ZnO/AlN/c-Al_2O_3. Thus, the *hf* region where the coupling coefficient is close to the maximum value broadens with increasing Al content, while K^2_{max} decreases slightly. When the thickness ratio of AlN to ZnO increases, the K^2_{max} and hf bandwidth of the second and third higher wave modes increases. The theoretical and experimental results of velocity dispersion in the $ZnO/Al_xGa_{1-x}N/c\text{-}Al_2O_3$ structures were found to be well matched.

13.4.3 Tailoring SAW properties using $Mg_xZn_{1-x}O$

$Mg_xZn_{1-x}O$ is a new material, and experimentally determined values for $Mg_xZn_{1-x}O$ material constants are not yet available. Previous studies have calculated these constants by using the material constants of the binary compounds ZnO and MgO. Martin's method was used to transform the stiffness coefficients of cubic MgO (3mm) to wurtzite phase (6mm). Martin first defines simplified stiffness coefficient vectors for the cubic (c^{ZB}) and wurtzite (c^{WZ}) phases:

$$c_1^{ZB} = c_{11}^{ZB}, \quad c_2^{ZB} = c_{12}^{ZB}, \quad c_3^{ZB} = c_{44}^{ZB} \tag{13.9a}$$

$$c_1^{WZ} = c_{11}^{WZ}, \quad c_2^{WZ} = c_{33}^{WZ}, \quad c_3^{WZ} = c_{12}^{WZ}, \quad c_4^{WZ} = c_{13}^{WZ},$$
$$c_5^{WZ} = c_{44}^{WZ}, \quad c_6^{WZ} = c_{66}^{WZ} \tag{13.9b}$$

The equations relating the two vectors are:

$$c_i^{WZ} = \bar{c}_i^{WZ} - D_i, \quad i = 1, ..., 6 \tag{13.10}$$

where $\bar{c}^{WZ}$ are the average stiffness coefficients related to the cubic phase coefficients with the transformation matrix P:

$$\bar{c}_i^{WZ} = \sum_{j=0}^{3} P_{ij} c_j^{ZB} \tag{13.11a}$$

$$P = \frac{1}{6} \begin{bmatrix} 3 & 3 & 6 \\ 2 & 4 & 8 \\ 1 & 5 & -2 \\ 2 & 4 & -4 \\ 2 & -2 & 2 \\ 1 & -1 & 4 \end{bmatrix} \tag{13.11b}$$

Table 13.2: MgO material constants.

Crystal structure	Rock salt
Stiffness coefficients [GPa]	$c_{11} = 286$
	$c_{22} = 87$
	$c_{44} = 148$
Dielectric coefficient (constant strain)	$\varepsilon_{11} = 9.6$
Density [kg/m^3]	3650

D is the internal strain contribution related to the average stiffness equations:

$$D = \left[\frac{\Delta^2}{\bar{c}_5^{WZ}}, 0, \frac{-\Delta^2}{\bar{c}_5^{WZ}}, 0, \frac{\Delta^2}{\bar{c}_6^{WZ}}, \frac{\Delta^2}{\bar{c}_5^{WZ}} \right]^T \tag{13.12a}$$

$$\Delta = \sum_{j=0}^{3} Q_j c_j^{ZB} \tag{13.12b}$$

$$Q = \frac{\sqrt{2}}{6} [1 \quad -1 \quad -2] \tag{13.12c}$$

This method has a $\pm 5\%$ error in determining the material constants. Then Vegard's law was applied to estimate $Mg_xZn_{1-x}O$ material constants:

$$C_{Mg_xZn_{1-x}O} = xC_{MgO} + (1 - x)C_{ZnO} \tag{13.13}$$

where C is the appropriate material constant. The material constants of ZnO are given in equations (13.1–13.5), and its density is 5665 kg/m^3. The MgO material constants are given in Table 13.2:

SAW propagation in the $Mg_xZn_{1-x}O$/r-Al_2O_3 structure was first investigated by N. W. Emanetoglu *et al.* [34]. Figure 13.13 shows the calculated velocity dispersion and piezoelectric coupling curves for the Rayleigh wave mode in the $Mg_xZn_{1-x}O$/r-Al_2O_3 system, with $x = 0$ (ZnO), 0.1, 0.2 and 0.3. At small values of hf, most of the energy of the SAW propagates in the sapphire substrate, and the velocity difference between the different compositions of $Mg_xZn_{1-x}O$ is quite small. At large hf values, the acoustic velocity difference becomes significant. At $hf = 5000$, the Rayleigh wave velocity in ZnO/r-Al_2O_3 is 2677.14 m/s, and 3398.56 m/s in $Mg_{0.3}Zn_{0.7}O$/r-Al_2O_3. The diminishing effect of the Mg content on the piezoelectric coupling is apparent even at low values of *hf*. The maximum coupling for the Rayleigh

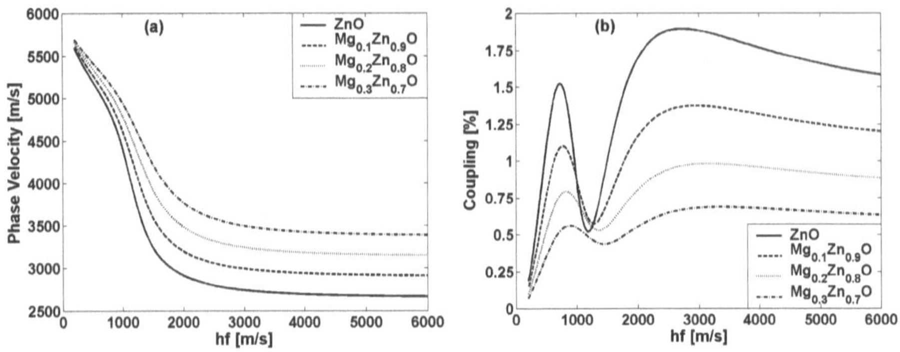

Figure 13.13: (a) Velocity dispersion and (b) piezoelectric coupling curves for the Rayleigh wave mode propagating in the $Mg_xZn_{1-x}O$/r-Al_2O_3 system.

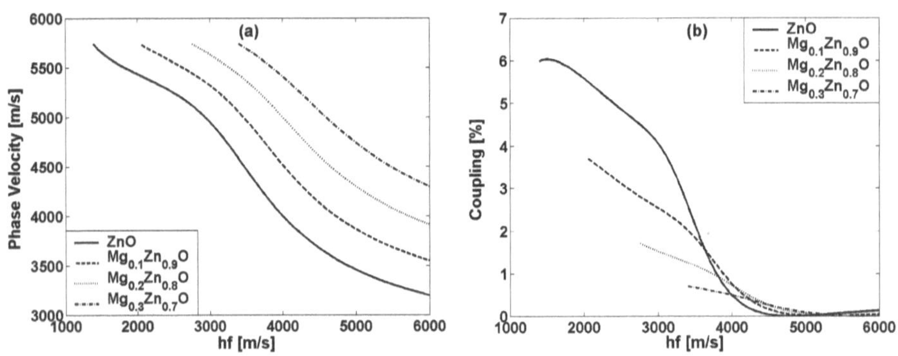

Figure 13.14: (a) Velocity dispersion and (b) piezoelectric coupling curves for the Sezawa wave mode propagating in the $Mg_xZn_{1-x}O$/r-Al_2O_3 system.

wave in ZnO/r-Al_2O_3 is 1.89% at $hf = 2730$, while the maximum coupling in $Mg_{0.3}Zn_{0.7}O$/r-Al_2O_3 is 0.68% at $hf = 3345$.

Figure 13.14 shows the calculated velocity dispersion and piezoelectric coupling curves for the Sezawa wave mode in the $Mg_xZn_{1-x}O$/r-Al_2O_3 system, with $x = 0$ (ZnO), 0.1, 0.2 and 0.3. At $hf = 5000$, the Sezawa wave velocity in ZnO/r-Al_2O_3 is 3455.15 m/s, and 4738.84 m/s in $Mg_{0.3}Zn_{0.7}O$/r-Al_2O_3. The maximum coupling for the Sezawa wave in ZnO/r-Al_2O_3 is 6.04% at $hf = 1500$, while the maximum coupling in $Mg_{0.3}Zn_{0.7}O$/r-Al_2O_3 is 0.76% at $hf = 3255$.

Figure 13.15 presents the calculated velocity dispersion and piezoelectric coupling curves for the Love wave mode in the same material systems. The Love wave mode behavior is similar to that of the Rayleigh wave mode. At $hf = 5000$, the Love wave velocity in ZnO/r-Al_2O_3 is 2869.04 m/s, and 3627.91 m/s in $Mg_{0.3}Zn_{0.7}O$/r-Al_2O_3. The maximum coupling for the Love wave in ZnO/r-Al_2O_3 is 6.46%

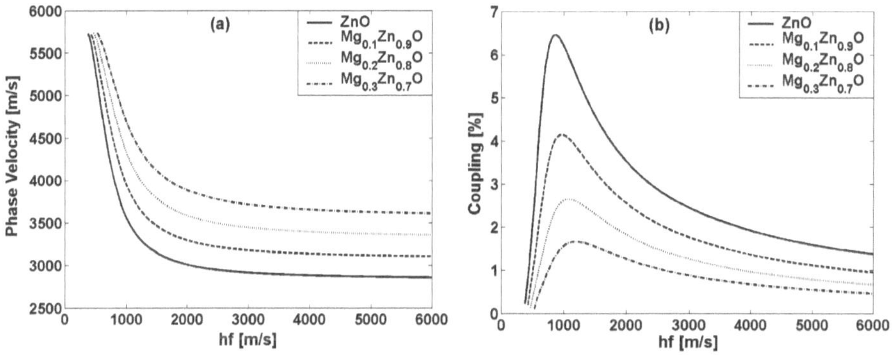

Figure 13.15: (a) Velocity dispersion and (b) piezoelectric coupling curves for the Love wave mode propagating in the $Mg_xZn_{1-x}O$/r-Al_2O_3 system.

at $hf = 870$, while the maximum coupling in $Mg_{0.3}Zn_{0.7}O$/r-Al_2O_3 is 1.66% at $hf = 1185$. The maximum coupling of the Sezawa mode is similar to that of the Love wave mode in ZnO/r-Al_2O_3, with a much higher velocity. However, the Sezawa wave mode's maximum coupling decreases much faster than that of the Love wave mode as the Mg content increases.

Further flexibility in tailoring piezoelectric properties is provided by the use of $Mg_xZn_{1-x}O$/ZnO multilayer structures. The SAW properties of two-layer and three-layer $Mg_xZn_{1-x}O$/ZnO structures on r-Al_2O_3 substrates were simulated using the transfer matrix method, with the IDT on top of the thin film structure. Piezoelectric properties can be tailored by changing the Mg composition and the thickness ratios in the structures. For example, an advantage of the ZnO/$Mg_{0.3}Zn_{0.7}O$/r-Al_2O_3 structure in comparison of the ZnO/r-Al_2O_3 structure for the Sezawa wave mode is the broadening of the hf region in which the piezoelectric coupling is close to the maximum value. This allows for wider bandwidth SAW devices to be fabricated using this structure. The piezoelectric tailoring effect is most apparent for the second Rayleigh type higher order wave mode (HOWM). A three-layer structure allows for greater flexibility in SAW property tailoring by providing additional parameters which can be modified. It is interesting to note that the maximum coupling of the second HOWM occurs for the two-layer structure with $h_{1,\mathrm{ZnO}} = h_{2,\mathrm{MgZnO}}$, while the maximum coupling of the third HOWM occurs for the three-layer structure with $h_{1,\mathrm{ZnO}} = h_{2,\mathrm{MgZnO}} = h_{3,\mathrm{ZnO}}$. Additionally, maximum coupling for any $(h_1 + h_3)$:h_2 ratio in the ZnO/$Mg_xZn_{1-x}O$/ZnO structure occurs when $h_1 = h_3$.

In summary, SAW properties can be tailored using $Mg_xZn_{1-x}O$/ZnO multilayer structures on r-Al_2O_3 substrates. The second and third Rayleigh type HOWMs benefit the most from the use of multilayer structures, with substantial increases in maximum coupling. The Sezawa wave mode also benefits by an increase in the hf range in which K^2 is close to maximum. This is analogous to reducing the gain of an amplifier circuit to increase its bandwidth. As for the Love wave modes, even though a benefit was not observed for the HOWMs, which have lower coupling than the base mode, the base Love wave mode benefits by a broadening of the maximum K^2 range, despite a lowering in the maximum K^2 value.

13.4.4 Integrated SAW devices

(i) Integration of SAW devices with Si ICs

A monolithic integration of SAW filters on Si substrates is highly desirable to miniaturize the outer dimensions of the cellular phones. The ZnO/SiO_2/Si layered structure is attractive for monolithic integration of SAW filters with electronic circuitry, because it combines attractive SAW propagation and transduction properties with a relatively simple fabrication process and a strong bonding of the ZnO on Si and SiO_2 substrates [1,52]. A. E. Corner *et al.* reported a ZnO SAW convolver fabricated on a silicon substrate [53]. The device employs a planar-processed, high-speed DMOS transistor as a mixer to implement the convolution cross terms, utilizing the nonlinear dependence of the drain current on gate voltage. J. C. Haartsen *et al.* reported a monolithic, programmable tapped delay line (PTDL) SAW filter in a ZnO/SiO_2/Si layered structure [54]. The electric input signal was converted into an acoustic wave by an IDT and subsequently sampled by the taps in the propagation path. The top gate electrode of the JFET had a p^+ implanted ring configuration, while the drain and source electrodes were made by n^+ diffused layers. The IDTs were located at the interface of ZnO and SiO_2 films. By controlling the DC bias, the JFET can be controlled. When a reverse bias was applied on the pn junction, a depletion region was created beneath the IDT. K. Tsubouchi *et al.* developed a compact spread-spectrum (SS) wireless modem using a SAW convolver [55–57]. The demodulation method of the SS modem was a completely asynchronous one, which is very useful for application to wireless communication under weak signals. The great merit of the ZnO-SiO_2-Si SAW convolver is its high efficiency. This convolver is suitable for a portable SS wireless modem which requires a moderate bandwidth-time product, corresponding to the process gain in the convolver.

(ii) Integration of piezoelectricity with photoconductivity

Photoconductive and photovoltaic UV detectors have been reported on polycrystalline and epitaxial ZnO films. These UV photodetectors use current or voltage as their output. Such a photodetector can be realized by using a semiconductor on, or in, the SAW device. The free carriers generated in the semiconductor layer

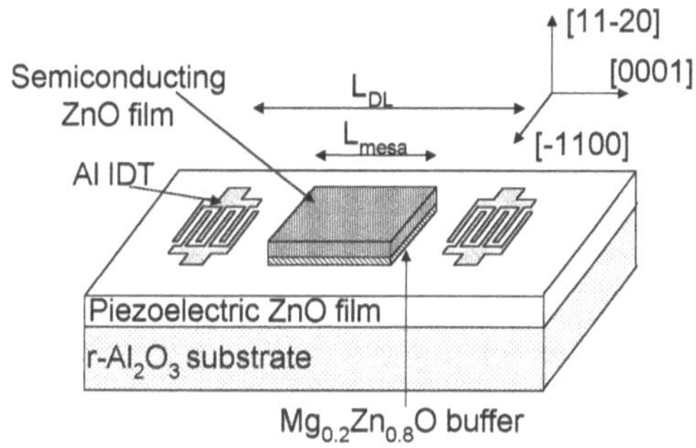

Figure 13.16: Schematic of UV-SAW optical sensor [59].

will interact with the electric field accompanying the propagating SAW, resulting in a velocity reduction and increased insertion loss. Sharma *et al.* reported a ZnO/$LiNbO_3$ SAW UV photodetector [58], which used the $LiNbO_3$ substrate for SAW excitation, and the sputter deposited ZnO as the photosensitive layer. This device was reported to result in a frequency shift of 170 kHz at 37 MHz, for a UV light intensity of 40 mW/cm^2. The SAW oscillator output showed a decrease in amplitude and a downshift shift in oscillation frequency under a UV light illumination due to acousto-electric interactions between the photo generated charge carriers and the electrical potential associated with SAW. The change in amplitude and the shift in frequency showed a linear variation with UV light intensity. A low level UV intensity ~ 450 nW/cm^2 was easily detectable. Results show its immense application in wireless UV light detection.

N. W. Emanetoglu *et al.* demonstrated a SAW UV detector using the optoelectronic-acoustic interaction in a ZnO/r-Al_2O_3 system [59]. The Sezawa wave mode, which is the first higher order acoustic wave mode generated in the dispersive ZnO/r-Al_2O_3 system, was used to achieve a larger photoresponse, as the Sezawa wave mode has higher acoustic velocity and larger effective coupling, leading to a larger photoresponse. The device structure is schematically shown in Figure 13.16. The semiconducting ZnO/$Mg_{0.2}Zn_{0.8}O$ buffer/piezoelectric ZnO structure was grown on r-Al_2O_3 using metalorganic chemical vapor deposition (MOCVD). The Li-doped piezoelectric ZnO was 2.0 μm thick. The 50 nm thick $Mg_{0.2}Zn_{0.8}O$ buffer was used to isolate the semiconducting ZnO layer from the piezoelectric ZnO layer, and as a diffusion barrier for Li. The as-grown, unintentionally doped MOCVD ZnO showed *n*-type conductivity, with a carrier concentration of about 10^{17} cm^{-3}. The piezoelectric ZnO layer was doped with Li to increase its resistivity to above 10^7 Ω cm, in order to ensure efficient SAW excitation. The SAW delay line devices were aligned parallel to the ZnO *c*-axis, which is in the surface plane of the $(11\bar{2}0)$ ZnO film, to generate the Rayleigh type wave modes.

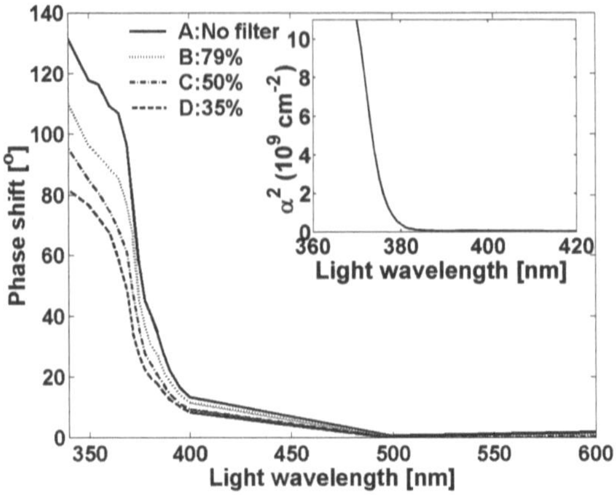

Figure 13.17: Phase shift vs. light wavelength for light power of 2.32 mW/cm^2, 1.83 mW/cm^2, 1.18 mW/cm^2, and 0.81 mW/cm^2. The responses have been normalized for constant power with respect to 365 nm [59].

The epitaxial piezoelectric ZnO layer thickness of 2.0 μm ensures the excitation of the Sezawa wave mode, in the test devices below 1 GHz. In the IDT region, the base Rayleigh wave mode had an estimated maximum coupling coefficient of 1.9% with a velocity of 2768 m/s, while the maximum coupling coefficient for the Sezawa wave mode was estimated to be 6% with an acoustic velocity of 5658 m/s.

A 75 W Xe lamp with a monochromator system was the UV source to illuminate the semiconducting ZnO mesa through a shadow mask. The light power was varied from 810 μW/cm^2 to 2.32 mW/cm^2 at $\lambda_{\text{light}} = 365$ nm. The phase shift at center frequency for the Sezawa wave mode with respect to the dark condition, as a function of light wavelength and power level, is shown in Figure 13.17. The inset shows the transmission spectrum of ZnO epilayer grown on r-Al_2O_3 for comparison. The SAW UV detector response closely follows the absorption spectrum. For light wavelengths above 400 nm, the phase shift is small. The phase shift increases rapidly as the light wavelength approaches the band edge. As the light wave approaches the band edge, at about 372 nm, the differential insertion loss rapidly increases, similar to the phase shift. The phase shift at 365 nm for a light power of 2.32 mW/cm^2 is 107°, corresponding to a frequency shift of 1.36 MHz in an oscillator circuit, calculated for the standard oscillator circuit configuration with the SAW in the feedback path:

$$\Delta f = \frac{1}{\tau} = \frac{1}{\frac{L_{\text{mesa}}}{v_{\text{SAW,mesa}}} + \frac{L_{\text{IDT}}}{v_{\text{SAW,IDT}}}} \frac{\Delta\varphi}{360} \qquad (13.14)$$

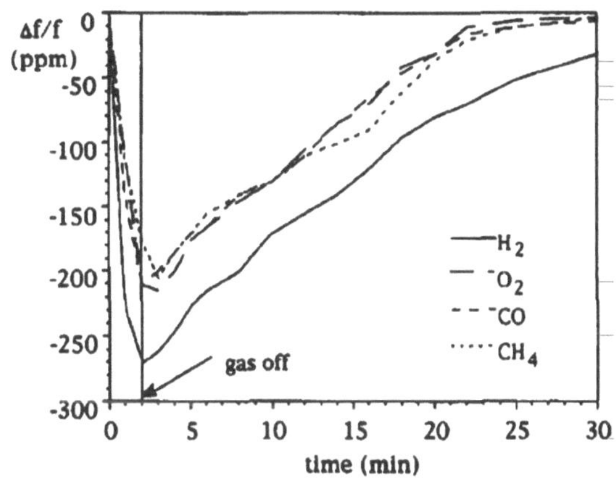

Figure 13.18: Acoustic responses of the sensors vs. time for different gases at room temperature and atmospheric pressure (ZnO:Li/Si) [60].

where τ is the delay time across the device, $v_{SAW,mesa}$ the acoustic velocity in the mesa region, $v_{SAW,IDT}$ the acoustic velocity in the IDT region, L_{IDT} the delay path length in the IDT region, and $\Delta\varphi$ the phase shift across the device. The Sezawa wave velocity in the mesa region, $v_{SAW,mesa}$, was estimated to be 5430 m/s. This 1.36 MHz frequency shift corresponded to a 0.19% relative shift for a light power of 2.32 mW/cm^2. The interaction of the SAW with the photogenerated carriers in the semiconducting ZnO layer results in a phase shift and an insertion loss change, as functions of light wavelength and power. The SAW UV detector can be used as a passive zero-power remote wireless sensor.

(iii) ZnO SAW sensors

ZnO-on-Si surface acoustic wave (SAW) delay lines can be used as detectors of combustible gases (CO, NO, H_2), reactive gas such as O_2 and other gases (CH_4, N_2, Ar) [60]. V. Z. Anisimkin *et al.* reported sputter deposited ZnO thin films on the acoustic propagation path ensure both electromechanical transduction and gas detection. The gas sensing was performed at temperatures of 20°C to 160°C, with operation frequencies of 100–200 MHz and atmospheric pressure. The SAW phase velocity change (acoustic response) upon absorption and desorption of a gas was demonstrated and compared with the change of electrical resistivity (electrical response) measured on the sensor in the same conditions. Figure 13.18 shows the the acoustic responses for H_2, O_2, CO and CH_4 versus time.

Recently, S. J. Ippolito *et al.* developed a layered surface acoustic wave (SAW) device for the monitoring of NO_2 and H_2 in synthetic air which is fabricated on XZ $LiNbO_3$ with a 1.2 μm ZnO guiding layer [61]. InO_x layers of 40 and 200 nm

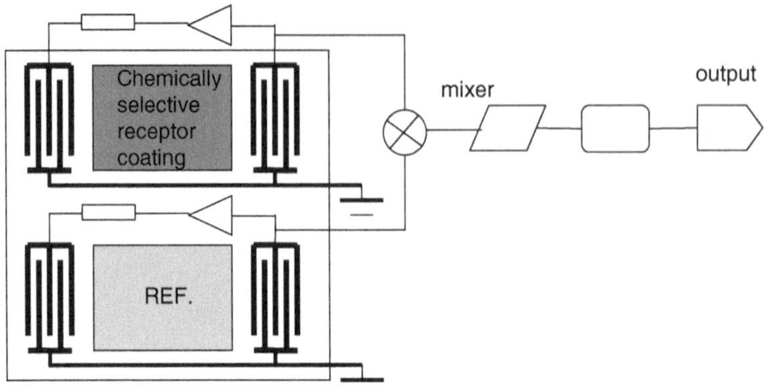

Figure 13.19: System block diagram of a SAW sensor chip.

thickness were employed to increase selectivity and sensitivity. The sensors were tested over a range of operating temperatures between 100 and 273°C. A large response magnitude with fast response and recovery time was observed. S. Ahmadi *et al.* developed inexpensive and miniaturized SAW gas sensors that are highly selective and sensitive [62]. These sensors were implemented with micro-electro-mechanical systems (MEMS) in CMOS technology. *c*-oriented thin film ZnO was grown on silicon based die. The frequency deviation due to gas absorbent was measured using the system presented with block diagram in Figure 13.19. The SAW sensor chip was composed of two SAW devices fabricated on the same substrate in CMOS technology, where one device was coated with a gas absorbent polymer and the second one was sealed to serve as a reference frequency source.

For layered SAW devices, the sensitivity depends on the physical properties of the films and the substrates. Love waves, which are mainly on SH-polarized guided waves, have high sensitivity in liquid. Love wave mode sensors based on ZnO/90° rotated ST-cut quartz structure have been studied [47,63]. The Love mode guiding layers in these sensors were ZnO and SiO_2 thin films. The substrates are 90° rotated ST-cut quartz crystals. Z. Wang *et al.* reported that Love mode devices with SiO_2 layers had a maximum sensitivity value of 400 cm^2/g and devices with ZnO layers had a significantly higher sensitivity of 950 cm^2/g [64]. S. Y. Chu *et al.* reported Love wave sensors made using the ZnO/90° rotated ST-cut quartz structure with high sensitivity to viscosity and conductivity in liquid environment. Additionally, the TCF of the quartz substrate was compensated by the ZnO film [65]. The high sensitivity of the Love wave sensor with ZnO/quartz structure was speculated to be due to the ZnO film's rough surface and the fact that the shear wave velocity of ZnO is much smaller than that of 90° rotated ST-cut quartz.

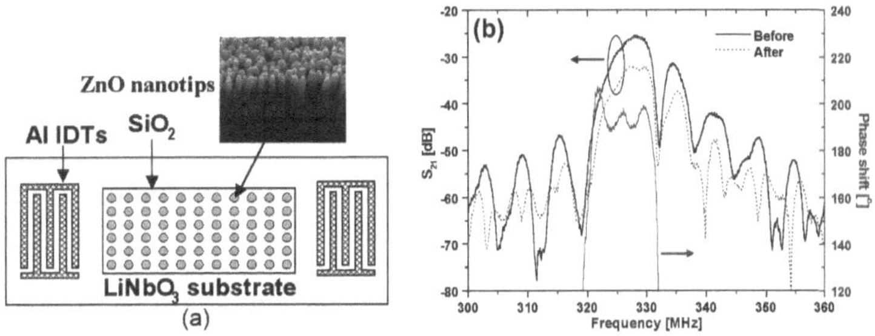

Figure 13.20: (a) Schematic of the ZnO nanotip-based SAW biosensor structures, (b) S_{21} (transmission) spectra before and after DNA oligonucleotide immobilization. The phase shift at the center frequency of 327.94 MHz is 191°. The circle indicates the frequency responses of the SAW sensor before and after immobilization, with respect to the left axis, the insertion loss [66].

A novel structure consisting of ZnO nanotips grown on a 128° Y-cut $LiNbO_3$ SAW delay line has been reported by Z. Zhang *et al.* [66]. The ZnO nanotips were grown by MOCVD on the top of a SiO_2 layer which was deposited and patterned on the $LiNbO_3$ SAW delay path, as shown in Figure 13.20. The ZnO nanotip arrays enhanced DNA immobilization by a factor of 200 compared to ZnO films with smooth surface. This device structure possesses the advantages of both traditional SAW sensors and ZnO nanostructures, and shows promising potential for portable sensors in biological and biomedical applications. During the DNA oligonucleotide immobilization test, when a 1 μg/μl DNA oligonucleotide solution was applied on the sensing channel, a 191° phase shift and 6.5 dB of additional insertion loss at the center frequency of 327.94 MHz was observed compared with the reference channel In the radioactive test, 4.32×10^2 ng/cm^2 APE and 7.33×10^2 ng/cm^2 complementary DNA oligonucleotide were detected. In the SAW device test, the sensitivity to immobilization was calculated to be 2.2×10^{-3} g/cm^2, while the sensitivity to the target second strand DNA oligonucleotide was 1.5×10^{-3} g/cm^2.

13.5 ZnO-based thin film bulk acoustic wave devices

13.5.1 Background

Thin film bulk acoustic wave resonators (TFBARs) and filters using bulk acoustic wave (BAW) have many advantages, such as small size, low insertion loss and lower power consumption than their bulk crystal based counter part. TFBARs allow the implementation of BAW devices from several hundred MHz up to 20 GHz. They

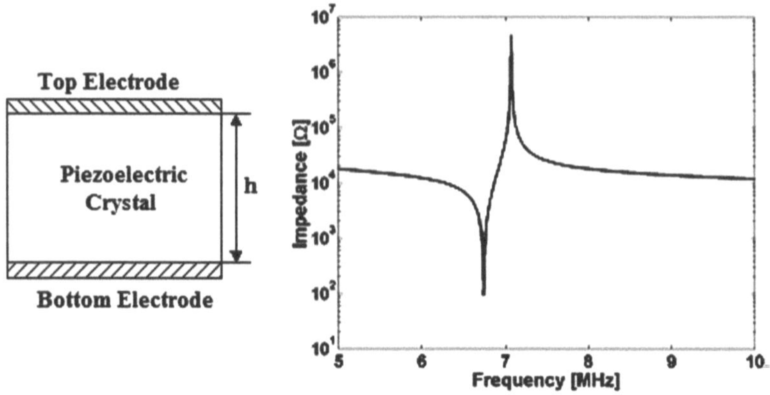

Figure 13.21: Frequency response of a bulk acoustic wave resonator.

are also attractive for the capability of monolithic integration of BAW resonators and filters with RF ICs, leading to miniaturization, and reducing cost.

Figure 13.21 shows the schematic of the simplest BAW resonator and a sample frequency response. The BAW resonator consists of a piezoelectric layer with thickness of h, a top and a bottom metal electrodes. In Figure 13.21, two acoustic resonances are observed. The first resonance, the series resonance f_s, is the acoustic resonance. Its frequency is determined by ratio of the phase velocity, v_a, and the thickness of the resonator, h:

$$f_s = \frac{v_a}{2h} \tag{13.15}$$

The second resonance, the parallel resonance f_p, is due to the interaction of the static capacitance of the BAW structure with the acoustic resonance. The series and parallel resonances are related to each other via the piezoelectric coupling coefficient, k_t^2. For a single piezoelectric layer without metallization:

$$k_t^2 = \frac{\pi}{2}\frac{f_s}{f_p}\tan\frac{\pi}{2}\left(1 - \frac{f_s}{f_p}\right) \tag{13.16}$$

A practical device is composed of a piezoelectric layer, and top and bottom electrodes. In such a case, an effective coupling coefficient, k_{eff}^2, is defined as:

$$k_{eff}^2 = 1 - \left(\frac{f_s}{f_p}\right)^2 \tag{13.17}$$

Several techniques are used to isolate the TFBAR structure from the substrate, as shown in Figure 13.22, in order to obtain a single resonance at the design frequency. As shown in Figure 13.22(a) and (b), the TFBAR structures require micromachining

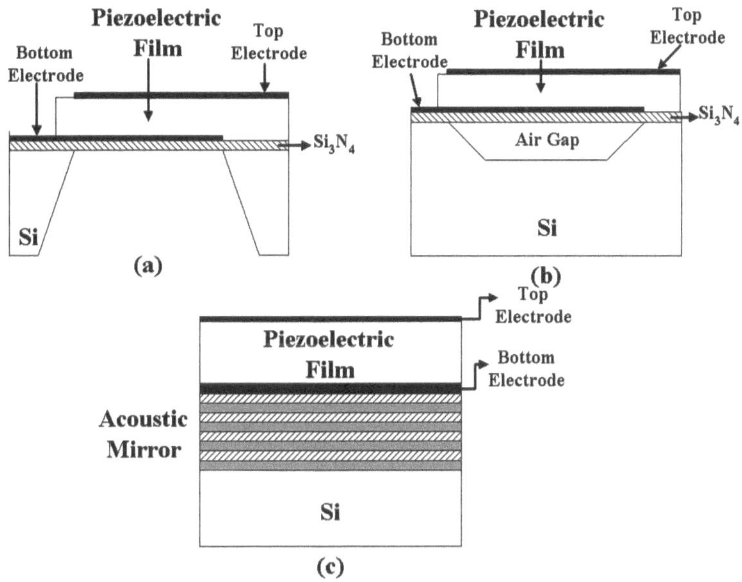

Figure 13.22: Three types of thin film bulk acoustic wave resonators: (a) film bulk acoustic resonator, (b) micromachined thin film resonator with air gap, and (c) solidly mounted resonator (SMR).

to fabricate an air gap, or use the deep backside etching to form a cavity structure. In contrast, the solidly mounted resonator (SMR) structure shown in Figure 13.22(c) uses an acoustic mirror to isolate the TFBAR from the substrate. The mirror structure is comprised of quarter wavelength reflectors, made of materials with a large acoustic impedance difference, such as SiO_2 and W. Therefore, the SMR is a low cost, easily manufacturable TFBAR structure and is attractive for monolithic integration with RF ICs.

A high frequency filter can be realized using piezoelectric ZnO thin film bulk acoustic wave resonators (TFBARs), as ZnO has a relatively large effective electromechanical coupling coefficient, k_{eff}. The bandwidth of BAW filters depends on the electromechanical coupling coefficient. ZnO TFBARs have been used in telecommunications and microwave applications, with resonators operating up to 18 GHz reported in the late-1970s [67]. TFBAR technology attracted increasing interest from the 1980s onwards [68,69], to meet the increasing demands of mobile communications. *c*-axis oriented piezoelectric ZnO thin films have been successfully used to develop TFBARs for telecommunications applications. K. M. Lakin *et al.* have demonstrated a series of ZnO TFBAR devices [70–75]. They demonstrated sputtered *c*-axis inclined piezoelectric ZnO films on Si substrates for shear wave resonators, which have coupling coefficients on the order of 17%. SMR miniature filters were also developed for integration into wireless PCS systems, achieved frequencies of operation in the 500 MHz to 6 GHz range with high Q factor (68,000 @ 1.6 GHz). Today, TFBAR becomes a promising technology, which has been wisely used due to its high feasibility of realizing monolithic integrated microwave circuits, miniaturizing the filters and duplexers for portable communication systems.

13.5.2 Solidly mounted thin film bulk acoustic wave resonators

A solid mounted thin film bulk acoustic resonator (TFBAR) consists of a piezoelectric film deposited on acoustic mirror on a substrate. These quarter-wavelength acoustic mirrors consist of alternating low- and high-acoustic-impedance quarter-wavelength layers. SMRs are now receiving attention for bulk acoustic wave devices operating in the GHz which can be monolithically integrated with ICs without the use of micromachining. C. W. Seabury *et al.* have reported on the fabrication of thin film ZnO based bulk acoustic mode filters [76], in which ZnO thin films were deposited on glass, semi-insulating GaAs, and high resistivity Si with an acoustic mirror. J. Kaitila *et al.* demonstrated SMR type ZnO TFBAR filters on glass substrates for the mobile Extended GSM (EGSM) R_x band centered at 942.5 MHz [77]. The resonators had an effective coupling coefficient of 0.236 and Q of approximately 800. Y. S. Park *et al.* reported SMR type ZnO TFBARs on Si substrates operating at 2.347 GHz for voltage controlled oscillator (VCO) applications [78]. This work reflects the important contribution ZnO has made for the 1–3 GHz filtering paradigms capable of handling greater than 2 watts of power with low loss. R. Kubo *et al.* demonstrated ZnO TFBARs at 5 GHz, with a quality factor (Q) of 1000 and effective electromechanical coupling coefficient (k_{eff}^2) of 6.7% [79]. M. Yim *et al.* demonstrated a ZnO TFBAR which can be monolithically integrated with Si IC processes to miniaturize the filters and duplexers for CDMA personal communication [80,81]. On a 4-inch p-type (100) Si wafer with an 600 nm thermal oxide layer, a five-layer Bragg reflector was formed by alternately depositing additional four-layer of silicon dioxide (SiO_2) and tungsten (W) using an RF magnetron sputtering technique. The 1.5 μm thick piezoelectric ZnO layer was sandwiched by two metal layers forming the top and bottom electrodes. TFBAR devices annealed at 400°C for 30 min showed excellent resonance characteristics. The return loss of TFBAR with Al top electrode was about 24–37 dB at a resonance frequency of ~1.92 GHz. The return loss of TFBAR with Co top electrode was about 38–46 dB at a resonance frequency of ~1.78 GHz. The Q-factor values of the TFBAR devices with Co electrodes ranged from 4000 to 6000, while those with Al electrodes ranged from 2000 to 5000.

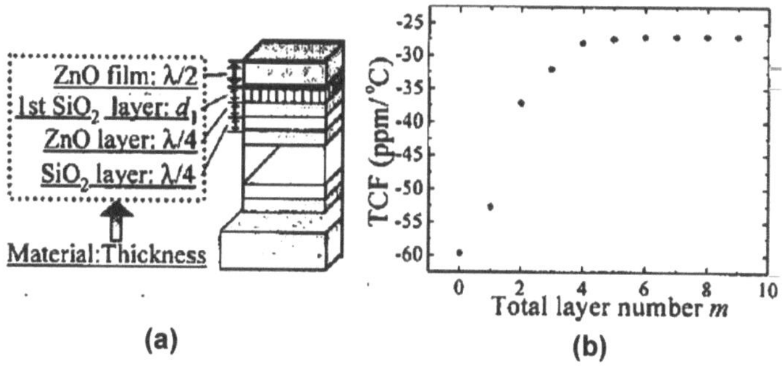

Figure 13.23: (a) Schematic of $ZnO/SiO_2/ZnO$-SMR, (b) TCF versus on the total layer number [82].

For practical applications, the temperature dependence of frequency (TCF) should be very low. S. Ohta *et al.* presented the theoretical analysis of the temperature characteristics of SMRs using ZnO as the piezoelectric material, SiO_2 as the low impedance material, and ZnO as the high impedance material [82]. The $ZnO/SiO_2/ZnO$-SMR is schematically presented on Figure 13.23(a). The dependence of TCF on the total layer number is presented on Figure 13.23(b). The calculated results showed that the temperature coefficient of frequency (TCF) varied from -60 ppm/°C to -27 ppm/°C with increasing the total layer number. This improvement of TCF results from the positive TCF of SiO_2. Also the temperature compensation of the SMR can be achieved by increasing the thickness of the first $\lambda/4$ SiO_2 layer just below the piezoelectric film or by depositing a thin SiO_2 capping layer on the piezoelectric film, although the effective electromechanical coupling factor K^2 and the electric Q decrease a little. A zero TCF can be obtained by properly choosing the SiO_2 layer thickness.

13.5.3 Edge supported ZnO TFBAR devices

Edge supported ZnO thin film bulk acoustic wave resonators include an air gap structure made by MEMS, or a backside substrate etched cavity structure by deep etching. Q. X. Su *et al.* reported a ZnO TFBAR sandwiched by Al and Au/Cr electrodes on high resistivity Si substrate with backside etched cavity [83]. High quality *c*-axis oriented 1.2 μm ZnO films were deposited on the bottom electrode using RF sputtering technology. A Si_3N_4 layer was used as the support membrane and for the protection of the Si etching area underneath the devices. Al should be a better electrode material than Au for TFBAR because of its smaller mass loading effect (2700 kg/m^3), larger acoustic velocity (6350 m/s) and lower acoustic attenuation

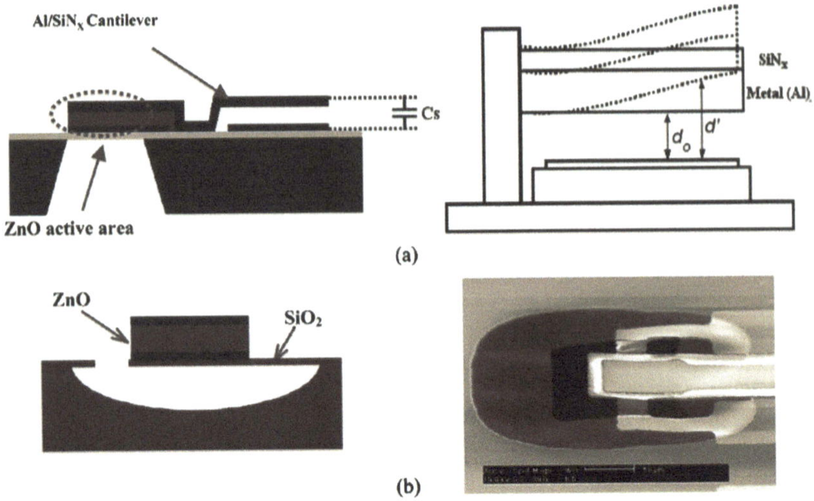

Figure 13.24: Schematic of (a) TC TFBAR and the bi-metal action of the series capacitor (C_S), (b) surface-micromachined TC TFBAR on SiO_2 cantilever [84].

(acoustic impedance: 1.7×10^7 kg/m^{2}s). However the resonance characteristics of ZnO TFBARs with Al bottom electrode were poorer than those with Au electrodes because of an amorphous layer which was formed at the ZnO-Al interface during sputtering, resulting in poor quality ZnO films. The Au/Cr bottom electrodes were 100 nm thick. A band-pass ladder filter was implemented by properly connecting two TFBARs in series and parallel. The bandwidth is determined mainly by the difference between the serial and parallel resonance frequency of the TFBAR, which are determined by the ZnO electromechanical coupling coefficient. R. Kubo *et al.* reported two kinds of 5 GHz TFBARs using 500 nm thick piezoelectric ZnO thin film: fundamental resonance and secondary harmonics [79]. For the fundamental mode resonator, the Q factor and k_{eff} were 1000 and 6.7%, respectively. These values for the secondary mode resonator were 700 and 4.7%, respectively. The filter bandwidth using the fundamental resonance at 3 dB was 191 MHz at 4.9 GHz and the minimum insertion loss was 1.3 dB, while the bandwidth of a filter using the secondary mode resonator was 130 MHz at 5.2 GHz and the minimum insertion loss was 1.6 dB. A temperature-compensated TFBAR has been recently reported by W. Pang *et al.* [84]. Figure 13.24(a) shows a cross-sectional view of the Type 1 temperature compensated (TC) TFBAR and the bi-metal action of the series capacitor (C_S). The series capacitor consisted of a bottom electrode and a cantilever made of an Al–SiN_x bi-layer. The two materials (forming the bi-layer) had a large difference in their thermal expansion coefficients, so that when the ambient temperature varied, the cantilever deflected, causing a variation in the air-gap capacitance. This compensated the frequency shift due to the stiffness change of the TFBAR layers with temperature. Figure 13.24(b) shows a cross-sectional view of Type 2, a surface-micromachined TC TFBAR on SiO_2 cantilever. The Type 1 TC TFBAR was built on the 0.6-μm-thick SiN_x diaphragm fabricated with micromachining. The submicrometer air-gap capacitor was formed after removing the silicon sacrificial layer. The TFBARs top-view active area was $30 \times 30\,\mu m^2$, while the top-view area of the cantilever-type bimorph capacitor was 50 μm wide and 240 μm long. For fabricating Type 2 TC TFBAR, a CMOS compatible surface micromachining process was developed, suitable for a post-CMOS fabrication of TFBAR for a monolithic integration of TFBAR and circuits on a single chip. After deposition and patterning the ZnO and Al electrodes, the TFBAR was released by etching the silicon substrate. When temperature increased from room temperature to 58°C, the cantilever bent upward, and lowered the series capacitance, resulting in a positive TCF. The TCF was around −19 ppm/°C between 33°C and 54°C, better than that of an uncompensated TFBAR. The Type 2 TFBAR composed of 0.1-μm-thick Al, 0.48-μm-thick ZnO, 0.1-μm-thick Al and 0.86-μm-thick SiO_2. It had a TCF of −0.45 ppm/°C at the second harmonic frequency of 4.4 GHz between 80°C and 105°C. The measured Q of Type 2 TC TFBAR was about 330.

13.5.4 $Mg_xZn_{1-x}O$ thin film resonators

R. H. Wittstruck *et al.* demonstrated a multi-mode $Mg_xZn_{1-x}O$ TFBAR on *r*-plane sapphire [85], consisting of the Al/$Mg_xZn_{1-x}O$/n^+ ZnO/r-sapphire structure. Al and n^+ ZnO served as the top and bottom electrode, respectively. The n^+ ZnO electrode was unique, as it preserved the crystalline relationship required for multi-mode coupling with the substrate while providing an electrode function. On r-Al_2O_3, the $Mg_xZn_{1-x}O$ film's primary axis of symmetry lies in the surface plane. This relationship promotes shear bulk wave propagation that affords sensing in liquid phase media without the dampening effects found in longitudinal wave mode BAW devices. When a signal is applied between the top and the bottom electrodes, an acoustic shear wave mode is excited. By varying the Mg composition in the $Mg_xZn_{1-x}O$ films, or using ZnO/$Mg_xZn_{1-x}O$ heterostructure with different thickness ratio, the frequency response can be tailored, which provides an alternative and complementary method for frequency tuning in addition to thickness variation.

Recently, a single mode TFBAR has been demonstrated by Y. Chen *et al.* using piezoelectric $Mg_xZn_{1-x}O$ layers on Si substrates [36]. A complete TFBAR device structure consisting of Al/$Mg_xZn_{1-x}O$/Au/acoustic mirror/Si is shown in Figure 13.25. Al and Au layers served as the top and bottom electrode, respectively.

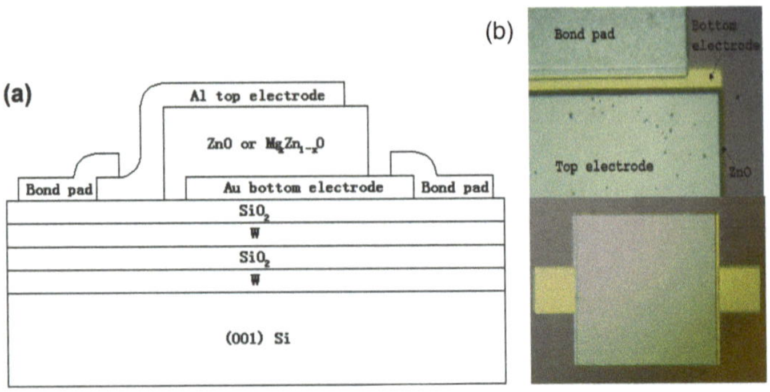

Figure 13.25: (a) A schematic diagram of a $Mg_xZn_{1-x}O/Si$ TFBAR structure, (b) optical microscope pictures and edge details of the TFBAR [36].

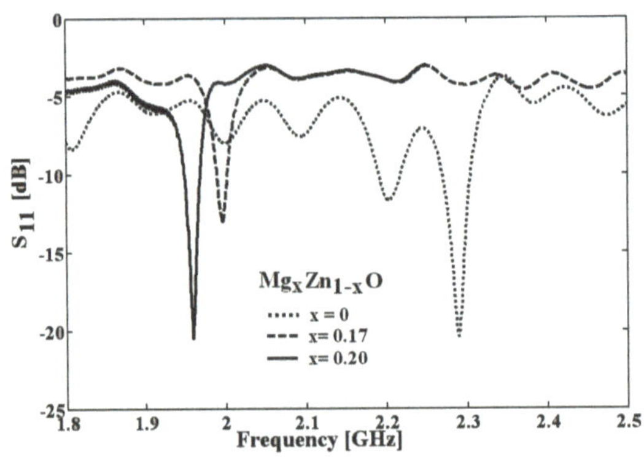

Figure 13.26: Measured S_{11} spectra of 150 μm × 150 μm $Mg_xZn_{1-x}O$ TFBARs on mirror/Si substrates, with $x = 0$, 0.17 and 0.20 [36].

The acoustic mirror, composed of two periods of alternating quarter-wavelength Bragg reflector, was used to isolate the resonator from the substrate. RF sputtered $Mg_xZn_{1-x}O$ thin films had a preferred *c*-axis orientation. Thus, when a signal was applied between the top and the bottom electrodes, a longitudinal acoustic wave mode was excited.

The frequency response of the thin film resonator can be tailored by using $Mg_xZn_{1-x}O$ instead of ZnO as the piezoelectric layer. Simulations indicated that for a TFBAR with 1.25 μm thick ZnO, $Mg_{0.1}Zn_{0.9}O$ and $Mg_{0.2}Zn_{0.8}O$ piezoelectric layers, the center frequency shifts from 1.976 GHz for ZnO to 2.040 GHz for $Mg_{0.1}Zn_{0.9}O$, and 2.114 GHz for $Mg_{0.2}Zn_{0.8}O$. Figure 13.26 shows the S_{11} spectra of $Mg_xZn_{1-x}O$ TFBAR whose device area was 150 μm × 150 μm. The solid line is from a TFBAR with a 1.41 μm thick $Mg_{0.2}Zn_{0.8}O$ piezoelectric layer.

The center frequency is at 1.959 GHz with $S_{11} = -20.5$ dB. The dashed line represents a TFBAR with a 1.30 μm thick $Mg_{0.17}Zn_{0.83}O$ piezoelectric layer. The center frequency is at 1.994 GHz with $S_{11} = -13.1$ dB. The dotted line represents a ZnO TFBAR with film thickness about 1.10 μm. The center frequency is at $f_0 = 2.289$ GHz and $S_{11} = -20.4$ dB. Considering the thickness difference of three devices, $Mg_{0.2}Zn_{0.8}O$ has a longitudinal acoustic velocity that is 6.6% larger than that of $Mg_{0.17}Zn_{0.83}O$, and 9.7% larger than that of ZnO.

13.5.5 BAW sensors

BAW sensors, such as the well known quartz crystal microbalance (QCM), are known to have a limitation in achieving an even higher sensitivity due to their low-frequency operation which is several hundred MHz, and they suffer from poor integration into arrays. On the other hand, the thin film bulk acoustic resonator (TFBAR) has recently emerged as an attractive sensor device, as TFBAR can operate at much higher frequencies. In addition, TFBAR type sensors are much smaller, and can be readily integrated as arrays. BAW sensors based on TFBARs with Bragg reflectors can be fabricated on silicon, glass or other substrates using a planar technology. In the case of the Si substrate, the sensors can be integrated with other Si-based electronic components on the same substrate. Another distinct advantage of BAW sensors is that they are miniature and compatible with small-size microwave aerials, and hence can be used for wireless distance probing in ecological applications.

The sensitivity of a BAW sensor due to the mass loading effect is calculated using the Sauerbrey's formula [86]:

$$\Delta f = -\frac{2f_0^2}{A\sqrt{c_{66}\rho}}\Delta m = -S_f\,\Delta m \tag{13.18}$$

where f_0 is the reference resonance frequency and $f_0 + \Delta f$ is the loaded frequency, c_{66} is a stiffness constant of the piezoelectric material, ρ is its density. In the case of ZnO, $c_{66} = 4.43 \times 10^{10}$ N/m^2 and $\rho = 5680$ kg/m^3.

Recently, Y. Chen *et al.* reported the resonance characteristics of a 150 μm × 150 μm ZnO TFBAR on mirror/Si substrate before and after mass-loading [36]. The piezoelectric ZnO film thickness was 1.20 μm. A 60 nm SiO_2 thin film was deposited on top electrode to investigate the mass-sensitivity. The density of PECVD deposited SiO_2 was 2.3 g/cm^3. $\Delta m/A$ was 13.8 μg/cm^2. As shown in Figure 13.27, the measured frequency shift was 23.7 MHz. Thus, a mass sensitivity S_f

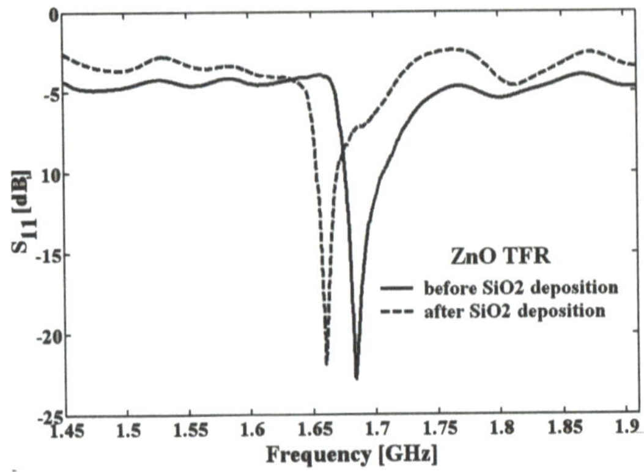

Figure 13.27: Measured S_{11} spectra of a 150 μm × 150 μm ZnO TFBAR on mirror/Si substrate, before and after SiO_2 deposition [36].

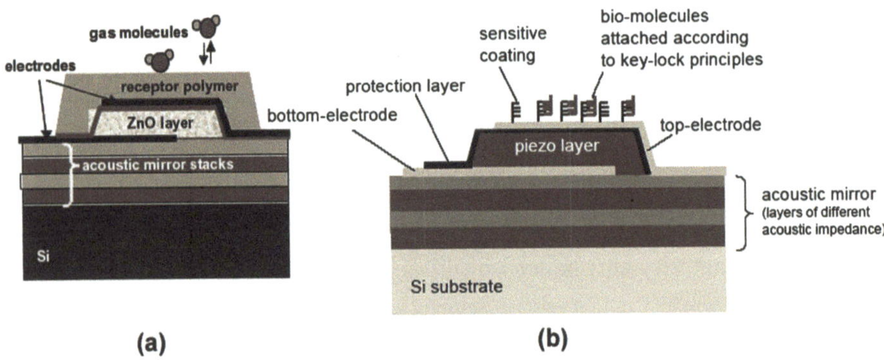

Figure 13.28: Schematic of the solidly mounted TFBAR used as (a) gas and (b) bio sensors [88,89].

of 1.7 Hz cm^2/pg was achieved. This sensitivity is useful for ultra-sensitive-mass loading chemical sensor, biosensor, and environmental sensor applications.

R. Gabl *et al.* reported thin film bulk acoustic resonators used for bio- and gas-detection [87–89]. Solidly mounted ZnO TFBARs with resonant frequency at 2 GHz were fabricated on Si substrates with 3-fold ZnO/Pt acoustic mirror. A schematic cross section of the ZnO TFBAR sensor is shown in Figure 13.28(a). It consists of a thin-film bulk acoustic resonator formed by a layer stack of bottom electrode, piezoelectric layer and top-electrode which are integrated onto a silicon substrate. The active vibrating region of the resonator is coated with a receptor layer which is sensitive to the adsorption or absorption of the molecules that are to be detected. The attachment of these molecules leads to an increase of the resonator mass load and thus to a frequency shift of the resonance frequency which can be electrically determined using state of the art RF-circuitry. The quarter wavelength thick bottom electrode acts acoustically as an efficient reflection mirror and ensures a high mass sensitivity as well as a low ohmic series resistance. The extracted electromechanical coupling coefficient k_{eff} was 19%. The sensitivity reached a maximum of 2.5 Hz cm^2/pg, a value which was 2500 times larger compared with typical 20 MHz QCM. To test the viability of TFBARs for gas sensing, a humidity absorbing polyimide from E + E Elektronik was spinned on the wafer surface. The measurements showed a nearly linear decrease in frequency with increasing humidity for TFBARs coated with a 50 nm polyimide layer. The shift was caused by an increase of the polymers' effective mass by absorbing water molecules. A sensitivity of 6.8 kHz/% r.h. was extracted. To test the feasibility of TFBARs for bio sensing after DNA immobilization and after protein coupling, TFBAR sensors with areas ranging from 150 μm × 150 μm to 500 μm × 500 μm were investigated, as shown in Figure 13.28(b). As a reference, several sensors were intentionally left untreated by the protein coupling. Both DNA immobilization and protein coupling resulted in clear signals. The results showed that the resolution of the sensor is sufficient for detection of DNA.

H. Zhang *et al.* [90] reported an implantable, resonant mass sensor for liquid mass sensing. The sensor consisted of a TFBAR with a piezoelectric 1.6 μm thick ZnO film and two 0.2 μm Al layers sitting on a 0.8 μm silicon nitride (Si_3N_4) diaphragm, which formed the tip portion of an implantable silicon probe for recording neuronal activities in a living biological tissue. When a TFBAR without Si_3N_4 supporting membrane was immersed into water, the quality factor Q dropped substantially from about 200 to about 10 at around 1 GHz. When a Si_3N_4 supporting membrane was used, an apparent second harmonic resonance appeared besides the fundamental one with its resonance frequency about twice that of the fundamental resonance. When this TFBAR was immersed into water, the Q dropped from 340 to 40 at 2 GHz. Although the mass sensitivity of the sensor was almost the same in air and water, the reduced Q in water affected the minimum detectable signal (mass change) adversely due to increased phase noise. With such a Q value, a 5 ppm resonant frequency shift was detected, corresponding to 10^{-8} g/cm^2 mass change on the sensor surface. It was found that the resonant frequency of TFBAR coated with TiO_2 was sensitive to the metal ion mass as the positive metal ion and O^- on the TiO_2 caused mass loading effect. This TFBAR sensor also was sensitive to the pH which was attributed to an increase in the number of anionic site on the TiO_2 surface.

Mai *et al.* [91] presented a feasibility study of ZnO-based TFBAR devices and their fabrication for the ultra-mass-sensitive sensor application. In this work, a considerable shift in the resonance frequency was observed due to the mass-loading effect by a mixture of ink and volatile methanol. A high sensitivity of 5 Hz cm^2/pg was obtained in the TFBAR-based device, which is ~5 orders of magnitude higher than that of 0.057 Hz cm^2/ng reported in the conventional 5-MHz quartz crystal microbalance. This approach seems useful for the ultra-high mass resolution chemical sensor or biosensor applications.

13.6 Conclusion

In this chapter, we have reviewed the progress in piezoelectric ZnO materials and devices. While piezoelectric ZnO films have been widely used in signal processing and frequency control, advances in the field have generated many new devices and applications. Piezoelectric ZnO films have been not only deposited on quartz, Si and sapphire substrates, but are now also integrated with the new wide band gap semiconductors, such as SiC and GaN, as well as diamond to develop high power, or high frequency (multi GHz) and zero temperature coefficient of frequency SAW devices. The new ternary piezoelectric material, $Mg_xZn_{1-x}O$ and its heterostructure $ZnO/Mg_xZn_{1-x}O$ provide design flexibility for multilayered piezoelectric devices to tailor the frequency response. Furthermore, as ZnO can be made multifunctional through proper doping, it offers the promise of integrating multifunctional devices in ZnO and $ZnO/Mg_xZn_{1-x}O$ multilayer structures. The UV-SAW detector is an example, in which a semiconductor ZnO layer is integrated with a piezoelectric ZnO layer to form a UV detector. This new device has its output in the frequency domain, suitable for a much desired wireless sensor, and also has low power operation. Another interesting new development is the combination of SAW sensors with ZnO nanostructures, which enhances DNA immobilization on the sensor surface. Piezoelectric ZnO has been used in SAW and BAW sensors. The new developments are based on the integration of TFBARs with MEMS using micromachining technology, or with a quarter-wavelength Bragg reflector acoustic mirror. Piezoelectric ZnO, its heterostructures and nanostructures have shown a great potential for future sensor technologies, including sensors on-a-chip, XYZ-on-a-chip, and laboratory on-a-chip.

Acknowledgements

We would like to thank many of our team members at Rutgers for their collaboration and assistances, especially, Richard H. Wittstruck, Sriram Muthukumar, Yimin Chen, Haifeng Sheng, Karen Wu, Zheng Zhang, Gaurav Saraf, Hanghong Chen, Pan Wu, Jian Zhong and Jun Zhu. The special thanks are due to Dr. Arthur

Ballato at US Army CECOM for his valuable help and discussion. Y. Lu gratefully acknowledges the financial support from NSF under the grants (ECS-008549, CCR-0103096), AFSOR under the grant (F49620-03-1-0292), New Jersey Commission of Science & Technology (NJCST) for the Research Excellent Center Program under the MUSE grant, and Rutgers University Academic Excellence Fund.

References

[1] J. H. Visser, M. J. Vellekoop, A. Venema, E. van der Drift, P. J. M. Rek, A. J. Nederlof, M. S. Nieuwenhuizen, and M. J. Vellekoop, *Proc. 1989 IEEE Ultrasonics Symp.*, pp. 195, 1989.

[2] Y. Kim, W. D. Hunt, F. S. Hickernell, R. J. Higgins, and C.-K. Jen, *IEEE Trans. on Ultrason., Ferroelect. and Freq. Contr.*, vol. 42, no. 3, pp. 351, May 1995.

[3] S. J. Chang, Y. K. Su, and Y. P. Shei, *J. Vac. Sci. Technol. A*, vol. 13, no. 2, pp. 385, 1995.

[4] K. Yamanouchi, N. Sakurai, and T. Satoh, *Proc. IEEE Ultrasonics Symp.*, pp. 351, 1989.

[5] H. Nakahata, A. Hachigo, S. Shikata, and N. Fujimori, *Proc. 1992 IEEE Ultrasonics Symp.*, pp. 377, 1992.

[6] T.-T. Wu, Y.-Y. Chen, *Proc. 2001 IEEE Ultrasonics Symp.*, pp. 107, 2001.

[7] T. Mitsuyu, S. Ono, and K. Wasa, *J. Appl. Phys.*, vol. 51, no. 5, pp. 2464, 1980.

[8] C. K. Lau, S. K. Tiku, and K. M. Lakin, *J. Electrochem. Soc.: Solid-State Sci. B Tech.*, vol. 127, no. 8, pp. 1843, 1980.

[9] J. Koike, H. Tanaka, and H. Ieki, *Jpn. J. Appl. Phys.*, vol. 34, pp. 2678, 1995.

[10] H. Ieki, H. Tanaka, J. Koike, and T. Nishikawa, *IEEE MTT-S Dag.*, pp. 409, 1996.

[11] N. W. Emanetoglu, C. Gorla, S. Liang, and Y. Lu, *Proc. 1998 IEEE Int. Freq. Contr. Symp.*, pp. 790, 1998.

[12] N. W. Emanetoglu, C. R. Gorla, S. Liang, Y. Liu, and Y. Lu, *Mat. Sci. Semiconductor Process.*, vol. 4, pp. 1, Sep. 1999.

[13] M. Kadota, *Proc. 1997 IEEE Ultrasonics Symp.*, vol. 1, pp. 261, 1997.

[14] M. Kadota, T. Kasanami, and M. Minakata, *Electronics Letters*, vol. 28, Issue 25, pp. 2315, 1992.

[15] M. Kadota and M. Minakata, *IEEE Trans. on Ultrason., Ferroelect. and Freq. Contr.*, vol. 42, Issue 3, pp. 345, 1995.

[16] R. D. Vispute, V. Talyansky, Z. Trajanovic, S. Choopun, M. Downes, R. P. Sharma, T. Venkatesan, M. C. Woods, R. T. Lareau, K. A. Jones, and A. A. Iliadis, *Appl. Phys. Lett.*, vol. 70, no. 20, pp. 2735, 1997.

[17] V. Craciun, J. Elders, J. G. E. Gardeniers, and I. W. Boyd, *Appl. Phys. Lett.*, vol. 65, no. 23, pp. 2963, 1994.

[18] R. Triboulet and J. Perrière, *Progress in Crystal Growth and Characterization of Materials*, vol. 47, Issues 2–3, pp. 65, 2003.

[19] F. Cracium, P. Verardi, M. Dinescu, and G. Guidarelli, *Thin Solid Film*, 343–344, pp. 90–93, 1999.
[20] G. H. Kim, B. D. Ahn, H. S. Kang, S. H. Lim, and S. Y. Lee, *Superlattices and Microstructures*, In Press, Corrected Proof, Available online 7 October 2005.
[21] R. Ondo-Ndong, F. Pascal-Delannoy, A. Boyer, A. Giani, and A. Foucaran, *Material Science and Engineering: B*, 97(1), pp. 68, 2003.
[22] J.-B. Lee, H.-J. Lee, and J.-S. Park, *Proc. 2003 IEEE Freq. Contr. Symp.*, pp. 868, 2003.
[23] S.-Y. Chu, W. Water, and J.-T. Liaw, *Journal of the European Ceramic Society*, 23, pp. 1593, 2003.
[24] Y. Yoshino, K. Inoue, M. Takeuchi, and K. Ohwada, *Vacuum*, vol. 5, no. 4, pp. 601, 1998.
[25] M. Kadota, T. Miura, and M. Minakata, *J. Cryst. Growth*, 237/239, pp. 523, 2002.
[26] S. J. Chang, Y. K. Su, and Y. P. Shei, *J. Vac. Sci. Technol. A*, vol. 13, no. 2, pp. 381, Mar./Apr. 1995.
[27] Y. Kashiwaba, K. Haga, H. Watanabe, B. P. Zhang, Y. Segawa, and K. Wakatsuki, *Phys. Status Solid. B*, 229, pp. 921, 2002.
[28] B. P. Zhang, Y. Segawa, K. Wakatsuki, Y. Kashiwaba, and K. Haga, *Appl. Phys. Lett.*, 79(24), pp. 3953, 2001.
[29] C. R. Gorla, N. W. Emanetoglu, S. Liang, W. E. Mayo, Y. Lu, M. Wraback, and H. Shen, *J. Appl. Phys.*, 85(5), pp. 2595, 1999.
[30] N. W. Emanetoglu, G. Patounakis, S. Liang, C. R. Gorla, R. Wittstruck, and Y. Lu, *IEEE Trans. on Ultrason., Ferroelect. and Freq. Contr.*, vol. 48, Issue 5, pp. 1389, 2001.
[31] J. B. Lee, H. J. Lee, S. H. Seo, and J. S. Park, *Thin Solid Film*, 398–399, pp. 641, 2001.
[32] A. Ohtomo, M. Kawasaki, T. Koida, K. Masubuchi, H. Koinuma, Y. Sakurai, Y. Yoshida, T. Yasuda, and Y. Segawa, *Appl. Phys. Lett.*, vol. 72, no. 19, pp. 2466, 1998.
[33] A. K. Sharma, J. Narayan, J. F. Muth, C. W. Teng, C. Jin, A. Kvit, R. M. Kolbas, and O. W. Holland, *Appl. Phys. Lett.*, vol. 75, no. 21, pp. 3327, 1999.
[34] N. W. Emanetoglu, S. Muthukumar, P. Wu, R. Wittstruck, Y. Chen, and Y. Lu, *IEEE Trans. on Ultrason., Ferroelect. and Freq. Contr.*, vol. 50, no. 5, pp. 537, 2003.
[35] H. Wu, N. W. Emanetoglu, G. Saraf, J. Zhu, P. Wu, Y. Chen, and Y. Lu, *Proc. 2003 IEEE Int. Ultrasonics Symp.*, pp. 897, 2003.
[36] Y. Chen, G. Saraf, R. H. Wittstruck, N. W. Emanetoglu, and Y. Lu, *Proc. 2005 IEEE Int. Freq. Contr. Symp.*, pp. 142, 2005.
[37] G. Saraf, Y. Chen, Y. Lu, T. Siegrist, and L. S. Wielunski, *2005 II–VI Workshop*, Sept. 20–22, 2005, Boston.
[38] A. H. Weber, G. Weiss, and S. Hunklinger, *Proc. 1991 IEEE Int. Ultrasonics Symp.*, pp. 363.
[39] P. Wu, N. W. Emanetoglu, X. Tong, and Y. Lu, *Proc. 2001 IEEE Int. Ultrasonics Symp.*, pp. 211.
[40] A. Haddoul, T. Grybal, J. E. Lefebvre, V. Sadaunel, V. Zhangl, and E. Cattanz, *Proc. 2001 IEEE Ultrasonics Symp.*, pp. 91, 2001.

[41] V. Y. Zhang, J. Z. Lefebvre, and T. Gryba, *Proc. 2001 IEEE Int. Ultrasonics Symp.*, pp. 261.
[42] M. Tomar, V. Gupta, and K. Sreenivas, *Proc. 2003 IEEE Ultrasonics Symp.*, pp. 901.
[43] M. Kadota and H. Kando, *IEEE Trans. on Ultrason., Ferroelect. and Freq. Contr.*, vol. 51, no. 4, pp. 464, 2004.
[44] M. Kadota, *Jpn. J. Appl. Phys.*, vol. 36, no. 5B, pp. 3076, 1997.
[45] T.-T. Wu and W.-S. Wang, *J. Appl. Phys.*, vol. 96, no. 9, pp. 5249, 2004.
[46] I. S. Didenko, F. S. Hickernell, and N. F. Naumenko, *IEEE Trans. on Ultrason., Ferroelect. and Freq. Contr.*, vol. 47, no. 1, pp. 179, January 2000.
[47] D. L. Dreifus and R. J. Higgins, *Proc. 1997 IEEE Ultrasonics Symp.*, pp. 191, 1997.
[48] K. Higaki, H. Nakahata, H. Kitabayashi, S. Fujii, K. Tanabe, Y. Seki, and S. Shikata, *1997 IEEE MTT-S Digest*, pp. 829, 1997.
[49] H. Nakahata, A. Hachigo, K. Itakura, and S. Shikata, *Proc. 2000 IEEE Ultrasonics Symp.*, pp. 349.
[50] V. U. Prechtel, V. Ziegler, S. Kolodzik, B. Plehn, H. Downar, J. Haering, R. Kunze, G. Martin, H. Schmidt, and M. Weihnacht, *2004 IEEE Ultrasonics Symp.*, pp. 199, 2004.
[51] Y. Chen, N. W. Emanetoglu, G. Saraf, P. Wu, Y. Lu, *IEEE Trans. on Ultrason., Ferroelect. and Freq. Contr.*, vol. 52, no. 7, pp. 1161, 2005.
[52] P. T. M. van Zeijl, J. H. Visser, and L. K. Nanver, *IEEE Trans. Consumer Electron.*, vol. CE-35, pp. 512, 1989.
[53] A. E. Corner and R. S. Muller, *IEEE Electron Device Letters*, vol. 3, Issue 5, May 1982.
[54] J. C. Haartsen, *1990 IEEE MTT-S Digest*, pp. 1115, 1990.
[55] K. Tsubouchi, T. Tomioka, T. Sato, C. Endo, N. Mikoshiba, *Proc. 1988 IEEE Ultrasonics Symp.*, vol. 1, pp. 213.
[56] K. Tsubouchi, N. Endo, N. Mikoshiba, *Proc. 1989 IEEE Ultrasonics Symp.*, vol. 1, pp. 165.
[57] K. Tsubouchi, H. Nakase, A. Namva, K. Masu, *IEEE Trans. on Ultrason., Ferroelect. and Freq. Contr.*, vol. 40, no. 5, pp. 478, 1993.
[58] P. Sharma, K. Sreenivas, *Appl. Phys. Lett.*, vol. 83, Issue 17, pp. 3617, 2003.
[59] N. W. Emanetoglu, J. Zhu, Y. Chen, J. Zhong, Y. Chen, and Y. Lu, *Appl. Phys. Lett.*, 85, pp. 3702, 2004.
[60] V. I. Anisimkin, M. Penza, A. Valentini, F. Quaranta, and L. Vasanelli, *Sensors and Actuators B*, 23, pp. 197, 1995.
[61] S. J. Ippolito, S. Kandasamy, K. Kalantar-zadeh, W. Wlodarski, K. Galatsis, G. Kiriakidis, N. Katsarakis, and M. Suchea, *Sensors and Actuators B*, 111–112, pp. 207–212, 2005.
[62] S. Ahmadi, F. Hassani, O. Tigli, S. Ahmadi, C. Korman, and M. Zaghloul, *Proc. 2003 IEEE Sensors*, pp. 1199
[63] K. K. Zadeh, W. Wlodarski, K. Galstsis, and A. Holland, *Proc. 2002 IEEE Int. Freq. Contr. Symp.*, pp. 74, 2002.

[64] Z. Wang and J. D. N. Cheeke, *Appl. Phys. Lett.*, 64, pp. 2940, 1994.
[65] S. Y. Chu, W. Water, and J. Liaw, *Ultrasonics*, 41, pp. 133, 2003.
[66] Z. Zhang, N. W. Emanetoglu, G. Saraf, Y. Chen, P. Wu, J. Zhong, Y. Lu, J. Chen, O. Mirochnitchenko, and M. Inouye, *IEEE Trans. on Ultrason., Ferroelect. and Freq. Contr.*, vol. 53, no. 4, pp. 786, 2006.
[67] K. A. Ingebrigtsen, *J. Appl. Phys.*, vol. 41, no. 2, pp. 454, 1970.
[68] F. S. Hickernell, *Proc. 1980 IEEE Int. Ultrasonics Symp.*, pp. 785, 1980.
[69] K. M. Lakin and J. S. Wang, *Appl. Phys. Lett.*, vol. 38, no. 3, pp. 125, 1981.
[70] G. R. Kline and K. M. Lakin, *Appl. Phys. Lett.*, vol. 43, no. 8, pp. 750, 1983.
[71] K. M. Lakin, G. R. Kline, R. S. Ketcham, A. R. Landin, W. A. Burkland, K. T. McCarron, S. D. Braymen, and S. G. Burns, *Proc. 1987 IEEE Freq. Contr. Symp.*, pp. 371, 1987.
[72] K. M. Lakin, *IEEE Freq. Contr. Symp.*, pp. 201, 1991.
[73] K. M. Lakin, G. R. Kline, and K. T. McCarron, *Proc. 1992 IEEE Ultrasonics Symp.*, pp. 471, 1992.
[74] K. M. Lakin, G. R. Kline, and K. T. McCarron, *IEEE Trans. Ultrason., Ferroelect. and Freq. Contr.*, vol. 41, no. 12, pp. 2139, 1993.
[75] K. M. Lakin, *IEEE 2003 FCS-EFTF Paper We1A-4 (Invited)*, May 5–8, 2003.
[76] C. W. Seabury, P. H. Kobrin, R. Addison, and D. P. Havens, *1997 IEEE MTT-S Digest*, pp. 181, 1997.
[77] J. Kaitila, M. Ylilammi, J. Molarius, J. Ella, and T. Makkonen, *Proc. 2001 IEEE Int. Ultrasonics Symp.*, pp. 803, 2001.
[78] Y. S. Park, S. Pinkett, J. S. Kenney, and W. D. Hunt, *Proc. 2001 IEEE Int. Ultrasonics Symp.*, pp. 839, 2001.
[79] R. Kubo, H. Fujii, H. Kawamura, M. Takeuchi, K. Inoue, Y. Yoshino, T. Makino, and S. Arai, *Proc. 2003 IEEE Int. Ultrasonics Symp.*, pp. 166, 2003.
[80] Munhyuk Yim, Dung-Hyun Kim, Dongkyu Chai, Junpil Jung, Jinseuk Park, and Giwan Yoon, *Proc. 2003 Radio and Wireless Conference*, pp. 389, 2003.
[81] Dong-Hyun Kim, Munhyuk Yim, Dongkyu Chai, and G. Yoon, *Electronics Letters*, vol. 39, Issue 13, pp. 962, 2003.
[82] Satoshi Ohta, Kiyoshi Nakamura, Arata Doi, and Yasuaki Ishida, *Proc. 2003 IEEE Ultrasonics Symp.*, pp. 2011, 2003.
[83] Q. X. Su, P. B. Kirbf, E. Komuro, and R. K. Whatmore, *Proc. 2000 IEEE/EIA Int. Freq. Contr. Symp.*, pp. 434, 2000.
[84] Wei Pang, Hongyu Yu, Hao Zhang, Eun Sok Kim, *IEEE Electron Device Letters*, vol. 26, no. 6, pp. 369, 2005.
[85] R. H. Wittstruck, X. Tong, N. W. Emanetoglu, P. Wu, Y. Chen, J. Zhu, S. Muthukumar, Y. Li, and A. Ballato, *IEEE Trans. on Ultrason., Ferroelect. and Freq. Contr.*, vol. 50, pp. 1272, 2003.
[86] G. Sauerbrey, Verwendung von Schwingquarzen zur Wagung dunner Schichten und zur Mikrowagung, *Z Phys.*, 155, pp. 206, 1959 (in German).
[87] R. Gabl, E. Green, M. Schreiter, H. D. Feucht, H. Zeininger, R. Primig, D. Piker, G. Eckstein, W. Wersing, W. Reichl, and J. Runck, *Proc. 2003 IEEE Sensors*, vol. 2, pp. 1184, 2003.

[88] R. Gabl, H.-D. Feucht, H. Zeininger, G. Eckstein, M. Schreiter, R. Primig, D. Pitzer, and W. Wersing, *Biosensors and Bioelectronics*, 19, pp. 615, 2004.
[89] W. Reichl, J. Runck, M. Schreiter, E. Green, and R. Gabl, *Proc. 2004 IEEE Sensors*, vol. 3, pp. 1504, 2004.
[90] Hao Zhang, Mong S. Marma, Eun Sok Kim, Charles E. McKenna, and Mark E. Thompson, *17th IEEE International Conference on Micro Electro Mechanical Systems*, pp. 347, 2004.
[91] Linh Mai, Dong-Hyun Kim, Munhyuk Yim, and Giwan Yoon, *Microwave and Optical Technology Letters*, vol. 42, no. 6, pp. 505, 2004.

Chapter 14

Gas, Chemical and Biological Sensing with ZnO

Young-Woo Heo

Department of Inorganic Materials and Engineering, Kyungpook National University, Daegu, 702-701, Korea

Fan Ren

Department of Chemical Engineering, University of Florida, Gainesville, FL 32611, USA

David P. Norton

Department of Materials Science and Engineering, University of Florida, Gainesville, FL 32611-6400, USA

14.1 Introduction

There is emerging interest in the use of wide bandgap semiconductors as sensitive gas and chemical sensors [1–10]. ZnO is a piezoelectric, transparent wide bandgap semiconductor used in surface acoustic wave devices. The bandgap can be increased by Mg doping. ZnO has been effectively used as a gas sensor material based on the near-surface modification of charge distribution with certain surface-absorbed species [11]. In addition, it is attractive for biosensors given that Zn and Mg are essential elements for neurotransmitter production and enzyme functioning [12,13].

The reversible chemisorption of reactive gases at the surface of these metal oxides can produce a large and reversible variation in the conductance of the material [14]. The gas-sensing mechanism suggested include the desorption of adsorbed surface oxygen and grain boundaries in poly-ZnO [15], exchange of charges between adsorbed gas species and the ZnO surface leading to changes in depletion depth [16] and changes in surface or grain-boundary conduction by gas adsorption/desorption [17].

Zinc Oxide Bulk, Thin Films and Nanostructures
C. Jagadish and S. Pearton (Editors)

The sensing mechanism for chemical adsorbates in piezoelectric materials originates from compensation of the polarization-induced bound surface charge by interaction with the polar molecules in the liquids [2]. Steinhoff et al. [4] suggested that the native oxide on the nitride surface was responsible for the pH sensitivity of the response of gateless GaN-based heterostructure transistors to electrolyte solutions. The pH response of metal oxide surfaces has been modeled by a number of groups in terms of formation of hydroxyl groups that lead to a pH dependent net surface change with a resulting change in voltage drop at the semiconductor/liquid interface [18–21]. Ambacher et al. [22–28] have shown a strong sensitivity of AlGaN/GaN heterostructures to ions, polar liquids, hydrogen gas and even biological materials. In particular they have shown that it is possible to distinguish liquid with different polarities.

The human immune system is extraordinarily complex. Antigen-antibody interactions are important for a biological sensing. Antibodies are protein molecules, which are composed of equal numbers of heavy and light polypeptide amino acid chains held together with disulfide bonds. These highly specialized proteins are able to recognize and bind only certain types of antigen molecules at receptor sites. Sensors based on using antibodies to detect certain specific antigens are called immunosensors. These immunosensors are useful for quantifying how well the human immune system is functioning, and could serve as valuable diagnostic tools. They also can be employed for identifying environmental contaminants and chemical or biological agents.

Silicon-based field effect transistors have been widely used as bio-sensors. For example, the silicon-based ion sensitive field effect transistor (ISFET) is already commercialized to replace conventional electrolyte pH meters. Silicon-based metal oxide semiconductor field effect transistors (MOSFETs) and ISFETs have also been used for immunosensors. However, silicon-based FETs are damaged by exposure to solutions containing ions. Si-FET based biosensors need an ion-sensitive membrane coated over their surface and a reference electrode (usually Ag/AgCl) is required to supply the bias voltage. Special precautions must also be taken to apply those biomembranes over the FETs so that they retain their enzymatic activity.

A large variety of ZnO one-dimensional structures have been demonstrated [22–38]. The large surface area of the nanorods and bio-safe characteristics of ZnO makes them attractive for gas and chemical sensing and biomedical applications, and the ability to control their nucleation sites makes them candidates for micro-lasers or memory arrays.

Ideal sensors have the ability to discriminate between different gases. Arrays that contain different metal oxides (e.g. SnO_2, ZnO, CuO, WO_3) on the same chip can be used to obtain this result. To date, most of the work on ZnO nano-structures has focused on the synthesis methods. There have been only a few reports of the electrical characteristics [22–26]. There is strong interest in developing gas, chemical, and biological sensors for use in both industry and domestic applications.

This chapter will discuss the gas, chemical, and biological sensing with ZnO. Session 2 introduces the response characteristics of ZnO in hydrogen, ethylene, CO, Ozone, UV, and pH. Session 3 provides the approaches to sense the biological agents with ZnO.

14.2 Gas and chemical sensing

14.2.1 Hydrogen sensing

There is currently great interest in the development of hydrogen sensors for applications involving leak detection in hydrogen fuel storage systems and fuel cells for space craft [39–42]. One of the main demands for such sensors is the ability to selectively detect hydrogen at room temperature in the presence of air. In addition, for most of these applications, the sensors should have very low power requirements and minimal weight. Nanostructures are natural candidates for this type of sensing. One important aspect is to increase their sensitivity for detecting gases such as hydrogen at low concentrations or temperatures, since typically an on-chip heater is used to increase the dissociation efficiency of molecular hydrogen to the atomic form and this adds complexity and power requirements.

Different metal coating layers on multiple ZnO nanorods are compared for enhancing the sensitivity to detection of hydrogen at room temperature. Pt is found to be the most effective catalyst, followed by Pd. The resulting sensors are shown to be capable of detecting hydrogen in the range of ppm at room temperature using very small current and voltage requirements and recover quickly after the source of hydrogen is removed.

ZnO nanorods were grown by nucleating on a Al_2O_3 substrate coated with Au islands [43]. For nominal Au film thicknesses of 20 Å, discontinuous Au islands are realized after annealing. The nanorods were deposited by Molecular Beam Epitaxy (MBE) with a base pressure of 5×10^{-8} mbar using high purity (99.9999%) Zn metal and an O_3/O_2 plasma discharge as the source chemicals. The Zn pressure was varied between 4×10^{-6} and 2×10^{-7} mbar, while the beam pressure of the O_3/O_2 mixture was varied between 5×10^{-6} and 5×10^{-4} mbar. The growth time was ~2 h at 600°C. The typical length of the resultant nanorods was 2~10 μm, with typical diameters in the range of 30–150 nm. Figure 14.1(top) shows a scanning electron micrograph of the as-grown rods. Selected area diffraction patterns showed the nanorods to be single-crystal. In some cases, the nanorods were coated with Pd, Pt, Au, Ni, Ag or Ti thin films (~100 Å thick) deposited by sputtering.

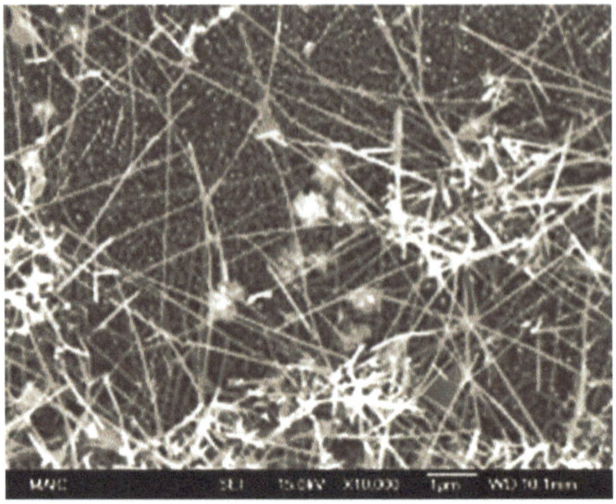

Figure 14.1: SEM of ZnO multiple nanorods.

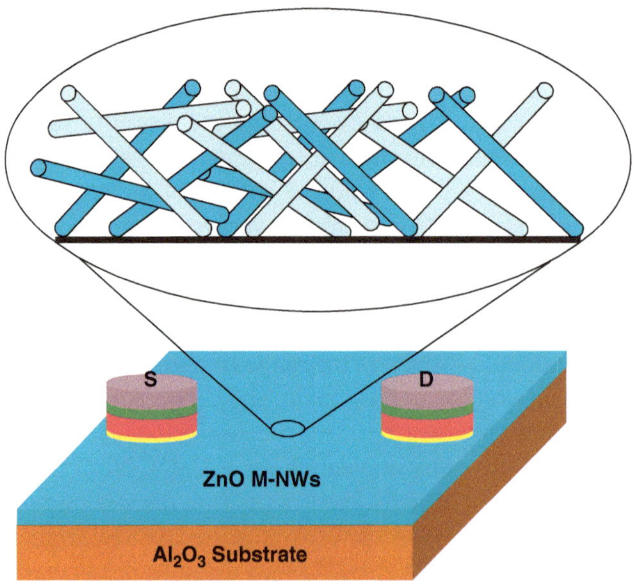

Figure 14.2: Schematic of contact geometry for multiple nanorod gas sensor.

Contacts to the multiple nanorods were formed using a shadow mask and sputtering of Al/Ti/Au electrodes. The separation of the electrodes was ~30 μm. A schematic of the resulting sensor is shown in Figure 14.2. Au wires were bonded to the contact pad for current-voltage (I-V) measurements performed at 25°C in a range of different ambients (N_2, O_2 or 10–500 ppm H_2 in N_2). Note that no currents

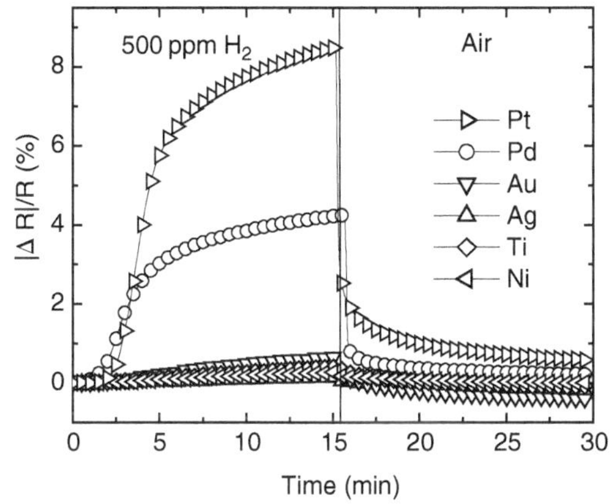

Figure 14.3: Time dependence of relative resistance response of metal-coated multiple ZnO nanorods as the gas ambient is switched from N_2 to 500 ppm of H_2 in air as time proceeds. There was no response to O_2.

were measured through the discontinuous Au islands and no thin film of ZnO on the sapphire substrate was observed with the growth condition for the nanorods. Therefore the measured currents are due to transport through the nanorods themselves. The I-V characteristics from the multiple nanorods were linear with typical currents of 0.8 mA at an applied bias of 0.5 V.

Figure 14.3 shows the time dependence of relative resistance change of either metal-coated or uncoated multiple ZnO nanorods as the gas ambient is switched from N_2 to 500 ppm of H_2 in air and then back to N_2 as time proceeds. These were measured a bias voltage of 0.5 V. The first point of note is that there is a strong increase (approximately a factor of 5) in the response of the Pt-coated nanorods to hydrogen relative to the uncoated devices. The maximum response was $\sim$8%. There is also a strong enhancement in response with Pd coatings, but the other metals produce little or no change. This is consistent with the known catalytic properties of these metals for hydrogen dissociation. Pd has a higher permeability than Pt but the solubility of H_2 is larger in the former [44]. Moreover, studies of the bonding of H to Ni, Pd and Pt surfaces have shown that the adsorption energy is lowest on Pt [45]. There was no response of either type of nanorod to the presence of O_2 in the ambient at room temperature. Once the hydrogen is removed from the ambient, the recovery of the initial resistance is rapid (<20 sec). By sharp contrast, upon introduction of the hydrogen, the effective nanorod resistance continues to change for periods of >15 min. This suggests that the kinetics of the chemisorption of molecular hydrogen onto the metal and its dissociation to atomic hydrogen are the rate-limiting steps in the resulting change in conductance of the ZnO [46]. An activation energy of 12 kJ/mol was calculated from a plot of the rate of change of

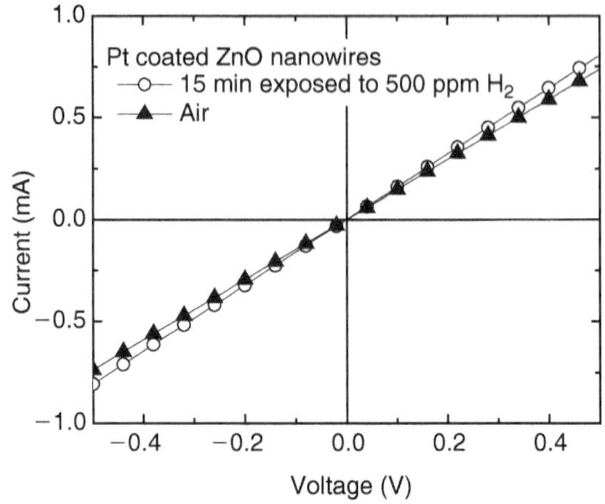

Figure 14.4: I-V characteristic of Pt-coated nanorods in air and after 15 min in 500 ppm H_2 in air.

nanorod resistance. This energy is somewhat larger than that of a typical diffusion process and suggests the rate-limiting step mechanism for this sensing process is more likely to be the chemisorption of hydrogen on the Pd surface. This reversible change in conductance of metal oxides upon chemisorption of reactive gases has been discussed previously [46]. The gas sensing mechanisms suggested in the past include the desorption of adsorbed surface hydrogen and grain boundaries in poly-ZnO [47], exchange of charges between adsorbed gas species and the ZnO surface leading to changes in depletion depth [48] and changes in surface or grain boundary conduction by gas adsorption/desorption [49]. Finally, Figure 14.3 shows an incubation time for response of the sensors to hydrogen. This could be due to some of the Pd becoming covered with native oxide which is removed by exposure to hydrogen. A potential solution is to use a bi-layer deposition of the Pd followed by a very thin Au layer to protect the Pd from oxidation. However, this adds to the complexity and cost of the process and, since the Pd is not a continuous film, the optimum coverage of Au would need to be determined. We should also point out that the I-V characteristics were the same when measured in vacuum as in air, indicating that the sensors are not sensitive to humidity.

The power requirements for the sensors were very low. Figure 14.4 shows the I-V characteristics measured at 25°C in both a pure N_2 ambient and after 15 min in a 500 ppm H_2 in N_2 ambient. Under these conditions, the resistance response is 8% and is achieved for a power requirement of only 0.4 mW. This compares well with competing technologies for hydrogen detection such as Pd-loaded carbon nanotubes [50,51]. Moreover, the 8% response compares very well to the existing SiC-based sensors, which operate at temperatures >100°C through an on-chip heater in order

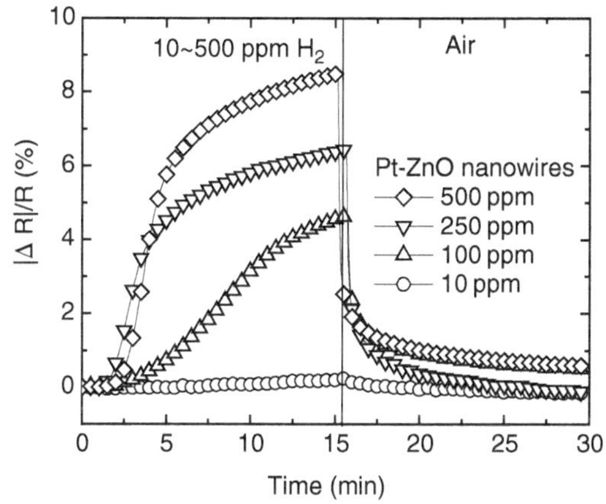

Figure 14.5: Time dependence of resistance change of Pd-coated multiple ZnO nanorods as the gas ambient is switched from N_2 to various concentrations of H_2 in air (10–500 ppm) and then back to N_2.

to enhance the hydrogen dissociation efficiency [50]. Figure 14.5 shows the sensors can detect 100 ppm H_2.

14.2.2 CO and C_2H_4 sensing

Pt/ZnO bulk Schottky diodes are capable of detection of low concentrations (1%) of carbon monoxide at temperatures >100°C and low concentrations ~10% of ethylene at temperatures between 50 and 300°C.

Figure 14.6(top) shows a schematic of the completed Pt/ZnO bulk Schottky diode, while Figure 14.6(bottom) shows a packaged sensor. The bulk ZnO crystals from Cermet, Inc. showed an electron concentration of $9 \times 10^{16}\,cm^{-3}$ and an electron mobility of $200\,cm^2/Vs$ at room temperature from van der Pauw measurements. The back (O-face) of the substrate was coated with full-area Ti (200 Å)/Al (800 Å)/Pt (400 Å)/Au (800 Å) by e-beam evaporation. After metal deposition, the samples were annealed in a Heatpulse 610 T system at 200°C for 1 min in N_2 ambient. The front face was deposited with plasma-enhanced chemical vapor deposited SiN_x at 100°C and windows opened by wet etching so that a thin (20 nm) layer of Pt could be deposited by e-beam evaporation. After the final metal of e-beam deposited Ti/Au (300 Å/1200 Å) interconnection contacts was deposited, the devices were bonded to electrical feedthroughs and exposed to different gas ambients in an environmental chamber while the diode current-voltage (I-V) characteristics were monitored, generally 8 min after introducing the CO and C_2H_4 partial pressure into

Figure 14.6: Schematic of ZnO Schottky rectifier (*top*) and photograph of packaged device (*bottom*).

the test chamber. Figure 14.6(top) shows a schematic of the completed device, while Figure 14.6(bottom) shows a packaged sensor.

The I-V characteristics at 150°C of the ZnO rectifiers are shown in Figure 14.7, for N_2 and 1% and 10% CO in N_2 ambients. The forward and reverse currents increase as the concentration of CO increases. In SnO_2 conductance devices [52], the detection of CO at 200–500°C was analyzed in terms of a surface reaction between absorbed CO and chemisorbed oxygen to produce CO_2. This leads to a charge transfer to SnO_2 and a change in the conductance. In the case of ZnO, the mechanism may be the same, but must occur by diffusion of oxygen through the Pt contact since the rest of the device is covered with SiN_x, which is impermeable to oxygen. The net effect is a reduction in effective barrier height, Φ_B, of ~20 meV at 150°C for 10% CO in the N_2 ambient.

The changes in forward bias at fixed current (20 mA) and forward current at fixed bias (1 V) from the rectifiers are shown in Figure 14.8 for different CO

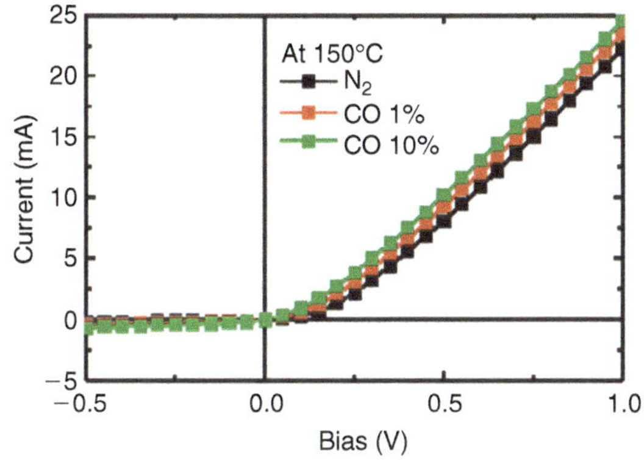

Figure 14.7: I-V characteristics of ZnO rectifier at 150°C in different ambients.

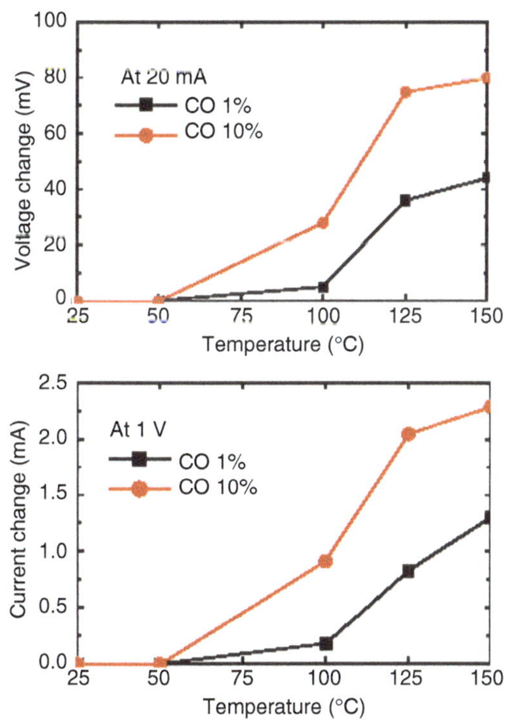

Figure 14.8: Change in voltage at fixed forward current (*top*) or change in current at fixed forward bias (*bottom*) for two different CO concentrations in N_2, as a function of measurement temperature.

concentrations in the ambient, as a function of temperature. The response is not detectable below ~90°C, but increases significantly with temperature. In the model of [52] for CO detection by SnO_2, if the temperature is too low then the CO product species will not be desorbed and therefore poisons the site for future adsorption of oxygen. The specific on-state resistance of the rectifier R_{ON} is given by

$$R_{ON} = (4V_B^2/\varepsilon\mu E_M^3) + \rho_s W_S + R_C,$$

where ε is the ZnO permittivity, μ the carrier mobility, ρ_S and W_S the substrate resistivity and thickness and R_C the contact resistance. Over a relatively limited range of CO partial pressures in the N_2 ambient, the on-state resistance decreased according to $R_{ON} = (R_O + A(P_{CO})^{0.5})^{-1}$, where A is a constant and R_O is the resistance in pure N_2 ambient. This is a similar dependence to the case of CO detection by SnO_2 conduction sensors, in which the effective conductance increased as the square root of CO partial pressure. The gas sensitivity can be calculated from the difference in resistances in CO containing ambients, divided by the resistance in pure N_2, i.e. $(R_{N2} - R_{CO})/R_{N2}$. At 150°C, the gas sensitivity was 4% for 1% CO in N_2 and 8% for 10% CO in N_2. These values are comparable to those reported for CO detection by SnO_2 conductance sensors [52].

Figure 14.9 shows the time dependence of the changes in forward current at fixed bias (0.4 V) for different temperatures. The increase in current shows a square-root dependence on time, suggesting that a diffusion process is occurring. The activation energy of this process was determined from the Arrhenius plot at the bottom of Figure 14.9, namely 40.7 kJ/mol. This may represent the effective activation energy needed to catalytically dissociate the CO on the Pt metal contact surface and then have diffusion of the atomic oxygen to the interface with the ZnO. It is much lower than the bond strength of the CO molecule, which is 1076.5 kJ/mol [53]. The detection mechanism is clearly still not firmly established in these devices and needs further study.

Figure 14.10 shows the I-V characteristics at 50 and 150°C of the Pt/ZnO diode both in pure N_2 and in ambients containing various concentrations of C_2H_4. At a given forward or reverse bias, the current increases upon introduction of the C_2H_4, through a lowering of the effective barrier height. One of the main mechanisms is once again the catalytic decomposition of the C_2H_4 on the Pt metallization, followed by diffusion to the underlying interface with the ZnO. In conventional semiconductor gas sensors, the hydrogen forms an interfacial dipole layer that can collapse the Schottky barrier and produce more Ohmic-like behavior for the Pt contact. The recovery of the rectifying nature of the Pt contact was many orders of magnitude longer than for Pt/GaN or Pt/SiC diodes measured under the same

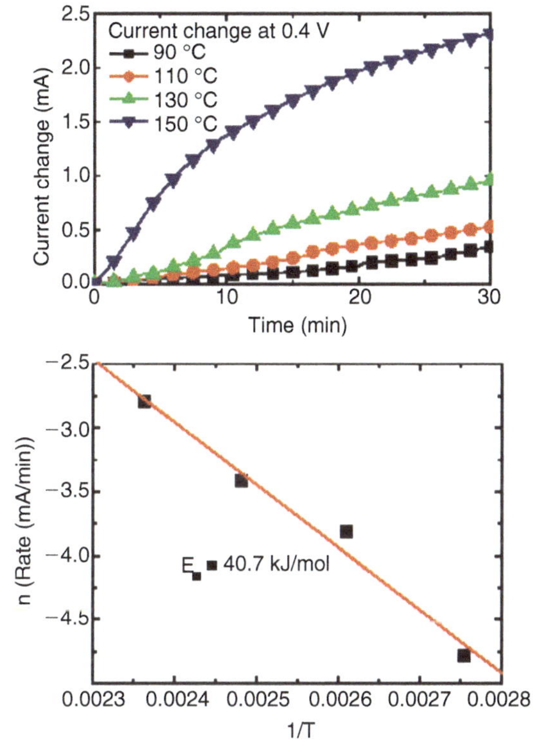

Figure 14.9: Time dependence of current change at fixed forward bias for measurement in 10% CO in N_2, for different measurement temperatures (*top*), and Arrhenius plot of time rate of change of current (*bottom*).

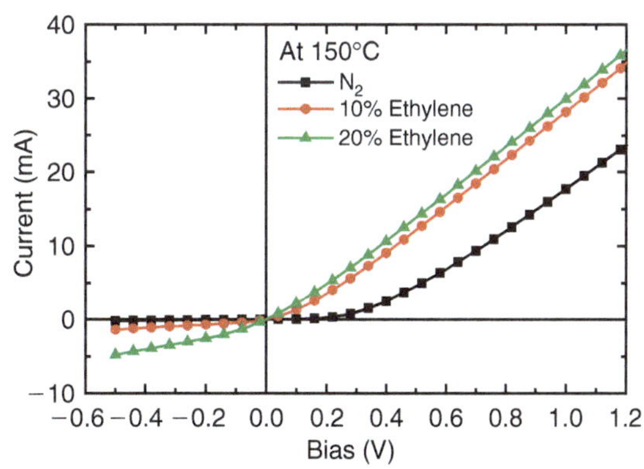

Figure 14.10: I-V characteristics at 150°C of Pt/ZnO diodes measured in different gas ambients.

conditions in the same chamber. For measurements over the temperature range 50–150°C, the activation energy for recovery of the rectification of the contact was estimated from the change in forward current at a fixed bias of 1.5 V. This was thermally activated through a relation of the type $I_F = I_O \exp(-E_a/kT)$ with a value for E_a of ~0.22 eV, comparable for the value of 0.17 eV obtained for the diffusivity of atomic deuterium in plasma exposed bulk ZnO. This suggests that at least some part of the change in current upon hydrogen gas exposure is due to in-diffusion of hydrogen shallow donors that increases the effective doping density in the near-surface region and reduces the effective barrier height.

14.2.3 Ozone sensing

Ozone gas ambient influences the current-voltage characteristics of multiple ZnO nanorods prepared by site-selective Molecular Beam Epitaxy (MBE). These structures can readily detect a few percent of ozone in N_2 at room temperature. Over a limited range of partial pressures of $O_3(P_{OZONE})$ in the ambient gas, the conductance G of the sensor at fixed bias voltage decreased according to the relation $G = (G_O + A(P_{OZONE})^{0.5})^{-1}$, where A is a constant and G_O the resistance in N_2.

The site selective growth of the nanorods has been described in a previous session. Selected area diffraction patterns showed the nanorods to be single-crystal. Figure 14.11 shows a transmission electron micrograph of a single ZnO nanorod.

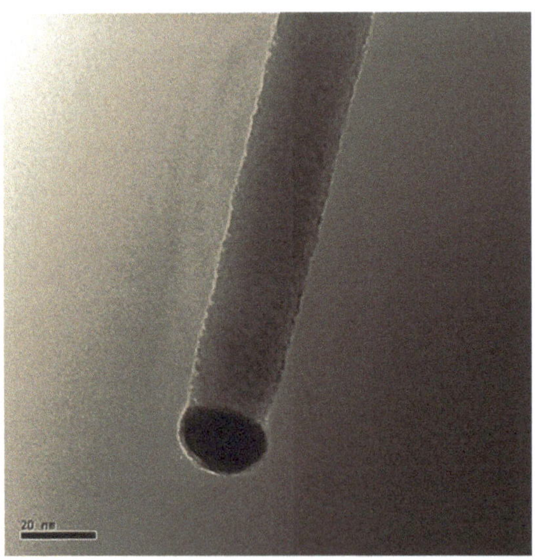

Figure 14.11: TEM of ZnO nanorod.

E-beam lithography was used to pattern sputtered Al/Ti/Au electrodes contacting both ends of multiple nanorods on Al_2O_3 using shadow mask. The separation of the electrodes was ~650 μm. A scanning electron micrograph of the completed device is shown in Figure 14.12. Au wires were bonded to the contact pad for current-voltage (I-V) measurements performed over the range 25–150°C in a range of different ambients (N_2, 3% O_3 in N_2). Note that no currents were measured through the discontinuous Au islands and no thin film of ZnO was observed with the growth condition of nanorods.

The nanorods were more sensitive to the presence of ozone in the measurement ambient. Figure 14.13 shows the room temperature I-V characteristics from the multiple nanorods measured in pure N_2 or 3% O_3 in O_2. The changes in current are much larger than for the case of hydrogen detection. Over a relatively limited range of O_3 partial pressures in the N_2 ambient, the conductance increased according

(a)
(b)
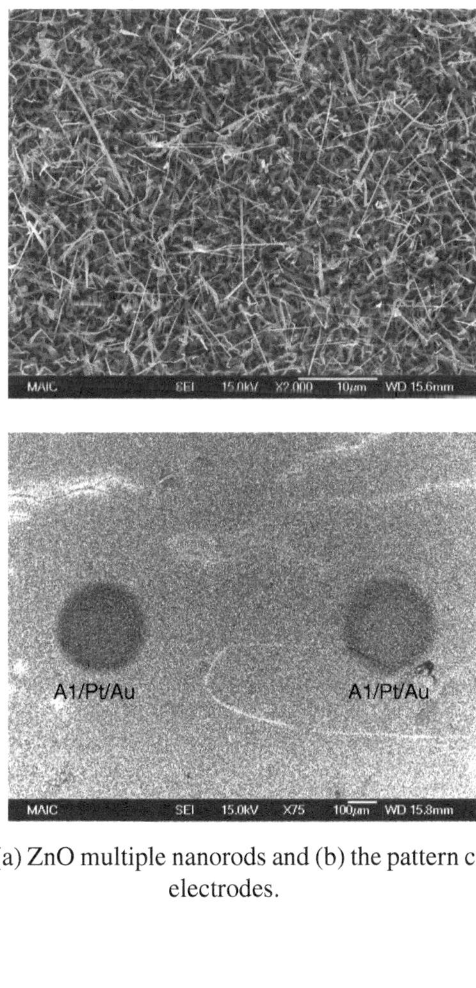

Figure 14.12: SEM of (a) ZnO multiple nanorods and (b) the pattern contacted by Al/Pt/Au electrodes.

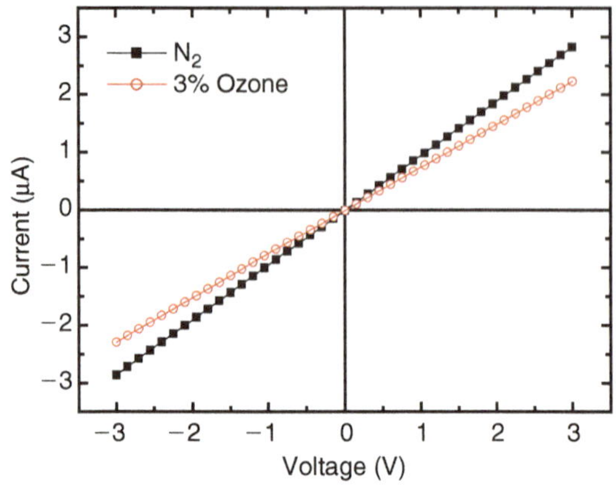

Figure 14.13: I-V characteristics at 25°C of ZnO multiple nanorods measured in either N_2 or 3% O_3 in O_2.

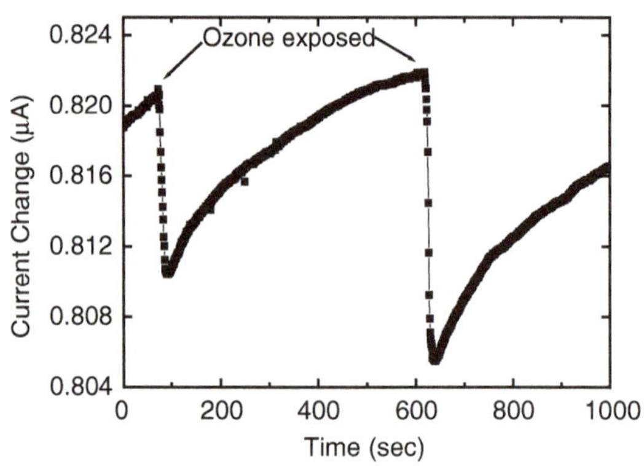

Figure 14.14: Time dependence of current at 1 V bias when switching back and forth from N_2 to 3% O_3 in O_2 ambients.

to $G = (G_O + A(P_{O_3})^{0.5})^{-1}$, where A is a constant and G_O the resistance in pure N_2 ambient. This is a similar dependence to the case of CO detection by SnO_2 conduction sensors, in which the effective conductance increased as the square root of CO partial pressure. The gas sensitivity can be calculated from the difference in conductance in O_3-containing ambients, divided by the conductance in pure N_2, i.e. $(G_{N_2} - G_O)/G_{N_2}$. At 25°C, the gas sensitivity was 21% for 3% O_3 in O_2.

Figure 14.14 shows the time dependence of change in current at a fixed voltage of 1 V when switching from back and forth from N_2 to 3% O_3 in O_2 ambients. The recovery time constant is long (>10 min) so that the nanorods are best suited to initial detection of ozone rather than to determining the actual time dependence of

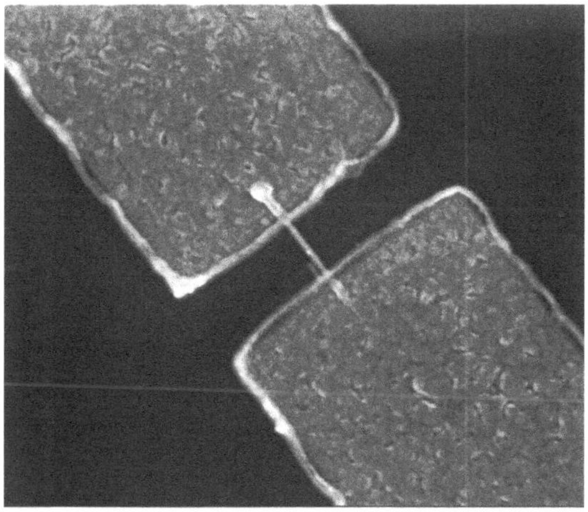

Figure 14.15: SEM micrograph of ZnO nanorod between two Ohmic contact pads of Al/Ti/Au.

change in concentration. In the latter case a much faster recovery time would be needed. The gas sweep-out times in our test chamber are relatively short (~a few seconds) and therefore the long recovery time is intrinsic to the nanorods.

14.2.4 UV sensing

E-beam lithography was used to pattern sputtered Al/Pt/Au electrodes contacting both ends of a single nanorod. The separation of the electrodes was ~0.3 μm. A scanning electron micrograph of the completed device is shown in Figure 14.15. Au wires were bonded to the contact pad for current-voltage (I-V) measurements performed at 25°C in air. In some cases, the nanorods were illuminated with above bandgap light at 254 or 366 nm from an Hg arc lamp with power density of $0.1\,\mathrm{W \cdot cm^{-2}}$.

Figure 14.16 shows the I-V characteristics of the nanorods in the dark and under illumination from 366 nm light. The conductivity is greatly increased as a result of the illumination, as evidenced by the higher current. No effect was observed for illumination with below bandgap light. The conductivity of the nanorod during illumination with 366 nm light was $0.2\,\Omega \cdot \mathrm{cm}$.

Figure 14.17 shows the photoresponse of the single ZnO nanorod at a bias of 0.25 V under pulsed illumination from a 366 nm wavelength Hg lamp. The photoresponse is much faster than that reported for ZnO nanorods grown by thermal evaporation from ball-milled ZnO powders [54] and likely is due to the reduced

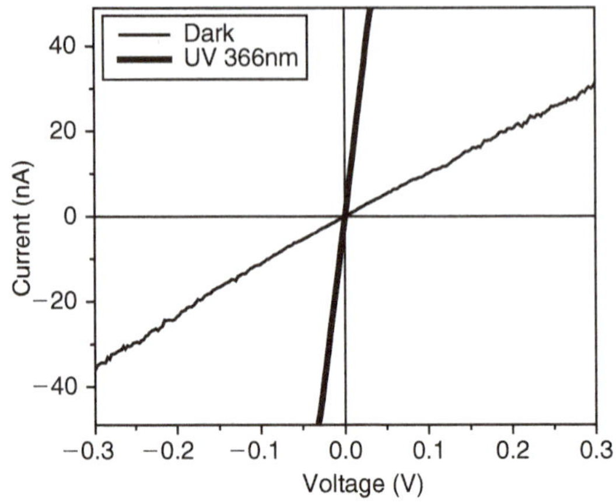

Figure 14.16: I-V characteristics from single ZnO nanorod measured at 25°C in the dark or under illumination from a 366 nm Hg lamp.

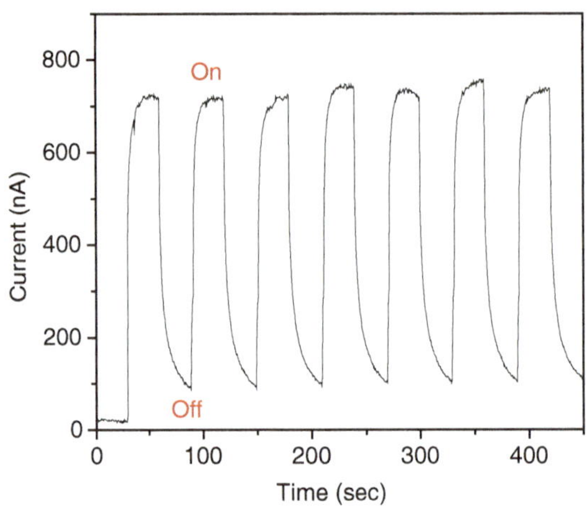

Figure 14.17: Time dependence of photocurrent at as the 366 nm light source is modulated.

influence of the surface states seen in that material. The generally quoted mechanism for the photoconduction is creation of holes by the illumination that discharge the negatively charged oxygen ions on the nanorod surface, with detrapping of electrons and transit to the electrodes [54–56]. The recombination times in high quality ZnO measured from time-resolved photoluminescence are short, on the order of tens of ps [57–59], while the photoresponse measures the electron trapping time

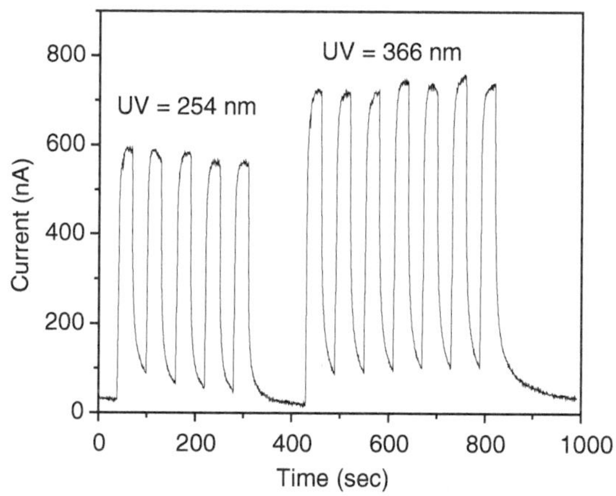

Figure 14.18: Time dependence of photocurrent in the nanorod as either a 254 nm or 366 nm light source is modulated.

[60]. There is also a direct correlation reported between the photoluminescence lifetime and the defect density in both bulk and epitaxial ZnO [59]. In our nanorods, the electron trapping times are on the order of tens of seconds and these trapping effects are only a small fraction of the total photoresponse recovery characteristic. Note also the fairly constant peak photocurrent as the lamp is switched on, showing that any traps present have discharged in the time frame of the measurement.

Figure 14.18 shows the photoresponse from the nanorods during pulsed illumination from either 254 or 366 nm light. The lower peak photocurrent in the former case may be related to the more efficient absorption near the surface of the nanorods.

E-beam lithography was used to pattern sputtered Al/Pt/Au electrodes contacting both ends of a single nanorod. The separation of the electrodes was ~3 μm. E-beam evaporated Pt/Au was used as the gate metallization by patterning a 1 μm wide strip orthogonal to the nanorod. A scanning electron micrograph of the completed device is shown in Figure 14.19. Au wires were bonded to the contact pads for current-voltage (I-V) measurements at 25°C. In some cases the diodes were illuminated with above band-edge ultra-violet (UV) light during the measurement.

Figure 14.20 shows I-V characteristics from the Pt/ZnO nanorod diode, measured both in the dark and with UV illumination. While the measurement in the dark shows rectifying behavior, the UV illumination produces an Ohmic characteristic. This is similar to the result reported by Keem et al. who showed that the above bandgap light lowered the potential barrier between the Schottky contact and the ZnO nanorod. In their case however, the photoresponse recovery times were very long (>10^4 sec), which was ascribed to the dominance of surface states, whereas our recovery times

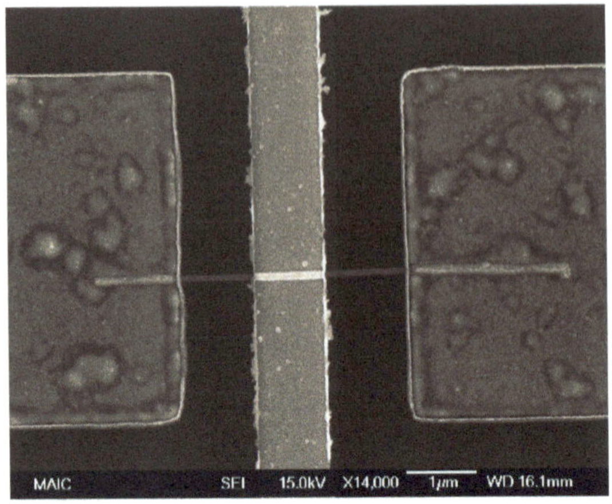

Figure 14.19: SEM micrograph of ZnO nanorod Schottky diode.

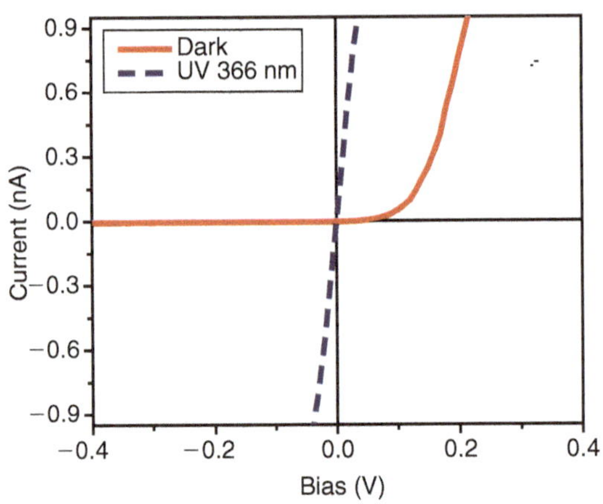

Figure 14.20: I-V characteristics from the ZnO Schottky nanorod diodes both in the dark and with UV illumination.

were limited by turn-off of the Hg lamp and suggest the conduction in our nanorods is dominated by bulk transport.

14.2.5 pH response

ZnO nanorod surfaces respond electrically to variations of the pH in electrolyte solutions introduced via an integrated microchannel. The ion-induced changes in

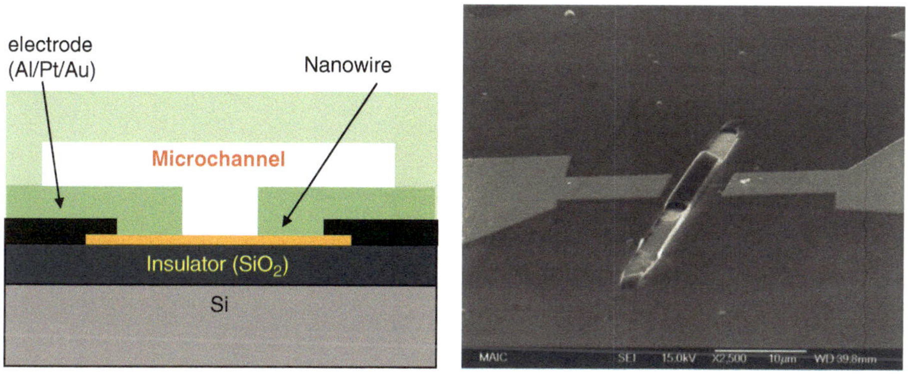

Figure 14.21: Schematic (*left*) and SEM (*right*) of ZnO nanorod with integrated microchannel (4 μm width).

surface potential are readily measured as a change in conductance of the single ZnO nanorods and suggest that these structures are very promising for a wide variety of sensor applications.

The preparation of the nanorods has been described in a previous section. They were released from the substrate by sonication in ethanol and then transferred to SiO_2-coated Si substrates. E-beam lithography was used to pattern sputtered Al/Pt/Au electrodes contacting both ends of a single nanorod. The separation of the electrodes was ~3.7 μm. Au wires were bonded to the contact pad for current-voltage (I-V) measurements to be performed. An integrated microchannel was made from SYLGARD@ 184 polymer from DOW CORNING. After mixing this silicone elastomer with curing agent using a weight ratio of 10:1 and mixing with for 5 min, the solution was evacuated for 30 min to remove residual air bubbles. It was then applied to the already etched Si wafer (channel length, 10–100 μm) in a cleaned and degreased container to make a molding pattern. Another vacuum de-airing for 5 min was used to remove air bubbles, followed by curing for 2 h at 90°C. After taking the sample from the oven, the film was peeled from the bottom of the container. Entry and exit holes in the ends of the channel were made with a small puncher (diameter less than 1 mm) and the film immediately applied to the nanorod sensor. The pH solution was applied using a syringe autopipette (2–20 μl). A schematic of the structure and a scanning electron micrograph of the completed device is shown in Figure 14.21.

Prior to the pH measurements, we used pH 4, 7, 10 buffer solutions from Fisher Scientific to calibrate the electrode and the measurements at 25°C were carried out in the dark or under ultra-violet (UV) illumination from 365 nm light using an Agilent 4156C parameter analyzer to avoid parasitic effects. The pH solution made

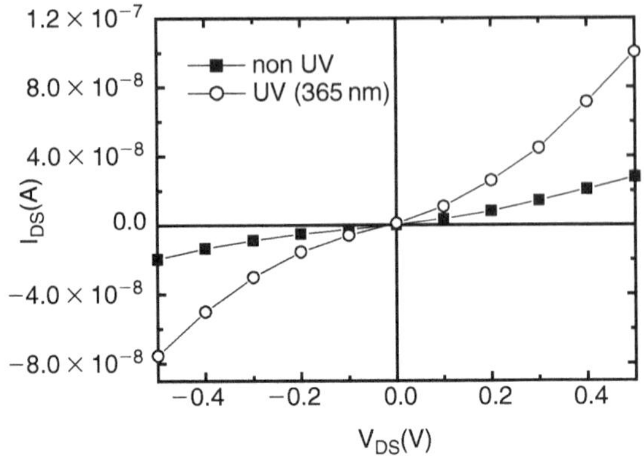

Figure 14.22: I-V characteristics of ZnO nanorod after wire-bonding, measured either with or without UV (365 nm) illumination.

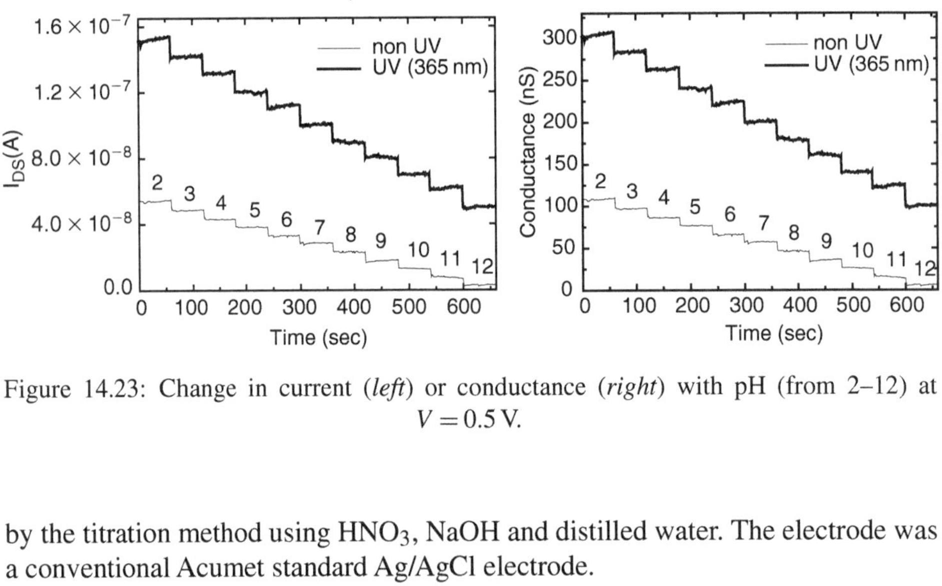

Figure 14.23: Change in current (*left*) or conductance (*right*) with pH (from 2–12) at $V = 0.5\,V$.

by the titration method using HNO_3, NaOH and distilled water. The electrode was a conventional Acumet standard Ag/AgCl electrode.

Figure 14.22 shows the I-V characteristics from the single ZnO nanorod after wire bonding, both in the dark and under UV illumination. The nanorods show a very strong photoresponse. The conductivity is greatly increased as a result of the illumination, as evidenced by the higher current. No effect was observed for illumination with below bandgap light. The photoconduction appears predominantly to originate in bulk conduction processes with only a minor surface trapping component [61].

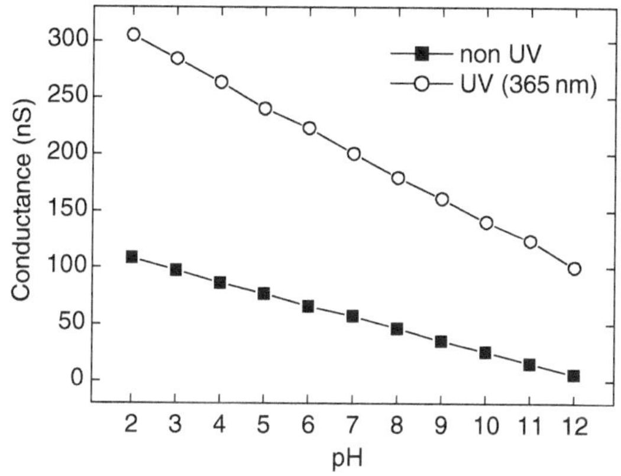

Figure 14.24: Relation between pH and conductance of ZnO nanorod either with or without UV (365 nm) illumination.

The adsorption of polar molecules on the surface of ZnO affects the surface potential and device characteristics. Figure 14.23(top) shows the current at a bias of 0.5 V as a function of time from nanorods exposed for 60 sec to a series of solutions whose pH was varied from 2 to 12. The current is significantly reduced upon exposure to these polar liquids as the pH is increased. The corresponding nanorod conductance during exposure to the solutions is shown at the bottom of Figure 14.23. The I-V in this figure was measured at pH at 7 (2.8×10^{-8} A without UV). The experiment was conducted starting at pH = 7 and went to pH = 2 or 12. The I-V measurement in the air was slightly higher that in the pH = 7 (10–20%). The data in Figure 14.23 shows that the HEMT sensor is sensitive to the concentration of the polar liquid and therefore could be used to differentiate between liquids into which a small amount of leakage of another substance has occurred. The conductance of the rods is higher under UV illumination but the percentage change in conductance is similar with and without illumination.

Figure 14.24 shows the conductance of the nanorods in either the dark or during UV illumination at a bias of 0.5 V for different pH values. The nanorods exhibit a linear change in conductance between pH 2–12 of 8.5 nS/pH in the dark and 20 nS/pH when illuminated with UV (365 nm) light. The nanorods show stable operation with a resolution of ~0.1 pH over the entire pH range, showing the remarkable sensitivity of the HEMT to relatively small changes in concentration of the liquid.

There is still much to understand about the mechanism of the current reduction in relation to the adsorption of the polar liquid molecules on the ZnO surface. It is clear that these molecules are bonded by van der-Waals type interactions and that they screen surface change that is induced by polarization in the ZnO. Different chemicals are likely to exhibit degrees of interaction with the ZnO surface. Steinhoff et al. [4] found a linear response to changes in the pH range 2–12 for ungated GaN-based transistor structures and suggested that the native metal oxide on the semiconductor surface is responsible.

14.3 Biological sensing

The development of advanced biological sensors could impact significantly the areas of genomics, proteomics, biomedical diagnostics, and drug discovery. In this regard, nanoscale sensors based on nanowires have received considerable recent attention. Nanowires can be used for label-free, direct real-time electrical detection of biomolecule binding and have the potential for very high sensitivity detection since the depletion or accumulation of charge carriers, which are caused by binding of charged macromolecules to receptors linked to the device surfaces, can affect the entire cross-sectional conduction pathway of these nanostructures and change conductance. In addition, electric measurements always have the convenience in making use of existing facility and equipment. In the following, we will discuss approaches where biosensors can be manufactured with the oxide nanostructures.

14.3.1 Surface modification of ZnO

The first step in biological sensing will be to identify those types of peptides which exhibit significant binding to the ZnO surface. An inorganic may be recognized physically or chemically through its surface composition, structure, crystallography or morphology. This effort focuses on the chemical ligand (amino acid) recognition of surfaces. Determination of this preferred recognition can be accomplished using a phage display library with multiple rounds of target binding, elution and amplification of the specifically bound phage. Analysis of the presence and position of certain amino acids should provide insight into the specific binding domains for a given inorganic or class of inorganics.

On metal surfaces, hydrophobic peptides with hydroxyl amino acids are prevalent. Tight binders on metal oxide surfaces, like ZnO, also show hydroxyl functional groups with a high concentration of basic amino acids such as arginine [62,63]. Similarly, GaAs was shown to possess selectivity for Lewis-base functional groups [64] Short peptide domains containing serine- and threonin-rich regions along with the presence of amine Lewis bases such as asparagine and glutamine were found to predominate. GaAs possesses a Lewis Acid surface and should also provide good interaction with basic amino acids. However, it also bears some similarity to ZnO which showed a surprising attraction for at least one peptide with no basic amino acids [62].

ZnO surfaces can be functionalized with bioreactive amine groups through a silanization reaction. Aminopropyl triethoxysilane (APS) is dissolved in water and reacted with clean ZnO and oxidized GaN surfaces. Condensation reactions between the silane and the hydroxyl groups on the metal oxide produce stable silicon-oxygen-metal bonds. The amine functionality on the silane allows a wide variety of bioconjugation reaction to be performed. Initial tests involved the addition of biotin to the surface with NHS chemistry. This immobilized biotin can undergo a nearly irreversible antigen-antibody bonding with streptavidin, a large protein (Figure 14.25, left).

To prove the effectiveness of the modification technique, a confocal microscope was used to image an amine coated GaN surface with regions of biotin modification. Solutions of ruthenium bipyridine (Rubpy) dye doped nanoparticles coated with either streptavidin or bovine serum albumin (BSA) were incubated on the GaN

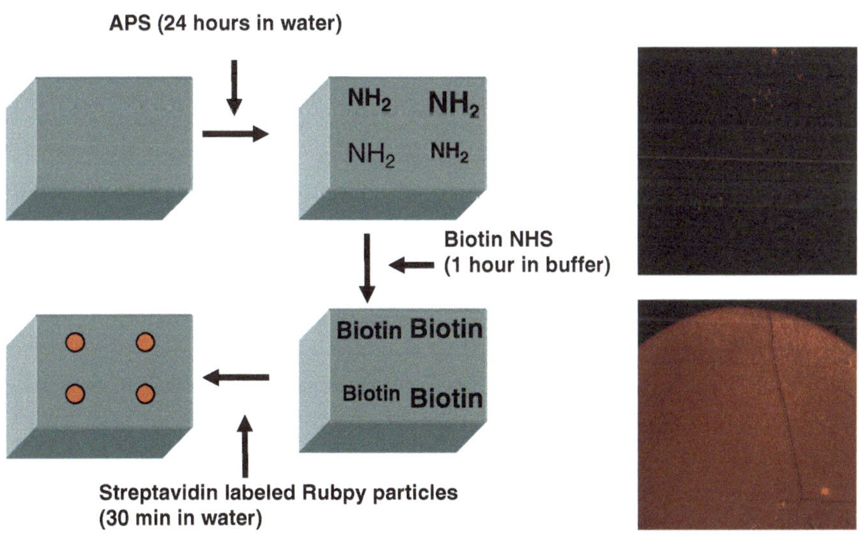

Figure 14.25: *Left*: Schematics of surface modification of nanowires for biomolecule immobilization for biosensor and biochip applications. *Right*: Preliminary results of surface modification: confocal images of a GaN and ZnO surface modified with avidin binding with Rubpy nanoparticles through biotin-avidin linkage (bright image); The same surface modified with BSA and showed no binding to the nanoparticles (dark image).

surface. To the right is an image of the streptavidin coated nanoparticles immobilized on the biotinylated surface and on the left is a biotinylated surface incubated with BSA coated particles. The biotinylated surface specifically reacted with the streptavidin and not with BSA, immobilizing the fluorescent particles. Non-biotinylated surfaces showed low background emission from the particles. After immobilization of the particles, washing in buffer solution did not reduce the fluorescence intensity. This indicates that modification with biotin creates a surface capable of specific and strong antigen-antibody conjugation.

The same chemistry will be applied to ZnO gates on FETs and on nanowires. The biotin-streptavidin system acts as a model for other antigen-antibody systems and can be extended to immobilization of oligonucleotide DNA sequences on to the surface [65–67]. Additional silane chemistries will be possible, including the attachment of thiol or carboxylic acid groups to the sensor surface [68]. With this chemistry, semiconductor sensors can be modified to specifically conjugate to a wide number of biomolecules of interest. Furthermore, attachment of the target analyte to another component, such as a molecule or a nanoparticle with a strong electron donating or accepting character, will allow the amplification of the conductivity change in the semiconductor.

14.3.2 Using protein immobilization on nanowire surface for ultrasensitive detection of proteins and single viruses

Viruses are among the most important causes of human disease and an increasing concern as agents for biological warfare and bioterrorism. Rapid, selective, and sensitive detection of viruses is central to implementing an effective response to viral infection, such as through medication or quarantine. Established methods for viral analysis include plaque assays, immunological assays, transmission electron microscopy, and PCR-based testing of viral nucleic acids [69,70]. These methods have not achieved rapid detection at single virus level and often require a relative high level of sample manipulation that is inconvenient for infectious materials. One promising approach for the direct electrical detection of biological macromolecules will be to use semiconducting nanowires.

A two-step procedure will be used to covalently link antibody receptors to the surfaces of the ZnO nanowire devices. First, the devices are reacted with a 1% ethanol solution of 3-(trimethoxysilyl) propyl aldehyde for 30 min, washed with ethanol, and heated at 120°C for 15 min. Second, mAb receptors are coupled to the aldehyde-terminated nanowire surfaces by reaction of 10–100 μg/ml antibody in a pH 8, 10 mM phosphate buffer solution containing 4 mM sodium cyanoborohydride.

The surface density of antibody is controlled by varying the reaction time from 10 min (low density) to 3 h (high density). Unreacted aldehyde surface groups are subsequently passivated by reaction with ethanolamine under similar conditions. Device arrays for multiplexed experiments are made in the same way except that distinct antibody solutions are spotted on different regions of the array. The antibody surface density vs. reaction time is quantified by reacting Au-labeled IgG antibodies with aldehyde-terminated nanowires on a transmission electron microscopy grid, and then imaging the modified nanowire by transmission electron microscopy, which enables the number of antibodies per unit length of nanowire to be counted.

Virus samples are delivered to the nanowire device arrays by using fluidic channels formed by either a flexible polymer channel or a 0.1-mm-thick glass coverslip sealed to the device chip. Virus samples are delivered through inlet/outlet connection in the polymer or holes made through the back of device chip in the case of the coverslip. Similar electrical results are obtained with both approaches, although the latter is used for all combined electrical/optical measurements.

The same strategy discussed above can be easily adopted for protein studies, where antibodies can be immobilized for target protein detection, making these nanowires suitable for proteomics applications. It is necessary to mention here that the protein modified nanowires will make multiplex analysis of proteins easy, producible, highly sensitive and quantitative.

14.3.3 Immobilization of nucleic acids on nanowires for biosensors for genes and mRNA

The ZnO nanowire devices are functionalized with DNA, PNA (peptide nucleic acid) or LNA (locked nucleic acid) probes using intervening avidin protein layer. This can be done through the surface modification discussed above. A variety of surface modification schemes have been developed for biomolecule immobilization [65–68]. PNA and LNA can be chosen as recognition sequences since they are known to bind to DNA/mRNA with much greater affinity and stability than corresponding DNA recognition sequences, and have shown binding with single base specificity. A microfluidic delivery system can be used to flow the complementary or mismatched DNA/mRNA samples. Direct comparison of the data will highlight the substantial net conductance change associated with hybridization of the DNA complementary to the receptor. This could enable high-throughput, highly sensitive DNA detection for basic biology research, disease diagnosis and genetic screening.

Figure 14.26: Schematic diagram of exposed nanowire and aptamer flow.

14.3.4 Direct immobilization of aptamers for ultrasensitive detection of proteins and drug molecules

Antibodies can be assembled onto nanowires as mentioned in previous part for protein recognition. Similarly, aptamers, designer DNA/RNA probes for target protein recognition and small molecular analysis [71,72], can also be immobilized. Upon the addition of target protein molecules, the device can function as ultrasensitive and selective detectors for protein recognition (Figure 14.26). Aptamers can also be used for analyzing small molecules like amino acids and cocaine [73]. If the aptamer for cocaine is immobilized onto the nanowire, an ultrasensitive biosensors for this substance will be possible.

14.3.5 ZnO nanowires with various doping and surface chemistry terminations

It will be a basic step to synthesize ZnO nanowires with varying transport properties (n-type to insulating to p-type) and surface termination. The former provides the opportunity to control the electrical response to surface absorption via depletion of the nanowire interior. The latter, surface termination will largely control both the chemistry of surface functionalization as well as electronic response to absorption events.

Electron doping will be accomplished via defects originating from Zn interstitials in the ZnO lattice [13]. The intrinsic defect levels that lead to n-type doping lie approximately 0.05 eV below the conduction band. Insulating material requires a reduction in oxygen vacancies in undoped material. One of the significant challenges with ZnO is the difficulty in producing p-type material. Recently, we have investigated the use of P as an acceptor dopant in ZnO. The capacitance-voltage properties of metal/insulator/P-doped (Zn,Mg)O diode structures were measured and found to exhibit a polarity consistent with the P-doped (Zn,Mg)O layer being p-type.

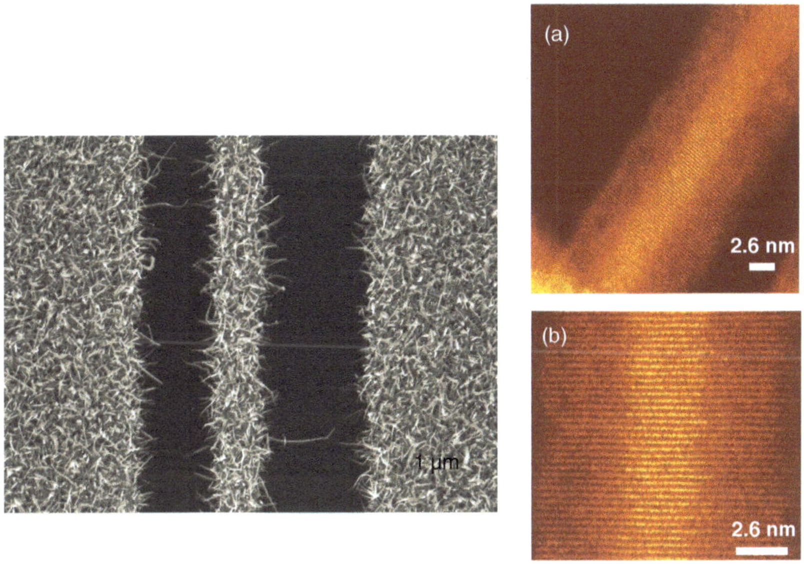

Figure 14.27: SEM (*left*) and Z-STEM (*right*) of cored ZnMgO nanorods grown on Ag-coated Si substrate.

For detectors that rely on surface absorption modulation of sub-surface transport, the ability to selectively terminate the surface of the nanowires with various cations is attractive. In thin-film semiconductor research, the formation of heteroepitaxial interfaces has proven to be enabling in the development of numerous device concepts, as well as in the investigation of low dimensional phenomena [74]. The synthesis of 1-D linear heterostructures is scientifically interesting and potentially useful [75], particularly if a technique is employed that allows for spatial selectivity in nanowire placement.

The heteroepitaxial cored nanostructures are based on the (Zn,Mg)O alloy system, and were synthesized using the catalysis-driven molecular beam epitaxy method shown in Figure 14.27. Doping ZnO with Mg widens the gap in this wurtzite structure compound. Site-selective nucleation and growth of cored nanorods was achieved by coating Si substrates with Ag islands. For a nominal Ag film thickness of 20 Å, discontinuous Ag islands are realized. On these small metal catalyst islands, (Zn,Mg)O nanorods were observed to grow. Under continuous Zn, Mg, and O flux, nanorod material nucleates on the catalysts particles. Close inspection of the nanorod structure indicates that the Zn and Mg concentrations are not uniformly distributed in the rods. Instead, there is a radial segregation into Zn-rich and Mg-rich regions, apparently reflecting differences in the solubility limits of bulk

ZnO-MgO or Zn-Mg-Ag-O versus epitaxial solid solutions. In bulk material, the solubility of Mg in ZnO is relatively low, on the order of 4 at.% [76]. In contrast, Mg content as high as $Zn_{0.67}Mg_{0.33}O$ has been reported to be metastable in the wurtzite structure for epitaxial thin films. For this composition, the bandgap of ZnO can be increased to ~3.8 eV. For (Zn,Mg)O nanorod growth, it appears that both growth modes are relevant, but for different regions in the rod. Under low temperature MBE growth conditions, a solubility-driven segregation occurs during the catalyst-driven core formation, with the core composition determined by bulk solid solubility. Subsequently, an epitaxial sheath grows with Mg content and crystal structure determined by epitaxial stabilization. The net result is the growth of (Zn,Mg)O nanorods that are not uniform in composition across the diameter, but distinctly cored. Figure 14.27(right) shows a high resolution Z-STEM image of a nanorod grown under the conditions described. The lattice image for the nanorod specimen indicates that the rod is crystalline with the wurtzite crystal structure maintained throughout the cross-section. The c axis is oriented along the long axis of the rod. The higher contrast for the center core region clearly indicates a higher cation atomic mass. The structures consist of a zinc-rich $Zn_{1-x}Mg_xO$ core (small x) surrounded by a $Zn_{1-y}Mg_yO$ (large y) sheath containing higher Mg content. Given the band offset between the ZnO core and (Zn,Mg)O sheath, the nanowire structure can easily function as an ISFET with the (Zn,Mg)O serving as the insulating dielectric. Using this approach, ZnO/(Zn,Mg)O sheathed nanowires can be synthesized for conductance measurements and surface functionalization.

14.3.6 A 3D self-consistent simulator for a biological sensor

Biomolecular sensing in the experiments is achieved by measuring change of electrical conductance through ZnO nanorods. In order to model the change of nanowire conductance due to the presence of the charge on biomolecules, a 3D Poisson equation is self-consistently solved with the carrier transport equation for the nanorod, in the presence of the charge on biomolecules.

Figure 14.28(left) shows simulation results. A three-dimensional Poisson equation is self-consistently solved with the carrier transport equation for a device structure as shown in Figure 14.21, in the presence of the biomolecular charge. Figure 14.10(right) plots the conduction band edge at equilibrium with different charge on a charge dot above the middle of the tube. It shows that when the charge dot is close to the nanorod surface (5 nm away), change of a single electron charge on the dot results in obvious change of the potential profile. The potential barrier created by the charge on the dot has significant effect on the carrier transport through a quasi-one-dimensional wire, which indicates that biological sensors with ultra-high

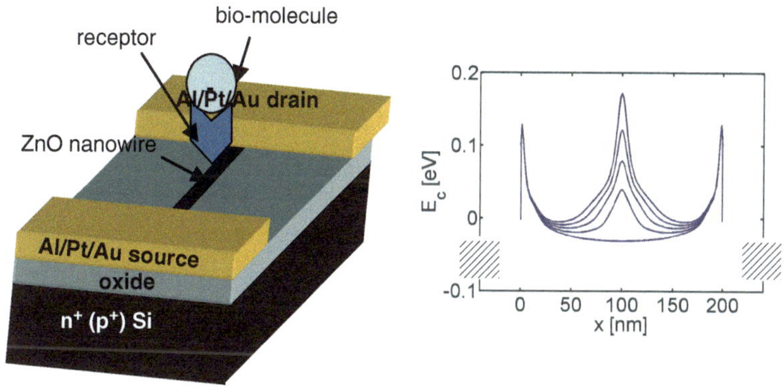

Figure 14.28: (*left*) The schematic picture of the modeled ZnO nanorod biological nanosensor, A device simulator that self-consistent solves 3D Poisson equation and the carrier transport equation is being developed to model the nanosensor. To consider electrostatic sensing, the biomolecule is modeled as a charge dot above the nanorod. (*right*) The conduction band profile for different amount of charges on the biomolecule. The channel length is 200 nm. A molecule with a uniform charge distribution with a diameter of 1 nm is placed 5 nm above the middle of the nanorod. The charge on the dot varies from 0 to $-4e$ at $-e$/step. The metal-nanorod Schottky barrier height is ~0.13 eV.

sensitivity (single-electron sensitivity) may be achieved by using a nanorod as the channel.

14.4 Conclusion

In this chapter, we reviewed recent progresses of ZnO in sensing the gas, chemical and biological agents. Pt-coated ZnO nanorods appear well suited to detection of ppm concentrations of hydrogen at room temperature. The recovery characteristics are fast upon removal of hydrogen from the ambient. Bulk ZnO Schottky diodes appear well suited to detection of CO and C_2H_4. They are sensitive at temperatures as low as 90°C for percentage levels of CO in N_2. The response is time dependent, with the characteristics of a diffusion-controlled process.

ZnO nanorods appear well-suited to detection of O_3. They are sensitive at temperatures as low as 25°C for percent levels of O_3 in O_2. The recovery characteristics are quite slow at room temperature, indicating that the nanorods can be used only for initial detection of ozone.

Single ZnO nanorods with excellent photocurrent response and current-voltage characteristics have been prepared by site selective MBE. The I-V characteristics are linear even when measured in the dark and the photoconduction appears predominantly to originate in bulk conduction processes with only a minor surface trapping component. These devices look very promising for UV detection.

ZnO nanorods show dramatic changes in conductance upon exposure to polar liquids. Bonding of polar liquid molecules appear to alters the polarization-induced positive surface change, leading to changes in the effective carrier density and hence the drain-source current in biased nanorods. This suggests the possibility of functionalizing the surface for application as biosensors, especially given the excellent biocompatibility of the ZnO surfaces which should minimize degradation of adsorbed cells.

The sensitivities of ZnO in gas, chemical, and biological agents make it attractive for senor applications. The ZnO nanorods can be placed on cheap transparent substrates such as glass, making them attractive for low-cost sensing applications, and can operate at very low power conditions. Of course, there are many issues still to be addressed, in particular regarding the reliability and long-term reproducibility of the sensor response before it can be considered for space-flight applications.

Acknowledgments

The authors would like to thank B.S. Kang, H.T. Wang, L.C. Tien, Suku Kim, C.R. Abernathy, B.P. Gila, K. Pruessner, K.N. Seibin, M. Kaufman, M.F. Chisholm, and S.J. Pearton.

References

[1] O. Ambacher, M. Eickhoff, G. Steinhoff, M. Hermann, L. Gorgens, V. Werss, et al., Proc ECS, 214, 27 (2002).

[2] R. Neuberger, G. Muller, O. Ambacher, M. Stutzmann, Phys. Stat. Solidi A 185, 85 (2001).

[3] J. Schalwig, G. Muller, O. Ambacher, M. Stutzmann, Phys. Stat. Solidi A 185, 39 (2001).

[4] G. Steinhoff, M. Hermann, W.J. Schaff, L.F. Eastmann, M. Stutzmann, M. Eickhoff, Appl. Phys. Lett. 83, 177 (2003).

[5] M. Eickhoff, R. Neuberger, G. Steinhoff, O. Ambacher, G. Muller, M. Stutzmann, Phys. Stat. Solidi B 228, 519 (2001).

[6] J. Schalwig, G. Muller, M. Eickhoff, O. Ambacher, M. Stutzmann, Sens. Actuat. B 81 425 (2002).

[7] M. Eickhoff, J. Schalwig, G. Steinhoff, R. Neuberger, M. Hermann, O. Ambacher, M. Stutzmann, Phys. Stat. Solidi C 6, 1908 (2003).

[8] R. Mehandru, B. Luo, B.S. Kang, J. Kim, F. Ren, S.J. Pearton, C.C. Fan, G.T. Chen, J.I. Chyi, Solid State Electronics 48, 351 (2004).
[9] O. Ambacher, M. Eickhoff, A. Link, M. Hermann, M. Stutzmann, F. Bernardini, V. Tilak, L.F. Eastmann, Phys. Stat. Solidi C 6, 1878 (2003).
[10] S.J. Pearton, B.S. Kang, S. Kim, F. Ren, B.P. Gila, C.R. Abernathy, J. Lin, S.N.G. Chu, J. Phys.: Condensed Matter 16, R961 (2004).
[11] W.P. Kang, C.K. Kim, Sens. Actuat. B 13–14, 682 (1993).
[12] B.G. Miler, T.W. Trautman, R. Wolfenden, J. Am. Chem. Soc. 120, 2666 (1998).
[13] D.C. Look, J.W. Hemsky, J.R. Sizelove, Phys. Rev. Lett. 82, 2552 (1999).
[14] H.L. Hartnagel, A.L. Dawar, A.K. Jain, C. Jagadish, *Semiconducting Transparent Thin Films* (IOP, Bristol, 1995).
[15] K.D. Mitzner, J. Sternhagen, D.W. Galipeau, Sens. Actuat. B 93, 92 (2003).
[16] P. Mitra, A.P. Chatterjee, H.S. Maiti, Mater. Lett. 35, 33 (1998).
[17] J.F. Chang, H.H. Kuo, I.C. Leu, M.H. Hon, Sens. Actuat. B 84, 258.
[18] D.E. Yales, S. Levine, T.W. Healy, J. Chem. Faraday Trans. 70, 1807 (1974).
[19] W.M. Siu, R.S.C. Collold, IEEE Trans. Electron. Dev. ED-26, 1805 (1979).
[20] L. Bousse, N.F. DeRooij, P. Bergbeld, IEEE Trans. Electron. Dev. ED-30, 1263 (1983).
[21] C.D. Fung, P.W. Cheng, W.H. Ho, IEEE Trans. Electron. Dev. ED-33, 8 (1986).
[22] Q.H. Li, Q. Wan, Y.X. Liang, T.H. Wang, Appl. Phys. Lett. 84, 4556 (2004).
[23] K. Keem, H. Kim, G.T. Kim, J.S. Lee, B. Min, K. Cho, M.Y. Sung, S. Kim, Appl. Phys. Lett. 84, 4376 (2004).
[24] B.H. Kind, H. Yan B. Messer, M. Law, P. Yang, Adv. Mater. 14, 158 (2002).
[25] C.H. Liu, W.C. Liu, F.C.K. Au, J.X. Ding, C.S. Lee, S.T. Lee, Appl. Phys. Lett. 83, 3168 (2003).
[26] W.I. Park, G.C. Yi, J.W. Kim, S.M. Park, Appl. Phys. Lett. 82, 4358 (2003).
[27] J.J. Wu, S.C. Liu, Adv. Mater. 14, 215 (2002).
[28] H.T. Ng, J. Li, M.K. Smith, P. Nguygen, A. Cassell, J. Han, M. Meyyappan, Science 300, 1249 (2003).
[29] M.H. Huang, S. Mao, H. Feick, H. Yan, Y. Wu, H. Kind, E. Weber, R. Russo, P. Yang, Science 292, 1897 (2001).
[30] J.Q. Hu, Y. Bando, Appl. Phys. Lett. 82, 1401 (2003).
[31] W.I. Park, G.C. Yi, M.Y. Kim, S.J. Pennycook, Adv. Mater. 15, 526 (2003).
[32] Y.W. Heo, V. Varadarjan, M. Kaufman, K. Kim, D.P. Norton, F. Ren, P.H. Fleming, Appl. Phys. Lett. 81, 3046 (2002).
[33] M. He, M.M.E. Fahmi, S. Noor Mohammad, R.N. Jacobs, L. Salamanca-Riba, F. Felt, M. Jah, A. Sharma, D. Lakins, Appl. Phys. Lett. 82, 3749 (2003).
[34] M.J. Zheng, L.D. Zhang, G.H. Li, X.Y. Zhang, X.F. Wang, Appl. Phys. Lett. 79, 839 (2001).
[35] S.C. Lyu, Y. Zhang, H. Ruh, H.J. Lee, H.W. Shim, E.K. Suh, C.J. Lee, Chem. Phys.
[36] B.P. Zhang, N.T. Binh, Y. Segawa, K. Wakatsuki, N. Usami, Appl. Phys. Lett. 83, 1635 (2003).
[37] Z.W. Pan, Z.R. Dai, Z.L. Wang, Science 291, 1947 (2001).
[38] J.Y. Lao, J.Y. Huang, D.Z. Wang, Z.F. Ren, Nano Lett. 3, 235 (2003).

[39] G.W. Hunter, P.G. Neudeck, R.S. Okojie, G.M. Beheim, V. Thomas, L. Chen, D. Lukco, C.C. Liu, B. Ward, D. Makel, Proc. ECS Vol. 01–02, 212 (2002).

[40] L.Y. Chen, G.W. Hunter, P.G. Neudeck, D.L. Knight, C.C. Liu, Q.H. Wu, Proceedings of the Third International Symposium on Ceramic Sensors, eds H.U. Anderson, M. Liu, and N. Yamazoe, Electrochemical Society Inc., Pennington, NJ, pp. 92–98 (1996).

[41] G.W. Hunter, C.C. Liu, D. Makel, MEMS Handbook, ed. M.G. Hak (CRC Press, Boca Raton, 2001).

[42] L. Chen, G.W. Hunter, P.G. Neudeck, J. Vac. Sci. Technol. A 15, 1228 (1997); J. Vac. Sci. Technol. A 16, 2890 (1998).

[43] Y.W. Heo, D.P. Norton, L.C. Tien, Y. Kwon, B.S. Kang, F. Ren, S.J. Pearton, J.R. LaRoche, Mat. Sci. Eng. R 47, 1 (2004).

[44] See the databases at http://www.rebresearch.com/H2perm2.htm and http://www.rebresearch.com/H2sol2.htm.

[45] W. Eberhardt, F. Greunter, E.W. Plummer, Phys. Rev. Lett. 46, 1085 (1981).

[46] A.A. Tomchenko, G.P. Harmer, B.T. Marquis, J.W. Allen, Sens. Actuat. B 93, 126 (2003).

[47] K.D. Mitzner, J. Sternhagen, D.W. Galipeau, Sens. Actuat. B 93, 92 (2003).

[48] P. Mitra, A.P. Chatterjee, H.S. Maiti, Mater. Lett. 35, 33 (1998).

[49] J.F. Chang, H.H. Kuo, I.C. Leu, M.H. Hon, Sens. Actuat. B 84, 258 (1994).

[50] Y. Lu, J. Li, H.T. Ng, C. Binder, C. Partridge, M. Meyyapan, Chem. Phys. Lett. 391, 344 (2004).

[51] I. Sayago, E. Terrado, E. Lafuente, M.C. Horillo, W.K. Maser, A.M. Benito, R. Navarro, E.P. Urriolabeita, M.T. Martinez, J. Gutierrez, Syn. Metals 148, 15 (2005).

[52] H.L. Hartnagel, A.L. Dawar, A.K. Jain, C. Jagadish, *Semiconducting Transparent Thin Films* (IOP, Bristol, 1995).

[53] H. Windischmann, P. Mark, J. Electrochem. Soc. 126, 627 (1979).

[54] K. Keem, H. Kim, G.T. Kim, J.S. Lee, B. Min, K. Cho, M.Y. Sung, S. Kim, Appl. Phys. Lett. 84, 4376 (2004).

[55] Y. Takahashi, M. Kanamori, A. Kondoh, H. Minoura, Y. Ohya, Jpn. J. Appl. Phys. 33, 6611 (1994).

[56] P. Sharma, K. Sreenivas, K.V. Rao, J. Appl. Phys. 93, 3963 (2003).

[57] U. Ozgur, A. Teke, C. Liu, S.J. Cho, H. Morkoc, H.O. Everitt, Appl. Phys. Lett. 84, 3223 (2004).

[58] B. Gou, Z.R. Qiu, K.S. Wong, Appl. Phys. Lett. 82, 2290 (2003).

[59] T. Koida, S.F. Chichibu, A. Uedono, A. Tsukazaki, M. Kawasaki, T. Sota, Y. Segewa, H. Koinuma, Appl. Phys. Lett. 82, 532 (2003).

[60] S.A. Studenikin, N. Golego, M. Cocivera, J. Appl. Phys. 87, 241 (2000).

[61] B.S. Kang, Y.W. Heo, L.C. Tien, F. Ren, D.P. Norton, S.J. Pearton, Appl. Phys. A 80, 497 (2005).

[62] C. Nguyen, J. Dai, M. Darikaya, D.T. Schwartz, F. Baneyx, J. Am. Chem. Soc. (2003).

[63] K. Kjaergaard, J.K. Sorensen, M.A. Schembri, P. Klemm, Appl. Environ. Microbiol. 66, 10 (2000).

[64] S.R. Whaley, D.S. English, E.L. Hu, P.F. Barbara, A.M. Belcher, Nature 405, 665 (2000).
[65] X. Fang, X. Liu, S. Schuster, W. Tan, J. Am. Chem. Soc. 121, 2921–2922 (1999a).
[66] W. Tan, K. Wang, T. Drake, Curr. Opin. Chem. Biol. 8 (5), 547–553 (2004).
[67] P. Kohli, C. Harrell, S. Miller, Z. Cao, R. Gasparac, W. Tan, C. Martin, Science 305, 984–986 (2004).
[68] W. Tan, K. Wang, X. He, J. Zhao, T. Drake, L. Wang, R. P. Bagwe, Medicinal Research Reviews 24 (5), 621–638 (2004).
[69] J.C. Mohle-Boetani, J.A. Farrar, S.B. Werner, D. Minassian, R. Bryant, S. Abbott, L. Slutsker, D.J. Vugia, Annals of Internal Medicine 135, 239–247 (2001).
[70] X. Zhao, L.R. Hilliard, S.J. Mechery, Y. Wang, R. Bagwe, S. Jin, W. Tan, Proceedings of National Academy of Sciences 101, 15027–15032 (2004).
[71] M.C. Vicens, A. Sen, A. Vanderlaan, T. Drake, W. Tan, ChemBioChem 5 (2005).
[72] M.N. Stojanovic, P. de Prada, D.W. Landry, J. Am. Chem. Soc. 123, 4928–4931 (2001).
[73] X. Fang, A. Sen, M. Vicens, W. Tan, ChemBioChem 4, 829–834 (2003).
[74] F. Capasso, Science 235, 172 (1987).
[75] L.J. Lauhon, M.S. Gudiksen, D. Wang, C.M. Lieber, Nature 420, 57 (2002).
[76] J.F. Sarver, F.L. Katnack, F.A. Hummel, J. Electrochem. Soc. 106, 960 (1959).

Chapter 15

ZnO-Based Light Emitters

Andrei Osinsky

SVT Associates, Inc., Eden Prairie, MN 55344, USA

Sergey Karpov

STR, Inc., P.O. Box 70604, Richmond, VA 23255-0604, USA;
Soft-Impact, Ltd., P.O. Box 83, St. Petersburg, 194156 Russia

Abstract: In this chapter, we review state-of-the-art ZnO-based light emitters with the primary focus on physical aspects of their operation and novel technical solutions. Since ZnO laser diodes have not been demonstrated to date, we confine our discussion to light-emitting diodes (LEDs). Both homojunction and hybrid LEDs that incorporate in their structure alternative materials such as III-nitrides or SiC are considered in the review. Due to special characteristics of the ZnO-based LEDs related to spontaneous electric polarization inherent in wurtzite semiconductors and type-II band alignment at ZnO/III-N interfaces, we employ numerical simulations to get a better understanding of physical mechanisms involved in the device operation and factors affecting the device characteristics.

We consider both single- and double-heterostructure LEDs and determine critical factors in device design. Among these factors, interfacial charges due to spontaneous electric polarization have a significant impact on the ZnO-based LED operation. In combination with the type-II band alignment at a ZnO/AlGaN heterojunction, the polarization charge may provide an effective carrier confinement near the ZnO/AlGaN interface that, in turn, may result in tunneling radiative recombination of electrons and holes on opposite sides of the interface. The tunneling emission spectra are found to correlate well with those observed in a number of ZnO-based LEDs.

From comparative analysis, some general problems facing ZnO-based light emitters are revealed: (i) a considerable imbalance between the electron and hole partial currents, resulting in additional carrier losses on the contacts, (ii) a high LED series resistance that is not necessarily related to a low conductivity of p-doped layers, (iii) poor control of the emission spectra in the devices, etc. Some ways of improving LED performance which address the above problems as well some still open issues are discussed in the review.

Zinc Oxide Bulk, Thin Films and Nanostructures
C. Jagadish and S. Pearton (Editors)

15.1 Introduction

The first efficient light emitters, LEDs and laser diodes (LDs) operating in the blue/green spectral range were fabricated from cubic II-VI (MgZnCdSSe) compounds [1]. Although these LEDs and LDs exhibited excellent optical and electrical characteristics of [1,2], they failed to achieve long-term operation due to strong degradation processes commonly attributed to the extremely high mobility of native point defects in the II-VI semiconductors. A great deal of effort was spent toward improving the lifetime of the II-VI light emitters, but the problem has not been overcome. Due to the breakthroughs in InGaN-LED technology achieved in the early 90s [3,4], III-nitride LEDs have replaced II-VI devices, and AlGaInN compounds have become the standard materials for light emitters in the blue/green spectral range. In contrast to II-Vs semiconductors, III-nitrides provided not only superior device performance, but also a lifetime long enough to be useful in industrial applications [5,6]. More recently, the spectral range of the III-nitride LEDs has been extended to the near-UV (340–400 nm) [7,8] and the deep-UV (250–300 nm) [9] regimes. Earlier investigations indicated that III-nitride epitaxial materials had an inherent problem with high dislocation density due to the lack of native GaN or AlN substrates [10]. By using specially designed buffer layers and GaN "templates" (i.e. thick epilayers normally grown by hydride vapor phase epitaxy) the typical threading dislocation density of $\sim 10^9\,cm^{-2}$ of heterostructures grown by MOCVD on commercial sapphire or SiC substrates can be reduced to $\sim 1\text{–}2 \times 10^8\,cm^{-2}$ and $\sim 10^7\,cm^{-2}$, respectively. A dislocation density of $\sim 10^7\text{–}10^8\,cm^{-2}$ is still too high, however, to eliminate their impact on the internal quantum efficiency (IQE) of III-nitride LEDs [11,12]. This problem is especially crucial for In-free UV LEDs which lack the beneficial effects that In-composition fluctuations have on radiative recombination efficiency. In addition to IQE, the threading dislocations affect the overall quality of epitaxial materials, hinder layer-by-layer growth, and reduce the lifetime of devices operating at high current density, i.e., of high-power LEDs and LDs. To date, the problem with high dislocation densities in III-nitride heterostructures has not yet been completely resolved.

A renaissance for II-VI semiconductors took place at the beginning of this century due to developments in ZnO technology and related MgZnO and CdZnO compounds. Zinc oxide has a bandgap of 3.37 eV at room temperatures, which is suitable for light emission in the near-UV (~370 nm) spectral range. MgZnO has a wider bandgap and a relatively low lattice constant mismatch with ZnO, e.g., ~0.4% for $Mg_{0.2}Zn_{0.8}O$ (Fig. 15.1a). Thus, MgZnO/ZnO/MgZnO heterostructures with effective carrier confinement in the ZnO active region can be fabricated. Moreover, by introducing a CdZnO active layer the emission spectral range may be extended into the visible regime.

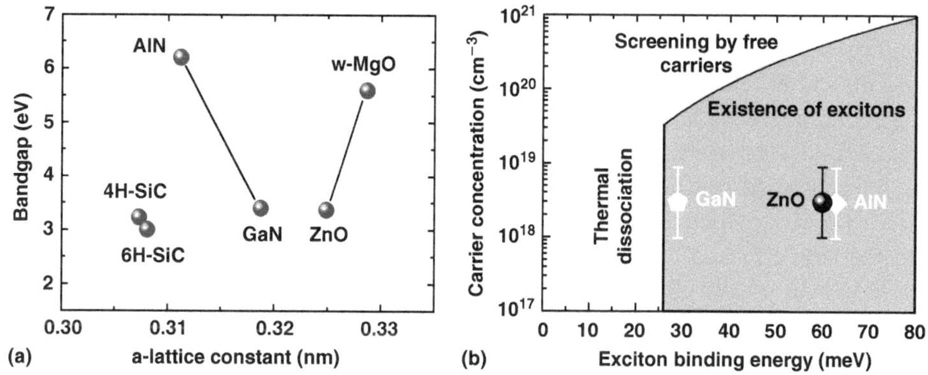

Figure 15.1: Bandgap as a function of a-lattice constant for wurtzite semiconductors (a) and diagram of A-exciton existence/dissociation at room temperature calculated for dielectric permittivity $\varepsilon = 8.8$ (b).

One important advantage offered by ZnO is its high exciton binding energy of ~60 meV, an advantage further enhanced by introducing ZnO/MgZnO quantum wells [13]. Using the well-known relationship between the exciton Bohr radius a_B and its binding energy E_b, $E_b = q^2/2\varepsilon a_B$, where q is the electron charge and ε is the dielectric permittivity of the material, we estimated the conditions for the exciton existence in ZnO at room temperature (see Fig. 15.1b). An exciton dissociates due to either thermal ionization, at $E_b < kT$, or screening by free carriers, at $na_B^3 < 1$, according to the Mott criterion. Here, k is the Boltzmann constant, T is the temperature, and n is the free carrier concentration. It can be seen from Fig. 15.1 that under typical LED operating conditions an exciton in ZnO ($n \sim 10^{18}$–10^{19} cm^{-3}) lies far from the dissociation boundaries. This means that the excitonic radiative recombination in ZnO is important at room temperature and can lead to the fabrication of highly efficient light emitters.

An additional advantage of ZnO is the availability of bulk ZnO crystals suitable for the fabrication of homoepitaxial ZnO substrates [14]. The history of III-nitride materials demonstrates clearly how critical such homoepitaxial substrates are for reducing the defect density in epitaxial materials. In contrast to group-III nitrides, ZnO and related compounds can be potentially grown on homoepitaxial substrates, leading to a low dislocation density in device heterostructures.

To date, however, the potential for ZnO has not been fully achieved, primarily because of two principal obstacles. One is related to the physical nature of the ternary MgZnO and CdZnO alloys. Among the binary oxides, ZnO, CdO, and MgO, zinc oxide is the only one with a thermodynamically stable wurtzite phase, while MgO and CdO crystallize in the cubic rock-salt lattice structure. This difference in the crystal symmetry results in phase separation for alloys with intermediate composition. This limits, in particular, the allowable composition range of wurtzite $Mg_xZn_{1-x}O$ to $x \approx 0.15$ [15] and, hence, the available conduction and valence band offsets at the ZnO/MgZnO interface. The second main obstacle is the lack of a reliable and stable p-doping for ZnO and, especially, MgZnO. Despite the successful demonstration of ZnO p-doping with P, As and N (see the review [16] and references therein), the hole concentration in the p-ZnO is still limited to ~2–3×10^{17} cm^{-3}, which is insufficiently high to provide efficient carrier injection into the active region of an LED or a LD. While the first obstacle can be partly resolved by a proper choice of the ternary alloy composition, the second one still remains a major roadblock in the development of the ZnO-based LEDs and LDs. One possible solution for overcoming the problem with ZnO p-doping is the use of hybrid ZnO/AlGaN or more complex II-O/III-N heterostructures as suggested recently by a number of research groups.

In this chapter we review the state-of-the-art ZnO-based light emitters with our primary focus on physical aspects of their operation and novel technical solutions. We concentrate, first of all, on LEDs, since ZnO LDs have not yet been demonstrated. Since the experimental results for ZnO LEDs are rather limited, we employ numerical simulations throughout the paper in order to interpret more clearly specific characteristics of device operation. The simulations follow one principal assumption. Namely, because fairly thick layers are normally used in most ZnO-based LEDs, the elastic strain in their heterostructures is neglected, i.e., all the epilayers are assumed to relax completely, and, hence, the piezoelectric charges at the structure interfaces are assumed to vanish. In contrast, interfacial charges arising from spontaneous polarization in wurtzite semiconductors are normally taken into account. This requires high quality heterostructure interfaces in order to avoid any influence of interfacial traps on the polarization charges.

The simulations are carried out with materials parameters close to those recommended in [17] for III-nitrides and in [18] for ZnO and MgZnO.

The chapter is organized as follows. Operation of homojunction ZnO LEDs is considered in Sec.15.2. In Sec.15.3 we will discuss the most frequent type of the ZnO-based emitters – hybrid single-heterostructure (SHS) LED implementing in their structure alternative materials, like III-nitrides or SiC. More complex, double-heterostructure LEDs, are reviewed in Sec.15.4. A summary of specific features of the ZnO-based LEDs and conclusions are given in Sec.15.5.

15.2 ZnO homojunction LEDs

The main problem faced in fabricating homojunction ZnO LEDs is the formation of p-doped contact layers. In early investigations, this was achieved by laser doping

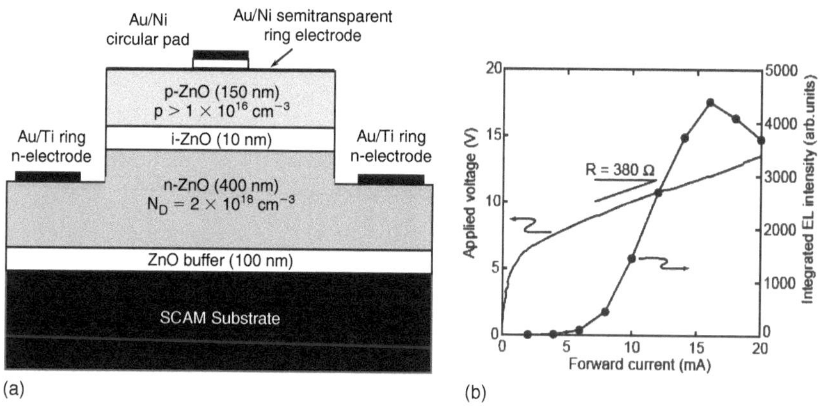

Figure 15.2: Schematic p-i-n ZnO LED (a) and its *I-V* and *L-I* characteristics (b) from [21].

an initially n-type ZnO [19] or plasma-enhanced pulsed laser reactive deposition [20]. In each of these cases, non-linear current-voltage (*I-V*) characteristics were observed, providing evidence for p-doped ZnO formation. The spectra of the LEDs consisted of wide peaks in the ~400–1000 nm spectral range, which could not be associated with emission from the n- or p-layers alone. The authors of Ref.[20] reported on changes in the emission spectra after many injection trials. The effect was attributed to instability of the nitrogen acceptors in ZnO, which indicated one more problem on the road to developing ZnO-based LEDs.

An advanced ZnO p-i-n LED was suggested in [21] (see Fig. 15.2a). This LED structure grown by molecular-beam epitaxy (MBE) consists of a ZnO buffer layer on a $ScAlMgO_4$ (SCAM) substrate and a 10 nm i-ZnO layer placed between an n-ZnO contact layer 400 nm thick and a p-ZnO contact layer 150 nm thick. The p- and n-doping parameters, summarized in Fig. 15.2a, were controlled independently in satellite samples.

The experimental *I-V* and output power-current (*L-I*) characteristics of the p-i-n LED are shown in Fig. 15.2b. The observed turn-on voltage of about 5.4 V is significantly higher than the value of ~3.0 V obtained by simulation of the carrier transport in this LED structure. The high turn-on voltage may be attributable to additional potential barriers in the structure related, possibly, to non-ohmic contact behavior. This, in turn, may be because of the p-contact since the reported hole concentration in p-ZnO is less than $10^{17}\,cm^{-3}$. The non-ohmic contact behavior manifested itself in an earlier report [22] via kinks observed in the *I-V* curve of a ZnO LED with a similar design.

The series resistance $R_S = 380\,\Omega$ determined from the slope of the *I-V* curve was ascribed in [21] to current spreading in the n-ZnO contact layer. Our estimates

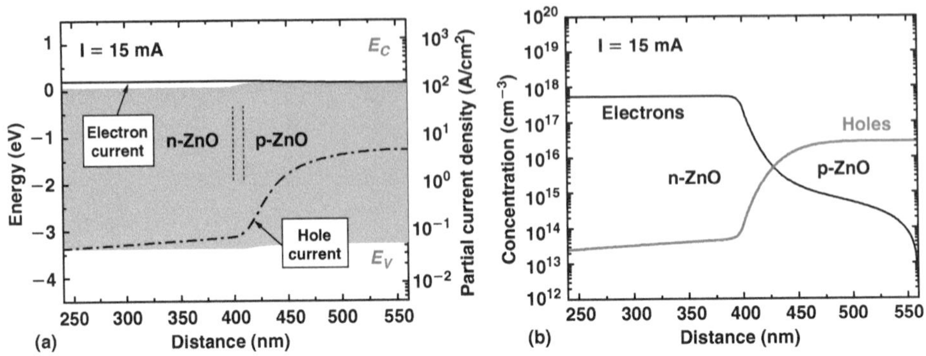

Figure 15.3: Simulated band diagram and partial electron/hole current densities (a) and carrier concentration distributions (b) in the p-i-n LED from [21]. The bandgap in the structure is marked in (a) by gray color.

made on the basis of the reported electron concentration and mobility show that the current-spreading resistance is at least an order of magnitude lower that the series resistance. In combination with the observed high turn-on voltage, this suggests that contact resistance is largely responsible for the high series resistance of the LED.

Our simulations also show that for the currents of interest, the p-n junction barrier almost disappears (Fig. 15.3a). As a result, the non-equilibrium electrons and holes are no longer confined near the p-n junction and can penetrate into the p- and n-regions, respectively (Fig. 15.3b). However, because of significantly different mobilities and concentrations of electrons and holes, there is a considerable imbalance between the partial currents of the carriers (see Fig. 15.3a), i.e., electrons easily pass through the p-layer and recombine non-radiatively at the p-contact electrode. These losses inherent to all homojunction LEDs normally result in a low device efficiency.

The spectra of the p-i-n LED exhibited a multi-peak structure caused by the interference-modulation of a wider peak centered at 440 nm and extending from 390 to 720 nm. This agrees well with the earlier reported results [19,20]. The wide spectral-range emission was attributed in [21] to a high concentration of the nitrogen acceptors incorporated into ZnO.

In general, the investigation of homojunction ZnO LEDs has demonstrated the feasibility of such devices, but revealed a number of serious drawbacks. In particular, those are (i) poor carrier confinement at the p-n junction under practical operating conditions, (ii) a significant imbalance between electron and hole partial currents caused by the difference in both carrier concentrations and mobilities, and (iii) a broad emission spectra related to the commonly used nitrogen acceptors.

Some of these problems can be partially resolved by increasing the hole concentration in p-ZnO or finding alternative acceptor impurities, while others require modifications in the basic LED design.

15.3 ZnO-based hybrid single-heterostructure LEDs

A natural solution that avoids the problems related to p-doping of ZnO, is to use materials which can be easily and reliably doped. A number of such materials have been investigated in previous studies, e.g., p-Cu_2O obtained by oxidation of a copper metal film deposited on vapor-phase-grown ZnO [23], p-ZnTe deposited on n-ZnO [24], and p-$SrCu_2O_2$ obtained by pulsed laser deposition [25]. Only for the latter case did the LED emission spectra peak at 382 nm, which is close to the band-to-band transition. For the remaining materials, wide emission bands centered in the green spectral range were observed. Moreover, for each of these p-type materials the LED efficiency was extremely low, a result attributed to high lattice constant and crystal symmetry mismatches with ZnO. Therefore, more appropriate p-doped semiconductors, GaN, AlGaN, and SiC, have been the subject of more recent investigations as discussed below.

15.3.1 n-ZnO/p-GaN single-heterostructure LED

As shown previously in Fig. 15.1a, GaN is a wurtzite material with a lattice constant close to that of ZnO (mismatch of ~1.9%). Consequently, III-nitrides were a natural first choice to consider for making a p-contact layer [26,27]. It is known that ZnO and GaN form a heterojunction with a type-II band alignment [28]. This fact as well as the existence of spontaneous electric polarization in both GaN and ZnO [29] determines a specific mechanism of carrier confinement near the ZnO/GaN interface, as described below.

The LEDs reported in [26,27] were based on a single n-ZnO/p-GaN heterostructure grown in the [0001] direction. Both the ZnO and GaN layers were ~1 μm thick and doped with electron and hole concentrations of $n = 4.5 \times 10^{18}$ cm^{-3} and $p = 3.5 \times 10^{17}$ cm^{-3}, respectively. Carrier mobilities as high as $\mu_p = 10\,\mathrm{cm^2/V \cdot s}$ and $\mu_n = 40\,\mathrm{cm^2/V \cdot s}$ were measured for holes in GaN and electrons in ZnO. We believe that the n-ZnO layer was completely relaxed because of its sufficiently large thickness.

The band diagrams, carrier concentration distributions, and radiative and non-radiative recombination rates calculated for different values of the forward bias U_b are plotted in Fig. 15.4. The calculations were in part based on a model for

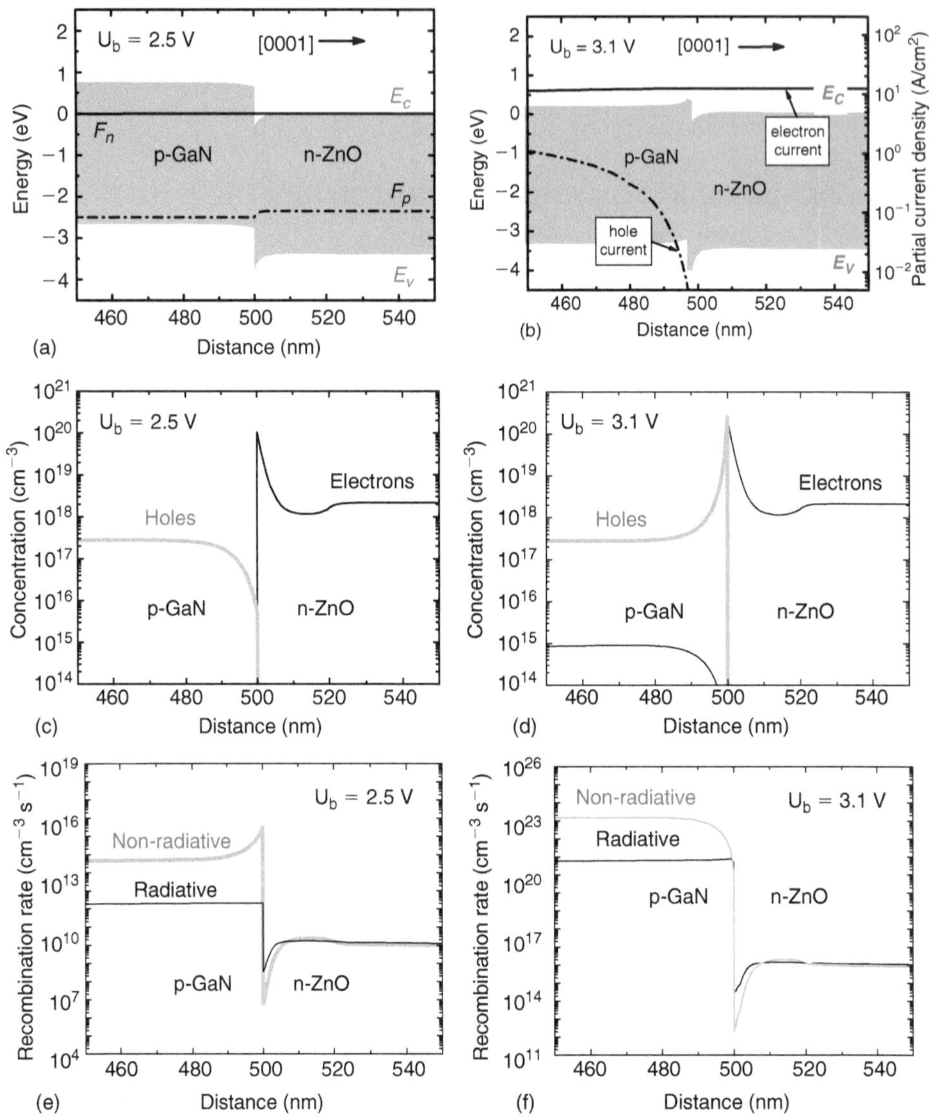

Figure 15.4: Simulated band diagram (a,b), carrier concentration distributions (b,c), and bulk radiative and non-radiative recombination rates (e,f) in the n-ZnO/p-GaN LED from [26]. The bandgap in the structure is marked in (a,b) by gray color.

the non-radiative carrier recombination at threading dislocations suggested in [12], and assuming a dislocation density of $\sim 10^9\ cm^{-2}$ in the heterostructure. Such a dislocation density is typical of the GaN layers grown by metalorganic vapor-phase epitaxy (MOVPE) and MBE. Since both the ZnO and GaN layers are considered relaxed, the interfacial polarization charge is entirely determined by the difference in the spontaneous polarization between ZnO and GaN.

Due to a positive polarization charge induced at the ZnO/GaN interface and the type-II band alignment, a local quantum well for electrons is formed in ZnO at the heterojunction. In contrast, a similar well for holes in GaN first appears only when a large enough forward bias is applied (compare Fig. 15.4a and Fig. 15.4b). The well formation results in a significant increase in the electron and hole concentrations near the ZnO/GaN interface, providing effective carrier confinement that is not typical of conventional single-heterostructures (Fig. 15.4d). At a low bias, the hole quantum well disappears, and the hole concentration in GaN at the interface falls down dramatically (Fig. 15.4c).

Figure 15.4b also demonstrates an imbalance between the partial electron and hole currents in the LED structure. The imbalance is caused by (i) the difference in the electron and hole mobilities and (ii) the difference in the heights of the potential barriers for electrons and holes in the space-charge regions formed at the ZnO/GaN interface. Indeed, a lower hole concentration in p-GaN provides a wider space-charge layer and, hence, a higher barrier height. As a result, electrons from n-ZnO penetrate easily into p-GaN, inducing in this layer intensive radiative/non-radiative carrier recombination (Fig. 15.4e,f).

By comparing the cathodoluminescence (CL) spectra from separate n-ZnO and p-GaN epilayers to the electroluminescence (EL) spectrum of the n-ZnO/p-GaN LED (Fig. 15.5) the authors of [26,27] have concluded that the emission from GaN provides a major contribution to the EL spectrum of the LED. This conclusion is supported by our simulation predicting a dominant carrier recombination in the p-GaN layer at any bias (Fig. 15.4e,f). However, an alternative interpretation of the spectrum position is also possible. Figure 15.5b compares the observed EL spectrum with that predicted for tunneling light emission that is discussed in Sec.15.3.2 below.

The measured I-V characteristic of the LED was remarkably different than what was expected from the Shokley's diode model. Moreover, the electric current I was found to be nearly proportional to the square of the forward voltage V. Such behavior is typical of space-charge limited current, as mentioned in [27], and is frequently observed in the ZnO-based LEDs. We will discuss this issue in more detail in the next section. The series resistance of the LED estimated from the I-V curve reported in [27] was $\sim$75 Ω.

15.3.2 Inverted p-ZnO/n-GaN single-heterostructure LED

An inverted p-ZnO/n-GaN SHS LED was studied in [30]. The LED heterostructure consisted of a 1.6 μm n-GaN:Si layer ($n = 1.2 \times 10^{18}$ cm^{-3}, $\mu_n = 200$ cm^2/V · s)

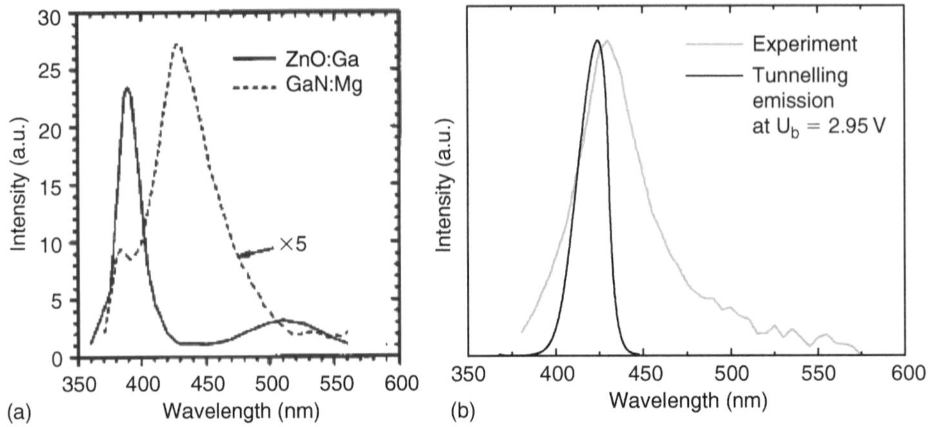

Figure 15.5: Room temperature CL spectra from n-ZnO and p-GaN epilayers (a) and EL spectrum of n-ZnO/p-GaN LED (b) [27]. For comparison, a spectrum of the tunneling light emission is plotted in (b).

grown by MOVPE, followed by a 0.4 μm p-ZnO:P layer ($p = 6.7 \times 10^{17}\,\text{cm}^{-3}$, $\mu_p = 1.4\,\text{cm}^2/\text{V}\cdot\text{s}$) deposited on the n-GaN by the RF-magnetron sputtering. The LED die $300 \times 300\,\mu\text{m}^2$ in area was fabricated with Ti/Al and NiO/Au ohmic contacts to n-GaN and p-ZnO, respectively.

As already mentioned in Sec.15.3.1, a positive surface charge is formed at the ZnO/GaN interface due to the difference in the spontaneous polarization of ZnO and GaN. (Again, we assume both ZnO and GaN layers to be relaxed because of their sufficiently large thicknesses, so that the piezoeffect does not give rise to a polarization charge.) In contrast to the n-ZnO/p-GaN SHS LED considered in the previous section, both electron and hole quantum wells adjacent to the ZnO/GaN interface (Fig. 15.6a) are found to exist at any, even a low, forward bias. This results in very atypical carrier concentration distributions in the LED structure in which electrons and holes are accumulated near the interface in the p-ZnO and n-GaN layers, respectively (Fig. 15.6b).

As in the case of the n-ZnO/p-GaN SHS LED, an imbalance between the partial electron and hole currents results in deep penetration of electrons into p-ZnO and, hence, in a dominate carrier recombination in the bulk of the layer (Fig. 15.6c).

Figure 15.6d compares the experimental and theoretical I-V characteristics of the LEDs, assuming the series resistance of the diode to be of 40 Ω. It is seen that the experimental curve differs from the simulated one by turn-on voltage and a general behavior that resembles the well-known dependence $I \propto V^2$ typical of space-charge limited current. A similar dependence was also observed in the n-ZnO/p-GaN LEDs [26,27] (see Sec.15.3.1). We believe, however, that the mechanism for the latter

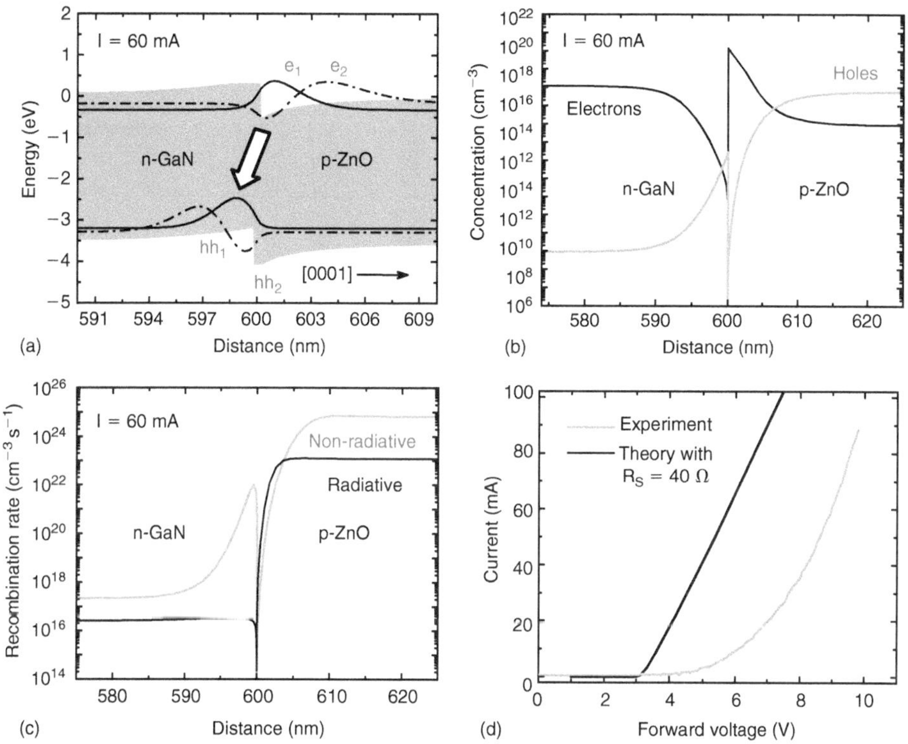

Figure 15.6: Computed band diagram and electron/hole wave functions (a), distributions of carrier concentrations (b), bulk radiative/non-radiative recombination rates (c), and comparison of theoretical and experimental *I-V* characteristics (d) for p-ZnO/n-GaN LED from [30].

LEDs is different than the space-charge current limitation, since the electron concentration in the p-ZnO layer is not high enough to break off the quasi-neutrality of the material (see Fig. 15.6b). A possible alternative mechanism leading to this behavior is self-heating of the diode. In fact, if the temperature of an LED increases with current, the electrical conductivity of either the p-ZnO or the p-GaN layers also rises due to more efficient acceptor activation. In this case, a superlinear *I-V* dependence would be observed at high currents.

The measured EL spectrum of the p-ZnO/n-GaN SHS LED shown in Fig. 15.7 peaks at 409 nm, while the PL emission from both p-ZnO and n-GaN is found to be centered at 385 nm and 365 nm, respectively. This disparity can be explained by tunnelling recombination between the electrons and holes confined in the wells adjacent to the ZnO/GaN interface (see Fig. 15.6a). As the electron and hole wave functions are extended into n-GaN and p-ZnO, respectively, their overlapping integral does not

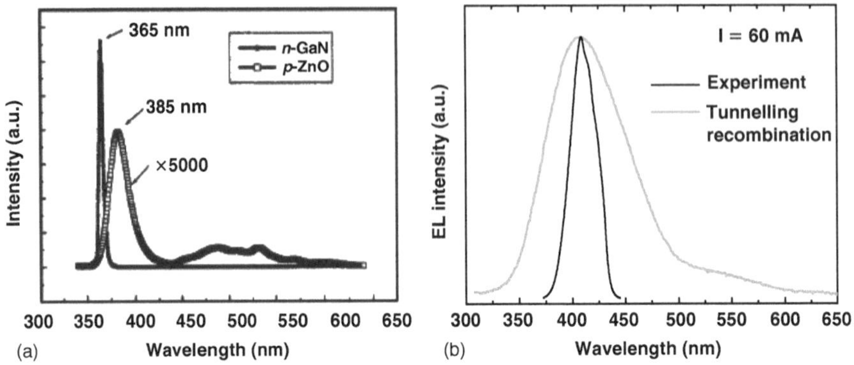

Figure 15.7: Measured room-temperature PL spectra from separate p-ZnO and n-GaN layers (a) and comparison of the predicted tunnelling recombination spectrum with the spectrum of EL from the p-ZnO/n-GaN LED [30] (b).

equal zero although it is rather small. Nevertheless, the tunneling radiative recombination between the electrons and holes confined on different sides of the ZnO/GaN interface may contribute significantly to the emission spectra, if the carrier concentrations at the interface exceed those of the bulk by a high enough value.

The spectrum corresponding to the tunneling emission is compared with the experimental one in Fig. 15.7. The peak position predicted by the theory is in excellent agreement with the position of the experimental spectrum, but the width is quite different. This may be caused by a non-uniformity of the ZnO/GaN interface formed during the heterostructure growth. Indeed, any imperfection may induce local strain at the interface which will affect the width of the adjacent potential wells via the piezoelectric effect and, thus, modulate the tunnelling emission spectrum.

Next, we would like to discuss in more detail the tunneling emission in the SHS ZnO/GaN LEDs. As mentioned above, this effect is expected to be important, if the electron and hole concentrations at the interface are high enough. This requirement can be satisfied by the formation of adjacent potential wells for electrons and holes in which the carrier concentrations are much higher than in the bulk material. The latter becomes possible under certain conditions dependent on the type-II band line-up in the LED heterostructure, the interface polarization charge, the doping levels of both ZnO and GaN, and the applied bias. If the polarization charge vanishes, for instance, because of interface defects, the tunneling light emission very likely becomes unobservable.

Simulations predict the tunnelling emission spectra in the ZnO/GaN LEDs to peak at 2.95–3.1 eV (400–420 nm) and to exhibit a pronounced blue shift with the electric current through the diode (Fig. 15.8). These predictions, however, do not account for both the temperature-dependent bandgap narrowing and the spectra

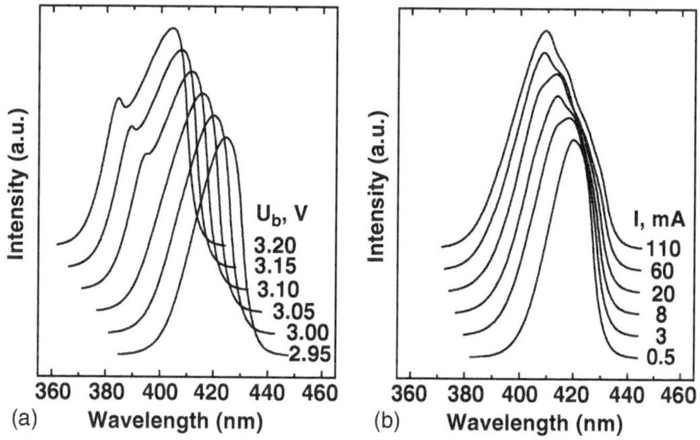

Figure 15.8: Tunnelling emission spectra of the n-ZnO/p-GaN LED [27] (a) and p-ZnO/n-GaN [30] (b) LED heterostructures computed for different values of the forward bias/current.

shift to the midgap due to heavy p-doping of ZnO or GaN. These factors result in a spectra broadening and a current dependent red shift that make it difficult to identify the tunneling emission experimentally.

15.3.3 n-ZnO/p-SiC single-heterostructure LED

An interesting attempt to utilize the possible heavy doping of ZnO with donors and SiC with acceptors for LED fabrication has been reported in [31]. In this study, the p-n junction was formed by a 100 nm n-ZnO layer ($n > 10^{20}$ cm^{-3}, $\mu_n \sim 7$ cm^2/V · s) deposited on a p-doped Si-faced 4H-SiC ($p \sim 10^{19}$ cm^{-3}, $\mu_n \sim 120$ cm^2/V · s) at 150°C using the Zn:Al target sputtering in an oxygen atmosphere (5×10^{-4} Torr). A 50 nm n$^+$-ZnO ($n > 10^{21}$ cm^{-3}) contact layer was then deposited on top of the n-ZnO layer. An Al/Ti ohmic contact was formed on the back surface of the SiC substrate by rapid thermal annealing. A separate 100 nm n$^+$-ZnO layer deposited on a quartz plate was used as the n-electrode by manual contacting with the n-side of the LED.

The I-V characteristic of the LED shown in Fig. 15.9a gives series resistance for the diode in the range of ~1.5 kΩ, which does not agree at all with the high carrier concentrations of the n-ZnO layer and p-SiC substrate reported in [31]. Apparently, the poor manually held n-contact is responsible for the extremely high resistance of the LED. The EL spectra of the n-ZnO/p-SiC LED exhibits a narrow peak at 385 nm (Fig. 15.9b) that has been attributed in [31] to the light emission in n-ZnO by comparing the EL spectra with the PL ones obtained from the LED structure and

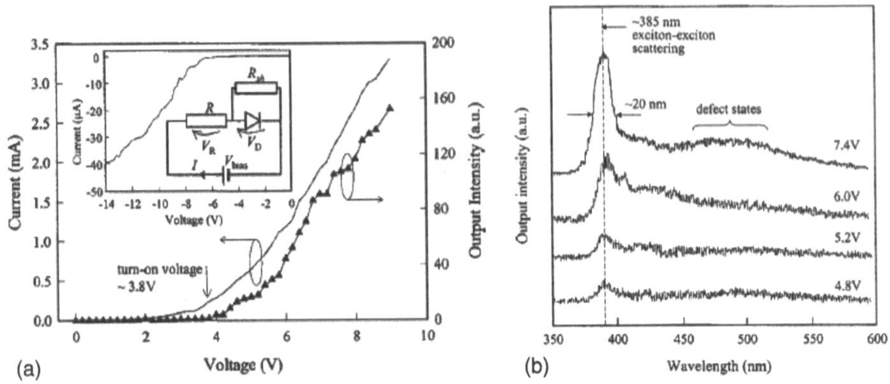

Figure 15.9: *I*-*V* and *L*-*I* characteristics of the n-ZnO/p-SiC LED from [31] (a) and its room temperature emission spectra at different forward voltages (b).

SiC substrate alone. Our simulations of the LED heterostructure predict, however, considerable electron penetration into the p-SiC substrate and a dominant radiative recombination in the SiC substrate despite a much lower recombination rate constant (non-direct transitions). The discrepancy between the experiment and theoretical predictions is not currently understood.

Another example of using p-SiC for LED fabrication has been recently reported in [32]. A 0.5 μm n-ZnO was grown on p-6H-SiC substrate by plasma-enhanced MBE at 650°C. The substrate was 400 μm thick and had a hole concentration of $4 \times 10^{17}\,cm^{-3}$. From *C*-*V* measurements, the electron concentration in the ZnO epilayer was found to be much higher than the hole concentration in SiC. Mesa-structures with a 250 μm diameter were then etched to form the ohmic contacts to the diode.

The *C*-*V* measurements have also revealed an increase in the carrier concentration on both sides of the ZnO/SiC interface, which correlates with our simulations assuming the Si/Zn polarity of the grown LED heterostructure. The density of surface states measured in [32] was in the range of $\sim 10^{11}$–$10^{13}\,cm^{-2}\,eV^{-1}$ (Fig. 15.10a). The upper limit of the surface state density produces an interfacial charge comparable with that induced by the difference in the spontaneous electric polarization of ZnO and SiC. It is not clear, however, whether the spontaneous polarization controls the ZnO/SiC interfacial charge or if it is due to the interfacial traps resulting from defect generation at the initial stage of ZnO growth on SiC.

The EL spectra of the ZnO/SiC LED (Fig. 15.10b) were determined in [32] to originate from SiC. This conclusion is in agreement with our theoretical predictions of a dominant radiative recombination in p-SiC caused by the asymmetry in the

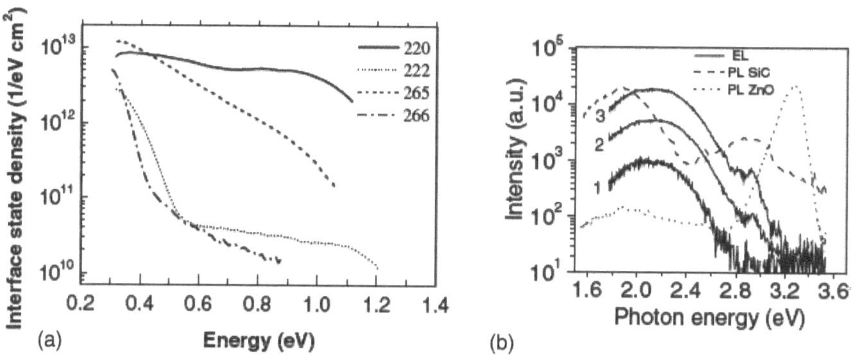

Figure 15.10: Distributions of interface state density counted from the valence band edge (a) and EL spectra (b) in hybrid n-ZnO/p-6H-SiC LED from [32]. PL spectra from separate ZnO and SiC are also plotted in (b) for comparison.

electron and hole partial currents (see the remark concerning the ZnO/SiC LED from Ref.[31]).

Generally, the lattice constant mismatch between SiC and ZnO (~5.4–5.7%) is much higher then that between GaN and ZnO (~1.9%), which should result in a higher defect density in the ZnO/SiC LEDs, especially adjacent to the interface when compared to those for ZnO/GaN. This fact, in combination with the electron/hole current imbalance shifting the light emission region into SiC, which provides indirect radiative transitions, makes, in our opinion, the ZnO/SiC LEDs less promising than hybrid ZnO/III-V devices.

15.3.4 n-ZnO/p-AlGaN single-heterostructure LED

The n-ZnO/p-AlGaN LED heterostructure has been proposed in [33] as one step in improving the performance of n-ZnO/p-GaN LED as suggested in [26,27]. The use of AlGaN as a p-contact layer instead of GaN enables an increased conduction band offset, which is beneficial for the electron confinement and reduction in the interfacial polarization charge. In particular, for spontaneous electric polarization values of $P^S_{\text{ZnO}} = -0.050\,\text{C/m}^2$, $P^S_{\text{GaN}} = -0.029\,\text{C/m}^2$, and $P^S_{\text{AlN}} = -0.081\,\text{C/m}^2$, one can derive from the Vegard law that the ZnO/$Al_xGa_{1-x}N$ interface becomes nearly free of charge at $x \cong 0.4$ (see also Fig. 15.11a).

The LED structure in [33] consisted of a 0.8 μm p-$Al_{0.12}Ga_{0.88}N$ layer grown by hydride vapor-phase epitaxy (HVPE) followed by a 1.0 μm n-ZnO grown by chemical vapor deposition stimulated by the RF-discharge plasma. The hole and electron concentrations in the layers were estimated as $\sim 5 \times 10^{17}\,\text{cm}^{-3}$ and

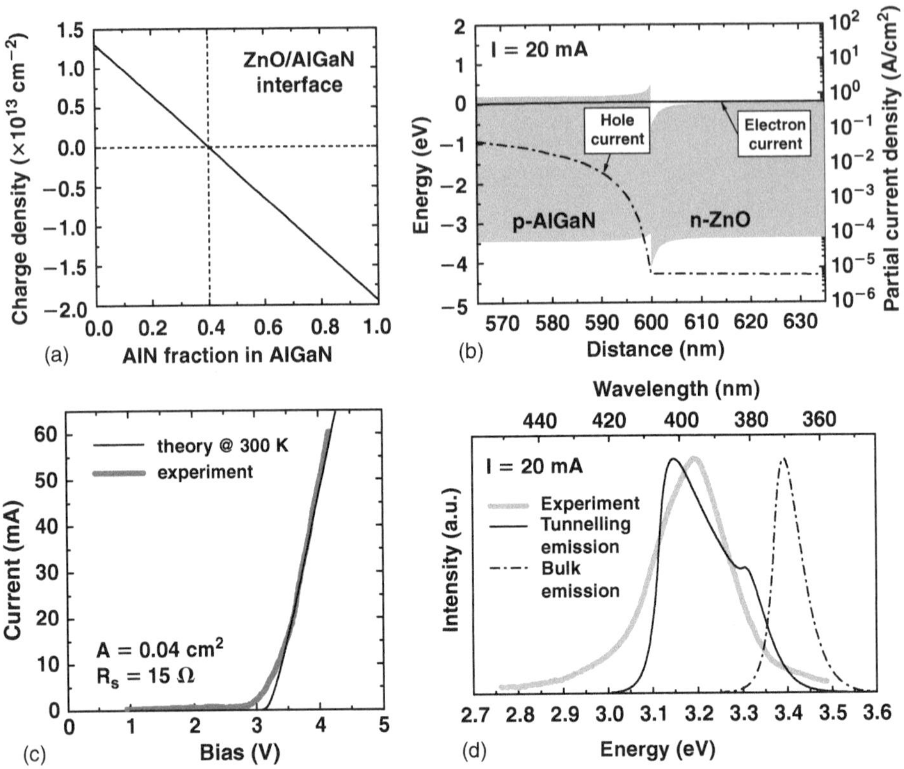

Figure 15.11: Interface polarization charge as a function of AlGaN composition (a), computed band diagram (b), *I*-*V* characteristics (c), and comparison of the measured emission spectrum with those predicted for tunnelling and bulk emission in n-ZnO/p-AlGaN LED from Ref.[33].

$\sim 7 \times 10^{17}$ cm^{-3}, respectively. Simulations of the LED structure (see also [34]) show that its behavior is quite similar to that of the n-ZnO/p-GaN LED considered in Sec.15.3.1 (compare the band diagrams in Fig. 15.11b and Fig. 15.4a). The large-area contacts formed on the LED die provide a comparably low series resistance of the device, ~15 Ω. A small difference is found between the measured and theoretically predicted turn-on voltage (Fig. 15.11c). It is interesting that the *I*-*V* curve of this LED is found to be much closer to the Shokley's characteristic of an ideal p-n junction diode than the characteristics reported for direct and inverted ZnO/GaN LEDs discussed in the previous sections. It is not clear at the present time whether this is related to the substitution of GaN by AlGaN or to improvements in the LED fabrication technology.

As in the case of the n-ZnO/p-GaN LED, the interpretation of the emission spectra measured in [33] is ambiguous. On the one hand, the observed EL spectra may be treated as the superposition of the light emission from both n-ZnO and p-AlGaN layers, which is independent of temperature for some obscure reasons. On the other hand, tunneling emission may also explain the observed spectra (Fig. 15.11d). More effort is required to clarify the nature of the emitted light in hybrid ZnO/AlGaN and similar heterostructures.

15.3.5 Resume on single-heterostructure LEDs

The review of the ZnO-based SHS LEDs carried out in the previous subsections allows us to point out some general operating features.

First, all the SHS LEDs exhibit a remarkable imbalance between the electron and hole partial currents resulting in electron losses at the p-contact electrode and enhanced carrier recombination in the p-region. This effect occurs irrespective of whether the p-layer is made from ZnO or from a III-nitride compound. The reasons for the imbalance are (i) the difference in the electron and hole mobilities, (ii) a lower (as a rule) hole concentration in p-doped layers, and (iii) the difference in the heights of the potential barriers for electrons and holes formed at the ZnO/III-N interface. The current imbalance may be suppressed by insertion of an electron-blocking layer into the p-doped region of the diode or by using a high Al-content p-AlGaN that provides a larger conduction band offset at the interface and, thus, may reduce the current imbalance.

Second, hybrid ZnO/III-N SHS LEDs frequently emit light with the energy well below the bandgaps of both ZnO and the III-nitride material employed. Although important, defect-mediated luminescence and light emission from a heavy-doped p-material with a broad spectrum extended to the midgap of the semiconductor does not explain all the currently observed results. An additional mechanism, tunneling radiative recombination of electrons and holes confined in adjacent potential wells on opposite sides of the ZnO/III-N heterojunction, is predicted to be essential if a certain polarization charge is induced at the interface. Even when the epilayers are relaxed, the interfacial charge can still arise due to the difference between the spontaneous polarization of ZnO and the III-nitride material. The latter, however, requires a sufficiently high interface quality, to avoid forming interfacial traps and reducing the polarization charge that contributes to the non-radiative carrier recombination. Thus, further effort is necessary to identify the mechanisms for light emission in hybrid LEDs and the factors controlling the emission spectrum position and width.

Normally the series resistance of hybrid ZnO/III-N LEDs varies in the range of 30–80 Ω. This is approximately an order of magnitude lower than the series resistance of the p-i-n junction ZnO LED (380 Ω) but it is an order of magnitude higher than typical series resistances of III-nitride LEDs. We have failed to find a correlation between the LED series resistance and the conductivity of the bottom contact layer that controls the current spreading in the LED die. Apparently this means that a considerable contribution to the series resistance comes from the contacts formed to both the p- and n-regions. Thus, resistance reduction is still an important task for improving the overall performance of the ZnO-based LEDs.

Among various hybrid SHS LEDs considered here, the ZnO/III-N ones seem to be the most promising due to a lower lattice mismatch between the constituent materials providing a lower defect density in the grown heterostructure.

15.4 ZnO-based double-heterostructure LEDs

The next natural step in the development of the ZnO-based LEDs is the use of double-heterostructures (DHS) which normally provide better carrier confinement in the active regions of the devices. In the case of conventional (cubic) III-V and II-VI compounds, the DHS always resulted in a better LED or LD performance. However, in wurtzite semiconductors, like II-O or III-N, possessing spontaneous polarization and strong piezoeffect, a built-in electric field introduces a remarkable asymmetry in the band diagrams of the LEDs with respect to the p- and n-regions. For instance, for the same heterostructure design the injection properties of GaN-based LEDs have been predicted to depend significantly on the crystal polarity, i.e., on the signs of the polarization charges induced at the structure interfaces [35]. A similar effect is also expected to be important for ZnO-based LEDs.

15.4.1 GaN/ZnO/GaN double-heterostructure LED

The n-GaN/n-ZnO/p-GaN DHS LED was proposed in [36] as an extension of previous studies on hybrid ZnO/III-N heterostructures [26,27,33]. The LED consisted of a 0.7 μm p-GaN layer ($p \sim 4 \times 10^{17}\ \mathrm{cm}^{-3}$) grown by MBE on a MOVPE-deposited unintentionally doped (UID) 2 μm GaN template, a 0.4 μm UID n-ZnO layer ($n \sim 10^{17}\ \mathrm{cm}^{-3}$) grown by plasma-enhanced MBE at 600°C, and a 0.4 μm UID n-GaN layer ($n < 10^{17}\ \mathrm{cm}^{-3}$) grown by plasma-enhanced MBE at 600°C. The scanning-electron micrograph of the DHS cross-section is shown in Fig. 15.12, indicating a much higher contrast non-uniformity in the upper n-GaN epilayer.

Since the LED heterostructure was grown on MOVPE-deposited GaN, one can expect the whole crystal to have a metal, Ga/Zn, polarity. Our simulations of the DHS LED made on this assumption predict, however, that the Ga/Zn-faced diode heterostructure is nonconductive. In fact, because of a particular distribution of the interface polarization charges the equilibrium band diagram of the diode

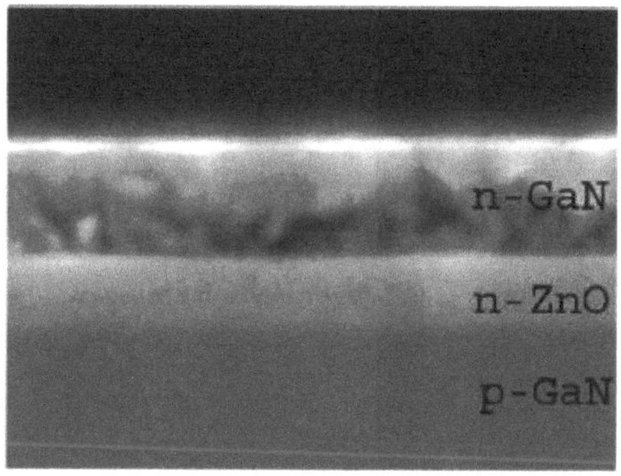

Figure 15.12: Scanning-electron micrograph of the n-GaN/n-ZnO/p-GaN DHS from Ref.[36].

(Fig. 15.13a) is lined up in such a way as to provide an alternating sequence of the electron and hole sheets in the diode (Fig. 15.13b). As a result, the diode is found to be incapable of conducting current at any forward voltage applied up to ~10–12 V.

Next we assumed that the DHS LED structure had an opposite, N/O, polarity that leads to the sign inversion of all the interfacial charges. In this case, simulations produce quite reasonable results presented in Fig. 15.13c,d,e. In particular, the calculated *I-V* characteristic fits the measured one well (Fig. 15.13f) when the series resistance of the LED is assumed to be as high as 95 Ω. The simulations also indicate the existence of a electron/hole partial current imbalance (Fig. 15.13c) and a domination of the carrier recombination in p-GaN (Fig. 15.13e).

Figure 15.14a compares the EL spectrum of the DHS LED with the PL spectra of separate epitaxial layers in the LED structure. The comparison shows that it is difficult to explain the observed EL spectrum by superposition of the emission spectra obtained from different layers. On the other hand, the calculated spectrum of the interface emission, i.e., the tunneling emission originating from both interfaces of the DHS LED structure, can be closely fitted with the peak position and the short-wavelength wing of the measured EL spectrum. The long-wavelength wing of the spectrum is consistent with light emission from p-GaN (see also Fig. 15.13e) and interface non-uniformity, as discussed in Sec.15.3.2.

The authors of [36] report also on a remarkable, ~140 meV, blue shift of the EL spectrum, when the forward current is varied from 48 to 155 mA. Since all the epitaxial layers in the LED structure are rather thick, ~400 nm and larger, this shift cannot be explained by the conduction band filling with electrons and the quantum-confined Stark effect, as in the case of the InGaN single/multiple quantum-well LEDs. Self-heating of the device caused by increasing the current would lead to the opposite trend. However, the observed blue shift agrees well, at

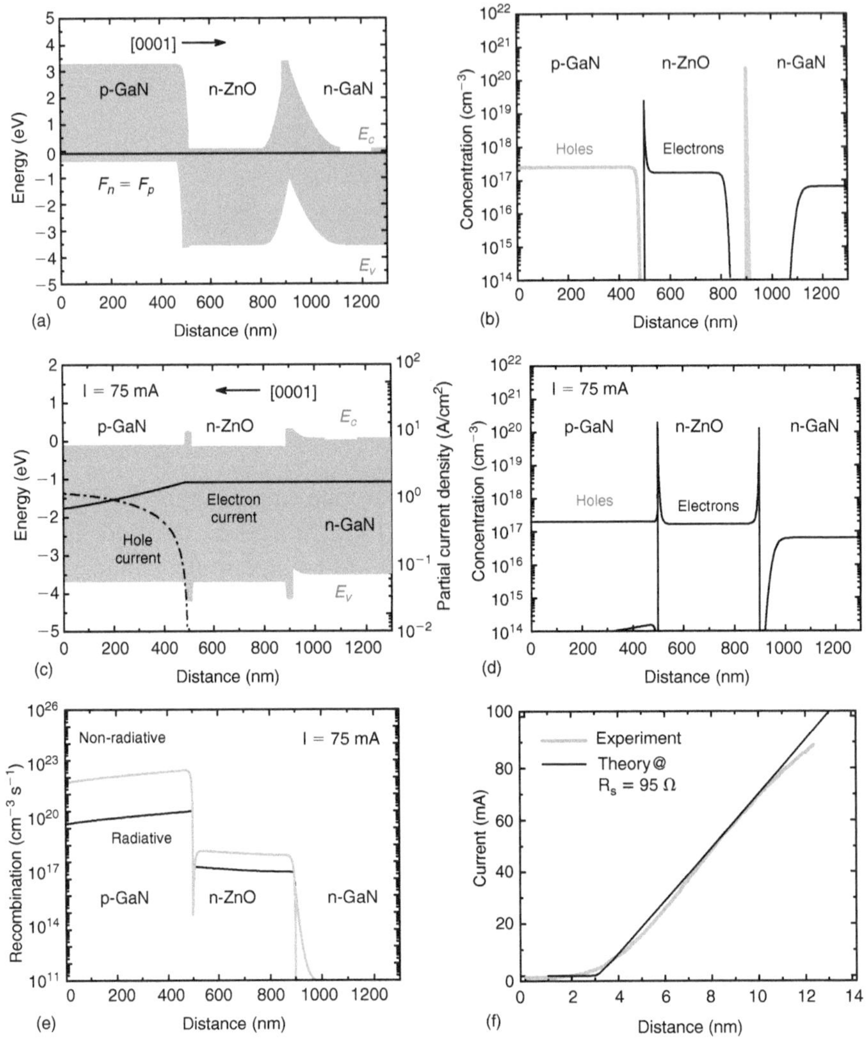

Figure 15.13: Band diagrams and carrier concentrations in Ga/Zn-faced (a,b) and N/O-faced (c,d) DHS LED. Bulk recombination rates (e) and comparison of measured and computed the *I*-*V* characteristics (f) for N/O-faced DHS LED from [36].

least qualitatively, with that predicted for the tunneling emission at the structure interfaces (see Sec.15.3.2 and Fig. 15.8). So, the blue shift of the spectrum observed in [36] may be evidence for the importance of the tunneling emission.

The above analysis shows that polarity control in ZnO-based DHS LEDs is a crucial factor affecting general device performance. It is known that MOVPE of GaN

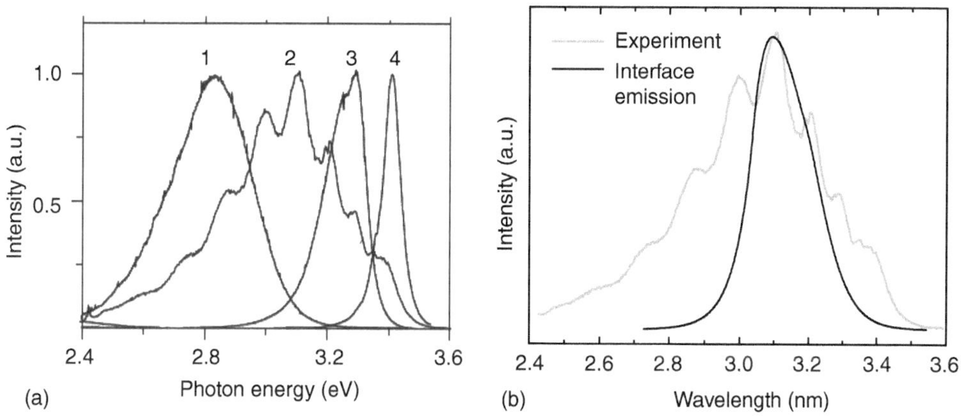

Figure 15.14: Room-temperature EL spectrum of the n-GaN/n-ZnO/p-GaN LED at the injection current of 155 mA (curve 2) and PL spectra of p-GaN (curve 1), ZnO (curve 3) and n-GaN (curve 4) layers [36] (a) and comparison of the measured EL spectrum with that predicted for interface luminescence (b).

normally provides a Ga-faced material, while the polarity of III-nitrides grown by MBE or HPVE can be controlled by a proper choice of the buffer layer growth conditions. Unfortunately, little is known about polarity control in the growth of ZnO and related materials. More effort is required in the investigation of this important issue.

Our calculations of the electron and hole partial currents and distributions of their concentrations carried out for the N/O-faced DHS LED also suggest that the use of the n-GaN layer for carrier confinement is ineffective, since the potential barrier formed at the n-GaN/n-ZnO interface hampers hole penetration into the ZnO active region. This problem may be overcome by using a higher n-doping of ZnO to provide hole tunneling through the barrier or by modification of the heterostructure design.

15.4.2 MgZnO/ZnO/AlGaN/GaN double-heterostructure LED

Figure 15.13c shows clearly that the carrier confinement in the ZnO active region of a GaN/ZnO/GaN DHS LED is created due to the potential barriers formed in the space-charge regions surrounding the active layer. The higher the applied forward voltage, the narrower the space-charge regions and, hence, the less the carrier confinement in such an LED. Confinement can be improved by using wide-bandgap MgZnO and AlGaN emitters for cladding the active layer. Such a DHS LED structure has been recently studied in a number of works [18,37,38,39].

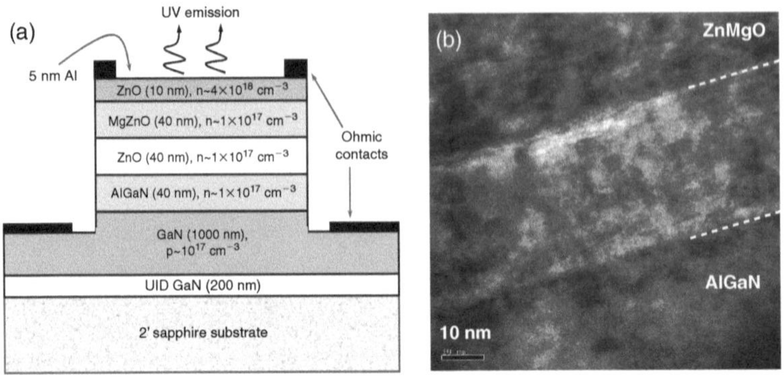

Figure 15.15: Schematic hybrid DHS LED (a) [38] and cross-sectional TEM image of its active region (b) [39].

A typical MgZnO/ZnO/AlGaN/GaN DHS LED consisted of a 0.8 μm p-GaN contact layer ($p \sim 10^{17}$ cm^{-3}), a 40 nm p-$Al_{0.16}Ga_{0.84}N$ emitter ($p \sim 10^{17}$ cm^{-3}), a 40 nm UID ZnO active layer ($n \sim 10^{17}$ cm^{-3}) followed by a 40 nm n-$Mg_{0.1}Zn_{0.9}O$ emitter ($n \sim 10^{17}$ cm^{-3}) and a 10 nm n-$Mg_{0.1}Zn_{0.9}O$ contact layer ($n \sim 4 \times 10^{18}$ cm^{-3}), all grown by MBE (Fig. 15.15a) [38]. The details of the growth procedure and structure characterization are described in [18,38,39]. The cross-sectional transmission-electron microscope (TEM) image of the active ZnO region and neighboring layers is shown in Fig. 15.15b. The LED dice with the square 200–400 μm mesa-isolated structures were then fabricated using a two-step etching technique.

The measured *I-V* characteristics of the LED show good rectifying behavior up to at least 400°C and turn-on voltages in the range of ~3.2–3.5 V [38,39]. The increase of the turn-on voltage to ~5 V (see Fig. 15.16a) was observed at room temperature [37]. The comparison of the theoretical and experimental *I-V* curves (Fig. 15.16) shows remarkable deviation of the experimental characteristic from those predicted on the basis of the drift-diffusion model. This may be caused by non-ohmic contact behavior and/or self-heating of the diode (see discussion in Sec.15.3.2). The series resistance of the LED decreases with temperature and varies in the range of ~100 (370°C)–150 (22°C) Ω. Estimates show that a considerable part of the series resistance can be associated with the p-contact resistance.

Figure 15.17 compares the predicted behavior of the DHS LED structure at different temperatures. The simulations have been carried out, assuming no strain relaxation in the heterostructure, as the AlGaN, ZnO, and MgZnO layers are rather thin. The temperature rise results in higher electron and hole concentrations in the LED structure due to more effective donor and acceptor ionization and, hence, to redistribution of the diode band diagram via widening of the space-charge regions

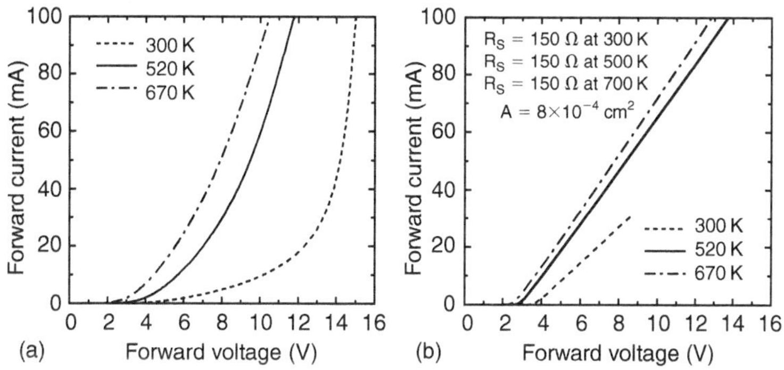

Figure 15.16: Experimental [37] (a) and simulated (b) *I-V* characteristics of the DHS LED.

(Fig. 15.17a,b). This leads, in particular, to a less efficient carrier confinement in the active region at high temperature and, consequently to a higher imbalance between the electron and hole partial currents (Fig. 15.17e,f).

Figure 15.17c demonstrates clearly that the radiative recombination of electrons and holes dominates in the ZnO active layer, which is quite different from what is predicted for the SHS GaN/ZnO/GaN DHS LEDs considered above. This effect is due to the beneficial influence of the wide-bandgap MgZnO and AlGaN emitters on the carrier confinement. At a higher temperature, the effect becomes less pronounced (Fig. 15.17d) because of the increasing imbalance between the electron and hole currents.

The comparison of the predicted and measured results for output power as a function of forward current at different temperatures is given in Fig. 15.18a. To get a better fit to experiment, we had to assume in the simulation a ~90 K self-heating of the LEDs with respect to the environment. In spite of the fact that this assumption is not quite correct, since the device self-heating depends generally on the current value, the theoretical predictions correlate well with the experimental data. In order to explain the strong temperature dependence of the output power, we plot in Fig. 15.18b the computed IQE of the DHS LED *versus* current density. It can be seen that the IQE reaches a maximum at a particular current density and then falls down dramatically. The drop is found to correlate with the increase in the electron flux passing through the ZnO active region and MgZnO emitter and arriving directly at the n-electrode. A rise in the potential barrier at the ZnO/AlGaN interface with temperature suppresses this effect, resulting in a higher IQE at a particular higher temperature. A further increase in temperature again leads to an IQE decrease caused by another mechanism – competition between the radiative and non-radiative recombination rates (compare Fig. 15.17c and Fig. 15.17d).

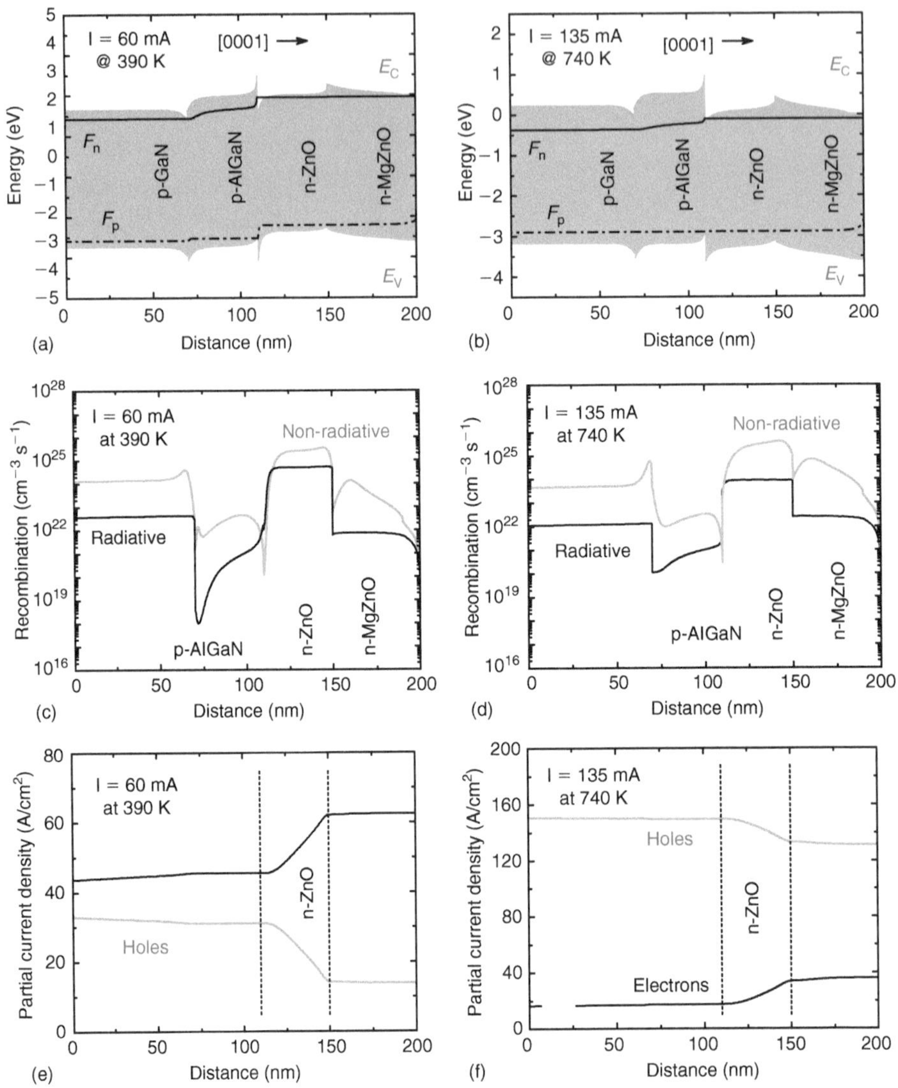

Figure 15.17: Comparison of band diagrams (a,b), bulk recombination rates (c,d), and partial electron and hole current densities (e,f) for the DHS LED at 390 K (a,c,e) and 740 K (b,d,f).

The calculated emission spectra (Fig. 15.19a,b) generally exhibit two peaks: one related to the band-to-band transition (~360 nm) and the other to the tunneling electron-hole recombination at the n-ZnO/p-AlGaN interface (~390–410 nm). The contribution of the tunneling recombination is found to be quite sensitive to both

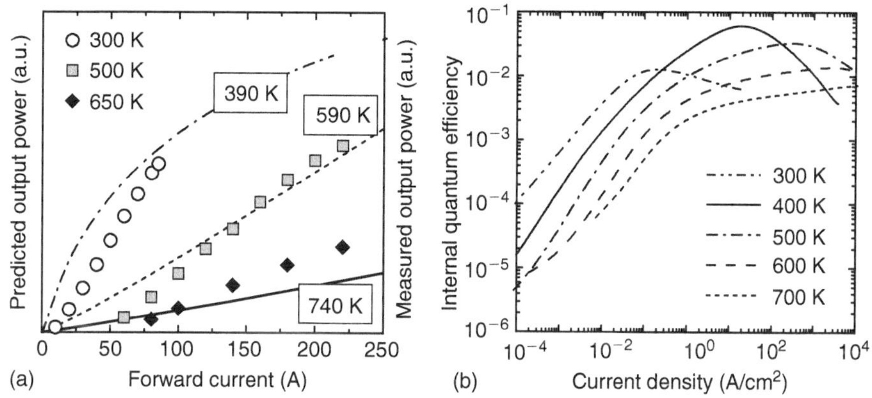

Figure 15.18: Comparison of predicted (lines) and measured in [37] (symbols) DHS LED output power as a function of forward current (a) and theoretical IQE versus current density (b) at different temperatures.

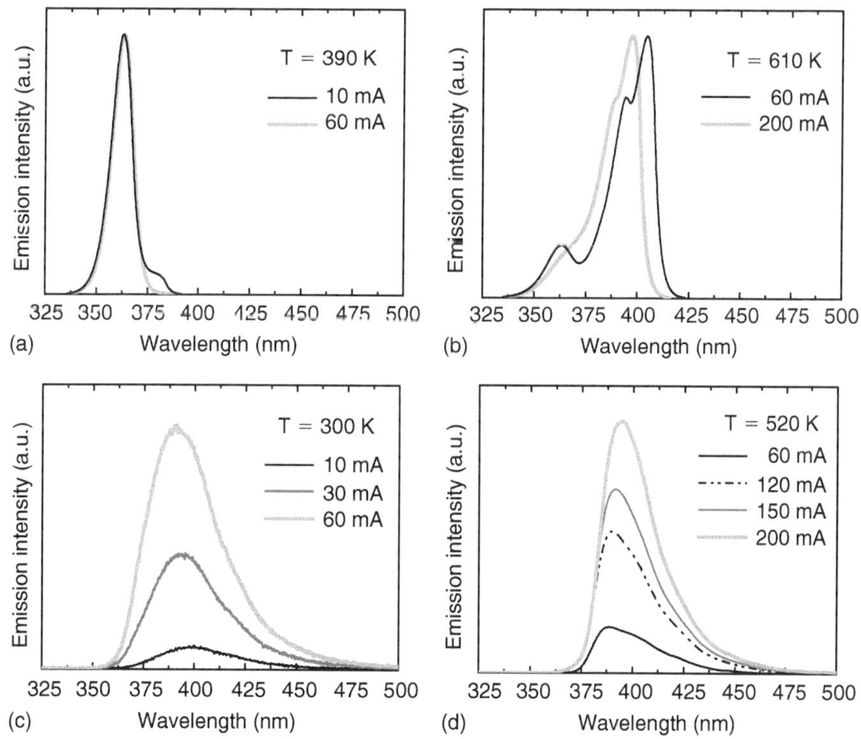

Figure 15.19: Emission spectra computed for different temperatures with account of possible self-heating of DHS LED (a,b) and its spectra measured at room temperature and 220°C (c,d) [39].

forward current and temperature. Simulations predict domination of emission arising from the band-to-band transitions at room temperature, while tunneling emission dominates at high temperature. Experimental spectra peak at ~380–400 nm (see Fig. 15.19c,d) and have a broad long-wavelength extended up to 450–470 nm. No qualitative difference between the room-temperature and high-temperature spectra is experimentally observed. The discrepancy is not yet understood and requires a more careful examination.

Since the series resistance of the LED is at least 100 Ω, the self-heating of the diode is quite likely. The red shift in the emission spectra with current seen in Fig. 15.19c may be the evidence for the self-heating of the DHS LED.

15.4.3 Double-heterostructure LED with CdZnO active layer

As it was already mentioned in Sec.15.1, the use of a CdZnO active layer allows extension of the LED emission spectrum into the visible regime. Up to now, there has been no experimental demonstration of such LEDs. Nevertheless, development of their heterostructures suitable for the device fabrication has been already started by simulation [40]. The simulations were based on the original data on the CdZnO bandgap *versus* composition and CdZnO/ZnO band offsets reported in [41]. It has been found, in particular, that the built-in polarization field in a symmetrical MgZnO/CdZnO/MgZnO DHS is systematically lower than in the hybrid ZnO/GaN structure. The parametric modeling of the I-V characteristics and emission spectra as a function of the active layer/cladding thickness, composition, and doping level provides a guideline for further development of LEDs with CdZnO active region.

15.4.4 Resume on double-heterostructure LEDs

To date, only a few papers have reported on investigations of ZnO-based DHS LEDs. Nevertheless, a number of general features have already been identified specific to this kind of LED heterostructure design.

First of all, DHS LEDs contain at least two heterointerfaces normally accumulating polarization charges of opposite signs. These charges induce in the structure an electric field that may be either co-directed or oppositely directed to the built-in p-n junction field. The latter leads eventually to a considerable difference in the injection properties of LEDs with different polarities. Therefore, the polarity control in the DHS heterostructures becomes a critical factor to their performance.

A GaN/ZnO/GaN DHS is likely ineffective from the point of view of carrier confinement in the ZnO active layer because of the similarity in the bandgaps of the constituent materials. The confinement is predicted to be created in such a structure entirely due to the potential barriers formed in the space-charge regions adjacent to the active layer interfaces. The barrier heights vanish, however, with emitter doping, resulting in dramatic minority carrier losses at the diode contacts. The use of wide bandgap emitters like AlGaN or MgZnO improves the carrier confinement in the DHS LEDs.

The use of the DHS with wide bandgap emitters also leads to more stable LED EL spectra due to improved localization of the radiative recombination in the ZnO active layer. Potentially this can enable better control of the spectrum position and width as well as the optimization of the LED IQE.

15.5 Conclusion

Presently, ZnO-based LEDs are in the early stage of research and development, prototypes for future devices are being considered and examining in detail, the properties of employed materials and related heterostructures are investigated, and various technological approaches to the device fabrication are being tested. The major problem hindering the development of the ZnO light emitters is still the lack of efficient, reliable, and stable p-doping of ZnO and MgZnO. This alone is the main reason that little progress has been achieved on the road to LEDs fabricated entirely from MgZnCdO materials. On the other hand, the unresolved problem of p-doping can be addressed through the alternative feasible approach based on incorporating p-doped III-nitride semiconductors in LED structures. Despite the more complex fabrication technology of such hybrid heterostructures, which requires a combination of quite different growth conditions for II-O and III-N materials, this approach looks, nevertheless, quite promising due to additional degrees of freedom provided for bandgap engineering of the ZnO-based LEDs.

Another general problem of the ZnO-based LEDs is poor control of the emission spectra. Frequently a broad emission spectrum extending into the visible range is generated in the LEDs rather than a narrow peak at 370–380 nm corresponding to the band-to-band transitions in ZnO. This phenomenon may result from a number of possible mechanisms. One is the defect-mediated luminescence caused by an insufficient quality of the ZnO. The defect density can be reduced by application of standard epitaxial growth techniques like MBE and CVD, use of the same reactor for growing both III-N and II-O compounds, and optimization of growth conditions.

Another reason frequently invoked is the emission spectrum broadening in a heavily p-doped wide bandgap material. In addition, as discussed in this review, the tunnelling radiative recombination at the heterostructure interfaces may also contribute to the emission spectra of hybrid II-O/III-N LEDs. The latter mechanism seems to be essential, if the II-O/III-N interface is of high quality, i.e., if the interfacial states and growth defects do not give noticeable rise to the interfacial charge and non-radiative carrier recombination. At present, poor emission spectra control is a factor hindering the utilization of the excitonic emission to improve the radiative recombination yield in the ZnO-based emitters. Therefore, a better understanding of the mechanisms responsible for the light emission in these devices and determining ways of improved emission spectra control are needed.

The third problem that has to be addressed is the high series resistance of almost all ZnO-based LEDs studied. As the major component of the device utilizes p-doped materials as the access layers in the planar LED dice, the high series resistance observed may be attributed to a low hole mobility/concentration in such a layer and correspondingly high ohmic contact resistance. In this respect, approach based on III-nitride LED technology might be helpful: the use of a short-period superlattice to enhance lateral current spreading in an LED die and the insertion of a thin narrow-bandgap layer to improve the p-contact resistance.

Our simulations demonstrate that the polarization charges have a significant effect on both electrical and optical properties of the ZnO-based LEDs. These predictions are based, however, on the poorly known polarization properties of the wurtzite semiconductors. Thus, a more detailed examination of the fundamental properties of these materials is required to complete development of the ZnO-based emitters. From the technological point of view, the control and management of crystal polarity during growth of the ZnO-based and hybrid heterostructure is fairly desirable. In our opinion, proper bandgap engineering of the LED heterostructures accounting for polarization factors may also be an important step leading to improvement of the ZnO-based LED performance.

In this review we intentionally did not consider technological aspects of the ZnO-based LED fabrication, which are discussed, e.g., in [42]. Nevertheless, one important trend deserves mention. We think that the transition from the frequently used deposition techniques like magnetron or laser-assisted sputtering to epitaxial growth of II-oxides by MBE or CVD has already demonstrated significant improvement in the material quality, which is necessary for fabrication of high-efficiency light emitters. Thus, further development of epitaxial techniques is required, including those enabling the growth of II-oxides and III-nitrides in the same reactor.

The analysis carried out in this chapter demonstrates the variety of coupled mechanisms involved in the ZnO-based LED operation, which make these devices quite different from those made from the cubic III-V and II-VI compounds. This complicates the intuitive design of ZnO-based LEDs based on analogy with conventional devices. Thus, device simulations at the stage of the development and LED die design may be quite helpful for understanding both the physics underlying the device operation and factors most critical to the device performance. However, our current knowledge of the basic properties of the MgCdZnO and hybrid

II-O/III-N materials and heterostructures is still incomplete. Further investigation into the materials properties will be an important contribution in the research and development of these quite promising light-emitting devices.

Acknowledgements

The simulations of the ZnO-based LEDs presented in this chapter are made with the SiLENSe 2.0 package provided by Semiconductor Technology Research, Inc. (http://www.semitech.us/products/SiLENSe/).

References

[1] L.A. Kolodziejski, R.L. Gunshor, and A.V. Nurmikko, Annual Review of Mat. Sci. 25 (1995) 711.
[2] A. Nurmikko and R.L. Gunshor, Semicond. Sci. Technol. 12 (1997) 1337.
[3] S. Nakamura, T. Mukai, and M. Senoh, Appl. Phys. Lett. 64 (1994) 1687.
[4] S. Nakamura, J. Vac. Sci. Technol. A 13 (1995) 705.
[5] S. Nakamura, Solid State Commun. 102 (1997) 237.
[6] I. Akasaki, Mat. Sci. Engineer. B 74 (2000) 101.
[7] T. Nishida, N. Kobayashi, T. Ban, Appl. Phys. Lett. 82 (2003) 1.
[8] M. Iwaya, S. Takanami, A. Miyazaki, Y. Watanabe, S. Kamiyama, H. Amano, and I. Akasaki, Jpn. J. Appl. Phys. 42 (2003) 400.
[9] J.P. Zhang, A. Chitnis, V. Adivarahan, S. Wu, V. Mandavilli, R. Pachipulusu, M. Shatalov, G. Simin, J.W. Yang, and M. Asif Khan, Appl. Phys. Lett. 81 (2002) 4910.
[10] O. Ambacher, J. Phys. D 31 (1998) 2653.
[11] H. Amano, S. Takanami, M. Iwaya, S. Kamiyama, and I. Akasaki, Phys. Stat. Solidi (a) 195 (2003) 491.
[12] S.Yu. Karpov and Yu.N. Makarov, Appl. Phys. Lett. 81 (2002) 4721.
[13] T. Makino, Y. Segawa, M. Kawasaki, and H. Koinuma, Semicond. Sci. Technol. 20 (2005) S78.
[14] K. Maeda, M. Sato, I. Niikura, and T. Fukuda, Semicond. Sci. Technol. 20 (2005) S49.
[15] A. Ohmoto and A. Tsukazaki, Semicond. Sci. Technol. 20 (2005) S1.
[16] D.C. Look, Semicond. Sci. Technol. 20 (2005) S55.
[17] I. Vurgaftman, J.R. Meyer, and L.R. Ram-Mohan, J. Appl. Phys. 89 (2001) 5815.
[18] A. Osinsky, J.W. Dong, M.Z. Kauser, B. Hertog, A.M. Dabiran, C. Plaut, P.P. Chow, S.J. Pearton, X.Y. Dong, and C.J. Palmstrøm, Electrochem. Soc. Proceedings 2004–6 (2004) 70.
[19] T. Aoki, Y. Hatanaka, and D.C. Look, Appl. Phys. Lett. 76 (2000) 3257.
[20] X.-L. Guo, J.-H. Choi, H. Tabata, and T. Kawai, Jpn. J. Appl. Phys. 40 (2001) L177.
[21] A. Tsukazaki, M. Kubota, A. Ohtomo, T. Onuma, K. Ohtani, H. Ohno, S.F. Chichibu, and M. Kawasaki, Jpn. J. Appl. Phys. 44 (2005) L643.

[22] A. Tsukazaki, A. Ohtomo, T. Onuma, M. Ohtani, T. Makino, M. Sumiya, K. Ohtani, S.F. Chichibu, S. Fuke, Y. Segawa, H. Ohno, H. Koinuma, and M. Kawasaki, Nature Mater. 4 (2005) 42.
[23] I.T. Drapak, Semiconductors 2 (1968) 624.
[24] A.E. Tsurkan, N.D. Fedotova, L.V. Kicherman, and P.G. Pas'ko, Semiconductors and Semimetals 6 (1975) 1183.
[25] H. Ohta, K. Kawamura, M. Orita, M. Hirano, N. Sarukura, and H. Hosono, Appl. Phys. Lett. 77 (2000) 475.
[26] B.M. Ataev, Ya.I. Alivov, V.A. Nikitenko, M.V. Chukichev, V.V. Mamedov, and S.Sh. Makhmudov, J. Optoelectronics and Advanced Mater. 5 (2003) 899.
[27] Ya.I. Alivov, J.E. Van Nostrand, D.C. Look, M.V. Chukichev, and B.M. Ataev, Appl. Phys. Lett. 83 (2003) 2943.
[28] S.-K. Hong, T. Hanada, H. Makino, Y. Chen, H.-J. Ko, T. Yao, A. Tanaka, H. Sasaki, and S. Sato, Appl. Phys. Lett. 78 (2001) 3349.
[29] F. Bernardini, V. Fiorentini, and D. Vanderbilt, Phys. Rev. B 63 (2001) 193201.
[30] D.-K. Hwang, S.-H. Kang, J.-H. Lim, E.-J. Yang, J.-Y. Oh, J.-H. Yang, and S.-J. Park, Appl. Phys. Lett. 86 (2005) 222101.
[31] C. Yuen, S.F. Yu, S.P. Lau, Rusli, and T.P. Chen, Appl. Phys. Lett. 86 (2005) 241111.
[32] Ya.I. Alivov, D. Johnstone, Ü. Özgür, V. Avrutin, Q. Fan, S.S. Akarca-Biyikli, and H. Morkoç, Jpn. J. Appl. Phys. 44 (2005) 7281.
[33] Ya.I. Alivov, E.V. Kalinina, A.E. Cherenkov, D.C. Look, B.M. Ataev, A.K. Omaev, M.V. Chukichev, and D.M. Bagnall, Appl. Phys. Lett. 83 (2003) 4719.
[34] K.A. Bulashevich, I. Yu. Evstratov, V.N. Nabokov, and S.Yu. Karpov, Appl. Phys. Lett. 87 (2005) 243502.
[35] S.Yu. Karpov, K.A. Bulashevich, I.A. Zhmakin, M.O. Nestoklon, V.F. Mymrin, and Yu.N. Makarov, Phys. Stat. Solidi (b) 241 (2004) 2668.
[36] Ya.I. Alivov, Ü. Özgür, S. Doğan, C. Liu, Y. Moon, X. Gu, V. Avrutin, Y. Fu, and H. Morkoç, Solid-State Electronics, to be published (2005).
[37] A. Osinsky, J.W. Dong, M.Z. Kauser, B. Hertog, A.M. Dabiran, P.P. Chow, S.J. Pearton, O. Lopatiuk, and L. Chernyak, Appl. Phys. Lett. 85 (2004) 4272.
[38] J.W. Dong, A. Osinsky, B. Hertog, A.M. Dabiran, P.P. Chow, Y.W. Heo, D.P. Norton, and S.J. Pearton, J. Electron. Mat. 34 (2005) 416.
[39] H.S. Yang, S.Y. Han, Y.W. Heo, K.H. Baik, D.P. Norton, S.J. Pearton, F. Ren, A. Osinsky, J.W. Dong, B. Hertog, A.M. Dabiran, P.P. Chow, L. Chernyak, T. Steiner, C.J. Kao, and G.C. Chi, Jpn. J. Appl. Phys. 44 (2005) 7296.
[40] S.Y. Ham, H. Yang, D.P. Norton, S.J. Pearton, F. Ren, A. Osinsky, J.W. Dong, B. Hertog, and P.P. Chow, J. Vac. Sci. Technol. B 23 (2005) to be published.
[41] J.-J. Chen, F. Ren, Y. Li, D.P. Norton, S.J. Pearton, A. Osinsky, J.W. Dong, P.P. Chow, and J.F. Weaver, Appl. Phys. Lett. 87 (2005) 192106.
[42] Ü. Özgür, Ya.I. Alivov, C. Liu, A. Teke, M.A. Reshchikov, S. Doğan, V. Avrutin, S.-J. Cho, and H. Morkoç, J. Appl. Phys. 98 (2005) 041301.

Chapter 16

Ferromagnetism in ZnO Doped with Transition Metal Ions

D. P. Norton and S. J. Pearton

Department of Materials Science and Engineering, University of Florida, Gainesville, FL 32611, USA

J. M. Zavada

Army Research Office, Research Triangle Park, NC 27709, USA

W. M. Chen and I. A. Buyanova

Department of Physics and Measurement Technology, Linköping University, 58183, Linköping, Sweden

Abstract: Spin-dependent phenomena in wide bandgap semiconductors such as ZnO may lead to devices with new or enhanced functionality, such as polarized solid-state light sources (spin light-emitting diodes), novel microprocessors and sensitive biological and chemical sensors. In this paper we review experimental results on transition metal doping of ZnO and the current state of theories for the ferromagnetism. It is important to re-examine some of the earlier concepts for spintronics devices, such as the spin field-effect transistor, to account for the presence of the strong magnetic field which has deleterious effects. In some of these cases, the spin device appears to have no advantage relative to the conventional charge-control electronic analog. The most promising materials for room temperature polarized light emission are expected to be GaN and ZnO, but progress on realizing devices has been disappointing. To date we have been unable to detect optical spin polarization in ZnO. The short spin-relaxation time observed likely results from the Rashba effect. Possible solutions involve either cubic phase ZnO or use of additional stressor layers to create a larger spin-splitting in order to get polarized light emission from these structures or to look at alternative semiconductors and fresh device approaches.

16.1 Introduction

There is high current interest in the development of dilute magnetic semiconductor materials (DMS) exhibiting ferromagnetic behavior for spin-based light-emitting diodes, sensors, and transistors.[1–3] These materials are formed through the

Zinc Oxide Bulk, Thin Films and Nanostructures
C. Jagadish and S. Pearton (Editors)

introduction of transition metal (TM) ions, such as Mn and Cr, or rare earth ions into the host semiconductors. Both types of ions have partially filled *d* and *f* shells, respectively, which give rise to unpaired electrons. The magnetic behavior of such materials depends upon the concentration of the TM ions in the crystal, the carrier density and the crystal quality. The term DMS refers to the fact that some fraction of the atoms in a non-magnetic semiconductor host like ZnO are replaced by magnetic ions. Generally, when 3*d* transition-metal ions are substituted for the cations of the host, their electronic structure is influenced by the strong 3*d* orbitals of the magnetic ion and the *p* orbitals of the neighboring anions. There are two interacting subsystems in DMS materials, namely the delocalized conduction band electrons and valence band holes and the random, diluted system of localized magnetic moments associated with the magnetic atoms.

The discovery of ferromagnetic ordering in Mn-doped narrow band gap semiconductors generated tremendous attention. However, DMS materials such as GaMnAs, InMnAs, and GaMnSb, have thus far shown relatively low magnetic ordering temperature (~170 K for GaMnAs), which limits their utility. Recently, many research groups have reported work directed towards achieving ferromagnetism at or above room temperature in wide bandgap materials, such as GaMnN and ZnMnO.[4–69] This property makes these materials attractive for use as non-volatile switching elements, where the bit state of the device is determined through control of electron spin. In addition, the integration of photonic (laser and light-emitting diodes), electronic (field-effect and bipolar transistors) and magnetic (information storage) devices on a single substrate, leading to a new class of electronic devices that offer multi-purpose functionality might be possible with the availability of DMS materials with high Curie temperatures. Figure 16.1 shows that the wider bandgap semiconductors, which tend to have smaller lattice constants, large *p*-*d*

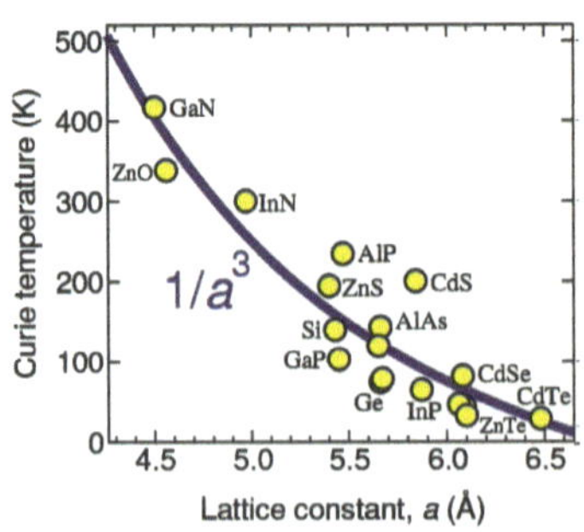

Figure 16.1: Predicted Curie temperature as a function of lattice constant for a variety of semiconductors (after S. C. Erwin (Naval Research Laboratory)). The materials predicted to have high T_C's have large *p-d* hybridization and small spin-orbit interaction.

hybridization and small spin-orbit interaction, are predicted to have higher Curie temperatures.[12]

16.2 Mechanisms for ferromagnetism

Mean field theory predicts that the Curie temperature for high carrier density DMS materials such as GaMnAs will be given by[1–3,12]

$$T_C = CN_{Mn}\beta^2 m^* p^{1/3}$$

where N_{Mn} is the concentration of uncompensated Mn spins, β is the coupling constant between localized Mn spins and the free holes (*p-d* coupling), m^* is the effective mass of the holes and p is the free hole concentration. It is now generally accepted that for low carrier density systems such as ZnO, mechanisms other than carrier-induced ferromagnetism are more likely. An example is shown in Figure 16.2, which shows a schematic of the formation of bound magnetic polarons in a system that also has direct anti-ferromagnetic coupling between closely spaced transition metal ions.[5] As the sample temperature is lowered, the polarons increase in radius and eventually overlap at the Curie temperature.

Much of the theory for ZnO DMS materials has come from Sato and Katayama-Yoshida.[6,13,33–37] An example is shown in Figure 16.3, where the ferromagnetic state is predicted to stabilize for most of the transition metal dopants except Mn. For ZnO, some of the theoretical predictions suggest that only *p*-type (Zn,Mn)O can lead to ferromagnetism (Figure 16.4). However most of the experimental observations of ferromagnetism have been for insulating (Zn,Mn)O, *n*-type (Zn,Mn)O, and undoped (Zn,Mn)O. Values of T_C above room temperature have been reported for Mn, Co, Sc, Ti and V-doped *n*-type ZnO powders and films deposited on variety of substrates (usually sapphire) by a variety of methods, including sintering and pulsed laser deposition. Table 16.1 shows a compilation of recent work in this area.[8,9]

The most recent model is that due to Coey et al.,[5] the so-called spin-split-orbit model shown schematically in Figure 16.5. This is consistent with the observed magnetization in *n*-type ZnO doped with transition metals. The ferromagnetic exchange is mediated by shallow donor electrons that form bound magnetic polarons, which in turn overlap to create a spin-split impurity band. Magnetic ions in different charge states couple by virtual hopping of the ‘extra’ electron from one ion to the other.[5] The 3*d* electrons in the partially occupied 3*d*-orbitals are allowed to hop to the 3*d*-orbitals of the neighboring TM, if neighboring TM ions have parallel magnetic moments. The ferromagnetic exchange is mediated by charge carriers in a spin-split

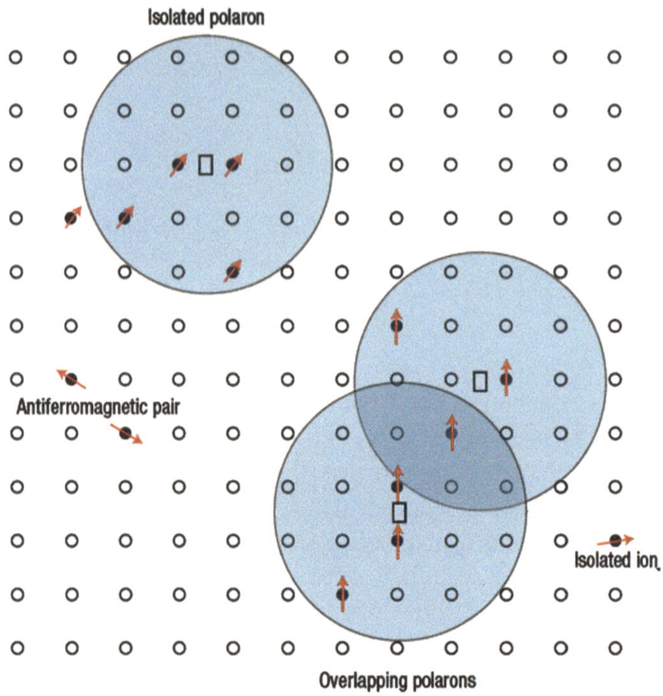

Figure 16.2: Representation of magnetic polarons. A donor electron in its hydrogenic orbit couples with its spin antiparallel to impurities with a $3d$ shell that is half-full or more than half-full. The figure is drawn for $x = 0.1$, $\gamma = 12$. Cation sites are represented by small circles. Oxygen is not shown; the unoccupied oxygen sites are represented by squares (after Coey et al.[5]).

impurity band formed by extended donor states. The impurity states hybridize with the d-orbits of the transition metal elements. It was shown that for Sc, Ti and V the spin-up states of the $3d$ transition metal element are aligned with the impurity levels, resulting in significant alignment. Similarly, for Fe, Co and Ni doping, the spin-down states perform the same function. It was suggested that Mn and Cr doping would not lead to strong magnetization due to a small hybridization,[5,50,51] although other reports suggest otherwise.

To be considered a true DMS, it is necessary to show that the carrier population is polarized. This can be done through transport measurements, especially the observation of the Anomalous Hall Effect (AHE) or by observation of polarized light emission from a quantum well using the DMS material as a spin injector. The latter has not yet been demonstrated for the ZnO system. The anomalous Hall effect is a ferromagnetic response of charge carriers in electrically conductive ferromagnets,

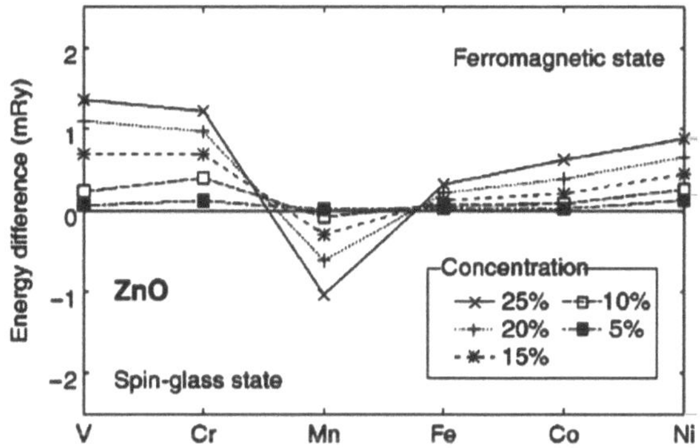

Figure 16.3: The stability of the ferromagnetic states in ZnO-based DMSs. V, Cr, Mn, Fe, Co, or Ni is doped as a magnetic impurity. The vertical axis is the energy difference per one formula unit between the ferromagnetic and the spin-glass state. A positive energy difference indicates that the ferromagnetic state is more stable than the spin-glass state (after K. Sato, and H. Katayama-Yoshida[34,37]).

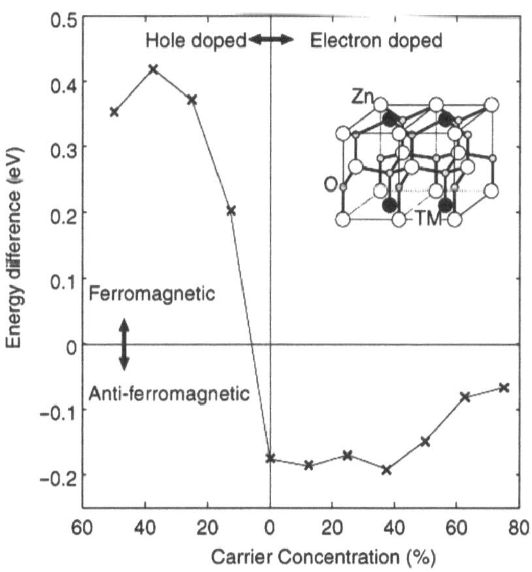

Figure 16.4: Stability of the ferromagnetic ordering of Mn magnetic moments in ZnO. The energy difference of $\Delta E = TE$ (anti-ferromagnetic ordering) − TE (ferromagnetic ordering) is plotted as a function of carrier concentration. The carrier concentration means N (hole doping) and Ga (electron doping) concentration in the supercell. The inset shows the simple orthorhombic supercell used in the preset calculations (after K. Sato, and H. Katayama-Yoshida[34,37]).

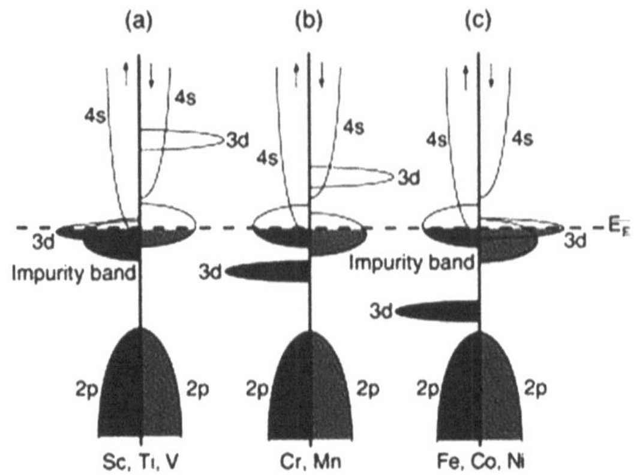

Figure 16.5: Schematic and structure of oxide with 3*d* impurities and spin-split donor impurity band. In (a) we show the position of the 3*d* level for low T_C when the splitting of the impurity band is small. In (b) and (c), respectively, we show the positions of the minority or majority spin 3d bands which lead to high T_C (after Coey et al.[5]).

producing a Hall voltage proportional to the magnetization. Its presence is generally attributed to asymmetric carrier scattering by magnetic impurities in the presence of spin-orbit interactions.[70,71] The Hall resistivity (ρ_{xy}) in ferromagnets is generally expressed as $\rho_{xy} = R_o B + R_s \mu_0 M$, where R_o is the ordinary Hall coefficient, B is the magnetic induction, R_s is the anomalous Hall coefficient, μ_0 is the magnetic permeability and M is the magnetization. The first term denotes the ordinary Hall effect, which is proportional to the inverse of the carrier density, while the second term denotes the anomalous Hall effect which is proportional to the magnetization.

The anomalous Hall effect represents ferromagnetic spin polarization of the charge carriers. Thus, the observation of the anomalous Hall effect was recognized as strong evidence for the intrinsic ferromagnetism of DMS. However, the AHE can be observed for nonmagnetic material embedded with magnetic clusters when their density is around the bulk percolation threshold.[70] The nonmagnetic material embedded with magnetic clusters may show an appreciable magnetoresistance in accordance with the saturation of the magnetization, implying the spin dependent scattering of the charge carriers between the magnetic clusters. Thus, even the presence of the AHE is not complete proof for the existence of a true DMS material.[70]

Ando et al.[24,27] reported magneto-optical effects in $Zn_{1-x}Co_xO$ thin films by magnetic circular dichroism (MCD), indicating $Zn_{1-x}Co_xO$ is a true DMS material with polarized carrier population. This method, which measures the optical transitions allowed under various optical polarizations involving split-off bands due to the Zeeman effect, which is enhanced by *sp-d* exchange interactions. ZnO films grown on lattice-matched $ScAlMgO_4$ substrates exhibited MCD structure at the

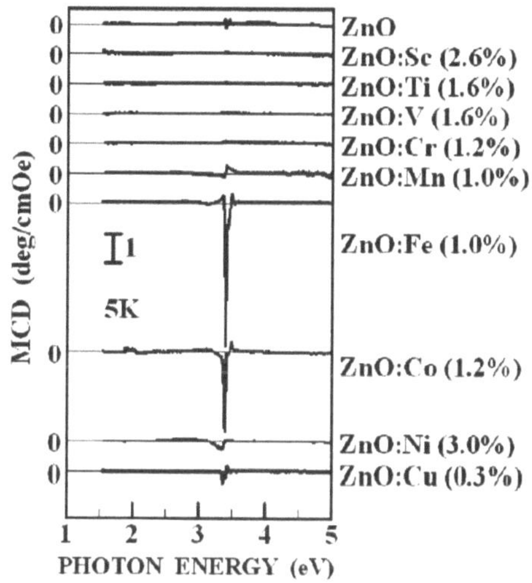

Figure 16.6: Transmission MCD spectra of ZnO and ZnO: transition metal (after Ando et al.[27]).

band gap of 3.4 eV. The MCD spectra of ZnO films doped with Sc, Ti, V, Cr, Mn, Co, Ni, and Cu using pulsed laser deposition were measured, as shown in Figure 16.6. Other films were grown on lattice-mismatched Al_2O_3 substrates. Those doped with Mn, Fe, Co, Ni, and Cu showed clear MCD structures near 3.4 eV. The films doped with Sc, Ti, V, and Cr did not exhibit any magneto-optical effect. The magnetic field and temperature dependence of the MCD spectra show that all of these latter films were paramagnetic. It is also clear from these data that factors such as the crystalline quality can influence the magnetic properties of DMS materials and this is certainly true of ZnO.[24,64]

16.3 Experimental results for ZnO DMS

Table 16.1 shows that there is still a wide variation reported for the magnetic properties of transition metal doped ZnO.[8,9,71] For example, ferromagnetism with $T_C > 300$ K was observed in $Zn_{1-x}(Co_{0.5}Fe_{0.5})_xO$ thin films prepared by magnetron co-sputtering and post annealed in vacuum. However, bulk $Zn_{1-x}Co_xO$ has been found to be antiferromagnetic in polycrystalline powder samples prepared by both solid-state and liquid-phase reactions. This antiferromagnetic behavior may result from Co clusters observed in the $Zn_{1-x}Co_xO$ powder, together with a population of interstitial Co atoms instead of substitutional Co.[71] Some experimental data show

Table 16.1: List of ZnO-based DMS recently reported (after refs. 8 and 9).

Compound	TM content	Substrate	Fabrication method	Growth temperature (°C)	Oxygen pressure (Torr)	Post-annealing	T_C (K)	Notes	Reference
ZnO:Mn	<0.35	*c*-Sapphire	PLD	600	5×10^{-5}		N/A		15
ZnO:Mn	0.36	*c*-Sapphire	PLD	600	5×10^{-5}		N/A	Spin-glass	14
$Zn_{1-x}TM_xO$		*c*-Sapphire	PLD	500–600	1×10^{-9} to 10^{-6}		N/A		16
ZnO:Co	0.02–0.5	*c*-Sapphire	PLD	300–700	1×10^{-6} to 10^{-1}			Spin-glass	17
ZnO:Mn	0.01–0.36	*c*-Sapphire	PLD	610	5×10^{-5}			Paramagnetic	18
ZnO:Mn	0.07	a-Sapphire	Sputtering	400	0.06			Paramagnetic	20
ZnO:Mn	0.03–0.2	GaAs(100)	Sputtering	500–600	8×10^{-4}				63
ZnO: (Co, Mn, Cr, or Ni)	0.05–0.25	*r*-Sapphire	PLD	350–600	$2\text{–}4 \times 10^{-5}$		280–300	$2\ \mu_B$/Co	28
ZnO:Ni	0.01–0.25	*c*-Sapphire	PLD	300–700	1×10^{-5}			Superpara- or ferro-magnetic	55
ZnO:V	0.05–0.15	*r*-Sapphire	PLD	300	10^{-5} to 10^{-3}		>350	$0.5\ \mu_B$/V	56

ZnO:(Co, Fe)	<0.15	SiO_2/Si	Magnetron sputtering	600	2×10^{-3}	600°C, 10 min, 1.0×10^{-5} Ton	>300	12–15 emu/cm^3	30
ZnO:Co	0.03–0.05	Bulk ZnO	Ion implantation			700°C, 5 min under O_2	>300	Oriented Co precipitates	57
ZnO:Co	0–0.25	*c*-Sapphire	Sol–gel	<350		700°C, 1 min	>350	0.56 μ_B/Co	58
ZnO:Mn	0–0.3	*c*-Sapphire	PLD				>30–45	0.15–0.17 μ_B/Mn	23
ZnO:Mn	<0.04		Sintered pellets	500–700°C	Air, atmospheric pressure		>425	0.006 emu/gm, single phase	38
ZnO:Mn	0.02	Fused quartz	PLD	400°C			>425	0.05 emu/gm, single phase	38,39
ZnO:(Fe, Cu)	0–0.1	Solid state reaction	897				550	0.75 μ_B/Fe	50
ZnO:Co	0.015		PLD	650°C	5×10^{-5}		>300	Ferromagnetic	60
ZnO:(Co, Al)	0.04–0.12	Glass	RF sputtering		1×10^{-2} in Ar		>350	0.21 μ_B/Co	61
ZnO:Mn	0.04–0.09	*c*-Sapphire	Reactive sputtering	200–380			>400	3 μ_B/Co	62
ZnO:(Mn, Sn)	0–0.3		Implantation			5 min, 700°C	250	Ferromagnetic	26

(*Continued*)

Table 16.1: (*Continued*)

Compound	TM content	Substrate	Fabrication method	Growth temperature (°C)	Oxygen pressure (Torr)	Post-annealing	T_C (K)	Notes	Reference
ZnO:Mn and Sn	Mn 0.03, Sn < 0.1	*c*-Sapphire	PLD	400–600°C	0.02		>300	Ferromagnetic	64
ZnO:Mn and Co	0.05–0.15		Crystalline precursor					Antiferromagnetic	65
ZnO:Mn and Co	<0.05	Bulk	Melt growth	1000°C				Paramagnetic	66
ZnO:Co	0.1	O-face ZnO	PLD					Antiferromagnetic	67
ZnO:Co	<0.35	*r*-Sapphire	MOCVD	350–600°C	40	20 min, 500°C in vacuum	>350	Ferromagnetic	68
ZnO:Co and Fe	<0.15	SiO_2/Si	Magnetron sputtering	600°C	2×10^{-3}	10 min, 600°C, 10^{-5} Torr	>300	Ferromagnetic, 12–15 emu/cm^3	69
ZnO:Mn	0.1	*r*-Sapphire	PLD	650°C	0.1		>300	0.075 μ_B/Mn	41
ZnO:Mn and Cu	0.05–0.1	*r*-Sapphire	PLD	650°C	0.1		400	0.1 μ_B/Mn	41
ZnO:Sc, Ti, V, Fe, Co, or Ni	0.05	*r*-Sapphire	PLD	600°C	0.1–750		>300	0.5 μ_B/Ti, 5.9 μ_B/Co, 0.3 μ_B/Sc	5,50
ZnO:Mn	0.02	Bulk pellets	Powder, pellets and laser-ablated films	500°C			>300	0.16 μ_B/Mn	38,39
ZnO:Cr		STO	PLD				>400	Ferromagnetic	40
ZnO:Mn	0.08	Tetrapods	Evaporation	600°C			43	(Zn,Mn)Mn_2O_4 phases	54
ZnO:Mn	0.05	ZnO sub	PLD	200–600°C			250	High T_C fpr lower T_G	51

that homogeneous films of $Zn_{1-x}Co_xO$ exhibit spin-glass behavior, whereas inhomogeneous $Zn_{1-x}Co_xO$ films are more likely to demonstrate room-temperature ferromagnetism.[8,9,71] This is evidence that Co clusters might be the source of the high T_C ferromagnetism in some $Zn_{1-x}Co_xO$. Norton et al. observed such an effect for ZnO films implanted with high doses of Co ions.[57] Given that the concentrations of TM ions needed to achieve high Curie temperatures are at or above the solid solubility, it is clear that small variations in the growth parameters will lead to a variety of outcomes, ranging from uniform alloys to clustering of TM atoms to precipitation and formation of second phases that may contribute to the observed magnetic properties.

Fukumura et al.[14–16] first reported ZnMnO grown by pulsed laser deposition (PLD). Up to 35% Mn was incorporated into ZnO without degrading the crystallographic quality. This is in sharp contrast to III-V DMS, where about 5–9% is tolerable before the crystallinity breaks down[71] and precipitation is detectable by x-ray diffraction, transmission electron microscopy and chemical profiling techniques. The high solubility of Mn in ZnO is due to the exactly half-filled $3d$ orbitals of Mn^{2+}. It costs a considerable amount of energy to add an electron with opposite spin to the $3d^5$ orbital, so that the Mn behaves similarly to Zn.[71] The reports of high solubility of Mn in ZnO led to many efforts on synthesis of TM-doped ZnO, as detailed in Table 16.1. The growth of these ferromagnetic materials by thin film techniques provides excellent control of the TM dopant concentration and the ability to grow single-phase layers under optimized conditions, as measured by conventional X-ray diffraction and transmission electron microscopy measurements. However, in some cases the results concerning the existence of ferromagnetism have been somewhat controversial, partially because of the sensitivity of the resulting magnetic properties to the exact growth conditions and also to incomplete characterization of the magnetic properties. While some groups have reported ferromagnetism in (Zn,TM)O systems, others report observations of antiferromagnetic or spin-glass behavior. The reported values of T_C scatter from 25 to >550 K. Similarly, conflicting results also exist concerning the distribution of transition metal ions such as Mn.[71] In RF magnetron sputter-deposited (Zn,Mn)O, Cheng et al.[20] found Mn to be distributed homogeneously, while by contrast, Jin et al.[16] reported clustering of Mn atoms. Once again, the exact growth conditions are crucial in determining the magnetic properties of the material at these high impurity concentrations.[8,9,70,71] A key issue in many of the published reports is whether the resulting material is indeed an alloy of transition-metal elements with the host material or whether it remains as the host material with clusters, precipitates or second phases that are responsible for the observed magnetic properties. A relatively complete characterization of the DMS would involve magnetic hysteresis measurements as well as field-cooled and zero field-cooled magnetization, magnetotransport, high resolution transmission electron microscopy, chemical bonding information obtained from X-Ray Photoelectron Spectroscopy (XPS) and lattice location measurements by ion channeling or Extended X-Ray Absorption Fine Structure (EXAFS). In most cases, such a detailed characterization is not carried out. The ZnO system provides an excellent platform for studying such effects, since most of the secondary phases that could form cannot explain the observed magnetism.

16.4 Theory for ferromagnetism in DMS

The mean-field Zener model proposed by Dietl et al.[12] has been successful in explaining the transition temperatures observed for III-V DMS such as *p*-(Ga,Mn)As and II-VI DMS such as (Zn,Mn)Te. The mean-field Zener theory is based on the original model of Zener and the Ruderman-Kittel-Kasuya-Yoshida (RKKY) interaction.[5,70,71] In the Zener model, the direct interaction between *d* shells of adjacent Mn atoms (super-exchange) leads to an antiferromagnetic configuration of the *d* shell spins because the Mn-*d* shell is half-filled. By contrast, the indirect coupling of spins through the conduction electrons tends to align the spins of the incomplete *d* shells in a ferromagnetic manner. It is only when this coupling dominates over the direct super-exchange coupling between adjacent *d* shells that ferromagnetism is present.[12,71] Accordingly, the mean-field approach assumes that the ferromagnetism occurs through interactions between the local moments of the Mn atoms mediated by free holes in the material. This approach works well for materials in which the transition metal dopants introduces a shallow level in the bandgap and therefore the material achieves a near-metallic conductivity when percent levels of the TM dopant are introduced on substitutional sites within the host.

The spin–spin coupling in the Zener approach is also assumed to be a long-range interaction, allowing the use of a mean-field approximation.[12,71] The mean-field model calculates the effective spin-density due to the Mn ion distribution within the host material. The direct Mn-Mn interactions are antiferromagnetic (AF) so that the Curie temperature T_C, for a given material with a specific Mn concentration and hole density (derived from Mn acceptors and/or intentional additional shallow level acceptor doping), is determined by a competition between the ferromagnetic and antiferromagnetic interactions.[12,71] As compared to the RKKY interaction, the mean-field Zener model takes into account the anisotropy of the carrier-mediated exchange interaction associated with the spin-orbit coupling in the host material. In the process it reveals the important effect of the spin-orbit coupling in the valence band in determining the magnitude of the T_C and the direction of the easy axis in DMS.[71] Based on this model, it was predicted that TM-doped *p*-type GaN and ZnO, as shown in Figure 16.1, are the most promising candidates for ferromagnetic DMS with high Curie temperature.[12] However, one aspect of these predictions that is usually overlooked is that they assumed very high hole concentrations ($>10^{20}\,cm^{-3}$) and large Mn contents (5 at.%). While the latter is possible in ZnO, the former is not possible and indeed the relatively low carrier densities in most wide bandgap semiconductor-based DMS precludes the achievement of the classical carrier-induced ferromagnetism.

To overcome the limitations of the Zener-type carrier-induced models for low carrier density materials where the TM dopant may not necessarily introduce additional carriers, Sato and Katayama-Yoshida[6,13,33–37] performed first principles *ab initio* calculations of the electronic structures of TM-doped ZnO and proposed the double exchange mechanism for the ferromagnetism. In the double exchange mechanism, also originally proposed by Zener, magnetic ions in different charge states couple with each other by virtual hopping of the extra electron from one ion to the other. In the DMS material, if neighboring TM magnetic moments are in the same direction, the TM-*d* band is widened by the hybridization between the spin-up states. Therefore, in the ferromagnetic configuration the band energy can be lowered by introducing carriers in the *d* band. In these cases, the 3*d* electron in the partially occupied 3*d*-orbitals of the TM is allowed to hop to the 3*d*-orbitals of the neighboring TM, if neighboring TM ions have parallel magnetic moments. As a result, the *d*-electron lowers its kinetic energy by hopping in the ferromagnetic state, or the so-called double exchange mechanism.

Sato and Katayama-Yoshida have reviewed their first principles approaches to theory of magnetism in DMS previously.[6,13,33–37] The basic approach is to calculate the total energy and electronic structures for the TM-doped material using density functional theory (DFT). The magnetic state of the DMSs can be investigated by calculating the electronic structure of a ferromagnetic DMS in which all the magnetic moments of the substitutional ions are parallel to each other and that of a spin-glass like state in which there are at least two components of transition metal ions whose magnetic moments are anti-parallel.[6,13,33–37,71] The total energy of the ferromagnetic state minus that of the spin-glass state is calculated as a function of transition-metal composition. This approach was used to predict that the TM-doped ZnO is a candidate for high-T_C ferromagnetic DMS.[6,13,33–37] The first principle calculations predicted that transition metals V, Cr, Fe, Co, and Ni-doped ZnO would exhibit ferromagnetism at TM concentrations from 5% to 25%, whereas the Mn-doped ZnO would be antiferromagnetic in the ground state because of the exact half-filled d^5 state of Mn ions. In these calculations, the ferromagnetic state in Mn-doped ZnO was found to be stabilized by hole doping due to the double exchange mechanism as shown in Figure 16.4. It is important to realize that T_C values well in excess of room temperature, which would be needed since devices operate in excess of this temperature due to self-heating, were only predicted for high TM concentrations where it is expected to be difficult to reproducibly incorporate such large dopant fractions.

Ferromagnetism in DMS has also been explained by the formation of bound magnetic polarons (BMP).[5,71] The bound magnetic polarons are formed by interaction of many transition-metal spins with that of much lower number of weakly bound carriers such as excitons. The localized holes of the polarons act on the transition-metal impurities surrounding them, creating an effective magnetic field and aligning all the spins. As temperature decreases the interaction distance increases and neighboring magnetic polarons overlap via magnetic impurities to form correlated clusters of polarons. The ferromagnetic transition occurs when the size of the overlapping clusters reaches that of the sample itself. This model is very attractive for low carrier density systems like ZnO. The polaron model is applicable to both *p*- and *n*-type host materials, which has long been a criticism of the competing models where generally holes are needed to stabilize the ferromagnetism. Experimentally, of course, all the reports of ferromagnetism in ZnO have come in *n*-type or insulating material. Even though the direct exchange interaction of the localized holes is antiferromagnetic, the interaction between bound magnetic polarons may be ferromagnetic at large enough concentrations of magnetic impurities.

The first-principle calculations also suggested that the *n*-type doping in ZnO can increase the Curie temperature of Fe-, Co- and Ni-doped samples when the effects of disorder are taken into account by what is called the coherent potential approximation.[6,13,33–37] Considering that *n*-type ZnO is readily available and the intrinsic defects such as O-vacancies and Zn-interstitials form donor states, it was concluded that (Zn,Fe)O, (Zn,Co)O and (Zn,Ni)O are promising candidates for high-T_C ferromagnets. It was also suggested that (Zn,Mn,Fe)O, (Zn,Mn,Co)O, or (Zn,Mn,Ni)O may show carrier-induced ferromagnetism under electron doping by tuning the ratio of Mn to Fe, Co or Ni.[6,13,33–37]

16.5 Spin relaxation

Detailed time-resolved magneto-PL results from undoped and Mn-doped ZnO spin-LED structures have recently been performed in an effort to detect polarized light emission. The preliminary conclusions can be summarized as follows:

1. In an applied magnetic field, the ZnO PL becomes polarized to about $P = (\sigma^+ - \sigma^-)/(\sigma^+ - \sigma^-) \approx 10\%$ at 6 T. Representative time-integrated PL spectra are shown in Figure 16.7. The polarization degree decreases with increasing

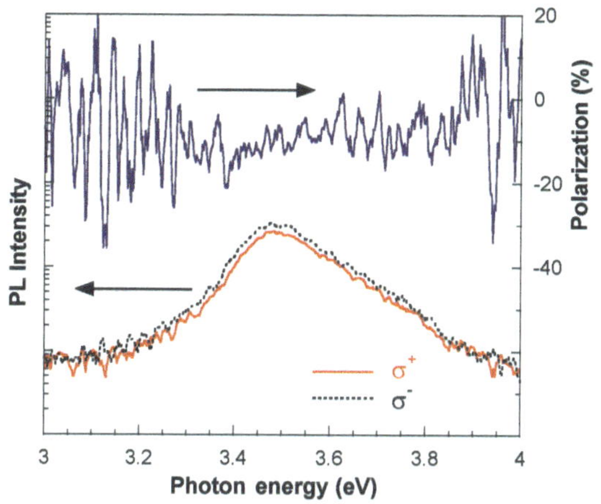

Figure 16.7: (top) PL spectra at 2 K from Zn(Mn,Mg)O/ZnO/AlGaN spin LEDs grown on sapphire, as well as their polarization properties at (bottom).

emission energy where contributions from MgZnO and possibly also AlGaN barriers dominate. The behavior is similar for both undoped and doped ZnO.

2. By monitoring the low-energy side of the PL emission from ZnO, PL polarization is zero immediately after the laser pulse but thereafter undergoes a rise within the first ~100 ps. This observation is similar to that seen in InGaN spin detectors,[72–74] except a change of the sign in PL polarization indicating a reverse in ordering of exciton spin states between ZnO and InGaN. Considering the similar band structure between the two materials, we follow the conclusion drawn for InGaN[72–74] and tentatively attribute the observed polarization rise to the spin relaxation near the bottom of the exciton band.
3. No information on presumably faster spin relaxation accompanying energy relaxation of hot carriers/excitons, which is relevant to spin injection from barrier materials, could be obtained from the present results. For that optical orientation experiments have to be performed, although given that the valence band structure of ZnO is rather similar to that of GaN, the outlook is not hopeful.[72–74] It appears the spin relaxation occurs mostly during energy relaxation processes at a high momentum that is beyond the reach of commonly available experimental techniques, as is true for the previously reported InGaN quantum wells.[72–74]

If one can design a resonant tunneling diode (RTD) such that the polarized spins carried by either electrons or holes can be injected directly to the ground state of the corresponding carriers in a spin detector without involving energy and momentum relaxation, there may have a hope to preserve the spin polarization much better.

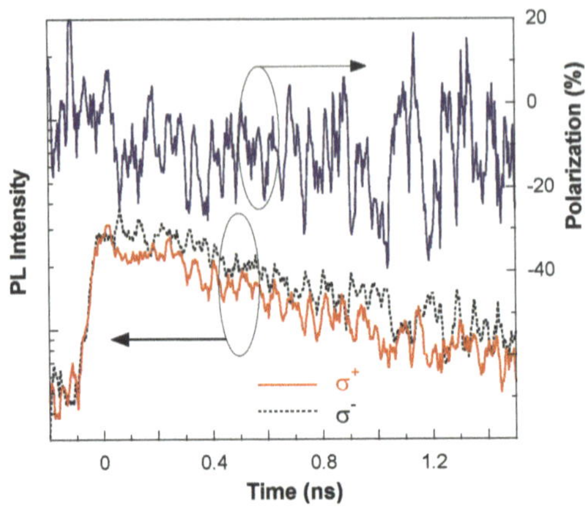

Figure 16.8: (top) Time resolved PL intensity measured from ZnO (detected within the spectral range of 365–432 nm) from (the same LED structures shown in Figure 16.7), as well as the polarization properties at a field of 6 T (bottom).

Spin relaxation within the spin sublevels of the ground state seems to be within the range that is detectable, based on past results from InGaN.[72–74] Viewing from the observed 50 ps spin relaxation time for the exciton ground state in InGaN QW, a spin LED or a device that utilizing ZnO-based materials for spin transport has to operate on a similar or shorter time scale to be effective even with a RTD-type of spin injection. For example, a pulsed operation mode should be employed.

16.6 Spintronic devices

In recent work, spin-dependent tunneling between two ferromagnetic films separated by an insulating (I) film shows junction magnetoresistance (JMR) of better than 20%.[75–78] This suggests that tunneling may be a more effective way of achieving spin injection than diffusive transport. A large magnetoresistance can also be obtained if the tunnel barrier is also ferromagnetic. A prototype device, shown in Figure 16.9, comprises a non-magnetic electrode, a ferromagnetic insulating tunnel barrier (the polarizer), and a ferromagnetic counterelectrode (the analyzer). Below T_C of the ferromagnetic insulator, the tunnel barrier is spin split, giving a highly polarized tunnel current as indicated in the schematic. If the moments of the ferromagnetic counterelectrode analyzer are parallel (antiparallel) to the moments of the spin polarized tunnel current, then the resistance is low (high). Magnetoresistance exceeding 100% has been obtained in an Al-EuS-Gd device,[79] where EuS, the ferromagnetic insulator, has a $T_C = 16.8$ K and Gd, the analyzer, a T_C near room temperature.

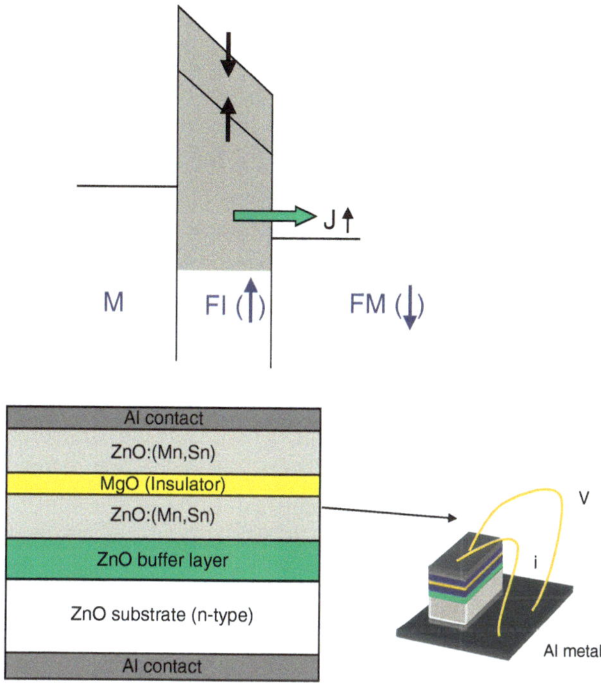

Figure 16.9: Spin tunnel device in which the insulating tunnel barrier is ferromagnetic (top) and possible embodiment in the ZnO materials system (bottom).

One key objective in the future will be to find materials that enable operation of these novel spin filter devices at room temperature. Promising barrier materials might be transition metal doped ZnO, insulating ferrites ($CoFe_2O_4$), or insulating DMS materials with localized carriers to provide the necessary magnetic interactions. For satisfactory operation the localization length ξ_L must be less than the sample size (barrier thickness) but greater than the length characterizing the magnetic interactions. An example of such a device based on ZnO is shown at the bottom of Figure 16.9.

A number of initial device configurations for GaMnAs have already been demonstrated at low temperatures. This list includes exchange-biased samples, spin-dependent resonant tunneling diodes, magnetic tunnel junctions and spin-polarized light-emitting diodes (spin LEDs).[75–82] The presence of tunable wavefunction overlap between magnetic ions and carriers confined to quantum structures has also been reported.[81] More controversial are reports of very large magneto-transport effects, the so-called "giant planar Hall effect".[83] There are also numerous theoretical analyses of new bipolar device configurations combining *n*-doped semiconductors with GaMnAs.

However, recent reexamination of the spinfet concept has revealed serious potential shortcomings. Cahay and Bandyopadhyay[84–87] modeled phase-coherent spin transport in the weakly disordered quasi-one-dimensional channel of a gate-controlled electron spin interferometer. When the effects of an axial magnetic field in the channel of the interferometer (caused by the ferromagnetic contacts), a Rashba spin-orbit interaction, and elastic (nonmagnetic) impurity scattering are all considered, it was shown that in the presence of an axial magnetic field, nonmagnetic impurities can cause spin relaxation in a manner similar to the Elliott-Yafet mechanism. Bandyopadhay and Cahay[84–87] suggest that is generally untrue that spin-based devices will be faster and consume less power than their electronic counterparts. They did an analysis of the switching voltages in a one-dimensional spinfet relative to a Si-based MOSFET and found that the former is actually not a lower power device. Their main conclusion was that unless materials with extremely strong spin-orbit interaction can be developed, the spintronic devices will not measure up to their electronic analogs. Several other manifestations of the spinfet have recently been proposed, but in each case, doubts have been raised whether these present versions show any advantage over the original Datta-Das design. Bandyopadhyay and Cahay[84–87] concluded that it is unlikely that currently proposed spinfets will play a significant role in digital, analog or mixed signal circuits but may be more suited to applications in memory, where high gain, high frequency, etc. are not necessary. They also concluded that spintronic devices may also have better noise margin since spin does not easily couple to stray electric fields (unless the host material has very strong spin-orbit interactions). Thus they concluded that it is also possible that spintronics may be able to outpace electronics in nonconventional applications such as single spin logic, spin neurons and using spin in a quantum dot to encode qubits. There is still hope that integration of spintronic devices with existing electronic and photonic devices and circuits may offer new functionalities and that this is where the DMS materials may have the most impact.

16.7 Future work

More detailed studies of the carrier type and concentration dependence of magnetization are needed in all of the transition metal doped oxides. In particular, methods such as Extended X-Ray Absorption Fine Structure (EXAFS) or X-Ray Photoelectron Spectroscopy (XPS) that establish lattice location or chemical state should be applied in order to give more insight into possible mechanisms for the observed ferromagnetism. Spin-sensitive local probe microscopy will also be highly desirable in order to understand spin interactions and magnetism on the nano and atomic scale. Spin polarized injection, transport and detection experiments in ZnO or

ZnCdO/ZnMgO heterostructures are lacking at present as are clear demonstrations of carrier- or light-induced ferromagnetism.[70,71] The ZnO system is very promising for demonstrating polarized light emission, given the recent breakthroughs in achieving *p*-type conduction. Alternatively, *p-n* oxide heterojunctions could be used for either electrical or optical detection of spin transport. Device demonstrations of the exploitation of spin effects in oxides are the key to advancing both the science and technology of these materials for spintronic applications.[8,8,70,71]

References

[1] P. Kacman, Semicond. Sci. Technol. **16**, R25 (2001).
[2] J. K. Furdyna, J. Appl. Phys. **64**, R29 (1988).
[3] S. A. Wolf, D. D. Awschalom, R. A. Buhrman, J. M. Daughton, S. von Molnár, M. L. Roukes, A. Y. Chtchelkanova, and D. M. Treger, Science **294**, 1488 (2001).
[4] S. J. Pearton, C. R. Abernathy, M. E. Overberg, G. T. Thaler, D. P. Norton, N. Theodoropoulou, A. F. Hebard, Y. D. Park, F. Ren, J. Kim, and L. A. Boatner, J. Appl. Phys. **93**, 1 (2003).
[5] See, for example, J. M. D. Coey, M. Venkatesan, and C. B. Fitzgerald, Nat. Mater. **4**, 173 (2005) and references therein; M. Venkatesan, C. B. Fitzgerald, and J. M. D. Coey, Nature **430**, 630 (2004); J. M. D. Coey, J. Appl. Phys. **97**, 10D313 (2005).
[6] K. Sato, and H. Katayama-Yoshida, Semicond. Sci. Technol. **17**, 367 (2002).
[7] W. Prellier, A. Fouchet, and B. Mercey, J. Phys.: Condens. Matter **15**, R1583 (2003).
[8] T. Fukumura, Y. Yamada, H. Toyosaki, T. Hasegawa, H. Koinuma, and M. Kawasaki, Appl. Surf. Sci. **223**, 62 (2004).
[9] S. J. Pearton, W. H. Heo, M. Ivill, D. P. Norton, and T. Steiner, Semicond. Sci. Technol. **19**, R59 (2004).
[10] S. J. Pearton, C. R. Abernathy, G. T. Thaler, R. M. Frazier, D. P. Norton, F. Ren, Y. D. Park, J. M. Zavada, I. A. Buyanova, W. M. Chen, and A. F. Hebard, J. Phys.: Condens. Matter **16**, R209 (2004).
[11] I. Zutic, J. Fabian, and S. D. Sarma, Rev. Mod. Phys. **76**, 323 (2004).
[12] T. Dietl, H. Ohno, F. Matsukura, J. Cibert, D. Ferrand, Science **287**, 1019 (2000).
[13] K. Sato, and H. Katayama-Yoshida, Jpn. J. Appl. Phys. **40**, L334 (2001).
[14] T. Fukumura, Z. Jin, M. Kawasaki, T. Shono, T. Hasegawa, S. Koshihara, and H. Koinuma, Appl. Phys. Lett. **78**, 958 (2001).
[15] T. Fukumura, Z. Jin, A. Ohtomo, H. Koinuma, and M. Kawasaki, Appl. Phys. Lett. **75**, 3366 (1999).
[16] Z. Jin, T. Fukumura, M. Kawasaki, K. Ando, H. Saito, T. Sekiguchi, Y. Z. Yoo, M. Murakami, Y. Matsumoto, T. Hasegawa, and H. Koinuma, Appl. Phys. Lett. **78**, 3824 (2001).
[17] J.-H. Kim, H. Kim, D. Kim, Y.-E. Ihm, and W.-K. Choo, J. Appl. Phys. **92**, 6066 (2002).

[18] A. Tiwari, C. Jin, A. Kvit, D. Kumar, J. F. Muth, and J. Narayan, Solid State Commun. **121**, 371 (2002).
[19] S. J. Han, T. H. Jang, Y. B. Kim, B. G. Park, J. H. Park, and Y. H. Jeong, Appl. Phys. Lett. **83**, 920 (2003).
[20] X. M. Cheng, and C. L. Chien, J. Appl. Phys. **93**, 7876 (2003).
[21] Z. Jin, Y.-Z. Yoo, T. Sekiguchi, T. Chikyow, H. Ofuchi, H. Fujioka, M. Oshima, and H. Koinuma, Appl. Phys. Lett. **83**, 39 (2003).
[22] K. Ando, H. Saito, V. Zayets, and M. C. Debnath, J. Phys.: Condens. Matter **16**, S5541 (2004).
[23] S. W. Jung, S.-J. An, G.-C. Yi, C. U. Jung, S.-I. Lee, and S. Cho, Appl. Phys. Lett. **80**, 4561 (2002).
[24] Y. W. Heo, M. P. Ivill, K. Ip, D. P. Norton, and S. J. Pearton, Appl. Phys. Lett. **84**, 2292 (2004).
[25] S. W. Lim, M. C. Jeong, M. H. Ham, and J. M. Hyoung, Jpn. J. Appl. Phys. **43**, L280 (2004).
[26] D. P. Norton, S. J. Pearton, A. F. Hebard, N. Theodoropoulou, L. A. Boatner, and R. G. Wilson, Appl. Phys. Lett. **82**, 239 (2003).
[27] K. Ando, H. Saito, Z. Jin, T. Fukumura, M. Kawasaki, Y. Matsumoto, and H. Koinuma, J. Appl. Phys. **89**, 7284 (2001).
[28] K. Ueda, H. Tabata, and T. Kawai, Appl. Phys. Lett. **79**, 988 (2001).
[29] K. Ando, H. Saito, Z. Jin, T. Fukumura, M. Kawasaki, Y. Matsumoto, and H. Koinuma, Appl. Phys. Lett. **78**, 2700 (2001).
[30] Y. M. Cho, W. K. Choo, H. Kim, D. Kim, and Y. E. Ihm, Appl. Phys. Lett. **80**, 3358 (2002).
[31] S. W. Yoon, S.-B. Cho, S. C. We, S. Yoon, B. J. Suh, H. K. Song, and Y. J. Shin, J. Appl. Phys. **93**, 7879 (2003).
[32] J. H. Kim, H. Kim, Y. E. Ihm, and W. K. Choo, J. Appl. Phys. **92**, 6066 (2002).
[33] K. Sato, and H. Katayama-Yoshida, Jpn J. Appl. Phys. **40**, L651 (2001).
[34] K. Sato, and H. Katayama-Yoshida, Physica E **10**, 251 (2001).
[35] T. Yamamoto, and H. Katayama-Yoshida, Jpn. J. Appl. Phys. **38**, L166 (1999).
[36] H. Katayama-Yoshida, and K. Sato, Physica B **327**, 337 (2003).
[37] K. Sato, and H. Katayama-Yoshida, Physica B **308**, 904 (2001).
[38] P. Sharma, A. Gupta, K. V. Rao, F. J. Owens, R. Sharma, R. Ahuja, J. M. Osorio Guillen, B. Johansson, and G. A. Gehring, Nat. Mater. **2**, 673 (2003).
[39] P. Sharma, A. Gupta, F. J. Owens, A. Inoue, and K. V. Rao, J. Mag. Magn. Mater. **282**, 115 (2004).
[40] I. Satoh, and T. Kobayashi, Appl. Surf. Sci. **216**, 603 (2003).
[41] N. H. Hong, V. Brize, and J. Sakai, Appl. Phys. Lett. **86**, 082505 (2005).
[42] N. H. Hong, J. Sakai, and A. Hassini, J. Phys.: Condens. Matter **17**, 199 (2005).
[43] Y. H. Jeong, S. J. Han, J. H. Park, and Y. H. Lee, J. Magn. Mag. Mater. **272–276**, 1976 (2004).
[44] D. A. Schwartz, K. R. Kittilstved, and D. R. Gamelin, Appl. Phys. Lett. **85**, 1395 (2004).

[45] P. V. Radovanovic, and D. R. Gamelin, Phys. Rev. Lett. **91**, 157202 (2003).
[46] T. Mizokawa, T. Nambu, A. Fujimori, T. Fukumura, and M. Kawasaki, Phys. Rev. B **65**, 085209 (2002).
[47] C.-H. Chien, S. H. Chiou, G. Y. Guo, and Y.-D. Yao, J. Magn. Mag. Mater. **282**, 275 (2004).
[48] X. Feng, J. Phys.: Condens. Matter **16**, 4251 (2004).
[49] S. Y. Yun, G.-B. Cha, Y. Kwon, S. Cho, S. C. Soon, and C. Hong, J. Magn. Mag. Mater. **272–276**, E1563 (2004).
[50] M. Venkatesan, C. B. Fitzgerald, J. G. Lunney, and J. M. D. Coey, Phys. Rev. Lett. **93**, 177206 (2004).
[51] K. W. Nielsen, J. B. Phillip, M. Opel, A. Erb, J. Simon, L. Alff, and R. Gross, Superlatt. Micro. (in press).
[52] M. Kunisu, F. Oba, H. Ikeno, I. Tananka, and T. Yamamoto, Appl. Phys. Lett. **86**, 121902 (2005).
[53] E. Rita, U. Wahl, J. G. Correia, E. Alves, and J. C. Soares, Appl. Phys. Lett. **85**, 4899 (2004).
[54] R. K. Zheng, H. Liu, X. X. Zhang, V. A. L. Roy, and A. B. Djurisic, Appl. Phys. Lett. **85**, 2589 (2004).
[55] T. Wakano, N. Fujimura, Y. Morinaga, N. Abe, A. Ashida, and T. Ito, Physics E **10**, 260 (2001).
[56] H. Saeki, H. Tabata, and T. Kawai, Solid State Commun. **120**, 439 (2001).
[57] D. P. Norton, M. E. Overberg, S. J. Pearton, K. Pruessner, J. D. Budai, L. A. Boatner, M. F. Chisholm, J. S. Lee, Z. G. Khim, Y. D. Park, and R. G. Wilson, Appl. Phys. Lett. **83** (2003).
[58] H. J. Lee, S. Y. Jeong, C. R. Cho, and C. H. Park, Appl. Phys. Lett. **81**, 4020 (2002).
[59] S. J. Hahn, J. W. Song, C. H. Yang, S. H. Park, J. H. Park, Y. H. Jeong, and K. W. Rhie, Appl. Phys. Lett. **81**, 4212 (2002).
[60] K. Rode, A. Anane, R. Mattana, J.-P. Contour, O. Durand, and R. LeBourgeois, J. Appl. Phys. **93**, 7676 (2003).
[61] S. G. Yang, A. B. Pakhomov, S. T. Hung, and C. Y. Wong, IEEE Trans. Magn. **38**, 2877 (2002).
[62] N. Theodoropoulou, G. P. Berera, V. Misra, P. LeCalir, J. Philip, J. S. Moodera, B. Satapi, and T. Som (to be published).
[63] S. Lim, M. Jeong, M. Ham, and J. Myoung, Jpn. J. Appl. Phys. 2B **43**, L280 (2004).
[64] M. Ivill, S. J. Pearton, D. P. Norton, J. Kelly, and A. F. Hebard, J. Appl. Phys. **97**, 053904 (2005).
[65] G. Lawes, A. S. Risbud, A. P. Ramirez, and R. Seshadri, Phys. Rev. B **71**, 045201 (2005).
[66] M. H. Kane, K. Shalini, C. J. Summers, R. Varatharajan, J. Nause, C. R. Vestal, Z. J. Zhang, and I. T. Ferguson, J. Appl. Phys. **97**, 023906 (2005).
[67] N. Jedrecy, H. J. von Bardeleben, Y. Zheng, and J.-L. Cantin, Phys. Rev. B **69**, 041308 (2004).

[68] A. C. Tuan, J. D. Bryan, A. B. Pakhomov, V. Shutthanandan, S. Thevuthasan, D. E. McCready, D. Gaspar, M. H. Engelhard, J. W. Rogers, Jr., K. Krishnan, D. R. Gamelin, and S. A. Chambers, Phys. Rev. B **70**, 054424 (2004).
[69] Y. M. Cho, W. K. Choo, H. Kim, D. Kim, and Y. E. Ihm, Appl. Phys. Lett. **80**, 3358 (2002).
[70] T. Fukumura, H. Toyosaki, and Y. Yamada, Semicond. Sci. Technol. **20**, S103 (2005).
[71] C. Liu, F. Yun, and H. Morkoç, J. Mater. Sci.: Mater. In Electronics (in press).
[72] I. A. Buyanova, M. Izadifard, L. Storasta, W. M. Chen, J. Kim, F. Ren, G. Thaler, C. R. Abernathy, S. J. Pearton, C.-C. Pan, G.-T. Chen, J.-I. Chyi, and J. M. Zavada, J. Electron. Mater. **33**, 467 (2004).
[73] I. A. Buyanova, M. Izadifard, W. M. Chen, J. Kim, F. Ren, G. Thaler, C. R. Abernathy, S. J. Pearton, C. Pan, G. Chen, J. Chyi, and J. M. Zavada, Appl. Phys. Lett. **84**, 2599 (2004).
[74] I. A. Buyanova, J. P. Bergman, W. M. Chen, G. Thaler, R. Frazier, C. R. Abernathy, S. J. Pearton, J. Kim, F. Ren, F. V. Kyrychenko, C. J. Stanton, C.-C. Pan, G.-T. Chen, J.-I. Chyi, and J. M. Zavada, J. Vac. Sci. Technol. B **22**, 2668 (2004); W. M. Chen, I. A. Buyanova, K. Nishibayashi, K. Kayanuma, K. Seo, A. Murayama, Y. Oka, G. Thaler, R. Frazier, C. R. Abernathy, F. Ren, S. J. Pearton, C.-C. Pan, G.-T. Chen, and J.-I. Chyi, Appl. Phys. Lett. **87**, 192107 (2005).
[75] R. Jansen, J. Phys. D **36**, R289–R308 (2003).
[76] P. Van Dorpe, V. F. Motsnyi, M. Nijboer, E. Goovaerts, V. I. Safarov, J. Das, W. Van Roy, G. Borghs, and J. De Boeck, Jpn. J. Appl. Phys. **42**, L502 (2003).
[77] H. J. Zhu, M. Ramsteiner, H. Kostial, M. Wassermeier, H. P. Schönherr, and K. H. Ploog, Phys. Rev. Lett. **87**, 016601 (2001).
[78] V. F. Motsnyi, J. De Boeck, J. Das, W. Van Roy, G. Borghs, E. Goovaerts, and V. Safarov, Phys. Lett. **81**, 265 (2002).
[79] P. E. LeClair, J. K. Ha, H. J. M. Swagten, J. T. Kohlhepp, C. H. van de Vin, and W. J. M. de Jonge, Appl. Phys. Lett. **80**, 625 (2002).
[80] M. Tanaka, and Y. Higo, Phys. Rev. Lett. **87**, 026602 (2001).
[81] R. Mattana, Phys. Rev. Lett. **90**, 166601 (2003).
[82] Y. Ohno, Nature **402**, 790 (1999).
[83] H. X. Tang, R. K. Kawakami, D. D. Awschalom, and M. L. Roukes, Phys. Rev. Lett. **90**, 107201 (2003).
[84] S. Bandyopadhyay, and M. Cahay, Appl. Phys. Lett. **85**, 1814 (2004).
[85] S. Bandyopadhyay, and M. Cahay, Appl. Phys. Lett. **85**, 1433 (2004).
[86] M. Cahay, and S. Bandyopadhyay, Phys. Rev. B **68**, 115316 (2003).
[87] M. Cahay, and S. Bandyopadhyay, Phys. Rev. B **69**, 045301 (2004).

Index

Note: Page numbers in italics refer to figures and tables.

Kolibri 360 GmbH
Ferdinand-Jühlke-Straße 7
99095 Erfurt, Deutschland
produktsicherheit@kolibri360.de